*Frontispiece by Ruth Weisberg*

# Foundations of Measurement

## VOLUME I
### Additive and Polynomial Representations

David H. Krantz
Columbia University

R. Duncan Luce
University of California, Irvine

Patrick Suppes
Stanford University

Amos Tversky

DOVER PUBLICATIONS, INC.
Mineola, New York

*Bibliographical Note*

This Dover edition, first published in 2007, is an unabridged republication
of the work originally published by Academic Press, Inc., San Diego and
London, in 1971.
The quotation on the opposite page is taken from Plato's *Protagoras* in
*The Dialogues of Plato* translated by Benjamin Jowett, 4th ed., 1953, Vol. 1,
pp. 183–184. Copyright © 1953 by the Clarendon Press, Oxford.

*International Standard Book Number*
*ISBN-13: 978-0-486-45314-9*
*ISBN-10: 0-486-45314-6*

Manufactured in the United States by LSC Communications
45314603    2023
www.doverpublications.com

*... What measure is there of the relations of pleasure to pain other than excess and defect, which means that they become greater and smaller, and more and fewer, and differ in degree? For if any one says: "Yes, Socrates, but immediate pleasure differs widely from future pleasure and pain"—to that I should reply: And do they differ in anything but pleasure and pain? There can be no other measure of them. And do you, like a skillful weigher, put into the balance the pleasures and the pains, and their nearness and distance, and weigh them, and then say which outweighs the other. If you weigh pleasures against pleasures, you of course take the more and greater; or if you weigh pains against pains, you take the fewer and the less; or if pleasures against pains, then you choose that course of action in which the painful is exceeded by the pleasant, whether the distant by the near or the near by the distant; and you avoid that course of action in which the pleasant is exceeded by the painful. Would you not admit, my friends, that this is true? ...*

# Summary of Contents

# Table of Contents

## 4.  Difference Measurement

## 5. Probability Representations

## 6. Additive Conjoint Measurement

# Preface

---

Scattered about the literatures of economics, mathematics, philosophy, physics, psychology, and statistics are axiom systems and theorems that are intended to explain why some attributes of objects, substances, and events can reasonably be represented numerically. These results constitute the mathematical foundations of measurement. Although such systems are of some mathematical interest, they warrant our attention primarily as empirical theories—as attempts to formulate properties that are observed to be true about certain qualitative attributes. Some of the theories appropriate to classical physics are so well accepted that they are usually considered in the province of philosophy rather than physics, but this should not be allowed to becloud the basic empirical character of any theory that purports, for example, to justify treating mass as an additive numerical property. From time to time, the empirical nature of basic measurement assumptions is forcibly brought to everyone's attention—for example, when the theory of relativity made clear that velocities do not combine additively; when quantum mechanics made clear that the probability theory of elementary particles is somewhat different from that appropriate to macroscopic events; and when we recall that the attribute of hardness still lacks any satisfactory measurement analysis.

In the nonphysical sciences, measurement has always been problematic, and it has become increasingly evident to nearly everyone concerned that we must devise theories somewhat different from those that have worked in physics. Because of the active—and we believe crucial—concern with measurement in these sciences, it is not terribly surprising that four behavioral scientists might attempt to summarize the field. The methodology of measurement has, of course, a long history in the physical sciences, and we have also tried to cover the major foundational problems there, ranging from the theory of extensive measurement to dimensional analysis.

We must emphasize that this book deals with Foundations and not with the history or current practice of measurement in any field. To the extent that we deal with empirical examples, it is to motivate theories and to illustrate their testing. The reader who is interested in specific empirical material will have to consult references in particular specialties. For example, an excellent history of weight measurement, with many illustrations of apparatus, is the volume by Kisch (1965).

As one explores the measurement literature it becomes clear that, in spite of the fact that each proof makes some peculiar use of the structure in question, many proofs are quite similar. Moreover, little is done to relate the particular theorem being presented to any others except those having exactly the same primitive concepts. Eventually, this becomes frustrating: results are reproved; it is extremely difficult to maintain a clear idea of the structure of the field; it is uncertain how many really basic ideas there are; and it is likely that the field gradually becomes inefficient in the sense that, for about the same effort, stronger theorems could be proved if other results were used instead of returning to first principles. We have attempted to organize the central results in a cumulative fashion.

As a result of our attempt, we have concluded that at present there are three distinct mathematical results (see Chapters 1 and 2) used to construct numerical representations of qualitative structures. With some minor exceptions, the remaining major theorems of this volume either reduce to one of these results directly or indirectly via some other representation theorem. By "reduce" we do not mean to suggest "reduce readily," for in many cases the proofs are quite lengthy and it is not always immediately obvious what to reduce to what. In organizing the material, we have undoubtedly made some arbitrary decisions about this, although as a matter of fact we suspect that there may be rather less flexibility than might first seem. In the process, we have usually arrived at theorems somewhat better than those previously published, and virtually all of our proofs of major results are different from those in the literature. (Some of our new results have been published independently of this book; any paper on measurement by the authors dated 1967–1970 is a byproduct of work on the book.)

Since the book includes both new results and new proofs it is a research monograph, but we hope that it is more than that. We have gone to pains to make it a textbook—in fact, two textbooks. If one deletes Chapter 2 and all of the sections headed "Proofs," what remains is a self-contained book that should serve as a comprehensive introduction to the mathematical theory of measurement for nonspecialists who have enough mathematics to understand the formulation of the problems (see below), but who have no particular reason for studying the proofs. For those with the requisite mathematical background, the proofs are there in separate sections. Rather more detail is provided than a trained mathematician is likely to want, but with his facility at skipping things that are routine to reconstruct, this should not really bother him. For the novice, the detail is important because he must learn to develop complete proofs, even though he may ultimately present them only in abbreviated form. The reason is that, more than in some areas of mathematics, it is all too easy to accept and use, as if it were proved, a familiar property that has not been proved and may, given the axioms, be exceedingly tricky or tedious to establish. The most "obvious" properties can be major stumbling blocks or the source of erroneous proofs.

To further the usefulness of the book as a text, we have included a number of problems which the student can use to exercise and test his developing skills. Many of these exercises are quite easy for a person familiar with the area, but a few are moderately difficult. In addition, a number of unsolved problems are mentioned throughout the book. Some of these are difficult and are suitable for dissertations; others, no doubt, will turn out to be easier than we have thought.

The material that we have elected to cover and our mode of presentation make it impossible to package it conveniently in a single volume. This first volume covers all of the major representation theorems in which the qualitative structure is reflected as some sort of a polynomial function of one or more numerical functions defined on the basic entities. The simplest and best-known examples are additive expressions of a single measure, such as the probability of disjoint events being the sum of their probabilities, and additive expressions of two measures, such as the logarithm of momentum being the sum of log mass and log velocity terms. The second volume will include representations in terms of distances in some sort of space, a treatment of the exceedingly perplexing problems raised by errors of measurement, and analyses of the philosophical issues that center about axiomatizability and meaningfulness.

## Mathematical Background

For a reader who plans to skip proofs, a modest background in mathematics and some diligence should suffice. He certainly must know well the elementary material on sets, relations, functions, and probability that is provided in introductory courses at many colleges and universities and is found in, among other books, Kershner and Wilcox (1950) and Suppes (1957). With no more than that, there will be some difficult spots and a few sections will be unintelligible because they depend upon calculus or topological concepts. For the reader who plans to follow the proofs, a background in calculus and in elementary abstract algebra is needed. In a few places, some elementary topological material that can be found in, for example, Kelley (1955) and some of the functional equations found in Aczél (1966) are also required.

**Selecting Among the Chapters**

Within the present volume, various courses of study are possible by selecting among the chapters. The main constraints are conceptual dependencies and the logical development of the proofs. These two are diagrammed separately in Figure A. Obviously, many more options are avail-

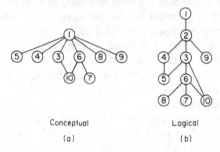

Conceptual                    Logical

(a)                              (b)

**FIGURE A.** (a) Conceptual dependencies of chapters. (b) Logical development of the proofs.

able to those who do not plan to read proofs, and so it may be useful for us to state what we believe to be the minimum core of information for physical and behavioral scientists. A physical scientist who plans to ignore proofs should cover Chapters 1, 3, 6, and 10; he will find that the initial sections of Chapters 4, 5, and parts of Chapter 9 are also valuable. To follow the proofs, Chapter 2 must be added. For a behavioral scientist, the essential chapters are 1, 4, 5, 6, 8, and 9, with Chapter 3 also being desirable. To follow the proofs, Chapters 2 and 3 must be added. In addition, to guide the reader further we have marked with a square (□) those sections of Chapters 3–10 that we consider to be the basic core material.

# Acknowledgments

Each of us has received substantial assistance while working on this book and we gratefully acknowledge it.

Krantz: Grants GB-4947 and 8181 from the National Science Foundation to the University of Michigan; Public Health Service Special Postdoctoral Fellowship, 1968; and the hospitality of the Center for Advanced Study in the Behavioral Sciences, Stanford, California, during the summer of 1967 and all of 1970–1971.

Luce: Grant GB-6536 from the National Science Foundation to the University of Pennsylvania; a National Science Foundation Senior Postdoctoral Fellowship, 1966–1967, which with a sabbatical leave from the University of Pennsylvania permitted him to spend that academic year at the Center for Advanced Study in the Behavioral Sciences, Stanford, California; an Organization of American States Professorship for the academic year 1968–1969 at the Pontificia Universidade Católica do Rio de Janeiro, Brazil; and a grant from the Alfred P. Sloan Foundation to The Institute for Advanced Study, Princeton, New Jersey.

Suppes: Grant GJ-443X from the National Science Foundation to Stanford University.

Tversky: Grant GB-6782 from the National Science Foundation to the University of Michigan and the hospitality of the Center for Advanced Study in the Behavioral Sciences, Stanford, California, during the summer of 1967 and all of 1970–1971.

Drafts of the chapters have been circulated among a small group of people whom we know to be especially interested in these matters. We have benefited greatly from their comments, some of which have been detailed and penetrating. We thank: J. Aczél, Maya Bar-Hillel, R. L. Causey, R. M. Dawes, Z. Domotor, P. C. Fishburn, R. Giles, E. W. Holman, M. V. Levine, K. R. MacCrimmon, A. A. J. Marley, F. S. Roberts, R. J. Titiev, J. W. Tukey, and T. S. Wallsten.

# Notational Conventions

Throughout the book we have attempted to use a consistent notation which adheres to standard mathematical conventions, although we have been forced to make departures in certain places.

We use I to denote the integers, Ra the rationals, and Re the real numbers. The positive and negative parts of these sets are denoted by using the superscripts $+$ and $-$. For example, $I^+$ is the set of positive integers.

We use the usual notation of set theory, with the Cartesian product notation $A_1 \times \cdots \times A_n = \mathsf{X}_{i=1}^{n} A_i$. Elements of the Cartesian products $A \times B$ or $\mathsf{X}_{i=1}^{n} A_i$ are usually denoted $(a, b)$ or $(a_1, \ldots, a_n)$ respectively. However, in certain places—notably in Chapters 4, 6, and 7—we omit parentheses and commas, writing $ab$ and $a_1 a_2 \cdots a_n$, whenever no ambiguity arises.

The symbols $\gtrsim$, $\succ$, and $\sim$ are used to denote binary relations on some (usually nonnumerical) set which correspond, respectively, to the usual relations $\geq$, $>$, and $=$ on numbers. If $\sim$ is an equivalence relation on $A$, then $A/\!\!\sim$ is a set of equivalence classes on $A$ under $\sim$, often denoted $\mathbf{A}$, and $\mathbf{a}$ denotes the equivalence class containing $a$. Thus, $a \sim b$ implies $\mathbf{a} = \mathbf{b}$.

We adhere to the following general conventions for the use of alphabet symbols except when dealing with specific subject matters, in which case we use the usual symbols of the field in question. Lower case letters are mostly elements of sets. In particular, integers are usually denoted by $i, j, k, l, m,$ and $n$ whereas other real numbers are usually written $\alpha, \beta, \gamma, \mu, \nu, \rho,$

and $\sigma$. The two major exceptions to this rule are that $\phi$, $\psi$, and $\theta$ are used for the functions that represent a structure in the real numbers, and $f$, $g$, and $h$ are sometimes used for other functions, often real-valued ones of real variables. Upper case Latin letters are usually sets; however, $F$, $G$, and $H$ are functions, often real-valued ones of two or more real variables. Upper case script letters are confined to families or collections of objects which, themselves, have structure, e.g. functions.

Chapters are divided into sections and some of these have subsections. All references are of the form Chapter A, Section A.B, or Section A.B.C, where A is the chapter number, B the section number, and C the subsection number. Within each chapter, each of the following categories is numbered separately and consecutively: definitions, theorems, lemmas, equations, figures, and tables. Intrachapter references omit the chapter number; interchapter references are prefixed by the chapter number. For example, the third definition in Chapter 6 is referred to in that chapter as Definition 3, and in other chapters as Definition 6.3. In the exercises at the end of each chapter, the directly relevant subsections for each exercise are listed.

In the proof sections the relevant theorems are restated in an abbreviated form. The end of each proof is marked by a $\diamondsuit$. Finally, the abbreviation "iff" is used in formal discussions and displays for "if and only if."

# Foundations of Measurement

---

## VOLUME I
### Additive and Polynomial Representations

# Chapter 1   Introduction

## 1.1   THREE BASIC PROCEDURES OF FUNDAMENTAL MEASUREMENT

When measuring some attribute of a class of objects or events, we associate numbers (or other familiar mathematical entities, such as vectors) with the objects in such a way that the properties of the attribute are faithfully represented as numerical properties. In this book we investigate various systems of formal properties of attributes that lead to measurement in this sense. Some commentators on physical measurement have claimed that an attribute must exhibit a more or less unique set of formal properties in order for it to have a "fundamental" measure—one that does not require the prior measurement of other quantities. For example, length can be measured fundamentally (see below), whereas density = mass/volume depends on the prior measurement of mass and volume. Not only is the intuitive distinction between fundamental and other sorts of measurement an elusive one (see Chapter 10), but whatever it may be, it is surely wrong to think that there is only one fundamental system of properties adequate to lead to numerical measurement. We present many quite different systems that are all fundamental by the intuitive criterion of independence of other measurement.

Despite the variety of systems that may lead to measurement, only a few basic procedures are known for assigning numbers to objects or events on

1

the basis of qualitative observations of attributes. In the rest of this section we sketch the three procedures that underlie the systems in Chapters 2–13. To make the discussion concrete, we formulate the ideas in terms of length measurement.

Suppose that we have a set of straight, rigid rods whose lengths are to be measured. If we place the rods $a$ and $b$ side by side and adjust them so that one is entirely beside the other and they coincide at one end, then either $a$ extends beyond $b$ at the other end, or $b$ beyond $a$, or they appear to coincide at that end also. We say, respectively, that $a$ is longer than $b$, $b$ is longer than $a$, or that $a$ and $b$ are equivalent in length. For brevity, we write, respectively, $a > b$, $b > a$, or $a \sim b$. Two or more rods can be *concatenated* by laying them end to end in a straight line, and so we can compare the qualitative length of one set of concatenated rods with that of another by placing them side by side, just as with single rods. The concatenation of $a$ and $b$ is denoted $a \circ b$, and the observation that $c$ is longer than $a \circ b$ is denoted $c > a \circ b$, etc. Many empirical properties of length comparison and of concatenation of rods can be formulated and listed, e.g., $>$ is transitive; $\circ$ is associative; if $a > b$, then $a \circ c > b$; etc. Although we do not systematically list these properties now, we will use them freely in the intuitive discussion that follows.

### 1.1.1 Ordinal Measurement

In measuring length ordinally, we confine our observations to comparisons between simple, unconcatenated rods, and we are concerned only with assigning numbers $\phi(a)$, $\phi(b)$, etc. to rods $a$, $b$, etc. so as to reflect the results of these comparisons. That is to say, *we require the numbers be assigned so that $a > b$ if and only if $\phi(a) > \phi(b)$*. Information arising from concatenation is not used at all.

A natural procedure for assigning numbers is this. We assign to the first rod selected any number whatsoever. If the second rod chosen exceeds the first, we assign it any larger number, whereas if the first exceeds the second, we assign the second rod any smaller number. We deal with the third rod similarly except if it is between the first two. Then we assign it any number between the numbers selected for the first two. This procedure can be continued indefinitely. The assignments of numbers already made need not be affected by subsequent observations because between any two distinct numbers others always exist.

The only difficulty that can arise in carrying out the above procedure is the existence of rods $a$ and $b$ such that neither $a > b$ nor $b > a$, i.e., $a \sim b$. If the comparison does not establish an order, we may be tempted to conclude that the lengths are actually equal and so we assign the same number to $a$

and $b$. However, if the comparison process for establishing relations is insensitive to very small disparities in length we may then find that $b \sim c$, $c \sim d$, and $b > d$. But to represent these relations by numbers, we require $\phi(b) = \phi(c)$, $\phi(c) = \phi(d)$, and $\phi(b) > \phi(d)$, which is impossible. Thus, this procedure for ordinal measurement is suitable only when the sensitivity of the comparison process exceeds the disparities among the rods under consideration. In the ideal case either $a > b$ or $b > a$ for any two rods $a$ and $b$, except for carefully prepared "perfect copies" (e.g., meter sticks or intervals on a meter stick); for the latter, $\sim$ holds between any two copies and thus is transitive.

Of course, really perfect copies cannot be prepared. Whenever physical differences become sufficiently small, any method for observing them ultimately deteriorates. In some cases, and perhaps in all, observations of two sufficiently similar entities are inconsistent when the same comparison is repeated several times. And when inconsistencies can occur, violations of transitivity may arise. But the ideal case described above can still be achieved if copies are prepared by "standard" methods that are much more sensitive than the "working" methods used to establish $>$ and if the set of rods is restricted so that either $a > b$ or $b > a$ except for standardized copies. Chapter 15 discusses the case where $\sim$ is not transitive, but $>$ is transitive.

Clearly, the above procedure of ordinal measurement can be applied to any attribute of objects—not just length of rods—provided that a suitable comparison process leads to relations $>$ and $\sim$ with the requisite properties and that the set of objects is finite. The same procedure can be used for an inductive definition of a scale $\varphi$ on a countable ordered set (see Section 2.1). What is less obvious is how to construct the scale when there are pairs of objects that cannot be compared directly. An interesting case, in which an ordering $>$ is inferred from "revealed preference" observations, has been treated in the economic literature (Arrow, 1959; Hansson, 1968b; Houthakker, 1950, 1965; Richter, 1966; Samuelson, 1938, 1947; and Uzawa, 1960). We do not deal with this problem; instead, this book is mainly about the ramifications and extensions of another procedure, to which we now turn.

### 1.1.2 Counting of Units

If we take into account the concatenation of rods as well as their ordering, then further constraints on the numerical assignments arise quite naturally. Suppose, for example, that $a', a'', a''',...$ are perfect copies of the rod $a$ (see Section 1.1.1). If $a \circ a' > b$ and $b > a$, then we not only want to assign numbers such that $\phi(a \circ a') > \phi(b) > \phi(a) = \phi(a')$, but we also want to represent that $a \circ a'$ is twice as long as $a$, i.e., $\phi(a \circ a') = 2\phi(a)$. Hence we make the assignment $\phi(b)$ so that it is between $\phi(a)$ and $2\phi(a)$. Similarly, if $a \circ a' \circ a'' \circ a''' \succ b$ and $b \succ a \circ a' \circ a''$, we place $\phi(b)$ between $3\phi(a)$ and $4\phi(a)$. And so on. The sequence $a$, $2a = a \circ a'$, $3a = (2a) \circ a''$, $4a$, $5a$,... is called a *standard sequence* based on $a$. A meter stick graded in millimeters provides, in convenient form, the first 1000 members of a standard sequence constructed from a one-millimeter rod. If we observe that rod $b$ falls between $na$ and $(n + 1)a$, say, between 480 and 481 mm, then we assign it a length between $n\phi(a)$ and $(n + 1)\,\phi(a)$ (in the present example, between $480\phi(a)$ and $481\phi(a)$, where $\phi(a)$ is the number assigned to a one-millimeter rod and its copies). The value of $\phi(a)$ depends on the selection of a particular rod (say, $e$) to have unit length. If $e \sim ma$, then $\phi(a) = 1/m$. Thus, if $e$ is the meter stick, then $m = 1000$ and the length assigned to $b$ must be between 0.480 and 0.481 meters; if $e$ is a centimeter rod, then $m = 10$ and $\phi(b)$ must be between 48.0 and 48.1 cm.

By choosing finer and finer standard sequences, keeping the unit of measurement fixed, of course, the value of $\phi(b)$ can be placed within an interval as small as we like. In Section 2.2 we prove a theorem establishing the convergence of these estimates as the standard sequence becomes arbitrarily fine.

Note that for purposes of ordinal measurement we could have disposed of the problem of transitivity of $\sim$ by restricting the set of rods so that either $a \succ b$ or $b \succ a$ for all rods $a$ and $b$, but for standard sequences copies are essential, and the discussion in Section 1.1.1 of perfect copies applies here as well.

Three remarks should be made about the procedure of counting using standard sequences.

1. *The numbers obtained form a satisfactory ordinal measure* (Section 1.1.1): If $b \succ c$, then for some sufficiently fine-grained standard sequence based on some $a$, we have $b \succ na$ and $na \succ c$, so $\phi(b) > n\phi(a) > \phi(c)$. It follows that $b \succ c$ if and only if $\phi(b) > \phi(c)$.

2. *The numbers assigned are additive with respect to concatenation*: $\phi(b \circ c) = \phi(b) + \phi(c)$. The reason is that if $n$ copies of $a$ must be concatenated to approximate $b$ and $n'$ copies to approximate $c$, then the concatenation of $n + n'$ copies of $a$ will approximate the concatenation of $b$ with $c$. The additivity equation holds only approximately for coarse standard sequences and approaches exactness for finer and finer sequences.

3. *Regardless of the choice of unit, ratios of numerical assignments are uniquely determined by the procedure:* For if $n$ copies of $a$ must be concatenated to approximate $b$ and $n'$ copies to approximate $c$, then $n/n'$ approximates $\phi(b)/\phi(c)$ more closely the finer the standard sequence.

The technique we have sketched is obviously applicable to any similar situation where the relation $>$ and the concatenation $\circ$ are both defined empirically. Whenever it is applied, we say that the attribute in question has been *extensively* measured (see Chapter 3). It turns out, however, that the same basic technique actually can be applied in many other, much less obvious ways (see Section 1.3.2 for an illustration and Chapters 4–8, 12, and 13 for the detailed exploitation of this idea).

### 1.1.3 Solving Inequalities

Suppose that five rods denoted $a_1$, $a_2$ ,..., $a_5$ , are found to satisfy

$$a_1 \circ a_5 > a_3 \circ a_4 > a_1 \circ a_2 > a_5 > a_4 > a_3 > a_2 > a_1. \qquad (1)$$

Data such as these can arise whenever a limited set of preselected objects and concatenations are compared and where it is impractical to go through the elaborate process of constructing standard sequences. Denote by $x_i$ the unknown value of the length of $a_i$, i.e., $x_i = \phi(a_i)$, $i = 1,...,5$. From the above observations, the unknown lengths $x_i$ must satisfy the following system of simultaneous linear inequalities.

$$
\begin{aligned}
x_1 + x_5 - x_3 - x_4 &> 0, \\
x_3 + x_4 - x_1 - x_2 &> 0, \\
x_1 + x_2 - x_5 &> 0, \\
x_5 - x_4 &> 0, \qquad (2) \\
x_4 - x_3 &> 0, \\
x_3 - x_2 &> 0, \\
x_2 - x_1 &> 0.
\end{aligned}
$$

Any solution to this set of seven inequalities in five unknowns gives a possible set of values for the lengths of $a_1$ ,..., $a_5$ . One can thus measure the five rods by finding a solution, if one exists. Alternatively, one can obtain bounds on certain ratios of numerical assignments from the inequalities; for example, it can be shown (with some manipulation) that the ratio $x_3/x_2$, or $\phi(a_3)/\phi(a_2)$, is between 1 and $\frac{3}{2}$ for any solution to the above inequalities (Exercise 1).

In setting up the inequalities in Equation (2) on the basis of the observations that are given by Equation (1), the concatenation operation $\circ$ is translated into addition $+$ of real numbers, and the observational order $>$ is translated into the order $>$ of real numbers. Thus, for example, $a_1 \circ a_2 > a_5$ is represented as $x_1 + x_2 > x_5$. This translation uses properties 1 and 2 of the previous section. In other words, measurement of length by counting units in a standard sequence is a procedure which assigns the sum $\phi(b) + \phi(c)$ to the concatenation $b \circ c$, and which assigns numbers whose numerical order preserves the observational order. To measure by solving inequalities, one *assumes* that these two properties are to be satisfied by the numerical assignment; this assumption allows one to set up the inequalities to be solved.

Solving inequalities to obtain numerical assignments has many applications other than the measurement of length. These are discussed in Chapter 9. In some cases more complicated numerical operations than simple addition are assumed to correspond to empirically defined notions, and this results in nonlinear inequalities.

## 1.2 THE PROBLEM OF FOUNDATIONS

### 1.2.1 Qualitative Assumptions: Axioms

When measuring the lengths of ordinary (as opposed to atomic-sized or interstellar) objects, the standard-sequence procedure of Section 1.1.2 is obviously the one of most interest. In analyzing its foundations, one is led to the following question: *What basic assumptions must be satisfied by $>$ and $\circ$ in order that the standard-sequence procedure can be carried through in a self-consistent manner?* In Section 1.1 some basic properties of $\circ$ and $>$, e.g., if $a > b$, then $a \circ c > b$, were cited, and in describing the counting-of-units procedure we tacitly invoked others. For example, we used the one just mentioned as follows: In finding elements $na$ and $(n + 1)a$, in the standard sequence based on $a$, such that $(n + 1)\,a > b$ and $b > na$, we need to be assured that the integer $n$ is unique. We infer this, because if $(n + 1)\,a > b$, then $(n + 2)\,a \circ a$ is also longer than $b$, etc., so that all subsequent members of the sequence are longer than $b$. The critical step, inferring that $(n + 1)\,a \circ a > b$ on the basis of $(n + 1)\,a > b$, depends on the truth of the above property (and on the transitivity of $>$). Similarly, another property tacitly assumed is the existence of some integer $n$ such that $(n + 1)\,a > b$; i.e., we need to know that $a$ is not "infinitesimally small" compared to $b$. For no really good reason, except mathematical tradition, this is called an Archimedean assumption.

To put our understanding of the counting-of-units procedure in good order, we must make all of these assumptions explicit. A measurement procedure certainly is not adequately understood if it depends on properties that are not explicitly recognized. Once they are explicit, deciding whether or not the same measurement procedure is applicable in a new domain reduces to testing whether or not the requisite properties are satisfied.

Still more orderly foundations are obtained if we can deduce most properties as theorems from a few basic properties. Applicability to a new situation is then judged according to whether those few properties are satisfied. This logical step is called the *axiomatization* of the measurement procedure. The few explicit properties from which all others are deduced are called the *axioms*. Axiomatization of any body of propositions can always be achieved in more than one way; some criteria for "good" axiomatizations are discussed later.

Geometry is a beautiful and far-reaching example of a foundational treatment of measurement. The science of geometry (i.e., earth measurement) was probably first developed as a set of practical procedures, either for the direct measurement of lengths and areas on the earth's surface or in connection with the astronomy devised to serve astrology. Eventually, the tacit assumptions of practice were formulated explicitly as theorems of geometry, and these were systematically organized and deduced from a few axioms and postulates by Euclid. Certain additional tacit assumptions, unrecognized by Euclid, were discovered later, and an orderly axiomatization of Euclidean geometry was finally devised by Hilbert (1899) and others. These axiomatic studies are known as the foundations of geometry. For further discussion of geometry as measurement, see Chapters 11–13.

Although it seems quite natural to view the task of the foundations of measurement to be the explication and systematization of the assumptions required by particular interesting procedures of measurement, doing so has actually led to some serious misunderstandings. These stem from the easy supposition that an empirical concatenation operation is *sine qua non* for the standard-sequence procedure. Campbell (1920, 1928), in his influential books on measurement, and some later philosophers (e.g., Cohen & Nagel, 1934; Ellis, 1966) treated fundamental measurement as practically synonymous with procedures involving empirically defined concatenation operations. (One of Campbell's remarks, p. 327 of the 1957 edition of 1920, makes clear that he was aware of a potential distinction: "Of course there may possibly be some other way of assigning numerals to represent properties differing in first principles from that described in Chapter X [which is devoted to the procedure of Section 1.1.2]; but until somebody suggests such a way, it is hardly worthwhile to discuss the possibility; it is certainly not employed in the actual physics of today.") The absence of appropriate, empirically defined, concatenation operations in psychology has even led some serious students of measurement to conclude that fundamental measurement is not possible there in the same sense that it is possible in physics (Guild, 1938). Reese (1943) discussed this in detail and attempted to provide psychological examples of scales based on concatenation. Many examples given in this book show that Campbell's viewpoint is untenable.

## 1.2.2 Homomorphisms of Relational Structures: Representation Theorems

We may view the foundations of measurement in a slightly different way by focusing on properties of the numerical assignment, rather than on the procedure for making the assignment. Specifically, for the standard-sequence procedure, we may pose the following question: Given a set of rods, a comparison relation $>$, and a concatenation $\circ$, what assumptions concerning $>$ and $\circ$ are necessary and/or sufficient to construct a real-valued function $\phi$ that is order preserving and additive—i.e., that satisfies properties 1 and 2 of Section 1.1.2? This question still asks for an axiomatization—the listing of certain properties of $>$ and $\circ$—however, the conclusion aimed for is not that a certain procedure is possible, but rather that a numerical function $\phi$ satisfying certain properties exists. The procedure to be used in assigning numbers (constructing $\phi$) is not specified in posing the problem; thus, quite distinct axiomatizations, which operate through different procedures—say, counting units and solving inequalities—may be expected.

The next step, a small but important one, recognizes that the numerical assignment $\phi$, satisfying properties 1 and 2 of Section 1.1.2 (order preserving and additive), is a homomorphism of an empirical relational structure into a numerical relational structure. To make our meaning clear we must first say what we mean by a relational structure and then what we mean by a homomorphism between two relational structures.

A *relational structure* is a set together with one or more relations on that set. If we denote the set of all the rods and all the finite concatenations of rods under consideration by $A$, then the empirical relational structure for the procedures of Sections 1.1.2 and 1.1.3 is denoted $\langle A, >, \circ \rangle$. (The concatenation operation is a ternary relation on $A$, holding among $a$, $b$, and $c = a \circ b$, whereas $>$ is a binary relation on $A$.) An appropriate numerical relational structure is $\langle \text{Re}, >, + \rangle$, where Re is the set of real numbers, $>$ is the usual greater than relation, and $+$ is the ordinary operation of addition. (Angle brackets $\langle \ \rangle$ rather than parentheses are used in giving an explicit listing of a relational structure.) The numerical assignment $\phi$ is a *homomorphism*[1] in the sense that it sends $A$ into Re, $>$ into $>$, and $\circ$ into $+$ in such a way that $>$ preserves the properties of $>$ (property 1, Section 1.1.2) and $+$ the properties of $\circ$ (property 2, Section 1.1.2).

This formulation generalizes naturally to other relational structures. Given an empirical relation $R$ on a set $A$ and a numerical relation $S$ on Re,

[1] We speak of a homomorphism (rather than an isomorphism) because $\phi$ is not usually one to one; in general $\phi(a) = \phi(b)$ does not mean that the rods $a$ and $b$ are identical, but merely of equal length. A one-to-one homomorphism is called an *isomorphism*.

a function $\phi$ from $A$ into Re takes $R$ into $S$ provided that the elements $a, b,...$ in $A$ stand in relation $R$ if and only if the corresponding numbers $\phi(a)$, $\phi(b),...$ stand in relation $S$. More generally, if $\langle A, R_1,..., R_m \rangle$ is an empirical relational structure and $\langle \text{Re}, S_1,..., S_m \rangle$ is a numerical relational structure, a real-valued function $\phi$ on $A$ is a *homomorphism* if it takes each $R_i$ into $S_i$, $i = 1,..., m$. Still more generally, we may have $n$ sets $A_1,..., A_n$, $m$ relations $R_1,..., R_m$ on $A_1 \times \cdots \times A_n$, and a vector-valued homomorphism $\phi$, whose components consist of $n$ real-valued functions $\phi_1,..., \phi_n$ with $\phi_j$ defined on $A_j$, such that $\phi$ takes each $R_i$ into relation $S_i$ on $\text{Re}^n$. A *representation theorem* asserts that if a given relational structure satisfies certain axioms, then a homomorphism into a certain numerical relational structure can be constructed. A homomorphism into the real numbers is often referred to as a scale in the psychological measurement literature.

From this standpoint, measurement may be regarded as the construction of homomorphisms (scales) from empirical relational structures of interest into numerical relational structures that are useful. Foundational analysis consists, in part, of clarifying (in the sense of axiomatizing) assumptions of such constructions.

This view of measurement would be entirely too abstract were it not for the fact that interesting examples of measurement exist that involve relational structures quite different from $\langle A, \succ, \circ \rangle$. The development of such additional examples of measurement spurred the formulation of this abstract viewpoint. Among the key examples were the axiomatizations of utility measurement by von Neumann and Morgenstern (1947), Savage (1954), Suppes and Winet (1955), and Davidson, Suppes, and Siegel (1957), and the axiomatization of semiorders by Luce (1956). An explicit statement, of the relational structure viewpoint was first given by Scott and Suppes (1958); also see Ducamp and Falmagne (1969) and Suppes and Zinnes (1963).

Despite the proliferation of measurement axiomatizations, the procedures for constructing numerical assignments remain about the same; the most important ones are those described in Section 1.1. One of our goals in this book is to show clearly just how the assignments of numbers in what are quite disparate representation theorems all reduce to one or another of these three basic procedures.

### 1.2.3 Uniqueness Theorems

In discussing the measurement of length based on the counting of units (Section 1.1.2), we pointed out that the number $\phi(a)$ assigned to rod $a$ depends on which rod $e$ is chosen as unit, i.e., $\phi(e) = 1$. This choice is entirely arbitrary. Moreover, as we noted, the ratio, $\phi(a)/\phi(e)$ is uniquely determined independent of whether $e$ or some other rod $e'$ is chosen as the unit. Thus, if $\phi$ is the numerical function constructed with $e$ as unit and if $\phi'$ is constructed with $e'$ as unit, we have

$$\phi(a)/\phi(e) = \phi'(a)/\phi'(e),$$

or, substituting $\phi(e) = 1$ and $\phi'(e) = \alpha$,

$$\phi'(a) = \alpha\phi(a). \tag{3}$$

Conversely, starting with $\phi$ based on $e$ as unit, we can select $e'$ with $\phi(e') = 1/\alpha$, and obtain a new scale (homomorphism) $\phi'$ satisfying Equation (3). These facts are usually expressed by saying that the *similarity* transformations

$$\phi \rightarrow \alpha\phi \doteq \phi', \quad \alpha > 0, \tag{4}$$

are *permissible* transformations of the scale $\phi$. A scale whose permissible transformations are only those of Equation (4) is called a *ratio scale*.

The term "ratio scale" comes from the fact that if $\phi \rightarrow \alpha\phi$ are the only permissible transformations, then the ratios of scale values are determined uniquely.

Other families of measurement procedures are related by different sets of admissible transformations. For example, Celsius temperature is related to Fahrenheit by $C = (5/9)(F - 32)$. Observe that in any ordinary temperature measurement, two arbitrary choices are made, the zero point and the unit. Varying these leads to the *affine* transformations which are of the form

$$\phi \rightarrow \alpha\phi + \beta, \quad \alpha > 0. \tag{5}$$

A scale whose permissible transformations are only those of Equation (5) is called an *interval scale* because ratios of intervals are invariant:

$$\frac{\phi'(a) - \phi'(b)}{\phi'(c) - \phi'(d)} = \frac{[\alpha\phi(a) + \beta] - [\alpha\phi(b) + \beta]}{[\alpha\phi(c) + \beta] - [\alpha\phi(d) + \beta]}$$

$$= \frac{\phi(a) - \phi(b)}{\phi(c) - \phi(d)}.$$

Two other classes of transformations play a key role in measurement. The *power* transformations are of the form

$$\phi \rightarrow \alpha\phi^\beta, \quad \alpha > 0, \quad \beta > 0, \tag{6}$$

and a scale whose permissible transformations are those of Equation (6) is called a *log-interval scale* because a logarithmic transformation of such a scale results in an interval scale. As we shall argue in Chapter 10, many of

the common physical scales that are usually said to be ratio scales are, in fact, log-interval scales. Density is an example.
   Finally, the *monotonic increasing* transformations are of the form

$$\phi \rightarrow f(\phi), \tag{7}$$

where $f$ is any strictly increasing real-valued function of a real variable, and a scale whose permissible transformations are those of Equation (7) is called an *ordinal scale*. The reason is that only order is preserved under these transformations.
   Stevens (1946, 1951) was the first to recognize and emphasize the importance of the type of uniqueness exhibited by a measurement homomorphism and he isolated these four types—ratio, interval, log-interval, and ordinal.[2]
   A classification of measurement in terms of permissible transformations is clear cut only so long as it is certain which transformations are permissible. Little ambiguity exists for length: there is a family of closely related procedures, described in Section 1.1.2, which differ from one another only in the rather trivial and arbitrary matter of which size rod is chosen as unit. Permissible transformations are precisely those produced' by variations in this matter of procedure. On closer examination, however, it becomes less clear which other choices in a measurement procedure are arbitrary and which are not. For example, in measuring length, we choose not only the unit, but we choose to count and record the number of copies of $a$ (say, $n$) that are needed to approximate $b$, rather than to record, for example, the square or exponential of that number, $n^2$ or $e^n$. Is this also an arbitrary choice? Would a procedure that recorded $n^2$ be closely related to one that recorded $n$, with the consequence that the transformation $\phi \rightarrow \phi^2$ is also permissible? If not, why not?

---

[2] Surprisingly, he later (Stevens, 1957, 1968) generated an ambiguity in the use of these terms by describing his magnitude estimation scale as a "ratio scale" of measurement. In this experimental procedure, observers are asked to assign numbers to stimuli "in proportion to the sensations evoked," and the resulting numbers are taken to be scale values. In the sense that subjects are asked to produce numbers that preserve subjective "ratios," one sees why this scale might be described as a ratio scale—except for the fact that he earlier introduced the term to refer to those theories in which any two homomorphisms are related by a similarity transformation. Stevens has not provided any argument showing that the procedure of magnitude estimation can be axiomatized so as to result in a ratio-scale representation; he has neither described the empirical relational structure, the numerical relational structure, nor the axioms which permit the construction of a homomorphism. In Chapter 4, we provide a set of plausible axioms for families of matching experiments (which generalize magnitude estimation) and, if the axioms are empirically valid, we have nearly justified Stevens' claim.

The reason for rejecting the $n^2$ or $e^n$ procedures is easily given: the resulting scales, related to the normal one by $\phi^2$ or $e^\phi$, are not additive. Instead, they satisfy other rules, namely,

$$\phi^2(a \circ b) = \phi^2(a) + 2[\phi^2(a)\,\phi^2(b)]^{1/2} + \phi^2(b),$$
$$e^\phi(a \circ b) = e^\phi(a)\,e^\phi(b).$$

Hence, they do not yield homomorphisms of $\langle A, >, \circ \rangle$ into $\langle \mathrm{Re}, >, + \rangle$; instead, they yield homomorphisms of $\langle A, >, \circ \rangle$ into numerical structures $\langle \mathrm{Re}, >, * \rangle$, where $*$ is a binary operation differing from addition. In the above examples $*$ is defined, respectively, by $x * y = x + 2(xy)^{1/2} + y$ and $x * y = xy$.

The concept of permissible transformations is much clearer from the standpoint of homomorphisms between relational structures than from the standpoint of arbitrary choices in measurement procedures. A transformation $\phi \to \phi'$ is permissible if and only if $\phi$ and $\phi'$ are both homomorphisms of $\langle A, R_1 ,..., R_n \rangle$ into the *same* numerical structure $\langle \mathrm{Re}, S_1 ,..., S_n \rangle$. Thus, if $\phi$ is order preserving and additive—a homomorphism of $\langle A, >, \circ \rangle$ into $\langle \mathrm{Re}, >, + \rangle$—the same is true for $\alpha\phi$, provided that $\alpha > 0$; moreover, if $\phi'$ is *any* homomorphism of $\langle A, >, \circ \rangle$ into $\langle \mathrm{Re}, >, + \rangle$, then $\phi' = \alpha\phi$ for some $\alpha > 0$. The latter result, which is the substance of what is proved in a *uniqueness theorem*, is not obvious.

We conclude, then, that an analysis into the foundations of measurement involves, for any particular empirical relational structure, the formulation of a set of axioms that is sufficient to establish two types of theorems: a representation theorem, which asserts the existence of a homomorphism $\phi$ into a particular numerical relational structure, and a uniqueness theorem, which sets forth the permissible transformations $\phi \to \phi'$ that also yield homomorphisms into the same numerical relational structure. A measurement procedure corresponds to the construction of a $\phi$ in the representation theorem.

It is important to note that every pair of representation and uniqueness theorems involves a choice of a numerical relational structure. This choice is essentially a matter of convention, although the conventions are strongly affected by considerations of computational convenience. For example, either of the $*$ operations proposed above is clumsy compared with addition. The subject of alternative numerical structures is considered in more detail in Sections 3.9, 4.4.2, 6.5.2, 7.2, 7.4.2, and in Chapter 19.

For an interesting attempt to treat the mathematics of measurement in an elementary way by focusing mostly on the uniqueness properties of various representations without, however, going deeply into the question of the existence of representations, see Blakers (1967).

### 1.2.4 Measurement Axioms as Empirical Laws

We have just emphasized that the numerical scales of measurement are subject to arbitrary conventions. There are permissible transformations, corresponding to arbitrary choices of unit, and the very representation and uniqueness theorems themselves depend on the conventional choice of a numerical relational structure. What is invariant, and so is not a matter of convention, is the empirical relational structure and its empirical properties, some of which are formulated as axioms. A set of axioms leading to representation and uniqueness theorems of fundamental measurement may be regarded as a set of qualitative (that is, nonnumerical) empirical laws. In some cases, as in the measurement of length, these laws are rather trivial, i.e., not intrinsically very interesting. In other empirical contexts, the axioms can be quite interesting and nonobvious. In such cases, the development of measurement scales is closely linked to the formulation and testing of appropriate qualitative laws. This viewpoint has been discussed by Krantz (1971).

We shall make an effort to point out the status of various axioms or classes of axioms as empirical laws. The type of consideration that arises is illustrated in Sections 1.3, 1.4.5, and 1.5.

### 1.2.5 Other Aspects of the Problem of Foundations

In analyzing the foundations of measurement, one of the main concerns is formalization: the choice of an empirical relational structure as an abstraction from the available data, the choice of an appropriate numerical relational structure, the discovery of suitable axioms, and the construction of numerical homomorphisms, i.e., proving the representation theorem and uniqueness theorem. However, this formalization process does not exhaust the problem of foundations by any means. The most important omission is an analysis of error of measurement. This involves difficult conceptual problems concerning the relation between detailed, inconsistent data and the abstraction derived from them, the empirical relational structure. We discuss these aspects of the foundations of measurement, which are poorly understood, as well as we can in Chapters 15–17.

## 1.3 ILLUSTRATIONS OF MEASUREMENT STRUCTURES

In the previous section, we presented a general statement of what is involved in a foundational analysis of measurement. Nevertheless, we refrained from trying to hammer out an acceptable definition of the concept of a formal theory of measurement. Experience suggests that after some exposure to paradigmatic theories of measurement, students have little difficulty in recognizing other examples. It is similar to recognizing grammaticalness of English sentences: with some borderline exceptions, sequences of words are readily identified as grammatical or not grammatical, even though no general definition is yet available. In this section, then, we try to make our previous generalities about relational structures, axioms, etc. more comprehensible by presenting two examples. These examples also motivate the discussion of axiom types in Section 1.4 and provide the material for a set of exercises.

### 1.3.1 Finite Weak Orders

It is useful to begin with a consideration of ordinal measurement. Most of the relational structures we shall consider involve an ordering relation, and the concepts needed to handle the ordering relation in the more complicated structures can be developed and presented in isolation here. Moreover, weak orders provide the simplest illustration of the ideas presented above.

DEFINITION 1. *Let $A$ be a set and $\gtrsim$ be a binary relation on $A$, i.e., $\gtrsim$ is a subset of $A \times A$. The relational structure $\langle A, \gtrsim \rangle$ is a weak order iff,*[3] *for all $a, b, c \in A$, the following two axioms are satisfied:*

1. *Connectedness: Either $a \gtrsim b$ or $b \gtrsim a$.*
2. *Transitivity: If $a \gtrsim b$ and $b \gtrsim c$, then $a \gtrsim c$.*

Definition 1 is typical of our format: we single out and name a class of relational structures that satisfy a particular set of axioms. In this case, the name "weak order" is well established, although "pre-order" is sometimes used.

A weak order is always *reflexive* since Axiom 1 implies $a \gtrsim a$ for all $a$.

Note that the numerical relational structure $\langle \text{Re}, \geqslant \rangle$, where Re denotes the set of all real numbers and $\geqslant$ is the usual ordering (i.e., $x \geqslant y$ if and only if $x$ is greater than or equal to $y$), is a weak order. However, this weak order is special in that it is *antisymmetric*: if both $x \geqslant y$ and $y \geqslant x$, then $x = y$. Such an anti-symmetric weak order is called a *simple* or *total order*. In general, a weak order is distinguished from a simple order because it is possible that $a \gtrsim b$ and $b \gtrsim a$, for distinct elements $a, b$ of $A$. As we shall see, every weak order is associated in a natural way with a simple order. Most empirical ordering operations yield weak orders, in the sense that there exist distinct elements that are equivalent, i.e., $a \gtrsim b$ and $b \gtrsim a$ hold.

---

[3] In all formal definitions, theorems, and proofs, we use "iff" to stand for "if and only if."

THEOREM 1. *Suppose that A is a finite nonempty set. If $\langle A, \gtrsim \rangle$ is a weak order, then there exists a real-valued function $\phi$ on A such that for all $a, b \in A$,*

$$a \gtrsim b \quad \text{iff} \quad \phi(a) \geqslant \phi(b).$$

*Moreover, $\phi'$ is another real-valued function on A with the same property iff there is a strictly increasing function f, with domain and range equal to* Re, *such that for all $a \in A$*

$$\phi'(a) = f[\phi(a)],$$

*i.e., $\phi$ is an ordinal scale.*

Theorem 1 formulates the representation and uniqueness results for ordinal measurement in finite sets. It asserts that if a relational structure $\langle A, \gtrsim \rangle$ satisfies Axioms 1 and 2 of Definition 1, then there is a homomorphism $\phi$ into the numerical structure $\langle \text{Re}, \geqslant \rangle$, and the permissible transformations consist of all strictly increasing functions from Re onto Re. We shall usually state representation and uniqueness theorems in the format of Theorem 1; the appropriate numerical structure and the properties corresponding to a homomorphism (in this case, $\phi$ carries $\gtrsim$ into $\geqslant$) will be apparent.

The proof of Theorem 1 is well worth presenting in detail here; it introduces concepts and notations that are essential in dealing with weak orders throughout the book (Definition 2 below) and it illustrates the relation between axioms and measurement procedures. A proof of Theorem 1 should provide two things: a definite method for constructing the order-preserving function $\phi$, and a method for constructing the function $f$, given $\phi$ and $\phi'$. The method of constructing $\phi$ is precisely the measurement procedure. This will be true in all our representation theorems.

DEFINITION 2. *If $\gtrsim$ is a binary relation on A, two new relations $\sim$ and $\succ$ are defined on A as follows:*

$$a \sim b \quad \text{iff} \quad a \gtrsim b \text{ and } b \gtrsim a,$$
$$a \succ b \quad \text{iff} \quad a \gtrsim b \text{ and not } (b \gtrsim a).$$

*These are referred to as the* symmetric *and* asymmetric *parts of $\gtrsim$, respectively. If $\langle A, \gtrsim \rangle$ is a weak order, it is easy to prove (Exercise 4) that $\sim$ is an equivalence relation on A (i.e., it is reflexive, symmetric, and transitive) and that $\succ$ is transitive and asymmetric [i.e., if $a \succ b$, then not $(b \succ a)$]. The set*

$$\mathbf{a} = \{b \mid b \in A, \quad b \sim a\}$$

*is called the* equivalence class *determined by a. It is well known that* $\mathbf{a} \cap \mathbf{b}$
*is nonnull iff* $b \in \mathbf{a}$, *in which case* $\mathbf{a} = \mathbf{b}$ (*i.e.,* $c \sim a$ *iff* $c \sim b$); *hence, the
distinct equivalence classes form a* partition *of A* (*i.e., they form a family of
pairwise disjoint subsets whose union is A*). *The set of equivalence classes is
denoted* $A/\sim$. *The weak order* $\gtrsim$ *induces a new ordering relation* $\gtrsim$ *on* $A/\sim$, .

$$\mathbf{a} \gtrsim \mathbf{b} \qquad iff \qquad a \gtrsim b.$$

*It is easy to show that* $\gtrsim$ *is a simple order* (*Exercise* 4).

Definition 2 gives a brief review of several basic concepts used throughout
the book. If you are unfamiliar with these ideas, you should study some
elementary theory of sets and relations found in, e.g., Kershner and Wilcox
(1950) or Suppes (1957).

The concepts of Definition 2 are important for the following reason.
If $\phi$ has been constructed as specified in Theorem 1 and if $a \sim b$, then both
$\phi(a) \geqslant \phi(b)$ and $\phi(b) \geqslant \phi(a)$ must hold. By the antisymmetric property of
$\langle \text{Re}, \geqslant \rangle$, we have $\phi(a) = \phi(b)$. In short, two elements of $A$ that lie in the
same equivalence class must have the same scale value; clearly, the converse
is also true. The proof of Theorem 1 thus reduces to constructing scale
values that preserve the order between different equivalence classes. That is,
it suffices to construct a real-valued function $\phi$ on $\langle A/\sim, \gtrsim \rangle$, such that

$$\mathbf{a} \gtrsim \mathbf{b} \qquad iff \qquad \phi(\mathbf{a}) \geqslant \phi(\mathbf{b}).$$

We then obtain $\phi$ on $A$ by setting $\phi(a) = \phi(\mathbf{a})$.

In short, we have reduced the conditions of the representation theorem
to the case where the weak order is a simple order. The uniqueness theorem
also reduces to this case, because $\phi$ and $\phi$ have the same range in Re;
therefore, the function $f$ will be exactly the same: $\phi'(a) = f[\phi(a)]$ iff $\phi'(\mathbf{a}) =
f[\phi(\mathbf{a})]$.

We now complete the proof of the representation theorem. For each
$\mathbf{a} \in A/\sim$, let $\phi(\mathbf{a})$ be the number of distinct equivalence classes $\mathbf{b}$ such that
$\mathbf{a} \gtrsim \mathbf{b}$. (Note that this counting process assigns the number 1 to the lowest
equivalence class, 2 to the next lowest, etc.) If $\mathbf{a} \gtrsim \mathbf{b}$ then for every $\mathbf{c}$, if
$\mathbf{b} \gtrsim \mathbf{c}$, then $\mathbf{a} \gtrsim \mathbf{c}$ (transitivity), so if $\mathbf{c}$ is counted for $\phi(\mathbf{b})$, it is also counted
for $\phi(\mathbf{a})$. Thus $\phi(\mathbf{a}) \geqslant \phi(\mathbf{b})$. Conversely, if not $(\mathbf{a} \gtrsim \mathbf{b})$, then $\mathbf{b} \gtrsim \mathbf{a}$ (connectedness), and also $b \succ a$. Thus, there is at least one $\mathbf{c}$ (namely, $\mathbf{b}$) counted in
$\phi(\mathbf{b})$ but not in $\phi(\mathbf{a})$, and we have $\phi(\mathbf{b}) > \phi(\mathbf{a})$. This completes the proof of
the representation theorem.

The proof of the uniqueness theorem is easy and not particularly instruc- .
tive; so we omit it (but see Exercise 7).

Note that the procedure used to construct $\phi$—counting the number of equivalence classes below a given one—only works for finite sets. A different proof of the representation theorem can be given using the procedure outlined in Section 1.1.1 (see also Exercise 6) which applies to *countable* sets (those in one-to-one correspondence with the positive integers). See Theorem 2.1 for the details.

Finally, we remark that the axioms of transitivity and connectedness were essential in the above construction: their uses were noted in the course of the proof. Both of these qualitative laws have been challenged on empirical grounds in social-science applications. For example, consider a binary relation $\gtrsim$ that reflects an individual's preferences for various objects. He may decide $a > b$ and $b > c$ by concentrating on one dimension (say, quality) and ignoring small differences in another (say, price); but the price difference between $a$ and $c$ may be more salient, leading to $c \gtrsim a$. Similarly, there may be some pairs for which there is neither strict preference ($a > b$ or $b > a$) nor indifference ($a \sim b$). Thus, these laws can be nontrivial from an empirical standpoint.

## 1.3.2 Finite, Equally Spaced, Additive Conjoint Structures

In the simplest relational structure, $\langle A, \gtrsim \rangle$, considered above, we cannot count units because there is no way of identifying which element of $A$ is the "sum" of two others and, hence, no way of deciding what constitutes two units. In order to count units, the structure must have some additional features. One of the simplest possibilities is for the set $A$ to be a cartesian product, $A = A_1 \times A_2$. Empirically, this amounts simply to saying that two factors determine the ordering $\gtrsim$. A given object $a$ corresponds to a level $a_1$ of the $A_1$-factor and a level $a_2$ of the $A_2$-factor. Such objects are denoted $a = (a_1, a_2)$, $b = (b_1, b_2)$, etc.

As an example, let the $A_1$-factor be temperature, the $A_2$-factor humidity, and the relation $\gtrsim$ be discomfort. Thus, $(a_1, a_2) > (b_1, b_2)$ if temperature $a_1$ together with humidity $a_2$ is less comfortable than temperature $b_1$ together with humidity $b_2$. This example suggests an attribute such as discomfort induces an ordering on each component separately—in this case, the ordering by temperature and by humidity. Formally, we mean the ordering obtained when the value of the other component is held fixed. Thus, we have $\gtrsim_1$ on $A_1$ defined by

$$a_1 \gtrsim_1 b_1 \text{ iff } (a_1, c_2) \gtrsim (b_1, c_2) \text{ for all } c_2 \text{ in } A_2 ;$$

and $\gtrsim_2$ on $A_2$ defined by

$$a_2 \gtrsim_2 b_2 \text{ iff } (c_1, a_2) \gtrsim (c_1, b_2) \text{ for all } c_1 \text{ in } A_1 .$$

Observe that, at the moment, we cannot assert that either $\succsim_1$ or $\succsim_2$ is a weak order because we cannot be sure that they are connected; it may happen, for all we know, that for some $c_2$ in $A_2$ we have $(a_1, c_2) > (b_1, c_2)$ and for some $d_2$ in $A_2$ $(b_1, d_2) > (a_1, d_2)$. We suppose, however, in the following discussion that both $\succsim_1$ and $\succsim_2$ are weak orders.

With a product structure of this sort, the entities that can be concatenated are *intervals within one factor*. By an interval in $A_1$, we simply mean the formal entity denoted $a_1 b_1$, where $a_1$, $b_1$ are in $A_1$ and are called the "end points" of the interval. It will prove convenient to adhere to the convention that when $a_1 \succsim_1 b_1$ we write the interval as $a_1 b_1$, not as $b_1 a_1$. Since the intervals $a_1 b_1$ and $b_1 c_1$ are adjacent, $a_1 c_1$ can be regarded as their "sum." Two intervals $a_1 b_1$ and $c_1 d_1$ can be regarded as equal if they are matched by the *same* interval $a_2 b_2$ on the second factor, where by $a_1 b_1$ matching $a_2 b_2$ we simply mean that $(a_1, b_2) \sim (b_1, a_2)$. This assumes *additivity* in the effects of the two factors: if the sum of $a_1$ and $b_2$ effects equals the sum of $b_1$ and $a_2$ effects, then the difference between $a_1$ and $b_1$ effects must equal the difference between $a_2$ and $b_2$ effects. The method of forming equal intervals and concatenating them to obtain a standard sequence (Section 1.1.2) on the $A_1$-factor is illustrated in Figure 1. In short, the presence of a second

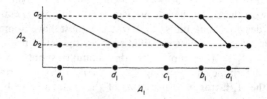

**FIGURE 1.** Use of an interval $a_2 b_2$ in $A_2$ to lay off equal adjacent intervals $a_1 b_1$, $b_1 c_1$ ,..., $d_1 e_1$ of a standard sequence on $A_1$. The points joined by straight lines are observed equivalences in the ordering $\succsim$ on $A_1 \times A_2$. Thus, $(a_1, b_2) \sim (b_1, a_2)$ implies $a_1 b_1$ matches $a_2 b_2$ ; ...; and $(d_1, b_2) \sim (e_1, a_2)$ implies $d_1 e_1$ matches $a_2 b_2$ .

factor together with the assumption of additivity of the effects of the two factors allows us both to calibrate equal units on the first factor and to combine adjacent equal units to form a standard sequence. This permits us to use a counting-of-units process formally the same as that of Section 1.1.2.

DEFINITION 3.   *Let $A_1$ and $A_2$ be nonempty sets and let $\succsim$ be a binary relation on $A = A_1 \times A_2$. The relational structure $\langle A_1 \times A_2, \succsim \rangle$ is called* *an* independent conjoint structure *iff, for all* $a, b, c \in A$ *and all* $a_i$, $b_i$, $c_i$, $d_i \in A_i$, $i = 1, 2$, *the following four axioms are satisfied*:

1. *Either* $a \gtrsim b$ *or* $b \gtrsim a$.
2. *If* $a \gtrsim b$ *and* $b \gtrsim c$, *then* $a \gtrsim c$.
3. *If* $(a_1, c_2) \gtrsim (b_1, c_2)$, *then* $(a_1, d_2) \gtrsim (b_1, d_2)$.
4. *If* $(c_1, a_2) \gtrsim (c_1, b_2)$, *then* $(d_1, a_2) \gtrsim (d_1, b_2)$.

Note that Axioms 1 and 2 can be combined into the statement that $\langle A, \gtrsim \rangle$ is a weak order.

DEFINITION 4. *Let* $\langle A_1 \times A_2, \gtrsim \rangle$ *be an independent conjoint structure. Define relations* $\gtrsim_1$ *on* $A_1$ *and* $\gtrsim_2$ *on* $A_2$ *by:*

$a_1 \gtrsim_1 b_1$ *iff there exists* $c_2$ *in* $A_2$ *with* $(a_1, c_2) \gtrsim (b_1, c_2)$;
$a_2 \gtrsim_2 b_2$ *iff there exists* $c_1$ *in* $A_1$ *with* $(c_1, a_2) \gtrsim (c_1, b_2)$.

It is easy to show (using Axioms 1–4) that $\langle A_i, \gtrsim_i \rangle$ is a weak order, $i = 1, 2$ (Exercise 11). Axiom 3 is precisely what is needed in order to prove that $\gtrsim_1$ on $A_1$ is a weak order, and Axiom 4 plays the parallel role in proving $\gtrsim_2$ on $A_2$ is a weak order. The term "independent" in Definition 3 refers to the fact that the induced order $\gtrsim_i$ on $A_i$ does not depend on the choice of $c_j$ or $d_j$ in the other factor $A_j$.

DEFINITION 5. *Define a relation* $J_i$ *on* $A_i$, $i = 1, 2$, *by* $a_i J_i b_i$ *iff, for all* $c_i \in A_i$, *exactly one of the following holds*: $c_i \gtrsim_i a_i$ *or* $b_i \gtrsim_i c_i$. *The structure* $\langle A_1 \times A_2, \gtrsim \rangle$ *is called an* equally spaced, additive conjoint structure, *if in addition to Axioms 1–4, the following axiom holds for all* $a_i$, $b_i \in A_i$, $i = 1, 2$:

5. *If* $a_1 J_1 b_1$ *and* $b_2 J_2 a_2$, *then* $(a_1, a_2) \sim (b_1, b_2)$.

The $J_i$ relation means that $a_i$ is strictly larger than $b_i$ (with respect to $\gtrsim_i$) and nothing lies between the two elements; thus, any $c_i$ is either $\gtrsim_i a_i$ or $\lesssim_i b_i$, but not both. Axiom 5 asserts that objects are equally spaced, meaning that any two $J$-intervals are equal in the intuitive sense of calibration of $A_1$-intervals against $A_2$-intervals as discussed above (Figure 1). The representation and uniqueness theorems are formulated as follows:

THEOREM 2. *Suppose that* $A_1$ *and* $A_2$ *are finite nonempty sets. If* $\langle A_1 \times A_2, \gtrsim \rangle$ *is an equally spaced, additive conjoint structure, then there exist real-valued functions* $\phi_i$ *on* $A_i$, $i = 1, 2$, *such that, for all* $a = (a_1, a_2)$, $b = (b_1, b_2) \in A$,

$$a \gtrsim b \quad iff \quad \phi_1(a_1) + \phi_2(a_2) \geq \phi_1(b_1) + \phi_2(b_2).$$

*Moreover, assuming that each $A_i/\sim_i$ contains at least two equivalence classes, then if $\phi_1'$, $\phi_2'$ are any other pair of real-valued functions with the above property there exist constants $\alpha$, $\beta_1$, $\beta_2$, with $\alpha > 0$, such that $\phi_i' = \alpha\phi_i + \beta_i$, $i = 1, 2$.*

The proof of this theorem is left as Exercises 12–14. All of the essential ideas for the proof were sketched above: the intervals in each factor form a finite standard sequence. Note that the representation theorem consists of a two-component vector homomorphism, namely $(\phi_1, \phi_2)$, between $\langle A_1 \times A_2, \gtrsim \rangle$ and $\langle \mathrm{Re} \times \mathrm{Re}, \gtrsim' \rangle$, where $\gtrsim'$ is defined on $\mathrm{Re} \times \mathrm{Re}$ by $(x, y) \gtrsim' (u, v)$ iff $x + y \geqslant u + v$. The uniqueness theorem asserts that the $\phi_i$ are interval scales with a common unit (constant $\alpha$) but independent zero points (constants $\beta_1$, $\beta_2$). This is not surprising. Since the counting-of-units process is applied to intervals on each factor, the ordinary ideas of extensive measurement lead one to expect invariance of ratios of intervals. But since only intervals in each factor are determined, the origins of the two scales are arbitrary.

In the previous section, we briefly discussed the empirical status of the weak order assumption (Axioms 1 and 2). Little needs to be added here except to note that the two factor aspect of the objects is more likely to lead to violations of transitivity. These will occur in any situation where attention is sometimes exclusively focused on one factor, sometimes on the other.

Axioms 3 and 4 are empirical laws of a very interesting type. We call them *independence laws*. Axiom 3 asserts that the ordering of $A_1$-effects is *independent of* the choice of a fixed level in $A_2$, which we abbreviate by saying that $A_1$ is independent of $A_2$. Axiom 4 asserts that $A_2$ is independent of $A_1$. Intuitively, independence is a qualitative, ordinal version of noninteraction between two variables. Of course, additivity (the conclusion of Theorem 2) asserts a quantitative noninteraction that is much stronger. Independence laws play a very prominent role in the discussion of additive and polynomial conjoint measurement (Chapters 6 and 7), in utility measurement (Chapter 8), and in multidimensional proximity measurement (Chapter 13).

The final axiom, 5, is of a very different nature. It is hardly ever satisfied by accident. If $A_1$ represents a finite set of levels of some factor and $A_2$ represents a different factor, there is no reason whatsoever to suppose that when we move from $(b_1, b_2)$ to the next higher level of $A_1$, say $(a_1, b_2)$, the effect is *exactly* the same as when we move to the next higher level of $A_2$, say $(b_1, a_2)$. What one might try to do, in practice, is to *select* subsets of levels of the two factors so as to satisfy this property, much as we select a standard sequence of weights and lengths. Thus, if we start by choosing high levels $a_1$ and $a_2$ and if we choose as the next highest level $b_1$ in $A_1$, then we are constrained to choose the next highest level $b_2$ in $A_2$ so that $(a_1, b_2) \sim (b_1, a_2)$. We are then forced to choose the next $A_1$ level $c_1$ to be such that $(c_1, a_2) \sim (b_1, b_2)$, since $b_1 J_1 c_1$ and $a_2 J_2 b_2$. Similarly, $c_2$ has to be chosen so that $(b_1, b_2) \sim (a_1, c_2)$. But now, with all degrees of freedom gone, we are forced to have $(b_1, c_2) \sim (c_1, b_2)$ which, empirically, could be false. Thus, it is not possible, in general, even if Axioms 1–4 are satisfied, to select elements satisfying Axiom 5. At the very least, the following law must hold (as was just shown):

5'.  If  $(a_1, b_2) \sim (b_1, a_2)$  and  $(c_1, a_2) \sim (b_1, b_2) \sim (a_1, c_2)$,  then  $(c_1, b_2) \sim (b_1, c_2)$.

If we continued the selection process, we would soon discover yet more laws of the same type that must hold in order to satisfy Axiom 5. So it is not satisfactory to propose as empirical laws either Axiom 5 or the simple statement that the sets $A_1$ and $A_2$ can be selected from some larger sets so that Axiom 5 holds. What is required for a satisfactory analysis is that one or more laws like 5', as simple as possible, be found that guarantee the internal consistency of constructed equally spaced sequences. This problem is solved in Chapter 6 in a surprisingly satisfying way.

As is probably obvious, the equal-spacing notion is identical to that of a standard sequence; both terminologies exist in the literature.

## 1.4  CHOOSING AN AXIOM SYSTEM

The following discussion of axiomatization neither exhibits the spirit of nor uses the highly developed technical apparatus of mathematical logic. Chapter 18 is devoted to such a formal treatment. Here we touch upon only a few of the best-known logical features of axiom systems (Section 1.4.5) and otherwise we describe in nontechnical terms some of the types of axioms typically found in measurement systems.

### 1.4.1  Necessary Axioms

The previous discussion suggests that much of the effort in analyzing measurement goes into finding a good axiom system. It should be clear by now that at least one axiom is required to construct a representation. Specifically, if $\gtrsim$ is an arbitrary binary relation on $A$, then there need not be any homomorphism of $\langle A, \gtrsim \rangle$ into $\langle \text{Re}, \geqslant \rangle$. Indeed, if we suppose that such a homomorphism $\phi$ exists, then it follows that $\gtrsim$ is transitive (for if $a \gtrsim b$ and $b \gtrsim c$, then $\phi(a) \geqslant \phi(b) \geqslant \phi(c)$; hence $\phi(a) \geqslant \phi(c)$, which implies $a \gtrsim c$). Thus $\gtrsim$ is not arbitrary at all. We express this by saying that transitivity is a *necessary* axiom. "Necessary" here means mathematical, not practical, necessity. An axiom is necessary if it is a consequence of the existence of the homomorphism which we are trying to establish. Reflexivity is also necessary in the sense of being a consequence of the representation of $\langle A, \gtrsim \rangle$ in $\langle \text{Re}, \geqslant \rangle$, but we did not need to assume it as an axiom to prove Theorem 1 because it follows from the other axioms. Connectedness is also a necessary axiom for Theorem 1, and it was needed in the proof of Theorem 1.

For the representation of additive conjoint structures (Theorem 2) Axioms 1–4 are all necessary. Axioms 1 and 2 are necessary for the same reason as in Theorem 1 since the mapping $\phi(a) = \phi_1(a_1) + \phi_2(a_2)$ is a homomorphism of $\langle A, \gtrsim \rangle$ into $\langle \text{Re}, \geqslant \rangle$. To show that Axiom 3 is necessary, suppose that $(a_1, c_2) \gtrsim (b_1, c_2)$ and that Theorem 2 holds. Then $\phi_1(a_1) + \phi_2(c_2) \geqslant \phi_1(b_1) + \phi_2(c_2)$. Adding $\phi_2(d_2) - \phi_2(c_2)$ to both sides of the above inequality in Re yields $\phi_1(a_1) + \phi_2(d_2) \geqslant \phi_1(b_1) + \phi_2(d_2)$, and this implies $(a_1, d_2) \gtrsim (b_1, d_2)$. Similarly, Axiom 4 is necessary.

In contrast, Axiom 5 is not necessary. This is shown by the following example. Let $A_1 = A_2 = \{0, 1, 2, 4\}$, and define $\gtrsim$ on $A = A_1 \times A_2$ by: $(a_1, a_2) \gtrsim (b_1, b_2)$ iff $a_1 + a_2 \geqslant b_1 + b_2$. It is obvious that the representation part of Theorem 2 holds with $\phi_1$ and $\phi_2$ just the identity functions. Slightly less obvious is the fact that the uniqueness part of Theorem 2 is true. But clearly it is true for $\{0, 1, 2\}$, since that part is equally spaced; and since for any $\phi_1'$, $\phi_2'$ providing a representation,

$$\phi_1'(0) + \phi_2'(4) = \phi_1'(2) + \phi_2'(2) = \phi_1'(4) + \phi_2'(0),$$

we have $\phi_1'(4)$ and $\phi_2'(4)$ as linear combinations of lower values, and this extends the uniqueness theorem to them. Obviously, Axiom 5 is not true in this structure.

Each of the axiom systems in this book contains several fairly simple necessary axioms. We usually present these axioms first, sometimes discussing their intuitive meanings and their roles as empirical laws. Almost always, the proof that they are necessary is very simple and either is given at once or is omitted altogether. We have no rule for selecting the right set of necessary axioms; in general, it is a matter of trial and error or of insight.

We do not try to keep the number of axioms used to a bare minimum. The number of axioms is a rather misleading quantity anyway, since they can always be reduced to one axiom by stringing them all together by conjunctions. More realistically, it is often possible to hide a rather complex property within apparently simple axioms. If that property can be seen to be wrong—i.e., contrary to empirical fact—then we see no point in burying it in an apparently more innocent system. If the proposed representation is wrong, then it needs to be altered. We try to state our axioms so that they are conceptually distinct, even at the cost of increasing their number.

### 1.4.2 Nonnecessary Axioms

Nonnecessary axioms are frequently referred to as *structural* because they limit the set of structures satisfying the axiom system to something less than the set determined by the representation theorem. Three main types of structural axioms occur. First, some demand that the system be nontrivial in one sense or another—that a certain set be nonempty, that there be at least two nonequivalent elements, etc. These do not really limit the applicability of the theory because the structures excluded are of no empirical interest. Second, we occasionally assume that certain sets are finite or countable. Both Theorems 1 and 2 above are of this character. The fact that in neither case did we list finiteness as a separate axiom, but included it instead as a hypothesis of the representation–uniqueness theorem, is purely a matter of style; in a fully formalized theory, all of the assumptions would be listed as axioms. This stylistic device is used occasionally throughout the book. Finiteness is a real limitation. In each case, however, we present alternative theorems which replace finiteness with other axioms (see Section 1.4.3).

Structural axioms of the third type assert that solutions exist to certain classes of equations or inequalities; these are known as *solvability* axioms. For example, in systems for length measurement, two different solvability axioms sometimes are used. One postulates that the set of rods $A$ is so "dense" that whenever $a > b$, then some $c$ in $A$ exists such that $a \gtrsim b \circ c$ (i.e., $c$ solves an inequality) or even that $a \sim b \circ c$ (i.e., $c$ solves an equation). It is easy to find examples of sets of rods where this does not hold, even though the representation and uniqueness theorems for extensive measurement of length are valid. Another type of solvability axiom asserts that the concatenation $a \circ b$ exists for, at least, certain pairs $a, b$ in $A$. This type of axiom is sometimes well concealed. For example, one of the primitive relations may be taken to be a binary operation, in which case by definition of an operation $a \circ b$ exists for every $a, b$. Nevertheless, in a formalization that makes explicit all existential assumptions, it would appear as follows: there is a primitive ternary relation, which can be written $a \circ b = c$, with the property that for every $a, b$ in $A$ exactly one $c$ exists in $A$ such that $a \circ b = c$ holds. In other cases, where $a \circ b$ is not defined for all $a$ and $b$, this kind of solvability is less well concealed. Nevertheless, it should always be considered as a specific assumption, for it plays much the same role in extensive measurement as does the other kind of solvability axiom in other measurement systems.

Axiom 5 for equally spaced, additive conjoint structures falls in none of the three classes, but in effect it is of the solvability type. If one were trying to *construct* a structure $\langle A_1 \times A_2, \gtrsim \rangle$ in which Axiom 5 held, one would proceed as indicated in Section 1.3.2, by choosing $a_1, a_2, b_1$, then selecting $b_2$ with $(a_1, b_2) \sim (b_1, a_2)$. To do so requires the assumption that $(a_1, b_2) \sim (b_1, a_2)$ can be solved for $b_2$. In Chapter 6, we replace Axiom 5 by solvability axioms of precisely that type together with necessary axioms similar to Axiom 5'.

We try to select structural properties that are as nonrestrictive as possible. Sometimes, alternative sets can be offered, covering different classes of structures. In particular, sometimes a trade can be effected in which the class of structures for which the representation is provable is enlarged (i.e., the structural axioms are weakened) at the expense of explicitly introducing additional necessary conditions which, in the presence of the former, stronger nonnecessary ones, had been deducible from the total axiom system. Such an exchange is deemed desirable when both an appreciable gain in applicability is effected and the added necessary conditions are neither too numerous nor too complex.

In many cases—especially in Chapters 3, 4, 5, 6, and 13, where we are dealing with additive representations of one kind or another—we have succeeded in limiting the structural properties to such an extent that we feel the remaining restrictions are of little practical import, i.e., they are quite likely to be acceptable empirically in many of the potential applications. In other cases—especially where nonlinear numerical structures are involved, as in Chapter 7—the structural restrictions are unsatisfactory, and much work needs to be done to weaken them. But even if strong structural axioms must be invoked, it is important to obtain axioms that are logically sufficient for the representation. If we only have necessary axioms, we remain unsure how to perform a thorough test of the representation.

### 1.4.3  Necessary and Sufficient Axiom Systems

The ubiquity of these nonnecessary restrictions may well seem puzzling. It seems far more desirable to find axiom systems composed entirely of necessary axioms that are also sufficient to prove the desired representation and uniqueness theorems. Such an axiom system is said to be *necessary and sufficient* for the representation. The advantage lies in the exclusion of examples such as on p. 22 where the additive conjoint representation holds, but Axiom 5 is violated.

As it happens, there are very few examples of what we consider satisfactory necessary and sufficient axiomatizations. Why is this? Loosely speaking, the reason is that the total set of structures admitting homomorphisms into a particular numerical structure is very heterogeneous and may include rather unusual and difficult-to-describe or pathological instances as well as more regular ones. Thus, the conditions which completely characterize such a set of structures are probably too complicated to be useful; in any event, they are not known. More systematic results, clarifying the above informal statement, are found in Chapter 9, Measurement Inequalities, and Chapter 18, Axiomatizability.

The requirement that the axiomatization be "satisfactory" is important (even though informal) because an unsatisfactory necessary and sufficient axiomatization always exists: take the representation and uniqueness theorems themselves as axioms. What criteria, then, do we impose on an axiomatization for it to be satisfactory? One demand is for the axioms to have a direct and easily understood meaning in terms of empirical operations, so simple that either they are evidently empirically true on intuitive grounds or it is evident how systematically to test them. In part, simplicity and clarity of meaning lie in the eye of the beholder. By the time you finish this book, some axioms may be clear which now might leave you aghast. Axiomatization is partly a search for simplicity and partly a restructuring of the axiomatizer's cognitive processes so that more things seem simple.

### 1.4.4 Archimedean Axioms

In addition to the types of axioms just described, a rather odd axiom is usually stated as part of each system. It is called *Archimedean* because it corresponds to the Archimedean property of real numbers: for any positive number $x$, no matter how small, and for any number $y$, no matter how large, there exists an integer $n$ such that $nx \geqslant y$. This simply means that any two positive numbers are comparable, i.e., their ratio is not infinite. Another way to say this, one which generalizes more readily to qualitative structures, is that the set of integers $n$ for which $y > nx$ is a finite set. For example, in extensive measurement, let $a$, $a \circ a = 2a$, $3a$,..., be a standard sequence. Then the Archimedean axiom says that for any $b$, the set of integers $n$ for which $b > na$ is finite. More generally, whenever we have defined a standard sequence, namely, entities having nonzero, equal spacing in the intended numerical representation, then we may always formulate the Archimedean property as: *every strictly bounded standard sequence is finite.*

It is evident that since the Archimedean property is true of the real numbers, it must also be true within the empirical relational system; it is a necessary axiom. What is surprising is that it is a needed axiom. In the few cases where the independence of axioms has been studied, the Archimedean axiom has been found to be independent of others; and no one seems to have suggested a more satisfactory substitute. It can be deleted if quite strong structural assumptions are made (see Section 6.11.1 and most of the systems in Pfanzagl, 1968), but with our relatively weak structural assumptions, we do not know how to eliminate it in favor of more desirable necessary axioms.

The objection to it as a necessary axiom is that either it is trivially true in a finite structure (that is why it was not stated in either Theorem 1 or 2) or it is unclear what constitutes empirical evidence against it since it may not be possible to exhibit an infinite standard sequence (see Section 1.5). Nonetheless, we can produce examples where it is violated, and these reveal something of the role it plays. Suppose that $\langle A_1 \times A_2, \succsim \rangle$ is a weak order in which any difference whatsoever on the first factor is decisive; the second factor matters only when a tie exists on the first. In such a case, any $A_1$-interval is infinitely large in comparison with any $A_2$-interval. If sufficiently large $A_2$ differences can compensate for small $A_1$ differences and vice versa, then the Archimedean axiom (for additive conjoint structures) seems reasonable. The difficulty in testing lies in deciding what evidence would be sufficient to conclude that a weak order on $A_1 \times A_2$ had the non-Archimedean character just described (see Section 1.5 and Chapters 17 and 18).

### 1.4.5  Consistency, Categoricalness, and Independence

All of the axiom systems we shall present have several nonisomorphic models in the real numbers. Therefore, the systems are consistent (i.e., something satisfies the axioms) and not categorical (i.e., two or more inherently different things satisfy them). The issue of independence of the axioms is more difficult. We have not knowingly included any axioms that are entirely derivable from the others in a system, but we may very well have done so inadvertently. In a few cases we establish independence formally (Chapters 3 and 6); but in many others, we are not sure that independence is met.

Axiom systems differ also in their general logical form and in the types of models they admit (cardinality, closure under submodels, etc.). There has been considerable work on such matters, including some recent results on the equivalence of different systems for finite models (Adams, Fagot, & Robinson, 1970); this work is discussed in Chapter 18.

### 1.5  EMPIRICAL TESTING OF A THEORY OF MEASUREMENT

Formal systems of measurement, although axiomatic, are not wholly or even largely evaluated on mathematical grounds. To be sure, we attempt to be explicit, precise, and consistent, and our proofs meet reasonable contemporary standards of rigor for informal set theory, but elegance and esthetics must give way, to a degree, to empirical criteria. The axioms purport to describe relations, perhaps idealized in some fashion, among certain potential observations, and adequacy of description is a more telling arbiter than beauty or simplicity. To carry out a satisfactory empirical evaluation of an axiom system is rather more difficult than it might first seem. Because some problems are very nearly universal, we sketch them briefly here.

### 1.5.1 Error of Measurement

The most pervasive problem is error—not human failure of one sort or another, but inherent features of the observational situation that cause us to fail to observe exactly what we wish to observe. As an example, suppose that we are judging qualitative weight by deflections of an equal-arm pan balance. When objects are placed in the two pans and do not cause a deflection of the arm from the horizontal, do we know that they have the same weight? In an operational sense relative to that balance, we do, but in some idealized sense we may doubt that we do. The contact between the knife edge and the arm exhibits some friction that makes the balance less than perfectly sensitive to what is placed in its pans. Moreover, the condition of the point of contact may vary over time as the result of movements and electrochemical effects, and so the amount of friction may fluctuate in some irregular way from observation to observation. Thus, we suspect that when two weights differ by an amount just at the edge of sensitivity of the balance, repeated observations may not yield the same results. Even when we avoid this boundary region of random error, we may still find evidence for systematic errors. For example, we may find that a sequence of weights has the property that each successive pair is judged equivalent in weight, but the first and last ones of the sequence are definitely not equivalent. Clearly then, the observed relation is not a weak order—in particular, the indifference relation $\sim$ is not transitive—and so the order cannot be represented in terms of $\geqslant$ in as simple a way as in Theorem 1. On the other hand, experience has shown that when we have such a sequence and when the observational conditions are improved by constructing a more sensitive balance, then at least one of the original equivalences is converted into a nonequivalence and, to a better approximation, the weak-order properties are satisfied. Of course, we can select a new set of objects that exhibits a refined version of the same phenomenon on the new balance. Nevertheless, the pattern of improved approximations is such that we elect to retain the assumption of an underlying weak ordering of weight and to say that, in any particular set of observations, there are systematic errors due to imperfections in the observational situation.

The existence of error also has implications for the construction of measurement scales. As we have indicated in Sections 1.1.2 and 1.3.2, such constructions most often involve a counting-of-units procedure based on some

appropriate definition of standard sequence. But any definition of standard
sequence involves finding exact replicas of a given object or interval. In the
presence of error, we are confronted with two choices. The first, which
often is the solution adopted, is to equate the replicas by a method that is
much more expensive, but much more precise, than the method that is
used to make comparisons in the field. For example, a good meter stick,
calibrated in millimeters, should have the property that, if a comparison
object falls clearly between the marks $x$ and $x + 1$ millimeters, then with
high probability its true length actually lies between $x$ and $x + 1$ millimeters.
This will happen if the procedure by which the millimeter steps are equated
yields a standard deviation not exceeding $10^{-3}$ mm or one micron.
The standard deviation of a sequence of $x$ steps, $x < 10^3$, will be about
$(10^3)^{1/2} \cdot 10^{-3} = 10^{-3/2}$ mm. Thus, an error of more than $10^{-1}$ mm will be
extremely improbable (three standard deviations); and so an object that is
clearly between the marks $x$ and $x + 1$, by visual comparison, is quite
likely to be actually between $x$ and $x + 1$ mm long. If an object falls
"right on" the $x$ mark, we will be quite uncertain whether its length is more
or less than $x$ mm, but with high probability it will lie in the interval
$(x - \frac{1}{2}, x + \frac{1}{2})$.

The second solution to the problem is to dispense with exact standard
sequences and to use only clear-cut observations of inequality to construct
approximate standard sequences. A good example is found in Section 4.4.4
on difference measurement. Some of the inequality observations may be
inferred rather than observed directly. For example, if $a$ is clearly greater
than $c$, but $b$ seems to be indifferent to $c$, we infer that $a$ is also greater
than $b$. In order that such inferences yield a weak order, the clear-cut
observations must satisfy the axioms of a semiorder (Chapter 15).

Obviously, a subtle interplay obtains among observations, theory, and
refined observations, in which the theory is both tested and used in a
normative fashion to define the existence and nature of error. It is probably
not possible at present to formulate generally the exact conditions that lead
us to attribute a discrepancy between theory and observation to error rather
than to an inadequacy in the theory. As explicit error theories are developed
to accompany measurement theories, it should become easier to make these
decisions more routine. Today, however, few error theories exist; what we
know about them is described in Chapters 15-17.

### 1.5.2 Selection of Objects in Tests of Axioms

A second ubiquitous problem of testing is that most theories are stated
for large, often infinite sets of elements, whereas empirical tests usually
involve small finite subsets. The general problem of inductive generalization

from limited data arises in testing all scientific theories, but there are some special problems connected with most measurement axiom systems.

First, some axioms may be easier to disconfirm than others. For example, transitivity can be disconfirmed if there is a single intransitive triple, $a \gtrsim b$, $b \gtrsim c$, $c > a$. Usually the empirical interpretation of $\gtrsim$ is such that it is possible to decide, for given $a$, $b$, whether $a \gtrsim b$, or at least, whether $a \gtrsim b$ is extremely probable. With such an empirical interpretation, transitivity can be unequivocally rejected or, at least, assigned a very low probability. On the other hand, the Archimedean axiom cannot be rejected merely because $b > a$, $b > 2a$,..., $b > ma$. There may be some $n > m$, for which $na \gtrsim b$. One may eventually consider it improbable that a large enough $n$ will ever be found and, thus, reject the Archimedean axiom—partly because it may seem not to make scientific sense to go on searching for $n$ large enough and partly because a more attractive and manageable theory may result from a non-Archimedean representation. Alternatively, there may be other empirical interpretations of $\gtrsim$, e.g., where underlying rules for generating $\gtrsim$ are directly observable, which enable us to reject the Archimedean axiom. One must keep in mind the fact that the refutability of axioms depends both on their mathematical form and on their empirical interpretation.

Nonnecessary axioms are usually not tested. If the elements are thought to exhibit some kind of fine grainedness, then one's belief in the existence of solutions to inequalities may be extremely strong. For example, if rod $b$ is longer than rod $c$, one assumes without much question that a small rod $d$ can be found such that $b \gtrsim c \circ d$. Frequently one also accepts, as an idealization, the "continuity" (mathematically, connectedness in the order topology) of the domain of objects, in which case solvability of equations is also accepted as a consequent idealization. However, if there is real doubt about the matter, solvability axioms can offer a difficulty similar to that sometimes encountered with Archimedean axioms: the mere fact that a solution has not yet been found may or may not lead one to believe that one never will be found, no matter what objects are tested.

Second, some axioms are more difficult to confirm than others. If we fail to disconfirm an axiom, we need to ask a question akin to that of the statistical power of the test: Did we select the elements in such a way that the data had some chance of showing the axiom wrong if, in fact, it is wrong? For example, if $a$ is much greater than $b$, and $b$ is much greater than $c$, it does not surprise us that $a$ is greater than $c$. A more convincing test of transitivity is obtained by selecting triples $a$, $b$, $c$ that are likely to violate it, if it is indeed false. Sometimes, an alternative theory can be a guide (see Tversky, 1969, for the use of an alternative theory to locate violations of transitivity). If we choose $a$ slightly greater than $b$ and $b$ slightly greater than $c$, we may be more gratified to find $a \succ c$; but then, there is more danger of falsely rejecting transitivity, as a result of errors of measurement.

Another manifestation of the problem of selection of objects in confirming axioms is the fact that, even though none of the necessary axioms of a system are disconfirmed by a given set of data, and even though the structural axioms of the system are assumed *a priori*, nevertheless, other necessary consequences of the representation, which were not needed to prove the representation theorem, may be disconfirmed by those same data. Because this seems almost contradictory, we amplify the point.

In most theories of measurement, several necessary axioms along with a few nonnecessary ones are shown to be sufficient for the numerical representation (homomorphism) to exist. From the representation, other necessary properties follow which, of course, do not need to be listed among the axioms, since they are deducible from them. Now, a particular set of data may not disconfirm any axiom of the theory; however, an axiom such as solvability may be false if attention is restricted just to that subset of objects tested: the solution to some inequality or equation may lie outside that subset. In fact, we may have accepted solvability to begin with because of the fine grainedness of the entire object set. Since the axiom system as a whole (including solvability) does not hold for the particular subset sampled, there is no mathematical reason why the unneeded necessary property must hold for that subset, even if none of the needed necessary axioms is disconfirmed for that subset. The disconfirmation of the necessary but unneeded property points, in fact, to an inadequate selection of objects for testing the necessary axioms. Since solvability presumably holds in the large structure, one of the tested necessary axioms would in fact be disconfirmed elsewhere, if suitable objects were selected. (For an example, see Section 9.1.)

These remarks are intended to point up the need for thoroughness in sampling objects before accepting any particular set of axioms as probably being satisfied. Another conclusion, which is tempting but overhasty, is that one should carry out all possible indirect tests of the axiom system by testing as many consequences of the system as possible in a given set of data. Since the ultimate consequence is the representation theorem itself, why not test whether a representation can be constructed for the sample at hand? There are two reasons why such a conclusion is unwarranted. The first has to do with fallibility of data. The more tests that are performed, the greater the chance that one of them will fail due to sampling error. This must be compensated for by relaxing the criterion for disconfirmation; but then, there may be an excellent chance of failing to reject the axiom system when it is in fact systematically wrong.

The existence of a numerical representation, for a fixed sample, generally corresponds to the existence of a solution to a large set of simultaneous inequalities (see Section 1.1.3 or Chapter 9). The sets of inequalities that arise in this way, in practice, rarely have solutions; but if one relaxes the criterion, accepting a "solution" that solves most of the inequalities, a "solution" may very well exist even when there is some systematic failure of one of the axioms.

The second danger in testing all necessary consequences of the representation by trying to construct the representation for a sample is that failure tends not to be instructive. Direct tests of particular axioms, on the other hand, are often very informative, since they can easily fail in a systematic way. Consider, for example, the independence axioms of additive conjoint measurement (Section 1.3.2). If both $(a_1, c_2) > (b_1, c_2)$ and $(a_1, d_2) < (b_1, d_2)$, we may be able to subclassify or order the $A_2$-factor so that the first inequality holds for values $c_2$ in certain classes or at one end of the dimension, and the reverse inequality holds in other classes or at the end of the dimension near $d_2$. This kind of systematic rejection greatly alleviates statistical problems in rejecting the axiom: we can be surer that the rejection is not due to sampling error if an alternative hypothesis is shown to fit the data very well. Moreover, such systematic rejection tells us a good deal about what is wrong, and it may suggest either other measurement schemes that will work or a different choice of basic factors.

Thus, one value of a satisfactory axiomatization is that it provides a set of relatively simple, conceptually distinct, empirically testable conditions to be tested. The problems of error and of selection of objects to be tested have no easy solutions, however; they must be tackled with whatever experimental and statistical tools are available.

## 1.6  ROLES OF THEORIES OF MEASUREMENT IN THE SCIENCES

As measurement surely plays an essential role in all science, one might anticipate great interest attaching to theories of measurement. This is not true, however, in much of contemporary physics. Some exists in applied physics—mechanics, thermodynamics, hydrodynamics, etc.—because the methods of dimensional analysis depend explicitly on properties of physical measures, and some also arises in connection with questions deep in the foundations of quantum theory and the theory of relativity. But for the most part, questions about physical measurement are regarded as being in the province of philosophy of physics, not in physics itself. Usually, the measurability of the variables of interest in physics is taken for granted and the actual measurements are reduced, via the elaborate superstructure of physical theory, to comparatively indirect observations. The construction and calibration of measuring devices is a major activity, but it lies rather far from the sorts of qualitative theories we examine here.

Other sciences, especially those having to do with human beings, approach measurement with considerably less confidence. In the behavioral and social sciences we are not entirely certain which variables can be measured nor which theories really apply to those we believe to be measurable; and we do not have a superstructure of well-established theory that can be used to devise practical schemes of measurement. For these reasons, the analysis of measurement and the construction of new systems of measurement have been an active preoccupation of some behavioral scientists. Included is some work that is highly sophisticated, and some that is remarkably naive.

A recurrent temptation when we need to measure an attribute of interest is to try to avoid the difficult theoretical and empirical issues posed by fundamental measurement by substituting some easily measured physical quantity that is believed to be strongly correlated with the attribute in question: hours of deprivation in lieu of hunger; skin resistance in lieu of anxiety; milliamperes of current in lieu of aversiveness, etc. Doubtless this is a sensible thing to do when no deep analysis is available, and in all likelihood some such indirect measures will one day serve very effectively when the basic attributes are well understood, but to treat them now as objective definitions of unanalyzed concepts is a form of misplaced operationalism.

Little seems possible in the way of a careful analysis of an attribute until means are devised to say which of two objects or events exhibits more of the attribute. Once we are able to order the objects in an acceptable way, we need to examine them for additional structure, for example, by selecting two or more factors that affect the ordering. Then begins the search for qualitative laws satisfied by the ordering and the additional structure. In contrast to fundamental physical measurement, which is typically one-dimensional (see, however, Chapter 10), many of the theories of measurement that appear applicable to behavioral problems are inherently multidimensional, and so the measurement theories deal simultaneously with several measures and the laws connecting them. These theories suggest new qualitative laws to be tested, and even when they are found to be wrong, much may be learned if the violations are systematic. Moreover, these theories lead to selection among the many factors that might be relevant by focusing attention on those variables that enter into simple qualitative laws.

The work on fundamental measurement representations, which is relatively recent in the behavioral sciences, contrasts with an older research field known as psychometrics and scaling theory. Most of the psychometric literature is based on numerical rather than qualitative relations (e.g., matrices of correlation coefficients, test profiles, choice probabilities), although there is a tradition, which has recently grown considerably, focusing on ordinal relations. This work aims to represent such relations by numerical relations, mostly of a geometric nature, that are more compact and more revealing than the input data. Among the unidimensional methods are Thurstonian scaling (Bock & Jones, 1968; Thurstone, 1959; Torgerson, 1958) and test theory (Lord and Novick, 1968); and among the multidimensional methods are the classic bilinear models of factor analysis (Harman, 1967; Spearman, 1927; Thurstone, 1947) and the ordinal procedures of Coombs (1964), Guttman (1944, 1968), and Shepard (1966). Most of these scaling procedures assume the validity of the proposed model and produce a best-fitting numerical representation of the data, whether or not the assumed model is really appropriate.

Here, by contrast, we are concerned almost exclusively with the qualitative conditions under which a particular representation holds. To some extent, therefore, theories of measurement may be regarded as complementary to the methods of scaling, with the former being concerned with empirical laws (axioms) that make a particular type of numerical representation appropriate and the latter with methods for finding a numerical representation of a particular type. This seeming complementarity is, however, somewhat illusory because the bulk of the scaling literature involves mapping one numerical structure into another one rather than a qualitative structure into a numerical one. For example, in scaling aptitude, intelligence, or social attitudes, test scores or numerical ratings are usually interpreted as measures of the attribute in question. But in the absence of a well-defined homomorphism between an empirical and a numerical relational structure, it is far from clear how to interpret such numbers. We return to this issue in Chapter 20.

The clearest complementarity exists with the ordinal scaling methods, which lately have become one of the main foci of scaling research (partly because the widespread availability of fast computers has made them practical). In fact, it was the earlier work on ordinal multidimensional scaling (e.g., Coombs, 1964; Shepard, 1966) that motivated the reworking of the foundations of geometry which is presented in Chapter 13. Such axiomatizations play the important role of showing how to test whether a particular scaling method is at all justified, and it invites the search for systematic departures from the axioms.

## 1.7  PLAN OF THE BOOK

Much of the book develops and demonstrates the theme stated in Sections 1.1 and 1.2 that, although many different empirical relational structures and many different axiom systems lead to measurement, the procedure for obtaining the numbers always reduces to one of three basic methods. Chapter 2 is the mathematical pivot which provides proofs, mainly using constructive methods, of a series of rigorous isomorphism theorems which amount to showing that, under suitable assumptions, the procedures outlined in Section 1.1 do give internally consistent numerical answers. In Chapters 3–9 the representation and uniqueness theorems are reduced to applications of the theorems in Chapter 2.

If you intend to skip proofs you need not read Chapter 2 in detail; however, you should go over Section 1.1 carefully to gain a good intuitive idea of how numerical scales are constructed. Then in reading a later chapter, you should try to see how the appropriate method of Section 1.1 is applied to the situation at hand. To understand this, you may find it useful to scan the statements of the theorems in Chapter 2.

Chapters 3–8 all depend on the counting-of-units procedure outlined in Section 1.1.2 and formulated more rigorously in Section 2.2. In Chapter 3, counting of units arises directly because the empirical relational structure contains a concatenation operation. This is extensive measurement. Some special variants arise in connection with problems of relativity and thermo-dynamics. The first half of the probability chapter (Chapter 5) may be considered as another variant of extensive measurement in which the union of disjoint events plays the role of concatenation. The latter half of Chapter 5 and Chapters 6 and 8 use the device discussed in Section 1.3.2, i.e., counting off equal units by laying equal intervals end to end, where equality of intervals is defined in terms of balancing by a single interval on another factor. Chapter 4 (Difference Measurement) studies this counting device in pure form; many results in later chapters reduce to those in Chapter 4. Chapter 6 deals with additive conjoint measurement, and Chapter 8 applies the results of Chapter 6 to expected-utility measurement. Still a third variant of the counting-of-units process is used for combined additive-multiplicative (polynomial) conjoint measurement studied in Chapter 7.

Some of the topics of Chapters 3–8 are reconsidered in Chapter 9 in terms of the solution-of-inequalities method (Sections 1.1.3 and 2.3) instead of the counting-of-units method. The results are primarily concerned with finite structures.

The final chapter of Volume I attempts to construct a bridge between fundamental measurement and dimensional analysis, and it includes formula-tions of the qualitative equivalents of numerical laws satisfied by funda-mentally measured variables.

In contrast to the relative unity of Volume I, the ten chapters of Volume II are more diverse. The first four deal with geometric representations. Since geometry is by far the earliest and most far-reaching example of measurement, we have included in Chapter 11 and 12 a general discussion of geometric structures and an overview of classical foundations of geometry, seen as measurement theory. A self-contained treatment of this classical theory is, of course, beyond the scope of this book. Chapter 13 presents a new approach to the foundations of geometry which is based on the ordering of distances as a primitive notion. The representation theorems rely on the theories of extensive, difference, and additive conjoint measurement developed in Chapters 3, 4, and 6, respectively, of Volume I. Chapter 14 deals with one of the best-developed measurement systems outside of the physical sciences, namely color measurement. The representation is geometric, but it is unlike any other in the book.

The next three chapters approach the problem of error of measurement in two very different ways. Chapters 15 and 16 deal with empirical relational structures in which error is incorporated directly and is dealt with axiomatically. Chapter 17 presents some statistical methods for testing theories of measurement of the kind discussed earlier in the book, where error is treated as an extraneous phenomenon.

Next we deal with two philosophical issues: the (logical) problem of axiomatizability (Chapter 18) and the relationship between uniqueness theorems and the meaningfulness of statements involving numerical measurements (Chapter 19).

Finally, Chapter 20 sums up the approach to measurement embodied in the rest of the book and compares it with other approaches.

## EXERCISES[4]

**1.** Show that the inequalities in Equation (2) imply $1 < x_3/x_2 < \frac{3}{2}$. (1.1.3)

**2.** Suppose that $P(x, y)$ denotes the proportion of times that $x$ is chosen over $y$ in a preference experiment. Define $x \gtrsim y$ iff $P(x, y) \geqslant \frac{1}{2}$. When does this yield a weak order (Definition 1)? (1.3.1)

**3.** Suppose that $\gtrsim$ is defined as in Exercise 2 and that $P(x, y)$ is given for all distinct pairs in the set $A = \{a, b, c, d\}$ by the values in the following matrix.

| $x$ \ $y$ | $a$ | $b$ | $c$ | $d$ |
|---|---|---|---|---|
| $a$ | — | .72 | .65 | .67 |
| $b$ | .28 | — | .39 | .32 |
| $c$ | .35 | .61 | — | .40 |
| $d$ | .33 | .68 | .60 | — |

---

[4] The directly relevant sections of Chapter 1 are listed in parentheses at the end of each exercise.

Verify that $\langle A, \gtrsim \rangle$ is a weak order (assuming $x \gtrsim x$ for all $x$). (1.3.1)

**4.** Let $\langle A, \gtrsim \rangle$ be a weak order. Show that the symmetric part $\sim$ (Definition 2) is an equivalence relation and that the asymmetric part $>$ is transitive and asymmetric. Show that $\gtrsim$ on $A/\sim$ is a simple order. (1.3.1)

**5.** Construct scale values $\phi(a)$, $\phi(b)$, $\phi(c)$, $\phi(d)$ successively, in that order, using the data in Exercise 3, by the method of Section 1.3.1.

**6.** Construct $\phi'(a)$, $\phi'(b)$, $\phi'(c)$, $\phi'(d)$ successively, in that order, using the data in Exercise 3, by the method sketched in Section 1.1.1.

**7.** Construct a strictly increasing real-valued function $h$ from Re onto Re such that $h[\phi(x)] = \phi'(x)$ for all $x$ in $A$, where $\phi$ is the scale of Exercise 5 and $\phi'$ is the scale of Exercise 6.

**8.** Suppose that the ordinal-scale values have been assigned to all elements of $A_1 \times A_2$, where $A_1 = \{a_1, b_1, c_1, d_1\}$, $A_2 = \{a_2, b_2, c_2, d_2\}$, as given in the following matrix.

| $A_1$ \ $A_2$ | $a_2$ | $b_2$ | $c_2$ | $d_2$ |
|---|---|---|---|---|
| $a_1$ | 5 | 1 | 13 | 29 |
| $b_1$ | 29 | 13 | 61 | 125 |
| $c_1$ | 61 | 29 | 125 | 253 |
| $d_1$ | 13 | 5 | 29 | 61 |

Verify that Axioms 3 and 4 of independent conjoint structures (Definition 3) are satisfied. Determine the weak orderings $\gtrsim_1$ and $\gtrsim_2$ on $A_1$ and $A_2$, respectively. (1.3.2)

**9.** Verify that Axiom 5 of Definition 5 holds for the matrix of Exercise 8. (1.3.2)

**10.** Construct functions $\phi_1$, $\phi_2$ on $A_1$, $A_2$ that satisfy the requirements of Theorem 2, for the data of Exercise 8. What is the relationship between the sums $\phi_1 + \phi_2$ and the numbers in the matrix? (1.3.2)

**11.** Let $\langle A_1 \times A_2, \gtrsim \rangle$ be an independent conjoint structure. Show that $\langle A_i, \gtrsim_i \rangle$, $i = 1, 2$, is a weak order (Definition 4). (1.3.2)

**12.** Suppose that $\langle A_1 \times A_2, \gtrsim \rangle$ is a finite, equally spaced, additive conjoint structure (Definition 5). Assume that the weak orders $\gtrsim_1$ and $\gtrsim_2$ are simple orders, and label the elements of $A_1$ as $a_1^{(i)}$, $i = 1,..., m$, with

$$a_1^{(m)} >_1 a_1^{(m-1)} >_1 \cdots >_1 a_1^{(1)}.$$

Similarly, let $\dot{A}_2 = \{a_2^{(j)} \mid j = 1,..., n\}$, with $a_2^{(j+1)} >_2 a_2^{(j)}$.

(a) Use Axiom 5, and mathematical induction, to prove that if $i + j = k + l$, then $(a_1^{(i)}, a_2^{(j)}) \sim (a_1^{(k)}, a_2^{(l)})$.

(b) Use the result of (a), plus Axioms 3 and 4, to show that if $i + j > k + l$, then $(a_1^{(i)}, a_2^{(j)}) > (a_1^{(k)}, a_2^{(l)})$.

(c) Show that $\phi_1(a_1^{(i)}) = i$, $\phi_2(a_2^{(j)}) = j$ satisfy the representation theorem (Theorem 2).     (1.3.2)

**13.** Extend the result of Exercise 12 to the case where $\succsim_1$, $\succsim_2$ need not be antisymmetric by using equivalence classes (Definition 2) with respect to $\sim_1$ and $\sim_2$.     (1.3.1, 1.3.2)

**14.** Prove the uniqueness theorem for finite, equally spaced, additive conjoint structures by showing that if $\phi_1'(a_1^{(1)}) = \sigma_1$, $\phi_2'(a_2^{(1)}) = \sigma_2$ and $\phi_1'(a_1^{(2)}) = \tau$, then

$$\phi_1' = (\tau - \sigma_1)(\phi_1 - 1) + \sigma_1,$$
$$\phi_2' = (\tau - \sigma_1)(\phi_2 - 1) + \sigma_2.$$

Do this by using the results of Exercise 12(a).     (1.3.2)

# Chapter 2  Construction of Numerical Functions

## 2.1  REAL-VALUED FUNCTIONS ON SIMPLY ORDERED SETS

In Chapter 1, we defined a *simple order* to consist of a set with a transitive, connected, antisymmetric binary relation. Theorem 1.1 established that any finite simple order can be represented by a finite set of real numbers together with their natural ordering. Such a representation is unique up to strictly increasing transformations of Re onto itself (ordinal scale). Here we prove similar representation and uniqueness theorems for certain infinite simple orders. The corresponding results for weak orders follow immediately from those for simple orders, factoring out equivalence classes of a weak order to obtain its associated simple order.

It is easy to show that not every simple order can be represented in $\langle \text{Re}, \geqslant \rangle$. Let $A = \text{Re} \times \text{Re}$, and define $\gtrsim$ on $A$ by

$$(x, y) \gtrsim (x', y')$$

if and only if either

  (i)  $x > x'$ or

  (ii)  $x = x'$ and $y \geqslant y'$.

Suppose that $\phi$ is an isomorphism of $\langle A, \gtrsim \rangle$ into $\langle \text{Re}, \geqslant \rangle$. For $x$ in Re, let $\phi_1(x) = \phi(x, 1)$ and let $\phi_0(x) = \phi(x, 0)$. Since $(x, 1) > (x, 0)$ and $\phi$ is order preserving, $\phi_1(x) > \phi_0(x)$. Since every open interval of Re contains a rational number, there is a rational $\psi(x)$ such that $\phi_1(x) > \psi(x) > \phi_0(x)$. Thus, $\psi$ is a function from Re into the set of rational numbers, denoted Ra. Moreover, $\psi$ is one to one, since if $x > x'$, $(x, 0) > (x', 1)$, so

$$\psi(x) > \phi_0(x) > \phi_1(x') > \psi(x').$$

However, it is well known that the rational numbers are countable (i.e., can be put into one-to-one correspondence with the set $I^+$ of positive integers) and the reals are not countable; therefore, a one-to-one mapping such as $\psi$ cannot exist, which implies that the isomorphism $\phi$ also cannot exist.

The previous example shows that some additional condition must be imposed on a simple order to obtain an isomorphism into $\langle \text{Re}, \geqslant \rangle$. We already know that the (structural) assumption of finiteness suffices. We next prove that it is also sufficient for the set $A$ to be countable.

THEOREM 1. *Let $\langle A, \succsim \rangle$ be a simple order. If $A$ is countable, then there exists a real-valued function $\phi$ on $A$ such that for all $a, b \in A$,*

$$a \succsim b \quad \text{iff} \quad \phi(a) \geqslant \phi(b).$$

Theorem 1 was proved by Cantor (1895).

Since Theorem 1 is merely a step toward a necessary and sufficient axiomatization, we do not formulate the corresponding uniqueness theorem at this point. The method of proof involves constructing $\phi$ precisely as was outlined in Section 1.1.1; after the values of $\phi$ have been found for any finite subset of $A$, we find the value for a new element of $A$ by locating the new element between its nearest neighbors in the finite set already considered and assigning a number between the numbers assigned to those neighbors. Since $A$ is countable, we construct the value of $\phi$ for the $n$th element of $A$ at the $n$th step; eventually, any given element of $A$ is reached. The function $\phi$ is thus considered to be defined over the whole of $A$, since given any particular element $a$ in $A$ we know precisely how to construct the number $\phi(a)$. (The function $\phi$ is said to be defined by *induction*.)

PROOF OF THEOREM 1. Let $a_n$ denote the element of $A$ that corresponds to $n$ in the given one-to-one correspondence between $A$ and $I^+$.
Define $\phi(a_1) = 0$. If $\phi(a_1),...,\phi(a_n)$ have been defined, where $n \geqslant 1$, define $\phi(a_{n+1})$ as follows:

(i) If $a_{n+1} > a_k$ for all $k$, $1 \leqslant k \leqslant n$, let $\phi(a_{n+1}) = n$.
(ii) If $a_{n+1} < a_k$ for all $k$, $1 \leqslant k \leqslant n$, let $\phi(a_{n+1}) = -n$.
(iii) If neither (i) nor (ii) applies, then there exist $i, j$, with $1 \leqslant i, j \leqslant n$,

such that $a_i > a_{n+1} > a_j$ and for any $k$, $1 \leqslant k \leqslant n$, either $a_k \gtrsim a_i$ or $a_j \gtrsim a_k$. Let $\phi(a_{n+1}) = \frac{1}{2}[\phi(a_i) + \phi(a_j)]$.

By induction, $\phi$ is defined on all of $A$. To prove that $\phi$ is order preserving, use mathematical induction: it is obviously order preserving on $\{a_1\}$, and if it is order preserving on $\{a_1, ..., a_n\}$, then, by the construction of $\phi(a_{n+1})$ just given, it is order preserving on $\{a_1, ..., a_{n+1}\}$. (See Exercise 1.)    $\Diamond$

There is a countable subset of the real numbers—the rationals—that is thoroughly interspersed in the reals in the sense that between every two reals there is a rational. If an uncountable simple order is to be represented in $\langle \text{Re}, \geqslant \rangle$, then one anticipates that it will exhibit the analogous property. In fact, the trouble in the example at the beginning of the chapter is precisely that there are uncountably many intervals of form $(x, 1) > (x, 0)$; therefore no countable set can be found such that it has a representative in every such interval. Conversely, if an uncountable set does have a countable subset thoroughly interspersed, then by Theorem 1 we can find a representation for that subset in $\langle \text{Re}, \geqslant \rangle$. We can then expect to extend the representation to the whole uncountable set by considering each element of the latter as a limit of elements in the countable subset. This suggests the feasibility of there being a simple necessary and sufficient condition for a simple order to be representable in $\langle \text{Re}, \geqslant \rangle$.

To make precise the meaning of "thoroughly interspersed" we introduce the following technical concept of *order dense*:

DEFINITION 1. *Let $\langle A, \gtrsim \rangle$ be a simple order and let $B$ be a subset of $A$. Then $B$ is order dense in $A$ iff for all $a, c \in A$ such that $a > c$, there exists $b \in B$ such that $a \gtrsim b \gtrsim c$.*

THEOREM 2. *If $\langle A, \gtrsim \rangle$ is a simple order, then the following two conditions are equivalent:*

(i)   *There is a finite or countable order-dense subset of $A$.*
(ii)  *There is an isomorphism of $\langle A, \gtrsim \rangle$ into $\langle \text{Re}, \geqslant \rangle$.*

This theorem seems first to have been stated in this generality by Birkhoff (1948, pp. 31–32), although a result almost as strong was proved by Cantor (1895). The proof sketched by Birkhoff is incomplete. Debreu (1954, Lemma II) proved that (i) implies (ii).

For the proof of this theorem, we remind the reader of three elementary facts about countable sets:

1. *Any infinite subset of a countable set is itself countable.*
2. *If $A$ is countable and $B$ is finite or countable, then $A \cup B$ is countable.*

3.  *The Cartesian product of finitely many countable sets is countable.*

(Since the rationals can be regarded as a subset of the Cartesian product $I \times I$, they are countable.)

If $\langle A, \gtrsim \rangle$ is a simple order and if $a, a' \in A$ are such that $a' > a$ and, for any $b \in A$, either $b \gtrsim a'$ or $a \gtrsim b$, then we say that $(a, a')$ is a *gap* and that $a, a'$ are *endpoints of a gap*. With this definition, we can formulate the following lemma.

LEMMA 1.  *Let $\langle A, \gtrsim \rangle$ be a simple order and let $A^*$ be the set of all endpoints of gaps. If either* (i) *or* (ii) *of Theorem 2 holds, then $A^*$ is either finite or countable.*

PROOF.  Let $A_1^*$ be the set of upper endpoints of gaps and $A_2^*$ be the set of lower endpoints.

Suppose that (i) holds, and let $B$ be a finite or countable, order-dense subset of $A$. By Definition 1, if $(a, a')$ is a gap, then either $a \in B$ or $a' \in B$. Hence, $A_1^* - B$ is in one-to-one correspondence with a subset of $B$ (each upper endpoint not in $B$ corresponds to its lower endpoint which is in $B$). Therefore, $A_1^* - B$ is finite or countable. Similarly, $A_2^* - B$ is finite or countable, and we know that $A^* \cap B$ is finite or countable. So $A^* = (A_1^* - B) \cup (A_2^* - B) \cup (A^* \cap B)$ is finite or countable.

Suppose that (ii) holds, and let $\phi$ be an isomorphism of $\langle A, \gtrsim \rangle$ into $\langle \mathrm{Re}, \geqslant \rangle$. If $(a, a')$ is a gap, then there exists a rational $\rho$ such that $\phi(a') > \rho > \phi(a)$. This leads to a one-to-one correspondence between $A_1^*$ and a subset of Ra and to one between $A_2^*$ and a subset of Ra, so $A^*$ is finite or countable.                                                                    $\diamondsuit$

PROOF OF THEOREM 2.  Suppose that $B$ is a finite or countable order-dense subset of $A$. Adjoin the greatest and least elements of $A$ (if such exist) to $B$. Let $A^*$ be the set of endpoints of gaps. By Lemma 1, $B^* = B \cup A^*$ is finite or countable. By either Theorem 1.1 or 1, there exists an order-preserving function $\phi'$ from $B^*$ to Re.

For $a \in A$, let $\phi(a)$ be the least upper bound of the set of numbers $\{\phi'(b) \mid b \in B^* \text{ and } a \gtrsim b\}$. For $a \in B^*$, obviously $\phi(a)$ exists and equals $\phi'(a)$. To show that $\phi(a)$ exists for $a \notin B^*$, note that there exist $a_1, a_2$ with $a_1 > a > a_2$. By order density, there exist $b_1, b_2 \in B$ such that

$$a_1 \gtrsim b_1 > a > b_2 \gtrsim a_2 .$$

The set $\{\phi'(b) \mid b \in B^* \text{ and } a \gtrsim b\}$ is nonempty [it contains $\phi'(b_2)$] and is bounded above [by $\phi'(b_1)$], so its least upper bound $\phi(a)$ exists.

We show that $\phi$ is order preserving. Suppose that $a' > a$. By the construction of $B^*$, there exist $b, b' \in B^*$ such that $a' \gtrsim b' > b \gtrsim a$. (If $a, a' \in B^*$, let $b = a$, $b' = a'$. If $a \notin B^*$, then by order density there exists $b' \in B$ with $a' \gtrsim b' \succ a$. Since $(a, b')$ is not a gap, there exists $a''$ with $b' \succ a'' \succ a$ and hence $b \in B$ with $a'' \gtrsim b \succ a$. A similar argument holds if $a' \notin B^*$.) Thus, $\phi(a') \geqslant \phi'(b') > \phi'(b) \geqslant \phi(a)$, as required.

Conversely, suppose that $\phi$ is an isomorphism of $\langle A, \gtrsim \rangle$ into $\langle \text{Re}, \geqslant \rangle$. Let $J$ be the set of pairs of rational numbers $(r, r')$ such that for some $a \in A$, $r' > \phi(a) > r$. For each $(r, r') \in J$, choose exactly one $a \in A$ such that $r' > \phi(a) > r$, and let $B_1$ be the set of elements so chosen.[1] Since $B_1$ is in one-to-one correspondence with a subset of Ra $\times$ Ra, $B_1$ is finite or countable. Let $A^*$ be the set of endpoints of gaps. By Lemma 1, $B = A^* \cup B_1$ is finite or countable.

We show that $B$ is order dense in $A$. Suppose that $a \succ c$. If $(c, a)$ is a gap, then $c$, $a \in A^*$ and we can take either of them as $b$. Otherwise, choose $b'$ with $a \succ b' \succ c$ and rationals $r$, $r'$ with $\phi(a) > r' > \phi(b') > r > \phi(c)$. Thus, $(r, r') \in J$, so for some $b \in B_1$, $r' > \phi(b) > r$. It follows that $a \succ b \succ c$, as required.          $\Diamond$

We next turn to the uniqueness theorem for ordinal measurement, which is very simple.

THEOREM 3.    *Let $\langle A, \gtrsim \rangle$ be a simple order and let $\phi$, $\phi'$ be two functions satisfying* (ii) *of Theorem 2. Let $R$, $R'$ be the respective ranges of $\phi$, $\phi'$. Then there exists a strictly increasing function $h$ from $R$ to $R'$, such that for all $a \in A$, $h[\phi(a)] = \phi'(a)$. Moreover, if $h$ is any strictly increasing function on the range $R$ of a representation $\phi$ of $\langle A, \gtrsim \rangle$, then $\phi'(a) = h[\phi(a)]$ defines another such representation.*

The proof is trivial, e.g., define $h$ on $R$ by $h[\phi(a)] = \phi'(a)$, and then show $h$ is well defined and strictly increasing.

Note that the class of permissible transformations includes all strictly increasing functions from Re onto Re, but includes others as well. For example, there are some strictly increasing functions on subsets of Re that cannot be extended to strictly increasing functions on Re. For finite sets $A$, any permissible transformation on $R$ can be extended to one from Re onto Re.

In the proof of Theorem 2, the definition of $\phi$ on $A$ (i.e., as the least upper bound of $\phi'(b)$, for $b \in B^*$ and $a \gtrsim b$) differs in an important way from the definition of $\phi'$ on $B^*$ (or of $\phi$ on a countable set, as in Theorem 1). Since $B^*$ is finite or countable, there is a method of obtaining the value of $\phi'$ for any given element of $B^*$ in a finite number of steps. But in order to

---

[1] Note the use of the Axiom of Choice. We try to point out its use, or the use of equivalents, throughout the book; we avoid using it when we know how.

obtain the value of $\phi$ for an element of $A - B^*$, one must first construct $\phi'$ for *every* element of $B^*$, which in the countable case requires infinitely many steps. In Section 2.2, we also use a limiting process to define $\phi$, but there we have the possibility of obtaining an approximation to the value of $\phi$, with any specified degree of accuracy, in a finite number of steps. The ratio-scale uniqueness theorem of Section 2.2 leads to a well-defined notion of accuracy of approximation (percentage error). With only the ordinal uniqueness of Theorem 3, however, there is no suitable notion of approxima-tion, and so the definition of $\phi$ is not very satisfactory.

Finally, we note that Debreu (1954, Lemma II) has shown that the order-preserving function $\phi$ of Theorem 2 can always be constructed so as to be continuous in every natural topology on the simple order $\langle A, \succsim \rangle$. This is easy to prove once all the required definitions are given, but we shall not do so since the ideas involved are not used elsewhere in this book. The essence of the matter is simply this. The function $\phi$ will be discontinuous if there are gaps in its range, so that, for instance, $a \in A$ is the least upper bound of $A_1 \subset A$, with respect to $\succsim$, but $\phi(a)$ is greater than the least upper bound of $\{\phi(a') \mid a' \in A_1\}$. Such gaps can be closed simply by modifying $\phi$ to close them; for example, if there were only one gap at $a$, one would simply subtract the appropriate constant from all values of $\phi(a'')$ for $a'' \succsim a$. Since the set of gaps can only be finite or countable, one can easily arrange to close them all at once by subtracting appropriate sums (with a finite or countable number of terms) of constants.

Because of the relatively nonunique measurement for simple orders, we make little use of the results of this section in the remainder of the book. Theorem 2 is used in Chapters 7 and 13.

## 2.2 ADDITIVE FUNCTIONS ON ORDERED ALGEBRAIC STRUCTURES

We next formulate and prove several extensions of a classical theorem of Hölder (1901). The first four subsections deal with the main isomorphism theorem, Theorem 4, due to Krantz (1968), which underlies the construction of real-valued functions throughout the book. We use the weakest structural conditions now known; the cost is some slight complications in the statement of the theorem. For example, we assume that the binary operation $\circ$ is defined only for a certain set $B$ of pairs $(a, b)$. Intuitively, $B$ should be thought of as the set of concatenable pairs. We do not assume that any equation can be solved, but only that if $a \succ b$ then for some $c, a \succsim b \circ c$.

Sections 2.2.5 and 2.2.7 deal with the special cases of Archimedean simply ordered groups and rings, respectively; Section 2.2.6 presents a brief comparison with other versions of Hölder's Theorem. A much more detailed treatment of the type of ordered algebraic structures considered here can be found in Fuchs (1963) and in a later survey by Vinogradov (1969).

### 2.2.1  Archimedean Ordered Semigroups

We start with the basic definition.

DEFINITION 2.  *Let $A$ be a nonempty set, $B$ and $\gtrsim$ nontrivial binary relations on $A$, and $\circ$ a binary operation from $B$ into $A$. The quadruple $\langle A, \gtrsim, B, \circ \rangle$ is an ordered local semigroup iff, for all $a$, $b$, $c$, $d \in A$, the following five axioms are satisfied:*

1.  $\langle A, \gtrsim \rangle$ *is a simple order.*

2.  *If $(a, b) \in B$, $a \gtrsim c$, and $b \gtrsim d$, then $(c, d) \in B$.*

3.  *If $(c, a) \in B$ and $a \gtrsim b$, then $c \circ a \gtrsim c \circ b$.*

4.  *If $(a, c) \in B$ and $a \gtrsim b$, then $a \circ c \gtrsim b \circ c$.*

5.  *$(a, b) \in B$ and $(a \circ b, c) \in B$ iff $(b, c) \in B$ and $(a, b \circ c) \in B$; and when both conditions hold $(a \circ b) \circ c = a \circ (b \circ c)$.*

*For the rest of this definition, assume that $\langle A, \gtrsim, B, \circ \rangle$ is an ordered local semigroup (Axioms 1–5 hold).*
*$\langle A, \gtrsim, B, \circ \rangle$ is called* positive *iff, for all $a, b \in A$,*

6.  *If $(a, b) \in B$, then $a \circ b \succ a$.*

*A positive semigroup $\langle A, \gtrsim, B, \circ \rangle$ is called* regular *iff, for all $a, b \in A$,*

7.  *If $a \succ b$, then there exists $c \in A$ such that $(b, c) \in B$ and $a \gtrsim b \circ c$.*

*For any $a \in A$, we define inductively a subset $N_a$ of $I^+$, and we define $na$ for each $n \in N_a$ by:*

(i)  $1 \in N_a$ *and* $1a = a$;

(ii)  *if $n - 1 \in N_a$ and $((n - 1) a, a) \in B$, then $n \in N_a$ and $na$ is defined to be $((n - 1) a) \circ a$;*

(iii)  *if $n - 1 \in N_a$ and $((n - 1) a, a) \notin B$, then for all $m \geqslant n$, $m \notin N_a$.*

*(Thus, $N_a$ is precisely the set of consecutive positive integers for which $na$ is defined.)*
*$\langle A, \gtrsim, B, \circ \rangle$ is called* Archimedean *iff for all $a, b \in A$:*

8.  $\{ n \mid n \in N_a$ *and* $b \succ na \}$ *is a finite set.*

An example of a structure satisfying Axioms 1–8 is provided by

$\langle \text{Re}^+, \geqslant, R_\Omega, + \rangle$, where $\geqslant$ is the usual order relation and $+$ the usual addition operation on $\text{Re}^+$, and

$$R_\Omega = \{(x, y) \mid x, y \in \text{Re}^+ \quad \text{and} \quad x + y < \Omega\},$$

where $0 < \Omega \leqslant +\infty$. That is, $+$ is defined for all pairs of positive reals whose usual sum is less than $\Omega$. By contrast, we do not obtain an Archimedean, regular, positive, ordered local semigroup if we define $+$ for just those pairs satisfying $x^2 + y^2 < \Omega$. In such a case, Axiom 5 is violated. For example, let $\Omega = 1$, $x = y = 0.2$, and $z = 0.9$. Then $x^2 + y^2 = 0.08$, so $(x, y)$ is in $R_\Omega$; and $(x + y)^2 + z^2 = 0.97$, so $(x + y, z)$ is again in $R_\Omega$; but $x^2 + (y + z)^2 = 1.25$, so $(x, y + z)$ is not in $R_\Omega$, violating part of the conclusion to Axiom 5. This example shows clearly that Axiom 5 has a strong structural significance.

The other structural axioms are 2 and 7, whereas 1, 3, 4, 6, 8, and part of 5 $[(a \circ b) \circ c = a \circ (b \circ c)]$ are deducible from the representation (Theorem 4 below).

We remark also that the axiom system used here has two solvability conditions. One is regularity (Axiom 7), which requires solvability of the inequality $a \gtrsim b \circ c$, for $c$, given $a > b$. The other solvability condition is Axiom 2, which requires solvability of the equation, $c \circ d = e$, for $e$, whenever $a \gtrsim c$, $b \gtrsim d$, and $(a, b) \in B$. [The assertion $(c, d) \in B$ is equivalent to the assertion that the required $e$ exists.]

We state two versions of the representation and uniqueness theorem. Theorem 4 gives the essential existence and uniqueness statements. A slight modification, Theorem 4', asserts that any structure $\langle A, \gtrsim, B, \circ \rangle$ satisfying Axioms 1–8 can be mapped isomorphically into a structure $\langle \text{Re}^+, \geqslant, R_\Omega, + \rangle$, where $R_\Omega$ is defined as above.

THEOREM 4.     Let $\langle A, \gtrsim, B, \circ \rangle$ be an Archimedean, regular, positive, ordered local semigroup (Axioms 1–8 of Definition 2 all hold). Then there is a function $\phi$ from $A$ to $\text{Re}^+$ such that for all $a, b \in A$,

(i)     $a \gtrsim b$ iff $\phi(a) \geqslant \phi(b)$;

(ii)    if $(a, b) \in B$, then $\phi(a \circ b) = \phi(a) + \phi(b)$.

Moreover, if $\phi$ and $\phi'$ are any two functions from $A$ to $\text{Re}^+$ satisfying conditions (i) and (ii), then there exists $\alpha > 0$ such that for any nonmaximal $a \in A$,

$$\phi'(a) = \alpha\phi(a).$$

The isomorphism version runs as follows:

THEOREM 4'.     Let the hypotheses of Theorem 4 hold and let $\phi$ be a function from $A$ to $\text{Re}^+$ satisfying (i) and (ii) of that theorem. Let $\Omega$ be the least upper bound of $\{\phi(a) \mid a \in A\}$. Let $A'$ be the set of nonmaximal elements of $A$ and let $B' = \{(a, b) \in B \mid a \circ b \in A'\}$. Then $\phi$ is an isomorphism of $\langle A', \gtrsim, B', \circ \rangle$ into $\langle \text{Re}^+, \geqslant, R_\Omega, + \rangle$.

The point of Theorem 4' is that $\phi$ not only carries $\gtrsim$ into $\geqslant$ and $\circ$ into $+$, but also carries $B'$ into $R_\Omega$, in the sense that

$$(a, b) \in B' \qquad \text{iff} \qquad (\phi(a), \phi(b)) \in R_\Omega.$$

If there is no maximal element in $A$, then $A' = A$ and $B' = B$. If there is a maximal element of form $a \circ b$, then the uniqueness statement of Theorem 4 and the isomorphism statement in Theorem 4' can easily be extended to it. The only case where the restriction to nonmaximal elements matters is when the maximal element is not of the form $a \circ b$ for any $(a, b) \in B$.

### 2.2.2  Proof of Theorem 4 (Outline)

Because of the importance of this theorem and because the construction of the function $\phi$ embodies the basic processes of additive measurement, we first give a detailed sketch of the proof and then a full proof.

To approximate $\phi(b)/\phi(c)$, we take a small $a$ and see how many copies of $a$ are required to approximate $b$ and how many to approximate $c$. If $ma \approx b$ and $na \approx c$ ($\approx$ means approximates), then $m/n \approx \phi(b)/\phi(c)$. This idea is justified by the desired properties of $\phi$ [(i) and (ii) of the theorem], since $ma \approx b$ should imply $m\phi(a) \approx \phi(b)$. Similarly, $n\phi(a) \approx \phi(c)$, so $m/n \approx \phi(b)/\phi(c)$. Thus, the first step of the proof is to define, for any $a, b$ such that $b \gtrsim a$, an integer $N(a, b)$ such that $N(a, b) a \approx b$. Specifically, $N(a, b)$ is the largest integer such that $[N(a, b) - 1]a$ is defined and $[N(a, b) - 1] a < b$. The Archimedean property guarantees that such a largest integer exists.

Two cases must now be distinguished. If there is a smallest element $a$ in $A$, then for every $b$, $N(a, b) a = b$. For if this were not true, we could use regularity (Axiom 7) to construct $a'$ such that $b \gtrsim [N(a, b) - 1] a \circ a'$, and it would follow that $a' < a$, contrary to minimality of $a$. In this case, $\phi(b) = N(a, b)$ gives the required isomorphism into $\langle \text{Re}^+, \geqslant, R_\Omega, + \rangle$.

In the second case, there is no least element of $A$, and the proof consists of showing that as $a$ is taken smaller and smaller, $N(a, b)/N(a, c)$ converges, for every $b, c$, to a limit in $\text{Re}^+$. This limit is defined to be $\phi(b)/\phi(c)$. The limit exists because, for $a'$ much smaller than $a$, $N(a', b) \approx N(a', a) N(a, b)$. This is intuitively obvious. If $N(a', a)$ copies of $a'$ approximate $a$, and $N(a, b)$ copies of $a$ approximate $b$, then $N(a', a) N(a, b)$ copies of $a'$ approximate $b$. Thus, $N(a', b)/N(a', c) \approx N(a, b)/N(a, c)$, since the common factor $N(a', a)$ drops out when the approximation of $b$ is divided by that of $c$. For example, if the gradation is changed from feet ($a$) to millimeters ($a'$), then all approximate measurements will be multiplied approximately by $N(a', a)$, the number of millimeters per foot, and ratios remain approximately the same.

Another important feature of the proof is the method for taking $a'$ sufficiently smaller than $a$. The trick is simple: take any $a' < a$, and then take $a''$ with $a \gtrsim a' \circ a''$ (regularity). The smaller of $a'$ and $a''$ is then less than "half" of $a$.

Once $\phi$ is constructed, the rest of the proof [that properties (i) and (ii) and uniqueness hold] is easy. Additivity of $\phi$, for example, follows using Axioms 3 and 4 to show that if $N(a, b) a \approx b$ and $N(a, c) a \approx c$, then $[N(a, b) + N(a, c)] a \approx b \circ c$. Ordering follows, using additivity. For if $b > c$, then for some $a$, $b \gtrsim c \circ a$. By construction of $\phi$, $\phi(b) \geqslant \phi(c) + \phi(a)$, where $\phi(a) > 0$; thus, $\phi(b) > \phi(c)$.

Note the role played by each axiom of Definition 2. Axioms 2 and 5 characterize the local semigroup: the operation $\circ$ is defined and associative for all sufficiently small $a$, $b$. We can thus generate $na$ from $a$, being unconcerned about the way that copies of $a$ are associated in forming $na$. Axiom 6 guarantees that the standard sequence $a, 2a, \ldots$ increases steadily, while Axiom 8 guarantees that eventually, such a sequence will approximate any $b$. The "increasing" and "approximate" notions are based on the ordering (Axiom 1). Regularity (Axiom 7) plays several roles, as was pointed out in the above sketch. Finally, Axioms 3 and 4 play an important role in additivity.

Note that the axioms do not explicitly assume commutativity, $a \circ b = b \circ a$. Since this property is implied by Theorem 4, it must follow from the axioms, but it is not used in the course of the proof.

### 2.2.3 Preliminary Lemmas

In this section we collect a group of simple results concerning ordered local semigroups. The hypothesis common to all the lemmas is that $\langle A, \gtrsim, B, \circ \rangle$ is an ordered local semigroup (Axioms 1–5 of Definition 2). Also, $N_a$ and $na$ are defined as in Definition 2.

LEMMA 2.    $m, n \in N_a$ and $(ma, na) \in B$ iff $m + n \in N_a$; and when both conditions hold $(ma) \circ (na) = (m + n)a$.

PROOF.    Exercise 3.                                                                        ◇

LEMMA 3.    If $a \gtrsim a'$, $b \gtrsim b'$, and $(a, b) \in B$, then $a \circ b \gtrsim a' \circ b'$.

PROOF.    Axioms 1–4.                                                                    ◇

LEMMA 4.    *If $a \gtrsim b$ and $n \in N_a$, then $n \in N_b$ and $na \gtrsim nb$.*

PROOF.    From Lemma 3, by induction.                    ◇

LEMMA 5.    *Let Axiom 6 hold. If $m, n \in N_a$, then $ma \gtrsim na$ iff $m \geqslant n$.*

PROOF.    Obvious.                                        ◇

LEMMA 6.    *If $m \in N_a$, then $n \in N_{ma}$ iff $nm \in N_a$; in which case $n(ma) = (nm)a$.*

PROOF.    The result is trivial for $n = 1$, $m \in I^+$. Suppose it holds for some $n \geqslant 1$, for all $m \in I^+$. If $n + 1 \in N_{ma}$, then

$$(n + 1)(ma) = [n(ma)] \circ (ma) \qquad \text{(Definition 2)}$$
$$= [(nm)a] \circ (ma) \qquad \text{(inductive hypothesis)}.$$

By Lemma 2, $nm + m = (n + 1) m \in N_a$, and $[(nm) a] \circ (ma) = [(n + 1) m]a$. Therefore, if $n + 1 \in N_{ma}$, then $(n + 1) m \in N_a$ and $(n + 1)(ma) = [(n + 1) m]a$.

Conversely, if $(n + 1) m \in N_a$, then by applying Lemma 2 we have $((nm)a, ma) \in B$. By the inductive hypothesis, $n \in N_{ma}$ and $(nm) a = n(ma)$. Hence, $(n(ma), ma) \in B$. It follows from Definition 2 that $n + 1 \in N_{ma}$, as required.

By induction, the result holds for all $n$.                    ◇

### 2.2.4  Proof of Theorems 4 and 4′ (Details)

If $b > a$, then by Axiom 8, there exists a largest positive integer, denoted $N(a, b)$, such that $N(a, b) - 1 \in N_a$ and $b > (N(a, b) - 1)a$. We can let $N(a, a) = 1$ by definition.

First, suppose that there exists a minimal element $a \in A$. We show that for every $b \in A$, $N(a, b) \in N_a$ and $N(a, b) a = b$. By Axiom 7, for any $b > a$ there exists $c$ such that $((N(a, b) - 1) a, c) \in B$ and $b \gtrsim [(N(a, b) - 1) a] \circ c$. By minimality of $a$, $c \gtrsim a$. By Axioms 1–3, we know that $N(a, b) \in N_a$ and $b \gtrsim N(a, b)a$. But then by maximality of $N(a, b)$, we cannot have $b > N(a, b)a$, hence, $b = N(a, b)a$. The same formula follows trivially if $b = a$.

Let $\phi(b) = N(a, b)$. We show that $\phi$ satisfies (i) and (ii) of Theorem 4. To prove (i), note that by Lemma 5, $N(a, b)a \gtrsim N(a, c)a$ iff $N(a, b) \geqslant N(a, c)$. Hence, $b \gtrsim c$ iff $\phi(b) \geqslant \phi(c)$. For (ii), note that

$$N(a, b \circ c) a = b \circ c$$
$$= [N(a, b) a] \circ [N(a, c) a]$$
$$= [N(a, b) + N(a, c)]a \qquad \text{(Lemma 2)}.$$

By Lemma 5, $N(a, b \circ c) = N(a, b) + N(a, c)$, or $\phi(b \circ c) = \phi(b) + \phi(c)$. Finally, if $\phi'$ is any other function satisfying (i) and (ii), then for any $b$ in $A$,

$$\begin{aligned}\phi'(b) &= \phi'(N(a, b)\, a)\\ &= N(a, b)\, \phi'(a) \qquad \text{[property (ii)]}\\ &= \alpha\phi(b),\end{aligned}$$

where $\alpha = \phi'(a) > 0$. This completes the proof of Theorem 4 for the case where a minimal element exists.

Next, suppose that $A$ has no minimal element. We shall construct a sequence $a_1, ..., a_m, ...$ in $A$ which converges to zero in the sense that for every $b \in A$, $N(a_m, b)$ is defined for sufficiently large $m$ and diverges to $+\infty$. We then show that for any such sequence, $\lim_{m\to\infty} N(a_m, b)/N(a_m, c)$ exists in $\mathrm{Re}^+$ for every $b, c \in A$. The limit is obviously independent of the sequence $a_m$, since any two sequences can be interleaved to form a third sequence with the same properties.

Let $a_1$ be arbitrary, and define the sequence inductively as follows. If $a_m$ has been defined, choose $a_m' < a_m$. By Axiom 7, choose $a_m''$ such that $(a_m', a_m'') \in B$ and $a_m \gtrsim a_m' \circ a_m''$. Define $a_{m+1} = \min\{a_m', a_m''\}$ (the minimum is taken with respect to the ordering $\gtrsim$). By Lemma 3, for $m \geq 1$,

$$a_m \gtrsim a_m' \circ a_m'' \gtrsim 2a_{m+1}.$$

Since $a_{m+1} \gtrsim 2a_{m+2}$, we can apply Lemma 3 again to obtain

$$a_m \gtrsim 2a_{m+1} \gtrsim (2a_{m+2}) \circ (2a_{m+2}).$$

By Lemma 2, $a_m \gtrsim 4a_{m+2}$. Continuing this argument, we have by induction

$$a_m \gtrsim 2^n a_{m+n}, \qquad m \geq 1, \quad n \geq 0.$$

It goes without saying that the inductive argument, as developed above, includes the proof that $2^n \in N_{a_{m+n}}$, $m \geq 1$, $n \geq 0$.

Having defined the sequence $a_m$ and established the basic property that $a_m \gtrsim 2^n a_{m+n}$, we now show that $N(a_m, b)$ is defined for sufficiently large $m$ and approaches $+\infty$. Suppose first that $a_2 > b$. By definition, $a_2 > [N(b, a_2) - 1]b$. Since $(a_2, a_2) \in B$, by Axiom 2, $((N(b, a_2) - 1)b, b) \in B$, so $N(b, a_2) \in N_b$. By definition, $N(b, a_2)\, b \gtrsim a_2$. Now choose $m$ so large that $2^m > N(b, a_2)$. We show that the supposition that $a_{m+2} \gtrsim b$ leads to a contradiction. In fact, if $a_{m+2} \gtrsim b$, then by Lemma 4, $2^m \in N_b$ and we have

$$N(b, a_2)\, b \gtrsim a_2 \gtrsim 2^m a_{m+2} \gtrsim 2^m b.$$

But by Lemma 5, $2^m > N(b, a_2)$ implies $2^m b > N(b, a_2)b$, a contradiction.

Thus, we have $b > a_{m+2}$. We established this formula assuming $a_2 > b$; but otherwise, the same formula holds with $m = 1$. Thus, for every $b$, there exists $m \geqslant 1$ such that for all $n \geqslant 0$, $b > a_{m+n+2}$. We now have

$$
\begin{aligned}
b &> [N(a_{m+2}, b) - 1]\, a_{m+2} && \text{(definition)} \\
&\gtrsim [N(a_{m+2}, b) - 1](2^n a_{m+n+2}) && \text{(Lemma 4)} \\
&= (2^n[N(a_{m+2}, b) - 1])\, a_{m+n+2} && \text{(Lemma 6).}
\end{aligned}
$$

By definition, therefore, $N(a_{m+n+2}, b) > 2^n[N(a_{m+2}, b) - 1]$. It follows that $N(a_{m+n+2}, b) \to +\infty$ as $n \to \infty$, as required.

To evaluate $N(a_m, b)/N(a_m, c)$, as $m \to \infty$, we need the following inequality:

*If $(a, a') \in B$ and $b \gtrsim a \gtrsim a'$, then*

$$
N(a', a)\, N(a, b) > N(a', b) - 1 \geqslant [N(a', a) - 1][N(a, b) - 1].
$$

For simplicity in proving this inequality, we denote $N(a', a)$ by $m + 1$ and $N(a, b)$ by $n + 1$. Note that if $b = a$ or $a = a'$ (or both), the inequality is trivial; so we can assume $b > a > a'$, hence, $m, n \geqslant 1$. We have $a > ma'$ and $(a, a') \in B$, so $m + 1 \in N_{a'}$ and $(m + 1)\, a' \gtrsim a$. Suppose that $(n + 1)(m + 1) \in N_{a'}$. By Lemma 6, $n + 1 \in N_{(m+1)a'}$; thus by Lemma 4, $n + 1 \in N_a$. By definition, $(n + 1)\, a \gtrsim b$, hence $(n + 1)(m + 1)\, a' \gtrsim b$. This shows that either $(n + 1)(m + 1) \notin N_{a'}$, or else $(n + 1)(m + 1)\, a' \gtrsim b$; in either case, we have $(n + 1)(m + 1) \geqslant N(a', b)$, which gives the left half of the inequality. From $b > na$ and $a > ma'$ we have $nm \in N_{a'}$ and $b > nma'$ (Lemmas 4 and 6). Thus, $N(a', b) - 1 \geqslant nm$, which is the right half of the inequality.

From this inequality, it follows that for any $b, c$, for all sufficiently large values of $m$, and for all $n \geqslant m$,

$$
\frac{[N(a_n, b) - 1]}{[N(a_n, c) - 1]} < \frac{N(a_n, a_m)\, N(a_m, b)}{[N(a_n, a_m) - 1][N(a_m, c) - 1]}.
$$

(Use the upper bound for the numerator and the lower bound for the denominator.) The first consequence of this is that for all $n \geqslant m$,

$$
\frac{[N(a_n, b) - 1]}{[N(a_n, c) - 1]} \leqslant 2 \frac{N(a_m, b)}{[N(a_m, c) - 1]}.
$$

Thus, the sequence $[N(a_n, b) - 1]/[N(a_n, c) - 1]$ is bounded above for arbitrary $b, c$. By interchanging the roles of $b$ and $c$, we prove that the sequence of inverses has a finite upper bound and, hence, the sequence itself has a positive lower bound. Let $L^*$ and $L_*$ be, respectively, the greatest and least limit points of the sequence; then $0 < L_* \leqslant L^* < \infty$. Holding $m$

fixed and letting $n \to \infty$ in our inequality, and noting that $N(a_n, a_m)/[N(a_n, a_m) - 1]$ converges to 1, we obtain

$$L^* \leqslant N(a_m, b)/[N(a_m, c) - 1].$$

Now letting $m \to \infty$ in this last inequality, we see that $L^* \leqslant L_*$, i.e., $L^* = L_*$ and the sequence $[N(a_n, b) - 1]/[N(a_n, c) - 1]$, hence also the sequence $N(a_n, b)/N(a_n, c)$ converges to a limit in $Re^+$.

For any $b \in A$, define

$$\phi(b) = \lim_{m \to \infty} N(a_m, b)/N(a_m, a_1).$$

Since $\phi(a_1) = 1$, $a_1$ has been taken as the unit of measurement; this choice is arbitrary.

We show that $\phi$ is additive (ii) and order preserving (i). Suppose that $(b, c) \in B$. For $m$ such that $b, c > a_m$, we have $b > [N(a_m, b) - 1] a_m$, $c > [N(a_m, c) - 1] a_m$, hence, $b \circ c \gtrsim [N(a_m, b) + N(a_m, c) - 2] a_m$ (Lemmas 2 and 3). If $N(a_m, b) + N(a_m, c) \in N_{a_m}$, then clearly

$$[N(a_m, b) + N(a_m, c)] a_m \gtrsim b \circ c.$$

Thus we have the inequality

$$N(a_m, b) + N(a_m, c) \geqslant N(a_m, b \circ c) > N(a_m, b) + N(a_m, c) - 2.$$

Dividing through by $N(a_m, a_1)$ and letting $m \to \infty$ yields $\phi(b) + \phi(c) = \phi(b \circ c)$. Thus, (ii) holds.

Suppose that $b > c$. By Axiom 7, there exists $c'$ with $(c, c')$ in $B$ and $b \gtrsim c \circ c'$. For each $m$, $N(a_m, b) \geqslant N(a_m, c \circ c')$; hence,

$$\begin{aligned}
\phi(b) &\geqslant \phi(c \circ c') \\
&\doteq \phi(c) + \phi(c') \qquad \text{[by (ii)]} \\
&> \phi(c) \qquad\qquad\quad [\phi(c') > 0].
\end{aligned}$$

It follows from this that $b \gtrsim c$ if and only if $\phi(b) \geqslant \phi(c)$, as required for (i).

Finally, we establish uniqueness. Suppose that $\phi'$ is any other function from $A$ to $Re^+$ satisfying (i) and (ii). If $b$ is nonmaximal in $A$, then by Axiom 7, there exists $c$ with $(b, c) \in B$. For $a_m \lesssim c$, $N(a_m, b) \in N_{a_m}$ and $N(a_m, b) a_m \gtrsim b$. By (i) and (ii),

$$N(a_m, b) \phi'(a_m) \geqslant \phi'(b) > [N(a_m, b) - 1] \phi'(a_m).$$

Since $a_2$ is nonmaximal, we have for $m > 2$

$$N(a_m, a_2)\, \phi'(a_m) \geqslant \phi'(a_2) > [N(a_m, a_2) - 1]\, \phi'(a_m).$$

Dividing these inequalities yields

$$N(a_m, b)/[N(a_m, a_2) - 1] > \phi'(b)/\phi'(a_2) > [N(a_m, b) - 1]/N(a_m, a_2).$$

Letting $m \to \infty$ yields $\phi'(b)/\phi'(a_2) = \phi(b)/\phi(a_2)$, or $\phi'(b) = \alpha\phi(b)$, where $\alpha = \phi'(a_2)/\phi(a_2) > 0$.

Note that the uniqueness statement can be extended to a maximal element $a \in A$ (if any) only if $a = b \circ c$ for some $b, c$. This is necessarily the case if there is a minimal element in $A$, but need not hold otherwise.

To prove Theorem 4', let $\Omega$ and $R_\Omega$ be defined as in that theorem. To show that $\phi$ is an isomorphism, it suffices to establish that

$$(a, b) \in B' \quad \text{iff} \quad (\phi(a), \phi(b)) \in R_\Omega\,.$$

One direction is obvious: if $(a, b) \in B'$, then $a \circ b \in A'$, and so $\phi(a) + \phi(b) < \Omega$. Conversely, suppose that $\phi(a) + \phi(b) < \Omega$. Since $\Omega$ is a least upper bound, there exists $c \in A$ such that $\phi(a) + \phi(b) < \phi(c)$. If there is a minimal element $a_1 \in A$, then $N(a_1, a) + N(a_1, b) \in N_{a_1}$, hence, by an application of Lemma 2, $(N(a_1, a)\, a_1\,, N(a_1, b)\, a_1) \in B'$, or $(a, b) \in B'$. If there is no minimal element in $A$, we know that $\phi(a_m) \to 0$, where $a_m$ is the sequence constructed in the proof of Theorem 4. Choose $a_m$ with $\phi(a_m) < \frac{1}{2}[\phi(c) - \phi(a) - \phi(b)]$. It follows readily that

$$N(a_m, a) + N(a_m, b) \in N_{a_m}\,,$$

and hence, that $(N(a_m, a)\, a_m\,, N(a_m, b)\, a_m) \in B$. Thus, $(a, b) \in B'$.   ◇

From the proof of the uniqueness theorem, we have a good idea of the precision of approximations to $\phi(b)$. In fact, the inequality used to prove uniqueness can be rewritten as follows.

$$\frac{N(a_m, b)}{N(a_m, a_2)} \cdot \frac{N(a_m, a_2)}{N(a_m, a_2) - 1} > \frac{\phi(b)}{\phi(a_2)} > \frac{N(a_m, b) - 1}{N(a_m, b)} \cdot \frac{N(a_m, b)}{N(a_m, a_2)}\,.$$

Hence, the proportion of error in estimating $\phi(b)/\phi(a_2)$ by $N(a_m, b)/N(a_m, a_2)$ is not greater than the larger of $1/N(a_m, a_2)$ and $1/N(a_m, b)$. These error limits are known, and approach zero at least as fast as $2^{-m}$ (by construction of the sequence $a_m$). Hence, we can specify precisely the finite observations required to obtain any preassigned accuracy. In this way, our proof does not have the undesirable nonconstructive features of the definition of $\phi$ in the proof of Theorem 2 (Section 2.1).

### 2.2.5 Archimedean Ordered Groups

As a corollary to Theorem 4, we obtain an isomorphism for the case of groups.

DEFINITION 3. *Let $\langle A, \gtrsim \rangle$ be a simple order and $\circ$ a binary operation on A such that $\langle A, \circ \rangle$ is a group. The triple $\langle A, \gtrsim, \circ \rangle$ is called a simply ordered group provided that for all $a, b, c \in A$, if $a \gtrsim b$, then $a \circ c \gtrsim b \circ c$ and $c \circ a \gtrsim c \circ b$. Let the group identity be denoted $e$. The group is Archimedean provided that if $a > e$ and $b \in A$, then $na > b$ for some $n \in I^+$.*

THEOREM 5. *Let $\langle A, \gtrsim, \circ \rangle$ be an Archimedean simply ordered group. Then $\langle A, \gtrsim, \circ \rangle$ is isomorphic to a subgroup of $\langle \text{Re}, \geqslant, + \rangle$, and if $\phi, \phi'$ are any two isomorphisms, $\phi' = \alpha\phi$ for some $\alpha > 0$.*

PROOF. Let $A^+ = \{a \mid a > e\}$, where $e$ is the identity. Let $B = A^+ \times A^+$, and let $\gtrsim^+$ be the restriction of $\gtrsim$ to $A^+$. If $a, b \in A^+$, then $a \circ b \gtrsim a \circ e = a > e$, so $a \circ b \in A^+$. Thus, $\circ$ induces a function $\circ^+$ from $B$ into $A^+$. We show that $\langle A^+, \gtrsim^+, B, \circ^+ \rangle$ satisfies Axioms 1–8 of Definition 2. In fact, Axioms 1–5 and 8 are immediate. To show positivity (Axiom 6), suppose that $c \in A^+$ and $a \gtrsim a \circ c$. Then

$$e = a^{-1} \circ a \gtrsim a^{-1} \circ (a \circ c) = c,$$

contradicting $c \in A^+$. Thus, $a \circ c > a$, as required. For regularity (Axiom 7), suppose that $c > a$ and let $b = c \circ a^{-1}$, where $a^{-1}$ is the inverse of $a$. If $e \gtrsim b$, then

$$a = e \circ a \gtrsim b \circ a = c,$$

a contradiction, so $b \in A^+$; and $c = b \circ a$.

We let $\phi^+$ be a function from $A^+$ to $\text{Re}^+$ satisfying (i) and (ii) of Theorem 4, and extend it to $A$ by noting that if $e > a$, then $a^{-1} \in A^+$ (by the same argument as for regularity, above). Therefore, define $\phi$ by

$$\phi(a) = \begin{cases} \phi^+(a), & a \in A^+, \\ 0, & a = e, \\ -\phi^+(a^{-1}), & e > a. \end{cases}$$

The remainder of the theorem is easy to verify. ◇

### 2.2.6 Note on Hölder's Theorem

Hölder (1901) presented a set of seven axioms that are necessary and sufficient for an isomorphism *onto* $\langle \text{Re}^+, \geqslant, + \rangle$. Thus, he also dealt with the case of an ordered semigroup. He used much stronger structural assumptions, including (in our notation) $B = A \times A$, no minimal element, solvability of $c = a \circ b$ for $a$ given $c > b$ and for $b$ given $c > a$, and the Dedekind property (if $A = C \cup D$, where $C$, $D$ are nonempty, $C \cap D = \varnothing$, and $c \in C$, $d \in D$ imply $c < d$, then either sup $C$ or inf $D$ exists). These are, of course, all necessary for an isomorphism *onto* Re$^+$. With these structural properties and a slightly stronger positivity assumption, he did not need our Axioms 2–4 and 8. Since the axioms follow easily from his assumptions, it is straightforward to prove his result as a corollary to Theorem 4.

The proofs given by Birkhoff (1967, p. 300) and Fuchs (1963, p. 45) of Theorem 5, which is now often called Hölder's theorem, are essentially the same as Hölder's proof. They all suffer from the disadvantage that the method of constructing $\phi$ could not be used in actual measurement, because it requires that $na$ be defined for every $n$. This means that the proof does not generalize to the case of more realistic structural assumptions.

### 2.2.7 Archimedean Ordered Semirings

Suppose that $\langle A, \gtrsim \rangle$ is a simple order, $B$, $B^*$ are two subsets of $A \times A$, and $\oplus$, $*$ are two binary operations from $B$ to $A$ and from $B^*$ to $A$, respectively, such that $\langle A, \gtrsim, B, \oplus \rangle$ and $\langle A, \gtrsim, B^*, * \rangle$ are both ordered local semigroups. If they are both Archimedean, regular, and positive, then two, possibly unrelated, isomorphisms $\phi$ and $\phi^*$ can be constructed into Re$^+$. If the two operations are linked, however, by the distributive laws of multiplication over addition, i.e.,

$$(a \oplus b) * c = (a * c) \oplus (b * c),$$
$$c * (a \oplus b) = (c * a) \oplus (c * b),$$

then it is desirable to construct a single isomorphism $\phi$ that is both additive and multiplicative. That is, in addition to satisfying (i) and (ii) of Theorem 4, $\phi$ must satisfy $\phi(a * b) = \phi(a) \phi(b)$, and thus yield an isomorphism of $\langle A, \gtrsim, B, B^*, \oplus, * \rangle$ into $\langle \text{Re}^+, \geqslant, R_\Omega, R_\Omega^*, +, \cdot \rangle$, where $+, \cdot$ are ordinary addition and multiplication in Re$^+$.

The requirement that $\phi$ be order preserving and additive already determines it up to ratio-scale transformations (Theorem 4). If $A$ has a multiplicative identity $e$, then from $a = a * e$ we have

$$\phi(a) = \phi(a * e) = \phi(a) \phi(e),$$

whence $\phi(e) = 1$. Thus, the additional requirement that $\phi$ be multiplicative determines the unit of measurement. We show below that even if $A$ has no multiplicative identity, there exists nevertheless one and only one additive representation that is also multiplicative. In most applications of this theorem, however, one still obtains ratio-scale, rather than absolute, measurement; the reason becomes clear in Chapter 7.

DEFINITION 4. *Suppose that $A$ is a set, $\gtrsim$, $B$, and $B^*$ are three binary relations on $A$, $\oplus$ is a function from $B$ into $A$, and $*$ is a function from $B^*$ into $A$. The sextuple $\langle A, \gtrsim, B, B^*, \oplus, * \rangle$ is an ordered local semiring iff the following four axioms are satisfied*:

1. $\langle A, \gtrsim, B, \oplus \rangle$ *is an ordered local semigroup (Definition 2, Axioms 1–5).*

2. $\langle A, \gtrsim, B^*, * \rangle$ *satisfies Axioms 1–4 of Definition 2 and the following modified version of Axiom 5*:
   5'. *If $(a, b)$, $(b, c) \in B^*$, then $(a * b, c) \in B^*$ iff $(a, b * c) \in B^*$; and if both conditions hold, then $(a * b) * c = a * (b * c)$.*

3. (i) *If $(b, c) \in B$ and $(a, b \oplus c) \in B^*$, then $(a, b)$, $(a, c) \in B^*$, $(a * b, a * c) \in B$, and $a * (b \oplus c) = (a * b) \oplus (a * c)$.*
   (ii) *If $(a, b) \in B$ and $(a \oplus b, c) \in B^*$, then $(a, c)$, $(b, c) \in B^*$, $(a * c, b * c) \in B$, and $(a \oplus b) * c = (a * c) \oplus (b * c)$.*

4. *For any $a \in A$, there exists $(b, c) \in B$ such that $(a, b \oplus c) \in B^*$.*

*The local semigroup mentioned in Axiom 1 is termed the additive $(\oplus)$ semigroup; the semiring is defined to be* positive, regular, *or* Archimedean *according to whether these properties are satisfied by the additive semigroup (see Definition 2, Axioms 6–8).*

The numerical ordered local semigroup described in Section 2.2.1 can be made into an ordered local semiring by putting

$$R_\Omega{}^* = \{(x, y) \mid x, y \in \mathrm{Re}^+ \text{ and } x \cdot y < \Omega\}.$$

It is easy to verify that $\langle \mathrm{Re}^+, \geqslant, R_\Omega, R_\Omega{}^*, +, \cdot \rangle$ satisfies Axioms 1–4 of Definition 4. Note that the replacement of Axiom 5 of Definition 2 by 5', in Axiom 2 of this definition, is necessary, since $x \cdot y < \Omega$ and $x \cdot y \cdot z < \Omega$ do not entail that $y \cdot z < \Omega$.

The representation and uniqueness theorem is given by the following:

THEOREM 6. *Let $\langle A, \gtrsim, B, B^*, \oplus, * \rangle$ be an Archimedean, regular, positive, ordered local semiring (see axioms of Definitions 4 and 2). Then there is a unique function $\phi$ from $A$ to $\mathrm{Re}^+$ such that, for all $a, b \in A$*:

   (i) $a \gtrsim b$ *iff* $\phi(a) \geqslant \phi(b)$;

   (ii) *if $(a, b) \in B$, then* $\phi(a \oplus b) = \phi(a) + \phi(b)$;

   (iii) *if $(a, b) \in B^*$, then* $\phi(a * b) = \phi(a)\, \phi(b)$.

*PROOF.* The proof is based on the method followed by Fuchs (1963, p. 126) to treat Archimedean ordered rings. A less abstract proof is sketched in Exercises 4–13. The actual construction of an isomorphism is probably better based on those exercises, although it could be carried out using the proof given here.

For any $a \in A$, let $A_a = \{b \mid (a, b) \in B^*\}$ and $B_a = \{(b, c) \mid (b, c) \in B$ and $b \oplus c \in A_a\}$. Let $\oplus_a$ and $\succsim_a$ be $\oplus$ and $\succsim$ restricted to $B_a$ and to $A_a$, respectively. Then it is easy to show that $\langle A_a, \succsim_a, B_a, \oplus_a \rangle$ is an Archimedean, regular, positive, ordered local semigroup.

First, note that Axiom 4 is precisely the statement that $A_a$ and $B_a$ are nonempty for any $a \in A$. Obviously, $\oplus_a$ is a function from $B_a$ to $A_a$.

Axioms 1, 3, 4, 6, and 8 of Definition 2 are immediate, since they hold for $\langle A, \succsim, B, \oplus \rangle$ and are true *a forteriori* for any subsets.

If $(b, c) \in B_a$, $b \succsim b'$, and $c \succsim c'$, then $b \oplus c \succsim b' \oplus c'$. Hence, by Axiom 2 of Definition 2, applied to $\langle A, \succsim, B^*, * \rangle$, $(a, b' \oplus c') \in B^*$, i.e., $(b', c') \in B_a$. This establishes Axiom 2 of Definition 2 for $\langle A_a, \succsim_a, B_a, \oplus_a \rangle$.

Axiom 5 of Definition 2 follows readily from the same axiom applied to $\langle A, \succsim, B, \oplus \rangle$: If $(b, b')$ and $(b \oplus b', b'') \in B_a$, then we have $(b', b'')$ and $(b, b' \oplus b'') \in B$, and $(b \oplus b') \oplus b'' = b \oplus (b' \oplus b'')$. Thus,

$$(a, b \oplus (b' \oplus b'')) \in B^*,$$

so $(b, b' \oplus b'') \in B_a$. By Axiom 6, $b \oplus (b' \oplus b'') \succ b' \oplus b''$, so

$$(a, b' \oplus b'') \in B^*,$$

hence $(b', b'') \in B_a$.

Finally, we show that Axiom 7 of Definition 2 holds. Suppose that $b, c \in A_a$ and $b \succ c$. We can find $c' \in A$ such that $b \succsim c \oplus c'$. Since $(a, c \oplus c') \in B^*$, $(c, c') \in B_a$, as required.

Note that, so far, Axiom 3 of Definition 4 has not been used.

By Theorem 4, there exists a function $\psi$ from $A$ to $\mathrm{Re}^+$ satisfying (i) and (ii) for $\langle A, \succsim, B, \oplus \rangle$. *A forteriori*, $\psi$ satisfies (i) and (ii) for each $\langle A_a, \succsim_a, B_a, \oplus_a \rangle$. However, we can define a new representation for the latter semigroup by

$$\psi_a(b) = \psi(a * b).$$

To prove that $\psi_a$ satisfies (ii), take $(b, c) \in B_a$. Then $(a, b \oplus c) \in B^*$, so by Axiom 3 of Definition 4, $(a, b)$ and $(a, c) \in B^*$, $(a * b, a * c) \in B$, and $a * (b \oplus c) = (a * b) \oplus (a * c)$. Hence,

$$\begin{aligned}
\psi_a(b \oplus c) &= \psi[a * (b \oplus c)] \\
&= \psi[(a * b) \oplus (a * c)] \\
&= \psi(a * b) + \psi(a * c) \\
&= \psi_a(b) + \psi_a(c).
\end{aligned}$$

To prove that $\psi_a$ satisfies (i), suppose that $b, c \in A_a$ and $b > c$. Choose $c' \in A_a$ such that $b \gtrsim c \oplus c'$. Then $a * b \gtrsim a * (c \oplus c')$. Hence,

$$\begin{aligned}
\psi_a(b) &= \psi(a * b) \\
&\geqslant \psi[a * (c \oplus c')] \\
&= \psi_a(c \oplus c') \\
&= \psi_a(c) + \psi_a(c') \\
&> \psi_a(c),
\end{aligned}$$

as required.

By the uniqueness part of Theorem 4, there is a positive constant, denoted $\phi(a)$, such that for all nonmaximal $b \in A_a$,

$$\psi_a(b) = \phi(a)\,\psi(b).$$

We now show that $\phi$ satisfies conditions (i)–(iii).

Suppose that $(a, a') \in B$. By Axiom 4 of Definition 4, we can choose a nonmaximal $b \in A_{a \oplus a'}$. Clearly, $b$ is also a nonmaximal element of $A_a$ and of $A_{a'}$. We have

$$\begin{aligned}
\phi(a \oplus a')\,\psi(b) &= \psi_{a \oplus a'}(b) \\
&= \psi[(a \oplus a') * b] \\
&= \psi[(a * b) \oplus (a' * b)] \\
&= \psi(a * b) + \psi(a' * b) \\
&= \psi_a(b) + \psi_{a'}(b) \\
&= \phi(a)\,\psi(b) + \phi(a')\,\psi(b) \\
&= [\phi(a) + \phi(a')]\,\psi(b).
\end{aligned}$$

Hence, $\phi$ satisfies (ii).

Suppose that $a > a'$; let $a \gtrsim a' \oplus a''$. Choose nonmaximal $b \in A_a$ and $A_{a' \oplus a''}$; then

$$\begin{aligned}
\phi(a)\,\psi(b) &= \psi_a(b) \\
&= \psi(a * b) \\
&\geqslant \psi[(a' \oplus a'') * b] \\
&= \phi(a' \oplus a'')\,\psi(b) \\
&= [\phi(a') + \phi(a'')]\,\psi(b).
\end{aligned}$$

Hence, $\phi(a) > \phi(a')$.

Finally, we prove (iii). Take $(a, a') \in B^*$. Choose nonmaximal $b' \in A_{a * a'}$ and nonmaximal $b'' \in A_a$. Let $b = \min\{b', b''\}$. Then $b$ is nonmaximal in both $A_{a*a'}$ and $A_{a'}$, and clearly, $a' * b$ is nonmaximal in $A_a$. We therefore have

$$\begin{aligned}
\phi(a * a')\,\psi(b) &= \psi_{a*a'}(b)\\
&= \psi[(a * a') * b]\\
&= \psi[a * (a' * b)]\\
&= \psi_a(a' * b)\\
&= \phi(a)\,\psi(a' * b)\\
&= \phi(a)\,\psi_{a'}(b)\\
&= \phi(a)\,\phi(a')\,\psi(b).
\end{aligned}$$

Thus, $\phi(a * a') = \phi(a)\,\phi(a')$.

For uniqueness, note that if $\alpha\phi$ satisfies (i)–(iii) of Theorem 6, then by (iii), $\alpha\phi(a * b) = [\alpha\phi(a)] \cdot [\alpha\phi(b)]$, whence $\alpha = \alpha^2$ or $\alpha = 1$. This argument applies except possibly for maximal elements. But if $a$ is maximal in $A$, we can choose $b$ such that $(a, b)$ is in $B^*$ and such that $b$, $a * b$ are nonmaximal. We now have $\phi(a) = \phi(a * b)/\phi(b)$, giving uniqueness of $\phi$ at $a$. $\diamond$

We next establish a corollary to Theorem 6 which is the classic result in this area (see Fuchs, 1963, p. 126); it stands to Theorem 6 as Hölder's Theorem 5 does to Theorem 4.

DEFINITION 5. *Suppose that $A$ is a set, $\succsim$ a binary relation on $A$, and $\oplus$ and $*$ binary operations on $A$. Then $\langle A, \succsim, \oplus, * \rangle$ is an* Archimedean ordered ring *provided that:*

(i)   $\langle A, \oplus, * \rangle$ *is a ring with zero element $\theta$;*

(ii)  $\langle A, \succsim, \oplus \rangle$ *is an Archimedean ordered group;*

(iii) *if $a \succ \theta$ and $b \succ c$, then $a * b \succ a * c$ and $b * a \succ c * a$.*

COROLLARY TO THEOREM 6.   *An Archimedean ordered ring is uniquely isomorphic to a subring of $\langle \mathrm{Re}, \geqslant, +, \cdot \rangle$.*

*PROOF.* By Theorem 5, there is an isomorphism $\phi'$ of $\langle A, \succsim, \oplus \rangle$ into a subgroup of $\langle \mathrm{Re}, \geqslant, + \rangle$. Let $A^+ = \{a \mid a \succ \theta\}$. According to Exercise 14, the restriction of $\succsim$, $\oplus$, and $*$ to $A^+$ forms an Archimedean, regular, positive, ordered, local semiring (with $B = B^* = A^+ \times A^+$). Let $\phi$ be the unique isomorphism of Theorem 6. By the uniqueness assertion of Theorem 5, there exists $\alpha > 0$ such that over $A^+$, $\phi = \alpha\phi'$. Extend $\phi$ to all of $A$ by defining $\phi = \alpha\phi'$. For all $a < \theta$, $\phi(a) = \alpha\phi'(a) = -\alpha\phi'(-a) = -\phi(-a)$. From this and the fact that $\phi$ preserves $*$ over $A^+$, it is trivial to show that it preserves $*$ over all of $A$. $\diamond$

## 2.3 FINITE SETS OF HOMOGENEOUS LINEAR INEQUALITIES

The general problem of this section is to find solutions to families of inequalities and equations of the form

$$\sum_{j=1}^{n} \alpha_{ij} x_j > 0, \qquad i = 1,\dots, m',$$

$$\sum_{j=1}^{n} \beta_{ij} x_j = 0, \qquad i = 1,\dots, m''. \tag{1}$$

Such a family of inequalities and equations arises in measurement contexts when the $x_1,\dots, x_n$ are unknown values of quantities to be measured, and each inequality or equation of System (1) is entailed, via a linear measurement model, by a corresponding observation of order or equality between two objects (see Section 1.1.3 for examples). Any solution to System (1) provides a measurement scale compatible with the given observations of order and equality. In all applications of System (1) in this book, the coefficients $\alpha_{ij}$, $\beta_{ij}$ are integers (see Chapter 9).

### 2.3.1 Intuitive Explanation of the Solution Criterion

We shall analyze the System (1) using vector methods. The desired solution to System (1), $(x_1,\dots, x_n)$, is a vector, and the coefficients of any inequality or equation of the system also form a vector, e.g., $(\alpha_{i1},\dots, \alpha_{in})$. For convenience, we abbreviate $(x_1,\dots, x_n)$ by $x$, $(\alpha_{i1},\dots, \alpha_{in})$ by $\alpha_i$, and $(\beta_{i1},\dots, \beta_{in})$ by $\beta_i$. Other abbreviations of the same sort are introduced later.

The *angle* between two vectors $x$ and $y$ is the angle whose vertex is the *origin* $\mathbf{0} = (0,\dots, 0)$, or $(0, 0, 0)$ in three dimensions, and whose sides are the lines from $\mathbf{0}$ through each vector. The *scalar product* of two vectors $x = (x_1,\dots, x_n)$ and $y = (y_1,\dots, y_n)$ is simply the sum of the products of their respective components, i.e., $\sum_{i=1}^{n} x_i y_i$. The reader can easily verify (in the two-dimensional case, for example) that the angle between $x$ and $y$ is acute, obtuse, or right depending on whether their scalar product is, respectively, positive, negative, or zero. In terms of these concepts, System (1) can be restated as

$$
\begin{aligned}
&x \text{ makes an acute angle with each } \alpha_i, && i = 1,\dots, m', \\
&x \text{ makes a right angle with each } \beta_i, && i = 1,\dots, m''.
\end{aligned}
\tag{1'}
$$

Suppose that $n = 3$, that $m'' = 2$, and that $\beta_1$, $\beta_2$ are distinct from $\mathbf{0}$ and have a nonzero angle between them. The three points, $\mathbf{0}$, $\beta_1$, $\beta_2$ lie in a unique plane, and if $x$ solves (1'), then the line from $0$ through $x$ must be at right angles to that plane. Thus, the direction of $x$ from $0$ is completely determined except for the choice of which side of the plane $x$ is on. If we choose $x$ on one side of the plane, then it makes an acute angle with any other point on the same side, a right angle with any point in the plane, and an obtuse angle with any point on the opposite side. Hence, it is obvious that in this case, *a solution to System* (1') *can be found iff all the points* $\alpha_i$ *lie on the same side of the plane determined by* $0, \beta_1, \beta_2$.

Let us keep $n = 3$ and consider values of $m''$ other than 2. If $m'' \geqslant 3$, there are two possibilities. If all the $\beta_i$ lie in a single plane (determined by two of them with $0$), then we take $x$ on a line perpendicular to this plane at $0$, and the situation is as before. Otherwise, they do not lie in a single plane, and so no $x$ can be found which forms a right angle with all the $\beta_i$. Hence the only vector $x$ which satisfies the equations is $0$, and this does not satisfy the strict inequalities. Thus, System (1') has no solution (unless there are no inequalities).

If $m'' = 1$, then $0$ and $\beta_1$ can be part of an infinite number of different planes. If all the $\alpha_i$ lie on the same side of just one of these planes, then an $x$ perpendicular to that plane solves System (1').

The extreme case is $m'' = 0$ in which there are no equalities. This case is common in the practice of measurement—it means that no two objects were accepted as exactly equal. Here, it suffices that there be any plane through $0$ such that all the $\alpha_i$ lie on the same side of it; there are no $\beta_i$ to constrain the position of the plane.

The analysis is identical for $n \geqslant 3$, except that the equivalent of a plane—a hyperplane—is determined by $0$ plus $n - 1$ independent $\beta_i$, if such exist.

And so we reduce the problem of solving System (1) or (1') to finding a plane (or a hyperplane) through $0$ and through all the $\beta_i$ (if any), such that all the $\alpha_i$ lie on the same side of it. A solution $x$ is just a vector such that the line through $0$ and $x$ is perpendicular to the plane and $x$ is on the same side as the $\alpha_i$.

To advance the problem further, we begin with the case $m'' = 0$. Is there a plane through $0$ such that all $\alpha_i$ are on one side of it? If so, consider the points $\alpha_i$ as the vertices of a polyhedron. This polyhedron lies all on one side of the plane, and so, in particular, $0$ is outside it. Conversely, if $0$ is not in the polyhedron with the $\alpha_i$ as vertices, then we can pass a plane through $0$ such that the whole polyhedron is on one side of it. It turns out that this is precisely the criterion for solvability of System (1') if $m'' = 0$: $0$ must be exterior to the polyhedron generated by the $\alpha_i$. In fact, in this case, one way to obtain a solution is to choose the point on the polyhedron that is closest to $0$. This point is a solution $x$. For through this point, we can pass a plane tangent to the polyhedron (in the sense that it does not enter it), and the parallel plane through **0** has the polyhedron on one side of it. The line from **0** to $x$ is perpendicular to both planes. This is illustrated, in a two-dimensional cross section, in Figure 1.

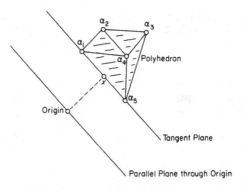

**FIGURE 1.** Polyhedron (shaded area) generated by five inequality vectors, $\alpha_1, ..., \alpha_5$, and the solution $x$ with planes perpendicular to $x$ (two-dimensional cross section).

The analysis for $m'' > 0$ is similar. If the plane through **0** is constrained to contain $\beta_1$ ($m'' = 1$), then we must require that the $\alpha$-polyhedron not intersect the (infinite) line through **0** and $\beta_1$. If the plane through **0** is determined by $\beta_1$, $\beta_2$, then the $\alpha$-polyhedron must not intersect this plane.

For $n > 3$, the analysis is analogous to the one just given. The only difference is that there are more possibilities for the number of independent $\beta_i$, other than 0, 1, or 2. If there are $n - 1$ independent equality vectors $\beta_i$, then they determine an $n - 1$ dimensional hyperplane, to which the solution $x$ must be perpendicular. All $\alpha_i$ must lie on the same side of this hyperplane. In general, any number of independent $\beta_i$, not exceeding $n - 1$, generate a subspace, and this subspace partially constrains the position for a hyperplane with the polyhedron of inequality vectors lying to one side of it. The general theorem we shall prove, then, is the following:

THEOREM 7 (Intuitive version). *System* (1′) *has a solution iff the polyhedron generated by the vectors* $\alpha_i$ *does not intersect the subspace generated by the vectors* $\beta_i$ *(when* $m'' = 0$, *the subspace in question is taken equal to* **0**).

### 2.3.2 Vector Formulation and Preliminary Lemmas

We now use vector concepts to give a precise formulation of Theorem 7, and as a preliminary to its proof, we state and prove a fundamental lemma of linear programming theory. Our treatment closely follows Gale (1960). Readers whose background in linear algebra is insufficient for them to follow this section can easily substitute pp. 28–47 of Gale's book for this section.

The *subspace* generated by vectors $\beta_1, ..., \beta_r$, where $\beta_i = (\beta_{i1}, ..., \beta_{in})$, is the set of all vectors of form

$$\sum_{i=1}^{r} \lambda_i \beta_i = \left( \sum_{i=1}^{r} \lambda_i \beta_{i1}, ..., \sum_{i=1}^{r} \lambda_i \beta_{in} \right),$$

where $\lambda_1, ..., \lambda_r$ are arbitrary real numbers.

The *convex hull* of vectors $\alpha_1, ..., \alpha_s$, where $\alpha_i = (\alpha_{i1}, ..., \alpha_{in})$, is the set of all vectors of form

$$\sum_{i=1}^{s} \lambda_i \alpha_i ,$$

where $\lambda_1, ..., \lambda_s$ are restricted to be greater than or equal to 0 and $\sum_{i=1}^{s} \lambda_i = 1$. This concept replaces the intuitive concept of polyhedron used above. We denote the scalar product of vectors $x, y$ by $x \cdot y = \sum_{i=1}^{n} x_i y_i$. We can now restate Theorem 7 as follows.

THEOREM 7   (Formal version).   *The system*

$$\begin{aligned} \alpha_i \cdot x > 0, \qquad i = 1,..., m', \\ \beta_i \cdot x = 0, \qquad i = 1,..., m'' \end{aligned} \tag{1''}$$

*has a solution iff the convex hull of $\alpha_1, ..., \alpha_{m'}$ does not intersect the subspace generated by $0, \beta_1, ..., \beta_{m''}$. That is, System (1) has a solution iff there is no solution $\lambda_1, ..., \lambda_{m'}, \mu_1, ..., \mu_{m''}$ to the system*

$$\sum_{i=1}^{m'} \lambda_i \alpha_{ij} = \sum_{i=1}^{m''} \mu_i \beta_{ij}, \qquad j = 1,..., n,$$

$$\lambda_i \geqslant 0, \qquad\qquad i = 1,..., m', \tag{2}$$

$$\sum_{i=1}^{m'} \lambda_i = 1.$$

The importance of this result depends partly on its constructive nature. One can start out to construct a solution to System (1''), but the procedure will eventually terminate in failure if and only if there is a solution to System (2).

An additional point is that the proof holds whether the numbers involved are real numbers or rationals. That is, if the $\alpha_{ij}$, $\beta_{ij}$ in System (1) are rational, then there is a rational solution to System (1) if and only if there is no rational solution to System (2). In System (2), the constraint $\sum \lambda_i = 1$ can just as well be replaced by $\sum \lambda_i > 0$, for then we could divide each value of $\lambda_i$ and of $\mu_i$ by $\sum \lambda_i$ to obtain $\sum \lambda_i = 1$. Let the revised system, with $\sum \lambda_i > 0$, be numbered (2'). Note that a rational solution to System (2) yields, by multiplying through by a common denominator, an integer solution to System (2'). The actual criterion thus amounts to the nonexistence of integer solutions to System (2'). This is the basis for measurement axiomatizations (see Chapter 9).

The proof is based solely on the following well-known result of linear algebra, which we formulate as Lemma 7, but do not prove (see Gale, 1960).

LEMMA 7. *Let* $\alpha_i = (\alpha_{i1}, ..., \alpha_{in})$, *and suppose that* $\alpha_1, ..., \alpha_m$ *are linearly independent* (*no one of them lies in the subspace generated by the other* $m - 1$). *Then for any numbers* $t_1, ..., t_m$ *the equations*

$$\alpha_i \cdot x = t_i, \qquad i = 1, ..., m$$

*have a solution* $x = (x_1, ..., x_n)$.

That is, it is always possible to solve $m$ linear equations $\alpha_i \cdot x = t_i$ in $n$ unknowns $x_1, ..., x_n$, provided that the $m$ linear expressions $\sum \alpha_{ij} x_j$ are independent of one another and so, in particular, $m \leqslant n$. Of course, the claim that the rest of the proof of Theorem 7 is constructive is based on the existence of a constructive proof to Lemma 7. The construction of a solution $(x_1, ..., x_n)$ is well known however. One merely uses the first equation to express one of the $x_i$ in terms of the other $n - 1$, substituting the expression for $x_i$ in the other $m - 1$ equations, reducing the problem to $m - 1$ equations in $n - 1$ unknowns. This process continues until either there are no equations left—in which case one can assign arbitrary values to the remaining unknowns—or until one of the substitutions results in a contradiction, in which case the process stops and it can be shown that the original $\alpha_i$ were not independent.

The next theorem belongs to the theory of linear inequalities, and since it serves as a basis for measurement axiomatizations we present its proof in detail.

LEMMA 8. *Let* $\{\alpha_{ij} \mid i = 1, ..., m, j = 1, ..., n\}$ *be an* $m \times n$ *matrix,* $\alpha_i = (\alpha_{i1}, ..., \alpha_{in})$, $\alpha^{(j)} = (\alpha_{1j}, ..., \alpha_{mj})$, *and* $z = (z_1, ..., z_n)$. *The system of inequalities*

$$\begin{aligned} \alpha_i \cdot x &\geqslant 0, \qquad i = 1, ..., m, \\ z \cdot x &< 0 \end{aligned} \tag{3}$$

*has a solution* $x = (x_1, ..., x_n)$ *iff the system*

$$y \cdot \alpha^{(j)} = z_j, \qquad j = 1, ..., n,$$
$$y_i \geqslant 0, \qquad i = 1, ..., m$$

(4)

*has no solution* $y = (y_1, ..., y_m)$.

PROOF. Suppose that $x$ is a solution to System (3) and $y$ is a solution to System (4). Then

$$0 > z \cdot x = \sum_{j=1}^{n} z_j x_j$$

$$= \sum_{j=1}^{n} [y \cdot \alpha^{(j)}] x_j$$

$$= \sum_{i=1}^{m} y_i [\alpha_i \cdot x] \geqslant 0,$$

since $y_i$, $\alpha_i \cdot x \geqslant 0$, $i = 1, ..., m$. This is impossible.

Suppose that System (4) has no solution. We show that System (3) has a solution. First, suppose that the equations $y \cdot \alpha^{(j)} = z_j$ have no solution at all, i.e., $z$ is not in the subspace generated by $\alpha_1, ..., \alpha_m$. Reindex the $\alpha_i$ so that $\alpha_1, ..., \alpha_r$ are an independent set and $\alpha_{r+1}, ..., \alpha_m$ are in the subspace generated by them. Then $\alpha_1, ..., \alpha_r$, $z$ are independent. By Lemma 7, we can find $x$ such that $\alpha_i \cdot x = 0$, $i = 1, ..., r$ and $z \cdot x < 0$. Since this implies $\alpha_i \cdot x = 0$, $i = r + 1, ..., m$, $x$ is a solution to (3).

Second, suppose that $y \cdot \alpha^{(j)} = z_j$, $j = 1, ..., n$, but that $y_i < 0$ for at least one $i$. We proceed by induction on $m$. If $m = 1$, then $y_1 \alpha_{1j} = z_j$, $j = 1, ..., n$, and $y_1 < 0$. Let $x = \alpha_1$. Now $\alpha_1 \neq \mathbf{0}$, otherwise $z = \mathbf{0}$, and $y_1 = 0$ would solve (4). Thus, $\alpha_1 \cdot x > 0$, $z \cdot x = y_1(\alpha_1 \cdot x) < 0$, and $x$ is a solution to System (3). Suppose now that Lemma 8 holds for $m \leqslant k$, where $k \geqslant 1$, and consider the case $m = k + 1$. Let $\beta^{(j)} = (\alpha_{1j}, ..., \alpha_{kj})$. There can be no solution to System (4) with $\beta^{(j)}$ substituted for $\alpha^{(j)}$, for if there were, we would let $y_{k+1} = 0$, obtaining a solution to the original system. By the inductive hypothesis, there is a vector $x' = (x_1', ..., x_n')$ with $\alpha_i \cdot x' \geqslant 0$, $i = 1, ..., k$ and $z \cdot x' < 0$. If $\alpha_{k+1} \cdot x' \geqslant 0$, then let $x = x'$ and we are done. So suppose that $\alpha_{k+1} \cdot x' < 0$. Define

$$\alpha_i' = (\alpha_i \cdot x') \alpha_{k+1} - (\alpha_{k+1} \cdot x') \alpha_i, \qquad i = 1, ..., k,$$
$$z' = (z \cdot x') \alpha_{k+1} - (\alpha_{k+1} \cdot x')z.$$

Consider the system

$$y' \cdot \alpha'^{(j)} = z_j', \qquad j = 1,\ldots, n,$$
$$y_i' \geqslant 0, \qquad i = 1,\ldots, k,$$

(4')

where $\alpha'^{(j)} = (\alpha'_{1j},\ldots, \alpha'_{kj})$, $\alpha_i' = (\alpha'_{i1},\ldots, \alpha'_{in})$. If $y'$ is a solution to System (4'), define $y$ by

$$y_i = y_i', \qquad i = 1,\ldots, k$$

$$y_{k+1} = -(\alpha_{k+1} \cdot x')^{-1} \left[ \sum_{i=1}^{k} y_i'(\alpha_i \cdot x') - (z \cdot x') \right].$$

We show that $y$ is a solution to System (4). In fact, $y_i = y_i' \geqslant 0$, $i = 1,\ldots, k$, and $y_{k+1} \geqslant 0$ since $-(\alpha_{k+1} \cdot x')^{-1} > 0$, $y_i' \geqslant 0$, $\alpha_i \cdot x' \geqslant 0$, and $-(z \cdot x') > 0$. Moreover, for $j = 1,\ldots, n$,

$$z_j = -(\alpha_{k+1} \cdot x')^{-1} \left[ z_j' - (z \cdot x') \alpha_{k+1,j} \right]$$

$$= -(\alpha_{k+1} \cdot x')^{-1} \left[ y' \cdot \alpha'^{(j)} - (z \cdot x') \alpha_{k+1,j} \right]$$

$$= -(\alpha_{k+1} \cdot x')^{-1} \left[ \sum_{i=1}^{k} y_i' \alpha'_{ij} - (z \cdot x') \alpha_{k+1,j} \right]$$

$$= -(\alpha_{k+1} \cdot x')^{-1} \left[ \sum_{i=1}^{k} y_i'(\alpha_i \cdot x') \alpha_{k+1,j} \right.$$

$$\left. - \sum_{i=1}^{k} y_i'(\alpha_{k+1} \cdot x') \alpha_{ij} - (z \cdot x') \alpha_{k+1,j} \right]$$

$$= \sum_{i=1}^{k} y_i' \alpha_{ij} + y_{k+1} \alpha_{k+1,j}$$

$$= y \cdot \alpha^{(j)}.$$

Thus, $y$ is a solution to System (4), contrary to hypothesis. We conclude that there exists no solution $y'$ to System (4'). By the inductive hypothesis, there exists a solution $x'' = (x_1'',\ldots, x_n'')$ to the system

$$\alpha_i' \cdot x'' \geqslant 0, \qquad i = 1,\ldots, k$$
$$z' \cdot x'' < 0.$$

(3')

Define

$$x = (\alpha_{k+1} \cdot x'') x' - (\alpha_{k+1} \cdot x') x''.$$

We show that $x$ is a solution to System (3). In fact, it is easy to verify, from the definition of $\alpha_i'$, $z'$, and $x$, that

$$\alpha_i \cdot x = \alpha_i' \cdot x'' \geq 0, \qquad i = 1,\ldots,k,$$
$$\alpha_{k+1} \cdot x = 0,$$
$$z \cdot x = z' \cdot x'' < 0.$$

By induction, the theorem follows for any $m$. $\qquad\qquad\qquad\qquad \diamondsuit$

The preceding proof allows us to construct the solution to a system of inequalities (3) whenever the solution exists. First, obtain a solution $x^{(1)}$ to $\alpha_1 \cdot x \geq 0, z \cdot x < 0$. If the equations $\alpha_1 \cdot x = 0$, $z \cdot x = -1$ have a simultaneous solution, then it will serve as $x^{(1)}$. If they do not, then $z = \lambda\alpha_1$ for some $\lambda$. If $\lambda \geq 0$, then no solution is possible, since $\alpha_1 \cdot x \geq 0$ implies $z \cdot x \geq 0$. If $\lambda < 0$, then choose $x^{(1)} = \alpha_1$. Now check the solution $x^{(1)}$ in successive inequalities $\alpha_2 \cdot x \geq 0,\ldots$ until, for some $k \geq 1$, $\alpha_{k+1} \cdot x^{(1)} < 0$. At this point, $x^{(1)}$ solves the first $k$ inequalities $\alpha_i \cdot x \geq 0$, as well as $z \cdot x < 0$, so relabel $x^{(1)}$ as $x^{(k)}$. To construct $x^{(k+1)}$, proceed as in the inductive step to the above proof, defining $\alpha_i'$, $i = 1,\ldots,k$ and $z'$, and solving the system $\alpha_i' \cdot x'' \geq 0$, $z' \cdot x'' < 0$. Since this system has fewer inequalities than System (3), a genuine reduction is achieved, although the method here described may have to be applied in full panoply to this lesser system. If the reduced system has no solution, then, as the lemma shows, the full system also has no solution. If $x''$ solves the reduced system, define $x^{(k+1)}$ from $x' = x^{(k)}$ and $x''$ as in the lemma. Now check $x^{(k+1)}$ in successive inequalities $\alpha_i \cdot x \geq 0, i \geq k + 2$, until it fails; then apply the method of the inductive step over again, etc. (see Exercise 15).

### 2.3.3  Proof of Theorem 7

First we show that Systems (1″) and (2) cannot both have solutions. For if they did, we would have

$$0 < \sum_{i=1}^{m'} \lambda_i(\alpha_i \cdot x) = \sum_{i=1}^{m''} \mu_i(\beta_i \cdot x) = 0.$$

Next, suppose that System (1″) has no solution. Define vectors $\gamma_i$ in $Re^{n+1}$ by

$$\gamma_i = (\alpha_{i1},\ldots,\alpha_{in},1), \qquad i = 1,\ldots,m',$$
$$\gamma_{m'+i} = (\beta_{i1},\ldots,\beta_{in},0), \qquad i = 1,\ldots,m'',$$
$$\gamma_{m'+m''+i} = (-\beta_{i1},\ldots,-\beta_{in},0), \qquad i = 1,\ldots,m''.$$

Let $z_j = 0$, $j = 1,..., n$, $z_{n+1} = 1$. If System (1″) has no solution, then neither does the system

$$\gamma_i \cdot \bar{x} \geqslant 0, \qquad i = 1,..., m' + 2m'',$$
$$z \cdot \bar{x} < 0. \tag{5}$$

For if System (5) had a solution $\bar{x} = (x_1 ,..., x_{n+1})$, $z \cdot \bar{x} < 0$ would imply $x_{n+1} < 0$. Thus, setting $x = (x_1 ,..., x_n)$, we have from $\gamma_i \cdot \bar{x} \geqslant 0$ the results

$$\alpha_i \cdot x + x_{n+1} \geqslant 0, \qquad i = 1,..., m',$$
$$\beta_i \cdot x \geqslant 0, \qquad i = 1,..., m'',$$
$$-\beta_i \cdot x \geqslant 0, \qquad i = 1,..., m''.$$

Thus, $\alpha_i \cdot x > 0$, $\beta_i \cdot x = 0$ as required in System (1″). Let $\gamma^{(j)} = (\gamma_{1j} ,..., \gamma_{mj})$, where $m = m' + 2m''$. By Lemma 8, the system

$$y \cdot \gamma^{(j)} = z_j , \qquad y_i \geqslant 0$$

has a solution. Let $\lambda_i = y_i$, $i = 1,..., m'$ and let $\mu_i = -y_{m'+i} + y_{m'+m''+i}$, $i = 1,..., m''$. Then

$$\sum_{i=1}^{m'} \lambda_i \alpha_{ij} - \sum_{i=1}^{m''} \mu_i \beta_{ij} = y \cdot \gamma^{(j)}$$

$$= z_j$$

$$= 0, \qquad j = 1,..., n,$$

and

$$\sum_{i=1}^{m'} \lambda_i = y \cdot \gamma^{(n+1)} = z_{n+1} = 1.$$

Thus, there is a solution to System (2).                                    $\Diamond$

Note that this proof leads to a construction of solutions to (1), by solving the associated System (5) using the method of Lemma 8. More efficient computational methods for solving System (5) are available, however, via linear programming (see Gale, 1960).

## 2.3.4 Topological Proof of Theorem 7

In this section, we give a shorter, more intuitive, but nonconstructive proof of Theorem 7, which fails if the base field is Ra rather than Re.

First, consider the case where there are no equalities in System (1), i.e., $m'' = 0$. In this case, the theorem states that a solution to System (1) exists if and only if the convex hull $K$ of $\{\alpha_1, ..., \alpha_{m'}\}$ does not intersect $\mathbf{0}$. The set $K$ is closed and bounded in $\mathrm{Re}^n$, hence, it is compact. For $y \in K$, $y \cdot y$ (the squared distance of $y$ from $\mathbf{0}$) is a continuous function of $y$, and thus is minimal for some $x \in K$. By convexity, $x$ is unique, but this result is not needed. (See Rudin, 1964, Sections 2.41, 4.11, 4.16 for the details of the argument just given.) We claim that $x$ is a solution to System (1). If not, then for some $i$, $\alpha_i \cdot x \leqslant 0$. For $\lambda \in \mathrm{Re}$, define

$$f(\lambda) = [\lambda \alpha_i + (1 - \lambda) x] \cdot [\lambda \alpha_i + (1 - \lambda) x].$$

Differentiating $f$ with respect to $\lambda$ and setting the derivative equal to zero yields that $f$ is minimum at

$$\lambda_0 = (x \cdot x - \alpha_i \cdot x)/(\alpha_i - x) \cdot (\alpha_i - x).$$

Since $x \cdot x$ and $(\alpha_i - x) \cdot (\alpha_i - x)$ are greater than 0 and $-\alpha_i \cdot x \geqslant 0$, $\lambda_0 > 0$. It is easily checked that $1 > \lambda_0$. Let $y = \lambda_0 \alpha_i + (1 - \lambda_0)x$. Then $y \in K$, $y \neq x$, and

$$y \cdot y = f(\lambda_0) < f(0) = x \cdot x,$$

contradicting minimality of $x \cdot x$. Therefore, $\alpha_i \cdot x > 0$ for $i = 1, ..., m'$.

Next, suppose that $m'' > 0$. Let $B$ be the subspace generated by $\beta_1, ..., \beta_{m''}$ and let $C$ be its orthogonal complement (see MacLane & Birkhoff, 1967, p. 240). Let $C$ be generated by vectors $\gamma_1, ..., \gamma_k$. Define new vectors $\alpha_i'$ by

$$\alpha_{ij}' = \alpha_i \cdot \gamma_j, \qquad i = 1, ..., m', \quad j = 1, ..., k.$$

If $x$ is any solution to System (1), then $x \in C$, so $x = \sum_{j=1}^{k} x_j' \gamma_j$, for some $x' = (x_1', ..., x_k')$. Thus,

$$\begin{aligned}
\alpha_i' \cdot x' &= \sum_{j=1}^{k} (\alpha_i \cdot \gamma_j) x_j' \\
&= \alpha_i \cdot x > 0, \qquad i = 1, ..., m'.
\end{aligned}$$

Conversely, if $x'$ is a solution of the system

$$\alpha_i' \cdot x' > 0, \qquad i = 1, ..., m', \tag{6}$$

then $x = \sum_{j=1}^{k} x_j' \gamma_j$ is a solution to System (1). Thus we can apply the criterion derived above for $m'' = 0$ to the reduced System (6). In fact, the convex hull of the $\alpha_i'$ intersects the origin (of $\mathrm{Re}^k$) if and only if there exist $\lambda_1, ..., \lambda_{m'} \geqslant 0$, with $\sum_{i=1}^{m'} \lambda_i = 1$, such that for $j = 1, ..., k$,

$$0 = \sum_{i=1}^{m'} \lambda_i \alpha'_{ij}$$

$$= \sum_{i=1}^{m'} \lambda_i (\alpha_i \cdot \gamma_j)$$

$$= \left[ \sum_{i=1}^{m'} \lambda_i \alpha_i \right] \cdot \gamma_j .$$

But this last equation means that $[\sum_{i=1}^{m'} \lambda_i \alpha_i] \cdot x = 0$ for every $x \in C$. A well-known result on orthogonal complements implies that $\sum_{i=1}^{m'} \lambda_i \alpha_i \in B$, and this is precisely the criterion for nonsolution of System (1).          ◇

## EXERCISES

**1.** Carry out the inductive step in the proof of Theorem 1, showing that if $\phi$ is order preserving on $\{a_1, ..., a_n\}$, then it is also order preserving on $\{a_1, ..., a_{n+1}\}$.     (2.1)

**2.** If you did not do Exercise 1.5, this would be a good time to do it.

**3.** Prove Lemma 2.     (2.2.3)

*In the following nine exercises, $\langle A, \gtrsim, \oplus, * \rangle$ is an Archimedean ordered ring (Definition 5) with additive identity (zero) $\theta$. Let $(n-1) a \oplus a = na$, $a^{n-1} * a = a^n$, and let $-a$ be the additive inverse of $a$.*     (2.2.7)

**4.** Prove the following. If $\phi$ is order preserving and multiplicative, then for any $a > \theta$, the following are equivalent:

 (i)  $\phi(a) < 1$,
 (ii) $a > a^2$.

**5.** Prove that if $\phi$ is order preserving, additive, and multiplicative, then for any $a > \theta$ and any $m, n > 0$, the following are equivalent:

 (i)  $\phi(a) < m/n$,
 (ii) $ma > na^2$.

**6.** Prove that for any $a > \theta$, the set of $m, n > 0$ such that $ma > na^2$ is nonempty.

Now *define* $\phi$ for $a > \theta$ by using Exercises 5 and 6:

$$\phi(a) = \inf\{m/n \mid m, n > 0 \text{ and } ma > na^2\}.$$

*The next six exercises show that $\phi$ is order preserving, additive, and multiplicative on $A^+ = \{a \mid a > \theta\}$, and hence, that $\phi$ yields an isomorphism into the ordered ring of real numbers. All $a$, $b$, $c$ below are understood to be $> \theta$.*

**7.** $m(a * b) = (ma) * b = a * (mb)$.

**8.** If $ma > na^2$, then for all $b$,

    (i)  $mb > na * b$,

    (ii)  $mb > nb * a$.

**9.** Suppose $a \gtrsim b$. If $ma > na^2$, then $mb > nb^2$; hence, $\phi(a) \geqslant \phi(b)$.

**10.** If $ma > na^2$, $m'b > n'b^2$, then $(mn' + m'n)(a \oplus b) > nn'(a \oplus b)^2$. Hence, $\phi(a \oplus b) \leqslant \phi(a) + \phi(b)$.

**11.** If $m/n < \phi(a)$, then $ma < na^2$. Hence, $0 < \phi(a) = \sup\{m/n \mid ma < na^2\}$, $\phi(a \oplus b) = \phi(a) + \phi(b)$, and $a > b$ implies $\phi(a) > \phi(b)$.

**12.** Prove that $\phi$ is multiplicative.

**13.** Modify Exercises 4–12 so as to generate a proof of Theorem 6 for ordered semirings.    (2.2.7)

**14.** In the proof of the corollary to Theorem 6 prove that the restriction of $\gtrsim$, $\oplus$, $*$ to $A^+$ forms an Archimedean, positive, regular, ordered, local semiring.    (2.2.7)

**15.** Use the method of Section 2.3.2 to solve the system

$$x_1 + x_2 \geqslant 0,$$
$$2x_1 + 3x_2 \geqslant 0,$$
$$2x_1 + 5x_2 < 0.$$

# Chapter 3 Extensive Measurement

□ 3.1 INTRODUCTION

Extensive attributes such as length and mass have been measured success-fully since antiquity. The modern theory of extensive measurement, however, originated less than a century ago when Helmholtz (1887) and Hölder (1901) developed the first axiomatic analysis of extensive attributes. Generally speaking, a theory of extensive measurement is a set of assumptions, or axioms, formulated in terms of an ordering $\gtrsim$ (of objects with respect to some property) and a concatenation operation $\circ$ (between objects) that permit the construction of a scale $\phi$ satisfying

(i)  $a \gtrsim b$ iff $\phi(a) \geqslant \phi(b)$,
(ii)  $\phi(a \circ b) = \phi(a) + \phi(b)$.

Since the additive representation of mass, length, and time duration have become a part of our daily experience we tend to take for granted the qualitative laws (e.g., $a \gtrsim b$ whenever $a \circ c \gtrsim b \circ c$) that underlie the numerical representation. Indeed, under the natural interpretations of $\gtrsim$ and $\circ$, these laws typically reduce to common physical truths. Nevertheless, the formulation of a set of axioms that are sufficient for the representation and acceptable from an empirical standpoint poses several problems to which the present chapter is devoted.

Several attempts to improve Hölder's (1901) theory have been made.

(See Huntington 1902, a, b; 1917; Suppes, 1951; Behrend, 1953, 1956; Hofmann, 1963). We begin the chapter (Sections 3.2 and 3.3) with a simple set of axioms, due to Roberts and Luce (1968), which are both necessary and sufficient for extensive measurement when the concatenation operation is closed, i.e., when any two objects can be concatenated. Other discussions of necessary and sufficient conditions for extensive measurement are found in Alimov, 1950, and Holman, 1969. As the closure requirement is clearly too restrictive, we turn (Sections 3.4 and 3.5) to a more general theory that was originally formulated by Luce and Marley (1969), which permits limited concatenation. This theory is of central importance not only because of its generality, but also because it is used as a base for the development of difference measurement (Chapter 4), probability representations (Chapter 5), and additive conjoint measurement (Chapter 6).

Several physical interpretations of the theory are discussed in Section 3.6, where the consideration of velocity leads to the investigation of structures that possess an essential maximum, such as the velocity of light, and to a simultaneous axiomatization of length and velocity (Sections 3.7 and 3.8). The attempt to axiomatize classical thermodynamics (Giles, 1964) leads to yet another generalization, obtained by weakening the assumption that $\gtrsim$ is connected. This development, due to Roberts and Luce (1968), is presented in Sections 3.12 and 3.13. The problem of alternative (nonadditive) numerical representations is explored in Section 3.9. In Sections 3.10 and 3.11 we study constructive scaling procedures. Finally, applications to the social sciences and the limitations of extensive measurement are discussed, respectively, in Sections 3.14 and 3.15.

## 3.2 NECESSARY AND SUFFICIENT CONDITIONS

### 3.2.1 Closed Extensive Structures

We start with three primitives: a nonempty set $A$, a binary relation $\gtrsim$ on $A$, and a closed binary operation $\circ$ that maps $A \times A$ into $A$. The interpretation is: $A$ is a set of objects or entities that exhibit the attribute in question; $a \gtrsim b$ holds if and only if $a$ exhibits, in some prescribed qualitative fashion, at least as much of the attribute as $b$; and $a \circ b$ is an object in $A$ that is obtained by concatenating (or composing) $a$ and $b$ in some prescribed, ordered fashion. An example is weight measurement (see Kisch, 1965, for a detailed history of methods and equipment) in which the elements of $A$ are material objects; $a \gtrsim b$ is established by placing $a$ and $b$ on the two pans of an equal-arm pan balance and observing which pan drops; and $a \circ b$ means that $a$ and $b$ are both placed in the same pan with $a$ beneath $b$.

As in Definition 1.2, we write $a \sim b$ if and only if $a \gtrsim b$ and $b \gtrsim a$; and $a > b$ if and only if $a \gtrsim b$ and not $(b \gtrsim a)$.

DEFINITION 1. *Let $A$ be a nonempty set, $\gtrsim$ a binary relation on $A$, and $\circ$ a closed binary operation on $A$. The triple $\langle A, \gtrsim, \circ \rangle$ is a closed extensive structure iff the following four axioms are satisfied for all $a, b, c, d \in A$:*

1. *Weak order:* $\langle A, \gtrsim \rangle$ *is a weak order, i.e., $\gtrsim$ is a transitive and connected relation.*

2. *Weak associativity:* $a \circ (b \circ c) \sim (a \circ b) \circ c$.

3. *Monotonicity:* $a \gtrsim b$ *iff* $a \circ c \gtrsim b \circ c$ *iff* $c \circ a \gtrsim c \circ b$.

4. *Archimedean:* *If $a > b$, then for any $c, d \in A$, there exists a positive integer $n$ such that $na \circ c \gtrsim nb \circ d$, where $na$ is defined inductively as: $1a = a$, $(n + 1) a = na \circ a$.*

*The structure is called* positive *if, in addition, it satisfies*

5. *Positivity:* $a \circ b > a$.

The monotonicity property, Axiom 3, is also called the *cancellation* or *homogeneity law.*

In a positive structure, the obvious analog of the Archimedean property of real numbers is simply: for $a, b$ in $A$, there exists a positive integer $n$ such that $na \gtrsim b$. This along with Axioms 1–3 and 5 is not sufficient to prove the desired additive representation. Such a structure must be supplemented with something like the following nonnecessary solvability condition: if $a > b$, then there is a $c$ in $A$ such that $a \sim b \circ c$. One may drop the positivity assumption, Axiom 5, provided that positive and negative elements are defined and the Archimedean property is stated separately for the two types of elements. (See Section 3.3.2.)

Alimov (1950) introduced the following idea to substitute for the usual statement of the Archimedean property. A pair of elements $a, b$ in $A$ is called *anomalous* if not $(a \sim b)$ and either for all $n$ in $I^+$

$$na > (n + 1)b \quad \text{and} \quad nb > (n + 1)a$$

or for all $n$ in $I^+$

$$(n + 1) b > na \quad \text{and} \quad (n + 1) a > nb.$$

Alimov showed that Axioms 1–3 and the assumption that there do not exist any anomalous pairs are necessary and sufficient conditions for the additive representation formulated in Theorem 1 below (Fuchs, 1963, p. 167). Thus, the prohibition of anomalous pairs plays the same role as our Axiom 4. In fact, in the presence of Axioms 1–3, the following statements are true:

1. Axiom 4 implies that no anomalous pairs exist (Lemma 1).

2. If no anomalous pair exists, then Axiom 4 is true. (Alimov proved the additive representation from the hypothesis, and Axiom 4 is easily seen to be a necessary consequence of that representation.)

3. The usual Archimedean property, together with solvability, implies both these properties, and it is implied by either one of them (Fuchs, 1963, p. 163, Lemma B).

4. Either of these properties implies that the semigroup is weakly commutative, i.e., for all $a, b$ in $A$, $a \circ b \sim b \circ a$ (Lemma 3).

It should be noted that Axiom 4 is, in fact, the ordinary Archimedean property for differences. For if we define $(a, b) \gtrsim (c, d)$ to mean $a \circ c \gtrsim b \circ d$, then Axiom 4 simply says that if $(a, b)$ is positive (i.e., $a \succ b$), then for some positive integer $n$, $n(a, b) = (na, nb) \gtrsim (c, d)$. Difference structures are studied in detail in Chapter 4. Holman (1969) gave still another version of the Archimedean axiom which is similar to Alimov's.

The empirical status of the Archimedean axiom was discussed in Sections 1.4.4 and 1.5.2. Of the remaining axioms, 1 and 3 are most interesting empirically. Violations of transitivity typically are due either to insensitivity to small differences or to shifts in the basis for comparison (see Section 1.3.1 and Chapter 15). Monotonicity might be violated because of variable sensitivity ($a \succ b$ but $a \circ c \sim b \circ c$ for large $c$) or for other reasons such as decreasing marginal utility (see p. 124 for an example). Associativity and commutativity ordinarily follow from the empirical interpretation of $\circ$. These properties are employed in the weak form ($\sim$ replaces $=$) since this makes them necessary consequences of the following representation.

THEOREM 1. *Let $A$ be a nonempty set, $\gtrsim$ a binary relation on $A$, and $\circ$ a closed binary operation on $A$. Then $\langle A, \gtrsim, \circ \rangle$ is a closed extensive structure iff there exists a real-valued function $\phi$ on $A$ ($\phi: A \to$ Re) such that for all $a, b \in A$*

(i) $a \gtrsim b$ *iff* $\phi(a) \geqslant \phi(b)$;
(ii) $\phi(a \circ b) = \phi(a) + \phi(b)$.

*Another function $\phi'$ satisfies* (i) *and* (ii) *iff there exists $\alpha > 0$ such that $\phi' = \alpha\phi$. The structure is positive iff for all $a \in A$, $\phi(a) > 0$.*

The empirical significance of the uniqueness assertion is that we may select any element in $A$ and make it our unit of measurement, for suppose that $e$ is in $A$ and that $\phi$ is any representation, then by choosing $\alpha = 1/\phi(e)$, we see that $\phi'(e) = \alpha\phi(e) = 1$. Although the choice of unit is wholly arbitrary, certain conveniences in the representation of data usually impose some constraints. First, it is generally convenient to select the unit somewhere in the midrange of the objects being measured so that the range of numbers employed runs roughly from $10^{-n}$ to $10^n$ rather than, say, from 1 to $10^{2n}$ or from $10^{-2n}$ to 1. Second, if three or more scales are related by some simple empirical law, such as $\theta = k\phi^p\psi^q$, then it is common practice to select the units of the three scales $\theta$, $\phi$, and $\psi$ so that $k = 1$. And third, if a scale has a natural upper bound—for example, the velocity of light or the probability of the universal event—it is sometimes convenient to take that bound as the unit. Clearly, nothing empirical compels any of these choices.

Before giving a formal proof of Theorem 1, it may be useful to sketch roughly what is involved, both mathematically and in empirical practice, in constructing the function $\phi$. Select any $e$ in $A$; this will be the unit. For any other $a$ in $A$, and for any positive integer $n$, the Archimedean axiom guarantees that there is an integer $m$ for which $me > na$. Let $m_n$ be the least integer for which this is true, namely, $m_n e > na \gtrsim (m_n - 1)e$. Thus, $m_n$ copies of $e$ are approximately equal to $n$ copies of $a$. As we select $n$ larger and larger, the approximation presumably gets closer and closer and, assuming that the limit exists, it is plausible to define

$$\phi(a) = \lim_{n\to\infty} m_n/n.$$

Such a proof then proceeds to show that this limit exists and that it has the properties asserted in Theorem 1.

The formal proof in Section 3.3 reduces Theorem 1 to its analog for strictly ordered groups, Theorem 2.5 (Hölder's Theorem).

### 3.2.2 The Periodic Case

Among the physical quantities that are considered extensive, several are basically periodic. The two most familiar examples are clock time, which has a period of either 12 or 24 hours, and angle, which has a period of $2\pi$ radians (or 1 cycle or 360 degrees, depending on the unit in use). In both cases, the periodic measure relates simply to an ordinary extensive measure. For example, the time duration of 76:35 hours is the same as 3 days and 4:35 hours, and so in terms of clock time it is simply equivalent to 4:35 hours, because the complete cycles (days) drop out. In a sense, therefore, there is no reason to deal with the periodic case explicitly; however, because these measures are so commonly used, it is interesting to see what is involved in axiomatizing them.

The two examples above do not suggest any changes in Axioms 1 and 2—weak ordering and weak associativity—of Definition 1, but a little reflection makes clear that 3 and 4 must be modified. The monotonicity axiom includes the assumption of weak commutativity (see Lemma 3), which we must retain (we add it into Axiom 2), and the property $a \gtrsim b$ iff $a \circ c \gtrsim b \circ c$, which we must modify since it is not true in a periodic structure. For example, on a 24-hour clock, $18 > 6$, but adding 12 hours yields $18 \circ 12 = 6 < 18 = 6 \circ 12$. The key to modifying the monotonicity property lies in deciding whether the concatenation causes both, one, or neither term to complete one cycle. In the "both" and "neither" case, the order is unchanged, but in the "one" case, the order is reversed. So the only question is how to formulate qualitatively whether a cycle is completed. It is easy to see that this corresponds to whether $a \circ c < c$ or $a \circ c \gtrsim c$. This, then, leads to the following monotonicity axiom:

$a \gtrsim b$ iff one of the following holds: (i) $a \circ c \gtrsim b \circ c \gtrsim c$, or (ii) $c > a \circ c \gtrsim b \circ c$, or (iii) $b \circ c \gtrsim c > a \circ c$.

Next we must modify the Archimedean axiom. The main idea is that when $a > b$ it is possible to find some integer $n$ so that $na$ just completes a cycle in the sense that $na = (n-1) a \circ a < a$ and $nb$ fails to complete a cycle in the sense that $nb = (n-1) b \circ b \gtrsim b$. In effect, this says that multiples of the angular difference between $a$ and $b$ can be made arbitrarily large.

Summarizing, we have the following notion (Luce, 1971a):

**DEFINITION 2.** *Let $A$ be a nonempty set, $\gtrsim$ a binary relation on $A$, and $\circ$ a closed binary operation on $A$. The triple $\langle A, \gtrsim, \circ \rangle$ is a* closed, periodic extensive structure *iff the following four axioms are satisfied for all $a, b, c, \in A$:*

1. *Weak order:* $\langle A, \gtrsim \rangle$ *is a weak order.*

2. *Weak associativity and weak commutativity:* $\langle A, \circ, \sim \rangle$ *satisfies* $(a \circ b) \circ c \sim a \circ (b \circ c)$ *and* $a \circ b \sim b \circ a$.

3. *Monotonicity:* $a \gtrsim b$ *iff either* (i) $a \circ c \gtrsim b \circ c \gtrsim c$, *or* (ii) $c > a \circ c \gtrsim b \circ c$, *or* (iii) $b \circ c \gtrsim c > a \circ c$.

4. *Archimedean:* *if* $a > b$, *then there exists a positive integer $n$ such that* $a > na$ *and* $nb \gtrsim b$, *where* $1a = a$ *and* $na = (n-1) a \circ a$.

As expected, the representation is similar to Theorem 1, except that it is periodic in character.

**THEOREM 2.** $\langle A, \gtrsim, \circ \rangle$ *is a periodic structure with identity $e$ (i.e.,* *for all $a \in A$, $a \circ e \sim e \circ a \sim a$) iff for any real $\alpha > 0$ there exists a unique function $\phi: A \rightarrow [0, \alpha]$ such that for all $a, b \in A$*

(i)   $a \gtrsim b$ iff $\phi(a) \geqslant \phi(b)$;
(ii)  $\phi(a \circ b) = \phi(a) + \phi(b) \pmod{\alpha}$;
(iii) $\phi(e) = 0$.

For other formulations of the measurement of angle, see Lenz (1967) and Zassenhaus (1954).

The proof of Theorem 2, which involves embedding the given structure in a structure satisfying Definition 1, is left as Exercises 5–10.


## 3.3  PROOFS

### 3.3.1  Consistency and Independence of the Axioms of Definition 1

The positive real numbers with the usual ordering $\geqslant$ and addition $+$ provide a model for the axioms, thereby establishing their consistency. To show independence, we must provide for each axiom an example of a structure that fails to satisfy that axiom but satisfies all of the others. These examples are numbered according to the axiom not satisfied; it is left to the reader to verify that they do what they are supposed to do.

1.   $A = I^+$; $a \gtrsim b$ iff $a + 1 \geqslant b$; $a \circ b = a + b + 2$.

2.   $A = Ra^+$, $a \gtrsim b$ iff $a \geqslant b$, and $a \circ b = \max(a, b) + \frac{1}{2}\min(a, b)$.

3.   $A = I^+ \times I^+$, $(a, b) \circ (c, d) = (a+c, b+d)$, $(a, b) \gtrsim (c, d)$ iff $ab \geqslant cd$.

4.   $A = \{(m, n) \mid m, n \in I^+\}$; $(a_1, a_2) \gtrsim (b_1, b_2)$ iff either $a_1 > b_1$ or $a_1 = b_1$ and $a_2 \geqslant b_2$; $(a_1, a_2) \circ (b_1, b_2) = (a_1 + b_1, a_2 + b_2)$.

5.   $A = I$; $\gtrsim$ is $\geqslant$; $\circ$ is $+$.


### 3.3.2  Preliminary Lemmas

The common hypothesis of the following six lemmas is that $\langle A, \gtrsim, \circ \rangle$ is a closed extensive structure (Definition 1, Section 3.2.1).

LEMMA 1.   *There is no anomalous pair.*

*PROOF.*  Suppose that $a, b \in A$ are anomalous and, with no loss of generality, that $a \succ b$. If for all $n \in I^+$, $na \succ (n + 1)b$ and $nb \succ (n + 1)a$, then by Axioms 2 and 3,

$$na \circ (a \circ b) \sim (n + 1)\, a \circ b \prec nb \circ b,$$

which, by Axiom 1, violates Axiom 4 with $c = a \circ b$ and $d = b$. If, on the other hand, $(n + 1) b > na$ and $(n + 1) a > nb$, then for all $n \in I^+$,

$$na \circ b < (n + 1) b \circ b \sim nb \circ (b \circ b),$$

which violates Axiom 4 with $c = b$ and $d = b \circ b$.                                          ◇

We say that $a \in A$ is *positive*, *null*, or *negative* iff, for all $b \in A$,

$$a \circ b > b \quad \text{and} \quad b \circ a > b,$$
$$a \circ b \sim b \quad \text{and} \quad b \circ a \sim b,$$

or

$$a \circ b < b \quad \text{and} \quad b \circ a < b,$$

respectively.

LEMMA 2.    *Every element is either positive, null, or negative.*

PROOF.    Suppose $a \in A$. By Axiom 1, either $a \circ a > a$, $a \circ a \sim a$ or $a \circ a < a$. Suppose $a \circ a > a$. For any $b \in A$, Axioms 2 and 3 yield

$$a \circ (a \circ b) \sim (a \circ a) \circ b > a \circ b,$$

and

$$(b \circ a) \circ a \sim b \circ (a \circ a) > b \circ a.$$

So by Axioms 1 and 3, $a \circ b > b$ and $b \circ a > b$, hence $a$ is positive. Similarly, if $a \circ a \sim a$, then $a$ is null, and if $a \circ a < a$, $a$ is negative.                    ◇

LEMMA 3.    $\langle A, \gtrsim, O \rangle$ *is weakly commutative.*

PROOF.    If one of $a$ and $b$ is null, say $a$, then by definition $a \circ b \sim b \sim b \circ a$, establishing commutativity in that case. Next, suppose that $a$ and $b$ are both nonnegative but not null, then it is easily shown that $a\square$ is positive. Now suppose $a \circ b > b \circ a$, then

$$(n + 1)(a \circ b) > n(a \circ b) > n(b \circ a),$$

and

$$(n + 1)(b \circ a) \sim b \circ n(a \circ b) \circ a > n(a \circ b),$$

which is impossible by Lemma 1. If $a \circ b < b \circ a$, then

$$(n + 1)(a \circ b) \sim a \circ n(b \circ a) \circ b > n(b \circ a),$$

and

$$(n + 1)(b \circ a) > (n + 1)(a \circ b) > n(a \circ b),$$

which is again impossible by Lemma 1, and so $a \circ b \sim b \circ a$, by Axiom 1. The argument is similar if both elements are nonpositive. If $a$ is positive and $b$ negative, then $a \circ b$ is either nonnegative or nonpositive. In the former case, by what has just been shown, $a \circ (a \circ b) \sim (a \circ b) \circ a \sim a \circ (b \circ a)$, and so by Axioms 1 and 3, $a \circ b \sim b \circ a$. The latter case is similar.   ◇

*Define the relation* $\approx$ *on* $A \times A$ *as follows: for all* $a, b, c, d \in A$,

$$(a, b) \approx (c, d) \quad \textit{iff} \quad a \circ d \sim b \circ c.$$

**LEMMA 4.**   *The relation* $\approx$ *on* $A \times A$ *is an equivalence relation.*

*PROOF.*   Clearly, $\approx$ is reflexive; by Lemma 3, it is symmetric; and to prove transitivity, suppose that $(a, b) \approx (c, d)$ and $(c, d) \approx (e, f)$. From the definition of $\approx$ and from Axioms 2 and 3,

$$(a \circ d) \circ f \sim (b \circ c) \circ f \sim b \circ (c \circ f) \sim b \circ (d \circ e).$$

Rearranging by Lemma 3 and canceling $d$, $(a, b) \approx (e, f)$.   ◇

*Define:*

$$[a, b] = \{(a', b') \mid a', b' \in A, \quad (a', b') \approx (a, b)\},$$
$$D = \{[a, b] \mid a, b \in A\} = A \times A / \approx,$$
$$P = \{[a, b] \mid a, b \in A, a \gtrsim b\},$$
$$* \; : \; [a, b] * [c, d] = [a \circ c, b \circ d],$$
$$[0] = [a, a], a \in A,$$
$$[a, b]^{-1} = [b, a]$$
$$\gtrsim_D \; : \; [a, b] \gtrsim_D [c, d] \quad \text{iff} \quad [a, b] * [c, d]^{-1} \in P.$$

By definition, $(a, b) \approx (c, d)$ iff $(a, b) \in [c, d]$ iff $[a, b] = [c, d]$ iff $a \circ d \sim b \circ c$.

**LEMMA 5.**   (i)   *The operation* $*$, *the identity* $[0]$, *the inverse, and the relation* $\gtrsim_D$ *on* $D$ *are well defined.*
   (ii)   $\langle D, * \rangle$ *is a commutative group.*
   (iii)   $\langle D, \gtrsim_D \rangle$ *is a simple order.*

*PROOF.*   (i)   If $(a, b) \approx (a', b')$ and $(c, d) \approx (c', d')$, then by definition, Axiom 2, and Lemma 3,

$$a \circ c \circ b' \circ d' \sim a \circ b' \circ c \circ d' \sim a' \circ b \circ c' \circ d \sim b \circ d \circ a' \circ c',$$

whence $[a, b] * [c, d] = [a', b'] * [c', d']$, so $*$ is well defined.

It can readily be shown that $[0]$ and $[a, b]^{-1}$ are well defined. If $a \gtrsim b$ and $(a, b) \approx (a', b')$, then $b \circ a' \circ a \gtrsim a \circ b' \circ b$ (Axioms 3 and 1) whence by Lemma 3 and Axioms 1–3, $a' \gtrsim b'$. Therefore $P$ is well defined. Since $*$, inverse, and $P$ are well defined, so is $\gtrsim_D$.

(ii)   The closure, associativity, and commutativity of $*$ follow immediately from the same properties of $\circ$ (Axiom 2 and Lemma 3). By Lemma 3 and Axiom 3, $[0]$ is an identity, since

$$[a, b] * [0] = [a, b] * [a, a] = [a \circ a, b \circ a] = [a, b].$$

Similarly, $[0] * [a, b] = [a, b]$. By Lemma 3,

$$[a, b] * [a, b]^{-1} = [a, b] * [b, a] = [a \circ b, b \circ a] = [a \circ b, a \circ b] = [0].$$

Similarly, $[a, b]^{-1} * [a, b] = [0]$. So $\langle D, * \rangle$ is a commutative group.

(iii)   Reflexivity:   Since $[a, b] * [a, b]^{-1} = [0] \in P$, $[a, b] \gtrsim_D [a, b]$.

Transitivity:   Suppose that $[a, b] \gtrsim_D [c, d]$ and $[c, d] \gtrsim_D [e, f]$. By Axiom 3, if $[u, v]$, $[x, y] \in P$, then $[u, v] * [x, y] = [u \circ x, v \circ y] \in P$, so $[a, b] * [e, f]^{-1} = [a, b] * [c, d]^{-1} * [c, d] * [e, f]^{-1} \in P$; hence $[a, b] \gtrsim_D [e, f]$. Finally, $\gtrsim_D$ is antisymmetric, since if $[a, b] \gtrsim_D [c, d]$ and $[c, d] \gtrsim_D [a, b]$, then $a \circ d \gtrsim b \circ c \sim c \circ b \gtrsim d \circ a \sim a \circ d$, and so $[a, b] = [c, d]$.   $\diamondsuit$

LEMMA 6.   $\langle D, \gtrsim_D, * \rangle$ is an Archimedean simply ordered group (Definition 2.3).

PROOF.   It is easy to see the structure satisfies monotonicity. To show that it is Archimedean, suppose that $[a, b]$, $[c, d] \in A$, and $[a, b] >_D [0]$. By definition and Lemma 5,

$$[a, b] = [a, b] * [0]^{-1} \in P,$$

and $[a, b] \neq [0]$, so $a > b$. By Axiom 4, there exists an $n \in I^+$ such that $na \circ d \gtrsim nb \circ c$. Therefore,

$$n[a, b] * [c, d]^{-1} = [na, nb] * [d, c] = [na \circ d, nb \circ c] \in P$$

whence $n[a, b] \gtrsim_D [c, d]$. Since $[a, b] >_D [0]$, it is immediate that $(n + 1)[a, b] >_D n[a, b]$, so $(n + 1)[a, b] >_D [c, d]$.   $\diamondsuit$

### 3.3.3   Theorem 1   (p. 74)

$\langle A, \gtrsim, \circ \rangle$ is a (positive) closed extensive structure iff it is homomorphic to a (positive) additive semigroup of real numbers. The homomorphism is unique up to multiplication by a positive constant.

*PROOF.*   Necessity.   It is routine to show that Axioms 1–4 are necessary, and, when the range of the homomorphism is positive, that the structure is positive.

   Sufficiency.   Lemma 6 and Theorem 2.5 prove the existence of a real-valued function $\psi$ on $D$ such that for all $[a, b]$, $[c, d] \in D$

   (i)  $[a, b] \gtrsim_D [c, d]$ iff $\psi([a, b]) \geq \psi([c, d])$,
   (ii) $\psi([a, b] * [c, d]) = \psi([a, b]) + \psi([c, d])$,

where $\psi$ is unique up to multiplication by a positive constant. It is easy to see that given (ii), (i) is equivalent to: $\psi([a, b]) \geq 0$ iff $[a, b] \in P$. Define $\phi$ on $A$ as follows: for all $a \in A$, $\phi(a) = \psi([2a, a])$. We verify that $\phi$ has the desired properties. First,

$$
\begin{aligned}
\phi(a) - \phi(b) &= \psi([2a, a]) - \psi([2b, b]) \\
&= \psi([2a, a]) + \psi([b, 2b]) \\
&= \psi([2a, a] * [b, 2b]) \\
&= \psi([2a \circ b, a \circ 2b]) \\
&= \psi([a \circ b, a \circ b]) + \psi([a, b]) \\
&= \psi([0]) + \psi([a, b]) \\
&= \psi([a, b]).
\end{aligned}
$$

Since $\psi([a, b]) \geq 0$ iff $a \gtrsim b$, $\phi(a) \geq \phi(b)$ iff $a \gtrsim b$. Second,

$$
\begin{aligned}
\phi(a \circ b) &= \psi([2(a \circ b), a \circ b]) \\
&= \psi([2a, a] * [2b, b]) \\
&= \psi([2a, a]) + \psi([2b, b]) \\
&= \phi(a) + \phi(b).
\end{aligned}
$$

   Uniqueness:   Note that if $\phi'$ is any function as in Theorem 1, then $\psi'([a, b]) = \phi'(a) - \phi'(b)$ is well defined, since $(a, b) \approx (c, d)$ if and only if $a \circ d \sim b \circ c$, and it satisfies the two properties of Theorem 2.5. Thus, from the uniqueness clause of that theorem, $\phi' = \alpha\phi + \beta$, where $\alpha > 0$. Since both $\phi$ and $\phi'$ are additive, $\beta = 0$.                                      ◇

## □ 3.4  SUFFICIENT CONDITIONS WHEN THE CONCATENATION OPERATION IS NOT CLOSED

   Although the notion of a closed extensive structure is simple and elegant, it is unrealistic as a theory of measurement. This is so not only because of the idealization involved in the assumption that $\langle A, \gtrsim \rangle$ is a weak order, but rather because $\circ$ is assumed to be a closed operation. The latter assumption entails both that $A$ is infinite and, with the Archimedean axiom, that we can construct arbitrarily large entities. But clearly, if $a$ and $b$ are rods, we can form $a \circ b$ only if we have enough room to do so. If our laboratory is not sufficiently large, we can move into the corridor, and when that fails there is the campus walk. But sooner or later we are forced to stop concatenating. The classical theory does not attempt to incorporate this practical limitation, and so it includes many imaginary constructions and comparisons that cannot actually be performed. We develop an alternative theory that embodies these limitations and provides a more realistic account of length and mass measurement. This theory allows us to subsume under extensive measurement probability measurement (Chapter 5) and other schemes in which there is no natural concatenation operation (Chapters 4 and 6).

### 3.4.1  Formulation of the Non-Archimedean Axioms

To embody the limitations on $\circ$, we proceed as follows. Let $A$ denote the set of all realizable entities (say, rods)—i.e., $A$ consists of all the primitive, indivisible rods together with those composite ones that can be formed by concatenations within the limitations of the space available. Since the cartesian product $A \times A$ contains all logically possible concatenations of the form $a \circ b$, where $a, b$ are in $A$, those that can actually be formed must constitute a subset $B$ of $A \times A$. Thus, if $(a, b)$ is in $B$, then $a$ and $b$ can be concatenated and so $a \circ b$ is in $A$, which means that the operation $\circ$ is a function from $B$ into $A$.

As would be expected, the axioms about $\langle A, \gtrsim, B, \circ \rangle$ must include the weak order, associativity, monotonicity, and Archimedean properties of Definition 1, but the last three must be somewhat modified to take into account the fact that not all concatenations are possible. In the process, the structure of $B$ is implicitly characterized. In addition, we must add axioms to compensate for the fact that $\circ$ is no longer closed. Specifically, we are led to assume that the structure is commutative, to impose a weak form of solvability, and, finally, to limit the discussion to positive structures.

Consider, first, the concatenation of three rods. Suppose that $a$ and $b$ can be concatenated and that, in turn, $a \circ b$ can be concatenated with $c$, i.e., $(a, b)$ is in $B$ and $(a \circ b, c)$ is in $B$. Then surely $b$ can be concatenated with $c$ (just remove $a$) and $a$ with $b \circ c$, i.e., $(b, c)$ and $(a, b \circ c)$ are in $B$. Moreover, as in the closed system, it is true under most empirical interpretations that the grouping used does not affect the end result. Actually, in the presence of the other axioms, it is sufficient to assume only that

$(a \circ b) \circ c \gtrsim a \circ (b \circ c)$. In Lemma 9 $(a \circ b) \circ c \lesssim a \circ (b \circ c)$ is proved. Consequently, parentheses that tell how successive concatenations are grouped can be omitted, because different groupings do not affect the result.

Next, we turn to monotonicity. Suppose that $a \gtrsim b$ and that $a$ can be concatenated with $c$; then since $b$ is no longer than $a$, $b$ can be concatenated with $c$ and $a \circ c \gtrsim b \circ c$. Of course, in practice, the latter comparison cannot be carried out directly since $c$ cannot be in two places at once, and so one must, for example, find (or construct) a $d$ with the properties $a \circ c \gtrsim d$ and $d \gtrsim b \circ c$. Because of the transitivity of $\gtrsim$, we may abbreviate this double comparison as $a \circ c \gtrsim b \circ c$.

Commutativity was deduced from the axioms of a closed structure, but as it cannot be deduced in the present structure, it must be assumed. Specifically, if $(a, b)$ is in $B$, then we assume that $(b, a)$ is also in $B$, as seems plausible, and that $a \circ b \sim b \circ a$. We combine monotonicity and commutativity in the following single axiom: if $a \gtrsim b$ and $(a, c)$ is in $B$, then $(c, b)$ is in $B$ and $a \circ c \gtrsim c \circ b$.

We next introduce an assumption that imposes a further limitation on our choice of $A$ and $B$. It says that if $a$ is longer than $b$, then there exists an element $c$ in $A$ which does not exceed the difference between $a$ and $b$. Formally if $a \succ b$, we assert the existence of a $c$ in $A$ such that $(c, b)$ is in $B$ and $a \gtrsim c \circ b$. This property means one of two things. If there is a shortest rod $b$ in $A$, then there is no rod intermediate in length between $b$ and $2b$. For if $a$ were such a rod and $c$ is chosen not to exceed the difference between $a$ and $b$, then $b \succ c$, which is impossible. Similarly, there is no rod intermediate between $2b$ and $3b$, etc. On the other hand, if there is no shortest rod in $A$, then the elements of $A$ must be indefinitely fine grained. Since the granular structure of matter is often extremely fine compared with the size of the measured entities, this assumption is quite reasonable except when we are concerned with lengths of the order of molecular dimensions. Obviously, the solvability postulate is not a necessary consequence of the desired additive representation, and so it is a candidate for weakening and modification.

### 3.4.2 Formulation of the Archimedean Axiom

As was noted earlier, if we define $na$ for $a$ in $A$ and $n$ in $I^+$ inductively as: $1a = a$, $2a = a \circ a,..., na = (n - 1) a \circ a,...$, then the immediate analog of the Archimedean property of real numbers is that for any $b$ in $A$ there exists an $n$ in $I^+$ such that $na \gtrsim b$. This formulation will not do because the limitations embodied in $B$ may prohibit us from forming $na$ for sufficiently large $n$. We can bypass this difficulty by requiring the finiteness of the set of positive integers $n$ for which $na$ is defined and $b \succ na$. Such a set can be finite either because, beyond a certain integer $n_0$, $na$ is no longer defined or because beyond a certain integer $n_0$, $na \gtrsim b$, for all $n \geqslant n_0$.

There is still another way to formulate the Archimedean axiom—one that will turn out to be a convenient standard formulation for all measurement systems that have a real-valued representation—which reads as follows:

*Every strictly bounded standard sequence is finite,*

where the precise definition of standard sequence is determined according to the particular structure under consideration.

Intuitively, a *standard sequence* is always an ordered sequence of entities in which successive elements are equally spaced. More explicitly, if $a_{n-1}$ and $a_n$ are both in the sequence and if $\phi$ is the desired representation, then $\phi(a_n) - \phi(a_{n-1})$ is a constant independent of $n$. The problem in formulating this axiom lies in finding a qualitative definition of equal spacing, i.e., a definition that does not presume the desired numerical representation. In the present case the definition is simply: $a_1, a_2, ..., a_n, ...$, is a standard sequence if, for $i = 2, 3, ..., a_n = a_{n-1} \circ a_1$.

When, as in this chapter, the sequence is of the form $a_1, ..., a_n, ...$, it is *strictly bounded* if there exists $b$ in $A$ such that for all $a_n$ in the sequence, $b > a_n$. In other cases, the natural definition of a standard sequence does not have a first element, in which case the subscript $n$ may be negative as well as positive and strictly bounded means that there exist $b$ and $c$ in $A$ such that $b > a_n > c$ for all $a_n$ in the sequence.

### 3.4.3   The Axiom System and Representation Theorem

We summarize the axioms compactly as follows.

DEFINITION 3.   *Let $A$ be a nonempty set, $\gtrsim$ a binary relation on $A$, $B$ a nonempty subset of $A \times A$, and $\circ$ a binary function from $B$ into $A$. The quadruple $\langle A, \gtrsim, B, \circ \rangle$ is an* extensive structure with no essential maximum *if the following six axioms are satisfied for all $a, b, c \in A$:*

1.   *$\langle A, \gtrsim \rangle$ is a weak order.*

2.   *If $(a, b) \in B$ and $(a \circ b, c) \in B$, then $(b, c) \in B$, $(a, b \circ c) \in B$, and $(a \circ b) \circ c \gtrsim a \circ (b \circ c)$.*

3.   *If $(a, c) \in B$ and $a \gtrsim b$, then $(c, b) \in B$ and $a \circ c \gtrsim c \circ b$.*

4.   *If $a > b$, then there exists $d \in A$ such that $(b, d) \in B$ and $a \gtrsim b \circ d$.*

5.   *If $(a, b) \in B$, then $a \circ b > a$.*

6.   *Every strictly bounded standard sequence is finite, where $a_1, ..., a_n, ...$ is a* standard sequence *if for $n = 2, ..., a_n = a_{n-1} \circ a_1$, and it is* strictly bounded *if for some $b \in A$ and for all $a_n$ in the sequence, $b > a_n$.*

The expression "with no essential maximum" is introduced as a restrictive term in the definition in anticipation of modifications introduced in Sections 3.6.4 and 3.7.

THEOREM 3.   *Let* $\langle A, \gtrsim, B, \circ \rangle$ *be an extensive structure with no essential maximum. Then there exists a function* $\phi: A \to \text{Re}^+$ *such that for all* $a, b \in A$

   (i)   $a \gtrsim b$ *iff* $\phi(a) \geqslant \phi(b)$,
*and*
   (ii)   *if* $(a, b) \in B$, *then* $\phi(a \circ b) = \phi(a) + \phi(b)$.

*If another function* $\phi'$ *satisfies* (i) *and* (ii), *then there exists* $\alpha > 0$ *such that, for all nonmaximal* $a \in A$, $\phi'(a) = \alpha\phi(a)$.

The gist of the uniqueness statement is that we can select any nonmaximal element of $A$ and make it the unit of our measurement. In the psychological literature at least, classes of representations that are unique up to multiplication by a positive constant have come to be called *ratio scales* (Stevens, 1946, 1951). The present representation is a ratio scale, in this classical sense, provided that there is no maximal element with respect to $\gtrsim$. It is also a ratio scale in this same sense if the maximal element $a$ is equivalent to $b \circ c$ for some $b, c$—for then $\phi(a) = \phi(b) + \phi(c)$ and $\phi'(a) = \alpha\phi(b) + \alpha\phi(c) = \alpha\phi(a)$. In cases where there is a maximal element that bears no relation (via concatenation) to the other elements of $A$, its scale value is not determined. It appears justifiable to extend the meaning of "ratio scale" to include this last case also.

In the paradigm examples of extensive measurement (length, mass) the axioms of Definition 3 are quite likely to be accepted, and in fact they exert a strong normative influence on procedures for determining $\gtrsim$ relations and concatenating objects. However, in those applications where there is strong empirical interest in testing these axioms, intricate problems arise. For example, the Archimedean axiom is always true in finite structures; to reject it, one might need to convince oneself that $na$ is approaching a limit much less than $b$, or one might need to obtain information about the underlying rules that govern the generation of the ordering $\gtrsim$. The reader is invited to contemplate also the problem of testing the existential clause of Axiom 2, i.e., that $b \circ c$ and $a \circ (b \circ c)$ exist.

## 3.5  PROOFS

### 3.5.1  Consistency and Independence of the Axioms of Definition 3

The consistency of the axioms of Definition 3 (Section 3.4.3) is obvious given that Theorem 3 is true; however, it is of some interest to prove their consistency by providing a very simple finite model. Let

$A = \{a, b, c, d, e\}$,

$\succsim$ : $a \sim b \succ c \sim d \succ e$ and all implications that follow from transitivity,

$B = \{(e, e), (d, e), (e, d), (c, e), (e, c)\}$,

$\circ$ : $e \circ e = c$, $\quad d \circ e = e \circ d = b$, $\quad c \circ e = e \circ c = a$.

The reader can check that this system fulfills the axioms.

The following examples, showing the independence of the axioms, are taken from Suppes (1951). They are numbered according to the axiom violated. Their verification is left as Exercise 14.

1.   $A = I^+$; $a \succsim b$ iff $a + 1 \geqslant b$; $B = A \times A$; $a \circ b = a + b - 1$.

2.   $A = \mathrm{Ra}^+$; $a \succsim b$ iff $a \geqslant b$; $B = A \times A$; $a \circ b = \max(a, b) + \frac{1}{2}\min(a, b)$.

3.   $A = \mathrm{Ra}^+$; $a \succsim b$ iff $a \geqslant b$; $B = A \times A$; $a \circ b = a + b/2$.

4.   $A$ is the positive integers greater than 1; $a \succsim b$ iff $a \geqslant b$; $B = A \times A$; $a \circ b = a + b$.

5.   $A = \{1\}$; $a \succsim b$ iff $a \geqslant b$; $B = A \times A$; $a \circ b = ab$.

6.   $A = \{(m, n) \mid m, n \in I$ and either $m > 0$ or $m = 0$ and $n > 0\}$, $(a_1, a_2) \succsim (b_1, b_2)$ iff $a_1 > b_1$ or $a_1 = b_1$ and $a_2 \geqslant b_2$; $B = A \times A$; $(a_1, a_2) \circ (b_1, b_2) = (a_1 + b_1, a_2 + b_2)$.

### 3.5.2  Preliminary Lemmas

LEMMA 7.   *If $(a, b) \in B$, then $(b, a) \in B$; and if, in addition, $a \succsim c$ and $b \succsim d$, then $(c, d) \in B$ and $a \circ b \succsim c \circ d$.*

PROOF.   Since $a \succsim a$ by Axiom 1, and $(a, b) \in B$, by Axiom 3, $(b, a) \in B$. Since $a \succsim c$, Axiom 3 yields $(b, c) \in B$ and $a \circ b \succsim b \circ c$. Similarly $(c, d) \in B$ and $b \circ c \succsim c \circ d$, so the result follows from Axiom 1.     $\diamondsuit$

LEMMA 8.   *If $(a, b) \in B$, then $a \circ b \sim b \circ a$.*

PROOF.   By Lemma 7, $(b, a) \in B$. By Axiom 1, $a \succsim a$, and so Axiom 3 yields $a \circ b \succsim b \circ a$. Similarly, $b \succsim b$, and so $b \circ a \succsim a \circ b$.     $\diamondsuit$

LEMMA 9.   *$(a, b)$ and $(a \circ b, c) \in B$ iff $(b, c)$ and $(a, b \circ c) \in B$; and if both conditions hold, then $a \circ (b \circ c) \sim (a \circ b) \circ c$.*

PROOF.   If $(a, b), (a \circ b, c) \in B$, then Axiom 2 gives $(b, c)$ and $(a, b \circ c) \in B$ and $(a \circ b) \circ c \succsim a \circ (b \circ c)$. By Lemmas 7 and 8, $(c \circ b, a) \in B$, which yields the reverse inequality, as required. If $(b, c)$ and $(a, b \circ c) \in B$, then Lemmas 7 and 8 yield $(c \circ b, a) \in B$, and Axiom 2 can be applied, etc.          $\Diamond$

Note that Lemmas 7–9 rely only on Axioms 1–4 of Definition 3.

### 3.5.3  Theorem 3  (p. 85)

*If $\langle A, \succsim, B, \circ \rangle$ is an extensive structure with no essential maximum, then there is an order-preserving ratio scale such that for $(a, b) \in B$, $\phi(a \circ b) = \phi(a) + \phi(b)$.*

*PROOF.*   Let **A** denote the set of equivalence classes of $A$ under $\sim$, i.e., $\mathbf{A} = A/\sim$, and let $\succsim$ be the simple order on **A** induced by $\succsim$. Define $\mathbf{B} = \{(\mathbf{a}, \mathbf{b}) \mid (a, b) \in B\}$, and for $(\mathbf{a}, \mathbf{b}) \in \mathbf{B}$ define $\mathbf{a} \circ \mathbf{b}$ to be the equivalence class of $a \circ b$. Observe that $\circ$ is well defined since if $a' \sim a$ and $b' \sim b$, then, by Lemma 7, $(a', b') \in B$ and by two applications of Axiom 3, $a \circ b \sim b \circ a' \sim a' \circ b'$, so $\mathbf{a'} \circ \mathbf{b'} = \mathbf{a} \circ \mathbf{b}$. We show that $\langle \mathbf{A}, \succsim, \mathbf{B}, \circ \rangle$ is an Archimedean, regular, positive, ordered local semigroup (Definition 2.2). Properties 1, 6, 7, and 8 follow respectively from Axioms 1, 5, 4, and 6 of Definition 3; properties 2–4 follow from Lemma 7; and property 5 follows from Lemma 9. Theorem 3 follows immediately from Theorem 2.4.          $\Diamond$

## ☐ 3.6  EMPIRICAL INTERPRETATIONS IN PHYSICS

In this section we discuss several classical empirical examples of extensive measurement in physics. In all these examples the primitives have a natural physical interpretation, and the axioms reduce to well-known physical principles.

Since the applications of extensive measurement in the social sciences are quite problematic, they are discussed separately in Section 3.14.

### 3.6.1  Length

In motivating the axioms of Definitions 1 and 3, we used the abutting of two rods along a straight line as the empirical interpretation of concatenation for length measurement. As Ellis (1966) pointed out, at least one other totally different interpretation of concatenation also satisfies the axioms and so leads to an additive representation; this measure of length is not linearly related to the usual one. Campbell (pp. 290–294 of 1957 edition) discussed other examples of a similar nature.

To present Ellis' interpretation we begin with a collection of rods. Let $a * b$ be the hypotenuse of the right triangle whose sides are $a$ and $b$. The comparison relation $\gtrsim$ is determined by placing two rods side by side, with one end coinciding, and observing which one extends at the other end. Using elementary properties of right triangles, it is easy to verify that the axioms of Definition 3 are satisfied. The only property that might present a slight difficulty is associativity. It is explained in Figure 1 where the lines

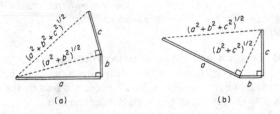

**FIGURE 1.** Orthogonal concatenation for length measurement illustrating left term (a) and right term (b) of the associative property.

are labeled by their lengths in the usual measure. Since Definition 3 is satisfied, Theorem 3 holds, hence there is a measure $\psi$ that is order preserving and additive over this new concatenation. Since the usual measure $\phi$ is also order preserving, $\psi$ and $\phi$ must be monotonically related, and by properties of triangles it is easy to see that $\psi$ is proportional to $\phi^2$.

To most people, the new interpretation seems much more artificial than the original one. In spite of this strong feeling, neither Ellis nor the authors know of any argument for favoring the first interpretation except familiarity, convention, and, perhaps, convenience. We are used to length being measured along straight lines, not along the hypotenuses of right triangles, but no empirical reasons appear to force that choice. Indeed, we could easily reconstruct the whole of physics in terms of $\psi$ by replacing all occurrences of $\phi$ by $\psi^{1/2}$. This would make some equations appear slightly more complicated; others would be simpler. In fact, when $\phi^2$ happens to be the more convenient measure, it is common to assign it a name and to treat it as the fundamental measure. Examples are the moment of inertia and the variance of a random variable. In the present case, if $a$ and $b$ are rods, the squares with side $a$ and with side $b$ can be concatenated by forming the square on the hypotenuse; $\phi^2$ will be an additive (area) measure for such concatenation of squares.

This conventional aspect of the empirical interpretation is rather closely related to the conventional aspect of the representation, which is discussed in Section 3.9.

### 3.6.2 Mass

A comparison relation for mass is often developed using an equal-arm balance. (When objects with densities less than air are to be judged the comparison is made in a vacuum.) The main requirements for such a balance are that it should have as little friction as possible and that if $a$ and $b$ balance when $a$ is in one pan and $b$ in the other, then they should also balance when their locations are reversed. Concatenation is simply interpreted as the positioning of both objects together in a pan. It is clear that, as with length, practical considerations, e.g., the strength and size of the balance, limit the possible concatenations. If we assume that the substances involved are noninteractive (among other things, chemically unreactive, at least over short periods of time), it is plausible that the axioms of Definition 3 should be met.

### 3.6.3 Time Duration

It has been said occasionally that time duration, unlike mass and length (the other two basic concepts of classical mechanics), cannot be fundamentally measured using the theory of extensive measurement. This is false, as can be seen in the following interpretation of the axioms (Campbell, 1957, pp. 550–553).

We begin with a set of pendulums, and our aim is to assign to each pendulum a number representing the time duration of its period. The comparison relation is determined by starting the two pendulums at the same instant and noting which pendulum first completes one period. To concatenate $a$ and $b$, we start a period of $b$ at the exact moment when $a$ completes one full period. The actual engineering of these comparisons and concatenations is not easy if any degree of precision is to be achieved. A possible procedure would involve a light beam at the low point of the swing which, when broken, triggers an extremely fast camera. By intelligent trial-and-error procedures, it can be arranged for the two pendulums to cross the beam at the same time, and then it is only necessary to see which returns to the beam first to make a comparison. This also serves to concatenate two periods. Actually, as will become apparent later (Section 3.10.1), we do not need to concatenate pendulums in this way. Under the above interpretation of the primitives it is plausible that the axioms will indeed hold. Thus, the theory of extensive measurement provides a justification for the usual time measures. Note that the limitations on concatenation (embodied in $B$) are indispensable in this case, because of lack of time rather than of space or strength.

### 3.6.4 Resistance

Physics presents several examples, such as electrical resistors and springs, in which both an ordering and its converse can be combined with concatenation operations to form extensive structures. Consider the electrical case. Suppose that $\gtrsim$ is a qualitative ordering by resistance of a set $A$ of objects. Thus, for $a$, $b$ in $A$, $a \gtrsim b$ means that $a$ exhibits as much or more electrical resistance as does $b$. We can wire resistors in series, denoted $a \circ b$, and it is not difficult to see that if classical physics is correct $\langle A, \gtrsim, \circ \rangle$ satisfies the axioms of an extensive structure, and there is an additive ratio scale representation $\phi$ called "resistance." Equally well, if we let $\gtrsim'$ denote the converse of $\gtrsim$, i.e., $a \gtrsim' b$ if and only if $b \gtrsim a$, and we wire resistors in parallel, denoted $a \parallel b$, then again it is not difficult to see that $\langle A, \gtrsim', \parallel \rangle$ is an extensive structure with an additive ratio scale $\psi$, which we may call "acceptance." Now, according to physics, acceptance is nothing but the reciprocal of resistance, i.e., $\psi = 1/\phi$. What qualitative property relating the two extensive structures implies this?

Clearly, such a property must formulate a close relation between the two operations $\circ$ and $\parallel$. Consider concatenations of $a$ with itself $n$ times. We denote the series one by

$$na = a \circ a \circ \cdots \circ a,$$

and the parallel one by

$$a^n = a \parallel a \parallel \cdots \parallel a.$$

The essence of the matter is that these two operations are inverses of one another in the following qualitative sense. For all $n$ in $I^+$ and all $a$ in $A$,

$$(na)^n \sim a.$$

From this we show that $\psi = \alpha/\phi$ (with $\alpha$ in $\mathrm{Re}^+$) follows provided that $\phi$ and $\psi$ are both onto the positive rationals. Since $\phi$ preserves the order of $\gtrsim$ and $\psi$ preserves the order of the converse of $\gtrsim$, there exists a strictly decreasing function $f$ such that $\psi = f\phi$. Applying this to the qualitative law and taking into account that $\phi$ is additive over $\circ$ and $\psi$ is additive over $\parallel$, we have for any $s$ in $\mathrm{Ra}^+$ and any $n$ in $I^+$,

$$f(s) = f[\phi(a)] = \psi(a) = \psi[(na)^n] = nf[\phi(na)] = nf[n\phi(a)] = nf(ns).$$

If we write $s = t/m$, where $t$ is rational and $m$ is a positive integer,

$$f(t/m) = mf(t),$$

and combining these two equations for any $r = n/m$ in Ra$^+$,

$$rf(rs) = \frac{n}{m} f\left(\frac{n}{m} s\right) = nf(ns) = f(s).$$

Setting $s = 1$, $r = f(1)/f(r)$, and so

$$f(rs)f(1) = f(r)f(s).$$

It is well known that the only strictly decreasing solution to this equation is $f(r) = \alpha r^{-\beta}$, where $\alpha = f(1) > 0$ and where $\beta > 0$. Substituting this into $rf(rs) = f(s)$ immediately yields $\beta = 1$, which we wished to prove.

In Section 10.7 a somewhat more difficult problem of multiple representations of the same ordering arises; we deal with it in much the same way by searching for a qualitative law that relates the two structures that have given rise to the alternative representations.

### 3.6.5  Velocity

Suppose that we have a set of bodies that are moving in a given direction at constant velocities. Ideally, the bodies are moving in a vacuum and subject to no forces. The comparison relation can be determined by comparing the distances traveled by two bodies in a given interval of time, provided that we accept the law that such a comparison is qualitatively independent of the time period chosen. The concatenation $a \circ b$ is, intuitively, the velocity $b$ superimposed on $a$. More precisely, $a \circ b$ is the velocity of a body that an observer on another body, moving at velocity $a$, would judge to be $b$. According to classical physics, i.e., prior to the end of the last century, these operations were believed to satisfy the axioms of Definition 3. This implies simple additivity of velocities, and so, in principle, any numerical value of velocity could be produced.

One of the basic postulates of the theory of relativity is that this is false; specifically, no body can move at a velocity exceeding that of light. Put another way, there is a natural upper bound to velocities which is quite distinct from the bounds that we have incorporated into our set $B$ to reflect practical limitations. Where, then, do the axioms fail? The most obvious place is Axiom 5, which asserts that $a \circ b$ is always strictly greater than $a$; according to the theory of relativity, with $a$ representing the velocity of light, this cannot be true. A second, less obvious, although important, failure occurs in the Archimedean Axiom 6. Suppose again that $a$ is the velocity of light. Then $nb$ can never exceed $a$, no matter how large $n$ is. So, it appears that if we wish to incorporate velocity measurement we should amend Axioms 5 and 6; to do so, we must first define abstractly

the notion of an essential maximum, e.g., the velocity of light in relativity
theory. For a general discussion of the measurement of the velocity of light,
see Dorsey (1944).

## 3.7  ESSENTIAL MAXIMA IN EXTENSIVE STRUCTURES

### 3.7.1  Nonadditive Representations

With relativistic velocity as our motivation, we introduce the following
two definitions:

DEFINITION 4.   *Let $A$ be a nonempty set, $\succsim$ a binary relation on $A$,
$B$ a nonempty subset of $A \times A$, and $\circ$ a binary function from $B$ into $A$.
An element $z \in A$ is an* essential maximum *(relative to $\succsim$ and $\circ$) iff it is
maximal, i.e., for all $a \in A$, $z \succsim a$, and there exists some $a \in A$ such that
$(z, a) \in B$.*

If Axioms 1 and 3 of Definition 3 hold and $z$ is an essential maximum, then
$z'$ is an essential maximum if and only if $z' \sim z$.

DEFINITION 5.   *Let $A$ be a nonempty set, $\succsim$ a binary relation on $A$,
$B$ a nonempty subset of $A \times A$, and $\circ$ a binary function from $B$ into $A$. The
quadruple $\langle A, \succsim, B, \circ \rangle$ is an* extensive structure *iff Axioms 1–4 of Definition 3
(Section 3.4.3) hold and Axioms 5 and 6 are replaced by*:

5′.   *For all $a, b \in A$, if $(a, b) \in B$ and $a$ is not an essential maximum, then
$a \circ b \succ a$.*

6′.   *If a standard sequence is strictly bounded by an element that is not an
essential maximum, then the sequence is finite.*

The following result gives a simple characterization of essential maxima.

THEOREM 4.   *Suppose that $\langle A, \succsim, B, \circ \rangle$ is an extensive structure, that
$a, z \in A$, and that $(z, a) \in B$. Then $z$ is an essential maximum iff $z \sim z \circ a$.*

From this result, we see that if Axiom 5 of Definition 3 holds, then there
can be no essential maximum, in which case Axioms 6 and 6′ are identical.
Put another way, Definition 3 is the special case of Definition 5 in which
no essential maximum exists, hence the phrase included in the original
definition.

The next result describes the relationship between extensive structures
with and without essential maxima.

**THEOREM 5.** *Suppose that $\langle A, \succsim, B, \circ \rangle$ is an extensive structure. Let $A'$ be the subset of nonmaximal elements of $A$; $\succsim'$ be the restriction of $\succsim$ to $A'$; $B' = \{(a, b) \mid a, b \in A', (a, b) \in B, \text{ and } a \circ b \in A'\}$; and $\circ'$ be the restriction of $\circ$ to $B'$. If $A'$ and $B'$ are nonempty, then $\langle A', \succsim', B', \circ' \rangle$ is an extensive structure with no essential maximum.*

It is clear from this theorem and from Theorem 3 that if we are willing to ignore the maximal elements, an additive representation for the remaining elements is available. However, this representation is unsatisfactory in two respects. First, when measuring velocity, we do not want to exclude light. Second, the additive representation does not have the essential feature that "velocity," measured in this way, be proportional to the distance traveled divided by the time elapsed. In fact, the additive quantity measured by concatenating velocities is not velocity in the usual sense, but rather what is called the "rapidity" of relative motion (Karapetoff, 1944).

To overcome these objections, we must abandon the additivity of the representation. If $\phi$ is the additive function on $A'$ derived from Theorems 3 and 5, and if $g$ is any strictly increasing function from $(0, 1)$ onto $(0, \infty)$, we can define a new function $\Phi$ on $A$ by

$$\Phi(a) = \begin{cases} g^{-1}[\phi(a)], & a \prec z, \\ 1, & a \sim z, \end{cases}$$

where $z$ is any essential maximum. If $a \circ b \prec z$, then $\Phi$ satisfies the addition formula

$$\Phi(a \circ b) = g^{-1}\{g[\Phi(a)] + g[\Phi(b)]\}.$$

This formula is valid even for $a \sim z$ if we introduce the conventions $g(1) = \infty$, $t + \infty = \infty$, and $g^{-1}(\infty) = 1$. To obtain the usual relativistic addition formula for velocities, choose $g(t) = \tanh^{-1}(t) = \frac{1}{2}\log_e(1 + t)/(1 - t)$. It is then easily verified that

$$\Phi(a \circ b) = [\Phi(a) + \Phi(b)]/[1 + \Phi(a)\,\Phi(b)].$$

This equation holds for $a \sim z$ with no special conventions.

This is not a satisfactory treatment of relativistic velocity. For one thing, we have not dealt with the interrelation of velocity, distance, and time. For another, the function $g$ is arbitrary, and choosing the inverse hyperbolic tangent, which yields the correct velocity formula, is completely *ad hoc* since it is not determined by the axioms. For a third, the uniqueness theorem for $\Phi$ clearly takes the form

$$\Phi'(a) = g^{-1}\{\alpha g[\Phi(a)]\},$$

where $\alpha > 0$. But $\Phi$ should be the absolute scale of relativistic velocity (i.e., the velocity of $a$ divided by the velocity of light) since $\Phi(z) = 1$, and it should not be subjected to the above transformations.

## 3.7.2 Simultaneous Axiomatization of Length and Velocity

Two of the above three problems can be overcome by axiomatizing the measurement of length and velocity simultaneously. We first introduce an extensive structure with no essential maximum, $\langle A, \gtrsim, B, \circ \rangle$, in which $\circ$ is to be interpreted as the usual concatenation of length. By Theorem 3, there is a ratio scale $\phi$ of length. We shall assume that velocities are identified by the distances traveled during a fixed interval of time. Specifically, let $z$ denote the distance traveled by light during the time interval denoted by $t_z$, i.e., if in some appropriate velocity units, the velocity of light is $c$, then $\phi(z) = ct_z$. Let $A_z$ denote all those lengths in $A$ which are shorter than or equal to $z$. According to the theory of relativity, each feasible velocity can be identified with an element of $A_z$, namely with the distance that a body moving at this velocity traverses in the time interval $t_z$. The concatenation of velocities induces a new concatenation $\circ_z$ on some appropriate subset $B_z$ of $A_z \times A_z$. The ordering of velocities is, for each $t_z$, simply the restriction of $\gtrsim$ to $A_z$, denoted $\gtrsim_z$. We shall suppose that the structure $\langle A_z, \gtrsim_z, B_z, \circ_z \rangle$ is an extensive one in which $z$ is an essential maximum.

If velocities are expressed as proportions of the velocity of light, e.g.,

$$u = \frac{\phi(a)/t_z}{\phi(z)/t_z} = \frac{\phi(a)}{\phi(z)} \quad \text{and} \quad v = \frac{\phi(b)}{\phi(z)},$$

then the concatenation of velocities is given by

$$u * v = \phi(a \circ_z b)/\phi(z).$$

As stated, $*$ depends upon the arbitrary choice of a time $t_z$ of observation. Since one of the main properties of velocity is the invariance of the ratio of distance traveled by the time elapsed, we must include an assumption to the effect that $*$ does not depend upon $z$. This is accomplished by formulating expressions for $u$ and $v$ in terms of any two time choices, $t_y$ and $t_z$, and then demanding that $u * v$ be independent of the choice of time.

The upshot of these assumptions is that absolute relativistic velocities satisfy an addition formula for concatenation of the form

$$u * v = f^{-1}[f(u) + f(v)],$$

where $f$ is determined up to multiplication by a positive constant. The problem of unwanted scale transformations for velocity is eliminated by linking the unit in which "finite" velocities are measured to that in which the velocity of light is measured; both are linked to the choice of a unit of length. The arbitrariness of the addition formula for the concatenation of velocities is also eliminated by considering two concatenations, $\circ$ and $*$, and requiring that the representation of $\circ$ be additive. The main weakness of this theory is that the function $f$, although fully determined, need not be the inverse hyperbolic tangent. We do not know how to axiomatize that specific result in any natural way using the present primitives.

We summarize the assumptions and result as follows.

THEOREM 6. *Suppose that $\langle A, \succsim, B, \circ \rangle$ is an extensive structure with no essential maximum and that any additive representation (Theorem 3) is onto $\mathrm{Re}^+$. Let $\phi$ be one such representation. For any $z \in A$, let*

$$A_z = \{a \mid a \in A \quad and \quad z \succsim a\}$$

*and let $\succsim_z$ be the restriction of $\succsim$ to $A_z$. Suppose that for each such $z$, there exist a nonempty $B_z \subset A_z \times A_z$ and $\circ_z : B_z \to A_z$ such that*

(i) $\langle A_z, \succsim_z, B_z, \circ_z \rangle$ *is an extensive structure in which $z$ is an essential maximum;*

(ii) *if $z \succ a, b$ and $(a, b) \in B_z$, then $z \succ a \circ_z b$;*

(iii) *if $a, b \in A_y$, $(a, b) \in B_y$, $c, d \in A_z$, $(c, d) \in B_z$, and if*

$$\phi(a)/\phi(y) = \phi(c)/\phi(z) \quad and \quad \phi(b)/\phi(y) = \phi(d)/\phi(z),$$

*then*

$$\phi(a \circ_y b)/\phi(y) = \phi(c \circ_z d)/\phi(z).$$

*Then there exists a strictly increasing function $f$: $(0, 1)$ onto $\mathrm{Re}^+$ such that for each $z \in A$ and for all $a, b \in A_z$ satisfying $(a, b) \in B_z$ and $z \succ a, b$,*

$$\phi(a \circ_z b)/\phi(z) = f^{-1}\{f[\phi(a)/\phi(z)] + f[\phi(b)/\phi(z)]\}.$$

*Moreover, $f$ is unique up to multiplication by a positive constant.*

In the above discussion and, in particular, in hypothesis (ii) of Theorem 6, we have assumed that if neither $a$ nor $b$ is an essential maximum, then their concatenation, $a \circ_z b$, is also not an essential maximum. Although we do not know of any empirical structure in which the concatenation of two such elements is an essential maximum, we include the following representation for such a case for the sake of completeness.

THEOREM 7. *Suppose that $\langle A, \succsim, B, \circ \rangle$ is an extensive structure that has at least one essential maximum $z$. Let $\langle A', \succsim', B', \circ' \rangle$ be defined* *as in Theorem 5. Suppose that $A'$ and $B'$ are nonempty and that there exist $x, y \in A'$ such that $(x, y) \in B$ and $z \sim x \circ y$. Then there exists $\phi \colon A \to \mathrm{Re}^+$ such that for all $a, b \in A$*

   (i)   *$a \gtrsim b$ iff $\phi(a) \geqslant \phi(b)$;*
   (ii)  *if $(a, b) \in B'$, then $\phi(a \circ b) = \phi(a) + \phi(b)$;*
   (iii)  *$\phi(z) = \inf\{\phi(x) + \phi(y) \mid (x, y) \in B, x \circ y \sim z\}$.*

*Moreover, $\phi$ is unique up to multiplication by a positive constant.*

The definitions and results in this section modify those given by Luce and Marley (1969).

## 3.8  PROOFS

### 3.8.1  Consistency and Independence of the Axioms of Definition 5

The example given in Section 3.5.1 establishes the consistency of the axioms of Definition 5 (Section 3.7.1); nonetheless, it is desirable to have a simple example of a structure that actually possesses an essential maximum.

$A = \{a, b, c\}$,
$\gtrsim : a \succ b, \quad b \succ c, \quad a \succ c,$
$B = A \times A$,
$\circ : a \circ a = a \circ b = b \circ a = a \circ c = c \circ a = b \circ b = b \circ c = c \circ b = a,$
    $c \circ c = b.$

The examples showing the independence of the axioms of Definition 3 (Section 3.5.1) are still appropriate.

The proofs of Theorems 4 and 5 are given as Exercises 17 and 18, respectively. The proofs of Theorems 6 and 7 appear in the following sections.

### 3.8.2  Theorem 6  (p. 95)

*Let $\langle A, \gtrsim, B, \circ \rangle$ be an extensive structure with no essential maximum and with an additive representation $\phi$ onto $\mathrm{Re}^+$. For any $z \in A$, suppose that $A_z = \{a \mid z \gtrsim a\}$, $\gtrsim_z$ is the restriction of $\gtrsim$ to $A_z$, and $\langle A_z, \gtrsim_z, B_z, \circ_z \rangle$ is an extensive structure with an essential maximum $z$; that $a \circ_z b$ is an essential maximum iff either $a$ or $b$ is an essential maximum; and that concatenation via $\circ_z$ is independent of the choice of $z$. Then there exists a strictly increasing ratio scale $f$ from $(0, 1)$ onto $\mathrm{Re}^+$ that renders $\circ_z$ additive over those elements that are not essential maxima.*

*PROOF.* If $g$ from $(0, 1)$ onto $Re^+$ is strictly increasing, then as we showed in Section 3.7, there exists $\Phi_z : A_z \to (0, 1)$ that is order preserving and for all $a, b < z, (a, b) \in B_z$,

$$\Phi_z(a \circ_z b) = g^{-1}[g\Phi_z(a) + g\Phi_z(b)].$$

Since both $\phi$ and $\Phi_z$ are order preserving on $A_z$ and $\phi$ is onto $(0, \phi(z)]$, there exists a strictly increasing function $h_z$ from $(0, \phi(z)]$ onto $(0, 1]$ such that for all $a \in A_z$, $\Phi_z(a) = h_z\phi(a)$. Define $f_z$ from $(0, 1)$ onto $Re^+$ by

$$f_z(t) = gh_z[t\phi(z)].$$

Since $g$ and $h_z$ are both strictly increasing, so is $f_z$. Suppose that $z > a, b$ and $(a, b) \in B_z$, then by the definition of $f_z$, $h_z$, and $\Phi_z$,

$$\begin{aligned}
f_z[\phi(a \circ_z b)/\phi(z)] &= gh_z\phi(a \circ_z b) \\
&= g\Phi_z(a \circ_z b) \\
&= g\Phi_z(a) + g\Phi_z(b) \\
&= gh_z\phi(a) + gh_z\phi(b) \\
&= f_z[\phi(a)/\phi(z)] + f_z[\phi(b)/\phi(z)].
\end{aligned}$$

Next we show that $f_z$ is unique up to multiplication by a positive constant. Suppose that $f_z'$ is another function with the same property. Since both $f_z$ and $f_z'$ are strictly increasing from $(0, 1)$ onto $Re^+$, there exists a strictly increasing function $F$ from $Re^+$ onto $Re^+$ such that $f_z' = Ff_z$. Let

$$\begin{aligned}
S_z &= \{\mu \mid \mu = f_z[\phi(a)/\phi(z)] \quad \text{for some} \quad a < z\}, \\
T_z &= \{(\mu, \nu) \mid \mu, \nu, \mu + \nu \in S_z\}.
\end{aligned}$$

As is easily seen, $\langle S_z, \geqslant, T_z, + \rangle$, where $\geqslant$ is the usual ordering and $+$ the addition of $Re^+$, is an extensive structure with no essential maximum. The function $F$, restricted to $S_z$, satisfies the representation stated in Theorem 3 since, if $(\mu, \nu) \in T_z$,

$$\begin{aligned}
F(\mu + \nu) &= F\{f_z[\phi(a)/\phi(z)] + f_z[\phi(b)/\phi(z)]\} \\
&= Ff_z[\phi(a \circ_z b)/\phi(z)] \\
&= f_z'[\phi(a \circ_z b)/\phi(z)] \\
&= f_z'[\phi(a)/\phi(z)] + f_z'[\phi(b)/\phi(z)] \\
&= F(\mu) + F(\nu).
\end{aligned}$$

Since the identity function is also a representation of $\langle S_z, \geqslant, T_z, + \rangle$, we know by Theorem 3 that for some $\alpha > 0$, $F(\mu) = \alpha\mu$ for all nonmaximal elements of $S_z$. But $S_z$ has no maximal elements by definition; thus, $f_z' = \alpha f_z$.

Finally, by hypothesis (iii), for any $x, y \in A$, $f_x$ and $f_y$ satisfy the above functional equation and are strictly increasing from $(0, 1)$ onto Re$^+$; consequently, they differ only by a positive multiplicative constant; hence the subscripts can be dropped.                                    ◇

### 3.8.3 Theorem 7 (p. 95)

Let $\langle A, \succsim, B, \circ \rangle$ be an extensive structure with an essential maximum $z$. Suppose $A'$ and $B'$ are nonempty (where $A'$ and $B'$ are as in Theorem 5), and there exist $x, y \in A$ such that $(x, y) \in B$, $z > x, y$, and $z \sim x \circ y$. Then there exists a ratio scale $\phi: A \to$ Re$^+$ such that

  (i)   $\phi$ is order preserving;
  (ii)  $\phi$ is additive over nonmaximal elements;
  (iii) $\phi(z) = \inf\{\phi(x) + \phi(y) \mid (x, y) \in B, x \circ y \sim z\}$.

PROOF.  By Theorems 3 and 5, there exists a ratio scale $\phi$ that is order preserving and additive over the nonmaximal elements $A'$. Extend $\phi$ to $A$ by defining $\phi(z)$ as in part (iii); obviously it remains a ratio scale. To complete the proof, we need only show that if $a \in A'$, then $\phi(a) < \phi(z)$. We distinguish two cases.

First, suppose that $A'$ has a minimal element $c$. For any $a \in A'$, there exists $m_a \in$ I$^+$ such that $a \sim m_a c$, for if not, then let $m$ be the maximum integer such that $mc \prec a$; $m$ exists by Axiom 6'. By Axiom 4, there exists $b \in A'$ such that $mc \circ b \precsim a$. Since $m$ is maximal, $b \prec c$, contrary to the choice of $c$. Suppose that $m_a \geqslant m_x + m_y$, where $(x, y) \in B$, $x \circ y \sim z$, then

$$a \sim m_a c \succsim m_x c \circ m_y c \sim x \circ y \sim z,$$

which by Axiom 1 is contrary to $a \in A'$. Thus, $m_a < m_x + m_y$ and so for $0 < \epsilon < 1$, since the $m$'s are integers, $m_a + \epsilon < m_x + m_y$. Choose $\phi$ for which $\phi(c) = 1$, then $\phi(a) = m_a$, $\phi(x) = m_x$, $\phi(y) = m_y$, and so

$$\phi(a) < m_a + \epsilon \leqslant \inf[m_x + m_y \mid (x, y) \in B, x \circ y \sim z] = \phi(z).$$

Second, suppose that there is no minimal element. For any $a \in A'$, there exists at least one $b \in A'$ with $(a, b) \in B'$. Let $\epsilon > 0$ be given. There exist $x, y \in A'$ such that $(x, y) \in B$, $x \circ y \sim z$, and

$$\phi(x) + \phi(y) \leqslant \phi(z) + \epsilon/2.$$

Since there is no minimal element, we may choose $c$ such that $\phi(c) < \epsilon/4$, and $a, b, x, y > c$. By Axiom 3 and Lemma 8, $(x, c), (y, c) \in B$ and $(a, c) \in B'$, so by Axiom 6 there are integers $m_x$, $m_y$, and $m_a$ such that

$$m_x c < x \lesssim (m_x + 1)c,$$
$$m_y c < y \lesssim (m_y + 1)c,$$
$$m_a c < a \lesssim (m_a + 1)c, \qquad (m_a + 1)\, c \in A'.$$

Thus, $\phi(a) \leqslant (m_a + 1)\, \phi(c)$ and, since $x \circ y \sim z$, $m_a + 1 \leqslant m_x + m_y + 2$. Therefore,

$$\phi(a) \leqslant (m_a + 1)\, \phi(c)$$
$$\leqslant (m_x + m_y + 2)\, \phi(c)$$
$$< \phi(x) + \phi(y) + 2\phi(c)$$
$$< \phi(z) + \epsilon.$$

Since $\epsilon > 0$ is arbitrary, $\phi(a) \leqslant \phi(z)$. But since $(a, c) \in B'$, $\phi(a \circ c) \leqslant \phi(z)$, so by Axiom 5

$$\phi(a) < \phi(a) + \phi(c) = \phi(a \circ c) \leqslant \phi(z). \qquad\qquad \diamondsuit$$

## □ 3.9 ALTERNATIVE NUMERICAL REPRESENTATIONS

We have encountered two examples of nonadditive representations for extensive structures. The first occurred when we considered two different empirical interpretations of concatenation for the measurement of length. With the natural linear abutting $\circ$, the scale $\phi$ is additive over $\circ$, and with the artificial orthogonal abutting $*$, the scale $\psi = \phi^2$ is additive over $*$, but in terms of $\circ$, the scale $\psi$ is nonadditive; in fact,

$$\psi(a \circ b) = \phi(a \circ b)^2$$
$$= [\phi(a) + \phi(b)]^2$$
$$= \psi(a) + \psi(b) + 2[\psi(a)\, \psi(b)]^{1/2}.$$

Suppose, however, that we define a new binary operation on the real numbers, denoted by $\oplus$, as follows: for all $x, y$ in $Re^+$,

$$x \oplus y = x + y + 2[xy]^{1/2},$$

then, clearly, $\psi$ is additive in $\oplus$ over $\circ$.

The second example occurred in Theorem 6 where, with relativistic velocity as our guide, we defined a function $f$ from $(0, 1)$ onto $\text{Re}^+$ such that the transformed velocity $f[\phi(a)/\phi(z)]$ is additive relative to $*$. The binary operation corresponding to this transformation is

$$x \oplus y = f^{-1}[f(x) + f(y)].$$

When $f = \tanh^{-1}$, we have

$$x \oplus y = (x + y)/(1 + xy).$$

The velocity measure $\phi(a)/\phi(z)$ is additive in $\oplus$ over $*$.

A third and similar example is frequently mentioned. Consider a positive extensive structure with a scale $\phi$ additive over $\circ$; then $\psi = \exp \phi$ is an alternative scale, which is multiplicative over $\circ$. Clearly, there is neither more nor less information contained in the multiplicative representation than in the additive one; the choice between $\phi$ and $\psi$ cannot be determined on experimental grounds. All of physics can obviously be rewritten in terms of multiplicative, rather than additive, scales. It would take some getting used to—for example, the multiplicative scales are unique up to positive powers rather than up to multiplication by positive constants—but that is only a matter of familiarity.

Since we have three unambiguous examples of nonadditive representations, it may be well to describe all such alternative representations. Let $\langle A, \succsim, B, \circ \rangle$ be an extensive structure with no essential maximum. It can be mapped additively into $\langle \text{Re}^+, \geqslant, + \rangle$ by some function $\phi$ (Theorem 3). Furthermore, let us suppose that there exists some different binary operation $\oplus$ on Re such that $\langle A, \succsim, B, \circ \rangle$ can be mapped additively into $\langle \text{Re}, \geqslant, \oplus \rangle$ by a function $\psi$. It is not difficult to see that the image of $\langle A, \succsim, B, \circ \rangle$ under $\psi$ must also satisfy the axioms of Definition 3, and so the image of $\psi$ can be mapped additively into $\langle \text{Re}^+, \geqslant, + \rangle$ by some function $f$ (see Figure 2). Thus,

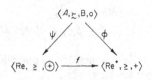

**FIGURE 2.**

by Theorem 3 and well-known properties of the composition of functions, there is a choice of units so that $\phi = f(\psi)$. Moreover, since both $\phi$ and $\psi$ are order preserving, $f$ is one to one, so its inverse exists on the range of $\phi$.

Observe that by additivity of both $\psi$ and $\phi$,

$$\begin{aligned}
\psi(a) \oplus \psi(b) &= \psi(a \circ b) \\
&= f^{-1}[\phi(a \circ b)] \\
&= f^{-1}[\phi(a) + \phi(b)] \\
&= f^{-1}\{f[\psi(a)] + f[\psi(b)]\},
\end{aligned}$$

which shows that the binary operation $\oplus$ is given by the following simple formula:

$$x \oplus y = f^{-1}[f(x) + f(y)]. \tag{1}$$

The converse is equally true: Suppose that $\langle A, \gtrsim, B, \circ \rangle$ is an extensive structure with no essential maximum, that $\phi$ is an order-preserving additive representation, and that $f$ is any strictly monotonic increasing function from a subset of the reals onto the range of $\phi$. Then if $\oplus$ is defined by Equation (1), $\psi = f^{-1}(\phi)$ is an order-preserving mapping from $\langle A, \gtrsim, B, \circ \rangle$ into $\langle \text{Re}, \geqslant, \oplus \rangle$ that is additive in $\oplus$ over $\circ$.

Equation (1) is essentially the well-known functional equation

$$g(x + y) = F[g(x), g(y)]$$

(see Aczel, 1966, Section 2.2). We have just shown that this equation, with $F$ specified, has a strictly increasing solution $g = f^{-1}$, whenever the operation $u \oplus v = F(u, v)$ is such that $\langle \text{Re}^+, \geqslant, \oplus \rangle$ satisfies Definition 3. This is closely related to the condition for the solvability of this equation given by Aczel.

In terms of Equation (1), the original three examples are:

1. Length: $f(x) = x^{1/2}, f^{-1}(x) = x^2$,

$$\begin{aligned}
x \oplus y &= f^{-1}[f(x) + f(y)] \\
&= (x^{1/2} + y^{1/2})^2 \\
&= x + y + 2(xy)^{1/2}.
\end{aligned}$$

2. Relativistic velocity: $f(x) = \frac{1}{2} \log[(1 + x)/(1 - x)], f^{-1}(x) = \tanh x$, and

$$\begin{aligned}
x \oplus y &= \tanh\left[\frac{1}{2}\log\left(\frac{1+x}{1-x}\right)\left(\frac{1+y}{1-y}\right)\right] \\
&= \frac{\exp\left[\frac{1}{2}\log\left(\frac{1+x}{1-x}\right)\left(\frac{1+y}{1-y}\right)\right] - \exp\left[-\frac{1}{2}\log\left(\frac{1+x}{1-x}\right)\left(\frac{1+y}{1-y}\right)\right]}{\exp\left[\frac{1}{2}\log\left(\frac{1+x}{1-x}\right)\left(\frac{1+y}{1-y}\right)\right] + \exp\left[-\frac{1}{2}\log\left(\frac{1+x}{1-x}\right)\left(\frac{1+y}{1-y}\right)\right]} \\
&= \frac{x + y}{1 + xy}.
\end{aligned}$$

3.  Multiplicative representation: $f(x) = \log x, f^{-1}(x) = \exp x$,

$$x \oplus y = \exp(\log x + \log y)$$
$$= \exp(\log xy)$$
$$= xy.$$

To arrive at the uniqueness of these alternative representations, suppose that $\phi$ and $\phi'$ are two additive representations and that $\psi$ and $\psi'$ are the corresponding representations under $f^{-1}$. By Theorems 1 and 3, there exists $\alpha > 0$ such that $\phi' = \alpha\phi$, so

$$\psi' = f^{-1}(\phi')$$
$$= f^{-1}(\alpha\phi)$$
$$= f^{-1}[\alpha f(f^{-1}(\phi))]$$
$$= f^{-1}[\alpha f(\psi)],$$

i.e., the typical transform is $F_\alpha = f^{-1}(\alpha f)$, where $\alpha > 0$. Furthermore, it is easy to show that any such transform generates a new representation of the same type.

This family of transformations is, in fact, a commutative group under composition since composition is associative and, as is easily seen, $F_\alpha(F_\beta) = F_\beta(F_\alpha) = F_{\alpha\beta}$, so $F_1$ is the identity and $F_\alpha$ and $F_{1/\alpha}$ are inverses.

In some discussions of measurement, great emphasis is placed upon a particular representation and its uniqueness properties—in the case of extensive measurement, the emphasis is on the additivity of the representation and its uniqueness up to multiplication by a positive constant. However, despite its great appeal and universal acceptance, the additive representation is just one of the infinitely many, equally adequate representations that are generated by the family of strictly monotonic increasing functions from the reals onto the positive reals. The essential fact about the uniqueness of a representation is not the particular group of admissible transformations, but that all groups are isomorphic and, in the case of extensive measurement, are all one-parameter groups; that is, there is exactly one degree of freedom in any particular representation.

## 3.10  CONSTRUCTIVE METHODS

The theory of extensive measurement justifies, at least in part, the representation of certain attributes by real numbers. These results are not, however, in a form that is especially suited to the practice of measurement. In this section we cite additional theorems which may be useful in the actual construction of scales of measurement.

### 3.10.1 Extensive Multiples

The proofs we have given, and most of those in the literature for the existence of extensive measures, rest in an essential way on the concatenation of an element with itself. And in at least one empirical realization, time duration, this is the only concatenation that can easily be carried out. Specifically, if $a$ denotes a particular pendulum, we generate $na$ merely by letting $a$ swing through $n$ periods. That this really generates $na$ rests upon the empirical fact that any two periods of a pendulum have the same duration, which can be shown as follows. Let $a$ and $b$ be two pendulums whose first periods match, i.e., $a \sim b$. Let us start $b$ at the instant when $a$ completes its $(n - 1)$st period, then we find that $b$ completes its first period at the same instant that $a$ completes its $n$th period. So, by the transitivity of $\sim$, $a$'s $n$th period is equivalent to its first. (This holds, of course, only for ideal, frictionless pendulums.)

These facts suggest that it might be worthwhile to see just how far we can go toward constructing the extensive scale using only concatenations of the form $na$ and excluding all those of the form $a \circ b$, where not $(a \sim b)$. As with our previous extensive structures and for the same reasons, we want to embody restrictions on the possible concatenations—in the case of time, to values of $n$ for which it is feasible to wait for $na$ to occur. It should be noted that we cannot specialize our previous results to handle the present idea because whenever $(a, a)$ is in $B$ and $a > b$, Lemma 7 states that $(a, b)$ is also in $B$ and so the unwanted concatenation $a \circ b$ is in the structure. We need a new axiom system, but if it is to lead to the same numerical measures as Definition 3, the elements $na$ of a general extensive structure should satisfy the new axioms.

DEFINITION 6. *Let $A$ be a nonempty set; $\succsim$ a binary relation on $A$; $I^+ = \{1, 2, 3,...\}$; $B$ a nonempty subset of $I^+ \times A$; and $f$ a function from $B$ into $A$, which will be abbreviated as $f(n, a) = na$. The quadruple $\langle A, \succsim, B, f \rangle$ is a structure of extensive multiples iff the following six axioms hold for all $a, b \in A$ and $m, n, p, q \in I^+$:*

1. *$\langle A, \succsim \rangle$ is a weak order.*

2. *$(1, a) \in B$ and $1a \sim a$.*

3. *Suppose that $(n, a) \in B$. Then, $n \geqslant m$ iff $(m, a) \in B$ and $na \succsim ma$.*

4. *If $(m, a)$, $(p, a)$, $(n + q, b) \in B$, $nb \succsim ma$, and $qb \succsim pa$, then $(m + p, a) \in B$ and $(n + q) b \succsim (m + p)a$.[1]*

5. *If $b > a$, then there exist $c \in A$ and $n \in I^+$ such that $(n + 1, c) \in B$, $b \succsim (n + 1) c > nc \succsim a$.*

---

[1] Note that, by Axiom 2, $(n + q, b) \in B$ implies $(n, b)$ and $(q, b) \in B$.

6. *Any strictly bounded standard sequence is finite, where* $a_1$, $a_2$,... *is a standard sequence if, for* $n = 2, 3,...,$ $a_n = na_1$.

Given the intended interpretation of $na$, Axioms 1–4 are clearly plausible. Axiom 5 is an existence statement which says that there are some comparatively small elements in the system. Axiom 6 is the usual Archimedean condition.

Two results are stated here and proved in Section 3.11. The first shows that the subset of multiples from an extensive structure with no essential maximum (Definition 3) fulfills Definition 6, and the second establishes that Definition 6 is sufficient to determine the extensive measure.

THEOREM 8. *Suppose that* $\langle A, \gtrsim, B, \circ \rangle$ *is an extensive structure with no essential maximum. Let* $B'$ *and* $na$ *be defined recursively by*:

(i) $(1, a) \in B'$ *and* $1a = a$, *for all* $a$;
(ii) *if* $(n - 1, a) \in B'$ *and* $((n - 1)a, a) \in B$,
    *then* $(n, a) \in B'$ *and* $na = (n - 1)a \circ a$,

*and, for* $(n, a) \in B'$, *let* $f(n, a) = na$. *Then* $\langle A, \gtrsim, B', f \rangle$ *is a structure of extensive multiples.*

COROLLARY. *If* $\phi$ *is any additive, order-preserving representation of an extensive structure with no essential maximum, then its restriction to the substructure of multiples is such that*

(i) $a \gtrsim b$ *iff* $\phi(a) \geqslant \phi(b)$;
*and*
(ii) *if* $(n, a) \in B'$, *then* $\phi(na) = n\phi(a)$.

THEOREM 9. *Suppose that* $\langle A, \gtrsim, B, f \rangle$ *is a structure of extensive multiples. Then there exists a positive, real-valued function* $\phi$ *on* $A$ *such that for all* $a, b \in A$ *and* $n \in I^+$

(i) $a \gtrsim b$ *iff* $\phi(a) \geqslant \phi(b)$;
(ii) *if* $(n, a) \in B$, *then* $\phi(na) = n\phi(a)$.

*If another function* $\phi'$ *satisfies* (i) *and* (ii), *then there exists* $\alpha > 0$ *with* $\phi'(a) = \alpha\phi(a)$ *for all nonmaximal* $a \in A$.

The uniqueness result is similar to that of Theorem 3.

The significance of these results is that we need only collect data about the subsystem of multiples in order to construct the desired extensive measure, since Theorem 9 insures that a representation, which is unique up to a similarity transformation (for all nonmaximal elements) can be constructed, and the Corollary to Theorem 8 insures that it coincides over the nonmaximal elements with one of the additive representations of the whole extensive structure.

These results appear to provide a natural way to measure time duration using pendulums, but they are of little help with mass and length. For these attributes, and for many others, the task of producing accurate multiples of form $na$ can be difficult, mainly because $\sim$ is not in practice a transitive relation (see Section 1.1.1). For these examples, it would be far better to know of a substructure that excludes all multiples and that determines the extensive scale. The idea of replacing concatenation by union of disjoint subsets was investigated by Adams (1965) and Krantz (1967); see also Section 5.2.3 and Chapter 15.

### 3.10.2  Standard Sequences

In the practice of measurement, it is usually convenient to agree upon some acceptable degree of precision to be achieved. This degree of precision can then be used as a basis for constructing what has been called a *standard series* (Campbell, 1920) or what in our discussion of the Archimedean axiom we have preferred to call a *standard sequence*. This allows us, using few qualitative observations, to assign to objects a numerical measure that approximates the extensive measure to within the preassigned limit of precision.

Let $e$ be an entity that defines the limit of precision in the sense that we are willing to fail to distinguish between two objects whose difference in the attribute is less than that $e$. Then we construct the sequence $ne$, where $n = 1, 2, \dots$ . Because of the inevitable imprecision in determining $\sim$ empirically, a number of internal cross checks are usually made in constructing a standard sequence. They are all based upon a simple consequence of the theory of extensive measurement: if $m, n, p, q$ in $I^+$ are such that $m + n = p + q$, then $me \circ ne \sim pe \circ qe$. Given the sequence and some $a$ that we wish to measure, we merely find that $n$ for which $ne > a \gtrsim (n - 1)e$. By Theorem 8, $n\phi(e) > \phi(a) \geqslant (n - 1)\,\phi(e)$, and so, aside from added errors introduced in determining $\sim$, $\phi(a)$ is determined to within an accuracy of $\phi(e)$.

It is evident that the whole sequence $e, 2e, 3e, \dots$ is redundant since, for example, if we have $e$ and $2e$, then we do not need $3e$ because it is equivalent to $e \circ 2e$. There are various ways to reduce the total number of objects actually constructed; perhaps the simplest is to include only those of the form $2^i e$, $i = 0, 1, 2, \dots$ . It is easy to see that this set suffices. If we wish to construct $ne$, we simply write $n$ in its binary expansion and include $2^i e$ in the concatenation if and only if the coefficient of the $i$th term of the expansion is 1.

The two best known examples of standard sequences are the sets of standard weights, which are a part of most laboratory balances, and rulers, which are marked in multiples of some amount, such as a millimeter.

## 3.11 PROOFS

### 3.11.1 Theorem 8 (p. 104)

*The substructure of the multiples of an extensive structure with no essential maximum forms a structure of extensive multiples.*

PROOF. We examine the six axioms of Definition 6 (Section 3.10.1) in order.

1. This axiom follows directly from Axiom 1 of Definition 3 (Section 3.4.3).

2. By definition $(1, a) \in B'$.

3. Suppose that $n \geqslant m$. If $n = m$, then $na = ma$. If $n > m$, then by the inductive definition of $na$, $na = ma \circ (n - m)a$, and so by Axiom 5 of Definition 3, $na > ma$. Obviously, the converse is also true.

4. By Lemma 7, $(ma, pa) \in B$, so by induction and associativity, $(m + p)a = ma \circ pa$, hence $(m + p, a) \in B'$. Also by Lemma 7, $(n + q)b = nb \circ qb \gtrsim ma \circ pa = (m + p)a$.

5. Suppose that $b > a$, then by Axiom 4 of Definition 3 there exists $d \in A$ with $(a, d) \in B$ and $b \gtrsim a \circ d$. If $d$ is minimal in $A$, then for some $n$, $a \sim nd$. (See proof of Theorem 2.4, Case A.) Thus,

$$b \gtrsim a \circ d \sim nd \circ d = (n + 1)d > nd \sim a$$

as required. If $d$ is not minimal, choose $e < d$ and by Axiom 4 let $f \in A$ be such that $(e, f) \in B$ and $d \gtrsim e \circ f$. Let $c = \min\{a, e, f\}$, and let $n$ be the least integer for which $nc \gtrsim a$ ($n$ exists, since when $a > mc$, then, since $d > c$, $(m + 1)c$ is defined). Then $(n + 1)c > b$ is impossible; otherwise, we would have

$$(n + 1)c > b \gtrsim a \circ d \gtrsim a \circ (e \circ f) \gtrsim a \circ 2c,$$

and it follows that $n \geqslant 2$ and $(n - 1)c \gtrsim a$, contradicting the minimality of $n$. Hence

$$b \gtrsim (n + 1)c > nc \gtrsim a$$

as required.

6. This axiom follows directly from Axiom 6 of Definition 3.

The corollary to Theorem 8 follows immediately from this theorem and Theorem 3.


### 3.11.2 Preliminary Lemmas

The common hypothesis of the lemmas of this section is that $\langle A, \gtrsim, B, f \rangle$ is a structure of extensive multiples (Definition 6, Section 3.10.1).

LEMMA 10.   *If in Axiom 4 nb $\sim$ ma and qb $\sim$ pa, then*

$$(n + q)\, b \sim (m + p)a.$$

*PROOF.*   Since Axiom 4 asserts $(m + p, a) \in B$, the axiom can be applied in both directions to yield the conclusion.                    $\Diamond$

LEMMA 11.   *For a, b $\in$ A and m, n $\in$ I$^+$, if b $\gtrsim$ a, n $\geqslant$ m, and (n, b) $\in$ B, then (m, a) $\in$ B and nb $\gtrsim$ ma.*

*PROOF.*   By Axiom 3, if $k \in$ I$^+$ and $n \geqslant k$, then $(k, b) \in B$. By Axiom 2, $1b \sim b \gtrsim a \sim 1a$. From these two facts, a finite induction using Axiom 4 yields $(m, a) \in B$ and $mb \gtrsim ma$. But, by Axiom 3, $nb \gtrsim mb$, and so by Axiom 1, $nb \gtrsim ma$.                    $\Diamond$

LEMMA 12.   *Suppose that a $\in$ A and m, n $\in$ I$^+$.*

  (i)   *(mn, a) $\in$ B iff (m, a) and (n, ma) $\in$ B.*
  (ii)  *If (mn, a) $\in$ B, then (mn) a $\sim$ n(ma).*

*PROOF.*   First, let us suppose that $(mn, a) \in B$. Since $n \geqslant 1$, Axiom 3 implies $(m, a) \in B$. For $n = 1$, both parts (i) and (ii) are true by Axiom 2. For $n > 1$, we proceed by induction. Suppose that both parts are true for $n - 1$, so $[m(n - 1)] a \sim (n - 1)(ma)$. By Axiom 2, $ma \sim 1(ma)$. Thus, by Axiom 4, $(n - 1 + 1, ma) = (n, ma) \in B$, and by Lemma 11,

$$(mn)\, a \sim [m(n - 1) + m]\, a \sim [(n - 1) + 1](ma) \sim n(ma).$$

The converse of part (i) is proved similarly.                    $\Diamond$

LEMMA 13.   *If u $\in$ A is minimal, then for each a $\in$ A there exists a unique $n_a \in$ I$^+$ such that $(n_a, u) \in B$ and a $\sim n_a u$.*

*PROOF.*   If $a \sim u$, by Axiom 2, $(1, u) \in B$ and $1u \sim a$, so let $n_a = 1$. If $a > u$, then by Axioms 2 and 6 there is a maximal integer $m_a$ such that

$(m_a, u) \in B$ and $a > m_a u$. By Axiom 5, there exist $n \in I^+$ and $c \in A$ such that $(n + 1, c) \in B$ and $a \gtrsim (n + 1) c > nc \gtrsim m_a u$. Since $c \gtrsim u$, Axioms 2 and 4 imply that $(m_a + 1, u) \in B$ and $a \gtrsim (n + 1) c \gtrsim (m_a + 1)u$. By the choice of $m_a$, $a \sim (m_a + 1)u$, so let $n_a = m_a + 1$. By Axiom 3, $n_a$ is clearly unique. ◇

LEMMA 14. *If $A$ has no minimal element, then there is a sequence, $a_1, ..., a_n, ...$, such that,*

(i)  *$(2^n, a_{n+1}) \in B$;*

(ii)  *$a_n \gtrsim 2a_{n+1}$;*

(iii)  *for $b \in A$, the set $\{n \mid n \in I^+$ and $a_n > b\}$ is finite.*

PROOF. Let $a_1$ be any element of $A$. Suppose that $a_1, ..., a_n$ have been constructed. Since $a_n$ is not minimal, Axiom 5 implies there exist $a_{n+1} \in A$ and $m_n \in I^+$ such that $(m_n + 1, a_{n+1}) \in B$ and $a_n \gtrsim (m_n + 1) a_{n+1}$. Since $m_n \geq 1$, Axiom 3 yields $(2, a_{n+1}) \in B$ and $a_n \gtrsim 2a_{n+1}$. By the induction hypothesis, $(2^{n-1}, a_n) \in B$, so by Lemma 11, $(2^{n-1}, 2a_{n+1}) \in B$. Thus, by Lemma 12, $(2^n, a_{n+1}) \in B$. So $a_1, ..., a_n, a_{n+1}$ fulfill parts (i) and (ii). We continue, by induction, to construct $a_n$ for all $n \in I^+$. It remains to show that part (iii) holds. By Axiom 6, for $b \in A$ there exists some $m_0 \in I^+$ such that for $m > m_0$ either $(m, b) \notin B$ or $mb \gtrsim a_1$. Choose $n$ such that $2^n > m_0$. If $a_{n+1} > b$, then since $(2^n, a_{n+1}) \in B$ it follows by Axiom 3 that $(2^n, b) \in B$, and so by a finite induction on part (ii) and Lemma 11,

$$a_1 \gtrsim 2^n a_{n+1} > 2^n b \gtrsim a_1,$$

which is impossible by Axiom 1. So, $b \gtrsim a_{n+1}$ for all $n$ such that $2^n > m_0$. ◇

LEMMA 15. *Suppose that $A$ has no minimal element and that $a_1, ..., a_n, ...$ is any sequence with the properties described in Lemma 14. For $a \in A$ and $n \in I^+$ for which $a \gtrsim a_n$, let $m_{n,a}$ be the maximal integer satisfying $a > ma_n$.[2] Then for $a, b \in A$, with $a$ not maximal,*

$$\lim_{n \to \infty} (m_{n,a}/m_{n,b})$$

*exists and is positive.*

PROOF. Since $a$ is not maximal, Axiom 5 guarantees that there exist $c \in A$ and $i \in I^+$ such that $(i + 1, c) \in B$ and $ic \gtrsim a$. By Lemma 14, for all

_____

[2] By Lemma 14 and Axioms 2 and 6, $m_{n,a}$ exists for all but finitely many $n$. .

sufficiently large $n \in I^+$, $c \gtrsim a_n$ and $a > m_{n,a}a_n$. Thus, by Axiom 4, $(m_{n,a} + 1, a_n) \in B$ and, by the choice of $m_{n,a}$,

$$(m_{n,a} + 1) a_n \gtrsim a > m_{n,a}a_n.$$

Moreover, for $n > 1$, $a_n$ is not maximal, so Lemma 14 and Axiom 6 insure that for all sufficiently large $k \in I^+$, there exists an integer $p_k$ such that $(p_k + 1, a_k) \in B$ and

$$(p_k + 1) a_k \gtrsim a_n > p_k a_k.$$

From these two inequalities and Lemmas 11 and 12, $(p_k + 1)(m_{n,a} + 1) a_k$ is either undefined or $\gtrsim a$. Moreover, $a > p_k m_{n,a} a_k$ and $b > p_k m_{n,b} a_k$. From the definitions of $m_{k,a}$ and $m_{k,b}$, we conclude that

$$p_k m_{n,a} + p_k + m_{n,a} \geq m_{k,a}$$

and

$$m_{k,b} \geq p_k m_{n,b}.$$

Dividing,

$$\frac{m_{k,a}}{m_{k,b}} \leq \frac{m_{n,a}}{m_{n,b}} + \frac{1}{m_{n,b}} + \frac{m_{n,a}}{p_k m_{n,b}}.$$

By Lemma 14, $p_k \geq 2^{k-n}$, so with $n$ fixed and $k \to \infty$, $p_k \to \infty$, hence

$$\limsup_{k \to \infty} \frac{m_{k,a}}{m_{k,b}} \leq \frac{m_{n,a}}{m_{n,b}} + \frac{1}{m_{n,b}} < \infty.$$

Since $m_{n,b} \to \infty$ as $n \to \infty$, we obtain

$$\limsup_{k \to \infty} \frac{m_{k,a}}{m_{k,b}} \leq \liminf_{n \to \infty} \frac{m_{n,a}}{m_{n,b}}.$$

This shows that the limit exists and is finite and nonnegative. Interchanging $a$, $b$, the sequence of reciprocals also has a finite limit, hence the limit of the original sequence is positive.                                                    ◇

### 3.11.3.  Theorem 9  (p. 104)

THEOREM.  *If $\langle A, \gtrsim, B, f \rangle$ is a structure of extensive multiples, then there exists a ratio scale $\phi$ on $A$ that is order preserving and, for $(n, a) \in B$, $\phi(na) = n\phi(a)$.*

*PROOF.*  Suppose, first, that $A$ has a minimal element $u$. For any $a \in A$, let $n_a$ be the unique positive integer shown to exist in Lemma 13, and define $\phi(a) = n_a$.

(i)  Since, by Axiom 3, $\phi(a) = n_a \geqslant n_b = \phi(b)$ iff $a \sim n_a u \gtrsim n_b u \sim b$, $\phi$ is order preserving.

(ii)  From $a \sim n_a u$ and $ma \sim n_{ma} u$, Lemma 10 implies $n_{ma} u \sim ma \sim mn_a u$. By the uniqueness of $n_{ma}$, $\phi(ma) = n_{ma} = mn_a = m\phi(a)$.

(iii)  Suppose that $\phi'$ satisfies (i) and (ii), then for $a \in A$,

$$\phi'(a)/\phi(a) = \phi'(n_a u)/\phi(n_a u) = \phi'(u)/\phi(u) = \alpha,$$

and so $\phi'(a) = \alpha \phi(a)$.

Now suppose $A$ has no minimal element. Let $a_n$ and $m_{n,a}$ be defined as in Lemma 15. Let $z$ be a maximal element of $A$ if there is one; otherwise, let it be any element. Define

$$\phi(a) = \lim_{n \to \infty} (m_{n,a}/m_{n,z}).$$

If $a$ is nonmaximal, the limit exists in $\mathrm{Re}^+$ by Lemma 15. Otherwise, $a \sim z$ and so $m_{n,a} = m_{n,z}$ for every $n \in I^+$ and the limit again exists and is 1.

(i)  Clearly, if $a \gtrsim b$, then $m_{n,a} \geqslant m_{n,b}$, so $\phi(a) \geqslant \phi(b)$. If $a > b$, by Axiom 5 there exist $c \in A$ and $n \in I^+$ such that $a \gtrsim (n+1)c > nc \gtrsim b$. If $\phi$ is additive, this implies $\phi(a) \geqslant (n+1)\phi(c) > n\phi(c) \geqslant \phi(b)$, hence $\phi(a) > \phi(b)$. We show that $\phi$ is additive.

(ii)  For sufficiently large $k$, $(m_{k,a}+1)a_k \gtrsim a > m_{k,a}a_k$, and $na > m_{k,na}a_k$, and so $n(m_{k,a}+1) > m_{k,na} \geqslant nm_{k,a}$. This implies, by definition of $\phi$, $n\phi(a) \geqslant \phi(na) \geqslant n\phi(a)$, hence $\phi(na) = n\phi(a)$.

(iii)  Suppose that $a, b \in A$ are nonmaximal and $\phi$ is any function satisfying properties (i) and (ii). Since for sufficiently large $n$,

$$(m_{n,a} + 1)a_n \gtrsim a > m_{n,a}a_n,$$

properties of $\phi$ yield

$$(m_{n,a} + 1)\phi(a_n) \geqslant \phi(a) > m_{n,a}\phi(a_n),$$
$$(m_{n,b} + 1)\phi(a_n) \geqslant \phi(b) > m_{n,b}\phi(a_n).$$

Thus,

$$(m_{n,a} + 1)/m_{n,b} > \phi(a)/\phi(b) > m_{n,a}/(m_{n,b} + 1),$$

and so

$$\phi(a)/\phi(b) = \lim_{n \to \infty} (m_{n,a}/m_{n,b}).$$

Therefore, $\phi$ is unique up to a positive constant for nonmaximal elements.

## 3.12  CONDITIONALLY CONNECTED EXTENSIVE STRUCTURES

### 3.12.1  Thermodynamic Motivation

This section provides an elementary axiomatization which can be interpreted as formulating an aspect of the second law of thermodynamics. This theory differs from the previous extensive theories in that the ordering is not connected and the representation involves at least two functions. In particular, the structure is represented by a family of real-valued additive functions, one of which does not decrease and the remainder do not change, i.e., they are conserved, when an isolated system passes from one state to another. It is natural to interpret the nondecreasing function as entropy and the conserved functions as volume, internal energy, and the numbers of units (molecules, atoms, etc.) of the relevant chemical and physical constituents (Callen, 1960). Whether or not this is more than a purely formal parallel cannot be decided without an analysis of specific systems. We show, as an example, that the entropy measure of statistical mechanics does satisfy the axiom system.

The basic ideas of this abstract approach to thermodynamics are due to Giles (1964). The reader interested in how this formulation relates to the classical ones and in the many ramifications of the theory beyond the basic theorems reported here should consult his book and the summary of the measurement results, their application to the study of energy and entropy, and some criticisms of Giles' formulation found in the excellent paper by Duistermaat (1968). See also Cooper (1967). As Giles and Duistermaat go far deeper into physics than is appropriate here, we present only a purely measurement portion of the theory as modified from Giles by Roberts and Luce (1968). This work ignores some of the constraints that are essential in an adequate theory of physical entropy.

There are three primitives that must be axiomatized and given a physical interpretation.

(i)  A set $A$ of states which a class of isolated physical systems can assume. Giles described states as follows:

> ... in a useful physical theory the *state* of a system represents its method of preparation: two systems are in the same state if they have been prepared in the same way, or more precisely if *our information about the method of preparation* is the same in each case. This conclusion may be modified slightly: two states or methods of preparation need not be distinguished if they are equivalent in respect of any prediction which might be made—that is, if they correspond to the same assertion concerning the result of any experiment which might be performed on the system (Giles, 1964, p. 17). (From *Mathematical Foundations of Thermodynamics* by R. Giles. Copyright © 1964, Macmillan.)

(ii) A binary operation ∘ which is interpreted as the union or sum of two states. In discussing this operation, Giles writes as follows (we substitute our symbol ∘ for Giles' +):

> Given two systems $\alpha$ and $\mathscr{B}$ we may if we wish consider them, taken together, as forming a single system. We call this single system the *union* or *sum* of $\alpha$ and $\mathscr{B}$ and denote it $\alpha \circ \mathscr{B}$. It is clear that this operation of addition is associative and commutative: it does not matter in what order we add a number of systems, the result is always the same.... It should be emphasized that the process of forming the union $\alpha \circ \mathscr{B}$ is purely a conceptual one; in particular, it does not imply any interaction between $\alpha$ and $\mathscr{B}$ .... We shall adopt the operation of addition of states as our second primitive concept. However, the above explanation is not a satisfactory rule of interpretation for this concept since it does not take cognisance of our interpretation of a state as a method of preparation. A better form would be as follows: *if a and b are two states then a ∘ b is the state whose method of preparation consists in the simultaneous and independent performance of the methods of preparation corresponding to the states a and b.* The associative and commutative laws of addition of states are evidently still applicable. (Giles, 1964, p. 22).

(iii) The last primitive is a binary relation $\gtrsim$ on the set of states $A$ (which Giles' denoted by →). The interpretation of $a \gtrsim b$ is either that a system $\mathscr{A}$ is in state $b$ at some time and evolves, in isolation, into state $a$ at a later time or that another system $\mathscr{K}$ can be found such that $\mathscr{A} \circ \mathscr{K}$ evolves, in isolation, from $b \circ k$ into $a \circ k$. Observe that for two arbitrary states $a$ and $b$—not necessarily of the same system—it may very well happen that neither $a \gtrsim b$ nor $b \gtrsim a$, i.e., the relation $\gtrsim$ is not necessarily connected.

> Let a and b be two states. We write $b \gtrsim a$ *if there exists a state k and a time interval* $\tau$ *such that a ∘ k evolves (in isolation) in the time* $\tau$ *into the state b ∘ k.* That is, ... such that the state whose method of preparation is "apply simultaneously and independently the methods of preparation corresponding to $a$ and $k$ and wait for a time $\tau$" is indistinguishable from the state whose method of preparation is "apply simultaneously and independently the methods of preparation corresponding to $b$ and $k$," in the sense that any experiment applied to these states will yield the same result (or rather the same statistical distribution of results) in each case (Giles, 1964, p. 24).

Assuming that the intended interpretation is relatively unambiguous—and we are by no means convinced that it is—our main problem is to provide an axiomatization of $\langle A, \gtrsim, \circ \rangle$ that imposes only physically plausible constraints and yields a suitable numerical representation. Since the relation $\gtrsim$ is not connected, the theory of extensive measurement is not applicable; no single order-preserving function will provide a suitable representation that will also distinguish between comparable and incomparable pairs of states. The solution to this difficulty employed in thermodynamics is to find, in addition to a function that reflects the order when a comparison is possible, a collection of additive functions all of which are conserved when $a$ and $b$

are comparable and at least one of which has different values on $a$ and $b$ when $a$ and $b$ are not comparable. This suggests a representation of the following form: There exists a nonnegative, additive function $\phi$ and a family $\mathscr{C}$ of additive functions on $A$ such that $a \gtrsim b$ if and only if $\phi(a) \geqslant \phi(b)$ and $\chi(a) = \chi(b)$ for all $\chi \in \mathscr{C}$. In discussing possible axioms, we refer to the function $\phi$ of the desired representation as the entropy function.

### 3.12.2 Formulation of the Axioms

With both the intended interpretation and the desired representation of $\langle A, \gtrsim, \circ \rangle$ in mind, some of the axioms can be arrived at easily. We certainly assume that $\gtrsim$ is reflexive (take $\tau = 0$) and that it is transitive (take $\tau_3 = \tau_1 + \tau_2$), but as was mentioned before we do not assume that it is connected, i.e., it is a quasiorder, but not a weak order. In lieu of connectedness we have another very strong property. Suppose that both $a \gtrsim b$ and $a \gtrsim c$, then according to the desired representation $b$ and $c$ both have the same value for every $\chi$ in $\mathscr{C}$ (namely, that of $a$) and so either $b \gtrsim c$ or $c \gtrsim b$ depending on which has the greater entropy. This important property, which Roberts and Luce called *conditional connectedness*, although a consequence of the representation we desire, is by no means a consequence of the proposed interpretation. It is a substantial physical law and, according to Giles, it fails for systems that exhibit hystersis. Giles credits Buchdahl (1958) and Falk and Jung (1959) with first recognizing the significance of conditional connectedness. Duistermaat (1968), who referred to it as a "causality principle," pointed out that "... its experimental verification may be a complicated matter."

As was noted, the interpretation of $\circ$ leads naturally to the assumption that $\circ$ is closed, associative, and commutative. In addition, the definition of $a \gtrsim b$ in terms of $a \circ k \gtrsim b \circ k$ strongly suggests imposing the monotonicity property: $a \gtrsim b$ if and only if $a \circ c \gtrsim b \circ c$.

Suppose that $n$ is a positive integer and that $a$ and $b$ are states such that $n$ copies of $b$ will pass into $n$ copies of $a$, i.e., $na \gtrsim nb$, then we assert that $b$ will pass into $a$, i.e., $a \gtrsim b$. Note that this seemingly reasonable property does not follow from the first four properties as the following example demonstrates. Let $A = I$; $\circ = +$; $a \gtrsim b$ if and only if $a - b$ is even. It is easy to see that the first four axioms hold; but the last does not since $4 \gtrsim 2$ whereas not $(2 \gtrsim 1)$.

As might be expected from our previous extensive structures, an Archimedean axiom must be included. In fact, since $\circ$ is closed, it is plausible to make essentially the same assumption as in a closed extensive structure (Definition 1). If we suppose that $a > b$, that $c$ and $d$ are comparable, and that $n$ is a positive integer, by conditional connectedness and the monotonicity property, $na \circ c$ and $nb \circ d$ are comparable and we assert that for some $n$, $na \circ c \gtrsim nb \circ d$.

These six properties are all necessary—i.e., follow from the desired representation. To formulate the remaining two nonnecessary assumptions, it is convenient to define a special class of states—Giles called them anti-equilibrium states—that have minimum entropy. In the presence of the other axioms, this minimum entropy has to be zero.

DEFINITION 7. *Let $\gtrsim$ be a binary relation on a set $A$. Define*

$$A_0 = \{x \mid x \in A \text{ and, for } a \in A, \text{ if } x \gtrsim a, \text{ then } a \gtrsim x\}.$$

The next two assumptions concern $A_0$. The first simply says that $A_0$ is closed under $\circ$, which insures that the elements of $A_0$ have zero "entropy" if the "entropy" function is additive. The second, which insures that $A_0$ is richly endowed with states, assumes that among the states that can evolve into a given state $a$, at least one is in $A_0$.

We may interpret $A_0$ to consist of the states of systems at absolute zero temperature. Although such states are exceedingly difficult to realize, they can be considered conceptually and the theory should apply to them. From this point of view, the two axioms seem physically acceptable.

### 3.12.3  The Axiom System and Representation Theorem

DEFINITION 8. *Let $A$ be a nonempty set; $\gtrsim$ a binary relation on $A$; $\circ$ a closed binary operation on $A$; and $A_0$ defined by Definition 7. The triple $\langle A, \gtrsim, \circ \rangle$ is an* entropy structure *iff the following eight axioms are satisfied for all $a, b, c, d, x, y \in A$:*

1. *Quasiordering:  The relation $\gtrsim$ is reflexive and transitive.*

2. *Conditional connectedness:  If $a \gtrsim b$ and $a \gtrsim c$, then either $b \gtrsim c$ or $c \gtrsim b$.*

3. *Weak associativity and weak commutativity: $(a \circ b) \circ c \sim a \circ (b \circ c)$ and $a \circ b \sim b \circ a$.*

4. *Monotonicity: $a \gtrsim b$ iff $a \circ c \gtrsim b \circ c$.*

5. *If for some $n \in I^+$, $na \gtrsim nb$, then $a \gtrsim b$. (Here, $1a = a$, $na = (n-1) a \circ a$.)*

6. *Archimedean:  If $a > b$ and either $c \gtrsim d$ or $d \gtrsim c$, then there exists $n \in I^+$ such that $na \circ c \gtrsim nb \circ d$.*

7. *If $x, y \in A_0$, then $x \circ y \in A_0$.*

8. *For each $a \in A$, there exists $z \in A_0$ such that $a \gtrsim z$.*

To formulate the representation, it is convenient to introduce the following notion:

DEFINITION 9.   *Suppose that* $\langle A, \gtrsim, \circ \rangle$ *is an entropy structure. A function* $\chi: A \rightarrow$ Re *is called a* component of content *iff, for all* $a, b \in A$,

   (i)   *if* $a \gtrsim b$, *then* $\chi(a) = \chi(b)$;
*and*

   (ii)   $\chi(a \circ b) = \chi(a) + \chi(b)$.

*It is called* nontrivial *if* $\chi \neq 0$.

THEOREM 10.   *Suppose that* $\langle A, \gtrsim, \circ \rangle$ *satisfies Axioms 1–4 and 6 of Definition 8. Then there exists* $\phi: A \rightarrow$ Re *such that for all* $a, b \in A$

   (i)   *if* $a$ *and* $b$ *are comparable, then* $a \gtrsim b$ *iff* $\phi(a) \geqslant \phi(b)$;
   (ii)   $\phi(a \circ b) = \phi(a) + \phi(b)$.

*If* $\phi'$ *is another function satisfying* (i) *and* (ii), *then for some real* $\alpha > 0$ *and some component of content* $\chi$, $\phi' = \alpha\phi + \chi$.

*If, in addition, Axioms 7 and 8 hold, then there exists* $\phi$ *satisfying* (i), (ii), *and*

   (iii)   $\phi \geqslant 0$;
   (iv)   $\phi(a) = 0$ *iff* $a \in A_0$.

*Another function* $\phi'$ *satisfies* (i)–(iv) *iff, for some real* $\alpha > 0$, $\phi' = \alpha\phi$.

The significance of this theorem is that when Axioms 1–4 and 6 of an entropy structure hold, there is a nonnegative extensive measure that does not decrease when one state passes into another. Moreover, this function is unique up to its unit. So it behaves as does entropy in classical thermodynamics.

It should be noted that Theorem 1 is really the special case of Theorem 10 when $\gtrsim$ is connected and, as we shall see, the proof of Theorem 10 is similar to that of Theorem 1.

The remaining problem is to characterize numerically whether or not $a$ and $b$ are comparable under $\gtrsim$, i.e., whether or not either $a$ can pass into $b$ or $b$ into $a$. This is formulated next.

THEOREM 11.   *Suppose that* $\langle A, \gtrsim, \circ \rangle$ *satisfies Axioms 1–5 of Definition 8. If* $\gtrsim$ *is connected, then there is no nontrivial component of content. If* $\gtrsim$ *is not connected, then for each incomparable pair* $a, b \in A$ *(i.e., neither* $a \gtrsim b$ *nor* $b \gtrsim a$*) there exists a component of content* $\chi$ *for which* $\chi(a) \neq \chi(b)$.

The significance of this theorem is that there is a class $\mathscr{C}$ of nontrivial components of content such that, for all $a, b$ in $A$, $a$ and $b$ are comparable if and only if $\chi(a) = \chi(b)$ for all $\chi$ in $\mathscr{C}$. In other words, there is a family of additive measures that are conserved during any change of state. As it stands this result is, unfortunately, of little physical interest. First, there is nothing to assure us that the class $\mathscr{C}$ is finite, which one would expect it to be in a physical theory. Second, there is nothing to assure us that the relevant components of content are well-behaved, nonpathological functions. Giles argued that it is sufficient to show that each $\chi$ in $\mathscr{C}$ is nonnegative, and he included a very complex axiom in his system which yields just this.

By combining Theorems 10 and 11, we have the following representation.

COROLLARY. *Suppose that* $\langle A, \gtrsim, \circ \rangle$ *is an entropy structure. Then there exists a family* $\mathscr{C}$ *of components of content and a function* $\phi: A \to \mathrm{Re}^+$ *such that, for all* $a, b \in A$,

(i)   $a \gtrsim b$ *iff* $\phi(a) \geqslant \phi(b)$ *and* $\chi(a) = \chi(b)$, *for all* $\chi \in \mathscr{C}$;
(ii)  $\phi(a \circ b) = \phi(a) + \phi(b)$;
(iii) $\phi(a) = 0$ *iff* $a \in A_0$.

*Another function* $\phi'$ *satisfies* (i)–(iii) *iff for some real* $\alpha > 0$, $\phi' = \alpha\phi$.

The key mathematical step in the proofs of both Theorems 10 and 11 is a group theoretic result which we first saw in Giles (1964) and Duistermaat (1968) states is well known. It says that if we have a commutative group with a real-valued additive function over a subgroup of that group, then this function can be extended additively over the whole group (Section 3.13.2).

### 3.12.4   Statistical Entropy

In statistical mechanics and in Shannon's theory of information, the state of a system is described by a probability distribution over a finite set. It is more convenient to treat as a state the equivalence class $p$ of vectors obtained by permutations from a vector $(p_1, ..., p_m)$, where, for $i = 1, ..., m$, $p_i \geqslant 0$ and $\sum_{i=1}^{m} p_i = 1$. Let $N(p)$ denote the number of components of a vector in $p$, and let $p \circ q$ be the equivalence class of vectors of the form $(p_1 q_1, ..., p_1 q_n, ..., p_m q_1, ..., p_m q_n)$. Observe that $N(p \circ q) = N(p) N(q)$. Let $A$ denote the set of all such equivalence classes of vectors, and let $A_0$ denote the set of equivalence classes of vectors of the form $(1, 0, ..., 0)$.

Now, suppose that $H$ is any real-valued function defined over $A$ for which:

(i)   $H(p) \geqslant 0$ and $H(p) = 0$ iff $p \in A_0$;
(ii)  $H$ is continuous;
(iii) $H(p \circ q) = H(p) + H(q)$.

For such an $H$, define

$$p \gtrsim q \quad \text{iff} \quad N(p) = N(q) \text{ and } H(p) \geqslant H(q).$$

It is easy to verify that $\langle A, \gtrsim, \circ \rangle$ is an entropy structure in the sense of Definition 8.

It is well known that Shannon's information measure, i.e., the entropy of statistical mechanics,

$$H_1(p) = -\sum_{i=1}^{m} p_i \log p_i$$

satisfies these three requirements, and so it is consistent with the axioms for an entropy structure. It does not follow, however, that any entropy system in which $A$ and $\circ$ are defined as above necessarily leads to the $H_1$ measure. As Rényi (1961) pointed out, the following class of functions

$$H_\alpha(p) = \frac{1}{1-\alpha} \log \sum_{i=1}^{m} p_i{}^\alpha, \quad \alpha \neq 1$$

also have the same three properties. Note that $\lim_{\alpha \to 1} H_\alpha(p) = H_1(p)$. This means that Definition 8 does not uniquely characterize physical entropy.

Rényi (1961) (also see Aczél, 1966, p. 153, the references given there, and Aczél, 1969) studied additional properties that characterize $H_1$ uniquely. In particular, he enlarged the space of distributions to include all those with $0 < \sum p_i \leqslant 1$ and introduced a further axiom concerning the composition $(p_1, ..., p_m, q_1, ..., q_n)$.

## 3.13 PROOFS

### 3.13.1 Preliminary Lemmas

The common hypothesis of the following four lemmas is that $\langle A, \gtrsim, \circ \rangle$ satisfies Axioms 1–4, Definition 8 , Section 3.12.3.

LEMMA 16. *If $b \gtrsim a$ and $c \gtrsim a$, then either $b \gtrsim c$ or $c \gtrsim b$.*

PROOF. By Axiom 4, $b \circ c \gtrsim a \circ c$ and $b \circ c \gtrsim b \circ a$, whence by Axioms 2 and 3, either $c \circ a \sim a \circ c \gtrsim b \circ a$ or $b \circ a \gtrsim a \circ c \sim c \circ a$. By Axioms 1 and 4, the conclusion follows. $\diamond$

As in the proof of Theorem 1 (Section 3.3.2), *define* $\approx$ on $A \times A$ as follows: for all $a, b, c, d \in A$,

$$(a, b) \approx (c, d) \quad \text{iff} \quad a \circ d \sim b \circ c.$$

LEMMA 17.   *The relation $\approx$ on $A \times A$ is an equivalence relation.*

PROOF.   The proof is the same as for Lemma 4 (Section 3.3.2) with the assumed commutativity of $\circ$ (Axiom 3) substituted for Lemma 3 (Section 3.3.2). $\diamondsuit$

Also, as in the proof of Theorem 1, define

$$[a, b] = \{(a', b') \mid a', b' \in A, \ (a', b') \approx (a, b)\}$$
$$D = \{[a, b] \mid a, b \in A\} = A \times A/\approx$$
$$P = \{[a, b] \mid a, b \in A, \ a \gtrsim b\}$$
$$* : [a, b] * [c, d] = [a \circ c, b \circ d]$$
$$[0] = [a, a], \ a \in A$$
$$[a, b]^{-1} = [b, a]$$
$$\gtrsim_D : [a, b] \gtrsim_D [c, d] \ \text{iff} \ [a, b] * [c, d]^{-1} \in P.$$

In addition, define

$$C = \{[a, b] \mid a \gtrsim b \ \text{or} \ b \gtrsim a\}$$

which is the set of comparable pairs modulo $\approx$ (called possible processes by Giles).

LEMMA 18.   (i)   *The operation $*$, the identity $[0]$, the inverse, and the relation $\gtrsim_D$ on $D$ are well defined.*

(ii)   $\langle D, * \rangle$ *and* $\langle C, * \rangle$ *are commutative groups.*

(iii)   *The relation $\gtrsim_D$ on $C$ is a simple order.*

PROOF.   Except for the two statements about $C$, the proof is the same as that for Lemma 5 (Section 3.3.2) with the assumed commutativity of $\circ$ substituted for Lemma 3.

The operation $*$ is closed in $C$. Suppose that $[a, b]$, $[c, d] \in C$. If $a \gtrsim b$, $d \gtrsim c$, then by Axioms 3 and 4, $a \circ c \gtrsim b \circ c$ and $b \circ d \gtrsim b \circ c$, and so by Lemma 16 either $a \circ c \gtrsim b \circ d$ or $b \circ d \gtrsim a \circ c$, and therefore $[a, b] * [c, d] \in C$. The other three cases are similar. If $[a, b] \in C$, then clearly $[a, b]^{-1} = [b, a] \in C$ and $[0] \in C$. Also $*$ is commutative since $\circ$ is. Thus, $\langle C, * \rangle$ is a commutative group.

Since by definition, $\gtrsim_D$ is connected over $C$, the proof of Lemma 5 that $\gtrsim_D$ is a simple order (over $D$) shows that here $\gtrsim_D$ is a simple order over $C$. $\diamondsuit$

LEMMA 19.   *If, in addition, Axiom 6 holds, then* $\langle C, \gtrsim_D, * \rangle$ *is an Archimedean simply ordered group (Definition 2.3).*

PROOF.   The proof is the same as for Lemma 6 (Section 3.3.2) with $C$ substituted for $D$ and Lemma 18 for Lemma 5. $\diamondsuit$

### 3.13.2  A Group-Theoretic Result

LEMMA 20.  *Suppose that* $\langle \mathcal{G}_0 , \circ \rangle$ *is a subgroup of commutative group* $\langle \mathcal{G}, \circ \rangle$ *and that* $\phi$ *is a real-valued additive function over* $\mathcal{G}_0$. *Then there exists a real-valued, additive extension* $\phi'$ *of* $\phi$ *over* $\mathcal{G}$, *i.e., for* $x \in \mathcal{G}_0$, $\phi'(x) = \phi(x)$.

PROOF.  Let $\mathcal{D}(\psi)$ denote the domain of a function $\psi$. Let $\mathcal{I}$ denote the set of additive, real-valued functions $\psi$ for which the following is true: $\mathcal{G}_0 \subset \mathcal{D}(\psi) \subset \mathcal{G}$; $\mathcal{D}(\psi)$ is a subgroup of $\mathcal{G}$; for $x \in \mathcal{G}_0$, $\psi(x) = \phi(x)$. Clearly $\mathcal{I}$ is nonempty since it includes $\phi$. We show that there exists $\phi' \in \mathcal{I}$ with $\mathcal{D}(\phi') = \mathcal{G}$.

Define $\lesssim$ on $\mathcal{I}$ as follows: if $\psi, \theta \in \mathcal{I}$, then $\psi \lesssim \theta$ iff $\mathcal{D}(\psi) \subset \mathcal{D}(\theta)$ and $\theta(x) = \psi(x)$ for $x \in \mathcal{D}(\psi)$. It is trivial to see that $\lesssim$ is a partial order. Let $\mathcal{L}$ be any simply ordered, nonempty subset of $\mathcal{I}$, then we show that $\mathcal{L}$ has an upper bound $\bar{\psi}$ in $\mathcal{I}$. Let $\mathcal{D}_{\mathcal{L}} = \bigcup_{\psi \in \mathcal{L}} \mathcal{D}(\psi)$, and define $\bar{\psi}$ as follows: For any $x \in \mathcal{D}_{\mathcal{L}}$ there exists at least one $\psi_x \in \mathcal{L}$ such that $x \in \mathcal{D}(\psi_x)$, and set $\bar{\psi}(x) = \psi_x(x)$. Observe that since $\mathcal{L}$ is simply ordered, it is immaterial which $\psi$ we choose so long as $x \in \mathcal{D}(\psi)$ because we obtain the same value for $\bar{\psi}(x)$. We show that $\bar{\psi} \in \mathcal{I}$. First, we establish that $\mathcal{D}_{\mathcal{L}}$ is a subgroup of $\mathcal{G}$. Obviously, it is sufficient to show that it is closed under $\circ$. If $x, y \in \mathcal{D}_{\mathcal{L}}$, then there exist $\psi, \theta \in \mathcal{L}$ such that $x \in \mathcal{D}(\psi)$ and $y \in \mathcal{D}(\theta)$. With no loss of generality, assume that $\psi \lesssim \theta$. Thus, $\mathcal{D}(\psi) \subset \mathcal{D}(\theta)$, and $\mathcal{D}(\theta)$ is closed, so $x, y, x \circ y \in \mathcal{D}(\theta) \subset \mathcal{D}_{\mathcal{L}}$. Second, $\mathcal{G}_0 \subset \mathcal{D}_{\mathcal{L}} \subset \mathcal{G}$ because $\mathcal{G}_0 \subset \mathcal{D}(\psi) \subset \mathcal{D}_{\mathcal{L}}$ for every $\psi \in \mathcal{L}$. Third, we show that $\bar{\psi}$ is additive: $\bar{\psi}(x \circ y) = \psi_{x \circ y}(x \circ y) = \psi_{x \circ y}(x) + \psi_{x \circ y}(y) = \psi_x(x) + \psi_y(y) = \bar{\psi}(x) + \bar{\psi}(y)$. And fourth, $\bar{\psi}(x) = \phi(x)$ for $x \in \mathcal{G}_0$ since $\psi(x) = \phi(x)$ for every $\psi \in \mathcal{L}$.

Therefore, by Zorn's Lemma, there exists a maximal element $\phi' \in \mathcal{I}$. We complete the proof by showing that $\mathcal{D}(\phi') = \mathcal{G}$. Suppose that, on the contrary, there exists $y \in \mathcal{G} - \mathcal{D}(\phi')$. Define $ny = (n - 1) y \circ y$, where $0y$ is the unit of $\mathcal{G}$ and $(-n)y$ is the inverse of $ny$. Let

$$\mathcal{D}'' = \{x \circ ny \mid x \in \mathcal{D}(\phi') \text{ and } n \text{ is an integer}\}.$$

Obviously, $\mathcal{D}(\phi')$ is a proper subgroup of $\mathcal{D}''$ (take $n = 0$). We distinguish two cases:

1.  For some $k > 0$, $ky \in \mathcal{D}(\phi')$. Thus, for any $z \in \mathcal{D}''$,

$$kz = kx \circ kny \in \mathcal{D}(\phi')$$

because $\mathcal{D}(\phi')$ is a commutative group. Define $\phi''(z) = \phi'(kz)/k$. If $z \in \mathcal{D}(\phi')$, then $\phi''(z) = \phi'(kz)/k = k\phi'(z)/k = \phi'(z)$ and, for $z, w \in \mathcal{D}''$, $\phi''(z \circ w) = \phi'[k(z \circ w)]/k = \phi'(kz \circ kw)/k = [\phi'(kz) + \phi'(kw)]/k = \phi''(z) + \phi''(w)$.

Hence, $\phi''$ is in $\mathscr{D}$ and $\phi'' > \phi'$, so $\phi'$ is not maximal in $\mathscr{I}$, contrary to choice.

2. For all $n > 0$, $ny \notin \mathscr{D}(\phi')$. It follows that each element of $\mathscr{D}''$ has a unique expression as $x \circ ny$, and we define $\phi''(x \circ ny) = \phi'(x)$. Obviously, if $x \in \mathscr{D}(\phi')$, $\phi''(x) = \phi'(x)$, and using the fact that $\circ$ is commutative,

$$
\begin{aligned}
\phi''[(x \circ ny) \circ (z \circ my)] &= \phi''[(x \circ z) \circ (n + m)y] \\
&= \phi'(x \circ z) \\
&= \phi'(x) + \phi'(z) \\
&= \phi''(x \circ ny) + \phi''(z \circ my),
\end{aligned}
$$

and again $\phi'$ is not maximal. Thus, $\mathscr{D}(\phi') = \mathscr{G}$.                     $\Diamond$

Note that the proof of Lemma 20 rests on Zorn's Lemma, which is, of course, equivalent to the axiom of choice.

### 3.13.3 Theorem 10  (p. 115)

*Suppose that $\langle A, \succsim, \circ \rangle$ satisfies Axioms 1–4 and 6 of Definition 8. Then there exists $\phi: A \to Re$ such that for all $a, b \in A$,*

(i) *if $a$ and $b$ are comparable, then $a \succsim b$ iff $\phi(a) \geqslant \phi(b)$;*

(ii) *$\phi(a \circ b) = \phi(a) + \phi(b)$.*

*This $\phi$ is unique up to multiplication by a positive constant and addition of a component of content. Adding Axioms 7 and 8 yields*

(iii) *$\phi \geqslant 0$;*

(iv) *$\phi(a) = 0$ iff $a \in A_0$.*

*In this case, $\phi$ is a ratio scale.*

PROOF.   By Lemma 19 and Theorem 2.4 (Section 2.2.1) there is an extensive measure $\psi$ on $C$ which is unique up to multiplication by a positive constant. By Lemmas 18 and 20, there is an additive extension of $\psi$ of $\langle D, * \rangle$. Define $\phi$ as in the proof of Theorem 1 (Section 3.3.3). The same argument as in Theorem 1 shows that it is an extensive measure [(i) and (ii) hold].

Any extensive measure $\phi'$ on $A$ defines a measure $\psi'$ on $D$ by $\psi'([a, b]) = \phi(a) - \phi(b)$. It is easy to show that $\psi'$ is an extensive measure on $\langle C, * \rangle$, and the uniqueness part of Theorem 2.4 establishes that $\psi = \alpha\psi'$ for some $\alpha > 0$. It is easy to show that $\chi = \phi - \alpha\phi'$ is a component of content, and so $\phi = \alpha\phi' + \chi$ for every pair of extensive measures $\phi$ and $\phi'$, with $\chi$ as above.

Next, suppose that Axioms 7 and 8 hold. For $a \in A$, let $x_a \in A_0$ be such that $a \gtrsim x_a$. At least one $x_a$ exists by Axiom 8. Let $\phi^*$ satisfy parts (i) and (ii) of the theorem. Define $\phi$ on $A$ as follows: $\phi(a) = \phi^*(a) - \phi^*(x_a)$. By Lemma 16 and definition of $A_0$, for all $x, y \ A_0$, if $a \gtrsim x$ and $a \gtrsim y$, then $x \sim y$, so $\phi(a)$ is invariant over the choice of $x_a$. Clearly, $\phi$ satisfies properties (iii) and (iv), and so we need only show (i) and (ii).

(i)   Suppose that $a$ and $b$ are comparable. Since $x_a \sim x_b$, then $\phi(a) \geqslant \phi(b)$ is equivalent to $\phi^*(a) \geqslant \phi^*(b)$ which, in turn, is equivalent to $a \gtrsim b$.

(ii)   By Axioms 1 and 4, $a \circ b \gtrsim x_a \circ x_b$, and by Axiom 7, $x_a \circ x_b \in A_0$, hence $x_a \circ x_b \sim x_{a \circ b}$. Therefore,

$$
\begin{aligned}
\phi(a \circ b) &= \phi^*(a \circ b) - \phi^*(x_{a \circ b}) \\
&= \phi^*(a \circ b) - \phi^*(x_a \circ x_b) \\
&= \phi^*(a) + \phi^*(b) - \phi^*(x_a) - \phi^*(x_b) \\
&= \phi(a) + \phi(b).
\end{aligned}
$$

Finally, suppose that $\phi'$ also satisfies (i)–(iv). We have already shown that $\phi = \alpha\phi' + \chi$, where $\alpha > 0$ and $\chi$ is a component of content. Since $\phi(x) = \phi'(x) = 0$ for $x \in A_0$, $\chi(x) = 0$. But $a \gtrsim x_a$ implies, by definition of $\chi$, that $\chi(a) = \chi(x_a) = 0$. Thus, $\chi = 0$.            $\diamondsuit$

### 3.13.4  Theorem 11  (p. 115)

*Suppose that $\langle A, \gtrsim, \circ \rangle$ satisfies Axioms 1–5 of Definition 8. If $\gtrsim$ is connected, then there is no nontrivial component of content. If $\gtrsim$ is not connected, then for any incomparable $a, b \in A$, there exists a component of content $\chi$ for which $\chi(a) \neq \chi(b)$.*

PROOF.   If $\gtrsim$ is connected, then any component of content is a constant and, by additivity, that constant must be 0.

Suppose that $a, b \in A$ are incomparable, i.e., $[a, b] \notin C$. For any integer $n$, define

$$
n[a, b] = \begin{cases} [na, nb], & \text{if } n > 0, \\ [0], & \text{if } n = 0, \\ [-nb, -na], & \text{if } n < 0, \end{cases}
$$

and let

$$
B[a, b] = B = \{n[a, b] * [c, d] \mid n \text{ an integer}, [c, d] \in C\}.
$$

We first show that $\langle B, * \rangle$ is a subgroup of $\langle D, * \rangle$. It suffices to show that $B$ is closed under $*$, that $[0] \in B$, and that the inverse of an element in $B$ is also in $B$.

*Closure*: Suppose that $n[a, b] * [c, d]$ and $m[a, b] * [e, f]$ are in $B$. By Axiom 3,

$$n[a, b] * [c, d] * m[a, b] * [e, f] = (n + m)[a, b] * [c, d] * [e, f] \in B$$

since, by Lemma 19, $[c, d] * [e, f] \in C$.

*Zero*:   $[0] = [0] * [0] = 0[a, b] * [a, a] \in B$.
*Inverse*:

$$\begin{aligned} n[a, b] * [c, d] * (-n)[a, b] * [c, d]^{-1} &= (n - n)[a, b] * [c, d] * [c, d]^{-1} \\ &= [0] * [0] \\ &= [0]. \end{aligned}$$

We show that $m = n$ when $n[a, b] * [c, d] = m[a, b] * [e, f]$. With no loss of generality, suppose that $m \geqslant n$. Premultiply the equality by $-m[a, b]$ and postmultiply it by $[d, c]$ to obtain $(n - m)[a, b] = [e, f] * [d, c] \in C$. By Axiom 5, $m = n$ since the other possibility, $[a, b] \in C$, is contrary to choice. This means that the function $\chi'$ on $B$ defined by $\chi'(n[a, b] * [c, d]) = n$ is well defined. Observe that for $[c, d], [e, f] \in C$,

$$\chi'([c, d]) = \chi'(0[a, b] * [c, d]) = 0,$$
$$\chi'([a, b]) = \chi'(1[a, b] * [0]) = 1,$$
$$\begin{aligned} \chi'(n[a, b] * [c, d] * m[a, b] * [e, f]) &= \chi'((n + m)[a, b] * [c, d] * [e, f]) \\ &= n + m \\ &= \chi'(n[a, b] * [c, d]) + \chi'(m[a, b] * [e, f]). \end{aligned}$$

Thus, $\chi'$ is additive over $\langle B, * \rangle$, a subgroup of the commutative group $\langle D, * \rangle$, whence, by Lemma 20, there is an additive extension $\chi''$ of $\chi'$ over $\langle D, * \rangle$. Because $\chi''$ is additive and by what we have shown, if $c, d \in A$,

$$\chi''([c, d]) + \chi''([d, c]) = \chi''([c, d] * [d, c]) = \chi''([0]) = 0.$$

For any $c \in A$, define $\chi$ by $\chi(c) = \chi''([c, 2c])$. We show that $\chi$ is a component of content for which $\chi(a) \neq \chi(b)$. First,

$$\begin{aligned} \chi(c) - \chi(d) &= \chi''([c, 2c]) - \chi''([d, 2d]) \\ &= \chi''([c, 2c]) + \chi''([2d, d]) \\ &= \chi''([c \circ 2d, 2c \circ d]) \\ &= \chi''([d, c]). \end{aligned}$$

If $[c, d] \in C$, then $\chi''([d, c]) = \chi'([d, c]) = 0$.

Second,

$$\chi(c \circ d) = \chi''([c \circ d, 2(c \circ d)])$$
$$= \chi''([c, 2c] * [d, 2d])$$
$$= \chi''([c, 2c]) + \chi''([d, 2d])$$
$$= \chi(c) + \chi(d).$$

So $\chi$ is a component of content, and

$$\chi(a) - \chi(b) = \chi''([b, a]) = -\chi''([a, b]) = -\chi'([a, b]) = -1,$$

so $\chi$ is nontrivial. $\diamondsuit$

## 3.14 EXTENSIVE MEASUREMENT IN THE SOCIAL SCIENCES

A major difficulty in most attempts to apply the theory of extensive measurement to nonphysical attributes such as utility, intelligence, or loudness is the lack of an adequate interpretation for the concatenation operation. Indeed, this lack has led some authors, such as Campbell (1920) and Guild (1938), to conclude that fundamental measurement is impossible in the social sciences. Their conclusion is incorrect on two counts. First, some attributes of the social sciences can be measured extensively; subjective probability is an outstanding example. See Reese (1943) for three not completely successful attempts. Second, and much more important, the existence of an empirical operation of concatenation, albeit valuable when it does exist, is *not* a necessary condition for fundamental measurement. Numerous examples of fundamental measurement, not based on an empirical operation of concatenation, are discussed in detail in later chapters.

It is instructive to spend a moment examining the difficulties involved in finding an adequate interpretation for an additive operation in psychology. Apparently, most entities to which psychological attributes, e.g., aesthetic value or intelligence, are associated cannot be concatenated in any satisfactory fashion. The intelligence of a *group* of people or the aesthetic value of a *set* of plays do not appear to be constructs that can be meaningfully ordered.

A few attributes, such as perceived weight, length, or time, are associated with entities that can be concatenated, but they pose a different problem. One may attempt to measure perceived weight, for example, by using the physical concatenation of objects and letting the judgment of a subject, rather than a pan balance, determine the ordering of the objects. When such an experiment is properly conducted, controlling for response biases, order effects, etc., a subject will certainly recover the weight ordering of the objects, within his limits of discrimination. Such a procedure, therefore, fails to yield psychological measurement of perceived weight different from physical measurement.

A third difficulty arises in connection with the measurement of the utility of composite entities such as commodity bundles. To illustrate, let $(x, y)$ denote a commodity bundle consisting of $x$ pairs of slacks and $y$ shirts. Using the natural concatenation of commodities, let $(x, y) \circ (z, w) = (x + z, y + w)$. For this interpretation, however, the monotonicity axiom is easily seen to be violated: an individual is very likely to prefer $(3, 0)$ over $(0, 3)$ and yet prefer $(3, 0) \circ (0, 3) = (3, 3)$ over $(3, 0) \circ (3, 0) = (6, 0)$, contrary to the axiom.

Thus, the attempt to apply extensive measurement to the social sciences is beset with serious difficulties. In some instances, no operation is available; in others, the available operation either leads to trivial results or to a violation of the axioms. These difficulties have led to the development of other axiom systems as a basis for fundamental measurement in the social sciences, such as difference measurement (Chapter 4), conjoint measurement (Chapters 6 and 7), and expected utility measurement (Chapter 8).

Two psychological attributes, subjective probability and risk, are exceptional in that they appear to be extensively measurable. The concatenation operation is interpreted as the union of disjoint events in the former and as the convolution of probability distributions in the latter. Chapter 5 is devoted to the measurement of subjective probability; the measurement of risk is discussed in the next section.

### 3.14.1  The Measurement of Risk

The notion of risk has been employed in the social sciences as a property of uncertain options, or lotteries, which affects decision making. For example, an economist may refer to one investment as riskier than another, or a psychologist may attribute to an individual a tendency to choose risky courses of action. In this section we present an analysis of the concept of risk from the viewpoint of measurement theory, as developed by Pollatsek and Tversky (1970). Following their approach we first show how the theory of extensive measurement can be applied to the measurement of risk, and then we demonstrate that, by adding several assumptions, the risk scale is essentially predetermined.

The theory is formulated in terms of a set $\mathcal{R} = \{f, g, h, ...\}$ of probability distributions over the real line. The elements of $\mathcal{R}$ are interpreted as options, or lotteries; that is, each element of $\mathcal{R}$ is a function that assigns to every monetary (or other numerical) outcome a probability of occurrence. For example, the lottery where one wins or loses $1 depending on whether a fair coin comes up heads or tails, respectively, is represented by the discrete probability distribution where each of the values $+1$ and $-1$ is obtained with probability of $\frac{1}{2}$.

The representation of uncertain options as probability distributions over the real line does not necessarily exclude lotteries with nonmonetary outcomes, e.g., the results of an election or the outcome of a game, provided they can be expressed in terms of appropriate numerical units.

Let $\circ$ denote convolution of probability distributions. Thus, if $f$ and $g$ are two discrete distributions where the values $x_1,...,x_m$ and $y_1,...,y_n$ are obtained with probabilities $f_1,...,f_m$ and $g_1,...,g_n$, respectively, then $f \circ g$ is the distribution where the values $x_i + y_j$ are obtained with probabilities $f_i g_j$, $i = 1,...,m, j = 1,...,n$. In the continuous case

$$f \circ g(t) = \int_{-\infty}^{\infty} f(t - x)g(x)\,dx.$$

It is well known that $\circ$ is associative and commutative.

The central construct of the theory is a binary relation of comparative risk, denoted $\gtrsim$, where $f \gtrsim g$ is interpreted as: $f$ is at least as risky as $g$. Such data can be obtained directly, for example, by asking an individual to judge which of two lotteries is riskier. Alternatively, one may infer risk judgments from other data (e.g., preferences) via an appropriate model.

DEFINITION 10. *Let $\mathscr{R}$ be a nonempty set of probability distributions over the real line that is closed under convolution, denoted $\circ$, and let $\gtrsim$ be a binary relation on $\mathscr{R}$. The triple $\langle \mathscr{R}, \circ, \gtrsim \rangle$ is a risk structure iff the following three axioms are satisfied for all $f, g, h, h' \in \mathscr{R}$.*

1. *Weak order:* $\langle \mathscr{R}, \gtrsim \rangle$ *is a weak order.*
2. *Monotonicity:* $f \gtrsim g$ *iff* $f \circ h \gtrsim g \circ h$.
3. *Archimedean: If $f > g$, then there exists a positive integer $n$ such that $f^n \circ h \gtrsim g^n \circ h'$, where $f^n = f^{n-1} \circ f$, $f^1 = f$.*

The above definition differs slightly from that of Pollatsek and Tversky (1970) in that it contains no explicit solvability axiom and the Archimedean axiom is formulated differently. In fact, the present definition of a risk structure is essentially equivalent to the definition of a closed extensive structure (Definition 1, Section 3.2.1). The difference between Definitions 1 and 10 stems from the fact that, because the convolution of distributions is, by definition, associative and commutative, there is no need to postulate associativity, and the monotonicity axiom assumes a simpler form.

Although the axioms of Definition 10 are all familiar, their application to a new domain calls for a brief discussion. The first axiom asserts that lotteries, conceived as probability distributions over the real line, can be weakly ordered with respect to risk. Although this assumption has been implicit in most discussions of risk, the possibility of obtaining a meaningful risk ordering should not be taken for granted. The monotonicity axiom expresses the compatibility between the risk ordering and the convolution operation. It is, clearly, equivalent to the assertion that $f \circ h \gtrsim g \circ h$ if and only if $f \circ h' \gtrsim g \circ h'$. Thus, it can be interpreted as saying that $f$ is judged riskier than $g$ in one context (i.e., when convoluted with $h$) if and only if $f$ is judged riskier than $g$ in any context. The validity of this axiom is not self-evident; nevertheless, it is theoretically appealing and empirically testable. Finally, the Archimedean axiom is introduced to ensure that no "risk difference" is infinitely larger than any other, i.e., any risk, however large, can always be exceeded by a finite number of copies of any other risk, however small.

**THEOREM 12.** *If $\langle \mathscr{R}, \circ, \gtrsim \rangle$ is a risk structure, then there exists a real-valued function R on $\mathscr{R}$ such that for any $f, g \in \mathscr{R}$*

   (i)  $f \gtrsim g$ iff $R(f) \geqslant R(g)$;
  (ii)  $R(f \circ g) = R(f) + R(g)$.

*Another function $R'$ satisfies (i) and (ii) iff there exists a real $\alpha > 0$ such that $R' = \alpha R$.*

It is readily seen that a risk structure is a closed extensive structure (Definition 1), hence Theorem 12 follows at once from Theorem 1. Moreover, Axioms 1, 2, and 3 are not only sufficient to establish Theorem 12, but they are also necessary conditions for the desired representation. Note that all the extensive measurement scales encountered thus far were positive. In contrast, the risk scale admits nonpositive values as well. In particular, if $0^*$ denotes the status quo of receiving 0 with probability 1, then it is easy to see that $f \circ 0^* = f$, and so $R(f) + R(0^*) = R(f \circ 0^*) = R(f)$, hence $R(0^*) = 0$. Thus, any option that is less risky than the status quo, if any, is assigned a negative risk value.

In order to formulate additional assumptions, four auxiliary concepts are introduced. First, generalizing from the notation $0^*$, for $\alpha$ in Re let $\alpha^*$ denote the sure-thing probability distribution for which

$$\alpha^*(t) = \begin{cases} 0, & \text{for } t \neq \alpha, \\ 1, & \text{for } t = \alpha. \end{cases}$$

Second, for $\alpha$ in Re, let $\alpha f$ denote the distribution after a scale change of multiplying by $\alpha$, i.e., $(\alpha f)(t) = f(t/\alpha)/\alpha$. Third, let $E(f)$ and $V(f)$ denote the expectation and variance, respectively, of the distribution $f$. Finally, a sequence $\{f_n\}$, $n = 1, 2,...$, of distributions is said to approach a limiting distribution $f$ provided that for all real $x$, $y$, with $x < y$,

$$\int_x^y f_n(t)\, dt \quad \text{approaches} \quad \int_x^y f(t)\, dt \quad \text{as} \quad n \to \infty.$$

DEFINITION 11. *A risk structure* $\langle \mathscr{R}, \circ, \gtrsim \rangle$ *is called* regular *iff the following four axioms are satisfied for all* $f$, $g \in \mathscr{R}$ *and for any* $\alpha \in \text{Re}$.

4.  *Completeness*: (i) $\alpha f \in \mathscr{R}$, (ii) $\alpha^* \in \mathscr{R}$;

5.  *Positivity*: *If* $\alpha > 0$, *then* $f \gtrsim f \circ \alpha^*$;

6.  *Scalar monotonicity*: *If* $E(f) = E(g) = 0$, *then* (i) $\alpha f > f$ *for* $\alpha > 1$, *and* (ii) $f > g$ *iff* $\alpha f > \alpha g$.

7.  *Continuity*: *If* $\{f_n\}$ *approaches* $f$ *as* $n \to \infty$, *then* $R(f_n)$ *approaches* $R(f)$, *provided that* $E(f_n) = E(f)$ *and* $V(f_n) = V(f)$, *for all* $n$.

The completeness axiom postulates that $\mathscr{R}$ includes all sure-thing lotteries and all changes of scale. The positivity axiom asserts that the convolution of a positive sure thing with a lottery cannot increase its risk. The scalar monotonicity axiom applies only to fair bets, i.e., to lotteries with zero expectation. It asserts that (i) scaling the lottery upward by a constant factor increases risk, and that (ii) changing scale of two lotteries preserves the risk ordering. In particular it implies that the risk ordering between lotteries (with zero expectation) is independent of the denomination of the outcomes, e.g., pennies, dollars. This is a strong, though testable, assumption. The continuity axiom is more technical in character. It asserts that if a sequence of distributions with common expectation and variance approach a limiting distribution, then their risks also approach that of the limiting distribution.

THEOREM 13. *If* $\langle \mathscr{R}, \circ, \gtrsim \rangle$ *is a regular risk structure, then there exist unique* $\mu > 0$ *and* $\eta > 0$, *such that for all* $f$, $g \in \mathscr{R}$ *(with finite expectations and variances)*

$$f \gtrsim g \quad \text{iff} \quad R(f) \geqslant R(g),$$

*where*

$$R(f) = \eta V(f) - \mu E(f)$$

The significance of Theorem 13 is in showing that in a regular risk structure the only possible orderings are those generated by linear combinations of expectation and variance. Indeed, expectation and variance have been considered major components of risk by both economists and psychologists. Further discussion of risk and its relation to preference can be found in Section 8.5.

Despite the formal similarity, the measurement of risk differs from other forms of extensive measurement in several important respects. First, the empirical adequacy of the axioms has not been established—indeed, there is no generally accepted procedure for obtaining the risk ordering. Second, the measurement of risk is not, strictly speaking, fundamental measurement because it presupposes the numerical representation of options as probability distributions on the real line. Finally, in a regular risk structure the risk scale is essentially predetermined, since the risk of any option can be readily computed once a single parameter $\theta$ is determined.

### 3.14.2 Proof of Theorem 13

The following two lemmas assume that $\langle \mathscr{R}, \circ, \succsim \rangle$ is a regular risk structure.

**LEMMA 21.** *There exists some real* $\mu > 0$ *such that for all real* $\alpha$, $R(\alpha^*) = -\mu\alpha$.

**PROOF.** Define $\phi(\alpha) = R(\alpha^*)$ for any real $\alpha$; it exists by part (ii) of Axiom 4. Hence, by definition and Theorem 12,

$$\phi(\alpha + \beta) = R(\alpha^* \circ \beta^*) = R(\alpha^*) + R(\beta^*) = \phi(\alpha) + \phi(\beta)$$

for any real $\alpha, \beta$. Furthermore, by Axioms 2 and 5, $\alpha \geqslant \beta$ implies $R(\alpha^*) \leqslant R(\beta^*)$, and hence $\phi$ is nonincreasing. Consequently, $\phi$ is linear and there exists some $\mu \geqslant 0$ such that for all $\alpha \in \text{Re}$, $R(\alpha^*) = \phi(\alpha) = -\mu\alpha$. $\Diamond$

**LEMMA 22.** *There exists some real* $\eta > 0$ *such that for all* $f \in \mathscr{R}$ *with* $E(f) = 0$ *and variance* $V(f) < \infty$, $R(f) = \eta V(f)$.

**PROOF.** Let $f, g \in \mathscr{R}$ such that $E(f) = E(g) = 0$, and $V(f) > V(g)$. Let $f_n = n^{-1/2}f^n$, $g_n = n^{-1/2}g^n$. Clearly, $f_n$, $g_n \in \mathscr{R}$. As $n \to \infty$, the sequences $\{f_n\}$ and $\{g_n\}$ approach, respectively, limiting distributions $f'$ and $g'$ which, by the central limit theorem, are normal with $E(f') = E(g') = 0$, and $V(f') = V(f) > V(g) = V(g')$. By properties of the normal distribution, $f' = \alpha g'$ for some real $\alpha > 1$. Hence, by part (i) of Axiom 6, $f' \succ g'$, and, by Theorem 12, $R(f') > R(g')$. By Axiom 7, however, there exists a positive integer $n$ such that $R(f_n) > R(g_n)$, and hence $f_n \succ g_n$. Consequently $f^n \succ g^n$, by part (ii) of Axiom 6, and $f \succ g$, by Axiom 2. Similarly, $V(f) = V(g)$ implies $f \sim g$. Hence, recalling Theorem 12, if $E(f) = E(g) = 0$, then

$f \gtrsim g$ iff $V(f) \geqslant V(g)$ iff $R(f) \geqslant R(g)$. Thus, there exists a strictly increasing function, $\phi$, such that for any $f \in \mathcal{R}$ with $E(f) = 0$, $R(f) = \phi[V(f)]$. Note that the set of distributions with zero expectation is closed under $\circ$, and that, by the definition of variance and Theorem 12, respectively, both $V$ and $R$ are additive over $\circ$, hence if $E(f) = E(g) = 0$, then

$$R(f \circ g) = \phi[V(f \circ g)] = \phi[V(f) + V(g)],$$

and

$$R(f \circ g) = R(f) + R(g) = \phi[V(f)] + \phi[V(g)].$$

It follows from the above equations that $\phi$ is linear; furthermore, by part (i) of Axiom 6 it is increasing. Hence, there exists some $\eta > 0$ such that $R(f) = \phi[V(f)] = \eta V(f)$. ◇

*If $\langle \mathcal{R}, \circ, \gtrsim \rangle$ is a regular risk structure, then there exists a unique $\theta$ in $(0, 1]$ such that for all $f, g \in \mathcal{R}$ (with finite expectations and variances)*

$$f \gtrsim g \quad iff \quad R(f) \geqslant R(g),$$

*where*

$$R(f) = \theta V(f) - (1 - \theta) E(f).$$

PROOF. For any $f \in \mathcal{R}$, define $f_0$ by the equation $f_0 = f \circ [-E(f)^*]$. By part (ii) of Axiom 4, $-E(f)^* \in \mathcal{R}$ and hence $f_0 \in \mathcal{R}$. Observe that by a property of expectation,

$$\begin{aligned} E(f_0) &= E\{f \circ [-E(f)]^*\} \\ &= E(f) + E(-E(f)^*) \\ &= E(f) - E(f) \\ &= 0. \end{aligned}$$

And by Theorem 12,

$$\begin{aligned} R(f_0) &= R\{f \circ [-E(f)]^*\} \\ &= R(f) + R\{[-E(f)]^*\}. \end{aligned}$$

Substituting from Lemmas 21 and 22,

$$\begin{aligned} R(f) &= \eta V(f_0) - \mu E(f) \\ &= \eta V(f) - \mu E(f). \end{aligned}$$ ◇

As E. Roskam has pointed out (unpublished manuscript), the original formulation of Theorem 13 incorporated a dimensional error.

## ☐ 3.15  LIMITATIONS OF EXTENSIVE MEASUREMENT

As an account of how numbers may be used to represent physical, psychological, and economic attributes, the theory of extensive measurement is not wholly adequate. Four limitations are cited here. Some of these are dealt with in later chapters.

First, even if we assume that a formal structure of the type $\langle A, \gtrsim, B, \circ \rangle$ is suitable, we may question the adequacy of the axioms given in Definition 3 and, in weakened form, in Definition 5. Probably the most vexing of the axioms is the Archimedean one which, as was pointed out, may be trivially true or may be difficult to test (see Section 3.4.2). Moreover, the axiom requires that we be able to construct multiples $na$ of a given element $a$. For many empirical realizations, multiples are not particularly easy to construct. Another vexing axiom is 4 which asserts that the sets $A$ and $B$ are sufficiently rich so that all inequalities of the form $a \gtrsim b \circ x$ can be solved for $x$ when $a > b$. The empirical significance of such an axiom is not transparent, and its falsity may be difficult to establish. We return to these problems in Chapter 9.

Second, as we saw in Section 3.6.1, there are alternative interpretations for $\circ$ that lead to nonlinearly related scales of length. The general problem of alternative representations was discussed in Section 3.9, and the particular problem of obtaining a nonadditive representation for velocity was partially answered in Section 3.7 by simultaneously axiomatizing length and velocity. In other empirical contexts, similar problems seem completely unresolved. Consider, for example, pure tones of a fixed frequency which are concatenated by physically adding them in phase. If we order the tones according to physical intensity (amplitude), then we expect the axioms of Definition 3 to be met and the additive representation simply recovers amplitude. If, however, the ordering is generated by a subject's judgments of relative loudness (ignore, for the moment, local inconsistencies in his judgments), then it is doubtful whether Axiom 5, $a \circ b > a$, would be satisfied; a maximal loudness probably exists. If so, we must use a bounded scale for loudness that is nonlinearly related to amplitude. But which representation should we select? Presumably, as with velocity, some other considerations about loudness are needed to determine the scale. No proposals have yet been made.

Third, is it acceptable to treat comparisons of an attribute as a weak order? In Chapter 1 we questioned not only the transitivity of indifference, but whether observed judgments should be treated as a relation at all. No more need be said about this now except to mention that attempts to modify the weak ordering assumption are described in Chapters 15 and 16.

Fourth, and finally, is the notion of concatenation satisfactory? When a more-or-less natural empirical concatenation exists, as with length and time, it seems sensible to use it in constructing a fundamental measure. But for many attributes no natural empirical interpretation of concatenation can be found. Besides the difficulties encountered in the social sciences, which were discussed in the previous section, we conclude by citing three physical attributes that cannot be adequately concatenated.

1.   Momentum of moving bodies is of basic importance, and it is given by $p(a) = m(a) \, v(a)$, where $m(a)$ is the mass of $a$ and $v(a)$ is its velocity. Although it is not difficult, in principle, to devise ways to decide qualitatively which of two objects has the greater momentum, concatenating two momentums is another matter if both the velocities and masses differ. In principal it can be done by locking the objects together after collision, but often this is impractical. For this reason, momentum is usually treated as a derived quantity that is simply calculated from the fundamental measures of mass and velocity. Actually, however, a fundamental approach is possible that does not require a concatenation operation provided that we take advantage of the fact that mass and velocity both contribute to momentum and that they do so independently. This construction is explored in detail in Chapter 6.

2.   Superficially, density seems somewhat like momentum. The relative densities of two homogeneous materials can be judged, in some cases, by finding a liquid in which the one material floats and the other sinks. Whenever this comparison can be made it turns out that the ratio of the mass to the volume for the more dense substance exceeds the corresponding ratio for the less dense one. Moreover—and this is essential—this ratio is independent of the volume of material used. It is called the density of the material, and it represents something about the material, not about the amount of it. Density also has been treated as an example of derived measurement. It is surely different from momentum. Although both express a new quantity as a simple function of two extensive measures, in the case of momentum we are free to manipulate the component variables independently and continuously, whereas, in the case of density, a change in the mass of the material automatically entails a proportionate change in volume. This is an empirical law, not a logical necessity. Furthermore, it is doubtful if all mass–volume combinations can be achieved by varying the material. Nonetheless, the methods of Chapter 6 can also be used to measure density, but they have to be employed somewhat differently from those for momentum, because different types of qualitative laws are involved. These laws are discussed in Chapter 10.

3.   Finally, the hardness of substances is yet a different case. Relative hardness can be evaluated by a scratch test. There does not appear to be any natural (or formal) concatenation operation, and so extensive measurement does not apply. Moreover, it has not been treated in a derived way,

like· density, because we do not know which variables have roles corresponding to mass, velocity, and volume in this example. Possibly the measurement of hardness can be treated as fundamental measurement of hardness differences, according to the methods developed in the next chapter. The differences can be determined by an observation that $x$ scratches $y$ deeper than $w$ scratches $z$.

## EXERCISES

**1.** Suppose that $\langle A, \succsim, \circ \rangle$ is a closed extensive structure (Definition 1). If $a \in A$ is idempotent ($a \circ a \sim a$) and $b \in A$, prove that $a \circ b \sim b \circ a \sim b$ and that if $a' \in A$ is also idempotent, then $a' \sim a$.        (3.2.1)

**2.** In Theorem 1, prove that

   (i)  $a$ is idempotent iff $\phi(a) = 0$;
   (ii) $a \circ b$ is idempotent iff $\phi(a) = -\phi(b)$.        (3.2.1)

**3.** Let $A$ consist of all elements of the form $(ma, nb)$, where $a$ and $b$ are arbitrary symbols and $m$ and $n$ are nonnegative integers. Define $(ma, nb) \circ (pa, qb) = ((m + p)a, (n + q)b)$  and  $(ma, nb) > (pa, qb)$  iff $m + n > p + q$ or $m + n = p + q$ and $m > p$. Prove that the pair $(1a, 0b)$ and $(0a, 1b)$ is anomalous (p. 73).        (3.2.1)

**4.** Suppose that $\langle A, \succsim, \circ \rangle$ satisfies Axioms 1–3 and 5 of Definition 1. Recall that the structure is *solvable* if whenever $a > b$, then there exists $c \in A$ such that $a \succsim b \circ c$ and that it satisfies the *usual Archimedean condition* if whenever $a > b$, then there exists a positive integer $n$ such that $nb \succsim a$. Prove that:

   (i)  If the structure is solvable and satisfies the usual Archimedean condition, then Axiom 4 holds.
   (ii) If Axiom 4 holds, then the structure satisfies the usual Archimedean condition.        (3.2.1)

*The following six exercises develop a theory of measurement of angle which takes into account that such measurements are periodic.*

**5.** Use Axioms 1–3 of Definition 2 to prove:

   (i)   $a \sim b$ iff $a \circ c \sim b \circ c$;
   (ii)  $a > b$ iff either (1) $a \circ c > b \circ c \succsim c$; (2) $c > a \circ c > b \circ c$; (3) $b \circ c \succsim c > a \circ c$;
   (iii) $a \circ b \succsim a$ iff $a \circ b \succsim b$.        (3.2.2)

**6.** Define

$$r(a, b) = \begin{cases} 1, & \text{if } a, b > a \circ b, \\ 0, & \text{if } a \circ b \gtrsim a, b. \end{cases}$$

Use Axioms 1–3 and Exercise 5 to prove

(i)   if $a \gtrsim b$, then $r(a, c) \geqslant r(b, c)$;

(ii)  $r(a, b) + r(a \circ b, c) = r(b, c) + r(a, b \circ c)$;

(iii) if $a \gtrsim b$ and $0 < k < n - 1$, then

$$r[(n - k - 1) a, a] + r[(n - k) a, kb] \geqslant r(b, kb) + r[(n - k - 1)a, (k + 1)b].$$

**7.** For $n \in I^+$, define

$$W(n, a) = \begin{cases} 0, & n = 1, \\ \sum_{i=1}^{n-1} r(a, ia), & n \geqslant 2. \end{cases}$$

Use Axioms 1–3 and Exercise 6 to prove, for $m, n \in I^+$,

(i)   $W(m + n, a) = W(m, a) + W(n, a) + r(ma, na)$;

(ii)  $W(mn, a) = mW(n, a) + W(m, na)$;

(iii) if $a \gtrsim b$, then $W(n, a) \geqslant W(n, b)$.

**8.** Use Axioms 1–4 and Exercise 7 to prove that if $a > b$ and $l \in I^+$, then there exists $k \in I^+$ such that $W(k, a) \geqslant W(k, b) + l$.

**9.** Let $I^0 = I^+ \cup \{0\}$ and define $A^* = I^0 \times A$; $\gtrsim^*$ on $A^*$ by: $(m, a) \gtrsim^* (n, b)$ iff $m > n$ or $m = n$ and $a \gtrsim b$; $*$ on $A^*$ by:

$$(m, a) * (n, b) = (m + n + r(a, b), a \circ b).$$

Use Exercises 5–8 to verify that if $\langle A, \gtrsim, \circ \rangle$ is a periodic extensive structure, then $\langle A^*, \gtrsim^*, * \rangle$ is an extensive structure that is nonnegative (i.e., for all $\alpha, \beta \in A^*$, $\alpha * \beta \gtrsim^* \alpha$).   (3.2.1 and 3.2.2)

**10.** Use Theorem 1 and Exercise 9 to prove Theorem 2.   (3.2.1 and 3.2.2)

**11.** Verify that the examples of Section 3.3.1 satisfy the axioms that they are suppose to.   (3.2 and 3.3.1)

**12.** Using the definitions given on p. 78, prove that:

(i)   If $a$ and $b$ are positive, then $a \circ b$ is positive.

(ii)  If $a$ and $b$ are negative, then $a \circ b$ is negative.

(iii) If $a$ is positive and $b$ is negative, then $a > b$.   (3.2.1 and 3.3.2)

**13.** Let $A = [0, 1]$, $B = \{(x, y) \mid x, y \in A\}$, $x \gtrsim y$ iff $x \geqslant y$, and $x \circ y = xy$.

Which axioms of Definition 3 does $\langle A, \gtrsim, B, \circ \rangle$ satisfy (in each case provide either a proof or a counterexample)?      (3.4.3)

**14.** Verify that the examples of Section 3.5.1 satisfy the axioms that they are supposed to.      (3.4.3 and 3.5.1)

**15.** Suppose that $\langle A, \gtrsim, B, \circ \rangle$ is an extensive structure with no essential maximum (Definition 3). Without using Theorem 3, prove that if $a \circ c \gtrsim b \circ c$, then $a \gtrsim b$.      (3.4.3 and 3.5.2)

**16.** Suppose that $\langle A, \gtrsim, A \times A, \circ \rangle$ is an extensive structure with no essential maximum. Without using Theorem 3, prove that it has no anomalous pair.      (3.2, 3.4.3, and 3.5.2)

**17.** Prove Theorem 4.      (3.7.1)

**18.** Prove Theorem 5.      (3.7.1)

**19.** Evaluate $x \oplus y$ for $f(x) = x/\alpha(1 + x)$.      (3.9)

**20.** If $x \oplus y = f^{-1}[f(x) + f(y)]$, then prove that

$$x_1 \oplus x_2 \oplus \cdots \oplus x_n = f^{-1}\left[\sum_{i=1}^{n} f(x_i)\right].$$

Using this, show that for relativistic velocity

$$nx = \frac{(1 + x)^n - (1 - x)^n}{(1 + x)^n + (1 - x)^n},$$

where $1x = x$ and $nx = (n - 1)x \oplus x$.      (3.9)

**21.** Without using Theorem 9, prove that in a structure of extensive multiples (Definition 6) there is no anomalous pair.      (3.2 and 3.10.1)

**22.** Complete the proof of the converse of part (i) of Lemma 12.
                                                                    (3.10.1 and 3.11.2)

**23.** In Definition 8 of an entropy structure, show that Axiom 5 does not follow from Axioms 1–4.      (3.12.3)

**24.** Verify that the structure defined on p. 117 is an entropy structure.
                                                                    (3.12.3 and 3.12.4)

**25.** Verify that the function

$$H_\alpha(p) = \frac{1}{1 - \alpha} \log \sum_{i=1}^{n} p_i^\alpha, \qquad \alpha \neq 1,$$

has the three properties listed on p. 116.      (3.12.4)

**26.** Suppose that $\langle A, \succsim, \circ \rangle$ is an entropy structure. Prove that:

  (i)  If $x, y \in A_0$ and if for some $a \in A$, $a \succsim x$ and $a \succsim y$, then $x \sim y$.
  (ii) If $x, y \in A$ and $x \circ y \in A_0$, then $x, y \in A_0$.     (3.12.3 and 3.13.1)

**27.** Suppose that $\langle A, \succsim, \circ \rangle$ is an entropy structure. Define $\succsim_E$ on $A$ as follows: for $a, b \in A$, $a \succsim_E b$ iff there exist $x, y \in A_0$ such that $a \circ x \succsim b \circ y$. Without using Theorem 10, prove that for all $a, b, c \in A$,

  (i)   $\langle A, \succsim_E \rangle$ is a weak order;
  (ii)  $a \succsim_E b$ iff $a \circ c \succsim_E b \circ c$;
  (iii) $A_0$ is an equivalence class of $\sim_E$ ;
  (iv)  if $x \in A_0$, then $a \circ x \sim_E a$.     (3.12.3 and 3.13.1)

**28.** Suppose that $\langle A, \succsim, \circ \rangle$ is an entropy structure. Define $\sim_C$ on $A$ as follows: for $a, b \in A$, $a \sim_C b$ iff either $a \succsim b$ or $b \succsim a$. Define $A^0 = \{p \mid p \in A$ and $p \circ p \succsim p\}$. Without using Theorems 10 and 11, prove that:

  (i)   $\sim_C$ is an equivalence relation on $A$;
  (ii)  $A^0$ is an equivalence class of $\sim_C$ ;
  (iii) if $p, q \in A^0$, then $p \circ q \in A^0$;
  (iv)  if $p, q \in A^0$ and $x, y \in A_0$, then $p \circ x \succsim q \circ y$ iff $p \succsim_E q$ and $x \sim_C y$, where $\succsim_E$ is defined in Exercise 27.     (3.12.3 and 3.13.1)

**29.** Let $\mathscr{R}$ be the set of all probability distributions on the real line with finite expectations and variances. For any $f, g \in \mathscr{R}$ define

$$f \succsim g \qquad \text{iff} \qquad \theta\sigma(f) - (1 - \theta) E(f) \geqslant \theta\sigma(g) - (1 - \theta) E(g),$$

where $0 < \theta < 1$, $\sigma(f) = [V(f)]^{1/2}$ is the standard deviation of $f$. Which of the axioms of a regular risk structure (Definition 11), are violated in this example?     (3.14.2)

**30.** Suppose an individual considers the following two lotteries equally risky. Lottery 1: A fair coin is tossed; if heads comes up he wins \$10; if tails comes up he loses \$8. Lottery 2: A fair die is rolled; if six comes up he wins \$33; otherwise he loses \$3. Compute the risk of the lotteries under the assumptions of a regular risk structure.     (3.14.2)

# Chapter 4  Difference Measurement

☐ **4.1  INTRODUCTION**

When a set $A$ is only ordered, measurement is unique only up to monotonic transformations (Section 2.1). If we also have a concatenation operation and measurement is required to represent it, then we have seen that much tighter representations result (Section 2.2; Chapter 3). The question arises whether similarly tight representations ever exist when there is no concatenation operation.

Recall that the key idea in constructing extensive measures is the existence of standard sequences of equally spaced elements. Since it is not essential that the standard sequences derive from a primitive, empirical concatenation, we should be able to devise other ways to construct them. The most obvious idea is to suppose that our primitive ordering is not of elements, but of "intervals" between elements, in which case indifference of the ordering establishes equal spacing. This chapter investigates that idea.

If the set $A$ consists of movable straight rods, we can compare them side by side and concatenate them end to end, employing extensive measurement. It is only a small step to consider a set $A$ consisting of immovable points on a straight line. If we want to measure intervals on $A$, we need an auxiliary set of movable rods which we can lay off against intervals of $A$. Now if these auxiliary rods happen to be graded according to a previous process of extensive measurement, this small step does not involve new fundamental measurement. An example is measuring a room with a meter stick. Suppose, however, that our auxiliary rods can be ordered, but have not been subjected to prior extensive measurement. In other words, we have a collection of uncalibrated rulers of different lengths. Then we can take a slightly larger step away from extensive measurement by constructing standard sequences directly in $A$ using the uncalibrated rods to lay off equal intervals. For example, even if one does not know the length of one's stride, one can still determine the approximate *ratio* of two distances by pacing them off.

Denoting elements of $A$ by $a$, $b$, $c$, $d$, we denote intervals in $A$ by $ab$, $cd$, etc. (We distinguish between $ab$ and $ba$.) Comparison with a set of movable rods generates an ordering on the intervals in $A$; e.g., $ab \gtrsim cd$ if some rod does not exceed $ab$ but exceeds or matches $cd$.

### 4.1.1 Direct Comparison of Intervals

In view of the availability of metersticks, why should one ever want to perform fundamental measurement of intervals in $A$ by using an ordering on the intervals? One reason is that, in numerous cases, pairs of elements $ab$, $cd$ can be compared, but no preexisting extensive measure can be applied. A second reason is that unless certain properties (axioms) are satisfied by the ordering of intervals, the measurement by matching with a calibrated extensive scale will not yield consistent results; and these basic axioms are best discovered by axiomatizing fundamental measurement based directly on ordering of intervals. We discuss both of these points in more detail, starting with the second one.

Suppose, for example, that one is measuring intervals along an arc of a large ellipse in terms of a 20-cm rule and a meter stick. The result from the rule is slightly larger than that from the meter stick, because 20-cm chords approximate the arc better than do 100-cm ones. Moreover, the change in measured length, passing from a meter stick to a 20-cm ruler, will be greater in the more curved part than in the flatter part of the ellipse. The point is that, if $ab$ in the flat part matches $a'b'$ in the curved part (using a 20-cm chord, say), and $bc$ in the flat part matches $b'c'$ in the curved part (using the 20-cm chord again), it will nevertheless turn out that $ac$ in the flat part is longer than $a'c'$ in the curved part (now using the meter stick). One cannot order intervals along an ellipse by comparison with straight rods and have things come out right; and it does not matter whether the rods are calibrated by previous extensive measurement or not. An assumption of additivity, namely, if $ab \sim a'b'$ and $bc \sim b'c'$, then $ac \sim a'c'$, is an essential property of fundamental measurement of intervals (see Axiom 4 of Definition 1 below). In a sense, this property helps to characterize "straightness" (or at any rate, constant curvature); we do not know that our interval measurements along a supposedly straight line are really valid unless we check this property, in one way or another.

A second reason for our interest in measurement of intervals is that cases arise in which we can order intervals using instruments which, in contrast to rods, cannot be extensively measured. Let us start with a somewhat contrived example from physics and then proceed to realistic examples from the social sciences.

Consider the problem of measuring the electrical resistance of intervals along a piece of wire (say, for purposes of calibrating a variable resistor). Suppose that the lengths of the intervals are either unknowable (the wire is coiled or inaccessible) or irrelevant (the resistivity is nonconstant if the thickness or composition of the wire varies from point to point). Moreover, suppose that the only instruments available are incandescent light bulbs, a power source of suitable output, other wires, and soldering tools. One solution is to connect one light bulb in series with an interval along the to-be-measured wire (using a pair of variable contacts to attach two leads to the endpoints of the interval) and a second light bulb to a fixed length of another (auxiliary) wire. Plug both bulbs into the power source. If the auxiliary wire has sufficiently small resistance, then the second bulb will have more current through it and be brighter than the first. Move one of the variable contacts along the to-be-measured wire toward the other until the resistance added by the interval is small enough so that the lights match in brightness. These two endpoints can then be taken as points $a_1$ and $a_2$ in a standard sequence. One next searches for a point $a_3$ such that the resistance of interval $a_2a_3$ is just sufficient for the two bulbs to match, keeping the same auxiliary wire in series with the second bulb. In this way, a sequence of intervals of equal resistance can be found. (Note that it is not assumed that the resistances of the two bulbs match exactly.) To obtain a finer or coarser sequence, shorter or longer pieces of auxiliary wire are used with the second bulb. It is necessary, of course, to check the appropriate axioms or, what amounts to the same thing, to show that the ratio of the resistances of any two intervals is invariant within error limits with the choice of standard sequence. The point of this example is to show that a standard sequence can be developed, using brightness matching to equate intervals, without any prior extensive measurement. (The reason the example seems contrived is that, ordinarily, one would obtain a calibrated electrical meter and measure resistance in terms of previous extensive measurement.)

We turn now to applications of difference measurement outside physics. One example is developed in detail in Chapter 5. Let $A$, $B$, $C$, and $D$ be events, with $A$ included in (implying) $B$ and $C$ included in $D$. If the conditional probability of $A$, given $B$, exceeds that of $C$, given $D$, then we say that the interval from $A$ to $B$ is smaller than that from $C$ to $D$. The scale of "position" given by this ordering of intervals is a scale of unconditional probability; ratios of unconditional probability yield the measures of the intervals, i.e., of conditional probability.

The most systematic empirical use of direct comparison of intervals has been the ordering of sensory intervals in psychophysical experiments. In the method of bisection, for example, an observer is asked to choose a stimulus whose brightness (or loudness, or some other psychological quality) is "halfway between" that of two given stimuli. That is, he is asked to judge which of two sensation "differences" is larger and to keep adjusting their common endpoint until they are equal.

In the method of cross-modality matching, used by S. S. Stevens and his collaborators (Stevens, Mack, & Stevens, 1960; Stevens, 1966), the observer is asked to produce a match between sensations, e.g., between the brightness and loudness of two stimuli. This method, in which only one stimulus from each modality is present at a time, can be fitted into our framework if we assume the observer uses remembered sensations for two of the endpoints. For example, if the brightness of light $a_1$ was previously matched to the loudness of sound $a_2$, and the observer is now given light $b_1$, we assume he chooses a sound $b_2$ such that the brightness "ratio" of $b_1$ and $a_1$ matches the loudness "ratio" of $b_2$ and $a_2$.

An even more common method is magnitude estimation in which, say, brightnesses are matched by numbers. The idea is that the numerical ratio is used as an index of the sensation "ratio"; thus, if lights $a$, $b$, $c$, and $d$ are assigned respective brightnesses 1, 5, 8, and 40, it is assumed that the sensation "ratios" $ba$ and $dc$, corresponding to five-to-one numerical ratios, are identical.

Other psychophysical methods employ some kind of objective index of the size of an interval, much as the brightness of a lamp wired in series is an index of electrical resistance. One of the most popular indices is the probability that the two endpoints of an interval are discriminated; this idea, dating back to Fechner, has spawned an enormous literature. Reaction time is believed by some to be a discriminability index that is applicable to large intervals, where the difference is always noted; many other indices are possible and are sometimes used in psychophysics.

Probability of choosing one alternative over another as an index of the size of the utility interval between them has been proposed by economists as well as by psychophysicists. This idea was studied by Block and Marschak (1960) and Debreu (1958). For some time previously, however, economists had been aware of the possibility of ordering utility intervals on one basis or another. In 1934, Lange (see also Phelps Brown, Bernadelli, & Lange, 1935) discussed what amounts to the uniqueness of a difference representation, pointing out that under fairly mild conditions the utility function must be an interval scale. Lange's interest in the question stemmed from the fact that Pareto and others had argued that only ordinal properties of utility were needed for much of economic theory; however, as Lange pointed out, these same authors frequently invoked marginal utility concepts which refer to orderings of differences of utilities. As he observed, such orderings remain invariant not under monotonic transformations, but only under positive linear transformations. Thus, implicitly, the utility functions entering these arguments must be interval scales, which was what Pareto had tried to avoid. A closer analysis of how utility intervals are actually ordered by use of marginal utility is given in Section 4.1.2.

So far we have mentioned the ordering of sensation intervals in psychophysics and of utility intervals in economics. Recent quantitative work in social sciences has led to many other relations of form $ab \gtrsim cd$. For example, $a$, $b$, $c$, and $d$ might be individuals in some social group, with pairs ordered by measuring the frequency of social contact; or they might be nations, with pairs ordered by the frequency with which members of the pair are included in lists of countries perceived (by respondents in a survey) as having a common political orientation; etc. However, these kinds of orderings of pairs are rarely analyzed by the methods of this chapter since it is extremely doubtful that the individual objects, $a$, $b$,..., are appropriately represented along a single continuum. On the contrary, the increased interest within the social sciences in orderings of pairs has come about because of new methods for representing entities as points in a multidimensional space or in some other mathematical structure in which the ordering of pairs is represented by some kind of distance measure. The measurement theories appropriate to such multidimensional representations are discussed in Chapters 11–13 (Volume II). We mention this here to make clear that, while direct comparisons of intervals can be attained in many ways, there is a sharp distinction between cases for which the ordering has the properties of intervals on a straight line (sensation continuum, conditional probability, utility continuum) and cases for which more complex structures are needed to provide a representation. This distinction is reflected in the formal properties given below (in particular, if $ab \sim a'b'$ and $bc \sim b'c'$, then $ac \sim a'c'$) and the axioms of Chapters 11–13.

Because of its widespread use in the social sciences, one last method of ordering intervals needs to be mentioned here: the rating scale. Suppose, for example, that respondents in a survey are asked to rate their reactions to noise from overflying jets on a scale from $+2$ (mildly pleasant) through 0 (indifferent) to $-10$ (extremely distressing). It is often said that rating-scale responses are merely ordinal, but this particular example suggests that rather more structure is implicit in the definition of the scale. Most people would think it absurd to rate the airplane noise on a scale from $+5$ (mildly pleasant) through 0 (indifferent) to $-5$ (extremely distressing). The affective interval (mild pleasantness, indifference) seems much smaller than the interval (indifference, extreme distress).

More typically, respondents are asked to rate a series of different items on a scale from $+10$ (extremely pleasurable) to $-10$ (extremely distressing). One item might be noise from overflying jets. If one respondent rates this noise at $+1$ and another, at $-5$, we generally feel it is safe to assume that the distress afforded the second man is greater than the pleasure afforded the first one. There are, however, many empirical and conceptual difficulties associated with this interpretation: some people have generalized tendencies (biases) toward extreme ratings or toward positive ones or toward negative ones; some have number preferences; and how is one man's pleasure to be compared with another's distress? Nevertheless, it may be possible to obtain orderings of intervals from a properly designed rating scale.

### 4.1.2 Indirect Comparison of Intervals

As we have indicated, one function of this chapter is to axiomatize measurement based on an ordering $\gtrsim$ of intervals; this is motivated by the many empirical instances of such orderings. Another function is to provide a possible basis for the development of additive conjoint measurement in Chapter 6. We have already noted in Chapter 1 that an ordering over a product $A_1 \times A_2$ can induce an ordering of intervals in each factor. If $(a_1, a_2) \gtrsim (b_1, b_2)$ and $(d_1, b_2) \gtrsim (c_1, a_2)$, then the $a_1b_1$ interval exceeds the $b_2a_2$ interval, while the latter exceeds the $c_1d_1$ interval; so we can conclude (indirectly) that $a_1b_1 \gtrsim_1 c_1d_1$. That is, each interval in $A_1$ is compared not with a movable rod, but with a fixed interval in $A_2$. The comparison with the fixed interval in $A_2$ is not made directly, but it is inferred from the ordering in $A_1 \times A_2$.

As an example, consider again marginal utility. If $a_1, b_1, \ldots$ denote amounts of some commodity or money and $a_2, b_2, \ldots$ of another commodity, then ordinal utility over commodity *vectors*—an ordering of $(a_1, a_2)$, etc.—leads to ordering of utility intervals within each commodity. If a man will pay \$20 for a new pair of shoes, but will not, even after being given \$20, spend it for an additional new pair of shoes, we can represent this by

$$(x + 1, y - 20) > (x, y),$$
$$(x + 1, y) > (x + 2, y - 20).$$

Here, $x$ is the number of pairs of shoes the man has to start with, and $y$ is his

wealth to start with. The first inequality represents his willingness to pay
$20 for one new pair of shoes; the second, his unwillingness to purchase a
second pair for $20 even if his wealth were restored to its original state.
We conclude that the interval between $x + 1$ and $x$ exceeds that between $y$
and $y - 20$, whereas the interval between $x + 2$ and $x + 1$ is smaller; we
say that the marginal utility of additional pairs of shoes is decreasing.

The translation of an order relation between $(a_1, a_2)$ and $(b_1, b_2)$ into one
between $a_1 b_1$ and $b_2 a_2$ depends, of course, on an additive rule of combination
for the contributions of $A_1$ and $A_2$, corresponding to subtraction of $a_1$, $b_1$
scale values.

Despite the interest in both marginal utility and in additive measurement
(see Chapter 6 for a historical review of the latter), the connection between
them dawned only gradually. The axiomatization of expected utility, by von
Neumann and Morgenstern (1947), revived the interest in interval scales of
utility. In 1955, Suppes and Winet presented the first axiomatization of
utility differences and also pointed out the possibility that utility differences
could be ordered indirectly, via additive structures of various kinds. Axiomatizations of additive measurement followed (see Chapter 6), but none of
these made explicit use of difference measurement on each factor.

As we demonstrate in detail in Exercises 11–18 of Chapter 6, following
the lines sketched in Chapter 1 and above, additive measurement can be
reduced to difference measurement on $A_1$ and $A_2$. The actual proof given in
the chapter reduces the result to extensive measurement on each component,
rather than to difference measurement because that proof is shorter. Our
deepest understanding of additive conjoint measurement lies in the reduction
to either extensive or difference measurement on each component separately.

Other empirical structures can also give rise, indirectly, to orderings of
intervals. One instance, discussed in Section 4.12, was introduced by Coombs
(1952), under the name *unfolding*. Given a family of preference orderings $\gtrsim_s$
generated by subjects $s$ in a set $S$, over a fixed set $A$, we can interpret this
family as a single ordering of intervals $as$, with $a$ in $A$ and $s$ in $S$. That is,
if $s$ prefers $a$ to $b$, the interval $as$ is smaller than the interval $bs$. Indirectly,
this induces order relations on intervals in $A$. For example, suppose that

$$b \gtrsim_s a \gtrsim_s c \quad \text{and} \quad b \gtrsim_{s'} c \gtrsim_{s'} a.$$

We interpret this to mean that $b$ is closer to $s$ than $a$ is to $s$, which is closer
than $c$ is to $s$; and $b$ is closer to $s'$ than $c$ is to $s'$, which is closer than $a$ is to $s'$.
If one tries to put $a$, $b$, $c$, $s$, and $s'$ on a line satisfying these requirements,
one easily finds that $b$ must lie between $a$ and $c$; hence (interpreting intervals
as absolute differences) $ac \gtrsim ab$, $bc$. Further discussion of this idea is given
in Section 4.12; see also Coombs (1964).

### 4.1.3 Axiomatization of Difference Measurement

In the second part of his classic paper of 1901, Hölder showed how the measurement of intervals between points on a line can be reduced to extensive measurement. The basic idea is very simple. In extensive measurement, a standard sequence has the form $a$, $a \circ a$, $a \circ a \circ a$,...; in difference measurement, a standard sequence has the form $a_1$, $a_2$, $a_3$,... where $a_2a_1 \sim a_3a_2 \sim \cdots$. Thus it is natural to identify $a_2a_1$ and $a_3a_2$ each with $a$, and the overall interval $a_3a_1$ with $a \circ a$. More generally, equivalent intervals are identified with a single element, their equivalence class; and equivalence classes are concatenated by selecting end-to-end intervals within the two classes and taking the equivalence class of the overall interval as the result of the concatenation.

In this method of reduction, the axioms for difference measurement are just those needed to establish that concatenation is well defined for equivalence classes and that the hypotheses of Hölder's Theorem are satisfied. We follow Hölder's method very closely; our axioms for positive differences (Section 4.2) and our proof are very similar to his. The main difference is that his version of extensive measurement used a closed operation, so his line had to be unbounded; we have the physically more realistic assumption of a bounded line, and we reduce difference measurement to the corresponding bounded extensive case (Theorem 2.4). Another difference is inherent in the different constructions of extensive scales: our method involves finer and finer standard sequences and is therefore more realistic physically (see discussion, Section 2.2.2).

One could just as well invert the order, proving a representation and uniqueness theorem for difference measurement directly and then reducing extensive measurement to difference measurement by, e.g., defining $a \circ d \gtrsim b \circ c$ if and only if $ab \gtrsim cd$. The basic method for constructing standard sequences and scales would remain the same.

Our first difference axiomatization deals with positive differences. The notion of direction of a difference is specified in the original relational structure, and if $ab$ is a positive difference, $ba$ is not. This is Hölder's method: he spoke of segments "von gleicher Richtung." The representation theorem states that if $ab$ and $cd$ are positive differences, then $ab \gtrsim cd$ if and only if $\phi(a) - \phi(b) \geq \phi(c) - \phi(d)$.

The next structure considered is very similar to the positive-difference structure. Sections 4.4 and 4.5 take up algebraic-difference structures, where both $ab$ and $ba$ are in the ordering and they are linked by the assumption that inequalities reverse when the signs of the differences are reversed: if $ab \gtrsim cd$, then $dc \gtrsim ba$. The representation asserts that $ab \gtrsim cd$ if and only if $\phi(a) - \phi(b) \geq \phi(c) - \phi(d)$; it applies to all pairs, regardless of sign. The theorem for algebraic-difference structures is reduced to the previous result for positive differences. These sections also deal with alternative numerical representations for difference structures (e.g., using ratios instead of differences) and with the interesting possibility of a double structure requiring both a difference and a ratio representation. We also show that an interval-scale representation for a difference structure can be obtained in the absence of equivalences, that is, when only strict inequalities between pairs are observed. This theory, which is based on approximate standard sequences, is applicable to situations where exact standard sequences cannot be constructed. An application of the algebraic-difference material to cross-modality matching in psychophysics is presented in Sections 4.6 and 4.7.

Sections 4.8 and 4.9 consider finite, equally spaced difference structures, which are similar to algebraic-difference structures, but can be treated by elementary methods. This has the important pedagogic feature that the proof (Section 4.9) is brief and self-contained.

A rather more subtle extension of positive differences to absolute differences is developed in Section 4.10. Observe that it is immaterial whether we write the representation for positive differences as we did or as: $ab \gtrsim cd$ if and only if $|\phi(a) - \phi(b)| \geqslant |\phi(c) - \phi(d)|$, $\phi(a) > \phi(b)$, and $\phi(c) > \phi(d)$. Of course, no one would write it this way except to point out a suggested generalization: $ab \gtrsim cd$ if and only if $|\phi(a) - \phi(b)| \geqslant |\phi(c) - \phi(d)|$. Not only is an axiomatization of absolute differences of interest in its own right, but it also plays an important role in Chapters 11–13 on multidimensional metric representations.

Another sort of absolute-difference representation is investigated in Section 4.12. Here, the ordering on $A \times A$ is conditionally connected (Section 3.12) in the strong sense that the only comparisons are between pairs whose second elements are identical, e.g., $ac, bc, dc,\ldots$. Many different empirical situations lead to data of this type. One example, discussed above in Section 4.1.2, is Coombs' theory of preference. If $c$ is the ideal point of a subject $s$, then the ordering $a \gtrsim_s b$ can be regarded as an ordering of absolute differences $bc \gtrsim ac$ with the second elements in common. Another situation is the method of triads in which a subject judges which of two comparison stimuli, $a$ or $b$, is more similar to a standard $c$; or the more general method of "cartwheels" (Coombs, 1964) in which subjects order $n$ stimuli with respect to their similarity to a standard. In the theory of preference, the standard is an internal ideal point; in "cartwheels" the standard is external. Still other applications are discussed in Section 4.12. We show how to use such conditionally connected data to infer comparisons between pairs $ac, bd$ with $c \neq d$, and we thereby reduce the analysis to the case of absolute differences.

## □ 4.2  POSITIVE-DIFFERENCE STRUCTURES

The basic entities of a positive-difference structure are a set $A$ (interpreted as the set of endpoints of intervals), a subset $A^*$ of $A \times A$ (interpreted as the set of positive intervals), and a binary relation $\gtrsim$ on $A^*$.

The first axiom merely asserts that $\gtrsim$ is a weak order on $A^*$.

The second axiom characterizes the set $A^*$ as a transitive relation on $A$. That is, if $ab$ and $bc$ are in $A^*$, so is $ac$. The example of points on a line would lead one to expect that $A^*$ is a total order, i.e., for $a \neq b$, either $ab$ or $ba$ is in $A^*$; but the example of conditional probability, where the interval between two events is positive only if one is a subevent of the other (conditioning) event, indicates the need for a more general theorem. In fact, we give two versions of the representation theorem. If $A^*$ is merely a partial order, a ratio scale $\psi$ on $A^*$ is obtained; if it is a total order, then we obtain the difference representation $\psi(ab) = \phi(a) - \phi(b)$, where $\phi$ is an interval scale on $A$.

The term "positive" connotes that $A^*$ is a strict (irreflexive) partial order, i.e., that $aa$ is not in $A^*$. However, this need not be assumed explicitly, since it follows from the third axiom. (Note, furthermore, that a transitive, irreflexive relation is always asymmetric: if $ab$ is in $A^*$, $ba$ is not.)

The third and fourth axioms are the heart of the system, since they concern the relation between $A^*$ and $\gtrsim$. Axiom 3 asserts that for $ab$, $bc$ in $A^*$, $ac$ is strictly greater ($\succ$) than both $ab$ and $bc$. (Hence, $aa$ in $A^*$ would imply $aa \succ aa$, contradicting the reflexivity of $\gtrsim$; so $A^*$ is irreflexive, as we claimed above.)

Axiom 4 is the so-called *weak monotonicity* condition: if $ab$, $bc$, $a'b'$, and $b'c'$ are in $A^*$, and if $ab \gtrsim a'b'$ and $bc \gtrsim b'c'$, then $ac \gtrsim a'c'$. This axiom and its logical relations with several similar conditions, were discussed by Block and Marschak (1960), who called it the sextuple condition because of the six elements involved. A geometric illustration is given in Figure 1.

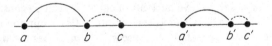

**FIGURE 1.**  A graphical illustration of weak monotonicity in which $ab \gtrsim a'b'$ and $bc \gtrsim b'c'$.

The role played by weak monotonicity can be explained in two different, but logically equivalent, ways. In the introduction, we discussed a closely related condition: if $ab \sim a'b'$ and $bc \sim b'c'$, then $ac \sim a'c'$. This condition fails on an ellipse when intervals are ordered by the lengths of chords. It is needed if the standard-sequence procedure is to yield consistent results, i.e., if adding up a sequence of adjacent intervals in one part of $A$ is to yield the same total interval as a sequence of equivalent intervals in another part of $A$. If one thinks of end-to-end intervals $ab$ and $bc$ as being concatenated to yield the interval $ac$, the assumption under discussion is a kind of homogeneity of concatenation with respect to location in $A$. (A circle or straight line is homogeneous in a way that an ellipse is not, because the curvature of the ellipse is variable.) Axiom 4 merely extends this condition, replacing $\sim$ by $\gtrsim$. The intuitive meaning is essentially unchanged.

The second explanation of Axiom 4 is in terms of the reduction of positive-difference measurement to extensive measurement. This is done by formalizing the informal notion of concatenation ($ab \circ bc = ac$). If $ab$ and $cd$ are any elements of $A^*$, find $a'$, $b'$, $c'$ in $A$ such that $ab \sim a'b'$ and $cd \sim b'c'$; then let $ab \circ cd = a'c'$. That is, two nonadjacent intervals are concatenated by mapping them into a pair of equivalent adjacent intervals. For this to make sense, concatenation must be unaltered if one interval is replaced by an equivalent one with different endpoints. Again, we need the weak mono-tonicity condition, at least in its $\sim$ form.

More formally, Definition 2 below defines concatenation of equivalence classes of intervals (elements of $A^*/\sim$ or, as denoted there, $\mathbf{A}^*$). The $\sim$ form of Axiom 4 asserts that this concatenation is well defined. We must then show that Axioms 1–8 of Definition 2.2 hold. The $\gtrsim$ form of Axiom 4 is required to show that the monotonicity conditions (Axioms 3 and 4 of Definition 2.2) are valid: if $ab \gtrsim a'b'$ and $bc \gtrsim b'c'$, then $ab \circ bc \gtrsim a'b' \circ b'c'$, etc. Axiom 4 also plays other roles in establishing the hypotheses of Hölder's theorem.

Axiom 5 is a solvability condition of the sort discussed in Section 1.4.2. It asserts that a given positive interval $cd$ can be "copied" within any larger positive interval $ab$, using either $a$ or $b$ as an endpoint of the copy. This allows boundedness. In contrast, Hölder assumed that $cd$ could be "copied" using an arbitrary point $a$ as an endpoint, which implies that an infinite sequence of adjacent positive intervals of equal size can be constructed.

Axiom 6, finally, is an Archimedean condition. Given our notion of end-to-end concatenation, a standard sequence $a_1, a_2, a_3, \ldots$ has the property that $a_3a_1 = a_3a_2 \circ a_2a_1 = 2(a_2a_1), \ldots; a_na_1 = (n-1)(a_2a_1)$. Thus, we assume that such a sequence is either finite or unbounded; this gives us the needed Archimedean axiom for extensive measurement. Hölder assumed a much more restrictive property (the so-called Dedekind property) in his treatment of difference measurement as well as in extensive measurement; he used the Dedekind property mainly to prove the needed Archimedean condition.

We remark that Axioms 1–4 and 6 are all obviously necessary for the desired representation theorem; Axiom 5 is the only structural restriction.

The formal definition of positive-difference structures, and the representation and uniqueness theorem, are as follows:

DEFINITION 1. *Suppose that $A$ is a nonempty set, $A^*$ a nonempty subset of $A \times A$, and $\succsim$ a binary relation on $A^*$. The triple $\langle A, A^*, \succsim \rangle$ is a positive-difference structure iff, for all $a$, $b$, $c$, $d$, $a'$, $b'$, and $c' \in A$, and all sequences $a_1, a_2, ..., a_i, ... \in A$, the following six axioms are satisfied:*

1. *$\langle A^*, \succsim \rangle$ is a weak order.*
2. *If $ab$, $bc \in A^*$, then $ac \in A^*$.*
3. *If $ab$, $bc \in A^*$, then $ac > ab$, $bc$.*
4. *If $ab$, $bc$, $a'b'$, $b'c' \in A^*$, $ab \succsim a'b'$, and $bc \succsim b'c'$, then $ac \succsim a'c'$.*
5. *If $ab$, $cd \in A^*$ and $ab > cd$, then there exist $d'$, $d'' \in A$ such that $ad'$, $d'b$, $ad''$, and $d''b \in A^*$ and $ad' \sim cd \sim d''b$.*
6. *If $a_1, a_2, ..., a_i, ...$ is a strictly bounded standard sequence ($a_{i+1}a_i \in A^*$ and $a_{i+1}a_i \sim a_2a_1$ for all $a_i$, $a_{i+1}$ in the sequence; and for some $d'd'' \in A^*$, $d'd'' > a_ia_1$ for all $a_i$ in the sequence), then it is finite.*

THEOREM 1. *If $\langle A, A^*, \succsim \rangle$ is a positive-difference structure, then there exists $\psi : A^* \to \text{Re}^+$ such that for all $a$, $b$, $c$, $d \in A$*

(i) *if $ab$, $cd \in A^*$, then $ab \succsim cd$ iff $\psi(ab) \geq \psi(cd)$;*
(ii) *if $ab$, $bc \in A^*$, then $\psi(ac) = \psi(ab) + \psi(bc)$.*

*If $\psi'$ also has these properties, then there exists $\alpha > 0$ such that $\psi' = \alpha\psi$. If, in addition, for all $a$, $b \in A$, $a \neq b$, either $ab$ or $ba \in A^*$, then there exists $\phi : A \to \text{Re}$ such that for $ab$ in $A^*$, $\psi(ab) = \phi(a) - \phi(b)$. If $\phi'$ has the same property, then there exists a constant $\beta$ such that $\phi' = \phi + \beta$.*

As we have indicated, the method of proof is to construct an extensive structure from the positive-difference one. Since we cannot concatenate $ab$ with $cd$, except by mapping $ab$, $cd$ onto equivalent intervals such that the lower endpoint of one interval coincides with the upper endpoint of the other, we need to introduce a defined notion of concatenation which applies to equivalence classes.

Denote by **ab** the $\sim$-equivalence class containing $ab$ and by $\mathbf{A}^*$, the set of all such equivalence classes. We will sometimes also use $\Gamma$, $\Delta$, $\Lambda$ to denote elements of $\mathbf{A}^*$. The induced order on $\mathbf{A}^*$ is denoted $\succsim$.

DEFINITION 2. *Let $\langle A, A^*, \succsim \rangle$ be a positive-difference structure. Define:*
$\mathbf{B} = \{(\Gamma, \Delta) \mid \Gamma, \Delta \in \mathbf{A}^* \text{ and there exist } a, b, c \in A \text{ such that } ab \in \Gamma, bc \in \Delta\}.$
$\mathbf{o} : \mathbf{B} \to \mathbf{A}^* \text{ by } \mathbf{ab} \circ \mathbf{bc} = \mathbf{ac}.$

It is immediate from Axiom 4 that o is well defined, and it will be easy to show that $\langle A^*, \gtrsim, B, o \rangle$ is an Archimedean, regular, positive, ordered local semigroup. Theorem 2.4 then leads to the construction of $\psi$ on $A^*$. The further construction of the scale of position $\phi$ follows readily. One fixes a zero point $a_0$ and defines $\phi(a)$ as $\psi(aa_0)$ or $\psi(a_0a)$, depending on whether $aa_0$ or $a_0a$ is a positive interval.

Note that $\phi$ can be constructed under slightly weaker conditions. Define $b$ and $b'$ to be equivalent ($b \, E \, b'$) if they form positive intervals in exactly the same way; i.e., for any $a$ in $A$ either $ab$, $ab'$ are in $A^*$ with $ab \sim ab'$, or $ba$, $b'a$ are in $A^*$ with $ba \sim b'a$, or none of $ab$, $ab'$, $ba$, $b'a$ is in $A^*$. Now assume that for all $a$, $b$, either $a \, E \, b$ or $ab$ is in $A^*$ or $ba$ is in $A^*$; $\phi$ can still be constructed. The proof is trivial: we can replace $A$ by $A/E$ and $A^*$ by $A^*/E$, where $A^*/E$ is the obvious subset of $(A/E) \times (A/E)$; the resulting structure satisfies all the hypotheses of Theorem 1.

## 4.3  PROOF OF THEOREM 1  (p. 147)

*If $\langle A, A^*, \gtrsim \rangle$ is a positive-difference structure, then there is an order-preserving ratio scale $\psi$ on $A^*$ such that for $ab$, $bc \in A^*$, $\psi(ac) = \psi(ab) + \psi(bc)$. If, in addition, $A^*$ is a strict simple order, then there is an interval scale $\phi$ on $A$ such that for $ab$, $cd \in A^*$, $ab \gtrsim cd$ iff $\phi(a) - \phi(b) \geqslant \phi(c) - \phi(d)$.*

It is convenient to have the following lemma:

LEMMA 1.  *Let $\langle A, A^*, \gtrsim \rangle$ be a positive-difference structure, and let **B** and o be as in Definition 2. If $(\Gamma, \Delta) \in B$ and $\Gamma \, o \, \Delta = ac$, then there exists $b \in A$ such that $ab$, $bc \in A^*$, $\mathbf{ab} = \Gamma$, and $\mathbf{bc} = \Delta$.*

PROOF.  By definition of **B**, there exist $a'$, $b'$, $c'$ with $\mathbf{a'b'} = \Gamma$ and $\mathbf{b'c'} = \Delta$. We have $ac \sim a'c' > a'b'$ (definition of o, Axiom 3, Axiom 1). By Axiom 5, we can find $b$ with $ab$, $bc \in A^*$ and $ab \sim a'b'$. To complete the proof, it suffices to show that $bc \sim b'c'$.

If $bc > b'c'$, choose $c''$ by Axiom 5 with $bc''$, $c''c \in A^*$ and $bc'' \sim b'c'$. By Axiom 4, $ac'' \sim a'c'$. But by Axiom 3, $ac > ac''$; this contradicts $ac \sim a'c'$. A symmetric argument disposes of $b'c' > bc$.                                                    ◇

We now show that if $\langle A, A^*, \gtrsim \rangle$ is a positive-difference structure, with $A^*$, $\gtrsim$, **B**, o defined as above, then $\langle A^*, \gtrsim B, o \rangle$ satisfies Axioms 1–8 of Definition 2.2. The proofs are divided into numbered paragraphs corresponding to those axioms.

1.  *Simple order:*  This is immediate from the fact that $\gtrsim$ is a weak order on $A^*$.

2. *Monotone closure of* **B**: Suppose that $(\Gamma, \Delta) \in \mathbf{B}$, $\Gamma \succsim \Gamma'$, and $\Delta \succsim \Delta'$. Let $\Gamma = \mathbf{ab}$, $\Delta = \mathbf{bc}$. If $\Gamma' = \Gamma$, let $b' = b$; otherwise, $\Gamma \succ \Gamma'$ so choose $b'$ with $ab'$, $b'b$ in $A^*$ and $ab' = \Gamma'$. For either choice of $b'$, $b'c \succsim bc$. Thus, we can choose $c'$ with $\mathbf{b}'\mathbf{c}' = \Delta'$. Hence, $(\Gamma', \Delta') \in \mathbf{B}$.

3 & 4. *Monotonicity*: Immediate, from Axiom 4.

5. *Associativity*: Suppose that $(\Gamma, \Delta)$, $(\Gamma \circ \Delta, \Lambda) \in \mathbf{B}$. By definition, we can choose $a$, $c$, $d \in A$ such that $\mathbf{ac} = \Gamma \circ \Delta$, $\mathbf{cd} = \Lambda$; we have $(\Gamma \circ \Delta) \circ \Lambda = \mathbf{ad}$. By Lemma 1, we can choose $b \in A$ such that $\mathbf{ab} = \Gamma$, $\mathbf{bc} = \Delta$. By definition, $(\Delta, \Lambda) \in \mathbf{B}$, $\Delta \circ \Lambda = \mathbf{bd}$, $(\Gamma, \Delta \circ \Lambda) \in \mathbf{B}$, and $\Gamma \circ (\Delta \circ \Lambda) = \mathbf{ad}$, as required. The proof for $(\Delta, \Lambda)$ and $(\Gamma, \Delta \circ \Lambda) \in \mathbf{B}$ is similar.

6. *Positivity*: Immediate from Axiom 3.

7. *Regularity*: Immediate from Axiom 5.

8. *Archimedean*: Suppose that Axiom 8 of Definition 2.2 is false. Then there exist $ab$, $cd \in A^*$, such that $\mathbf{cd} \succ n(\mathbf{ab})$ for every positive integer $n$. We now construct a sequence $a_1$, $a_2$,..., $a_n$,... in $A$ that violates Axiom 6 of Definition 1.

Let $a_1 = d$. Suppose that $a_1$,..., $a_n$ have been constructed such that (i) $ca_n \succsim ab$, and (ii) if $n \geqslant 2$, $a_n a_1$ is in the equivalence class $(n - 1)(\mathbf{ab})$. Using Axiom 5, we can construct $a_{n+1}$ such that $a_{n+1}a_n \in A^*$ and $a_{n+1}a_n \sim ab$; also, either $a_{n+1} = c$ or $ca_{n+1} \in A^*$. We have $a_{n+1}a_1 \in n(\mathbf{ab})$, so (ii) holds for $a_{n+1}$. If $a_{n+1} = c$, then $\mathbf{cd} = n(\mathbf{ab})$, contrary to hypothesis. If $ca_{n+1} \in A^*$, with $ab \succsim ca_{n+1}$, then $(n + 1)(\mathbf{ab}) \succsim \mathbf{cd}$, again, contrary to hypothesis (use monotonicity). Hence, we have $ca_{n+1} \succ ab$, so (i) holds for $a_{n+1}$. Thus, the construction can be continued indefinitely. Since $cd \succ a_n a_1$ for every $n$, the sequence $a_1$, $a_2$,..., $a_n$,... violates Axiom 6.

By Theorem 2.4, there exists a function $\psi$ from $A^*$ to $\mathrm{Re}^+$ which preserves order and satisfies $\psi(\Gamma \circ \Delta) = \psi(\Gamma) + \psi(\Delta)$. Define $\psi$ on $A^*$ by $\psi(ab) = \psi(\mathbf{ab})$; this obviously satisfies properties (i) and (ii) of Theorem 1. Uniqueness follows from Theorem 2.4, except that here the equation $\psi' = \alpha\psi$ also holds for a maximal element (if any exists) of $A^*$, since by Axiom 5, such an element can always be expressed as the concatenation of nonmaximal elements (unless $A^*$ contains only one element; in that case, uniqueness is trivial).

Assuming that $A^*$ is a strict simple order on $A$, fix an arbitrary $a_0$ in $A$ and define $\phi$ on $A$ by

$$\phi(a) = \begin{cases} \psi(aa_0), & \text{if } \quad aa_0 \in A^*, \\ 0, & \text{if } \quad a = a_0, \\ -\psi(a_0 a), & \text{if } \quad a_0 a \in A^*. \end{cases}$$

By hypothesis, $\phi$ is defined on all of $A$. To show that $\psi(ab) = \phi(a) - \phi(b)$, there are several cases to consider. For example, suppose that $aa_0$ and $a_0b \in A^*$. Then $\psi(ab) = \psi(aa_0) + \psi(a_0b) = \phi(a) - \phi(b)$. Similarly, if $ba_0 \in A^*$, then from $ab \in A^*$ and Axiom 2, we have $aa_0 \in A^*$; moreover, $\phi(a) = \psi(aa_0) = \psi(ab) + \psi(ba_0) = \psi(ab) + \phi(b)$. Finally, $a_0a$, $ab$, and $a_0b \in A^*$, the only other possibility, leads to $-\phi(b) = \psi(a_0b) = \psi(a_0a) + \psi(ab) = -\phi(a) + \psi(ab)$.

For uniqueness, note that any $\phi'$ with the same property satisfies

$$\phi'(a) - \phi'(a_0) = \psi(aa_0) = \phi(a)$$

for $aa_0 \in A^*$. The same equation holds for $a_0a \in A^*$, with all signs reversed. Thus, $\phi' = \phi + \beta$, where $\beta = \phi'(a_0)$.                                    ◇

The relation between $\phi$ and $\psi$ in this proof is similar to the solution to Sincov's functional equation $F(x, y) + F(y, z) = F(x, z)$, where

$$F : \text{Re} \times \text{Re} \to \text{Re},$$

which is $F(x, y) = f(x) - f(y)$. See Aczel (1966, p. 223).


☐ **4.4 ALGEBRAIC-DIFFERENCE STRUCTURES**

We now apply the results just obtained to structures in which both positive and negative intervals appear. The representation of such structures by algebraic differences $\phi(a) - \phi(b)$ was axiomatized by Debreu (1958), Scott and Suppes (1958), Suppes and Zinnes (1963), and Kristof (1967).[1,2] Our version differs slightly, in that it uses less severe structural restrictions (solvability) and uses the weak monotonicity condition instead of the quadruple condition.[3] The proof proceeds by defining a subset of positive intervals $A^*$— those for which $ab > aa$—and then showing that the axioms of Definition 1 hold. The representation extends to negative intervals via the usual sign-reversal axiom: if $ab \gtrsim cd$, then $dc \gtrsim ba$. Still another axiomatization can be obtained from the axiomatization of additive conjoint measurement in Chapter 6; it is noted at that point.

The difference representation can be replaced by an equivalent ratio representation $\phi(a)/\phi(b)$, as well as by many other alternative representations, as we note in Section 4.4.2.


[1] Hölder (1901) also considered a structure with both kinds of intervals, but it seems more logical to classify it with absolute-difference structures (Section 4.10).

[2] Closely related (higher-order metric) structures were considered by Hurst and Siegel (1956) and by Siegel (1956), but no numerical representation can be established for these structures.

[3] See Block and Marschak (1960) or Luce and Suppes (1965).

In Section 4.4.3 we pose and solve the following problem. Given *two* distinct orderings of intervals on the same set $A$, when can a single scale be constructed that gives a difference representation for one ordering and a ratio representation for the other? This has applications to sensory judgments with "difference" or "ratio" instructions.

### 4.4.1 Axiom System and Representation Theorem

We start with the basic definition:

DEFINITION 3. *Suppose $A$ is a nonempty set and $\gtrsim$ a quaternary relation on $A$, i.e., a binary relation on $A \times A$. The pair $\langle A \times A, \gtrsim \rangle$ is an* algebraic-difference structure *iff, for all $a$, $b$, $c$, $d$, $a'$, $b'$, and $c' \in A$ and all sequences $a_1, a_2, ..., a_i, ... \in A$, the following five axioms are satisfied:*

1. *$\langle A \times A, \gtrsim \rangle$ is a weak order.*
2. *If $ab \gtrsim cd$, then $dc \gtrsim ba$.*
3. *If $ab \gtrsim a'b'$ and $bc \gtrsim b'c'$, then $ac \gtrsim a'c'$.*
4. *If $ab \gtrsim cd \gtrsim aa$, then there exist $d'$, $d'' \in A$, such that $ad' \sim cd \sim d''b$.*
5. *If $a_1, a_2, ..., a_i, ...$ is a strictly bounded standard sequence ($a_{i+1}a_i \sim a_2a_1$ for every $a_i$, $a_{i+1}$ in the sequence; not $a_2a_1 \sim a_1a_1$; and there exist $d'$, $d'' \in A$ such that $d'd'' > a_ia_1 > d''d'$ for all $a_i$ in the sequence), then it is finite.*

Axioms 1 and 3–5 correspond respectively to Axioms 1 and 4–6 of Definition 1. The main change lies in Axiom 2, which is wholly different from the axioms of Definition 1 but plays an obvious role here. The properties expressed in Axioms 2 and 3 of Definition 1 are not needed here, since they are easily provable once $A^*$ is defined (see Lemmas 3 and 4 below).

THEOREM 2. *If $\langle A \times A, \gtrsim \rangle$ is an algebraic-difference structure, then there exists a real-valued function $\phi$ on $A$ such that, for all $a$, $b$, $c$, $d \in A$,*

$$ab \gtrsim cd \quad \textit{iff} \quad \phi(a) - \phi(b) \geq \phi(c) - \phi(d).$$

*Moreover, $\phi$ is unique up to a positive linear transformation, i.e., if $\phi'$ has the same property as $\phi$, then there are real constants $\alpha$, $\beta$, with $\alpha > 0$, such that $\phi' = \alpha\phi + \beta$.*

The proof is quite simple. First we define $A^*$ and two other relations:

DEFINITION 4. *Let $\langle A \times A, \gtrsim \rangle$ be an algebraic-difference structure (Definition 3). Define $A^*$ by*

$$ab \in A^* \quad \textit{iff} \quad ab > aa.$$

Let $\gtrsim^*$ be the restriction of $\gtrsim$ to $A^*$. Define $E = \{ab \mid neither\ ab\ nor\ ba \in A^*\}$.

We shall establish that $\langle A, A^*, \gtrsim^* \rangle$ is a positive-difference structure and that $E$ is an equivalence relation of the kind mentioned at the end of Section 4.2, i.e., $ab$ in $E$ means that $a, b$ are equivalent with respect to formation of positive intervals. This gives us a new positive-difference structure, composed of $E$-equivalence classes, for which the interval scale $\phi$ of Theorem 1 can be constructed. It follows almost immediately that $\phi$ satisfies Theorem 2 also. The proofs of these statements are given in Sections 4.5.1 and 4.5.2.

## 4.4.2 Alternative Numerical Representations

If $\phi$ satisfies the conclusion of Theorem 2, then $\psi = e^\phi$ gives a ratio representation: $ab \gtrsim cd$ if and only if $\psi(a)/\psi(b) \geq \psi(c)/\psi(d)$. Conversely, any positive-valued $\psi$ satisfying the above property yields a difference representation $\phi$ by taking logarithms.

More generally, let $L$ be an interval in Re and let $f$ be any strictly increasing functions from Re onto $L$. Define $\psi$ and $\ominus$ by

$$\psi(a) = f[\phi(a)], \qquad a \in A,$$
$$x \ominus y = f[f^{-1}(x) - f^{-1}(y)], \qquad x, y \in L.$$

It follows that

$$\psi(a) \ominus \psi(b) = f[\phi(a) - \phi(b)].$$

Thus, $\psi$ is a homomorphism that carries $\gtrsim$ into the numerical relation defined by ordering of $\ominus$ values. Hence, we see that there is a wide choice of equivalent numerical representations. The most convenient alternative is the ratio representation, with $L = \mathrm{Re}^+$, $f = \exp$.

It is not difficult to see that the $\psi$ representation is invariant under transformations of the form

$$T_{\alpha,\beta} = f(\alpha f^{-1} + \beta), \qquad \alpha, \beta \ \text{real}, \qquad \alpha > 0.$$

The set of all such transformations is a two-parameter group with $T_{1,0}$ the identity and $T_{1/\alpha, -\beta/\alpha}$ the inverse of $T_{\alpha,\beta}$.

## 4.4.3 Difference-and-Ratio Structures

Although the choice of a difference or a ratio or other representation is a matter of convenience, when two different orderings $\gtrsim_1$ and $\gtrsim_2$ exist on $A \times A$ the situation may be more constrained. Suppose that $\langle A \times A, \gtrsim_1 \rangle$

and $\langle A \times A, \gtrsim_2 \rangle$ are both algebraic-difference structures. It is interesting to determine the circumstances under which it is possible to construct a single scale $\phi$ on $A$ such that $\phi$ gives *both* a difference representation for $\gtrsim_1$ and a ratio representation for $\gtrsim_2$.

Let $\phi_1$ be a difference representation for $\gtrsim_1$ and $\phi_2$ a ratio representation for $\gtrsim_2$, then $\alpha_1(\phi_1 + \beta)$ and $\alpha_2\phi_2{}^\gamma$ are also difference and ratio representations of $\gtrsim_1$, $\gtrsim_2$, respectively. So the question raised in the preceding paragraph really is: Under what circumstances can $\alpha_1$, $\alpha_2$, $\beta$, and $\gamma$ be chosen such that

$$\alpha_1(\phi_1 + \beta) = \phi = \alpha_2\phi_2{}^\gamma? \tag{1}$$

It is obvious that $\beta$, $\gamma$, and $\alpha_1/\alpha_2$ are going to be uniquely determined by such a requirement if, indeed, the requirement can be satisfied at all. Thus, $\phi$ will be a ratio scale.

Assuming $\phi_1$ is a strictly increasing function of $\phi_2$, it is easy to see that the following two conditions are necessary if the required function $\phi$ exists: First,

$$ab \gtrsim_1 aa \quad \text{iff} \quad ab \gtrsim_2 aa. \tag{2}$$

This condition states that the relations $A_1{}^*$, $A_2{}^*$ (Definition 4) are identical, and it follows from Equation (1) because $\phi_1$ is a strictly increasing function of $\phi_2$. We could also admit the case where $A_1{}^*$ and $A_2{}^*$ are converse orders, but this can be bypassed by reversing the direction of $\gtrsim_1$ or $\gtrsim_2$. We shall assume that such a reversal has been performed, if necessary, so that we can concentrate only on the case where $A_1{}^*$, $A_2{}^*$ coincide.

Second, if both $aa' \gtrsim_2 bb'$ and $cc' \gtrsim_2 bb'$, then

$$\text{if } a'b' \gtrsim_1 b'c', \text{ then } ab \gtrsim_1 bc. \tag{3}$$

This follows from Equation (1) by letting $\phi(b) = t\phi(b')$, where $t > 0$. For then $\phi(a) \gtrsim t\phi(a')$ and $\phi(c) \gtrsim \phi(c')$. Subtracting inequalities leads to

$$\phi(a) - \phi(b) \gtrsim t[\phi(a' - \phi(b')] \text{ and } t[\phi(b') - \phi(c')] \gtrsim \phi(b) - \phi(c),$$

whence (3).

Note that if $\sim_2$ obtains in both antecedents, then "if . . . then" becomes "iff" in (3).*

---

*In the first edition, we used only the condition with $\sim_2$ and iff. As pointed out by Miyamoto (1983), the resulting axioms do not differentiate between a difference-and-ratio representation for $\gtrsim_1, \gtrsim_2$ and such a representation for the converses, $\gtrsim_1{}^*, \gtrsim_2{}^*$. The present stronger condition breaks this symmetry. Miyamoto's paper contains an interesting generalization of our Theorem 3 below, as well as a careful discussion of our error in the first edition and a brief review of empirical literature on difference-and-ratio representations.

It turns out that the converse is also true: given that $\langle A \times A, \succsim_1 \rangle$ and $\langle A \times A, \succsim_2 \rangle$ are distinct algebraic-difference structures and that conditions (2) and (3) hold, then there exists a ratio scale $\phi$ that gives simultaneously a difference representation for $\succsim_1$ and a ratio representation for $\succsim_2$.

THEOREM 3. *Let* $\succsim_1$ *and* $\succsim_2$ *be distinct quaternary relations on a set* $A$ *such that* $\langle A \times A, \succsim_1 \rangle$ *and* $\langle A \times A, \succsim_2 \rangle$ *are algebraic-difference structures. Suppose that for all* $a, b, c, a', b', c' \in A$:

$$ab \succsim_1 aa \qquad iff \qquad ab \succsim_2 aa; \qquad (2)$$

*if* $aa' \succsim_2 bb'$ *and* $cc' \succsim_2 bb'$, *then*

$$if\ a'b' \succsim_1 b'c',\ then\ ab \succsim_1 bc. \qquad (3)$$

*Then there exists* $\phi : A \to \mathrm{Re}$ *such that for all* $a, b, c, d \in A$:

(i)   $ab \succsim_1 cd$ *iff* $\phi(a) - \phi(b) \geq \phi(c) - \phi(d)$;

(ii)  $ab \succsim_2 cd$ *iff* $\phi(a)/\phi(b) \geq \phi(c)/\phi(d)$.

*Moreover, if* $\phi'$ *is any other function with the same properties, then* $\phi' = \alpha\phi$, $\alpha > 0$.

A similar theorem was proved by Pfanzagl (1968, p. 103).

An empirical application of the above theorem arises from the possibility that subjects may generate different orderings on a set of stimuli $A$ when they are instructed to rate sensation "differences" or sensation "ratios." There are two extreme possibilities. If the two sets of instructions generate the same ordering $\succsim$, and their only effect is to cause the experimenter to use difference or ratio numerical representations, then, of course, two different scales are obtained and they are related by $\phi_2 = e^{\phi_1}$. This is just the case of alternative numerical representations, discussed in Section 4.4.2. The instructions affect the experimenter's numerical relation, but not the subject's empirical relation. Torgerson (1961) argued that this is what in fact happens in psychophysics. Birnbaum (1980) presents rather convincing evidence that this is the case, at least in some domains. If, however, subjects generate *distinct* relations $\succsim_1$ and $\succsim_2$ (as argued e.g. by Rule, Curtis, & Mullin, 1981), one may ask whether the two relations are interlocked so as to satisfy the conditions of Theorem 3. If they are, then the subjects really do act *as though they are judging numerical differences and ratios* of sensations, when requested to judge "differences" and "ratios."

### 4.4.4 Strict Inequalities and Approximate Standard Sequences

We next show how to obtain interval-scale measurement for difference structures using just strict inequalities. The method is similar to that of Krantz (1967). This development is important because observations of equality or indifference are often either unreliable or difficult to obtain. Unreliability is a necessary consequence of imperfectly sensitive methods of comparison. For example, whenever two objects are judged equal in weight, it is practically certain that a more refined balance will reveal a difference. Moreover, in many situations, strict inequalities are considerably simpler to obtain and interpret than equations. Consider a study of savings behavior in which $ab \succ cd$ denotes the observation that the subject prefers to spend \$$a$ and save \$$b$ rather than spend \$$c$ and save \$$d$. Such an observation may be interpreted as showing that the utility difference (for spending) between \$$a$ and \$$c$ exceeds the utility difference (for saving) between \$$d$ and \$$b$. From the field observation that a person has chosen to spend \$$a$ and to save \$$b$, we may elect to infer that, for all $x$, $y$ satisfying $x + y = a + b$, $ab \succ xy$. But there is no simple way of inferring any *indifference* from such an observation. In an experimental (rather than a field) setting, we can instruct the subject to choose an amount $x$ such that $ab \sim cx$, but how should he interpret our instructions? In the case of strict preference, he simply has to judge which of two offers he would rather have. There is no comparable simple act which reflects indifference. Thus, a theory that does not require indifference judgments is more satisfactory from both theoretical and practical viewpoints.

If we are to bypass the use of exact equivalences in measurement, we should examine the use to which they have been put in the construction of numerical representations. Once that is ascertained, perhaps we can see how to modify our constructions. The primary role of solvability (Axiom 5 of Definition 1 and Axiom 4 of Definition 3) is to show that there exist standard sequences whose "mesh" corresponds to the interval $cd$ and which we can use to approximate any other interval $ab$. Thus, if $ab \succsim cd \succsim aa$, by solvability $b'$ exists for which $b'b \sim cd$. If it is still true that $ab' \succsim cd$, then by solvability there exists $b''b' \sim cd$. The process terminates with $b^{(n)}$ such that $cd \succ ab^{(n)}$, and $b, b', b'',..., b^{(n)}$ is a standard sequence. The whole point of this is merely to discover how many copies of $cd$ approximate $ab$.

The idea of the present approach is that one can answer this question without being able to solve exactly any equivalence. Of course, one can no longer construct exact standard sequences, but only approximate ones. Specifically, for any $cd \succ cc$, we let $\mathscr{G}(cd)$ denote the family of sequences, with mesh at least $cd$, by which we mean that if $\{a_i\}$ is in $\mathscr{G}(cd)$, then for all $a_n$, $a_{n+1}$ in $\{a_i\}$, $a_{n+1}a_n \succsim cd$. Similarly, $\mathscr{L}(cd)$ is the family of sequences with mesh less than $cd$, i.e., $cd \succ a_{n+1}a_n$. Clearly, any exact standard sequence, if one exists, is in $\mathscr{G}(cd)$.

We now modify the solvability and Archimedean axioms (4 and 5) of Definition 3 as follows:

4'. *If $ab \succsim cd \succ c'd' \succsim aa$, then there exist $x$, $y$ in $A$ such that both*

$$cd \succ ay \succ c'd';$$
$$cd \succ xb \succ c'd'.$$

5'. *If $cd \succ cc$, the sequence $a_1$, $a_2$,..., $a_i$,... is in $\mathscr{G}(cd)$, and $ab \succ a_ia_1$ for all $a_i$ in the sequence, then the sequence is finite.*

Axiom 4' asserts that two strict inequalities, rather than the equalities $cd \sim ay$ and $cd \sim xb$, are solvable for $x$ and $y$. This implies that $A$ is dense in the sense that between any two nonequivalent points there exists a third that is not equivalent to either of them. When inequalities are easier to observe than equalities, this axiom is easier to verify than Axiom 4. In addition, Axiom 4' may be considerably easier to refute. If one samples a large set of points, $x_1,..., x_n$, within the $ab$ interval and one finds that for all $x_i$, either $ax_i \succ cd$ or $c'd' \succ ax_i$, this provides evidence against Axiom 4', suggesting that there is no $x$ satisfying $cd \succ ax \succ c'd'$. It is considerably more difficult to interpret such a finding as evidence against the original solvability axiom (Axiom 4), unless the selection of points was rather exhaustive. A more detailed discussion of the problem of testing such axioms is deferred to Chapter 17.

The new Archimedean axiom, 5', is of course slightly stronger than Axiom 5, since it asserts the same conclusion for a wider class of sequences, but the intuitions behind it, its necessity for a numerical representation, and its use, remain the same as for Axiom 5.

If we change the definition of algebraic-difference structure by replacing Axioms 4 and 5 by 4' and 5', then Theorem 2 remains valid. We call this result Theorem 2'. Its proof involves the construction of approximate standard sequences. If $a_1,..., a_{m+1}$ is a sequence in $\mathscr{L}(cd)$ such that $a_{m+1}a_1 \succsim ab$, and if $b_1,..., b_m$ is a sequence in $\mathscr{G}(cd)$ such that $ab \succsim b_mb_1$, then we have

$$m[\phi(c) - \phi(d)] \geqslant \phi(a) - \phi(b) \geqslant (m - 1)[\phi(c) - \phi(d)].$$

Thus, the sequences $a_1,..., a_{m+1}$ and $b_1,..., b_m$ may be considered to approximate the interval $ab$ in terms of the "mesh" $cd$. We can employ Axioms 4' and 5' to show that such sequences exist. The construction of $\phi$ then proceeds much as in the proof of Hölder's Theorem (Theorem 2.4). Moreover, it can be shown that $\phi$ can be approximated, to any prespecified accuracy, on the

basis of finitely many inequality observations. The proofs of these statements
are left for the reader in Exercises 9–15. Unusually full hints are provided.

## 4.5 PROOFS

### 4.5.1 Preliminary Lemmas

The common hypothesis of the following five lemmas is that $\langle A \times A, \gtrsim \rangle$
satisfies Axioms 1–3 of Definition 3; $A^*$ and $E$ are defined by Definition 4.

**LEMMA 2.** *For $a, b \in A$, $aa \sim bb$.*

*PROOF.* By Axiom 2, $aa \gtrsim bb$ implies $bb \gtrsim aa$. By Axiom 1 (connectedness of $\gtrsim$), it follows that both $aa \gtrsim bb$ and $bb \gtrsim aa$.  ◇

**LEMMA 3.** *If $ab \in A^*$, then for all $c \in A$, $ac \succ bc$ and $cb \succ ca$.*

*PROOF.* Note that by Axiom 2, $ab \succ cd$ iff $dc \succ ba$. Hence, the two
conclusions of Lemma 3 are equivalent. Also, if $ab \in A^*$, then by Definition 4,
Lemma 2, and Axiom 1 (transitivity), $ab \succ bb$, whence, by Axiom 2, $bb \succ ba$.

Now suppose that $ab \in A^*$ but that for some $c$ the conclusions of the
lemma are false, i.e., $bc \gtrsim ac$ and $ca \gtrsim cb$. Applying Axiom 3 gives
$ba \gtrsim ab$. Thus, $ba \succ bb$, contrary to the previous paragraph.  ◇

**LEMMA 4.** *$A^*$ is transitive and irreflexive.*

*PROOF.* If $ab, bc \in A^*$, then using Lemma 3, Definition 4, Lemma 2, and
Axiom 1 (transitivity) in succession gives the chain

$$ac \succ bc \succ bb \sim aa.$$

Hence, $ac \in A^*$, so $A^*$ is transitive.

If $aa \in A^*$, then Lemma 3 gives $ab \succ ab$, contradicting Axiom 1 (reflexivity).  ◇

**LEMMA 5.** *If $ab \in E$, then for all $c \in A$, $ac \sim bc$ and $ca \sim cb$.*

*PROOF.* By Axiom 2, the conclusions of the Lemma are equivalent.
If $ab \in E$, then by Definition 4, $aa \gtrsim ab$ and $bb \gtrsim ba$. By Axioms 1 and 2
and Lemma 2, $ab \sim ba$.

Now suppose that $ab \in E$ but $ac \succ bc$. Using $ba \sim ab$ and $ac \succ bc$,
Axiom 3 yields $bc \gtrsim ac$, contradicting $ac \succ bc$. Similarly, $bc \succ ac$ leads to
a contradiction.  ◇

LEMMA 6.  *E is an equivalence relation.*

*PROOF.*  Reflexivity follows from irreflexivity of $A^*$ (Lemma 4) and symmetry is true by definition. To show transitivity, note that $ab, bc \in E$ yields, by Lemma 5, $ac \sim bc \sim cc$, hence, $ac \sim aa$.              ◇

### 4.5.2  Theorem 2  (p. 151)

*If $\langle A \times A, \succsim \rangle$ is an algebraic-difference structure, then there is an interval scale $\phi$ such that $ab \succsim cd$ iff $\phi(a) - \phi(b) \geq \phi(c) - \phi(d)$.*

*PROOF.*  We first show that $\langle A, A^*, \succsim^* \rangle$ is a positive-difference structure, where $A^*$ and $\succsim^*$ are defined by Definition 4. The six axioms of Definition 1 are established almost immediately as follows:

1. Immediate from the present Axiom 1.
2. Lemma 4.
3. Follows from Lemma 3.
4. Immediate from the present Axiom 3.
5. Immediate from the present Axiom 4.

6. Let $a_1, a_2, ..., a_i, ...$ be a strictly bounded standard sequence in the sense of Definition 1; thus, $d'd'' \succ a_i a_1$, $a_{i+1} a_i \sim a_2 a_1$, and $a_{i+1} a_i \in A^*$. By definition of $A^*$, transitivity of $\succsim$, and Axiom 2, we have

$$d'd'' \succ a_i a_1 \succ a_1 a_1 \succ d''d';$$

thus, $a_1, a_2, ..., a_i, ...$ is also a strictly bounded standard sequence in the sense of Definition 3. Therefore, the Archimedean property for positive-difference structures follows from the present one.

Next, by Lemmas 5 and 6, $E$ is an equivalence relation such that equivalence classes preserve the structure of $\langle A, A^*, \succsim^* \rangle$. That is, if $ab \in A^*$, $aa' \in E$, and $bb' \in E$, then $a'b' \in A^*$ and $ab \sim^* a'b'$. Thus, we have a new positive-difference structure $\langle A/E, A^*/E, \succsim^*/E \rangle$ on the equivalence classes under $E$. By definition of $A^*$ and $E$, and Lemma 4, $A^*/E$ is a strict simple order of $A/E$, so Theorem 1 yields an interval scale $\phi_E$ on $A/E$, which induces $\phi$ on $A$. It is clear that for $ab, cd \in A^*$, $ab \succsim cd$ iff $\phi(a) - \phi(b) \geq \phi(c) - \phi(d)$. The extension to all of $A \times A$ is immediate via Axiom 2. The interval-scale property of $\phi$ follows from that of $\phi_E$.              ◇

### 4.5.3  Theorem 3  (p. 154)

*If $\langle A \times A, \succsim_1 \rangle$ and $\langle A \times A, \succsim_2 \rangle$ are algebraic-difference structures, with $\succsim_1 \neq \succsim_2$, $A_1^* = A_2^*$, and such that, when $aa', cc' \succsim_2 bb'$, then $a'b' \succsim_1 b'c'$*

*implies ab $\gtrsim_1$ bc, then there is a ratio scale $\phi$ on A giving a difference representation for $\gtrsim_1$ and a ratio representation for $\gtrsim_2$.*

PROOF.    Let $\phi_1$, $\phi_2'$ be functions constructed on $A$ by Theorem 2, applied to $\langle A \times A, \gtrsim_1 \rangle$, $\langle A \times A, \gtrsim_2 \rangle$, respectively. Let $\phi_2 = \exp \phi_2'$. (Thus, $\phi_2$ gives a ratio representation for $\langle A \times A, \gtrsim_2 \rangle$.) Since $A_1^* = A_2^*$, we have $\phi_1(a) = h[\phi_2(a)]$ for all $a$ in $A$, where $h$ is a strictly increasing function. We show that $h$ has the form

$$h(x) = \alpha x^\nu - \beta, \tag{4}$$

which immediately implies that $\phi = \phi_1 + \beta$ gives both representations demanded by the theorem. To do this, we use the main assumption [condition (3)] to derive a functional equation for $h$, and then show that the strictly increasing solutions to this functional equation are those given by Equation (4).

To set up the functional equation, we first consider the nature of the domain and range of the function $h$. Its domain is the range of $\phi_2$, denoted $R_2$ (a subset of Re$^+$); its range is the range of $\phi_1$, denoted $R_1$. We prove that there are intervals $I_1$ in Re and $I_2$ in Re$^+$ ($I_1$ may be unbounded below and/or above; $I_2$ may be unbounded above) such that $R_i$ is dense[4] in $I_i$, $i = 1, 2$. For suppose there is an interval $(x, y)$ such that $x = \phi_1(a)$, $y = \phi_1(b)$, but for $z$ in the open interval $(x, y)$, $z \notin R_1$. Therefore, $ba$ is minimal in $A_1^*$ since if $ba \succ_1 dc \succ_1 dd$, we use solvability (Axiom 4) to obtain $d'$ with $bd' \sim_1 dc$; whence $\phi_1(b) - \phi_1(a) > \phi_1(b) - \phi_1(d') > 0$, and $\phi_1(d')$ is in $(x, y)$. Returning to the proof of Hölder's Theorem (Section 2.2.4), we are in the case where there is a minimal element in the semigroup, with every element a multiple of it. It is easy to see that this means that $R_1$ consists of a finite or countable sequence with constant spacing equal to $y - x$. Let the sequence be $..., x_0, x_1, ..., x_n, ...$ and let $a_n$ be the preimage of $x_n$, i.e., $\phi_1(a_n) = x_n$. Then $R_2$ consists of the elements $y_n = \phi_2(a_n)$. But by the same arguments as above, applied to $A_2^*$ and $\phi_2$, the elements $y_n$ must be equally spaced in the ratio sense [the elements $y_n' = \phi_2'(a_n) = \log y_n$ must be equally spaced in the difference sense]. This shows that $\phi_2'$ is linear with $\phi_2$, and the relations $\gtrsim_1$ and $\gtrsim_2$ coincide, contrary to the assumption that they are distinct. Thus, the assumption that $R_1$ contains a gap leads to a contradiction. A parallel argument applies to $R_2$. We therefore have intervals $I_1$ and $I_2$ with $R_i$ dense in $I_i$.

Since $h$ is strictly increasing, we can extend it to a mapping from $I_2$ onto $I_1$. For $x \in I_2$ we let $\bar{h}(x)$ be the supremum of $h(y)$ such that $y \in R_2$ and $y \leqslant x$ (this works except for a minimal $x \in I_2$, where we take the infimum of

---

[4] That is, the complement $I_i - R_i$ contains no open interval. Consequently, any point in $I_i$ is the limit of a monotone sequence of points in $R_i$.

$h(y)$, $y \in R_2$, $y \geqslant x$). It is straightforward to prove that $\bar{h}$ is a strictly increasing function from $I_2$ onto $I_1$ (and, hence, continuous), which coincides with $h$ on $R_2$. We now set up our functional equation using $\bar{h}$ instead of $h$.

Consider any $x, z \in R_2$, $x < z$, and let $t$ be any (positive) number such that either $t < 1$ and $tx \in R_2$ or $t > 1$ and $tz \in R_2$. Then for any $y \in R_2$ with $x < y < z$, we assert that $tx$, $ty$, $tz$ are all in $R_2$. To prove this, we use solvability. For example, if $tx < x$, let $a', a, b, c \in A$ be such that $\phi_2(a') = tx$, $\phi_2(a) = x$, $\phi_2(b) = y$, and $\phi_2(c) = z$. Then $ca', ba' \succsim_2 aa' \succsim_2 a'a'$, so we can choose $c', b'$ with $cc' \sim_2 bb' \sim_2 aa'$; and $\phi_2(b') = ty$, $\phi_2(c') = tz$. The proof is analogous for $tz > z$.

For $x$, $y$, $z$, and $t$ as above, let $\phi_2(a) = x$, $\phi_2(a') = tx,..., \phi_2(c') = tz$. By Equation (3), applied with $\sim_2$, $cb \succsim_1 ba$ iff $c'b' \succsim_1 b'a'$. Since $\phi_1(a) = h(x)$, $\phi_1(a') = h(tx)$, etc., this implies

$$h(z) - h(y) \geqslant h(y) - h(x) \qquad \text{iff} \qquad h(tz) - h(ty) \geqslant h(ty) - h(tx).$$

In other words,

$$h(y) \leqslant \tfrac{1}{2}[h(x) + h(z)] \qquad \text{iff} \qquad h(ty) \leqslant \tfrac{1}{2}[h(tx) + h(tz)].$$

For fixed $x$, $z \in R_2$ and for fixed $t$ such that $tx$, $tz \in R_2$, the above result extends by continuity to all $y \in I_2$ such that $x < y < z$. In particular, for $y$ such that

$$\bar{h}(y) = \tfrac{1}{2}[h(x) + h(z)]$$

we have

$$\bar{h}(ty) = \tfrac{1}{2}[h(tx) + h(tz)].$$

Eliminating $y$ between these two equations gives the fundamental equation

$$\bar{h}^{-1}\left[\frac{\bar{h}(tx) + \bar{h}(tz)}{2}\right] = t \cdot \bar{h}^{-1}\left[\frac{\bar{h}(x) + \bar{h}(z)}{2}\right]. \tag{5}$$

By the above argument, this is valid for all $x$, $z \in R_2$, $x < z$, and all $t$ such that either $t > 1$ and $tz \in R_2$ or $t < 1$ and $tx \in R_2$. But by continuity, the same equation holds for all $x$, $z \in I_2$, for all $t$ such that $tx$, $tz \in I_2$.

A simple substitution shows that any $\bar{h}$ of the form of Equation (4) satisfies Equation (5). In addition,

$$\bar{h}(x) = \alpha + \gamma \log x \tag{6}$$

also satisfies Equation (5), but this leads to

$$\phi_1(a) = \alpha + \gamma \log \phi_2(a) = \alpha + \gamma \phi_2'(a),$$

whence $\succsim_1 = \succsim_2$ contrary to assumption. Thus, solutions of the form of Equation (6) can be excluded,

We now show that the only solutions to Equation (5) are those given by Equations (4) and (6). For $I_2 = \mathrm{Re}^+$, this is a standard result (Aczel, 1966, Section 3.1.3). We sketch a proof for $I_2$ an arbitrary subinterval of $\mathrm{Re}^+$.

For $t \in \mathrm{Re}^+$, let $I_2(t)$ be the subinterval of $I_2$ consisting of all $x \in I_2$ such that $tx \in I_2$ also. Let $I_1(t)$ be the subinterval of $I_1$ containing all $\bar{h}(x)$ with $x \in I_2(t)$. Let $T$ be the interval in $\mathrm{Re}^+$ containing all $t$ for which $I_2(t)$ is nonempty.

For $t \in T$ and $u \in I_1(t)$, define

$$f_t(u) = \bar{h}[t\bar{h}^{-1}(u)]. \tag{7}$$

Thus, $f_t$ is a continuous function from $I_1(t)$ to $I_1$. For any $u, v \in I_1(t)$, let $x = \bar{h}^{-1}(u)$ and $z = \bar{h}^{-1}(v)$; then $x, z, tx$, and $tz \in I_2$, so Equation (5) holds. We have, from (5) and (7),

$$f_t \left( \frac{u + v}{2} \right) = \frac{f_t(u) + f_t(v)}{2}. \tag{8}$$

Equation (8) is called *Jensen's equation*; its only continuous solutions on an interval $I_1(t)$ are linear functions $\gamma_1 u + \gamma_2$ (Aczel, 1966, Section 2.1.4). Since $\gamma_1$, $\gamma_2$ may depend on $t$, we have

$$f_t(u) = \gamma_1(t) u + \gamma_2(t), \tag{9}$$

where $\gamma_1$, $\gamma_2$ are functions from $T$ to $\mathrm{Re}$, $\gamma_1$ being positive valued because $f_t$ is increasing. Combining Equations (7) and (9) and letting $u = \bar{h}(x)$, $x \in I_2(t)$, we have

$$\bar{h}(tx) = \gamma_1(t)\,\bar{h}(x) + \gamma_2(t). \tag{10}$$

This holds for all $t \in T$, $x \in I_2(t)$. We must therefore solve Equation (10). This is another well-known equation, with solutions given by Equations (4) and (6) (Aczel, 1966, Section 3.1.3), but, again, the standard treatment is for $t$, $x \in \mathrm{Re}^+$. Here, we must deal with the much more complicated domain just described.

For $s$, $t$ sufficiently near unity, $s$, $t \in T$, and there exist distinct $x, y$ such that $x, y \in I_2(s) \cap I_2(t)$, and $tx, ty \in I_2(s)$. Thus, we have from Equation (10),

$$
\begin{aligned}
\bar{h}(sty) - \bar{h}(stx) &= \gamma_1(st)[\bar{h}(y) - \bar{h}(x)] \\
&= \gamma_1(s)[\bar{h}(ty) - \bar{h}(tx)] \\
&= \gamma_1(s)\,\gamma_1(t)[\bar{h}(y) - \bar{h}(x)].
\end{aligned}
$$

Thus, for $s$, $t$ in a neighborhood of unity,

$$\gamma_1(st) = \gamma_1(s)\,\gamma_1(t). \tag{11}$$

Also, we can assume for suitable $x$ and for $s$, $t$ near unity, that $sx \in I_2(t)$ and $tx \in I_2(s)$. We obtain

$$\begin{aligned}
\bar{h}(stx) &= \gamma_1(s)[\gamma_1(t)\,\bar{h}(x) + \gamma_2(t)] + \gamma_2(s) \\
&= \gamma_1(t)[\gamma_2(s)\,\bar{h}(x) + \gamma_2(s)] + \gamma_2(t).
\end{aligned}$$

The latter part of this equation simplifies into

$$\gamma_2(t)[\gamma_1(s) - 1] = \gamma_2(s)[\gamma_1(t) - 1]. \tag{12}$$

If $\gamma_1(t) = 1$ and $\gamma_1(s) \neq 1$, then by Equation (12), $\gamma_2(t) = 0$. So, we have two cases to consider. If $\gamma_1$ is identically equal to 1 in the neighborhood of unity for which Equation (12) holds, then that equation is trivial. But then we have from Equation (10),

$$\bar{h}(tx) = \bar{h}(x) + \gamma_2(t). \tag{13}$$

This is true for $t$ in a neighborhood of unity, $x \in I_2(t)$. Applying this twice, we obtain

$$\begin{aligned}
\bar{h}(stx) &= \bar{h}(x) + \gamma_2(st) \\
&= \bar{h}(x) + \gamma_2(s) + \gamma_2(t).
\end{aligned}$$

Hence, we have

$$\gamma_2(st) = \gamma_2(s) + \gamma_2(t), \tag{14}$$

in that neighborhood of unity. In the second case, $\gamma_1$ is not identically equal to 1, but then $\gamma_1(t) = 1$ iff $\gamma_2(t) = 0$. For values of $s$, $t$ with $\gamma_1(s)$, $\gamma_1(t) \neq 1$, we can separate variables in Equation (12), obtaining

$$\gamma_2(t)/[\gamma_1(t) - 1] = \gamma_2(s)/[\gamma_1(s) - 1] = \beta \neq 0.$$

Thus, we have

$$\gamma_2(t) = \beta[\gamma_1(t) - 1]$$

for all $t$ (including those $t$ for which $\gamma_1(t) = 1$) in the neighborhood of unity for which Equation (12) holds. Substituting this in Equation (10) gives us

$$\bar{h}(tx) + \beta = \gamma_1(t)[\bar{h}(x) + \beta], \tag{15}$$

for $t$ in a neighborhood of unity, $x \in I_2(t)$.

Now Equations (11) and (14) are well-known variants of Cauchy's equation; in the domain Re+, their monotonic solutions are the power function

$$\gamma_1(t) = t^\nu, \tag{16}$$

and the logarithm,

$$\gamma_2(t) = \gamma \log t. \tag{17}$$

Since Cauchy's equation has the same solutions in any neighborhood of zero (see Aczel, 1966, Section 2.14; or the uniqueness part of Theorem 2.4 in this book), the variants, Equations (11) and (14), have, respectively, solutions Equations (16) and (17) in any neighborhood of unity. Note also that in case $\gamma_1 \equiv 1$, $\gamma_2$ must be strictly increasing [see Equation (13)] while in the alternative case, $\gamma_1$ is strictly monotonic [see Equation (15)]. Therefore, Equations (16) and (17) are applicable in these two cases, and we have either

$$\bar{h}(tx) = \bar{h}(x) + \gamma \log t \tag{18}$$

or

$$\bar{h}(tx) + \beta = t^\nu[\bar{h}(x) + \beta], \tag{19}$$

for $t$ near unity, $x \in I_2(t)$. But these equations lead immediately to Equations (6) and (4), respectively, for all $x \in I_2$. For example, suppose case 2 holds, so Equation (19) is valid. Let $x_0$ be arbitrary in the interior of $I_2$. For $x$ in a neighborhood of $x_0$, we put $x = tx_0$, giving

$$\bar{h}(x) + \beta = (x/x_0)^\nu[\bar{h}(x_0) + \beta] = \alpha x^\nu,$$

where $\alpha = x_0^{-\nu}[\bar{h}(x_0) + \beta]$. Suppose that $x_1$ is the least upper bound of all $x$ for which this is valid. If $x_1$ is not the least upper bound of $I_2$, then we can choose $y > x_1 > x$ such that

$$\bar{h}(y) + \beta = (y/x)^\nu[\bar{h}(x) + \beta],$$
$$\bar{h}(x) + \beta = \alpha x^\nu.$$

Hence, $\bar{h}(y) + \beta = \alpha y^\nu$, contrary to assumption. Similarly, the greatest lower bound for which the above equation holds must also be the greatest lower bound on $I_2$. Case 1, with Equation (18), is treated similarly.

Since the logarithmic solution has been excluded by the assumption that $\succsim_1$ and $\succsim_2$ are distinct, we are left with $\phi_1(a) + \beta = \alpha \phi_2(a)^\gamma$. Hence, $\phi(a) = \phi_1(a) + \beta = \alpha \phi_2(a)^\gamma$ is the desired difference-and-ratio representation.

The logarithmic solution is excluded because $\succsim_1 \neq \succsim_2$. This asymmetry of Equation (3), which has been unused thus far, excludes $\gamma \leq 0$. (See Miyamoto, 1983, for another approach.) Thus $\phi(a) = \phi_1(a) + \beta = \alpha \phi_2(a)^\gamma$ is the desired difference-and-ratio representation.  ◇

## 4.6  CROSS-MODALITY ORDERING

A very simple extension of algebraic-difference structures to the case where $\gtrsim$ is defined on the set of all pairs from any of $A_1 \times A_1, ..., A_m \times A_m$, i.e., on the set

$$\bigcup_{i=1}^{m} A_i \times A_i,$$

has an interesting empirical interpretation. Let $A_1, ..., A_m$ be different sensory continua (different modalities, or submodalities). For instance, $A_1$ might be a set of line segments of fixed orientation and different lengths; $A_2$, a set of white lights of fixed spectral composition and different radiances; $A_3$, a set of pure tones of fixed frequency and different sound pressures; etc. Observers are asked to compare the relation (the sensation "ratio") of a pair $a_1 b_1$ from $A_1 \times A_1$ with the relation (sensation "ratio") of a pair $c_2 d_2$ from $A_2 \times A_2$, etc. This can also be done by asking them to rate the sensation "ratios" $a_1 b_1$, $c_2 d_2$, etc. on some numerical rating scale.

In actual experimental practice, the somewhat different method of cross-modality matching is used: observers match, say, the brightness sensation of a light from $A_2$ by choosing an appropriate loudness sensation of a tone from $A_3$, etc. The transition from such cross-modality matching data to the empirical relation to be discussed in this section, namely, an ordering $\gtrsim$ on $\bigcup_{i=1}^{m} A_i \times A_i$, rests on a particular theoretical interpretation, according to which observers actually perform the matching task by matching sensation "ratios." More explicitly, if the observer matches $a_2$ to $c_3$ and $b_2$ to $d_3$, this is taken not as evidence that the stimulus $a_2$ produces a sensation equivalent to that of $c_3$, etc., but rather that the perceived relation, or sensation "ratio," of $a_2$ to $b_2$ is equivalent to that of $c_3$ to $d_3$, i.e., $a_2 b_2 \sim c_3 d_3$. According to this interpretation, cross-modality matching establishes an equivalence relation $\sim$ on $\bigcup_{i=1}^{m} A_i \times A_i$. The extension of $\sim$ to an ordering $\gtrsim$ may be made empirically in many ways. One is to assume that subjects order pairs within one modality (say, $A_1$) and that all others are ordered by matching with elements of $A_1 \times A_1$. For a discussion of this theoretical interpretation of cross-modality matching, in terms of an ordering of $\bigcup_{i=1}^{m} A_i \times A_i$, and of alternative theories, see Krantz (1972a).

The five axioms assumed here parallel those of Definition 3: weak ordering, sign reversal, weak monotonicity, solvability, and Archimedean. The key change involves casting the weak monotonicity condition in cross-modality form: if $a_i b_i \gtrsim a_j' b_j'$ and $b_i c_i \gtrsim b_j' c_j'$, then $a_i c_i \gtrsim a_j' c_j'$. In terms of the interpretation of cross modality matching given above, this condition (in its $\sim$ form) has an interesting empirical meaning. If $b_i$ is matched to $b_j'$ using $c_i$ and $c_j'$ as the reference levels, i.e., using $b_i c_i \sim b_j' c_j'$, then it does not matter in any other match whether $b_i$, $b_j'$, or $c_i$, $c_j'$, are used as reference levels.

The solvability axiom contains two parts: one asserts that each pair in $A_i \times A_i$ can be matched by a pair in $A_1 \times A_1$; the other is the usual solvability assumption of algebraic-difference structures applied to $A_1$. The Archimedean axiom is also formulated only for $A_1$. Thus, we assume that $\langle A_1 \times A_1, \gtrsim \rangle$ is an algebraic-difference structure and that each $\langle A_i \times A_i, \gtrsim \rangle$ can be mapped into $\langle A_1 \times A_1, \gtrsim \rangle$ in a suitable fashion.

The representation theorem involves functions $\phi_1, ..., \phi_m$ from $A_1, ..., A_m$ to $\mathrm{Re}^+$, with a ratio representation:

$$a_i b_i \gtrsim a_j' b_j' \qquad \text{iff} \qquad \phi_i(a_i)/\phi_i(b_i) \geq \phi_j(a_j')/\phi_j(b_j').$$

From a logical standpoint, the use of the ratio, rather than the difference, representation is arbitrary, but it agrees with established practice and may reflect the use of sensation "ratio" instructions (see Section 4.4.3).

The uniqueness theorem involves permissible transformations

$$\phi_i \rightarrow \alpha_i \phi_i{}^\gamma, \qquad i = 1, ..., m,$$

where $\gamma$ is independent of $i$. Thus, if the scale is fixed on one continuum, the other scales are ratio scales. In practice, one finds that judgments of length ratios are nearly veridical, so it is tempting to standardize the length or number scale as the physical scale, making all the others ratio scales.

DEFINITION 5. *Suppose $A_1, ..., A_m$ are nonempty sets and $\gtrsim$ is a binary relation on $\bigcup_{i=1}^{m} A_i \times A_i$. Then $\langle A_1, ..., A_m, \gtrsim \rangle$ is called a* cross-modality ordering structure *iff, for all $i$, $j$ with $1 \leq i, j \leq m$ and all $a_i$, $b_i$, $c_i \in A_i$, $a_j'$, $b_j'$, and $c_j' \in A_j$ the following five axioms hold:*

1. $\langle \bigcup_{i=1}^{m} A_i \times A_i, \gtrsim \rangle$ *is a weak order.*
2. *If $a_i b_i \gtrsim a_j' b_j'$, then $b_j' a_j' \gtrsim b_i a_i$.*
3. *If $a_i b_i \gtrsim a_j' b_j'$ and $b_i c_i \gtrsim b_j' c_j'$, then $a_i c_i \gtrsim a_j' c_j'$.*
4. (i) *There exist $d_1$, $e_1 \in A_1$ such that $a_i b_i \sim d_1 e_1$.*
   (ii) *$\langle A_1 \times A_1, \gtrsim \rangle$ satisfies Axiom 4 of Definition 3.*[5]
5. *$\langle A_1 \times A_1, \gtrsim \rangle$ satisfies Axiom 5 of Definition 3.*

THEOREM 4. *If $\langle A_1, ..., A_m, \gtrsim \rangle$ is a cross-modality ordering structure (Definition 5), then there exist functions $\phi_i$ from $A_i$ to $\mathrm{Re}^+$, $i = 1, ..., m$, such that for all $i$, $j$, $1 \leq i, j \leq m$ and all $a_i$, $b_i \in A_i$, $a_j'$, $b_j' \in A_j$,*

$$a_i b_i \gtrsim a_j' b_j' \qquad \text{iff} \qquad \phi_i(a_i)/\phi_i(b_i) \geq \phi_j(a_j')/\phi_j(b_j').$$

[5] It is understood that $\langle A_1 \times A_1, \gtrsim \rangle$ is the restriction of $\langle \bigcup_{i=1}^{m} A_i \times A_i, \gtrsim \rangle$ to $A_1 \times A_1$.

If $\phi_i'$ are any other such functions, $i = 1,..., m$, then there exist positive numbers $\alpha_1 ,..., \alpha_m$, and $\gamma$ such that $\phi_i' = \alpha_i \phi_i{}^\gamma$, $i = 1,..., m$.

## 4.7 PROOF OF THEOREM 4 (p. 165)

If $\langle A_1 ,..., A_m , \gtrsim \rangle$ is a cross-modality ordering structure, then there are positive-valued functions $\phi_i$, unique up to $\phi_i \rightarrow \alpha_i \phi_i{}^\gamma$, such that $a_i b_i \gtrsim a_j' b_j'$ iff $\phi_i(a_i)/\phi_i(b_i) \geq \phi_j(a_j')/\phi_j(b_j')$.

PROOF. Axioms 1–5 of Definition 5 applied to the structure $\langle A_1 \times A_1 , \gtrsim \rangle$ are the same as Axioms 1–5 of Definition 3. Thus, by Theorem 2, there exists a function $\phi_1$ from $A_1$ to Re$^+$, such that for all $a_1 , b_1 , a_1', b_1' \in A_1$, $a_1 b_1 \gtrsim a_1' b_1'$ iff $\phi_1(a_1)/\phi_1(b_1) \geq \phi_1(a_1')/\phi_1(b_1')$. Moreover, $\phi_1$ is unique up to $\phi_1 \rightarrow \alpha_1 \phi_1{}^\gamma$, where $\alpha_1 , \gamma$ are positive.

Fix $a_i^* \in A_i$, $i = 2,..., m$. For any $a_i \in A_i$, $2 \leq i \leq m$, we can find $d_1 , e_1 \in A_1$ [by Axiom 4(i)] such that $d_1 e_1 \sim a_i a_i^*$. Define $\phi_i(a_i)$ by

$$\phi_i(a_i) = \phi_1(d_1)/\phi_1(e_1).$$

Since $d_1 e_1 \sim d_1' e_1'$ implies that $\phi_1(d_1)/\phi_1(e_1) = \phi_1(d_1')/\phi_1(e_1')$, the $\phi_i$ do not depend on the particular values of $d_1 , e_1$ chosen, so long as $d_1 e_1 \sim a_i a_i^*$. Of course, the $\phi_i$ do depend on the choice of the $a_i^*$; in fact, $\phi_i(a_i^*) = 1$, since by Axiom 2 (see Lemma 2, Section 4.5.1), $a_i^* a_i^* \sim d_1 d_1$. So $a_i^*$ is the unit element for $\phi_i$.

To show that the $\phi_i$ have the desired property, it suffices to show that if $a_i b_i \sim a_1 b_1$, then $\phi_i(a_i)/\phi_i(b_i) = \phi_1(a_1)/\phi_1(b_1)$. For then we can take any $a_i b_i , a_j' b_j'$ and map them into $a_1 b_1 , a_1' b_1'$, etc.

Without loss of generality, suppose that $a_i b_i \gtrsim a_i^* a_i^*$. [In the alternative case, use Axiom 2 and show that $b_i a_i \sim b_1 a_1$ implies $\phi_i(b_i)/\phi_i(a_i) = \phi_1(b_1)/\phi_1(a_1)$.] We need to consider three cases: (I) $b_i a_i^* \gtrsim a_i^* a_i^*$; (II) $a_i a_i^* \gtrsim a_i^* a_i^* \gtrsim b_i a_i^*$; (III) $a_i^* a_i^* \gtrsim a_i a_i^*$. These are the only possible cases because $a_i b_i \gtrsim b_i b_i$ and $b_i a_i^* \sim b_i a_i^*$ yield $a_i a_i^* \gtrsim b_i a_i^*$.

(I) Let $a_i a_i^* \sim d_1 e_1$, $b_i a_i^* \sim d_1' e_1'$. Applying Axiom 4(ii) to $d_1 e_1 \gtrsim d_1' e_1' \gtrsim d_1 d_1$, we have $d_1''$ such that $d_1'' e_1 \sim d_1' e_1'$. Thus, $\phi_i(a_i) = \phi_1(d_1)/\phi_1(e_1)$ and $\phi_i(b_i) = \phi_1(d_1'')/\phi_1(e_1)$; so $\phi_i(a_i)/\phi_i(b_i) = \phi_1(d_1)/\phi_1(d_1'')$. It therefore suffices to prove that $d_1 d'' \sim a_i b_i$, which follows by Axiom 3 from $a_i a_i^* \sim d_1 e_1$, $a_i^* b_i \sim e_1 d_1''$ (the latter premise following via Axiom 2).

(II) We have $a_i b_i \gtrsim a_i a_i^*$, $a_i b_i \gtrsim a_i^* a_i^*$. Thus we can take $a_1 , a_1^*, b_1$, with $a_i a_i^* \sim a_1 a_1^*$, $a_i^* b_i \sim a_1^* b_1$, $a_i b_i \sim a_1 b_1$. Therefore,

$$\frac{\phi_i(a_i)}{\phi_i(b_i)} = \frac{\phi_1(a_1)/\phi_1(a_1^*)}{\phi_1(b_1)/\phi_1(a_1^*)} = \frac{\phi_1(a_1)}{\phi_1(b_1)}.$$

(III) This case is similar to (I). We start, however, with $a_i*b_i \sim d_1e_1$, $a_i*a_i \sim d_1'e_1'$; then since $d_1e_1 \gtrsim d_1'e_1' \gtrsim d_1d_1$, Axiom 4(ii) can be applied.

This concludes the proof that the $\phi_i$ have the requisite property. To show uniqueness, note merely that any $\phi_i'$ with the same property would satisfy

$$\phi_i'(a_i) = \phi_i'(a_i*) \cdot \frac{\phi_1'(d_1)}{\phi_1'(e_1)}$$

whenever $a_ia_i* \sim d_1e_1$. But by uniqueness for $\phi_1$,

$$\frac{\phi_1'(d_1)}{\phi_1'(e_1)} = \left[\frac{\phi_1(d_1)}{\phi_1(e_1)}\right]^\nu = \phi_i(a_i)^\nu;$$

thus, $\phi_i' = \alpha_i\phi_i{}^\nu$, where $\alpha_i = \phi_i'(a_i*)$. $\Diamond$

## 4.8 FINITE, EQUALLY SPACED DIFFERENCE STRUCTURES

A finite, equally spaced structure, of whatever type, corresponds to a single finite standard sequence of the corresponding general structure. The interest in such finite structures stems, partly, from the fact that empirical finite standard sequences are often constructed in order to carry out approximate measurement of other objects and, partly, because such simple self-contained proofs of the representation and uniqueness theorems are possible in these cases. Our less mathematically oriented readers may wish to follow the proofs for the equally spaced structures, even if they avoid all other proofs.

A method for developing axioms for a finite, equally spaced structure is to retain all the necessary conditions of the corresponding general structure, with the exception of the Archimedean one, which is automatically true in the finite case, and to replace the solvability condition by the requirement that successive pairs of elements are equally spaced. Sometimes, although apparently not in the present case, the necessary conditions can be simplified. So we retain Axioms 1–3 of Definition 3, and to formulate the idea of equal spacing, we introduce a defined concept of immediate successor. We say that $a$ is the immediate successor of $b$ if $ab$ is in $A^*$ and there is no $c$ such that $ac$, $cb$ are in $A^*$. The equal spacing axiom is very simple: if $a$ is the immediate successor of $b$, and $c$ of $d$, then $ab \sim cd$. Formally, we have:

DEFINITION 6. *Let $\langle A \times A, \gtrsim \rangle$ satisfy Axioms 1–3 of Definition 3, and let $A^*$ be defined by Definition 4. Define a relation $J$ on $A$: for all $a, b \in A$, $a J b$ iff $ab \in A^*$ (i.e., $ab \succ aa$) and there exists no $c \in A$ such that both $ac$, $cb \in A^*$.*

Note that $J$ may be vacuous in a general structure satisfying Axioms 1–3, but in a finite structure, every element has an immediate successor (predecessor) except for those that are maximal (minimal) with respect to the ordering $A^*$.

DEFINITION 7.   *Suppose $A$ is a finite nonempty set and $\gtrsim$ is a binary relation on $A \times A$. The pair $\langle A \times A, \gtrsim \rangle$ is a finite, equally spaced difference structure iff, for all $a$, $b$, $c$, $d$, $a'$, $b'$, $c' \in A$, the following four axioms hold:*

1. $\langle A \times A, \gtrsim \rangle$ *is a weak order.*
2. *If $ab \gtrsim cd$, then $dc \gtrsim ba$.*
3. *If $ab \gtrsim a'b'$ and $bc \gtrsim b'c'$, then $ac \gtrsim a'c'$.*
4. *If $a\, J\, b$ and $c\, J\, d$, then $ab \sim cd$.*

The new testable condition is Axiom 4; obviously, this is a strong structural condition which usually will not be met except when the objects of $A$ are specifically selected so that it is.

THEOREM 5.   *If $\langle A \times A, \gtrsim \rangle$ is a finite, equally spaced difference structure, then the conclusion of Theorem 2 holds: there exists $\phi : A \to \mathrm{Re}$ such that, for all $a$, $b$, $c$, $d \in A$,*

$$ab \gtrsim cd \qquad \text{iff} \qquad \phi(a) - \phi(b) \geq \phi(c) - \phi(d),$$

*and $\phi$ is unique up to a positive linear transformation. Furthermore, there exist functions $\phi$ satisfying these conditions that are integer valued.*

A closely related theorem may be found in Suppes (1957, p. 267–274); see also Davidson, Suppes, and Siegel (1957) and Luce and Suppes (1965).

The proof of Theorem 5 can be obtained from Theorem 2, since we can prove that the solvability axiom (Axiom 4 of Definition 3) follows from the present axioms. However, we give a self-contained proof of Theorem 5 in the next section.

## 4.9   PROOFS

### 4.9.1   Preliminary Lemmas

Since Axioms 1–3 of Definitions 3 and 7 are the same, Lemmas 2–6 of Section 4.5.1 are still true: $aa \sim bb$; if $ab \in A^*$, then $ac > bc$ for all $c$; $A^*$ is transitive and irreflexive; if $ab \in E$ (Definition 4), then $ac \sim bc$ for all $c$; and $E$ is an equivalence relation.

The remaining lemmas of this section concern properties of the $n$th power of the relation $J$ (Definition 6). This is defined recursively as

$a\,J^0\,b$     iff     $ab \in E$,

$a\,J^1\,b$     iff     $a\,J\,b$,

$a\,J^n\,b$     iff     there exists $c \in A$ such that $a\,J^{n-1}\,c$ and $c\,J\,b$.

Intuitively, for $n \geqslant 1$, $a\,J^n\,b$ means that there are exactly $n - 1$ intermediate elements between $b$ and $a$.

The common hypothesis of the following lemmas is that Axioms 1–4 of Definition 7 hold.

LEMMA 7.    $ab \in A^*$ iff there exists $n \geqslant 1$ such that $a\,J^n\,b$.

PROOF.    If $a\,J^n\,b$, $n \geqslant 1$, proceed by induction. By Definition 6, $a\,J^1\,b$ entails $ab \in A^*$. If $a\,J^{n-1}\,c$ and $c\,J^1\,b$, then by the inductive hypothesis, $ac$ is in $A^*$. Since $cb \in A^*$, and $A^*$ is transitive, we have $ab \in A^*$.

Conversely, if $ab \in A^*$ and not $(a\,J\,b)$, then there exists $c_1$ such that $ac_1$, $c_1 b \in A^*$. Since $A$ is finite, we choose $c_1$ minimal with respect to $A^*$ with this property; thus, $c_1\,J\,b$. Now repeat on $ac_1$, constructing $c_2$ with $ac_2 \in A^*$ and $c_2\,J\,c_1$. This stops only when $a\,J\,c_{n-1}$; but since $A$ is finite, it must stop. Thus, if not $(a\,J^1\,b)$, we have for some $n \geqslant 2$, $a\,J\,c_{n-1}$, $c_{n-1}\,J\,c_{n-2}$,..., $c_1\,J\,b$, i.e., $a\,J^n\,b$.      $\Diamond$

LEMMA 8.    If $ab$, $cd \in A^*$, then $ab \gtrsim cd$ iff there exist $m$, $n$ with $n \geqslant m \geqslant 1$, such that $a\,J^n\,b$, $c\,J^m\,d$.

PROOF.    First, suppose that such $m$, $n$ exist. We can take $a'$ with $a\,J^{n-m}\,a'$ $a'\,J^m\,b$. By Axioms 3 and 4 and induction, we have $a'b \sim cd$. Since $aa' \gtrsim cc$, we have $ab \gtrsim cd$.

If, in the above argument, $n > m$, then $aa' \in A^*$, hence, by Lemma 3, $ab > a'b \sim cd$. Thus, to prove the converse, suppose that $ab$, $cd \in A^*$. By Lemma 7, there exist $n$, $m \geqslant 1$ with $a\,J^n\,b$ and $c\,J^m\,d$. If $ab \gtrsim cd$, then $n \geqslant m$, since $m > n$ would, by the argument just given, entail $cd > ab$.      $\Diamond$

### 4.9.2   Theorem 5 (p. 168)

If $\langle A \times A, \gtrsim \rangle$ is a finite, equally spaced difference structure, then it has an interval-scale representation as an equally spaced, numerical difference structure.

PROOF.    Since $A$ is finite, there exists a minimal element $e$ with respect to the order $A^*$. For $ae \in E$, let $\phi(a) = 0$; otherwise, $ae \in A^*$ so, by Lemmas 7 and 8, there exists a unique $n \geqslant 1$ with $a\,J^n\,e$; let $\phi(a) = n$. By Lemmas 3 and 8, $ab \in A^*$ iff $\phi(a) > \phi(b)$.

Suppose that $ab$, $cd \in A^*$. If $ab \succ cd$, then $a\,J^n\,b$, $c\,J^m\,d$, with $n > m$. If $b\,J^p\,e$, $d\,J^q\,e$, $p, q \geqslant 0$, then $a\,J^{n+p}\,e$, $c\,J^{m+q}\,e$, hence, $\phi(a) - \phi(b) = n > m = \phi(c) - \phi(d)$. Similarly, $ab \sim cd$ implies $\phi(a) - \phi(b) = \phi(c) - \phi(d)$. The case where $ba$, $dc \in A^*$, with $ab \gtrsim cd$, is reduced to the above case by Axiom 2. Finally, if $ab \in A^* \cup E$, and $dc \in A^* \cup E$, then

$$\phi(a) - \phi(b) \geqslant 0 \geqslant \phi(c) - \phi(d).$$

Thus, $\phi$ has the desired properties.

If $\phi'$ is any other representation, and if there exist $a$, $b$ with $a\,J\,b$, let $\alpha = \phi'(a) - \phi'(b) > 0$. Let $\beta = \phi'(e)$. If $a\,J^n\,e$, then obviously (by induction),

$$\phi'(a) - \beta = n\alpha = \alpha\phi(a).$$

Thus, we have the desired uniqueness.                                              $\Diamond$

## 4.10   ABSOLUTE-DIFFERENCE STRUCTURES

We next develop a representation by means of absolute differences along a single dimension. Once again, we have a weak ordering $\gtrsim$ on $A \times A$, but now we want to obtain the representation

$$ab \gtrsim cd \qquad \text{iff} \qquad |\,\phi(a) - \phi(b)\,| \geqslant |\,\phi(c) - \phi(d)\,|.$$

Thus, instead of having the sign-reversal axiom that $ab \gtrsim cd$ implies $dc \gtrsim ba$, with its accompanying implication $ab \gtrsim aa \sim bb \gtrsim ba$ (or the reverse), our second axiom below (Definition 8) asserts that $ab \sim ba \succ aa \sim bb$ (for $b \neq a$).

A major impetus for including material on absolute differences is that we need it in proving the existence of multidimensional representations (Chapters 11–13). Objects that vary along several dimensions may not have any empirically useful weak ordering, and comparisons of differences between pairs of such objects are likely to be comparisons of overall, unordered difference, which one frequently seeks to represent as a function of the absolute differences on each of several dimensions.

The first representation theorem for absolute differences was given by Hölder (1901). He assumed that the nondiagonal elements of $A \times A$ could be divided into sets of "like intervals" and "opposite intervals,"[6] each of which behaved according to the properties of a positive-difference structure

---

[6] "Strecken von gleicher Richtung" and "Strecken von entgegengesetzter Richtung," literally, segments of like direction and segments of the opposed direction.

(in his sense, which included stronger structural restrictions than Definition 1 above). In addition, a "like interval" $ab$ could be compared with an "opposite interval." Thus, he treated his ordering of intervals as an ordering of absolute values. The isomorphism between the two kinds of intervals was guaranteed by the axiom: if $ab \gtrsim cd$, then $ba \gtrsim dc$; where $ab$, $cd$ are both "like intervals" or both "opposite intervals." Note that this is a natural adaptation of the sign-reversal axiom to absolute values. However, he did not assume that $ab \sim ba$. Thus, his representation included the possibility that the representations of the two structures were related by a multiplicative factor different from unity: $\psi(ab) = \alpha \cdot \psi(ba)$ if $ab$ is a "like interval" and $ba$ an "opposite interval."

Another axiomatization of absolute differences was given by Suppes and Winet (1955). They assumed, as did Hölder, that the directions of intervals are specified. They used as primitives an ordering of $A$ as well as an ordering of $A \times A$. As we have seen in Section 4.2, assuming an ordering of $A$ and assuming a special set of positive or "like" intervals is essentially the same thing.

In algebraic-difference structures, it is not necessary to give the ordering of $A$ separately because it can be reconstructed immediately from the ordering of $A \times A$: $ab$ is in $A^*$ if $ab > aa$. This works because of the sign-reversal axiom for algebraic differences: if $ab$ is in $A^*$, then $ba$ is not. In the absolute difference case, this device cannot be used to define an order on $A$ because $ab > aa$ for every $a \neq b$. Nonetheless, Tversky and Krantz (1970) showed how to recover an order on $A$ from the absolute-difference ordering. This is most easily accomplished by introducing the concept of *betweenness*. We say $b$ is *between* $a$ and $c$ if $ac \gtrsim ab$, $bc$. Betweenness, intuitively, leads to ordering: if $b$ is between $a$ and $c$ and if we arbitrarily define $a$ to be above $b$ in the ordering, then $b$ must be above $c$ in the ordering.

The results of this section incorporate this idea of obtaining the ordering on $A$ from that on $A \times A$. This requires a special axiom (Axiom 3 of Definition 8 below) to guarantee that betweenness is well behaved. This axiom asserts, in part, that if $b$ is between $a$ and $c$ and if $c$ is between $b$ and $d$, then $b$ and $c$ are both between $a$ and $d$.

Apart from the subtleties involved in recovering the notions of betweenness and unidimensional order from an ordering of differences, the axioms are quite similar to those for positive differences and algebraic differences. Axiom 4 is a variant of the weak monotonicity condition (corresponding to Axiom 4 of Definition 1 or Axiom 3 of Definitions 3, 5, and 7). If $ab \gtrsim a'b'$ and $bc \gtrsim b'c'$, then we assert that $ac \gtrsim a'c'$, provided only that we know that $b$ is between $a$ and $c$. This is analogous to the proviso that $ab$ and $bc$ are in $A^*$ in the positive-difference case. In addition to the usual monotonicity assumption, we add a further clause to the effect that if either antecedent inequality is strict, then so is the conclusion. This clause is a partial replacement for the missing ordering on $A$; it serves somewhat the same function as Axiom 3 of Definition 1. Axioms 5 and 6 are solvability and Archimedean properties, which require no further comment.

After defining an appropriate notion of order and demonstrating that it has the required properties, the proofs, in Section 4.11, reduce the problem to one of positive differences.

DEFINITION 8. *Suppose $A$ is a set with at least two elements and $\succsim$ is a binary relation on $A \times A$. The pair $\langle A \times A, \succsim \rangle$ is an* absolute-difference *structure iff, for all $a, b, c, d, a', b', c' \in A$, and all sequences $a_1, a_2, ..., a_i, ...$ of elements of $A$, the following six axioms hold:*

1. *$\langle A \times A, \succsim \rangle$ is a weak order.*

2. *If $a \neq b$, then $ab \sim ba \succ aa \sim bb$.*

3. (i) *If $b \neq c$, $ac \succsim ab$, $bc$, and $bd \succsim bc$, $cd$, then $ad \succsim ac$, $bd$.*
   (ii) *If $ac \succsim ab$, $bc$ and $ad \succsim ac$, $cd$, then $ad \succsim bd$.*

4. *Suppose that $ac \succsim ab$, $bc$. If $ab \succsim a'b'$ and $bc \succsim b'c'$, then $ac \succsim a'c'$; moreover if either $ab \succ a'b'$ or $bc \succ b'c'$, then $ac \succ a'c'$.*

5. *If $ab \succsim cd$, then there exists $d' \in A$ such that $ab \succsim d'b$ and $ad' \sim cd$.*

6. *If $a_1, a_2, ..., a_i, ...$ is a strictly bounded standard sequence (i.e., there exist $d', d'' \in A$, such that for all $i = 1, 2, ...,$ $d'd'' \succ a_{i+1}a_1 \succsim a_i a_1$ and $a_{i+1}a_i \sim a_2 a_1 \succ a_1 a_1$), then the sequence is finite.*

Axioms 3–6 become more transparent by rewriting them in terms of the defined notion of betweenness.

DEFINITION 9. *Suppose $\langle A \times A, \succsim \rangle$ satisfies Axioms 1 and 2 of Definition 8. We say that $b$ is* between *$a$ and $c$ (denoted $a \mid b \mid c$) iff $ac \succsim ab$, $bc$.*

Note that, by Axiom 2, betweenness is symmetric in the first and third variables: $a \mid b \mid c$ iff $c \mid b \mid a$. Furthermore, for any $a, b, c$, at least one of $a \mid b \mid c$, $a \mid c \mid b$, or $b \mid a \mid c$ must hold.

Axiom 3 asserts the following:

(i) *If $b \neq c$, $a \mid b \mid c$, and $b \mid c \mid d$, then both $a \mid b \mid d$ and $a \mid c \mid d$.*

(ii) *If $a \mid b \mid c$ and $a \mid c \mid d$, then $a \mid b \mid d$.*

Note that the conclusion to Axiom 3(i) in Definition 8 is $ad \succsim ac$, $bd$. Since the hypothesis includes $ac \succsim ab$ and $bd \succsim cd$, we can conclude that $a \mid b \mid d$ and $a \mid c \mid d$ as asserted here. Actually, in this form, it suffices to assert the conclusion $a \mid b \mid d$ because $a \mid c \mid d$ follows by symmetry (interchanging $a$ with $d$ and $b$ with $c$ throughout). The restriction $b \neq c$ is needed in 3(i);

otherwise, letting $b = c$, we obtain $a \mid b \mid d$ for all $a$, $b$, and $d$. Also, in 3(ii), the conclusion $a \mid b \mid d$ follows from $ad \gtrsim bd$, since the hypothesis already includes $ad \gtrsim ac \gtrsim ab$.

We also need the fact that, under the hypotheses of Axiom 3(ii), $b \mid c \mid d$; but this can be proved using Axioms 1, 2, 4, and 5 (Lemma 12).

Axiom 4 can be restated, using betweenness, as follows:

If $a \mid b \mid c$, $a' \mid b' \mid c'$, and $ab \sim a'b'$, then $bc \gtrsim b'c'$ iff $ac \gtrsim a'c'$. This statement obviously follows from the former one, using the strict inequality clause; and conversely, it is easy to prove the version in Definition 8, using the present statement and Axioms 1, 2, and 5.

Axiom 5 becomes:

If $ab \gtrsim cd$, then there exists $d'$ with $a \mid d' \mid b$ and $ad' \sim cd$.

Finally, note that the chain of inequalities in the hypothesis of Axiom 6 tells us that $a_{i+1} \mid a_i \mid a_1$ for all $i$, in addition to stating that successive intervals are equal and nonnull and that $a_i a_1$ is strictly bounded.

The ordering on $A$ is defined by choosing distinct fixed elements $x$ and $y$ and arbitrarily designating $xy$ as a positive interval. Then, other positive intervals are defined relative to $xy$, using betweenness.

DEFINITION 10.   *Suppose $\langle A \times A, \gtrsim \rangle$ is an absolute-difference structure. Choose distinct elements[7] $x$, $y \in A$. Define $A^*$ to be the set of all $ab \in A \times A$ such that at least one of the following holds:*

(i)   $a \mid x \mid y$,  $b \mid x \mid y$, and $ax > bx$;

(ii)   not $(a \mid x \mid y)$, not $(b \mid x \mid y)$, and $bx > ax$;

(iii)   $a \mid x \mid y$ and not $(b \mid x \mid y)$.

*Let $\gtrsim^*$ be the restriction of $\gtrsim$ to $A^*$.*

The essential result to be developed is that $\langle A, A^*, \gtrsim^* \rangle$ is a positive-difference structure, with $ab$ or $ba$ in $A^*$, for $a \neq b$. This leads, via Theorem 1, to the following representation and uniqueness theorem:

THEOREM 6.   *If $\langle A \times A, \gtrsim \rangle$ is an absolute-difference structure, then there exists a function $\phi : A \to Re$ such that for all $a$, $b$, $c$, $d \in A$,*

$$ab \gtrsim cd \qquad iff \qquad |\phi(a) - \phi(b)| \geq |\phi(c) - \phi(d)|.$$

*If $\phi'$ is any other function with the same property, then $\phi' = \alpha\phi + \beta$, where $\alpha$, $\beta$ are real, $\alpha \neq 0$.*

---

[7] Note that $A$ has at least two elements by Definition 8.

## 4.11 PROOFS

### 4.11.1 Preliminary Lemmas

Throughout this section, we assume that $\langle A \times A, \gtrsim \rangle$ is an absolute-difference structure and that $A^*$ is defined as in Definition 10.

LEMMA 9. *If* $ab \sim bc$ *and* $a \neq c$, *then* $ac \succ ab, bc$.

*PROOF.* By Axiom 2, $ac \succ aa$. By Axiom 1, $ba \sim ba$. If $bc \gtrsim ac$, then $b \mid a \mid c$, so Axiom 4 can be applied to the two preceding inequalities, obtaining $bc \succ ba$. This contradicts $ab \sim bc$.                    $\diamond$

LEMMA 10. *If* $ab \sim bc$, $ab' \sim b'c$, *and* $a \neq c$, *then* $b = b'$.

*PROOF.* By Lemma 9, $a \mid b \mid c$ and $a \mid b' \mid c$. Thus, both $ab \succ ab'$ and the converse inequality lead, by Axiom 4, to $ac \succ ac$, a contradiction. Therefore, $ab \sim ab'$.

Now suppose that $b \neq b'$. Then we have $bb' \succ b'b'$. If we apply Axiom 4 to this inequality together with $ab \sim ab'$, then $ab' \succ ab'$, a contradiction. Therefore $a \mid b \mid b'$ is impossible; the same applies for $a \mid b' \mid b$. Hence, $b' \mid a \mid b$. But this, with $a \mid b \mid c$, yields [Axiom 3(i)] $b' \mid a \mid c$, contrary to the result $ac \succ b'c$ from Lemma 9. Thus, $b = b'$.                    $\diamond$

LEMMA 11. *$A^*$ is a strict simple order.*

*PROOF.* We need to show transitivity and trichotomy (exactly one of $ab \in A^*$, $ba \in A^*$, or $a = b$ holds).

Suppose that $ab$, $bc \in A^*$. This can arise from the nine possible combinations of (i), (ii), (iii) of Definition 10. However, five combinations require both $b \mid x \mid y$ and not $(b \mid x \mid y)$. Thus, we are left with four cases to consider.

(i), (i):   $a \mid x \mid y$, $b \mid x \mid y$, $c \mid x \mid y$, $ax \succ bx \succ cx$. Then $ac \in A^*$ follows by clause (i) of Definition 10.

(ii), (ii):   Similar, using clause (ii).

(i), (iii):   $a \mid x \mid y$, $b \mid x \mid y$, and not $(c \mid x \mid y)$. Then $ac \in A^*$ follows by clause (iii) of Definition 10.

(iii), (ii):   Similar, by clause (iii).

To prove trichotomy, note first $a = b$ excludes both $ab \in A^*$ and $ba \in A^*$ (see Definition 10). Also, $ab \in A^*$ and $ba \in A^*$ are clearly contradictory. Thus, we need only show that if $a \neq b$, then either $ab$ or $ba \in A^*$. If neither $ab$ nor $ba \in A^*$, then by Definition 10, we must have one of the following two cases:

   I.  $a \mid x \mid y$, $b \mid x \mid y$, and $ax \sim bx$.
   II. Not $(a \mid x \mid y)$, not $(b \mid x \mid y)$, and $ax \sim bx$.

In Case I, $ax \sim bx$ and $xy \sim xy$ yield, by Axiom 4, $ay \sim by$. Since $x \neq y$, Lemma 10 yields $a = b$, as desired. So we need only consider Case II.

If $x \mid y \mid a$ and $x \mid y \mid b$ or if $x \mid a \mid y$ and $x \mid b \mid y$, then Axiom 4 with $ax \sim bx$ and $xy \sim xy$ yields, again, $ay \sim by$, whence, by Lemma 10, $a = b$. The only other possible subcases are $x \mid b \mid y$ and $x \mid y \mid a$ or $x \mid a \mid y$ and $x \mid y \mid b$. But these, by Axiom 3(ii), yield, respectively, $x \mid b \mid a$ or $x \mid a \mid b$; thus, either $ax \gtrsim ab$ or $bx \gtrsim ab$, contradicting $ab \succ ax$, $bx$ from Lemma 9, unless $a = b$. $\diamondsuit$

Lemma 11 establishes Axiom 2 for positive-difference structures, as well as the needed property that either $ab$ or $ba \in A^*$, unless $a = b$. (See hypotheses of Theorem 1.) The next five lemmas establish Axiom 3 for positive-difference structures: if $ab$, $bc \in A^*$, then $ac \succ ab$, $bc$.

**LEMMA 12.** *If $a \mid b \mid c$ and $a \mid c \mid d$, then $b \mid c \mid d$.*

**PROOF.** If $dc \succ db$, then since $ca \gtrsim ba$ and $d \mid c \mid a$, Axiom 4 yields $da \succ da$, a contradiction. Hence, $bd \gtrsim cd$.

If $bc \succ bd$, then $ab \sim ab$ and $a \mid b \mid c$ yield $ac \succ ad$, contradicting $a \mid c \mid d$. Hence, $bd \gtrsim bc$. $\diamondsuit$

**LEMMA 13.** *If $a$, $b$ are in the same region with respect to $x$, $y$ (i.e., either $a \mid x \mid y$, $b \mid x \mid y$; or $x \mid a \mid y$, $x \mid b \mid y$; or $x \mid y \mid a$, $x \mid y \mid b$), then either $a \mid b \mid x$ or $b \mid a \mid x$.*

**PROOF.** Suppose that $a \mid x \mid y$, $b \mid x \mid y$, and $ax \gtrsim bx$. By Axiom 5, choose $b'$ with $a \mid b' \mid x$ and $b'x \sim bx$. By Lemma 12, from $a \mid b' \mid x$ and $a \mid x \mid y$, we have $b' \mid x \mid y$. Applying Axiom 4 to $b'x \sim bx$ and $xy \sim xy$, we have $b'y \sim by$. By Lemma 10, since $x \neq y$, $b = b'$; hence, $a \mid b \mid x$. Similarly, if $a \mid x \mid y$, $b \mid x \mid y$, and $bx \gtrsim ax$, then $b \mid a \mid x$.

The proofs for the other two cases are analogous (using Axiom 3 instead of Lemma 12 where appropriate). $\diamondsuit$

**LEMMA 14.** *If $x \mid a \mid y$ and $x \mid y \mid b$, then $ba \notin A^*$.*

**PROOF.** By Axiom 3(ii), $x \mid a \mid b$, thus, $bx \gtrsim ax$. $\diamondsuit$

**LEMMA 15.** *If $a$, $b$, $c$ are distinct and $a \mid b \mid c$, then $ac \succ ab$, $bc$.*

**PROOF.** If $ab \sim ac$, then by Lemma 9, $bc \succ ac$; or if $bc \sim ac$, then by Lemma 9, $ab \succ ac$. $\diamondsuit$

We can now establish the result we want. Given $ab$, $bc \in A^*$, we use

Lemmas 13 and 14 to establish that, for each possible case under Definition 10, $a \mid b \mid c$. Then $ac > ab, bc$ follows by Lemma 15, in each case.

LEMMA 16.   *If* $ab, bc \in A^*$, *then* $ac > ab, bc$.

*PROOF.*   As in the proof of transitivity of $A^*$, there are four cases to consider.

(i), (i):   $a \mid x \mid y, b \mid x \mid y, c \mid x \mid y, ax > bx > cx$. By Lemma 13, $a \mid b \mid x$ and $b \mid c \mid x$; so we have $a \mid b \mid c$ from Lemma 12. Thus, $ac > ab, bc$, by Lemma 15.

(ii), (ii):   The cases where $x \mid a \mid y$, $x \mid b \mid y$, $x \mid c \mid y$ or $x \mid y \mid a$, $x \mid y \mid b$, $x \mid y \mid c$ follow from Lemmas 13 and 15, as in case (i), (i). By Lemma 14, the only other subcases to consider are: first, $x \mid a \mid y$, $x \mid b \mid y$, $x \mid y \mid c$, and second, $x \mid a \mid y, x \mid y \mid b, x \mid y \mid c$.

In the first case, Lemma 13 yields $x \mid a \mid b$; from $x \mid a \mid b$, $x \mid b \mid y$, $x \mid y \mid c$ we have $a \mid b \mid y$, $b \mid y \mid c$ (Lemma 12); thus, by Axiom 3(i), $a \mid b \mid c$. (Unless $b = y$, when $x \mid a \mid b$, $x \mid b \mid c$ yield $a \mid b \mid c$ directly by Lemma 12.) As above, $ac > ab, bc$ follows. The second case follows similarly, using Lemma 13 to establish $x \mid b \mid c$.

(i), (iii):   $a \mid x \mid y$, $b \mid x \mid y$, $ax > bx$, not $(c \mid x \mid y)$. Here, Lemma 13 gives $a \mid b \mid x$, and $b \mid x \mid y$ with $x \mid c \mid y$ or $x \mid y \mid c$ gives $b \mid x \mid c$. Therefore we have $a \mid b \mid c$, and thence, $ac > ab, bc$.

(iii), (ii):   Similar to (i), (iii).                                        ◇

## 4.11.2   Theorem 6   (p. 173)

*If* $\langle A \times A, \gtrsim \rangle$ *is an absolute-difference structure, then there is an interval scale* $\phi$ *on* $A$ *such that* $ab \gtrsim cd$ *iff* $\mid \phi(a) - \phi(b) \mid \geq \mid \phi(c) - \phi(d) \mid$.

*PROOF.*   We show that $\langle A, A^*, \gtrsim^* \rangle$ is a positive-difference structure by verifying the six axioms of Definition 1.

1.   Weak ordering: immediate.

2.   Transivity of $A^*$: Lemma 11.

3.   Lemma 15.

4.   Suppose that $ab, bc, a'b'$, and $b'c' \in A^*$ and $ab \gtrsim^* a'b'$, $bc \gtrsim^* b'c'$. By Lemma 16, $a \mid b \mid c$, hence $ac \gtrsim^* a'c'$.

5.   Suppose that $ab, cd \in A^*$ with $ab >^* cd$. Since we have $ab, ba \gtrsim^* cd$, by Axiom 5 of Definition 8, there exist $d', d'' \in A$ with $a \mid d' \mid b$, $b \mid d'' \mid a$, and $ad' \sim cd \sim d''b$. We necessarily have $ad', d'b \in A^*$. For suppose $ad' \notin A^*$. By Lemma 11, $a = d'$ or $d'a \in A^*$. In either case $d'b \gtrsim ab$ leads to a contradiction. Similarly, $d'b \in A^*$. In the same way, $ad'', d''b \in A^*$. This establishes Axiom 5 of Definition 1.

6. If $a_1$, $a_2$,..., $a_i$,... is a strictly bounded standard sequence, in the sense of Definition 1, then $a_{i+1}a_i \sim {}^*a_2a_1$ in $A^*$, $i = 1, 2,...$. By Lemmas 11 and 16, $a_{i+1}a_i$, $a_ia_1 \in A^*$ implies $a_{i+1}a_1 \in A^*$ and $a_{i+1}a_1 \succ a_ia_1$. Moreover, $a_2a_1 \in A^*$ implies $a_2a_1 \succ a_1a_1$. Thus, we also have a strictly bounded standard sequence in the sense of Definition 8.

Furthermore, we have (Lemma 11) that $ab$ or $ba \in A^*$, provided that $a \neq b$. Therefore, by Theorem 1, there exists $\phi$ on $A$ such that, for $ab$, $cd \in A^*$, $ab \succsim^* cd$ iff $\phi(a) - \phi(b) \geq \phi(c) - \phi(d)$. Moreover, $ab \in A^*$ iff $\phi(a) > \phi(b)$; so for $ab \in A^*$, $\phi(a) - \phi(b) = |\phi(a) - \phi(b)|$. This, with $ab \sim ba$, gives the absolute-value representation on all of $A \times A$.

If $\phi'$ is any other representation for $\langle A \times A, \succsim \rangle$, satisfying the theorem, then either $\phi'$ or $-\phi'$ yields a representation for $\langle A, A^*, \succsim^* \rangle$ (depending on whether $\phi'(x) > \phi'(y)$, in which case $\phi'$ preserves $A^*$, or $\phi'(x) < \phi'(y)$, when $-\phi'$ preserves $A^*$). Uniqueness therefore follows from the uniqueness part of Theorem 1.                    $\diamond$

## 4.12 STRONGLY CONDITIONAL DIFFERENCE STRUCTURES

In Section 4.1.3, we cited examples where the empirical ordering of intervals is conditionally connected in the strong sense that only intervals with a common endpoint are compared. Such orderings arise from the method of triads in which the similarity of two comparison stimuli is judged relative to a standard, and more generally in the method of cartwheels, in which the similarity of $n$ stimuli is judged relative to a common standard. In this instance, one could in fact generate a complete ordering by shifting to an alternative method in which similarities of distinct pairs are judged. But in another situation described before in which a preference ordering is interpreted as an ordering of absolute differences of the stimuli from the subject's ideal point, no change of method rectifies the situation—there is no direct way of comparing subject $s$'s preference for stimulus $a$ with subject $t$'s preference for stimulus $b$.

Another broad and important class of conditionally connected difference orderings arises in experiments for which an intrinsic asymmetry exists in the method used to order the pairs. A typical example is the complete-identification experiment in which distinct stimuli $s_1$,..., $s_m$ are assigned distinct responses, $r_1$,..., $r_m$, and the subject is asked to respond with $r_i$ when $s_i$ is presented. Under many circumstances, errors occur (either while learning the $s$–$r$ pairs or, in the case of intensity variation on a single stimulus continuum, even in asymptotic performance). The frequency with which $r_k$ occurs when $s_i$ is presented is sometimes interpreted as a measure of the similarity of $s_i$ to $s_k$. More precisely, if $r_k$ occurs more often with $s_i$ than with $s_j$, we may be tempted to conclude that the $s_i s_k$ interval is smaller than $s_j s_k$ interval, i.e.,

$$s_j s_k > s_i s_k .$$

However, if $r_k$ occurs more often with $s_i$ than $r_l$ does with $s_j$, we may not want to conclude that $s_j s_l > s_i s_k$, because such a difference in relative frequency may simply reflect an overall bias toward response $r_k$ rather than $r_l$. In particular, if either $r_k$ is used more for $s_i$ than $r_l$ for $s_k$ or if the total usage of $r_k$ exceeds that of $r_l$, a bias interpretation is very plausible.

One tack that can be taken is to formulate a theory of complete-identification behavior which takes bias into account and to use the theory to extract an ordering of stimulus pairs $s_i s_k$, etc.; see, for example, Luce (1963). Such theories may be difficult to check and may even be extraneous to the problem at hand. A different tack is to accept the incompleteness and work with a conditionally connected ordering. If the frequency of $r_k$ to $s_i$ is represented as the $(i, k)$ entry of a matrix, then we use only the ordering of entries *within each column* of the matrix. The number of examples of this general sort is very large (see Coombs, 1964, Chapter 19).

In this section, then, we assume as our primitives a set $A$ and a binary relation $\succsim$ on $A \times A$, which is connected only for pairs whose second elements are common. Thus, our first axiom states that $ac$ is comparable to $bd$ if and only if $c = d$. This is the property of *strong conditional connectedness*. Observe that such a relation is conditionally connected in the sense of Section 3.12.2. (This axiom implies, of course, that $\succsim$ is reflexive; in fact, $ab \sim ab$.) The representation to be established is

$$ac \succsim bc \qquad \text{iff} \qquad |\phi(a) - \phi(c)| \geq |\phi(b) - \phi(c)|.$$

We could also consider strongly conditional algebraic-difference structures, but the absolute-difference case appears to be more interesting since it arises naturally in Coombs' theory of preference. Note, however, that in the latter theory, the natural relation is a binary one on $A \times S$, where $as \succsim bs$ if subject $s$ prefers $b$ to $a$, i.e., if $b \succsim_s a$. We can only consider the relation to be on $A \times A$ if each subject is identified with an element of $A$ (his ideal point) and each element of $A$ is the ideal point for some subject. Thus, we do not have a completely satisfactory axiomatization of Coomb's theory.

The strategy we pursue in axiomatizing the strongly conditional difference representation is to extend the given relation $\succsim$ to a connected ordering $\succsim^*$ on $A \times A$ which satisfies the axioms of an absolute-difference structure (Definition 8). This strategy is feasible because a strongly conditionally connected ordering of pairs can contain some information about the ordering of pairs that have no common element. In order to see this, we introduce three defined notions: midpoint, betweenness, and extreme point.

DEFINITION 11. *Let $\sim$ be a symmetric binary relation on $A \times A$. For $a$, $b \in A$, we say that $c$ is a* midpoint *of $a$ and $b$ iff*

(i)  $a = b = c$, *or*

(ii)  $a \neq b$ and $ac \sim bc$.

In the present application (Definition 14), the symmetric relation is, of course, the conjunction of $\gtrsim$ and $\lesssim$. Axiom 5 of Definition 14 states that for any two elements of $A$, a midpoint exists; and, according to Lemma 21, the midpoint is unique. We therefore use the operation notation $a \circ b$ for the midpoint of $a$ and $b$. In view of the symmetry of $\sim$, $a \circ b = b \circ a$.

DEFINITION 12. *Suppose $\gtrsim$ is a binary relation on $A \times A$. We say that $b$ is* between *$a$ and $c$, denoted $a \mid b \mid c$, iff both $ac \gtrsim bc$ and $ca \gtrsim ba$. If $X \subset A$, then $a \in A$ is an* extreme point *of $X$ iff, for all $x, y \in X$, either $x \mid y \mid a$ or $y \mid x \mid a$.*

Note that the first part reformulates Definition 9 in a suitable way for strongly conditionally connected relations. Since interchanging $a$ and $c$ leaves the two defining conditions the same, we have that $a \mid b \mid c$ is equivalent to $c \mid b \mid a$ as before. However, it is no longer true, merely from the definition, that given any $a$, $b$, $c$, one of them is between the other two. For example, we could have $ac \succ bc$, $ba \succ ca$, and $cb \succ ab$. We specifically exclude this possibility in Axiom 4, by asserting a form of transitivity [Axiom 3(ii)], namely, that if $ac \gtrsim bc$ and $cb \gtrsim ab$, then $ca \gtrsim ba$ and (Axiom 2) that $bb \gtrsim ab$ implies $a = b$. It follows immediately from this that one of $a \mid b \mid c$, $b \mid a \mid c$, or $a \mid c \mid b$ holds.

Our Axiom 4 is a two-part one which is essentially the same as Axiom 3 for absolute-difference structures. If $b \neq c$, $a \mid b \mid c$ and $b \mid c \mid d$, then $a \mid b \mid c$ and $a \mid c \mid d$; and if $a \mid b \mid c$ and $a \mid c \mid d$, then $b \mid c \mid d$ and $a \mid b \mid d$.

Let $a_1 \mid a_2 \mid \cdots \mid a_n$ denote that all ternary betweenness relations hold in the indicated order, i.e., $a_i \mid a_j \mid a_k$ whenever $1 \leqslant i \leqslant j \leqslant k \leqslant n$. Then Axioms 2–4 can be used to show that for any $n$ elements, some $n$-ary betweenness relation obtains (Lemma 19).

We can now explain how the connected extension $\gtrsim^*$ is defined. The problem is to infer $ab \gtrsim^* cd$ from $\gtrsim$ relations. One case is obvious: if $c$, $d$ are nested between $a$ and $b$, i.e., if either $a \mid c \mid d \mid b$ or $a \mid d \mid c \mid b$.

If neither $cd$ is nested inside $ab$ nor $ab$ in $cd$, then the two intervals are either disjoint or interlocking. After exchanging the labels on $a$ and $b$ if necessary, and likewise for $c$ and $d$, we have two possible cases:

1.  $a \mid b \mid c \mid d$,
2.  $a \mid c \mid b \mid d$.

Writing the points in these orders on a line segment gives rise to the two cases of Figure 2. There we see that the *ab* interval exceeds the *cd* interval if and only if the distance from *a* to the midpoint *b* ∘ *c* exceeds the distance from *d* to *b* ∘ *c*. Since this latter is a meaningful ≳ statement, we can use it to define ≳*. Formally, we have:

**DEFINITION 13.** *Suppose* ≳ *is a binary relation on A × A, and let* ∼, *midpoint* ∘, *ternary betweenness, and quaternary betweenness be defined as above (Definitions 11 and 12 and discussion following). For a, b, c, d ∈ A, define ab* ≳* *cd iff one of the following holds:*

   (i)   *a | c | d | b  or  a | d | c | b;*

   (ii)  *either a | b | c | d or a | c | b | d, and (a, b ∘ c)* ≳ *(d, b ∘ c);*

   (iii) *condition (ii) applies after interchanging a and b and/or c and d.*

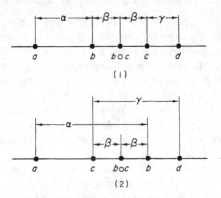

**FIGURE 2.** The two possible cases of *ab* ≳* *cd* when neither interval is wholly within the other. In case (1) $\alpha + \beta > \beta + \gamma$ iff $\alpha > \gamma$ and in case (2) $\alpha - \beta > \gamma - \beta$ iff $\alpha > \gamma$.

In practice, one does not necessarily have to determine the midpoint *b* ∘ *c* in order to determine whether *ab* ≳* *cd*. Suppose, for example, that *a | b | c | d*, and that some *e* is found with *b | e | c*, *ce* ≳ *be*, and *ae* ≳ *de*. It follows that *e* is on the *b* side of the midpoint *b* ∘ *c*, hence, since *ae* ≳ *de*, we must certainly have (*a, b ∘ c*) ≳ (*d, b ∘ c*). Therefore, *ab* ≳* *cd*. We state the definition using the midpoint equality, (*c, b ∘ c*) ∼ (*b, b ∘ c*), instead of the inequality, *ce* ≳ *be*, only for the sake of clarity. The importance of this remark, of course, is that in a finite sample of data (e.g., a strongly conditionally connected data matrix) there may be no midpoints; thus, inferences about the ≳* ordering must be based on inequalities.

Another characterization of the relation ≳* is given in Lemma 23 (Section 4.13.1). Intuitively, if *ab* ≳* *cd*, then (*a ∘ b, b*) ≳* (*c, c ∘ d*)—just divide by two. Thus, instead of $(a, b \circ c) \gtrsim (d, b \circ c)$, we can use $(a \circ b, b \circ c) \gtrsim (c \circ d, b \circ c)$.

Still a third characterization, based on Lemma 22, may be given. That lemma shows that if $a \mid b \mid c$, then $ab \gtrsim cb$ if and only if $a \mid a \circ c \mid b \mid c$. Applying this to $a \mid b \circ c \mid d$, we see that clause (ii) of Definition 13 could be replaced by $a \mid a \circ d \mid b \circ c \mid d$. That is, the order of intervals $ab$ and $cd$ is reflected in the order of the midpoints $a \circ d$ and $b \circ c$. This is the basis of Coombs' unfolding method. Once we know that $a \mid b \mid c \mid d$, then if there is a subject $s$ who prefers $b$ to $c$ and $d$ to $a$, we know that $ab \gtrsim^* cd$.

What additional axioms are needed in order to establish that $\langle A \times A, \gtrsim^* \rangle$ is an absolute-difference structure? So far, we are committed to strong conditional connectedness of $\gtrsim$ (Axiom 1), positiveness (Axiom 2), a limited form of transitivity [Axiom 3(ii)], a betweenness axiom (4), and the existence of midpoints (Axiom 5). The betweenness axiom corresponds to that of absolute-difference structures. Obviously, we shall also need solvability and Archimedean axioms. These are introduced as Axioms 8 and 9 below, respectively, and require little discussion.

The main problem will be to show that $\gtrsim^*$ is a weak order and that it satisfies the requisite version of weak monotonicity (Axiom 4 of Definition 8). Connectedness of $\gtrsim^*$ follows readily from Definition 13 and the properties of quaternary betweenness. But the transitivity of $\gtrsim^*$ is a problem, because it does not follow merely from transitivity of $\gtrsim$. We seem to need further axioms in order to prove it. The first allows us to make some of the inferences that could be made easily from transitivity of $\gtrsim$ were the symmetry property, $ab \sim ba$, true. One such inference was already incorporated in Axiom 3(ii): if $ac \gtrsim bc$ and $cb \gtrsim ab$, then $ca \gtrsim ba$. This would be obvious from the transitivity of $\gtrsim$ if we had the "forbidden" symmetry property. Nevertheless, even without symmetry, this inference is a necessary condition for the representation, since we do have the numerical chain

$$
\begin{aligned}
\mid \phi(c) - \phi(a) \mid &= \mid \phi(a) - \phi(c) \mid \\
&\geqslant \mid \phi(b) - \phi(c) \mid \\
&= \mid \phi(c) - \phi(b) \mid \\
&\geqslant \mid \phi(a) - \phi(b) \mid \\
&= \mid \phi(b) - \phi(a) \mid.
\end{aligned}
$$

Of course, not all inferences that can be made using transitivity of $\gtrsim$ plus "forbidden" symmetry do follow from the representation. For example, from $ba \gtrsim ca$ and $ac \gtrsim dc$, the same sort of absolute-value chain just given leads to

$$\mid \phi(a) - \phi(b) \mid \geqslant \mid \phi(c) - \phi(d) \mid;$$

but we cannot infer $ab \gtrsim cd$ because $b \neq d$. Thus, we want to make inferences only when the intervals at the ends of the chain have an endpoint in common. In particular, we assume, in addition to Axiom 4, the following two properties [Axiom 3(i) and (iii)]:

(i)   If $ad \gtrsim bd$ and $bd \gtrsim cd$, then $ad \gtrsim cd$.

(iii)  If $ad \gtrsim bd$, $db \gtrsim cb$, and $bc \gtrsim ac$, then $da \gtrsim ca$.

The first of these is simply transitivity when there is a common second element. The second involves two reversals such that the first and last terms have a common element. It is easy to see that it is also necessary.

These two are most frequently used in the proof together in the following situation. Suppose $a \mid b \mid c \mid d \mid e \mid f$ and that $bc \gtrsim fc$. Intuitively, if we expand $bc$ at both ends, to $ad$, and simultaneously contract $fc$ at both ends to $ed$, we will have the expanded $ad \gtrsim$ the contracted $ed$. Note that this inference in fact follows from 3(i) and 3(iii) if $bc \gtrsim fc$. For, by betweenness, $ac \gtrsim bc$, $fc \gtrsim ec$, so by 3(i) $ac \gtrsim ec$. And by betweenness $da \gtrsim ca$ and $ce \gtrsim de$, whence $ad \gtrsim ed$ by Axiom 3(iii). These assumptions will be employed most often in just this form: if $a \mid b \mid c \mid d \mid e \mid f$ and $bc \gtrsim fc$, then $ad \gtrsim ed$. Moreover, it is clear that the conclusion must be a strict inequality when any of the antecedents is strict. This, too, is made part of the axiom. (In our special case, if $a \neq b$, or $c \neq d$, or $e \neq f$, or $bc > fc$, then $ad > ed$; this uses Lemma 18 to obtain strict inequalities from betweenness of distinct elements.)

Axiom 3, which was just discussed, does not seem strong enough to establish transitivity of $\gtrsim^*$ since Definition 13 involves midpoints as well as betweenness. For this purpose, two assumptions seem needed; both are necessary conditions. The first, Axiom 6, is known as bisymmetry and asserts an internal consistency of midpoints, namely

$$(a \circ b) \circ (c \circ d) = (a \circ d) \circ (b \circ c).$$

(This property is studied more fully in Section 6.10) Observe that if $\phi(a \circ b) = [\phi(a) + \phi(b)]/2$ holds in the representation, then bisymmetry must follow. The second, Axiom 7, is a version of (very) weak monotonicity for this structure. It says that if $c$ is an extreme point of $\{a, a', b, b'\}$ (intuitively, $c$ is to either the left or the right of all of these points) and if $ac \gtrsim bc$ and $a'c \gtrsim b'c$, then $(a \circ a', c) \gtrsim (b \circ b', c)$. It is easy to see that this condition is necessary. From the extremity of $c$ and the hypothesis, either $\phi(a) - \phi(c) \geqslant \phi(b) - \phi(c)$ and $\phi(a') - \phi(c) \geqslant \phi(b') - \phi(c)$, from which

$$\frac{\phi(a) + \phi(a')}{2} - \phi(c) \geqslant \frac{\phi(b) + \phi(b')}{2} - \phi(c)$$

follows; or the same holds with all of the inequalities reversed.

From these and other of the axioms we derive the following property (Lemma 25). Suppose that $a \mid b \mid c \mid g$ and $a \mid e \mid f \mid g$ and that $d$ is a point whose distance from the midpoint $b \circ f$ is equal and opposite to the distance of $c \circ e$ from $b \circ f$, i.e., $d \circ (c \circ e) = b \circ f$. Then if $ab \gtrsim cb$ and $ef \gtrsim gf$, it follows that $ad \gtrsim gd$ (and strict inequalities lead to strict inequalities). Actually, Axiom 7 appears in the proof only via this lemma. Moreover, one can derive Axiom 6 from this lemma and Axioms 1–5. Thus, we could delete Axioms 6 and 7 and replace them by Lemma 25. The advantage of Lemma 25 as an axiom is that it does not appear to impose on $\circ$ such an elaborate constraint as bisymmetry, which would be difficult to test. The overriding disadvantage is that the property stated in Lemma 25 is highly nonintuitive.

We have now discussed all the axioms, and we summarize them as follows:

DEFINITION 14.   *Suppose that $A$ is a set with at least two elements and $\gtrsim$ is a binary relation on $A \times A$. Let $\sim$, $\circ$, and betweenness be as above (Definitions 11 and 12 and discussion following them). The pair $\langle A \times A, \gtrsim \rangle$ is a* strongly conditional absolute-difference structure *if for all $a, a', b, b', c$, and $d \in A$ and all sequences $a_1, a_2, ..., a_i, ...$ of elements of $A$, the following nine axioms hold:*

1.   *Either $ac \gtrsim bc$ or $bc \gtrsim ac$; and if $ac \gtrsim bd$, then $c = d$.*

2.   *If $aa \gtrsim ba$, then $a = b$.*

3.   *Transitivity:*
   (i)   *If $ad \gtrsim bd$ and $bd \gtrsim cd$, then $ad \gtrsim cd$.*
   (ii)   *If $ac \gtrsim bc$ and $cb \gtrsim ab$, then $ca \gtrsim ba$.*
   (iii)   *If $ad \gtrsim bd$, $db \gtrsim cb$, and $bc \gtrsim ac$, then $da \gtrsim ca$.*
*Furthermore, $>$ holds in each conclusion except if $\sim$ holds in every antecedent.*

4.   (i)   *If $b \neq c$, $a \mid b \mid c$, and $b \mid c \mid d$, then $a \mid b \mid c \mid d$.*
   (ii)   *If $a \mid b \mid c$ and $a \mid c \mid d$, then $a \mid b \mid c \mid d$.*

5.   *A midpoint $a \circ b$ exists.*

6.   *Bisymmetry:   $(a \circ b) \circ (c \circ d) = (a \circ d) \circ (b \circ c)$.*

7.   *Weak monotonicity:   If $c$ is an extreme point of $\{a, a', b, b'\}$, $ac \gtrsim bc$, and $a'c \gtrsim b'c$, then $(a \circ b, c) \gtrsim (a' \circ b', c)$. Furthermore, $>$ holds in the conclusion unless $\sim$ holds in both antecedents.*

8.   *Solvability:   If $a \mid b \mid c$, then there exists $e \in A$ such that $b \circ e = a \circ c$.*

9.   *If $a_1, a_2, ..., a_i, ...$ is a strictly bounded standard sequence ($a_2 \neq a_1$; for any $a_{i+1}, a_i, a_{i-1}$ in the sequence, $a_{i+1} \circ a_{i-1} = a_i$; and there exist $e, f \in A$ such that for every $a_i$ in the sequence, $e \mid a_i \mid f$), then it is finite.*

We have the following theorem:

THEOREM 7. *Suppose that $\langle A \times A, \succsim \rangle$ is a strongly conditional absolute-difference structure. Then there exists a function $\phi : A \rightarrow \text{Re}$ such that for all $a, b, c \in A$*

$$ac \succsim bc \qquad \text{iff} \qquad |\phi(a) - \phi(c)| \geqslant |\phi(b) - \phi(c)|.$$

*If $\phi'$ is any other real-valued function with the same property, then $\phi' = \alpha\phi + \beta$, where $\alpha, \beta$ are real, with $\alpha \neq 0$.*

## 4.13 PROOFS

### 4.13.1 Preliminary Lemmas

The common hypothesis of the following lemmas is that $\langle A \times A, \succsim \rangle$ satisfies the first five axioms of a strongly conditional absolute-difference structure. The use of any other axioms is stated explicitly.

LEMMA 17. *If $ac \succsim bc$, $cb \succsim ab$, and $ba \succsim ca$, then $a = b = c$.*

PROOF. By Axiom 3(ii), all $\succsim$ are actually $\sim$. Thus, $a \mid b \mid c$, $a \mid c \mid b$, and $b \mid a \mid c$. Axiom 4(ii) applied to the first two yields $a \mid b \mid c \mid b$, whence $bb \succsim cb$ and so, by Axiom 2, $b = c$. The proof that $a = b$ is similar.   $\diamondsuit$

LEMMA 18. *If $a, b, c$ are pairwise distinct, and $a \mid b \mid c$, then $ac > bc$ and $ca > ba$.*

PROOF. Suppose that $ac \sim bc$. By Lemma 17, $ab > cb$ cannot hold, otherwise, $a = b = c$. Thus, $cb \succsim ab$ (Axiom 1). Since also $bc \sim ac$, we have $c \mid a \mid b$. From the hypotheses $c \mid b \mid a$ and $c \mid a \mid b$ we obtain [Axiom 4 (ii)] $b \mid a \mid b$, i.e., $bb \succsim ab$. Thus, by Axiom 2, $a = b$, contrary to hypothesis. So, we must have $ac > bc$. A similar proof yields $ca > ba$.   $\diamondsuit$

LEMMA 19. *For any $n$ elements, $n \geqslant 3$, there is a way of relabeling them as $a_1, \ldots, a_n$, such that $a_1 \mid a_2 \mid \cdots \mid a_n$.*

PROOF. For $n = 3$, let the elements be $a, b, c$. By Axiom 1, $ac \succsim bc$ or $bc \succsim ac$. Relabel $a$ and $b$, if necessary, so that $ac \succsim bc$. If $ca \succsim ba$, then $a \mid b \mid c$. Otherwise, by Axiom 1, $ba > ca$ (and thus, $b \neq c$). By Lemma 17, $cb \succsim ab$ cannot hold; thus $ab > cb$ (Axiom 1), so $a \mid c \mid b$. This establishes the Lemma for $n = 3$.

Now we proceed by induction. Suppose that the lemma holds for any $n$

elements; we show that it holds for any $n + 1$ elements. We can assume that the $n + 1$ elements are all distinct, otherwise, the result is trivial from the induction hypothesis. Let $n$ of the elements be chosen arbitrarily and relabeled $b_1,..., b_n$ such that $b_1 \mid b_2 \mid \cdots \mid b_n$. Let the $(n + 1)$st element be $b$.

By the proof for $n = 3$, we must have either $b \mid b_1 \mid b_n$, $b_1 \mid b \mid b_n$, or $b_1 \mid b_n \mid b$. In the first case, $b \mid b_1 \mid b_n$, relabel $b$ as $a_1$ and $b_i$ as $a_{i+1}$, $i = 1,..., n$. We must show that $a_1 \mid a_i \mid a_j$ for $1 \leqslant i \leqslant j \leqslant n + 1$. From $a_1 \mid a_2 \mid a_{n+1}$ and $a_2 \mid a_i \mid a_{n+1}$, $2 \leqslant i \leqslant n + 1$, Axiom 4 (ii) gives $a_1 \mid a_2 \mid a_i \mid a_{n+1}$. If $2 \leqslant i \leqslant j$, we have $a_1 \mid a_i \mid a_{n+1}, a_i \mid a_j \mid a_{n+1}$, so by Axiom 4 (ii) $a_1 \mid a_i \mid a_j \mid a_{n+1}$. The proof for $b_1 \mid b_n \mid b$ is analogous, letting $b_i = a_i$, $i = 1,..., n$ and $b = a_{n+1}$. So we need only consider $b_1 \mid b \mid b_n$.

Let $m$ be maximal such that $b_m \mid b \mid b_n$. We have $m < n$ (otherwise, by Axiom 2, $b = b_n$). Our claim is that $b_m \mid b \mid b_{m+1}$. If $n = m + 1$, this is immediate. If $m + 1 < n$, then by assumption, $b_{m+1} \neq b_n$. Thus, we cannot have $b \mid b_n \mid b_{m+1}$, otherwise, $b_n \mid b_{m+1} \mid b_1$ yields $b \mid b_n \mid b_1$, by Axiom 4 (i). Also, we do not have $b_n \mid b \mid b_{m+1}$ ($m$ is maximal). Therefore, by elimination (using the proof for $n = 3$) we have $b_n \mid b_{m+1} \mid b$; and this, with $b_n \mid b \mid b_m$ yields $b_{m+1} \mid b \mid b_m$, by Axiom 4 (ii).

We now relabel $a_i = b_i$, $i = 1,..., m$, $a_{m+1} = b$, $a_i = b_{i-1}$, $i = m + 2,..., n + 1$. It is entirely straightforward to show that $a_1 \mid a_2 \mid \cdots \mid a_{n+1}$, using $a_m \mid a_{m+1} \mid a_{m+2}$ (i.e., $b_m \mid b \mid b_{m+1}$), established above, together with Axiom 4.    $\diamondsuit$

LEMMA 20.  $a \mid a \circ b \mid b$, and if $a \mid b \mid c$, then $a \mid a \circ b \mid a \circ c \mid b \circ c \mid c$.

PROOF. If $a \circ b \mid a \mid b$, then by Lemma 18, $(b, a \circ b) \sim (a, a \circ b)$ implies that $a \circ b, a, b$ cannot be pairwise distinct. It follows that $a \mid a \circ b \mid b$ [note that if $a = b$, then $a \circ b = a = b$, by Definition 11, part (i)]. Similarly, $a \mid b \mid a \circ b$ leads to $a \mid a \circ b \mid b$. By Lemma 19, however, if neither $a \circ b \mid a \mid b$ nor $a \mid b \mid a \circ b$, then $a \mid a \circ b \mid b$. This proves the first statement.

Now let $a \mid b \mid c$ hold. We have $a \mid a \circ b \mid b$ and $b \mid b \circ c \mid c$. Thus, by Axiom 4 (ii), we have

$$a \mid a \circ b \mid b \mid b \circ c \mid c.$$

By Lemma 19, we must have some betweenness relation for $a \circ b$, $a \circ c$, $b \circ c$. If $a \circ c \mid a \circ b \mid b \circ c$, then we have the 6-ary relation,

$$a \mid a \circ c \mid a \circ b \mid b \mid b \circ c \mid c.$$

But in this case, we can apply Axiom 3 to

$$a \mid a \mid a \circ c \mid a \circ b \mid b \mid c,$$

with $(a, a \circ c) \sim (c, a \circ c)$, obtaining[8] that either $(a, a \circ b) \succ (b, a \circ b)$ or $b = c$. Since the former possibility is false, the latter holds, whence $a \circ c = a \circ b$, and we have $a \circ b \mid a \circ c \mid b \circ c$. A similar proof leads from $a \circ b \mid b \circ c \mid a \circ c$ to $a \circ b \mid a \circ c \mid b \circ c$. Thus, $a \circ b \mid b \circ c \mid a \circ c$ holds in any case.                                                                                    $\diamond$

LEMMA 21.  *The midpoint of two elements is unique.*

PROOF.  Let $b$, $b'$ be midpoints of $a$, $c$. Either $a = c$ (whence, by Definition 11, $a = b = b' = c$) or $a \neq c$, in which latter case, Definition 11 and Axiom 2 yield that $a, b, c$ are pairwise distinct, as are $a, b', c$. We assume, then, that $a \neq c$.

By Lemma 20, $a \mid b \mid c$ and $a \mid b' \mid c$. Now $b \mid c \mid b'$ is impossible; otherwise, with $b \neq c$, Axiom 4 (i) leads to $a \mid b \mid c \mid b'$, and $a \mid c \mid b'$, $a \mid b' \mid c$ imply $b' = c$ by Lemma 18. By Lemma 19, we must have either $b \mid b' \mid c$ or $b' \mid b \mid c$. Relabeling $b$, $b'$, if necessary, assume that $b \mid b' \mid c$ holds. By Axiom 4 (ii), $a \mid b \mid b' \mid c$. But now we can apply Axiom 3 to

$$a \mid a \mid b \mid b' \mid c \mid c \qquad \text{and} \qquad ab \sim cb,$$

obtaining $ab' \succsim cb'$, with the inequality strict unless $b = b'$. Thus, $b = b'$.   $\diamond$

LEMMA 22.  *If $a \mid b \mid c$, then $ab \succsim cb$ iff $a \mid a \circ c \mid b$.*

PROOF.  If $a \mid b \mid c$ and $a \mid a \circ c \mid b$, then we have

$$a \mid a \mid a \circ c \mid b \mid c \mid c.$$

Since $(a, a \circ c) \sim (c, a \circ c)$, Axiom 3 yields $ab \succsim cb$, and furthermore, inequality is strict unless $b = a \circ c$. Thus, for the converse, if $a \mid b \mid c$ and not $a \mid a \circ c \mid b$, then Lemma 20 yields $a \mid b \mid a \circ c \mid c$, with $b \neq a \circ c$; and applying the result just proved to $c \mid a \circ c \mid b$, we have $cb \succ ab$.          $\diamond$

The following lemma is fundamental to the proof of Theorem 7, since it gives an alternative characterization of the relation $ab \succsim^* cd$; it also extends Axiom 7 to certain nonextreme points.

LEMMA 23.  *Suppose Axioms 1–6 hold. If either $a \mid b \mid c \mid d$ or $a \mid c \mid b \mid d$, then the following are pairwise equivalent:*

---

[8] A more formal proof is as follows: By Axiom 3(i), $(a, a \circ c) \sim (c, a \circ c) \succsim (b, a \circ c)$ implies $(a, a \circ c) \succsim (b, a \circ c)$. By Axiom 3(iii), $(a \circ b, a) \succsim (a \circ c, a)$, $(a, a \circ c) \succsim (b, a \circ c)$, $(a \circ c, b) \succsim (a \circ b, b)$ implies $(a, a \circ b) \succsim (b, a \circ b)$, with the strict inequality unless all antecedents are $\sim$. But in the latter case, $(c, a \circ c) \sim (b, a \circ c)$ yields (Lemma 18) $b = c$ or $a \circ b$. If $b = a \circ b$, then $(a \circ c, b) \sim (a \circ b, b)$ yields $b = a \circ c = a \circ b$. Therefore, in any case $a \circ c = a \circ b$.

(i)   $(a, b \circ c) \gtrsim (d, b \circ c)$;

(ii)  $(a \circ b, b \circ c) \gtrsim (c \circ d, b \circ c)$;

(iii) $(a \circ c, b \circ c) \gtrsim (b \circ d, b \circ c)$.

*PROOF.* Since the hypothesis is symmetrical in $b$, $c$ and since (iii) is obtained from (ii) by interchanging $b$ and $c$, it suffices to prove that (i) and (ii) are equivalent, given either alternative in the hypothesis.

With either alternative in the hypothesis, we have $a \mid b \circ c \mid d$. Therefore, by Lemma 22, (i) is equivalent to $a \mid a \circ d \mid b \circ c \mid d$.

Suppose (i) holds. By repeated use of Lemma 20, Axiom 4 (ii), and the observation just made above, we have

$$a \circ b \mid a \circ d \mid (a \circ d) \circ (b \circ c) \mid b \circ c \mid c \circ d.$$

By Axiom 6, $(a \circ d) \circ (b \circ c) = (a \circ b) \circ (c \circ d)$, thus, we have

$$a \circ b \mid (a \circ b) \circ (c \circ d) \mid b \circ c \mid c \circ d.$$

By Lemma 22, this is equivalent to (ii). Conversely, if (i) is false, we have

$$a \circ b \mid b \circ c \mid (a \circ d) \circ (b \circ c) \mid a \circ d \mid c \circ d,$$

with $b \circ c \neq a \circ d$. Axiom 6 now yields

$$a \circ b \mid b \circ c \mid (a \circ b) \circ (c \circ d) \mid c \circ d,$$

with $b \circ c \neq (a \circ b) \circ (c \circ d)$ (using Axiom 2). Therefore, (ii) is false, by Lemma 22.                                                                                    $\diamondsuit$

LEMMA 24.  *If $a \circ c = b \circ c$, then $a = b$.*

*PROOF.*  First suppose that $a \mid b \mid c$. By Lemma 20 $a \mid b \mid b \circ c \mid c$. Thus we have

$$a \mid b \mid b \circ c \mid a \circ c \mid c \mid c,$$

and Axiom 3 yields $(b, b \circ c) \succ (c, b \circ c)$ unless $a = b$. The case $b \mid a \mid c$ is similar.

Second, suppose that $a \mid c \mid b$. Then by Lemma 20, $a \mid a \circ c \mid c \mid b \circ c \mid b$, hence, $a \circ c = c = b \circ c$, which is impossible unless $a = b = c$.     $\diamondsuit$

LEMMA 25.  *Suppose that Axioms 1–7 hold. Suppose $a \mid b \mid c \mid g, a \mid e \mid f \mid g$, and that $d$ is such that $d \circ (c \circ e) = b \circ f$. If $ab \gtrsim cb$ and $ef \gtrsim gf$, then $ad \gtrsim gd$.*

*PROOF.*  By Lemma 22, $ef \gtrsim gf$ implies $e \mid e \circ g \mid f$, and so by Lemma 19,

$a \mid e \mid e \circ g \mid f \mid g$. Thus we have $fa \gtrsim (e \circ g, a)$. Now, by Axiom 7 applied to this and $ba \gtrsim (a \circ c, a)$ (Lemma 22),

$$[d \circ (c \circ e), a] = (b \circ f, a)$$
$$\gtrsim [(a \circ c) \circ (e \circ g), a].$$

But by Axiom 6,

$$[d \circ (c \circ e), a] \gtrsim [(a \circ g) \circ (c \circ e), a].$$

From the fact that Axiom 7 holds with strict inequalities, cancellation is possible, so

$$da \gtrsim (a \circ g, a),$$

which by Lemma 22 implies $ad \gtrsim gd$.                                        ◇

### 4.13.2  Theorem 7 (p. 184)

*If $\langle A \times A, \gtrsim \rangle$ is a strongly conditional absolute-difference structure, then there is an interval scale $\phi$ on $A$ such that $ac \gtrsim bc$ iff $\mid \phi(a) - \phi(c) \mid \geqslant \mid \phi(b) - \phi(c) \mid$.*

*PROOF.* We first show that $\gtrsim^*$ is an extension of $\gtrsim$, in the sense that $ac \gtrsim bc$ implies $ac \gtrsim^* bc$, and that it introduces no incorrect equivalences, i.e., if $ac > bc$, then $ac >^* bc$.

Suppose that $ac \gtrsim bc$. By Lemma 19, either $a \mid b \mid c$ or $a \mid c \mid b$. In the former case, $a \mid b \mid c \mid c$ yields $ac \gtrsim^* bc$ by Definition 13 (i); while in the latter case, $a \mid c \mid c \mid b$ and $(a, c \circ c) \gtrsim (b, c \circ c)$ yield $ac \gtrsim^* bc$ [Definition 13 (ii)].

Conversely, suppose that $ac \gtrsim^* bc$. This can arise from Definition 13 in six ways:

(i)$_1$  $a \mid b \mid c \mid c$;

(i)$_2$  $a \mid c \mid b \mid c$;

(ii)$_1$  $a \mid c \mid b \mid c$ and $(a, b \circ c) \gtrsim (c, b \circ c)$;

(ii)$_2$  $c \mid a \mid b \mid c$ and $(c, a \circ b) \gtrsim (c, a \circ b)$;

(ii)$_3$  $c \mid a \mid c \mid b$ and $(c, a \circ c) \gtrsim (b, a \circ c)$;

(ii)$_4$  $a \mid c \mid c \mid b$ and $(a, c \circ c) \gtrsim (b, c \circ c)$.

These all yield $ac \gtrsim bc$: the first three by $a \mid b \mid c$, the next two because they imply $a = b = c$ (via Axiom 2), and the last because $(a, c \circ c) \gtrsim (b, c \circ c)$ is the same as $ac \gtrsim bc$. This completes the proof that $ac \gtrsim bc$ iff $ac \gtrsim^* bc$.

Theorem 7 will now follow if we can show that $\langle A \times A, \gtrsim^* \rangle$ is an absolute-difference structure, since the representation $\phi$ of Theorem 6 yields precisely the representation claimed here because $\gtrsim$ and $\gtrsim^*$ are equivalent on pairs whose second elements are the same. We therefore verify in corresponding numbered sections that the six axioms of Definition 8 hold. The proofs are long, due to the necessity for considering many separate cases.

1. $\gtrsim^*$ *is a weak order.*

By Axiom 1, $ab \gtrsim ab$. Therefore, $ab \gtrsim^* ab$, and $\gtrsim^*$ is reflexive.

To show that $\gtrsim^*$ is connected, we use Lemma 19. Let $a$, $b$, $c$, $d$ be arbitrary. By that lemma, some quaternary betweenness relation holds. If $cd$ is nested within $ab$, then $ab \gtrsim^* cd$, by (i) of Definition 13; if $ab$ is nested within $cd$, then $cd \gtrsim^* ab$, similarly. If neither interval is nested within the other, then they are either disjoint or interlocking. Relabel $a$ and $b$ either as $a'$, $b'$ or as $b'$, $a'$, and similarly, relabel $c$, $d$ as either $c'$, $d'$ or $d'$, $c'$, such that $a' \mid b' \mid c' \mid d'$ (disjoint) or $a' \mid c' \mid b' \mid d'$ (interlocking). Since $(a', b' \circ c')$ and $(d', b' \circ c')$ are comparable, by Axiom 1, we have either $a'b' \gtrsim^* c'd'$ or $d'c' \gtrsim^* b'a'$, by (ii) of Definition 13. And then, by (iii) of Definition 13, it follows that either $ab \gtrsim^* cd$ or $cd \gtrsim^* ab$.

Third, we must show transitivity. Suppose that $ab \gtrsim^* cd$ and $cd \gtrsim^* ef$. There are four major cases to consider, since each of the two $\gtrsim^*$ statements can be based on nesting [(i) of Definition 13)] or on midpoints [(ii) of Definition 13].

*Case* (i), (i): $cd$ nested inside $ab$ and $ef$ nested inside $cd$. Axiom 4 leads easily to $ef$ nested inside $ab$, hence $ab \gtrsim^* ef$ by (i) of Definition 13.

*Case* (i), (ii): Exchange the labels on $c$ and $d$, if necessary, and also on $e$ and $f$, if necessary, so that either $c \mid d \mid e \mid f$ or $c \mid e \mid d \mid f$, and $(c, d \circ e) \gtrsim (f, d \circ e)$. Then exchange labels on $a$, $b$ if necessary so that $a \mid c \mid d \mid b$. There are now only five possible 6-ary betweenness relations for $a,...,f$:

1. $a \mid c \mid d \mid b \mid e \mid f$;
2. $a \mid c \mid d \mid e \mid b \mid f$;
3. $a \mid c \mid d \mid e \mid f \mid b$;
4. $a \mid c \mid e \mid d \mid b \mid f$;
5. $a \mid c \mid e \mid d \mid f \mid b$.

Subcases 3 and 5 have $ef$ nested inside $ab$, so $ab \gtrsim^* ef$ by (i) of Definition 13. Subcases 1, 2, and 4 all lead, by Lemma 19, to

$$a \mid c \mid d \circ e \mid b \circ e \mid f \mid f.$$

By Axiom 3, $(c, d \circ e) \gtrsim (f, d \circ e)$ implies that $(a, b \circ e) \gtrsim (f, b \circ e)$. Since these subcases all have $a \mid b \mid e \mid f$ or $a \mid e \mid b \mid f$, we have $ab \gtrsim^* ef$ by (ii) of Definition 13.

*Case* (ii), (i):   The proof is similar to that in case (i), (ii).

*Case* (ii), (ii):   Up to now, we have only used Axioms 1–6. (Note that Lemmas 17–21 use only Axioms 1–5.) This case is the principal one, and we use Axioms 6 and 7 (in part via Lemma 25).

Exchange labels on $a$, $b$ and/or on $c$, $d$, if necessary, so that $a \mid b \mid c \mid d$ or $a \mid c \mid b \mid d$. Furthermore, exchange labels on $e$, $f$ if necessary so that one of the following holds: $c \mid d \mid e \mid f$, $c \mid e \mid d \mid f$, $d \mid c \mid e \mid f$, or $d \mid e \mid c \mid f$. (The latter two possibilities could be eliminated by exchanging labels on $c$ and $d$, but that would upset the $a$, $b$, $c$, $d$ relation.)

We need to consider seven separate subcases. Subcase 1 covers all situations where $c \mid d \mid e \mid f$ or $c \mid e \mid d \mid f$, for in all these, we also have $a \mid b \mid e \mid f$ or $a \mid e \mid b \mid f$. (Use Axiom 4 (i), with $c \neq d$; if $c = d$, then necessarily $e = f$ and the result is trivial.) By Lemma 20 and Axiom 4 we have

$$a \circ b \mid b \circ c \mid c \circ d \mid c \circ d \mid d \circ e \mid e \circ f.$$

By Lemma 23, $ab \gtrsim^* cd$ yields $(a \circ b, b \circ c) \gtrsim (c \circ d, b \circ c)$ and $cd \gtrsim^* ef$ yields $(c \circ d, d \circ e) \gtrsim (e \circ f, d \circ e)$. By Axiom 6, $(b \circ e) \circ (c \circ d) = (b \circ c) \circ (d \circ e)$. Thus, we use Lemma 25 to infer that

$$(a \circ b, b \circ e) \gtrsim (e \circ f, b \circ e).$$

By Lemma 23, this yields $ab \gtrsim^* ef$.

The situations where $d \mid c \mid e \mid f$ or $d \mid e \mid c \mid f$ hold divide into six more subcases, according to the betweenness relations of $a$, $b$, $e$, $f$. By symmetry, we can always write $a$ to the left of $b$. Axiom 4 then requires that $f$ be written to the left of $e$. For example, $e \mid f \mid a \mid b$, with $a \mid b \mid d$, would lead to $e \mid f \mid d$, contradicting the assumed $d \mid e \mid f$ (unless $e = f$); etc. The six cases are:

2. $f \mid e \mid a \mid b$;
3. $f \mid a \mid e \mid b$;
4. $f \mid a \mid b \mid e$;
5. $a \mid f \mid e \mid b$;
6. $a \mid f \mid b \mid e$;
7. $a \mid b \mid f \mid e$.

The arguments in subcases 2, 3, 6, and 7 are all analogous to that of subcase 1 above.

In subcases 2 and 3, we have from Lemma 20,

$$f \circ e \mid e \circ a \mid a \circ b \mid a \circ b \mid b \circ c \mid c \circ d.$$

If not $ab \gtrsim^* ef$, then by Lemma 23, $(f \circ e, e \circ a) \succ (a \circ b, e \circ a)$. Also,

$ab \gtrsim^* cd$ yields $(a \circ b, b \circ c) \gtrsim (c \circ d, b \circ c)$. Axiom 6 yields

$$(e \circ c) \circ (a \circ b) = (e \circ a) \circ (b \circ c).$$

Therefore, by the strict inequality clause in Lemma 25, we have

$$(f \circ e, e \circ c) > (c \circ d, e \circ c).$$

By Lemma 23, $ef >^* cd$, which is a contradiction. Therefore, $ab \gtrsim^* ef$.

For subcases 6 and 7 we use Lemma 20 to obtain

$$d \circ c \mid c \circ e \mid e \circ f \mid e \circ f \mid f \circ b \mid b \circ a.$$

Now, $cd \gtrsim^* ef$ and $ef >^* ab$ yield, by Lemma 23, that $(d \circ c, c \circ e) \gtrsim (e \circ f, c \circ e)$ and $(e \circ f, f \circ b) > (b \circ a, f \circ b)$. Axiom 6 yields

$$(c \circ b) \circ (e \circ f) = (c \circ e) \circ (f \circ b),$$

hence, by the strict clause of Lemma 25, we have $(d \circ c, c \circ b) > (b \circ a, c \circ b)$, or $cd >^* ab$. So we conclude that $ab \gtrsim^* ef$.

Finally, we turn to subcases 4 and 5. In subcase 5, $ab \gtrsim^* ef$ is immediate by nesting [(i) of Definition 13]. In subcase 4, $ef \gtrsim^* ab$ follows by nesting, with strict inequality unless $f = a, b = e$. Thus, we need to show that subcase 4 cannot hold, except for that degenerate case. In fact, we must have $f \mid a \mid b \mid c \mid e \mid d$ or $f \mid a \mid b \mid e \mid c \mid d$ or $f \mid a \mid c \mid b \mid e \mid d$, all of which lead, by Lemma 20, to

$$f \mid a \mid b \circ c \mid e \circ c \mid d \mid d.$$

Since $ab \gtrsim^* cd$ implies $(a, b \circ c) \gtrsim (d, b \circ c)$, Axiom 3 yields $(f, e \circ c) \gtrsim (d, e \circ c)$, with inequality strict unless $f = a$ and $b \circ c = e \circ c$. But strict inequality contradicts $cd \gtrsim^* ef$; thus only the degenerate situation $f = a$, $b = e$ is possible (note Lemma 24). This completes the proof of transitivity.

2.  If $a \neq b$, then $ab \sim^* ba >^* aa \sim^* bb$.

Since $a \mid a \mid b \mid b$ holds, with $(a, a \circ b) \sim (b, a \circ b)$, $ab \sim^* ba$ and $aa \sim^* bb$ follow immediately, for any $a, b$. If $a \neq b$, then by Axiom 2, $ab > bb$, thus, by extension, $ab >^* bb$.

3.  (i)  If $b \neq c$, $ac \gtrsim^* ab$, $bc$, and $bd \gtrsim^* bc$, $cd$, then $ad \gtrsim^* ac$, $bd$.
    (ii)  If $ac \gtrsim^* ab$, $bc$ and $ad \gtrsim^* ac$, $cd$, then $ad \gtrsim^* bd$.

This follows immediately from Axiom 4; e.g., the hypotheses of 3 (i) imply $ac \gtrsim bc$, $ca \gtrsim ba$, or $a \mid b \mid c$; similarly, $b \mid c \mid d$; hence, by 4 (i), $a \mid b \mid c \mid d$, so $ad \gtrsim^* ac$, $bd$. Similarly, 4 (ii) implies 3 (ii).

4.  If $a \mid b \mid c$, $ab \gtrsim^* a'b'$, and $bc \gtrsim^* b'c'$, then $ac \gtrsim^* a'c'$, with strict inequality if either antecedent is strict.

We first show that this is relatively trivial unless $a' \mid b' \mid c'$. For example, if $a' \mid c' \mid b'$, then we have

$$ca \gtrsim ba \gtrsim^* b'a' \gtrsim c'a'.$$

It follows that $ac \gtrsim^* a'c'$. (Use the extension to replace $\gtrsim$ by $\gtrsim^*$ and use transitivity of $\gtrsim^*$ and $ab \sim^* ba$.) Moreover, if $ac \sim^* a'c'$ holds, then by transitivity of $\gtrsim^*$ and the extension property, we have

$$ca \sim ba, \qquad b'a' \sim c'a'.$$

Lemma 18 now yields $b = c$, $b' = c'$. From this it is immediate that $\sim^*$ holds in both antecedent hypotheses. A similar proof applies in case $b' \mid a' \mid c'$. So by Lemma 19, we are reduced to the situation where $a' \mid b' \mid c'$.

There are two cases to consider, depending on whether the triples $a$, $b$, $c$ and $a'$, $b'$, $c'$ are ordered in the same sense or oppositely in the 6-ary betweenness relation involving $a$, $b$, $c$, $a'$, $b'$, and $c'$. Subcases in which they are ordered in the same sense include $a \mid b \mid c \mid a' \mid b' \mid c'$, $a \mid a' \mid b \mid b' \mid c \mid c'$, etc. In all such subcases, $a$ or $c$ must be an exterior element in the 6-ary betweenness, for otherwise, we would have $a' \mid a \mid b \mid c \mid c'$, with $a \neq a'$, $c \neq c'$; and any location for $b'$ in the 6-ary betweenness would lead, by (i) of Definition 13, to either $a'b' >^* ab$ or to $b'c' >^* bc$. Without loss of generality we can always write $a$ on the left of the 6-ary betweenness. We then write $c'$ on the right, since otherwise, $a \mid a' \mid c' \mid c$, with $c \neq c'$, gives $ac >^* a'c'$ directly. Thus, the only subcases involve the ordering of $b$, $c$, $a'$, and $b$ .

In all these subcases, we have the midpoint relation

$$a \circ a' \mid b \circ a' \mid b \circ b' \mid b \circ b' \mid c \circ b' \mid c \circ c'.$$

(Use Lemma 20 and Axiom 4.) Moreover, in all these subcases, $ab \gtrsim^* a'b'$ leads to

$$(a \circ a', b \circ a') \gtrsim (b \circ b', b \circ a'),$$

and likewise, $bc \gtrsim^* b'c'$ leads to

$$(b \circ b', c \circ b') \gtrsim (c \circ c', c \circ b').$$

To see this, note that if either $a \mid a' \mid b \mid b'$ or $a \mid b \mid a' \mid b'$, then $ab \gtrsim^* a'b'$ implies $(a \circ a', b \circ a') \gtrsim (b \circ b', b \circ a')$ by Lemma 23; whereas if $a \mid a' \mid b' \mid b$, then we have $a \circ a' \mid a \circ b' \mid b \circ a' \mid b \circ b'$ and $(a \circ b') \circ (b \circ a') = (a \circ a') \circ (b \circ b')$ (Lemma 20 and Axiom 6), and hence,

$$a \circ a' \mid (a \circ a') \circ (b \circ b') \mid b \circ a' \mid b \circ b'$$

holds, and the desired inequality follows by Lemma 22. Similarly, $bc \gtrsim^* b'c'$ yields the desired inequality.

We also have $(a' \circ c) \circ (b \circ b') = (b \circ a') \circ (c \circ b')$, thus, Lemma 25 can be applied, yielding

$$(a \circ a', a' \circ c) \gtrsim (c \circ c', a' \circ c).$$

Lemma 23 then gives $ac \gtrsim^* a'c'$. Moreover, if either antecedent inequality is strict, then the strict case of Lemma 25 yields a strict conclusion. (The "if and only if" clause of Lemma 23 is used repeatedly here.)

The second main case, with opposite ordering of $a$, $b$, $c$ and $a'$, $b'$, $c'$, must now be considered. By arguments like those in the first case, we can restrict our attention to 6-ary betweenness relations with $a$ on the left and $a'$ on the right; thus, our subcases involve the betweenness relations of $b$, $c$, $b'$, and $c'$. There are three such subcases:

1. $b \mid c \mid c' \mid b'$ or $b \mid c' \mid c \mid b'$;
2. $c' \mid b \mid b' \mid c$ or $c' \mid b' \mid b \mid c$;
3. $b \mid c' \mid b' \mid c$.

These come from the 24 orderings of $b$, $c$, $b'$, and $c'$, by restricting them so that $b$ is to the left of $c$ and $c'$ to the left of $b'$, and by eliminating the case $c' \mid b \mid c \mid b'$, which entails $b'c' >^* bc$.

In subcase 1, $bc \gtrsim^* b'c'$ and Lemma 22 yield $b \mid b \circ b' \mid c \circ c' \mid b'$, while $ab \gtrsim^* a'b'$ and Lemma 22 yield $a \mid a \circ a' \mid b \circ b' \mid a'$. Combining these yields $a \mid a \circ a' \mid c \circ c' \mid a'$, or $ac \gtrsim^* a'c'$. Furthermore, by Lemma 22, $ac \sim^* a'c'$ entails $a \circ a' = b \circ b' = c \circ c'$, hence, both hypotheses are $\sim^*$.

Subcase 2 is similar: $ab \gtrsim^* a'b'$ gives $a \mid a \circ a' \mid b \circ b' \mid a'$, and $bc \gtrsim^* b'c'$ gives $c' \mid b \circ b' \mid c \circ c' \mid c$, whence, $a \mid a \circ a' \mid c \circ c' \mid a'$, etc.

Finally, in subcase 3, we have $a \mid a \circ a' \mid b \circ b' \mid a'$ and $b \mid b \circ b' \mid c \circ c' \mid c$, whence, $a \mid a \circ a' \mid c \circ c' \mid a'$ etc. This completes the proof.

5. *If* $ab \gtrsim^* cd$, *then there exists* $d' \in A$ *such that* $ab \gtrsim^* d'b$ *and* $ad' \sim^* cd$.

Since the statement to be proved is symmetric in $c$, $d$, though not in $a$, $b$, we need to consider five cases:

1. $a \mid b \mid c \mid d$;
2. $a \mid c \mid b \mid d$;
3. $b \mid a \mid c \mid d$;
4. $b \mid c \mid a \mid d$;
5. $a \mid c \mid d \mid b$.

Cases 1, 2, and 5 allow a simple construction of $d'$: since $a \mid c \mid d$, choose $d'$ by Axiom 8 such that

$$d' \circ c = a \circ d.$$

Intuitively, this should give the desired result: since $d' + c = a + d$, then $a - d' = c - d$, etc. More rigorously, we proceed to consider each case separately.

In case 1, $ab \gtrsim^* cd$ leads, by Lemma 22, to $a \mid a \circ d \mid b \circ c \mid d$. There are two subcases, depending on whether $a \mid b \mid a \circ d$ or $a \mid a \circ d \mid b$. In the former subcase, we have

$$a \mid d' \mid b \mid \begin{Bmatrix} a \circ d \\ d' \circ c \end{Bmatrix} \mid b \circ c \mid c \mid d.$$

Here, the position of $d'$ follows from Lemma 20, since $b \mid d' \mid c$ would imply $b \mid b \circ c \mid d' \circ c$. Since $(a, d' \circ c) \sim (d, d' \circ c)$ and $ab \gtrsim d'b$, the conclusions follow.

The second subcase of case 1 leads to

$$a \mid d' \mid \begin{Bmatrix} a \circ d \\ d' \circ c \end{Bmatrix} \mid b \mid b \circ c \mid c \mid d,$$

and the conclusions follow as above.

The argument for case 2 is similar, except that the subcases are defined by the possibilities $a \mid c \mid a \circ d$ and $a \mid a \circ d \mid c$. The former leads to

$$a \mid c \mid \begin{Bmatrix} a \circ d \\ d' \circ c \end{Bmatrix} \mid d' \mid b \mid d$$

(with the position of $d'$ relative to $b$ inferred from Lemma 20, since $c \mid d' \circ c \mid b \circ c$), while the latter leads to

$$a \mid d' \mid \begin{Bmatrix} a \circ d \\ d' \circ c \end{Bmatrix} \mid c \mid b \mid d.$$

Case 5 is even simpler, since $a \mid d' \mid b$ follows from $a \mid d' \mid d$, and $ad' \sim^* cd$ follows from

$$a \mid \begin{Bmatrix} a \circ d \\ d' \circ c \end{Bmatrix} \mid d.$$

In cases 3 and 4 we rely on the results from cases 1 and 2, respectively, first constructing $d''$ such that

$$d'' \circ c = b \circ d.$$

In each case, we have $b \mid d'' \mid a$ and $bd'' \sim^* cd$. Then we apply the result from case 3, to the quadruple

$$a \mid d'' \mid b \mid b, \qquad ab \gtrsim^* d''b,$$

constructing $d'$ to satisfy

$$d' \circ d'' = a \circ b.$$

We have $a \mid d' \mid b$ and $ad' \sim^* d''b \sim^* cd$, as required. This completes the proof.

6. *The Archimedean axiom.*

Suppose that $a_1$, $a_2$,..., $a_i$,... is a strictly bounded standard sequence, in the sense of Definition 8, i.e., there exist $d'$, $d''$ such that

$$d'd'' >^* a_{i+1}a_1 \gtrsim^* a_i a_1 \qquad \text{and} \qquad a_{i+1}a_i \sim^* a_2 a_1 >^* a_1 a_1$$

for all $a_{i+1}$, $a_i$ in the sequence. We use this to construct a strictly bounded standard sequence in the sense of Definition 14.

Note first that the $a_i$ are distinct (this follows from Axiom 2 of Definition 8, established above) and that $a_1 \mid a_2 \mid ... \mid a_i \mid ...$. The betweenness follows by induction, since we have $a_{i+1} \mid a_i \mid a_1$.

Now let $b_1 = d'$. Suppose that $b_1$,..., $b_k$ have been constructed, with $b_{i+1}b_i \sim^* a_2 a_1$, $1 \leqslant i < k$, $b_k b_1 \sim^* a_k a_1$, and $d' \mid b_i \mid d''$, $1 \leqslant i \leqslant k$. Suppose that there exists $a_{k+1}$ in the sequence $a_1$, $a_2$,.... We cannot have $a_2 a_1 \gtrsim^* d''b_k$, for if this were so, the inequalities $a_{k+1}a_k \gtrsim^* d''b_k$ and $a_k a_1 \sim^* b_k b_1$, together with $a_{k+1} \mid a_k \mid a_1$ and Axiom 4 of Definition 8 (established above) would yield $a_{k+1}a_1 \gtrsim^* d''b_1 = d'd'$. By Axiom 5 of Definition 8 (established above) there exists $b_{k+1}$ with $d'' \mid b_{k+1} \mid b_k$ and $b_{k+1}b_k \sim^* a_2 a_1$. By Axiom 4 (ii), we have $d'' \mid b_{k+1} \mid d'$, $b_{k+1} \mid b_k \mid b_1$. By Axiom 4 of Definition 8, $b_{k+1}b_1 \sim^* a_{k+1}a_1$. Thus, the construction can be continued to $b_{k+1}$, as long as there remains an $a_{k+1}$. Now $b_1$, $b_2$,..., $b_i$,... is a strictly bounded standard sequence, in the sense of Definition 14. The only thing that needs to be established for this is $b_{i+1} \circ b_{i-1} = b_i$; and this follows from $b_{i+1} \neq b_{i-1}$, $b_{i+1}b_i \sim b_{i-1}b_i$. The $b_i$ sequence is finite; therefore the $a_i$ sequence must be also.                                                                                          $\diamondsuit$

## EXERCISES

**1.** Give a model of a positive-difference structure (that is, specify how $A^*$, $\gtrsim$, may be defined so that Axioms 1–6 of Definition 1 hold) in which $A$ is the set of all integers.     (4.2)

**2.** Suppose that $A = \text{Re}$, $A^* = \{(x, y) \mid x, y \in \text{Re}, x > y\}$, and $\gtrsim$ on $A^*$ is defined by $(x, y) \gtrsim (x', y')$ iff either

(i)  $x > x'$ or
(ii) $x = x'$, $y \leqslant y'$.

Which axioms of Definition 1 does $\langle A, A^*, \gtrsim \rangle$ satisfy? In each case, provide either a proof or a counterexample.     (4.2)

**3.** Assuming that the representation given in Theorem 1 is valid, show that each axiom of Definition 1 except Axiom 5 is necessary.     (4.2)

**4.** Without using Theorem 1, prove that in a positive-difference structure, if $\Gamma \succsim \Gamma'$, $\Delta \succsim \Delta'$, and $(\Gamma, \Delta) \in \mathbf{B}$, then $\Gamma \circ \Delta \succsim \Gamma' \circ \Delta'$ and if either hypothesis is strict, so is the conclusion.     (4.2, 4.3)

**5.** Give a model of an algebraic-difference structure (Definition 3), in which

(i)   $A = \mathrm{Re} \cup \{w\}$, where $w \notin \mathrm{Re}$;

(ii)   $(x, y) \succsim (y, y)$ iff $x > y$, if $x, y \in \mathrm{Re}$;

(iii)   $(w, x) > (x, x)$ for all $x \in \mathrm{Re}$.     (4.4.1)

**6.** Suppose that $\succsim$ is the lexicographic ordering of $\mathrm{Re}^2$, i.e., for $x, x', y, y' \in \mathrm{Re}$, $(x, y) \succsim (x', y')$ iff either

(i)   $x > x'$, or

(ii)   $x = x'$ and $y \geqslant y'$.

Which axioms of an algebraic-difference structure are satisfied (proof or counterexample)?     (4.4.1)

**7.** Let $\langle A \times A, \succsim \rangle$ be an algebraic-difference structure. Show that in general there can be no representation of the form

$$ab \succsim cd \qquad \text{iff} \qquad g[\phi(a), \phi(b)] \geqslant g[\phi(c), \phi(d)],$$

where $g(x, y) = (x/y) + (x - y)$.     (4.4.1, 4.4.2)

**8.** Reaxiomatize algebraic-difference structures in terms of a relation $\succsim^*$, defined from $\succsim$, and interpreted as

$$ab \succsim^* cd \qquad \text{iff} \qquad \phi(a) + \phi(b) \geqslant \phi(c) + \phi(d).$$

$$(4.4.1, 6.5.4)$$

*The following seven exercises outline a proof of Theorem 2′. Note that Lemmas 2–6 of Section 4.5.1 are valid because they only depend on Axioms 1–3 of Definition 3 (Section 4.4.1). Throughout, the intervals ab, cd, etc. are assumed to be in $A^*$.*

**9.** Suppose $ab \succsim cd$. Construct a finite sequence $a_1, ..., a_{m+1}$ in $\mathscr{L}(cd)$ for which $a_1 = b$ and $a_{m+1} = a$.

**10.** (i) Let $M(cd, ab)$ be the least integer such that there exists a sequence $a_1, ..., a_{m+1}$ in $\mathscr{L}(cd)$ for which $a_1 = b$ and $a_{m+1}a_1 \succsim ab$. Show that if $a_1, ..., a_{m+1}$ is in $\mathscr{L}(cd)$, $b_1, ..., b_m$ is in $\mathscr{G}(cd)$, $a_1 = b = b_1$, and $a_{m+1}a_1 \succsim ab \succsim b_m b_1$, then $M(cd, ab) = m$.

(ii)   Construct a set of inequalities that, if observed, would guarantee that

$$1.20 < \frac{\phi(a) - \phi(b)}{\phi(a') - \phi(b')} < 1.75.$$

**11.**   (i)   If $ab \gtrsim a'b' \gtrsim cd$, then $M(cd, ab) \geqslant M(cd, a'b')$;

(ii)   If $ab \gtrsim cd \gtrsim ef$, then

$$M(ef, cd)\, M(cd, ab) \geqslant M(ef, ab) > [M(ef, cd) - 1][M(cd, ab) - 1].$$

**12.**   Construct a sequence $c_i d_i$, $i = 1, 2, \ldots$, for which $M(c_i d_i, ab) \to \infty$.

**13.**   For any sequence fulfilling the property of Exercise 12, prove that for all $ab, a'b'$, $M(c_i d_i, ab)/M(c_i d_i, a'b')$ converges to a limit in $\mathrm{Re}^+$.

**14.**   Prove that

(i)   $M(c_i d_i, ab) + M(c_i d_i, bc) \geqslant M(c_i d_i, ac)$

$$\geqslant M(c_i d_i, ab) + M(c_i d_i, bc) - 2.$$

(ii)   For fixed $a'b'$,

$$\psi(ab) = \lim_{i \to \infty} M(c_i d_i, ab)/M(c_i d_i, a'b')$$

is additive and, therefore, for $ab_0 \in A^*$, $\phi(a) = \psi(ab_0)$ satisfies the representation of Theorem 2'.

**15.**   Show that the representation in Theorem 2' is unique up to a positive linear transformation.

**16.**   Suppose that $\gtrsim_1$ and $\gtrsim_2$ are two cross-modality ordering relations (see Definition 5, Section 4.6) and that $\langle A_1 \times A_1, \gtrsim_1, \gtrsim_2 \rangle$ satisfies the interlocking conditions for a difference-and-ratio representation (Theorem 3, Section 4.4.3). What can be said about $\langle A_i \times A_i, \gtrsim_1, \gtrsim_2 \rangle$, for $i \neq 1$? A necessary condition, for the existence of a set of functions $\phi_1, \ldots, \phi_m$ that give difference-and-ratio representations for $\gtrsim_1, \gtrsim_2$ across modalities, is the following: if

$$a_i a_i' \sim_2 b_i b_i' \sim_2 c_j c_j' \sim_2 d_j d_j',$$

then $a_i b_i \gtrsim_1 c_j d_j$ if and only if $a_i' b_i' \gtrsim_1 c_j' d_j'$. Is this condition also sufficient?   (4.4.3, 4.6)

**17.**   Prove that a finite, equally spaced difference structure (Definition 7, Section 4.8) is an algebraic-difference structure.   (4.4.1, 4.8).

**18.**   Suppose that $A$ is the circumference of a circle of unit radius, i.e.,

$$A = \{(x, y) \mid x, y \in \mathrm{Re}, \; x^2 + y^2 = 1\}.$$

Order $A \times A$ by the ordering of the lengths of chords joining pairs of points in $A$. Which axioms of an absolute-difference structure (Definition 8) are satisfied (proof or counterexample)?     (4.10)

**19.** Show that the axioms of an absolute-difference structure (Definition 8) are all necessary for the representation of Theorem 6, except for the solvability axiom.     (4.10)

**20.** Suppose that dissimilarities of the 15 pairs of six stimuli, $a, b, c, d, e, f$, are ranked (from 15, highest dissimilarity, to 1, lowest) as in the following matrix.

|   | a | b | c | d | e | f |
|---|---|---|---|---|---|---|
| a |   | 4 | 7 | 8 | 14 | 15 |
| b |   |   | 1 | 5 | 11 | 13 |
| c |   |   |   | $2\frac{1}{2}$ | $9\frac{1}{2}$ | 12 |
| d |   |   |   |   | 6 | $9\frac{1}{2}$ |
| e |   |   |   |   |   | $2\frac{1}{2}$ |

Show that no absolute-difference representation is possible. Show that with only one inversion of order, a solution is possible. In the resulting representation, the $cf$ distance must be more than three times as large as the $cd$ distance.     (4.10)

**21.** Suppose that $\mathrm{Re}^2$ is ordered as follows:

$$(x, y) \gtrsim (x', y') \qquad \text{iff} \qquad |x - y| \geqslant |x' - y'|.$$

Does every pair of points in Re have a midpoint (Definition 11)? If so, what is it? Answer the same questions with respect to the lexicographic ordering of the plane (Exercise 6).     (4.12)

**22.** Suppose that four subjects, $s, t, u, v$, exhibit the following preference orderings over objects $a, b, c, d$, where S is subject and R is rank.

| S \ R | 1 | 2 | 3 | 4 |
|---|---|---|---|---|
| s | a | b | c | d |
| t | d | c | b | a |
| u | c | b | a | d |
| v | b | c | d | a |

Show that no one-dimensional unfolding representation is possible.     (4.12)

# Chapter 5  Probability Representations

The debate about what probability is and about how probabilities shall be calculated has been prolonged and is, in many respects, unresolved; nonetheless few disagreements exist about the mathematical properties of numerical probability. It is accepted that probabilities are numbers between 0 and 1 and that they are attached to entities called events. Moreover, it is widely agreed that events in a given context are subsets—although not necessarily all subsets—of a set known as a sample space. Sample spaces are intended to represent all possible observations that one might make in particular situations. The classical assumptions about events are embodied in the following definition:

DEFINITION 1. *Suppose that $X$ is a nonempty set (sample space) and that $\mathscr{E}$ is a nonempty family of subsets of $X$. Then $\mathscr{E}$ is an* algebra of sets *on $X$ iff, for every $A, B \in \mathscr{E}$:*

1. $-A \in \mathscr{E}$.
2. $A \cup B \in \mathscr{E}$.

*Furthermore, if $\mathscr{E}$ is closed under countable unions, i.e., whenever $A_i \in \mathscr{E}$, $i = 1, 2,...$, it follows that $\bigcup_{i=1}^{\infty} A_i \in \mathscr{E}$, then $\mathscr{E}$ is called a $\sigma$-algebra on $X$. The elements of $\mathscr{E}$ are called* events.

199

In words, to be called an algebra of sets, $\mathscr{E}$ must have the two properties of being closed under complementation and unions. As is shown in Section 5.3.1 (Lemma 1), it follows that $\mathscr{E}$ is also closed under intersections and differences.

Observe that $\{\varnothing, X\}$ is an algebra of sets and that it is a subalgebra of every algebra on $X$ since if $A$ is in $\mathscr{E}$, then $-A$ is in $\mathscr{E}$ and so $X = A \cup -A$ is in $\mathscr{E}$ and $\varnothing = -X$ is in $\mathscr{E}$. Only more complex algebras are usually of interest.

Later, in Section 5.4.1, we shall have occasion to consider a somewhat weaker definition in which we are assured that $A \cup B$ is in $\mathscr{E}$ only when $A \cap B = \varnothing$ and $A, B$ are in $\mathscr{E}$.

The following axiomatic definition of numerical probability is the basis of most current work in probability theory. It was first stated explicitly by Kolmogorov (1933).

DEFINITION 2.  *Suppose that $X$ is a nonempty set, that $\mathscr{E}$ is an algebra of sets on $X$, and that $P$ is a function from $\mathscr{E}$ into the real numbers. The triple $\langle X, \mathscr{E}, P \rangle$ is a* (finitely additive) probability space *iff, for every $A, B \in \mathscr{E}$:*

 1.  $P(A) \geqslant 0$.
 2.  $P(X) = 1$.
 3.  *If $A \cap B = \varnothing$, then $P(A \cup B) = P(A) + P(B)$.*

*It is a* countably additive probability space *if in addition*:

 4.  $\mathscr{E}$ *is a $\sigma$-algebra on $X$.*
 5.  *If $A_i \in \mathscr{E}$ and $A_i \cap A_j = \varnothing$, $i \neq j$, $i, j = 1, 2, \ldots$, then*

$$P\left(\bigcup_{i=1}^{\infty} A_i\right) = \sum_{i=1}^{\infty} P(A_i).$$

We do not study the surprising consequences of Definition 2; the interested reader should consult a good book on probability theory, e.g., Feller (1957, 1966). Our plan, instead, is to treat Definition 2 as (part of) a representation theorem; specifically, we inquire into conditions under which an ordering $\gtrsim$ of $\mathscr{E}$ has an order-preserving function $P$ that satisfies Definition 2. Obviously, the ordering is to be interpreted empirically as meaning "qualitatively at least as probable as." Put another way, we shall attempt to treat the assignment of probabilities to events as a measurement problem of the same fundamental character as the measurement of, e.g., mass or momentum. From this point of view, the debates about the meaning of probability are, in reality, about acceptable empirical methods to determine $\gtrsim$. It is not evident why the measurement of probability should have been the focus of more philosophic controversy than the measurement of mass, of length, or of any other scientifically significant attribute; but it has been. We are not suggesting that the controversies over probability have been unjustified, but merely that other controversial issues in the theory of measurement may have been neglected to a degree.

To those familiar with the debate about probability (see, for example, Carnap, 1950; de Finetti, 1937; Keynes, 1921; Nagel, 1939; Savage, 1954, 1961) these last remarks may seem slightly strange. Theories about the representation of a qualitative ordering of events have generally been classed as subjective (de Finetti, 1937), intuitive (Koopman, 1940a,b; 1941), or personal (Savage, 1954), with the intent of emphasizing that the ordering relation $\gtrsim$ *may* be peculiar to an individual and that he *may* determine it by any means at his disposal, including his personal judgment.[1] But these "mays" in no way preclude orderings that are determined by well-defined, public, and scientifically agreed upon procedures, such as counting relative frequencies under well-specified conditions. Even these objective procedures often contain elements of personal judgment; for example, counting the relative frequency of heads depends on our judgment that the events "heads on trial 1" and "heads on trial 2" are equiprobable. This equiprobability judgment is part of a partly "objective," partly "subjective" ordering of events.

Presumably, as science progresses, objective procedures will come to be developed in domains for which we now have little alternative but to accept the considered judgments of informed and experienced individuals.

Ellis pointed out that the development of a probability ordering is

> ... analogous to that of finding a thermometric property, which... was the first step towards devising a temperature scale.
>
> The comparison between probability and temperature may be illuminating in other ways. The first thermometers were useful mainly for comparing atmospheric temperatures. The air thermometers of the seventeenth century, for example, were not adaptable for comparing or measuring the temperatures of small solid objects. Consequently, in the early history of thermometry, there were many things which possessed temperature which could not be fitted into an objective temperature order. Similarly, then, we should not necessarily expect to find any single objective procedure capable of ordering all propositions in respect of probability, even if we assume that all propositions possess probability. Rather, we should expect there to be certain kinds of propositions that are much easier to fit into an objective probability order than others... (Ellis, 1966, p. 172).

Since virtually all representation theorems yield a unique probability measure, agreement about the probability ordering of an algebra of events

---

[1] Many of the relevant papers are collected together in Kyburg and Smokler (1964).

or, as is more usual, agreement about a method to determine that ordering is sufficient to define a unique numerical probability—an objective probability. Various authors have studied conditions under which two related probability measures on the same algebra of events, one thought of as objective and the other subjective, must agree. Perhaps the most interesting results are those included in Edwards (1962); also see Section 8.4.2.

In any case, the difficulties in measuring probability do not appear to be inherently different from those that arise when we apply extensive and other measurement methods elsewhere. In measuring lengths, for example, rods are practical only for relatively short distances; certainly other methods must be used in astronomical research, and considerable disagreement exists among astronomers about which of the alternatives is appropriate (for a general discussion, see Chapter 15 of North, 1965).

Besides the methods of ordering events by relative frequencies or by direct human judgments, there are indirect methods of inferring the ordering. Savage's (1954) book, *The Foundations of Statistics*, introduced an ordering of "personal probability" inferred from decisions among acts with uncertain outcomes. Briefly, if $a$, $b$ are outcomes and $A$ is an event, we denote by $a_A \cup b_{-A}$ an act whose outcome is $a$ if $A$ occurs and $b$ if $A$ does not occur. (The reason for the union notation will become clear in Chapter 8.) If $a$ is preferred to $b$, and if the act $a_A \cup b_{-A}$ is preferred to the act $a_B \cup b_{-B}$, then we may infer that $A$ is more probable than $B$ (for the decision maker whose preferences are studied). This idea is discussed further in Section 5.2.4, where more complete references are given. Usually, the idea is introduced in conjunction with another: the measurement of both probability and utility such that acts are ordered by their expected utilities. This was Savage's approach, and we devote a separate chapter, 8, to a new version of it. A fine survey of work on this topic is available in the book by Fishburn (1970).

Our concern in this chapter is with conditions under which an ordering of events can be represented by a probability measure. Whether or not these conditions are satisfied by event orderings established by one or another experimental procedure is not discussed here; see Luce and Suppes (1965, pp. 321–327).

## ☐ 5.2  A REPRESENTATION BY UNCONDITIONAL PROBABILITY

### 5.2.1  Necessary Conditions: Qualitative Probability

Since we know the representation theorem that we want to prove, we may proceed as in Chapters 3 and 4 to derive necessary conditions on the assumption that the representation exists.

As always, if $P$ is to be order preserving, $\succsim$ must be a weak ordering of $\mathscr{E}$. Since this property is familiar and its difficulties as an empirical law have already been discussed earlier (Section 1.3.1), albeit in other contexts, we need not repeat ourselves; slight rewording suits those comments to this context.

We next show that the certain event $X$ is strictly more probable than the impossible event $\varnothing$, the empty set, and that any event $A$ is at least as probable as $\varnothing$. Since, by Axiom 2 of Definition 2, $P(X) = 1$ and since $X \cup \varnothing = X$ and $X \cap \varnothing = \varnothing$, it follows from Axiom 3 that

$$1 = P(X) = P(X \cup \varnothing) = P(X) + P(\varnothing) = 1 + P(\varnothing),$$

and so $P(\varnothing) = 0$. Since $P$ is order preserving and $P(X) = 1 > 0 = P(\varnothing)$, $X \succ \varnothing$. Moreover, by Axiom 1, $P(A) \geqslant 0 = P(\varnothing)$, so $A \succsim \varnothing$. Observe that $A \sim \varnothing$ does *not* imply $A = \varnothing$. That is to say, nonempty events may be equivalent in probability to the impossible event, and these *null events*, as they are called, correspond to sets with zero probability. We can also derive that $X \succsim A$ and a number of other properties, but it is unnecessary to state them separately because they turn out to be direct consequences of the properties that we list (see Section 5.3.1).

Next, we wish to consider how the order relation and the operation of union relate. Since Axiom 3 of Definition 2 is our main tool and since it is conditional on the two events being disjoint, we may anticipate disjointness playing a role in the corresponding qualitative property. Consider three events $A$, $B$, $C$, with $A$ disjoint from $B$ and $C$. If $B \succsim C$, then by the order-preserving property and Axiom 3,

$$P(A \cup B) = P(A) + P(B) \geqslant P(A) + P(C) = P(A \cup C),$$

whence $A \cup B \succsim A \cup C$. Conversely, if $A \cup B \succsim A \cup C$,

$$P(A) + P(B) = P(A \cup B) \geqslant P(A \cup C) = P(A) + P(C),$$

and so subtracting $P(A)$, $P(B) \geqslant P(C)$, whence $B \succsim C$. Summarizing, if $A \cap B = A \cap C = \varnothing$, then $B \succsim C$ if and only if $A \cup B \succsim A \cup C$. Intuitively, this condition seems plausible. If $B$ is at least as probable as $C$, then adjoining to each the same disjoint event should not alter the ordering, and conversely, deleting such an event also should not alter it.

The primary relation between complementation and ordering is: If $A \succsim B$, then $-B \succsim -A$. This is not stated as a separate axiom because it follows from the others (Lemma 4, Section 5.3.1).

The final necessary condition that we derive in this section is an Archimedean property. We first derive a special case of it. Suppose that $A_1, \ldots, A_i, \ldots$

are mutually disjoint events (i.e., for $i \neq j$, $A_i \cap A_j = \varnothing$), each of which is equivalent in probability to some fixed event $A$, i.e., for $i = 1, 2, ..., A_i \sim A$. By a finite induction on Axiom 2 of Definition 1, $\bigcup_{i=1}^{n} A_i$ is an event. If a probability measure exists, then by a finite induction on Axiom 3 of Definition 2,

$$P\left(\bigcup_{i=1}^{n} A_i\right) = \sum_{i=1}^{n} P(A_i) = nP(A).$$

If $P(A) > 0$, i.e., if $A \succ \varnothing$, then since $P \leqslant 1$ it follows that $n$ cannot exceed $1/P(A)$. Thus, there can be at most finitely many mutually disjoint events each of which is equivalent in probability to a nonnull event. Actually, we need a slightly stronger form of this property which is formulated as follows:

DEFINITION 3.   *Suppose that $\mathscr{E}$ is an algebra of sets and that $\sim$ is an equivalence relation on $\mathscr{E}$. A sequence $A_1, ..., A_i, ...,$ where $A_i \in \mathscr{E}$, is a* standard sequence *relative to $A \in \mathscr{E}$ iff for $i = 1, 2, ...,$ there exist $B_i$, $C_i \in \mathscr{E}$ such that:*

   (i)    $A_1 = B_1$ and $B_1 \sim A$;
   (ii)   $B_i \cap C_i = \varnothing$;
   (iii)  $B_i \sim A_i$;
   (iv)   $C_i \sim A$;
   (v)    $A_{i+1} = B_i \cup C_i$.

We see that if event $A_i$ is assumed to correspond to $i$ disjoint copies of $A$, then it follows that $A_{i+1}$—which equals the disjoint union of $B_i$, which in turn is equivalent to $A_i$, and of $C_i$, which is equivalent to $A$—corresponds to $i + 1$ disjoint copies of $A$. So this is a sensible inductive definition which generalizes our original notion. Since by induction $P(A_i) = iP(A)$, we must assume the finiteness of any standard sequence relative to an $A \succ \varnothing$. Note that our usual formulation of the Archimedean property, namely that every strictly bounded standard sequence be finite, is somewhat simplified here since every standard sequence is bounded by $X$.

We summarize these properties as:

DEFINITION 4.   *Suppose that $X$ is a nonempty set, that $\mathscr{E}$ is an algebra of sets on $X$, and that $\succsim$ is a relation on $\mathscr{E}$. The triple $\langle X, \mathscr{E}, \succsim \rangle$ is a* structure of qualitative probability *iff for every $A, B, C \in \mathscr{E}$:*

1. *$\langle \mathscr{E}, \succsim \rangle$ is a weak order.*
2. *$X \succ \varnothing$ and $A \succsim \varnothing$.*
3. *Suppose that $A \cap B = A \cap C = \varnothing$. Then $B \succsim C$ iff $A \cup B \succsim A \cup C$.*

*Furthermore, the structure is* Archimedean *iff*

4. *For every $A > \varnothing$, any standard sequence relative to $A$ is finite.*

### 5.2.2 The Nonsufficiency of Qualitative Probability

The question we now ask and answer in the negative is: Does every (finite) structure of qualitative probability have a finitely additive, order-preserving probability representation? The following ingenious counterexample is due to Kraft, Pratt, and Seidenberg (1959). Let $X = \{a, b, c, d, e\}$ and let $\mathscr{E}$ be all subsets of $X$. Consider any order for which

$$\{a\} > \{b, c\}, \qquad \{c, d\} > \{a, b\} \quad \text{and} \quad \{b, e\} > \{a, c\}. \tag{1}$$

If an order-preserving representation $P$ exists, then

$$P(a) > P(b) + P(c), \qquad P(c) + P(d) > P(a) + P(b),$$
$$P(b) + P(e) > P(a) + P(c).$$

Adding these three inequalities and canceling $P(a) + P(b) + P(c)$, we obtain $P(d) + P(e) > P(a) + P(b) + P(c)$, whence $\{d, e\} > \{a, b, c\}$. Therefore, if we can construct an order that simultaneously satisfies inequality (1), $\{a, b, c\} > \{d, e\}$, and the axioms of qualitative probability, then we know that this order cannot possibly be represented by a numerical probability. The trick is to find a way to avoid the tedious task of exhaustively verifying Axiom 3 for the example.

Suppose we can find a probability measure $P$ on $X$ such that inequality (1) holds (and necessarily $\{d, e\} > \{a, b, c\}$) and such that all subsets $A$ of $X$ other than $\{d, e\}$ and $\{a, b, c\}$ are either strictly more probable than $\{d, e\}$ or strictly less probable than $\{a, b, c\}$, i.e., not[$P(\{d, e\}) \geqslant P(A) \geqslant P(\{a, b, c\})$]. Of course the ordering induced by $P$ automatically satisfies the Axioms of Definition 4. Now, change that order only to the extent of changing $\{d, e\} > \{a, b, c\}$ to $\{a, b, c\} > \{d, e\}$. Since no set lies between these two, their relations to all other sets are unaffected by the inversion, and so the new ordering, which cannot possibly have a probability representation, satisfies Definition 4. Therefore, we need only construct a measure $P$ for which inequality (1) holds and no event separates $\{d, e\}$ and $\{a, b, c\}$.

Let $0 < \epsilon < \frac{1}{3}$, and ignoring the normalizing factor $P(X) = 16 - 3\epsilon$, let

$$P(a) = 4 - \epsilon, \qquad P(b) = 1 - \epsilon, \qquad P(c) = 3 - \epsilon,$$
$$P(d) = 2, \qquad P(e) = 6.$$

Using additivity, it is easy to verify that inequality (1) holds, that $P(\{d, e\}) = 8$, and that $P(\{a, b, c\}) = 8 - 3\epsilon$. Since $3\epsilon < 1$, the only possibility for finding a distinct set $A$ such that $P(\{d, e\}) \geqslant P(A) \geqslant P(\{a, b, c\})$ is for $P(A)$ to be of the form $8 - i\epsilon$, $i = 0, 1, 2, 3$. However, the number 8 can be expressed as a sum of distinct elements from $\{1, 2, 3, 4, 6\}$ in only two ways, $1 + 3 + 4$ or $2 + 6$. These correspond to $\{a, b, c\}$ and $\{d, e\}$, respectively. Thus, no $A$ distinct from those two sets can have probability between 7 and 8 inclusive. This completes the example.

So further properties are needed in order to construct a representation.

### 5.2.3 Sufficient Conditions

Quite a number of conditions are known which, along with the properties of qualitative probability (Definition 4), are each sufficient to prove the existence of an additive probability representation. As would be expected, each condition postulates the existence of events with certain (strong) properties, and therefore limits the algebras of sets for which the theory establishes a representation. We present three conditions here which lead to unique, finitely additive measures, another in Section 5.4.2 which leads to a unique, countably additive measure, and a fifth in Section 5.4.3 which leads to a unique measure when $X$ is finite.

AXIOM 5″.  *Suppose $\langle X, \mathscr{E}, \succsim \rangle$ is a structure of qualitative probability. If $A, B \in \mathscr{E}$ and $A \succ B$, then there exists a partition $\{C_1, ..., C_n\}$ of $X$ such that, for $i = 1, ..., n$, $C_i \in \mathscr{E}$ and $A \succ B \cup C_i$.*

Savage (1954) pointed out that if $\mathscr{E}$ does not seem to fulfill Axiom 5″, one can always enlarge it to an algebra that does by adjoining all finite sequences of heads and tails generated by a coin of one's own choice. For any $n$, this leads to a partition into $2^n$ events, and, as he said,

> It seems to me that you could easily choose such a coin and choose $n$ sufficiently large so that you would continue to prefer to stake your gain on $A$, rather than on the union of $B$ and any particular sequence of $n$ heads and tails. For you to be able to do so, you need by no means consider every sequence of heads and tails equally probable (Savage, 1954, p. 38).

Savage established a reformulation of his axiom which is interesting and which we shall need again in Section 5.4.2. Suppose that $\langle X, \mathscr{E}, \succsim \rangle$ is a system of qualitative probability. It is called *fine* if, for every $A \succ \varnothing$, there exists a partition $\{C_1, ..., C_n\}$ of $X$ such that $A \succsim C_i$ for $i = 1, ..., n$. For $A, B$ in $\mathscr{E}$, define $A \sim^* B$ if for all $C, D \succ \varnothing$ such that $A \cap C = B \cap D = \varnothing$, then $A \cup C \succsim B$ and $B \cup D \succsim A$. The system is called *tight* if whenever $A \sim^* B$, then $A \sim B$.

**THEOREM 1.** *Suppose that $\langle X, \mathcal{E}, \succsim \rangle$ is a structure of qualitative probability. Axiom $5''$ is true iff the structure is both fine and tight.*

This is Savage's Theorem 4, p. 38.

Although Savage's axiom is, in the presence of the axioms for qualitative probability, actually stronger than the next condition, which was invoked both by de Finetti (1937) and Koopman (1940a,b), Savage felt that his was the easier to justify intuitively.

**AXIOM $5'$.** *Suppose that $\langle X, \mathcal{E}, \succsim \rangle$ is a structure of qualitative probability. For every positive integer $n$, there exists a partition $C_1, ..., C_n$ of $X$ such that, for $i, j = 1, ..., n$, $C_i \in \mathcal{E}$ and $C_i \sim C_j$.*

If $\mathcal{E}$ includes all finite sequences of heads and tails generated by what you believe to be independent tosses of a fair coin, i.e., a head is just as probable as a tail, then Axiom $5'$ must be fulfilled for you.

A common drawback of both axioms is that they force $X$ to be infinite—the illustrative coins suggest why. The following condition, which is due to Luce (1967), is fulfilled in many infinite structures and in some finite ones, although not in most.

**AXIOM 5.** *Suppose that $\langle X, \mathcal{E}, \succsim \rangle$ is a structure of qualitative probability. If $A, B, C, D \in \mathcal{E}$ are such that $A \cap B = \varnothing$, $A \succ C$, and $B \succsim D$, then there exist $C', D', E \in \mathcal{E}$ such that:*

(i)   $E \sim A \cup B$;

(ii)  $C' \cap D' = \varnothing$;

(iii) $E \supset C' \cup D'$;

(iv) $C' \sim C$ and $D' \sim D$.

It is difficult to say just what this means other than to restate it in words. If $A$ and $B$ are disjoint, if $A$ is more probable than $C$, and if $B$ is at least as probable as $D$, then the axiom asserts that somewhere in the structure there is an event $E$ that is equivalent in probability to $A \cup B$ such that $E$ includes disjoint subsets $C'$ and $D'$ that are equivalent, respectively, to $C$ and $D$. This axiom can be satisfied by some finite probability spaces, e.g., let $X = \{a, b, c, d\}$, let $P(a) = P(b) = P(c) = 0.2$, $P(d) = 0.4$, and $\mathcal{E}$ is all subsets. In some sense, however, most finite structures that have a probability representation violate the axiom, e.g., any structure whose equivalence classes fail to form a single standard sequence.

As was mentioned, Savage showed that, in the presence of the axioms for qualitative probability, Axiom $5''$ is strictly stronger than $5'$, and under the same conditions Luce (1967) showed that $5''$ is strictly stronger than 5. Fine (1971a,b) used a still weaker structural condition—the existence, for every $n$, of an $n$-fold "almost uniform" partition—in conjunction with the (necessary) condition that the order topology on $\mathscr{E}$ have a countable base. None of these axiom systems permits finite models, so none is weaker than Axioms 1–5. Fine (1971a) proved that all are strictly stronger than Axioms 1–5.

Assuming that Axioms 1–5 are weaker than the other systems that have been mentioned and that adding the (necessary) Archimedean condition to those of qualitative probability is not objectionable, the appropriate result to prove is the following.

THEOREM 2.  *Suppose that* $\langle X, \mathscr{E}, \succsim \rangle$ *is an Archimedean structure of qualitative probability for which Axiom 5 holds, then there exists a unique order-preserving function P from $\mathscr{E}$ into the unit interval* $[0, 1]$ *such that* $\langle X, \mathscr{E}, P \rangle$ *is a finitely additive probability space.*

The proof, which is given in Section 5.3.2, involves reducing this structure to one of extensive measurement. It is evident that the crucial property, if $A \cap B = \varnothing$, then $P(A \cup B) = P(A) + P(B)$, is formally similar to the corresponding one in extensive measurement provided that $\cup$ is treated as a concatenation operation $\circ$. Clearly the condition $A \cap B = \varnothing$ will have to be rephrased as a restriction on the concatenation operation of the form: $(A, B)$ in $\mathscr{B}$, where $\mathscr{B} \subset \mathscr{E} \times \mathscr{E}$.

In this connection, it should be noted that an ordering of an algebra of sets can be interpreted as a type of extensive measurement provided that the sets are collections of objects rather than events. Such an interpretation is especially natural for masses, in which case any collection of objects placed on one pan of a balance is an element of the algebra. Such an interpretation is less natural, though nonetheless possible, for length measurement. Observe that when we take one of the primitives to be an algebra of sets, we automatically assume the commutativity and associativity of concatenation, since both properties hold for unions of sets; these properties had to be listed more or less explicitly as axioms in the theories of Chapter 3. This is, of course, a general mathematical phenomena: the more structured the primitives, the weaker the axiomatic system need be in order to describe a given structure.

### 5.2.4  Preference Axioms for Qualitative Probability

If the ordering $\succsim$ on $\mathscr{E}$ is inferred from choices among acts, then the axioms of qualitative probability can be reformulated as assumptions concerning observable choices. For simplicity, we restrict attention to acts that have either two or three uncertain outcomes.

Let $\mathscr{E}$ be an algebra of sets on $X$ and $\mathscr{C}$ a set of outcomes or consequences; let $a_A \cup b_B \cup c_C$ be the act that has consequence $a$ if $A$ occurs, $b$ if $B$ occurs, and $c$ if $C$ occurs, where it is understood that $a$, $b$, $c$ are in $\mathscr{C}$, $A$, $B$, $C$ are in $\mathscr{E}$, and $A$, $B$, $C$ partition $X$. Denote the preference relation as $\succsim^*$. This is a weak ordering of both $\mathscr{C}$ and of the set $\mathcal{O}$ of all acts, i.e., formally, a subset of $(\mathscr{C} \times \mathscr{C}) \cup (\mathcal{O} \times \mathcal{O})$. We do not need to assume that a consequence and an act are compared.

We also identify acts that have the same event–outcome interpretation, e.g., permutations do not matter, $a_A \cup b_B \cup c_C = b_B \cup c_C \cup a_A$, and a three-outcome act with two identical outcomes reduces to a two-outcome act, $a_A \cup b_B \cup b_C = a_A \cup b_{B \cup C} = a_A \cup b_{-A}$.

We define a relation $\succsim$ on $\mathscr{E}$ as follows:

$A \succsim B$ iff for all $a, b \in \mathscr{C}$, if $a \succsim^* b$, then $a_A \cup b_{-A} \succsim^* a_B \cup b_{-B}$.

The following three axioms are sufficient (in the presence of the assumptions made informally above) to make $\langle X, \mathscr{E}, \succsim \rangle$ a structure of qualitative probability:

1.  *If $a \succ^* b$, $c \succsim^* d$, and $a_A \cup b_{-A} \succsim^* a_B \cup b_{-B}$, then*

$$c_A \cup d_{-A} \succsim^* c_B \cup d_{-B}.$$

2.  *If $a \succ^* b$, then $a_X \cup b_\phi \succ^* a_\phi \cup b_X$ and $a_A \cup b_{-A} \succsim^* a_\phi \cup b_X$.*

3.  *If $a_A \cup b_B \cup b_C \succsim^* a_{A'} \cup b_B \cup b_{C'}$, then*

$$a_A \cup a_B \cup b_C \succsim^* a_{A'} \cup a_B \cup b_{C'}.$$

We also need to assume that $\succsim^*$ is nontrivial on $\mathscr{C}$, i.e., there exist $a$, $b$ with $a \succ^* b$.

The first axiom implies that $\succsim$ is connected: for take $a \succ^* b$, then for any $A$, $B$, the connectedness of $\succsim^*$ yields some relation between $a_A \cup b_{-A}$ and $a_B \cup b_{-B}$, and this same relation then holds with $c$, $d$ substituted for $a$, $b$, provided that $c \succsim^* d$. Transitivity of $\succsim$ follows from transitivity of $\succsim^*$, so $\succsim$ is a weak order.

The second axiom obviously implies $X \succ \varnothing$ and $A \succsim \varnothing$.

The third axiom is based on the idea that if two acts have a common consequence $b$ when $B$ occurs, then the contingency $b_B$ is irrelevant to the decision between them, and so $b$ can just as well be replaced by $a$. The axiom only assumes this in a very restricted situation: when the other two outcomes are $b$ and $a$. The general principle, which this axiom exemplifies, is called the extended sure-thing principle. It leads to an expected utility representation (see Chapter 8, Savage's book, or Fishburn's book, referred to above). In this highly restricted form, however, it is already sufficient to yield Axiom 3 of qualitative probability: If $A$ is disjoint from $B \cup C$, then $B \succsim C$ if and only if $A \cup B \succsim A \cup C$.

The proof is fairly obvious: Letting $a \succ^* b$, $B \succsim C$ translates into

$$a_B \cup b_A \cup b_{-(A \cup B)} \succsim^* a_C \cup b_A \cup b_{-(A \cup C)}$$

while $A \cup B \gtrsim A \cup C$ translates into

$$a_B \cup a_A \cup b_{-(A \cup B)} \gtrsim^* a_C \cup a_A \cup b_{-(A \cup C)} .$$

Thus the axiom (which is equivalent to its own converse) gives the desired result.

Axioms 1 and 3 are highly interesting propositions about choices. If Axiom 1 is violated, there would seem to be some kind of special interaction between outcomes and events, e.g., such that even though $a >^* b$, $c \gtrsim^* d$, and $a_A \cup b_{-A} \gtrsim^* a_B \cup b_{-B}$, the combination $c_A$ is somehow much less valuable than $c_B$.

Violations of the extended sure-thing principle have been much discussed (see Chapter 8). From the present standpoint, the interesting thing to note is the light that Axiom 3 sheds on the principle of additivity of probabilities. If $a >^* b$ and $A \gtrsim A'$, then we must have

$$a_A \cup b_B \cup b_C \gtrsim^* a_{A'} \cup b_B \cup b_{C'} .$$

But if, somehow, $B$ adds more to $A'$ than it does to $A$, we could have the reversal,

$$a_{A'} \cup a_B \cup b_{C'} >^* a_A \cup a_B \cup b_C.$$

An example of a possible violation is offered by the following game (adapting an idea of Ellsberg, 1961). Consider three urns, containing 200 white balls, 200 black balls, and 100 red balls, respectively. One of the first two urns is selected by tossing a fair coin, and without informing the player which one was selected, white or black, its 200 balls are thoroughly mixed with the 100 red balls. Then a single ball is drawn from the mixture, and a prize is awarded or not depending on its color. Let $a$ denote a valuable prize and $b$ denote no prize. The player may prefer the payoff scheme

$$a_{\text{red}} \cup b_{\text{black}} \cup b_{\text{white}}$$

to the scheme

$$a_{\text{white}} \cup b_{\text{black}} \cup b_{\text{red}} .$$

In each case, the objective probability of winning the prize $a$ is exactly one-third. The difference is that in the first case, he knows there are one-third red balls, whereas, in the second case, depending on the coin toss, there may be no white balls at all.

Now suppose that we change $b$ to $a$ on black, i.e., compare

$$a_{\text{red}} \cup a_{\text{black}} \cup b_{\text{white}}$$

to

$$a_{\text{white}} \cup a_{\text{black}} \cup b_{\text{red}} .$$

The objective probability of winning $a$ is now two-thirds in each case, but the tables are turned. In the first scheme, the probability could be as low as one-third, depending on the coin toss, but in the second scheme, the probability is surely two-thirds. The point is that black adds a great deal more unambiguity to white than it does to red. Only if such considerations play no role—either because of the constancy of ambiguity over events or because of sophisticated decision-makers—may we expect the probability ordering inferred from decisions under uncertainty to lead to an additive representation.

Other discussions of the inference of a qualitative probability from decisions are found in Section 8.7.1 and in Anscombe and Aumann (1963), Davidson and Suppes (1956), Edwards (1962), Fishburn (1967g), Pratt, Raiffa, and Schlaifer (1964), Ramsey (1931), Savage (1954), and Suppes (1956).

## 5.3 PROOFS

### 5.3.1 Preliminary Lemmas

LEMMA 1. *If $\mathscr{E}$ is an algebra of sets on $X$ (Definition 1, Section 5.1), then $\varnothing$, $X \in \mathscr{E}$ and $\mathscr{E}$ is closed under set difference and intersection. If $\mathscr{E}$ is a $\sigma$-algebra, then it is closed under countable intersections.*

*PROOF.* This is a standard result, easily proved using the formulas $X = A \cup -A$, $A \cap B = -(-A \cup -B)$, etc.                                $\diamondsuit$

The common hypotheses of Lemmas 2–4 is that $\langle X, \mathscr{E}, \gtrsim \rangle$ is a structure of qualitative probability (Definition 4, Section 5.2.1) and that $A, B, C, D \in \mathscr{E}$.

LEMMA 2. *Suppose that $A \cap B = C \cap D = \varnothing$. If $A \gtrsim C$ and $B \gtrsim D$, then $A \cup B \gtrsim C \cup D$; moreover, if either hypothesis is $>$, then the conclusion is $>$.*

*PROOF.* Let $A' = A - D$, $D' = D - A$. Note that $A'$, $D'$ are in $\mathscr{E}$ (Lemma 1) and that

$$A' \cap B = A' \cap D = A \cap D' = C \cap D' = \varnothing,$$

and that $A' \cup D = A \cup D'$.
Using Axiom 3 twice,

$$A' \cup B \gtrsim A' \cup D = A \cup D' \gtrsim C \cup D'.$$

If either hypothesis is strict, then by the "if" part of Axiom 3,

$$A' \cup B > C \cup D'.$$

Observe that $A \cap D$ is disjoint from both $A' \cup B$ and $C \cup D'$; thus, by Axiom 3,

$$A \cup B = (A' \cup B) \cup (A \cap D) \gtrsim (C \cup D') \cup (A \cap D) = C \cup D;$$

and $>$ holds if $A' \cup B > C \cup D'$.                                      ◇

COROLLARY. *If* $A \cap B = C \cap D = \varnothing$, $A \sim C$, *and* $B \sim D$, *then* $A \cup B \sim C \cup D$.

LEMMA 3.  *If* $A \supset B$, *then* $A \gtrsim B$.

*PROOF.* Use Axiom 3 on $A - B \gtrsim \varnothing$ (Axiom 2), noting that $B$ is disjoint from $A - B$ and from $\varnothing$ and that $(A - B) \cup B = A$.                ◇

COROLLARY 1.  $X \gtrsim A$.

COROLLARY 2.  *If* $A \supset B$, *then* $A > B$ *iff* $A - B > \varnothing$.

LEMMA 4.  *If* $A \gtrsim B$, *then* $-B \gtrsim -A$.

*PROOF.* If $A \gtrsim B$, $-A > -B$, then by Lemma 2, $X > X$, contradicting Axiom 1.                                                          ◇

### 5.3.2  Theorem 2 (p. 208)

*If* $\langle X, \mathscr{E}, \gtrsim \rangle$ *is an Archimedean structure of qualitative probability for which Axiom 5 is true, then it has a unique, finitely additive probability representation.*

*PROOF.* Let **A** denote the equivalence class that includes $A$, and let $\mathscr{E}$ be the set of all equivalence classes, excluding $\varnothing$. By Axiom 2, $\mathbf{X} \in \mathscr{E}$. If $\mathbf{X}$ is the only element of $\mathscr{E}$, then we let $P(A) = 0$ if $A \sim \varnothing$, $P(A) = 1$ if $A \sim X$, and $P$ fulfills the assertions of the theorem (note that $\varnothing$ is closed under disjoint union, by Lemma 2). We assume, therefore, that $\mathscr{E}$ contains $\mathbf{A} \neq \mathbf{X}$.

We construct an extensive structure on $\mathscr{E}$ letting

$$\mathscr{B} = \{(\mathbf{A}, \mathbf{B}) \mid A > \varnothing, B > \varnothing, \text{ and there exist}$$
$$A' \in \mathbf{A}, \quad B' \in \mathbf{B} \quad \text{with} \quad A' \cap B' = \varnothing\}.$$

Thus $\mathscr{B}$ is nonempty because if $\mathbf{A} \neq$ both $\mathbf{X}$ and $\varnothing$, then $-A > \varnothing$ (Lemma 4), so the pair $(\mathbf{A}, -\mathbf{A}) \in \mathscr{B}$. We define ∘ on $\mathbf{B}$ by letting $\mathbf{A} \circ \mathbf{B} =$

$A \cup B$, if $A \cap B = \varnothing$. By the corollary of Lemma 2, $\circ$ is well defined. We let $\succsim$ be the induced simple order on $\mathscr{E}$.

We shall now prove that $\langle \mathscr{E}, \succsim, \mathscr{B}, \circ \rangle$ is an extensive structure with no essential maximum (Definition 3.3, Section 3.4.3). The six axioms are established in corresponding numbered paragraphs. It is convenient in some places to use boldface Greek capitals for elements of $\mathscr{E}$.

1. $\succsim$ is a weak order; in fact, it is a simple order.

5. Positivity follows from Corollary 2 of Lemma 3.

2. Associativity. Suppose that $(\mathbf{\Gamma}, \mathbf{\Delta})$, $(\mathbf{\Gamma} \circ \mathbf{\Delta}, \mathbf{\Lambda}) \in \mathscr{B}$. The essence of the proof is to construct $A, B, C$ pairwise disjoint such that $\mathbf{A} = \mathbf{\Gamma}$, $\mathbf{B} = \mathbf{\Delta}$, $\mathbf{C} = \mathbf{\Lambda}$. For then, the rest will follow via the associativity of the union operation. The construction is as follows. Take disjoint sets $D' \in \mathbf{\Gamma} \circ \mathbf{\Delta}$, $C' \in \mathbf{\Lambda}$. By positivity $\mathbf{D'} \succ \mathbf{\Delta}$, thus, by Axiom 5 we can take $E \sim D' \cup C'$ and $B, C$ disjoint with $E \supset B \cup C$ and $\mathbf{B} \in \mathbf{\Delta}$, $\mathbf{C} \in \mathbf{\Lambda}$. Then let $A = E - (B \cup C)$. It remains only to show that $A \in \mathbf{\Gamma}$; but since $(\mathbf{A} \circ \mathbf{\Delta}) \circ \mathbf{\Lambda} = \mathbf{E} = (\mathbf{\Gamma} \circ \mathbf{\Delta}) \circ \mathbf{\Lambda}$, this follows from Lemma 2, applied twice.

3. We need to show that if $(\mathbf{A}, \mathbf{C}) \in \mathscr{B}$ and $\mathbf{A} \succsim \mathbf{B}$, then $(\mathbf{C}, \mathbf{B}) \in \mathscr{B}$; the rest follows by the obvious commutativity of $\circ$ and by Lemma 2. If $\mathbf{A} = \mathbf{B}$, there is nothing to show; if $\mathbf{A} \succ \mathbf{B}$, apply Axiom 5.

4. Solvability. If $\mathbf{A} \succ \mathbf{B}$, apply Axiom 5 to $A > B$ and $\varnothing \sim \varnothing$ to obtain $A', B'$ with $A' \supset B'$, $A \sim A'$, $B \sim B'$, and let $C = A' - B'$; then $\mathbf{A} = \mathbf{B} \circ \mathbf{C}$.

6. Finally, we show that $\{n \mid \mathbf{B} \succ n\mathbf{A}\}$ is finite, for $A \in \mathscr{E}$, where $1\mathbf{A} = \mathbf{A}$, $n\mathbf{A} = (n - 1)\,\mathbf{A} \circ \mathbf{A}$. We do this by showing that the existence of $n\mathbf{A}$ implies the existence of an $n$-term standard sequence relative to $A > \varnothing$. For suppose that $n\mathbf{A}$ exists and that we have a $k$-term standard sequence relative to $A$, $A_1, \ldots, A_k$, with $k < n$, and with $A_i \in i\mathbf{A}$ for $i \leqslant k$. Since $(k\mathbf{A}, \mathbf{A}) \in \mathscr{B}$, we have $B_k \in k\mathbf{A}$ and $C_k \in \mathbf{A}$ such that $B_k \cap C_k = \varnothing$. Let $A_{k+1} = B_k \cup C_k$. Then clearly, $A_1, \ldots, A_{k+1}$ is a $(k + 1)$-term standard sequence relative to $A$, and $A_{k+1} \in (k + 1)\,\mathbf{A}$. Since a 1-term standard sequence can be obtained by $A_1 = A$, we have the required result by induction.

Now, by Theorem 3.3 (Section 3.4.3), there is a positive-valued ratio scale $\phi$ on $\mathscr{E}$ such that

$$\mathbf{A} \succsim \mathbf{B} \quad \text{iff} \quad \phi(\mathbf{A}) \geqslant \phi(\mathbf{B}),$$

and

$$\text{for} \quad (\mathbf{A}, \mathbf{B}) \in \mathscr{B}, \quad \phi(\mathbf{A} \circ \mathbf{B}) = \phi(\mathbf{A}) + \phi(\mathbf{B}).$$

Choose the unit so $\phi(\mathbf{X}) = 1$. For $A \in \mathscr{E}$, let

$$P(A) = \begin{cases} \phi(\mathbf{A}), & \text{if } A > \varnothing, \\ 0, & \text{if } A \sim \varnothing. \end{cases}$$

It is easy to see that $P$ fulfills the assertions of the theorem. Moreover, $P$ is unique, since if another such function $P'$ existed, then $\phi'(A) = P'(A)$ would be a representation of $\langle \mathscr{E}, \succsim, \mathscr{B}, \circ \rangle$. Since $\phi' = \alpha\phi$ and $\phi'(X) = \phi(X) = 1$, $\phi' = \phi$ and $P' = P$.                                                    $\diamondsuit$

## 5.4 MODIFICATIONS OF THE AXIOM SYSTEM

### 5.4.1 QM-Algebra of Sets

The notions of event and probability given in Definitions 1 and 2 have proved satisfactory for almost all scientific purposes. The one outstanding exception is quantum mechanics. In that theory both $P(A)$ and $P(B)$ may exist and yet $P(A \cap B)$ need not. For example, suppose that $A$ is the event that, at some time $t$, a certain elementary particle is located in some region $\alpha$ of space and $B$ is the event that, at the same time $t$, this same particle has a momentum whose value lies in some interval $\beta$. Then the event $A \cap B$ means that, at time $t$, the particle is both in the region $\alpha$ and has a momentum in $\beta$. One basic feature of quantum mechanics is that we may be able to observe whether event $A$ occurred or whether event $B$ occurred, but not necessarily be able to observe whether both events $A$ and $B$, i.e., event $A \cap B$, occurred. In the theory this means that no probability is assignable to $A \cap B$ even when $P(A)$ and $P(B)$ are specified.

Such an incompatibility between two very fundamental sets of ideas must be resolved, presumably by modifying probability theory in some way. Those interested in a full understanding of the physical issues involved and in some of the proposed modifications of both classical logic and probability theory should consult Birkhoff and von Neumann (1936), Kochen and Specker (1965), Reichenbach (1944), Suppes (1961, 1963, 1965, 1966), and Varadarajan (1962). Suffice it to say that the simplest proposal (Suppes, 1966) is to restrict further our notion of an algebra of sets (Definition 1) to:

DEFINITION 5. *Suppose that $X$ is a nonempty set and that $\mathscr{E}$ is a nonempty family of subsets of $X$. Then $\mathscr{E}$ is a* QM-algebra *of sets on $X$ iff, for every $A, B \in \mathscr{E}$:*

1. *$-A \in \mathscr{E}$;*
2. *If $A \cap B = \varnothing$, then $A \cup B \in \mathscr{E}$.*

*Furthermore, if $\mathscr{E}$ is closed under countable unions of mutually disjoint sets, then $\mathscr{E}$ is called a* QM $\sigma$-algebra .

Since the definition is still stated in terms of unions, and the difficulty seemed to be about intersections, it is not clear that we have coped with the problem; however, as is shown in Lemma 5 (Section 5.5.1), all is well in the sense that when $A$, $B$ are in $\mathscr{E}$, $A \cap B$ is in $\mathscr{E}$ if and only if $A \cup B$ is in $\mathscr{E}$, and if $A \supset B$, then $A - B$ is in $\mathscr{E}$.

As a simple example, if $\langle X, \mathscr{E}, P \rangle$ is a finitely additive probability space, then for any $A$ in $\mathscr{E}$, the set of all events probabilistically independent of $A$ is a QM-algebra of sets on $X$; see Section 5.8.

What is not clear about this suggestion is just how badly probability theory suffers. For one thing, how many of the important theorems of probability theory cannot now be proved? This is not the place to enter into that discussion. For another, to what extent have we impaired our ability to construct a numerical probability representation from a qualitative ordering? If one carefully reexamines Axioms 1–4 of Definition 4 and Axiom 5 of Section 5.2.3, the proofs of the lemmas used in the proof of Theorem 2, and the proof of that theorem, then one finds that, with the exception of Lemmas 1 and 2, all unions are of disjoint sets and that all differences are of one set that includes another. As we mentioned above, Lemma 5 (Section 5.5.1) replaces Lemma 1. As for Lemma 2, we note that only disjoint unions enter into its statement, that it is a necessary condition for the representation (the argument for necessity is very similar to the argument for Axiom 3), and that it is strictly stronger than Axiom 3. Therefore, we adopt it as Axiom 3′, in place of Axiom 3.

AXIOM 3′. *Suppose that* $A \cap B = C \cap D = \varnothing$. *If* $A \gtrsim C$ *and* $B \gtrsim D$, *then* $A \cup B \gtrsim C \cup D$; *moreover, if either hypothesis is* $\succ$, *then the conclusion is* $\succ$.

We shall encounter this axiom again in the treatment of qualitative conditional probability (Sections 5.6 and 5.7). Except for the proof of Lemma 5, we have established the following:

THEOREM 3. *If* $\mathscr{E}$ *is a* QM-*algebra of sets on* $X$ *and if* $\langle X, \mathscr{E}, \gtrsim \rangle$ *satisfies Axioms* 1, 2, 3′, 4, *and* 5, *then there is a unique function* $P$ *on* $\mathscr{E}$ *that satisfies the Kolmogorov axioms (Definition 2) and preserves the order* $\gtrsim$ *on* $\mathscr{E}$.

### 5.4.2 Countable Additivity

Theorem 2, and others like it, establish conditions under which finite additivity hold, but nothing has been said about countable additivity (see Definition 2), which is needed in many parts of probability theory. Countably additive representations of qualitative probability structures have been studied by Villegas (1964, 1967) and Fine (1971a). The main result (from our

point of view) is embodied in the following definition and theorem. The
definition follows Villegas (1964).

DEFINITION 6. *Suppose that* $\langle X, \mathscr{E}, \succsim \rangle$ *is a structure of qualitative
probability and that $\mathscr{E}$ is a $\sigma$-algebra. We say that $\succsim$ is monotonically continuous on $\mathscr{E}$ iff for any sequence $A_1, A_2, \ldots$ in $\mathscr{E}$ and any B in $\mathscr{E}$, if $A_i \subset A_{i+1}$
and $B \succsim A_i$ for all i, then $B \succsim \bigcup_{i=1}^{\infty} A_i$.*

THEOREM 4. *A finitely additive probability representation of a structure of
qualitative probability, on a $\sigma$-algebra, is countably additive iff the structure is
monotonically continuous.*

This theorem is proved in Section 5.5.2. An immediate corollary is that
Theorem 2 gives a set of sufficient conditions for a countably additive
probability representation for a binary relation $\succsim$ on a $\sigma$-algebra: Axioms 1–5
and monotone continuity.
   Given Axioms 1–3 (qualitative probability) and monotone continuity,
Axioms 4 and 5 can be dispensed with if a different structural condition is
used. To formulate the condition, we need a qualitative definition of a
probability *atom*. The numerical definition is that $A$ is an atom if $P(A) > 0$
and if $A \supset B$, then $P(B) = P(A)$ or $P(B) = 0$. Corresponding to this, we have:

DEFINITION 7. *Let $\succsim$ be a weak ordering of an algebra of sets $\mathscr{E}$. An
event $A \in \mathscr{E}$ is an atom iff $A > \varnothing$ and for any $B \in \mathscr{E}$, if $A \supset B$, then $A \sim B$
or $B \sim \varnothing$.*

The structural condition which replaces Axioms 4 and 5 is that $\mathscr{E}$ has no
atoms. Recall the definitions of *fine* and *tight* (Section 5.2.3).

THEOREM 5. *Suppose that $\langle X, \mathscr{E}, \succsim \rangle$ is a structure of qualitative
probability, $\mathscr{E}$ is a $\sigma$-algebra, $\succsim$ is monotonically continuous on $\mathscr{E}$, and there are
no atoms. Then the structure is both fine and tight, there is a unique order-
preserving probability, and it is countably additive.*

For the proof of this and other related results, we refer to Villegas (1964).
This structural condition turns out, of course, to be a special case of Axiom 5
(given $\sigma$-algebras and monotone continuity), since a structure of qualitative
probability that is both fine and tight satisfies Axiom 5 (see Section 5.2.3).

### 5.4.3 Finite Probability Structures with Equivalent Atoms

As in the chapters on extensive measurement and difference measurement,
it is possible here to give a separate and elementary axiomatization of a

single finite standard sequence. To do so, we replace the structural restriction (Axiom 5) by some sort of equal-spacing axiom.

In the case of probability structures, it is impossible for a single standard sequence to exhaust $\mathscr{E} - \{\varnothing\}$. For if $A_1, A_2, ..., A_n$ is a standard sequence, then by Definition 3, we have $A_2 \sim B_1 \cup C_1$, where $B_1, C_1$ are disjoint and both are $\sim A_1$; thus, either $B_1$ or $C_1$ must be outside the sequence. The most that one can expect is for a single standard sequence to exhaust the set of nonnull equivalence classes $\mathscr{E}$. However, there are a great variety of structures of $\mathscr{E}$ compatible with a single standard sequence in $\mathscr{E}$, e.g., any finite structure satisfying Axiom 5. Perhaps the simplest structure is one with a finite number of equivalent atoms—corresponding to a uniform distribution on a finite set of categories. This structure can be characterized by the following axiom:

AXIOM 6.    *If $A \gtrsim B$, then there exists $C \in \mathscr{E}$ such that $A \sim B \cup C$.*
This axiom, and the following theorem, are due to Suppes (1969, pp. 5–8).

THEOREM 6.    *Suppose that $\langle X, \mathscr{E}, \gtrsim \rangle$ is a structure of qualitative probability (Axioms 1–3 of Definition 4) such that $X$ is finite and Axiom 6 holds. Then there exists a unique order-preserving function $P$ on $\mathscr{E}$ such that $\langle X, \mathscr{E}, P \rangle$ is a finitely additive probability space. Moreover, all atoms in $\mathscr{E}$ are equivalent.*

The proof is found in Section 5.5.3.

Since any event in a finite probability space is a union of disjoint atoms and a null event, the probabilities are all multiples of the atomic probability.

## 5.5 PROOFS

### 5.5.1 Structure of QM-Algebras of Sets

LEMMA 5.    *Suppose that $\mathscr{E}$ is a QM-algebra of sets on $X$ (Definition 5), then for all $A, B \in \mathscr{E}$:*

(i)   *$\varnothing$ and $X \in \mathscr{E}$;*

(ii)  *if $A \supset B$, then $A - B \in \mathscr{E}$;*

(iii) *$A \cup B \in \mathscr{E}$ iff $A \cap B \in \mathscr{E}$.*

PROOF.    (i)    The proof is the same as Lemma 1, since $A \cap -A = \varnothing$.

(ii)    Since $A \supset B$ implies $-A \cap B = \varnothing$, and since $-A \in \mathscr{E}$, it follows that $-A \cup B \in \mathscr{E}$. So $A - B = -(-A \cup B) \in \mathscr{E}$.

(iii)    If $A \cup B \in \mathscr{E}$, then by part (ii), $A - B = (A \cup B) - B$ and

$B - A = (A \cup B) - A \in \mathscr{E}$. Thus, the symmetric difference,

$$(A - B) \cup (B - A) \in \mathscr{E};$$

and again, by part (ii), the intersection, which is the union minus the symmetric difference, is in $\mathscr{E}$. Conversely, if the intersection is in $\mathscr{E}$, then $A - B = A - (A \cap B)$ and $B - A = B - (A \cap B) \in \mathscr{E}$ by part (ii), so $A \cup B$, which is the disjoint union of $A \cap B$, $A - B$, and $B - A$, is in $\mathscr{E}$. $\qquad\qquad \diamondsuit$

### 5.5.2 Theorem 4 (p. 216)

*Suppose that a structure of qualitative probability $\langle X, \mathscr{E}, \gtrsim \rangle$ has an order-preserving probability measure $P$ and that $\mathscr{E}$ is a $\sigma$-algebra. Then $\langle X, \mathscr{E}, P \rangle$ is countably additive iff $\gtrsim$ is monotonically continuous on $\mathscr{E}$.*

PROOF. If $P$ is countably additive, let $\{A_i\}$ satisfy $A_i \subset A_{i+1}$ and $A_i \lesssim B$. By countable additivity,

$$P\left(\bigcup_{i=1}^{\infty} A_i\right) = \sum_{i=1}^{\infty} P(A_{i+1} - A_i).$$

Since the partial sums of the right-hand expression are $P(A_{i+1})$, they are bounded by $P(B)$, and so $P(\bigcup_{i=1}^{\infty} A_i) \leqslant P(B)$, implying $B \gtrsim \bigcup_{i=1}^{\infty} A_i$.

For the converse, note first that by finite additivity alone, if $\{A_i\}$ are pairwise disjoint, then

$$\sum_{i=1}^{\infty} P(A_i) \leqslant P\left(\bigcup_{i=1}^{\infty} A_i\right).$$

This is true because each partial sum on the left is $P(\bigcup_{i=1}^{n} A_i) \leqslant P(\bigcup_{i=1}^{\infty} A_i)$. Suppose that for some disjoint union, strict inequality holds, in violation of countable additivity, i.e., let

$$P\left(\bigcup_{i=1}^{\infty} A_i\right) - \sum_{i=1}^{\infty} P(A_i) = \epsilon > 0.$$

We consider two cases. First, suppose that for some $A_m$ in the sequence,

$$\epsilon \geqslant P(A_m) > 0.$$

Let

$$B_k = \bigcup_{i=1}^{k} A_i, \qquad B = \left(\bigcup_{i=1}^{\infty} A_i\right) - A_m.$$

For each $k$, $B_k \subset B_{k+1}$, and

$$P(B_k) \leqslant \sum_{i=1}^{\infty} P(A_i)$$

$$= P\left(\bigcup_{i=1}^{\infty} A_i\right) - \epsilon$$

$$\leqslant P\left(\bigcup_{i=1}^{\infty} A_i\right) - P(A_m)$$

$$= P(B).$$

Thus, $B \gtrsim B_k$. But since $A_m > \varnothing$, we have

$$\bigcup_{k=1}^{\infty} B_k = \bigcup_{i=1}^{\infty} A_i = B \cup A_m > B.$$

Thus, monotone continuity is contradicted.

The other case is where no such $A_m$ exists. Then clearly, $P(A_i) = 0$ for all but finitely many $i$. Let $J$ be the subset of integers $i$ for which $P(A_i) = 0$, and let

$$A = \bigcup_{i \in J} A_i.$$

By finite additivity,

$$P\left[\left(\bigcup_{i=1}^{\infty} A_i\right) - A\right] = P\left(\bigcup_{i \notin J} A_i\right)$$

$$= \sum_{i \notin J} P(A_i)$$

$$= \sum_{i=1}^{\infty} P(A_i)$$

$$= P\left(\bigcup_{i=1}^{\infty} A_i\right) - \epsilon.$$

Therefore, $P(A) = \epsilon$. Let

$$C_k = \bigcup_{\substack{i \in J \\ i \leqslant k}} A_i.$$

Then $P(C_k) = 0$ by finite additivity; so we have

$$C_k \subset C_{k+1}, \qquad \varnothing \gtrsim C_k,$$

but

$$\bigcup_{k=1}^{\infty} C_k = A \succ \varnothing,$$

violating monotone continuity.

### 5.5.3  Theorem 6  (p. 217)

*A structure $\langle X, \mathscr{E}, \succsim \rangle$ of qualitative probability for which $X$ is finite and which satisfies Axiom 6 (Section 5.4.3) has a unique probability representation; moreover, all atoms are equivalent.*

*PROOF.*  Define a new structure $\langle X', \mathscr{E}', \succsim' \rangle$ in which the nonempty null events are eliminated. (If $N$ is the union of all null events, $N \sim \varnothing$; put $X' = X - N$, $\mathscr{E}' = \{A - N \mid A \in \mathscr{E}\}$, $A - N \succsim' B - N$ iff $A \succsim B$.) It is easy to show that $\langle X', \mathscr{E}', \succsim' \rangle$ also satisfies the hypotheses of the theorem; moreover, any nonempty event in $\mathscr{E}'$ is expressible in a unique way as the union of finitely many pairwise disjoint atoms.

Let $A_1$ be a minimal atom with respect to $\succsim'$ (such exists by the finiteness of $X$) and let $\{A_1,..., A_n\}$ be the set of distinct atoms equivalent to $A_1$. We show there are no other atoms in $\mathscr{E}'$. If there are others, let $A$ be minimal among the atoms $\succ' A_1$. Define $B$ and $C$ by

$$B = A \cup (A_2 \cup \cdots \cup A_n),$$
$$C = A_1 \cup (A_2 \cup \cdots \cup A_n).$$

(If $n = 1$, $B = A$, $C = A_1$.) By Axiom 3, $B \succ' C$. By Axiom 6, there exists $D \in \mathscr{E}'$ such that $B \sim' C \cup D$. With no loss in generality, take $D$ disjoint from $C$. Since $D$ contains no minimal atoms, $D \succsim' A$; but then

$$B \sim' C \cup D \succsim' C \cup A = B \cup A_1 \succ' B,$$

a contradiction. Therefore $\{A_1,..., A_n\}$ is the set of all atoms.

Now construct $\mu$ on $\mathscr{E}$ by letting $\mu(A)$ be the cardinality of the set

$$\{A_i \mid A_i \subset A - N\}.$$

Let $P(A) = \mu(A)/n$. It is easy to verify that $P$ is an order-preserving probability and that $P$ is unique.                                          $\diamondsuit$

## 5.6  A REPRESENTATION BY CONDITIONAL PROBABILITY

Most work in the theory of probability requires the important defined notion of conditional probability. Intuitively, $P(A \mid B)$ is the probability of the event $A$ when we have the added information that the outcome is one of the elements of the event $B$—in other words, when we know that the event $B$ occurred. Thus, the outcome must be in both $A$ and $B$, that is, in $A \cap B$. This immediately suggests $P(A \cap B) = P(A \mid B)$; however, this cannot be correct since, given $B$, either $A$ or $-A$ must certainly occur so $P(A \mid B) + P(-A \mid B) = 1$, whereas, $P(A \cap B) + P(-A \cap B) = P(B)$. However, if we set

$$P(A \mid B) = P(A \cap B)/P(B), \tag{2}$$

all is well, provided that $P(B) > 0$. This is the accepted definition of *conditional probability.*

Because this concept is so important and because ordinary probability is the special case of it in which $B = X$, several authors have treated it as the basic notion to be axiomatized and have given generalizations of Kolmogorov's axioms (Definition 2). Copeland (1941, 1956) and Rényi (1955) are of particular interest. Császár (1955) studied conditions under which a real-valued function of two set-valued arguments, which is what conditional probability is, can be expressed in the quotient form of Equation (2) in terms of a function of one set-valued argument, which is what unconditional probability is.

Our concerns are somewhat different. We wish to know conditions under which a qualitative relation of the form $A \mid B \gtrsim C \mid D$, meaning "$A$ given $B$ is qualitatively at least as probable as $C$ given $D$," can be represented by a probability measure of the form of Equation (2) in the sense that

$$A \mid B \gtrsim C \mid D \qquad \text{iff} \qquad P(A \cap B)/P(B) \geqslant P(C \cap D)/P(D).$$

To our knowledge, the only published attempts to do this are Koopman (1940 a,b), Aczél (1961, 1966, p. 319), Luce (1968), and Domotor (1969). Domoter gave necessary and sufficient conditions in the finite case similar to the axiom systems in Chapter 9. The other three systems have much in common. Koopman treated $\varnothing$ as the only null event and did not assume that every pair $A \mid B$ and $C \mid D$ are necessarily comparable; whereas Aczél and Luce admitted other null events and did require comparability of all pairs. The lack of comparability forced Koopman to introduce some extra axioms. Aczél and Luce postulated an additivity requirement, similar to Axiom 3 of Definition 4; whereas, Koopman invoked the property that if $A \mid B \gtrsim C \mid D$, then $-C \mid D \gtrsim -A \mid B$, to serve an analogous function. The major differences lie in the choice of sufficient conditions and in the proofs. Aczél postulated a real-valued function on the pairs $A \mid B$, which gives the ordering, and imposed continuity conditions. His proof used results in functional equations. Koopman used a kind of uniform-partition postulate and constructed the probability function by a limiting process. Luce, whose necessary conditions were essentially the same as those of Aczél, reduced the result to known results in measurement theory, using a solvability axiom. He also used a functional-equation argument. Here, we present a modification of Luce's system.

In addition to the above studies, Copeland (1956) stated a system of axioms that is a good deal more transparent than Koopman's, but he did not construct a numerical representation. Moreover, he assumed that his basic system of elements is closed under the conditioning operation, which is to say, if $A$ and $B$ are events, where $B$ is nonnull (see below), then $A \mid B$ is also treated as an event. This seems a trifle odd intuitively, although, as a matter of fact, our structural restriction entails that every $A \mid B$ is equivalent in probability to some event.

### 5.6.1 Necessary Conditions: Qualitative Conditional Probability

At first glance, one might expect to begin with an algebra of sets $\mathscr{E}$ and a relation $\succsim$ on $\mathscr{E} \times \mathscr{E}$; however, the proviso that $P(B) > 0$ in Equation (2) makes it clear that matters are not quite this simple. Events with probability 0—null events—must be excluded from the conditioning position. Such events form a subset $\mathscr{N}$ of $\mathscr{E}$, and the relation $\succsim$ is on $\mathscr{E} \times (\mathscr{E} - \mathscr{N})$. We use the suggestive notion $A \mid B$ for a typical element of $\mathscr{E} \times (\mathscr{E} - \mathscr{N})$. Since we are familiar with the general procedure for finding necessary conditions, we first state the axioms in the form of a definition and then discuss those that are novel to this context.

DEFINITION 8. *Suppose that $X$ is a nonempty set, $\mathscr{E}$ is an algebra of sets on $X$, $\mathscr{N}$ is a subset of $\mathscr{E}$, and $\succsim$ is a binary relation on $\mathscr{E} \times (\mathscr{E} - \mathscr{N})$. The quadruple $\langle X, \mathscr{E}, \mathscr{N}, \succsim \rangle$ is a* structure of qualitative conditional probability *iff for every $A$, $B$, $C$, $A'$, $B'$, and $C' \in \mathscr{E}$ (or $\in \mathscr{E} - \mathscr{N}$, whenever the symbol appears to the right of $\mid$) the following six axioms hold:*

1. $\langle \mathscr{E} \times (\mathscr{E} - \mathscr{N}), \succsim \rangle$ *is a weak order.*

2. $X \in \mathscr{E} - \mathscr{N}$; *and $A \in \mathscr{N}$ iff $A \mid X \sim \varnothing \mid X$.*

3. $X \mid X \sim A \mid A$ *and $X \mid X \succsim A \mid B$.*

4. $A \mid B \sim A \cap B \mid B$.

5. *Suppose that $A \cap B = A' \cap B' = \varnothing$. If $A \mid C \succsim A' \mid C'$ and $B \mid C \succsim B' \mid C'$, then $A \cup B \mid C \succsim A' \cup B' \mid C'$; moreover, if either hypothesis is $\succ$, then the conclusion is $\succ$.*

6. *Suppose that $A \supset B \supset C$ and $A' \supset B' \supset C'$. If $B \mid A \succsim C' \mid B'$ and $C \mid B \succsim B' \mid A'$, then $C \mid A \succsim C' \mid A'$; moreover, if either hypothesis is $\succ$, then the conclusion is $\succ$.*

*The structure is* Archimedean *provided that:*

7.  *Every standard sequence is finite, where* $\{A_i\}$ *is a* standard sequence *iff for all i,* $A_i \in \mathcal{E} - \mathcal{N}$, $A_{i+1} \supset A_i$, *and* $X \mid X \succ A_i \mid A_{i+1} \sim A_1 \mid A_2$.

Of course, as in all our axiom systems, the choice of necessary axioms depends somewhat on the nature of the nonnecessary structural restrictions. In the case of Definition 4 (qualitative probability), the three axioms are extremely natural and have long been accepted as the definition of qualitative probability. We added only the Archimedean axiom and a structural restriction. Here, we must be somewhat more tentative about the choice of axioms. Lemmas 9–12 (Section 5.7.1) and their corollaries are also necessary conditions; they are derived from Axioms 1–6 only with the help of the structural axiom introduced below. Some of these properties could well be included in the concept of qualitative conditional probability. Indeed, in Section 5.6.3 we shall relabel Lemma 12 as Axiom 6' and use it instead of Axiom 6 in the development of a nonadditive conditional probability representation (Section 5.6.4).

Some properties, which it would seem natural to include in qualitative conditional probability, follow from Axioms 1–6 without any structural restriction. For example, Axiom 3 could be expanded into

$$X \mid X \succsim A \mid B \succsim \varnothing \mid X,$$

but this can be proved (Corollary 1 to Lemma 6).

The proof that Axioms 1–5 are necessary for the desired representation is quite easy. Note, for example, that $P(A \mid B) = P(A \cap B \mid B)$, which yields Axiom 4. The derivation of Axiom 5 is similar to that of its analog, Axiom 3 of Definition 4.

To show that Axiom 6 is necessary, note that if $B \mid A \succsim C' \mid B'$, with $A \supset B$ and $B' \supset C'$, then

$$P(B)/P(A) \geqslant P(C')/P(B').$$

Similarly, $C \mid B \succsim B' \mid A'$, $B \supset C$, and $A' \supset B'$ yield

$$P(C)/P(B) \geqslant P(B')/P(A').$$

Since the $P$-values are nonnegative, these inequalities can be multiplied to yield

$$P(C)/P(A) \geqslant P(C')/P(A'),$$

or $C \mid A \succsim C' \mid A'$, as required. Moreover, we know that $P(B)$ and $P(B')$ are positive, hence, by the second inequality, $P(C)$ is positive. Thus, if either inequality is strict, multiplication preserves the strict inequality.

The Archimedean property, 7, is also derived by multiplying fractions: Since $P(A_n) \leqslant 1$ and $P(A_1) > 0$,

$$0 < P(A_1) \leqslant \frac{P(A_1)}{P(A_n)}$$

$$= \frac{P(A_1)}{P(A_2)} \frac{P(A_2)}{P(A_3)} \cdots \frac{P(A_{n-1})}{P(A_n)}$$

$$= \left[ \frac{P(A_1)}{P(A_2)} \right]^{n-1}.$$

Since $X \mid X \succ A_1 \mid A_2$, it follows that $P(A_1)/P(A_2) < 1$. Thus, the inequality $P(A_1) \leqslant [P(A_1)/P(A_2)]^{n-1}$ can hold for only finitely many $n$.

See also the further discussion of Axiom 6 (as contrasted with Axiom 6′) in Sections 5.6.3 and 5.6.4 (also, following Lemma 12 in Section 5.7.1) and of Axiom 7 in Section 5.7.2.

### 5.6.2 Sufficient Conditions

Continuing the numbering of Definition 8, we add the nonnecessary property:

AXIOM 8.    *If $A \mid B \succsim C \mid D$, then there exists $C'$ in $\mathscr{E}$ such that $C \cap D \subset C'$ and $A \mid B \sim C' \mid D$.*

This simply says that $\mathscr{E}$ is sufficiently rich so that whenever $A \mid B \succ C \mid D$, we can add just enough to $C$ to make $C' \mid D$ equivalent to $A \mid B$. This is a rather strong structural restriction. In particular, Axioms 1–6 and 8 yield Axiom 5 of Section 5.2.3 (see 5.7.2), and moreover, it can be shown that, apart from trivial cases, Axiom 5′ of Section 5.2.3 holds. That is, $X$ can be partitioned into arbitrarily fine equivalent events. Koopman's system included the nonnecessary postulate that for every positive integer $n$ there is a nonnull event that can be partitioned into $n$ equiprobable events. It would, of course, be desirable to replace Axiom 8 by something weaker, even if additional necessary axioms had to be added.

THEOREM 7.    *Suppose that $\langle X, \mathscr{E}, \mathscr{N}, \succsim \rangle$ is an Archimedean structure of qualitative conditional probability (Definition 8) for which Axiom 8 holds. Then there exists a unique real-valued function $P$ on $\mathscr{E}$, such that for all $A, C \in \mathscr{E}$ and all $B, D \in \mathscr{E} - \mathscr{N}$:*

(i)    *$\langle X, \mathscr{E}, P \rangle$ is a finitely additive probability space;*

(ii)    *$A \in \mathscr{N}$ iff $P(A) = 0$;*

(iii)    *$A \mid B \succsim C \mid D$ iff $P(A \cap B)/P(B) \geqslant P(C \cap D)/P(D)$.*

The proof (Section 5.7) involves three major steps. First, a number of elementary properties, such as $A \mid B \gtrsim C \mid D$ implies $-C \mid D \gtrsim -A \mid B$, are shown to follow from the axioms (Section 5.7.1). Second, we introduce an ordering $\gtrsim'$ on $\mathscr{E}$, by letting $A \gtrsim' B$ if and only if $A \mid X \gtrsim B \mid X$ (Section 5.7.2). The structure $\langle X, \mathscr{E}, \gtrsim' \rangle$ is shown to satisfy the hypotheses of Theorem 2. By the proofs of Theorems 2 and 3.3 (reduction to extensive measurement, and thence, to Hölder's theorem) we obtain an Archimedean, positive, regular, ordered local semigroup (Definition 2.2), whose elements are equivalence classes in $\mathscr{E} - \mathscr{N}$.

Third, we obtain a semiring structure by using the conditional structure to define a multiplication operation (Section 5.7.3). If $A \mid X \sim C \mid B$, with $B \supset C$, then the representation that we aim for gives $P(A) P(B) = P(C)$. Therefore we define $\mathbf{A} * \mathbf{B} = \mathbf{C}$ (where boldface letters denote equivalence classes). It is then easy to show that the axioms of Definition 2.4 hold. The required representation follows from the semiring theorem, 2.6.

### 5.6.3  Further Discussion of Definition 8 and Theorem 7

In this subsection we discuss three topics: the role of various axioms in the proof of Theorem 7; alternative proofs and axiomatizations; and the testability of the axioms.

Axioms 1–3 of Definition 8 are obvious ground rules for qualitative conditional probability, and are often used in the proof without explicit reference. For example, we may say, "since $A \mid A \sim B \mid B...$," without noting that this uses Axioms 3 ($X \mid X \sim A \mid A$) and 1 (transitivity).

By restricting the relation $\gtrsim$ to hold between $A \mid B$ and $C \mid D$ only if $B \supset A$ and $D \supset C$, we could bypass Axiom 4 altogether. In effect, this is what we do during the proof, establishing the representation conditions for $B \supset A$, $D \supset C$:

(iii)′   $A \mid B \gtrsim C \mid D$ iff $P(A)/P(B) \geqslant P(C)/P(D)$.

We then use $A \mid B \sim A \cap B \mid B$ only to pass from this to the more usual representation, involving condition (iii) of Theorem 7.

Axiom 5 is, of course, the key assumption for an additive probability measure. It is analogous to Lemma 2 of Section 5.3.1 and Axiom 3′ of Section 5.4.1. From the standpoint of semirings, it has two principal uses. First, it is used to show that $\oplus$ is well defined and monotone, where if $A \cap B = \varnothing$, $\mathbf{A} \oplus \mathbf{B} = \mathbf{A} \cup \mathbf{B}$. This only uses part of the strength of the axiom, i.e., if $A \mid X \gtrsim A' \mid X$ and $B \mid X \gtrsim B' \mid X$, then $A \cup B \mid X \gtrsim A' \cup B' \mid X$ (of course, $A \cap B = A' \cap B' = \varnothing$). This use of Axiom 5 is not really explicit, being buried in the reductions to Theorems 2 and 3.3. Rather, 5 is used heavily in the preliminary lemmas to derive other needed features of qualitative conditional probability. The second major use in the semiring development is to prove right distributivity of multiplication $*$ over addition $\oplus$. Here, we use the form of the axiom with $\mid C$ instead of $\mid X$. The structure of the axiom suggests right distributivity, i.e., if $A \mid X \sim A' \mid C$ $(A * C = A')$ and $B \mid X \sim B' \mid C$ $(B * C = B')$, then $A \cup B \mid X \sim A' \cup B' \mid C$ $[(A \oplus B) * C = A' \oplus B' = (A * C) \oplus (B * C)]$.

Axiom 6 also plays two main roles in the semiring development, besides its use in some preliminary lemmas. First, it is used to show that multiplication $*$ is commutative. This leads immediately to the left distributive law. Second, Axiom 6 leads to Lemma 12 and its corollary (Section 5.7.1), which are used to prove left and right monotonicity of multiplication.

Except for commutativity of multiplication, we would be better off using Lemma 12 as an axiom, instead of Axiom 6. To further this discussion, we state Lemma 12 here as an alternative axiom, 6'.

AXIOM 6'. *Suppose that $A \supset B \supset C$ and $A' \supset B' \supset C'$. If $B \mid A \gtrsim B' \mid A'$ and $C \mid B \gtrsim C' \mid B'$, then $C \mid A \gtrsim C' \mid A'$; moreover, if either hypothesis is $>$ and $C \in \mathscr{E} - \mathscr{N}$, then the conclusion is $>$.*

It can be seen that Axiom 6' is very similar to Axiom 6. The hypotheses of Axiom 6' are "uncrossed," with $B \mid A$ dominating $B' \mid A'$, etc. Axiom 6' is very analogous to the weak monotonicity condition in difference structures (Chapter 4), whereas Axiom 6 is analogous to what has been called the strong 6-tuple condition (Block & Marschak, 1960). Here, we use Axiom 6, plus other conditions, including Axiom 8, to derive 6'; then 6' does most of the work, except for commutativity of multiplication. We could use 6' as an axiom, if we were willing to add another axiom to guarantee left distributivity. For example, we could assume:

*Suppose that $A \supset A'$, $B \supset B'$, and $A \cap B = \varnothing$. If $A' \mid A \sim B' \mid B$, then $A' \cup B' \mid A \cup B \sim A' \mid A \sim B' \mid B$.*

Alternatively, we could try to push through the analogy with difference structures, deriving 6 from 6' via a representation theorem from Chapter 4. This is done in the next subsection (proof in 5.7.4), but we seem to require a stronger structural condition (see Axiom 8', below) in order to obtain the solvability axiom of positive-difference structures.

Luce's (1968) proof relied on the positive-difference structure of conditional probability. The exponential of a difference representation gave a ratio representation $Q$. Normalizing so that $Q(X) = 1$, $Q$ fulfilled all the assertions of Theorem 7, except possibly additivity. The normalized $Q$ was unique up to $Q \to Q^\alpha$, and Luce used a functional-equation argument to show that $\alpha$ could be chosen such that additivity also held. Our use of the ordered local semiring incorporates the essence of a slightly different functional-equation argument, since the proof of Theorem 2.6 is based precisely on finding the unique additive representation for a semiring that is also multiplicative.

Archimedean axioms are familiar to the reader by this time, so we need not comment on the role of Axiom 7; but we do note that we have defined *standard sequence* in terms of the ratio structure, rather than in terms of the additive structure. We do so because the resulting definition is much simpler (compare Definition 3) and it leads directly to the Archimedean axiom for positive differences (see Section 5.7.4). However, the proof that Axioms 1–8 imply the Archimedean axiom for the additive structure is a bit tricky (5.7.2).

The structural condition, 8, plays many different roles. We note here that it is used to define multiplication: $\mathbf{A} * \mathbf{B} = \mathbf{C}$ where $C$ is the solution to $A \mid X \sim C \mid B$, with $B \supset C$. The solution exists by Axiom 8 (note Axiom 4 is also used), since $A \mid X \gtrsim \varnothing \mid B$ (Corollaries 1 and 2 of Lemma 6, below).

This completes the discussion of the roles of the various axioms, and of alternative axiomatizations and proofs. We now comment briefly on the testability of the axioms.

The status of the various axioms depends critically on the method used to establish the empirical ordering over $\mathscr{E} \times (\mathscr{E} - \mathscr{N})$. If the ordering is estimated from relative frequency counts—by the proportion of the times that $A$ occurs when $B$ occurs—we would most likely attribute any failure of the axioms to error. If the ordering is obtained by having a subject rate the likelihood of $A$, given $B$, then it would seem that the most interesting axioms to test are 5 and 6. Axiom 4 could fail if the subject did not perceive set relations correctly, but this would not bear strongly on the issue of a probability representation. The possibility that Axiom 5 might fail while 6 holds motivates our development, in the next section, of a nonadditive conditional representation. If Axiom 6 fails, then the properties of "subjective conditional probability" are indeed very far removed from the representation here axiomatized.

In the case where the relation $\gtrsim$ is inferred from decisions under uncertainty, Axiom 1 also comes into question, for the same reasons as in unconditional representations (see Section 5.2.4 and Chapter 8).

It seems hard to envisage situations in which tests of Axioms 2, 3, 7, or 8 would be empirically interesting. In most, if not all, practical situations, $\mathscr{N} = \{\varnothing\}$, which simplifies matters considerably.

### 5.6.4 A Nonadditive Conditional Representation

As we mentioned in the preceding subsection, there may be cases where an ordering of "conditional probability" satisfies the multiplicative property expressed by Axiom 6 or 6', but does not satisfy the additive property of Axiom 5. We therefore want a nonadditive representation to be obtained without using Axiom 5.

We retain Axioms 1–4 and 7 of Definition 8; we replace Axiom 6 by
Axiom 6′, introduced in the preceding subsection; and we replace Axioms 5
and 8 by the following:

5′.  *If $A \in \mathcal{N}$ and $A \supset B$, then $B \in \mathcal{N}$.*

8′.  *If $A \mid B \gtrsim C \mid D$, then there exists $C' \in \mathcal{E}$ such that $C \cap D \subset C'$ and
$A \mid B \sim C' \mid D$; if, in addition, $C \in \mathcal{E} - \mathcal{N}$, then there exists $D'$ such that
$D \supset D' \supset C \cap D$ and $A \mid B \sim C \mid D'$.*

Axiom 5′ corresponds to Lemma 8 (iii) below, i.e., it can be proved from
Axioms 1–6 of Definition 8. The first part of Axiom 8′ is just a restatement
of Axiom 8; the second part has the additional proviso that just enough can
be removed from $D - C$ to obtain a subset $D'$ such that $A \mid B \sim C \mid D'$.
Naturally, this must be restricted to the case where $C$ is not in $\mathcal{N}$.

**THEOREM 8.**  *Let $\langle X, \mathcal{E}, \mathcal{N}, \gtrsim \rangle$ satisfy the conditions of Theorem 7,
except that Axioms 5, 6, and 8 are replaced by 5′, 6′, and 8′, respectively.
Then there exists a function $Q$ from $\mathcal{E} - \mathcal{N}$ to $(0, 1]$, such that:*

(i)  $Q(X) = 1$;

(ii)  *if $A \cap B, C \cap D \in \mathcal{E} - \mathcal{N}$, then $A \mid B \gtrsim C \mid D$ iff*

$$Q(A \cap B)/Q(B) \geqslant Q(C \cap D)/Q(D).$$

*If $Q'$ is any other function with the same properties, then $Q' = Q^\alpha$, $\alpha > 0$.*

Note that this theorem deals entirely with the restriction of $\gtrsim$ to
$(\mathcal{E} - \mathcal{N}) \times (\mathcal{E} - \mathcal{N})$; we do not have $Q$ defined on elements of $\mathcal{N}$. In
order to extend $Q$ to $\mathcal{N}$, so that (ii) still holds, and so that $A$ is in $\mathcal{N}$ if and
only if $Q(A) = 0$, we would need to add certain additional conditions,
which are provable with the help of Axiom 5, but not provable from the
present axioms. For example, we would need properties such as
$A \mid B \gtrsim \varnothing \mid X$, $\varnothing \mid X \sim \varnothing \mid B$, and $A \mid B \sim \varnothing \mid B$ if and only if $A$ is in $\mathcal{N}$.
Since the extension to $\mathcal{N}$ complicates matters and seems of little interest,
we omit it.

Theorem 8 is proved in Section 5.7.4.

## 5.7  PROOFS

### 5.7.1  Preliminary Lemmas

The common hypothesis of the following four lemmas is that $\langle X, \mathcal{E}, \mathcal{N}, \gtrsim \rangle$
is a structure of qualitative conditional probability (Definition 8, Axioms
1–6). Axiom 7 is not used, and Axiom 8 is assumed only in Lemma 9.

LEMMA 6. *If $A \mid B \gtrsim C \mid D$, then $-C \mid D \gtrsim -A \mid B$.*

*PROOF.* Suppose that, on the contrary, $-A \mid B > -C \mid D$. By the strict inequality clause of Axiom 5, $(A \cup -A) \mid B > (C \cup -C) \mid D$. By Axioms 1 and 4, $B \mid B > D \mid D$, contradicting $B \mid B \sim D \mid D$ derived from Axioms 1 and 3.                                                      ◇

COROLLARY 1. $A \mid B \gtrsim \varnothing \mid X$.

*PROOF.* Axiom 3 and the lemma.                                             ◇

COROLLARY 2. $\varnothing \mid A \sim \varnothing \mid B$.

*PROOF.* By Axioms 1 and 3, $A \mid A \sim B \mid B$; by the lemma, $-A \mid A \sim -B \mid B$; and the conclusion follows by Axioms 4 and 1.    ◇

COROLLARY 3. *If $A \mid C \sim -A \mid C$ and $B \mid D \sim -B \mid D$, then*
$$A \mid C \sim B \mid D.$$

*PROOF.* If $A \mid C > B \mid D$, we can use Axiom 1 to derive
$$-A \mid C > -B \mid D;$$
this contradicts Lemma 6.                                                   ◇

Of course, in the representation, $A \mid C \sim -A \mid C$ will yield $P(A \mid C) = \frac{1}{2}$. Corollary 3 is the qualitative version of that result.

LEMMA 7. *If $A \supset B$, then $A \mid C \gtrsim B \mid C$.*

*PROOF.* By Corollaries 1 and 2 of Lemma 6, $(A - B) \mid C \gtrsim \varnothing \mid C$. Also, $B \mid C \sim B \mid C$. By Axiom 5, $A \mid C \gtrsim B \mid C$.            ◇

We collect together a number of properties of the set $\mathcal{N}$ in one lemma:

LEMMA 8.

   (i)   $\varnothing \in \mathcal{N}$.

  (ii)  *If $A \in \mathcal{N}$, then $-A \in \mathscr{E} - \mathcal{N}$.*

 (iii)  *If $A \in \mathcal{N}$, and $A \supset B$, then $B \in \mathcal{N}$.*

 (iv)  *If $A, B \in \mathcal{N}$, then $A \cup B \in \mathcal{N}$.*

*PROOF.*

   (i)   $\varnothing \mid X \sim \varnothing \mid X$.

  (ii)  If $A, -A \in \mathcal{N}$, then $A \mid X \sim \varnothing \mid X \sim -A \mid X$. By Lemma 6, $-A \mid X \sim X \mid X \sim A \mid X$, i.e., $X \mid X \sim \varnothing \mid X$, contradicting Axiom 2.

(iii)  By Lemma 7, $A \mid X \sim \varnothing \mid X$, $A \supset B$ imply $\varnothing \mid X \succsim B \mid X$. By Corollary 1 to Lemma 6, $B \mid X \sim \varnothing \mid X$, or $B$ is in $\mathcal{N}$.

(iv)  By part (iii), $(A - B) \mid X \sim \varnothing \mid X$; this, with $B \mid X \sim \varnothing \mid X$, yields $(A \cup B) \mid X \sim \varnothing \mid X$, by Axiom 5.    $\diamondsuit$

**LEMMA 9.**  *If $A \mid B \sim \varnothing \mid B$ and $B \supset A$, then $A \in \mathcal{N}$. Conversely, if Axiom 8 holds, $B \in \mathcal{E} - \mathcal{N}$, and $A \in \mathcal{N}$, then $A \mid B \sim \varnothing \mid B$.*

**PROOF.**  Suppose that $A \mid B \sim \varnothing \mid B$, $B \supset A$, but $A \in \mathcal{E} - \mathcal{N}$. By Corollary 2 of Lemma 6, $\varnothing \mid A \sim \varnothing \mid X$. Since $B \in \mathcal{E} - \mathcal{N}$, Axiom 2 and Corollary 1 of Lemma 6 yield $\varnothing \mid X < B \mid X$. We have

$$\varnothing \mid A < B \mid X, \qquad A \mid B \sim \varnothing \mid B,$$

whence, by Axiom 6, $\varnothing \mid B < \varnothing \mid X$, contradicting Corollary 2 of Lemma 6.

For the converse, suppose that $B \in \mathcal{E} - \mathcal{N}$ and $A \mid B \succ \varnothing \mid B$. We apply Axiom 8 to $A \mid B \succ \varnothing \mid X$ to obtain

$$A \mid B \sim C \mid X \succ \varnothing \mid X.$$

Thus, $C \in \mathcal{E} - \mathcal{N}$. Since also $B \in \mathcal{E} - \mathcal{N}$, we have $B \mid X \succ \varnothing \mid C$. By Axiom 4, $A \cap B \mid B \sim C \mid X$. By Axiom 6, $A \cap B \mid X \succ \varnothing \mid X$, so $A \cap B \in \mathcal{E} - \mathcal{N}$, and by Lemma 8 (iii), $A \in \mathcal{E} - \mathcal{N}$.    $\diamondsuit$

For the remaining lemmas, we assume that $\langle X, \mathcal{E}, \mathcal{N}, \succsim \rangle$ satisfies Axioms 1–6 and Axiom 8.

The next two lemmas have the content of Axiom 3 of positive-difference structures (Definition 4.1): if $ab$, $bc$ are positive intervals, then $ac$ is strictly greater than either of them. Here, $AB$ will be a positive interval if $A \supset B$ and $X \mid X \succ B \mid A$. Thus, we want to establish that if $A \supset B \supset C$, with $AB$, $BC$ positive, then $C \mid A$ is strictly smaller than both $B \mid A$ and $C \mid B$.

**LEMMA 10.**  *If $A \supset B \supset C$ and $B \in \mathcal{E} - \mathcal{N}$, then $C \mid B \succsim C \mid A$, and $\succ$ holds unless either $C$ or $A - B \in \mathcal{N}$.*

**PROOF.**  By Lemma 8 (iii), $A \in \mathcal{E} - \mathcal{N}$. If $C \in \mathcal{N}$, then Lemma 9 gives

$$C \mid B \sim \varnothing \mid B \sim \varnothing \mid A \sim C \mid A,$$

as required. So assume that $C \in \mathcal{E} - \mathcal{N}$.

We have $C \mid B \sim C \mid B$ and $C \mid C \succsim B \mid A$, whence, Axiom 6 yields $C \mid B \succsim C \mid A$. Moreover, $\succ$ holds unless $C \mid C \sim B \mid A$; but in that case, Lemma 6 gives $\varnothing \mid C \sim (A - B) \mid A$, and now, from Lemma 9 and $(A - B) \mid A \sim \varnothing \mid A$, we have $A - B \in \mathcal{N}$.    $\diamondsuit$

LEMMA 11.  *If* $A \supset B \supset C$ *and* $A \in \mathcal{E} - \mathcal{N}$, *then* $B \mid A \succsim C \mid A$, *and* $\succ$ *holds unless either* $B$ *or* $B - C \in \mathcal{N}$.

PROOF.  If $B \in \mathcal{N}$, then $C \in \mathcal{N}$ and $B \mid A \sim \varnothing \mid A \sim C \mid A$, as required. So assume that $B \in \mathcal{E} - \mathcal{N}$.

We have $A \mid A \succsim C \mid B$, $B \mid A \sim B \mid A$, so Axiom 6 yields $B \mid A \succsim C \mid A$, with $\succ$ unless $A \mid A \sim C \mid B$. In the latter case, $B - C \in \mathcal{N}$. $\quad\quad\diamondsuit$

The next lemma (Axiom 6′, Section 5.6.3) corresponds to Axiom 4 of positive differences, the weak monotonicity condition, with an additional strict inequality proviso.

LEMMA 12.  *If* $A \supset B \supset C$, $A' \supset B' \supset C'$, $B \mid A \succsim B' \mid A'$, *and* $C \mid B \succsim C' \mid B'$, *then* $C \mid A \succsim C' \mid A'$. *Moreover, the conclusion is* $\succ$ *unless either* $C \in \mathcal{N}$ *or both hypotheses are* $\sim$.

PROOF.  Suppose that the hypotheses of the lemma hold, but that $C' \mid A' \succsim C \mid A$. We must show that in fact $C' \mid A' \sim C \mid A$ and that either $C \in \mathcal{N}$ or both hypotheses are $\sim$.

By Lemma 10, we have $C' \mid B' \succsim C \mid A$. By Axiom 8, choose $B''$ with $A \supset B'' \supset C$ and $C' \mid B' \sim B'' \mid A$. If $B'' \in \mathcal{N}$, then Lemma 9 yields that $C$, $C' \in \mathcal{N}$ and $C' \mid A' \sim C \mid A \sim \varnothing \mid X$, so we are done. Assume $B'' \in \mathcal{E} - \mathcal{N}$. We show that $C \mid B'' \succsim B \mid A$. Otherwise, $B \mid A \succ C \mid B''$ and

$$C \mid B \succsim C' \mid B' \sim B'' \mid A$$

yield, by Axiom 6, that $C \mid A \succ C \mid A$. But now, $C \mid B'' \succsim B' \mid A'$ and $B'' \mid A \sim C' \mid B'$ yield $C \mid A \succsim C' \mid A'$, by Axiom 6. Thus, we have $C \mid A \sim C' \mid A'$, and by the strict clause of Axiom 6, $C \mid B'' \sim B' \mid A'$. We now have $C \mid B'' \sim B \mid A$, and can now infer, by the strict clause of Axiom 6, that $B'' \mid A \sim C \mid B$ (otherwise, $C \mid A \succ C \mid A$). Hence, $\sim$ holds in both hypotheses. $\quad\quad\diamondsuit$

COROLLARY.  *Suppose that* $C$, $D \in \mathcal{E} - \mathcal{N}$ *and that* $D \supset C \supset A \cup B$. *Then* $A \mid C \succsim B \mid C$ *iff* $A \mid D \succsim B \mid D$.

PROOF.  If $A \mid C \succsim B \mid C$, then since $C \mid D \sim C \mid D$, the lemma implies $A \mid D \succsim B \mid D$.

Conversely, if $B \mid C \succ A \mid C$, then the strict clause in the lemma yields $B \mid D \succ A \mid D$. $\quad\quad\diamondsuit$

This corollary is one of the main results to be used in the proof of

Theorem 7. It is used first to establish the Archimedean axiom for a structure of qualitative (unconditional) probability, in Section 5.7.2, on the basis of the Archimedean axiom of Definition 8. Subsequently, it is used to show monotonicity of multiplication in the ordered local semiring. The main line of development thus consists of Lemmas 6, 7, 8 (iii), 9, 10, and 12, with corollaries.

One more lemma is required, that will be used to obtain the solvability axiom for unconditional probability.

**LEMMA 13.**  *If  $A \mid X \gtrsim B \mid X$ , then there exists  $B' \subset A$  such that  $B' \mid X \sim B \mid X$ .*

**PROOF.**  By Lemma 6,  $-B \mid X \gtrsim -A \mid X$ . By Axiom 8, there exists  $-B' \supset -A$  such that  $-B \mid X \sim -B' \mid X$ . We have  $B' \subset A$  and, by Lemma 6,  $B' \mid X \sim B \mid X$ .                                             ◇

## 5.7.2  An Additive Unconditional Representation

**THEOREM 9.**  *Suppose that  $\langle X, \mathscr{E}, \mathscr{N}, \gtrsim \rangle$  is an Archimedean structure of qualitative conditional probability (Definition 8) for which Axiom 8 holds. Let  $\gtrsim'$  on  $\mathscr{E}$  be defined by*

$$A \gtrsim' B \quad iff \quad A \mid X \gtrsim B \mid X.$$

*Then  $\langle X, \mathscr{E}, \gtrsim' \rangle$  is an Archimedean structure of qualitative probability for which Axiom 5 of Section 5.2.3 holds.*

**PROOF.**  We verify the five axioms.

1.   $\gtrsim'$  is a weak order on  $\mathscr{E}$ , since  $\gtrsim$  is a weak order.

2.   $A \gtrsim' \varnothing$  by Corollary 1 of Lemma 6;  $X \succ' \varnothing$  by Axiom 2 (if  $X \sim' \varnothing$ , then  $X \in \mathscr{N}$ , but  $X \in \mathscr{E} - \mathscr{N}$ ).

3.   This follows as a special case of Axiom 5 of Definition 8 (note that the strict clause of that Axiom corresponds to the "if" clause in Axiom 3 of Definition 4).

4.   Suppose that  $A_1', A_2',...$  is a standard sequence in  $\langle X, \mathscr{E}, \gtrsim' \rangle$  relative to some  $A \succ' \varnothing$  (see Definition 3). We first construct another standard sequence  $A_1, A_2,...$ , with the properties that, for all  $i$ ,  $A_i \sim A_i'$  and  $A_i \subset A_{i+1}$ . To do this, let  $A_1 = A$  and assuming that  $A_{i-1}$  has been defined, choose  $A_i$  by applying Axiom 8. We have  $A_i' \succ' A_{i-1}'$  (use Axiom 5, Definition 3, and  $A \succ' \varnothing$ ), hence,  $A_i' \mid X > A_{i-1} \mid X$ . By Axiom 8, there exists  $A_i \supset A_{i-1}$  with  $A_i' \mid X \sim A_i \mid X$ .

Next we show that the subsequence  $A_{2^i}$ ,  $i = 0, 1,...$ , is a standard sequence (in the sense of Definition 8) of $\langle X, \mathscr{E}, \mathscr{N}, \gtrsim \rangle$. Let $C_1 = A_1$ and for $i > 1$, $C_i = A_i - A_{i-1}$. The $C_i$ are pairwise disjoint and

$$A_{2^i} = \bigcup_{k=1}^{2^i} C_k,$$

$$A_{2^{i+1}} - A_{2^i} = \bigcup_{k=2^i+1}^{2^{i+1}} C_k.$$

By Definition 3 and Axiom 5, we have $C_i \sim' A$ for all $i$. Thus, by Axiom 5 and induction, any two unions of $m$ $C_i$'s are $\sim'$, and in particular,

$$A_{2^i} \sim' A_{2^{i+1}} - A_{2^i}.$$

By Lemma 8, each $A_i \in \mathscr{E} - \mathscr{N}$. By the corollary of Lemma 12, and Axiom 4,

$$A_{2^i} \mid A_{2^{i+1}} \sim (A_{2^{i+1}} - A_{2^i}) \mid A_{2^{i+1}} \sim -A_{2^i} \mid A_{2^{i+1}}.$$

By Corollary 3 to Lemma 6, we have $A_{2^i} \mid A_{2^{i+1}} \sim A_1 \mid A_2$, all $i$. By Axioms 3 and 5, we have $X \mid X \sim A_2 \mid A_2 > A_1 \mid A_2$ (note that by Lemma 9, $C_2 \mid A_2 > \varnothing \mid A_2$). Thus, $\{A_{2^i}\}$ is indeed a standard sequence (Definition 8) and since it is finite, so must be the sequences $\{A_i\}$ and $\{A_i'\}$.

5.  Suppose that $A \cap B = \varnothing$, $A \succ' C$, and $B \gtrsim' D$. Applying Lemma 13 to $A \mid X > C \mid X$, we have $C' \subset A$ with $C' \mid X \sim C \mid X$. Similarly, we have $D' \subset B$ with $D' \mid X \sim D \mid X$. Since $A \cap B = \varnothing$, $C' \cap D' = \varnothing$ also, and so $C'$, $D'$, and $E = A \cup B$ fulfill the condition.                                  $\diamondsuit$

COROLLARY.  *Suppose that the hypotheses of Theorem 9 hold. Let $\mathscr{E}$ be the set of $\sim'$ equivalence classes, excluding $\mathscr{N}$; and for $A \in \mathscr{E} - \mathscr{N}$, let $\mathbf{A}$ denote the equivalence class of $A$. Let $\gtrsim$ be the induced simple order on $\mathscr{E}$; and if $\mathbf{A}, \mathbf{B} \in \mathscr{E}$, with $A \cap B = \varnothing$, let $\mathbf{A} \oplus \mathbf{B}$ be $\mathbf{A} \cup \mathbf{B}$. Let $\mathscr{B}$ be the set of $(\mathbf{A}, \mathbf{B})$ with $A \cap B = \varnothing$. Then $\langle \mathscr{E}, \gtrsim, \mathscr{B}, \oplus \rangle$ is an Archimedean, positive, regular, ordered local semigroup (Definition 2.2).*

PROOF.  From Section 5.3.2 and Theorem 9 we know that $\langle \mathscr{E}, \gtrsim, \mathscr{B}, \oplus \rangle$ is an extensive structure with no essential maximum (Definition 3.2); but since $\gtrsim$ is already a simple order, the argument in Section 3.5.3 shows that the structure is an Archimedean, positive, regular, ordered local semigroup.                                                                            $\diamondsuit$

### 5.7.3  Theorem 7  (p. 224)

*An Archimedean structure of qualitative probability $\langle X, \mathscr{E}, \mathscr{N}, \gtrsim \rangle$ has a unique representation $P$ such that $\langle X, \mathscr{E}, P \rangle$ is a finitely additive probability*

space, $\mathcal{N}$ is the class of events with probability zero, and $A \mid B \gtrsim C \mid D$ iff
$P(A \cap B)/P(B) \geqslant P(C \cap D)/P(D)$.

*PROOF.* We start by introducing a semiring operation into the semi-
group $\langle \mathscr{E}, \gtrsim, \mathscr{B}, \oplus \rangle$ defined in the corollary to Theorem 9, in the preceding
subsection. Take $\mathbf{A}, \mathbf{B} \in \mathscr{E}$. We have $A \mid X > \varnothing \mid B$ (since $\mathcal{N} \notin \mathscr{E}$ and since
$\varnothing \mid X \sim \varnothing \mid B$ by Corollary 2 to Lemma 6). By Axioms 8 and 4, there
exists $C$ with $B \supset C$ and $A \mid X \sim C \mid B$. We now define $\mathbf{A} * \mathbf{B} = \mathbf{C}$.

To show that $*$ is well defined, suppose that $A \mid X \sim A' \mid X, B \mid X \sim B' \mid X$,
$C \subset B, C' \subset B', A \mid X \sim C \mid B$, and $A' \mid X \sim C' \mid B'$. Applying Lemma 12 to
$C \subset B \subset X, C' \subset B' \subset X$, we have $C \mid X \sim C' \mid X$; thus, $\mathbf{C} = \mathbf{C}'$. By Lemma 9,
$\mathbf{A} * \mathbf{B} \in \mathscr{E}$, since in the above definition, $C \mid B > \varnothing \mid X \sim \varnothing \mid B$.

The set $\mathscr{B}^*$ for which $*$ is defined consists of all of $\mathscr{E} \times \mathscr{E}$ (this is because
the product of two numbers less than or equal to 1 is again less than or
equal to 1; $X$ is a natural identity for $*$). This greatly simplifies the proof of
the other axioms of Definition 2.4.

To show that $*$ is associative, choose $\mathbf{A}, \mathbf{B}, \mathbf{C} \in \mathscr{E}$. Let $\mathbf{A} * \mathbf{B} = \mathbf{C}'$,
$C' \subset B$; and let $\mathbf{C}' * \mathbf{C} = \mathbf{D}, D \subset C$. Thus, $(\mathbf{A} * \mathbf{B}) * \mathbf{C} = \mathbf{D}$. Similarly,
choose $\mathbf{A}' = \mathbf{B} * \mathbf{C}, A' \subset C$, and $\mathbf{D}' = \mathbf{A} * \mathbf{A}', D' \subset A'$, so that $\mathbf{A} * (\mathbf{B} * \mathbf{C}) =$
$\mathbf{D}'$. We have $A \mid X \sim C' \mid B$ and $A \mid X \sim D' \mid A'$, hence $D' \mid A' \sim C' \mid B$.
Also $A' \mid C \sim B \mid X$. Applying Lemma 12 to $D' \subset A' \subset C$ and $C' \subset B \subset X$,
we have $D' \mid C \sim C' \mid X$. Since $C' \mid X \sim D \mid C$, we have $D' \mid C \sim D \mid C$. By
the corollary to Lemma 12, this implies $\mathbf{D}' = \mathbf{D}$, as required.

Before establishing that the other axioms of Definition 2.4 are satisfied,
we note that $*$ is commutative: if $A \mid X \sim C \mid B$, with $C \subset B$, and
$B \mid X \sim D \mid A$, with $D \subset A$, then by Axiom 6 applied to $D \subset A \subset X$,
$C \subset B \subset X$, we have $D \mid X \sim C \mid X$, or $\mathbf{D} = \mathbf{C}$.

We now show that $*$ is right distributive over $\oplus$. Left distributivity then
follows by commutativity of $*$. Let $A \cap B = \varnothing, A \cup B \mid X \sim C' \mid C$,
$C' \subset C$. Then $(\mathbf{A} \oplus \mathbf{B}) * \mathbf{C} = \mathbf{C}'$. By Axiom 8, choose $A' \subset C'$ such that
$A \mid A \cup B \sim A' \mid C'$. (Note that $A \cup B, C' \in \mathscr{E} - \mathcal{N}$.) Let $B' = C' - A'$.
By Lemma 6, $B \mid A \cup B \sim B' \mid C'$. Now apply Lemma 12 to $A \subset A \cup B \subset X$,
$A' \subset C' \subset C$, obtaining $A \mid X \sim A' \mid C$; and similarly, by Lemma 12,
$B \mid X \sim B' \mid C$. Hence, $\mathbf{A} * \mathbf{C} = \mathbf{A}', \mathbf{B} * \mathbf{C} = \mathbf{B}'$. By construction, $A' \cap B' =$
$\varnothing$, therefore,

$$(\mathbf{A} * \mathbf{C}) \oplus (\mathbf{B} * \mathbf{C}) = \mathbf{A}' \cup \mathbf{B}'$$
$$= \mathbf{C}'$$
$$= (\mathbf{A} \oplus \mathbf{B}) * \mathbf{C}.$$

Finally, we note that $\mathbf{A} \gtrsim \mathbf{B}$ implies $\mathbf{A} * \mathbf{C} \gtrsim \mathbf{B} * \mathbf{C}$ follows immediately
from the corollary of Lemma 12; left monotonicity follows by commutativity;
also it could easily be proved directly from Lemma 12.

We now assert that the first three axioms of Definition 2.4 hold. Axiom 1, that $\langle \mathscr{E}, \succsim, \mathscr{B}, \oplus \rangle$ is an ordered local semigroup, was proved in the previous section; Axiom 2, that $\langle \mathscr{E}, \succsim, \mathscr{E} \times \mathscr{E}, * \rangle$ is an ordered local semigroup, has just been established (the assertions about pairs in $\mathscr{E} \times \mathscr{E}$ are all trivial, since $*$ is defined everywhere); and Axiom 3, distributivity, was also just proved. The final axiom asserts that for any $\mathbf{A} \in \mathscr{E}$, there exists $(\mathbf{B}, \mathbf{C})$ in $\mathscr{B}$ such that $(\mathbf{A}, \mathbf{B} \oplus \mathbf{C}) \in \mathscr{E} \times \mathscr{E}$. This is trivial unless $\mathscr{B}$ is empty. But $\mathscr{B}$ can be empty only if $\mathscr{E}$ contains exactly one element, $\mathbf{X}$; in which case, Theorem 7 is trivially true, letting $P(A) = 1$ or 0 accordingly as $A \sim' X$ or $A \sim' \varnothing$.

If $\mathscr{B}$ is nonempty, then $\langle \mathscr{E}, \succsim, \mathscr{B}, \mathscr{E} \times \mathscr{E}, \oplus, * \rangle$ is an Archimedean, positive, regular, ordered local semiring. By Theorem 2.6, there exists a unique function $\phi$ from $\mathscr{E}$ to $\text{Re}^+$ which is order preserving, additive, and multiplicative. Define

$$P(A) = \begin{cases} \phi(\mathbf{A}), & \text{if} \quad A \in \mathscr{E} - \mathscr{N}, \\ 0, & \text{if} \quad A \in \mathscr{N}. \end{cases}$$

We must show that $P$ satisfies the conclusions of Theorem 7.

Suppose that $A \mid B \succsim C \mid D$. We want to show that $P(A \cap B)/P(B) \geq P(C \cap D)/P(D)$. Without loss of generality, suppose $A \cap B, C \cap D \in \mathscr{E} - \mathscr{N}$ (since $C \cap D \in \mathscr{N}$ is trivial, and $A \cap B \in \mathscr{N}$ implies $C \cap D \in \mathscr{N}$, using Lemma 9). Choose $E, F \in \mathscr{E} - \mathscr{N}$ such that

$$E \mid X \sim A \mid B \succsim C \mid D \sim F \mid X.$$

We have $\mathbf{B} * \mathbf{E} = \mathbf{A} \cap \mathbf{B}$ and $\mathbf{F} * \mathbf{D} = \mathbf{C} \cap \mathbf{D}$. Therefore, since $\phi$ is multiplicative and order preserving and since $\mathbf{E} \succsim \mathbf{F}$, we have

$$\begin{aligned} \phi(\mathbf{A} \cap \mathbf{B})/\phi(\mathbf{B}) &= \phi(\mathbf{E}) \\ &\geq \phi(\mathbf{F}) \\ &= \phi(\mathbf{C} \cap \mathbf{D})/\phi(\mathbf{D}). \end{aligned}$$

The required inequality for $P$ follows immediately. The strict case $>$ similarly leads to $>$ in the $P$-inequality; so $A \mid B \succsim C \mid D$ if and only if $P(A \cap B)/P(B) \geq P(C \cap D)/P(D)$.

We need only show now that $\langle X, \mathscr{E}, P \rangle$ is a finitely additive probability space. Since $\mathbf{X}$ is a multiplicative identity, we have $P(X) = 1$; clearly $P \geq 0$; and additivity of $P$ follows from additivity of $\phi$ on $\langle \mathscr{E}, \mathscr{B}, \oplus \rangle$.

For uniqueness, note that another representation $P'$ would lead to another representation $\phi'$ of $\langle \mathscr{E}, \succsim, \mathscr{B}, \mathscr{E} \times \mathscr{E}, \oplus, * \rangle$, by letting $\phi'(\mathbf{A}) = P'(\mathbf{A})$. For example, to show that $\phi'$ is multiplicative, let $\mathbf{A} * \mathbf{B} = \mathbf{C}$. We can choose $C$ such that $A \mid X \sim C \mid B$ and $C \subset B$; and then, from the representation, $P'(A) \, P'(B) = P'(C)$, implying $\phi'(\mathbf{A} * \mathbf{B}) = \phi'(\mathbf{A})\phi'(\mathbf{B})$. $\diamond$

**5.7.4 Theorem 8** (p. 228)

*If* $\langle X, \mathscr{E}, \mathcal{N}, \gtrsim \rangle$ *satisfies Axioms 1–4, 5′, 6′, 7, and 8′ (Section 5.6), then there exists* $Q : (\mathscr{E} - \mathcal{N}) \to (0, 1]$, *such that* (i) $Q(X) = 1$ *and* (ii) $A \mid B \gtrsim C \mid D$ *iff* $Q(A \cap B)/Q(B) \geqslant Q(C \cap D)/Q(D)$. *This* $Q$ *is unique up to* $Q \to Q^{\alpha}$, $\alpha > 0$.

*PROOF.* We construct a positive-difference structure $\langle \mathscr{E} - \mathcal{N}, \mathcal{O}^*, \gtrsim^* \rangle$ (see Definition 4.1) as follows. Let $AB \in \mathcal{O}^*$ iff $A, B \in \mathscr{E} - \mathcal{N}$, $A \supset B$, and $X \mid X \succ B \mid A$. For $AB, CD \in \mathcal{O}^*$, let $AB \gtrsim^* CD$ iff $D \mid C \gtrsim B \mid A$. We verify the six axioms of Definition 4.1.

1.  Weak ordering is immediate.

2 & 3.  If $AB, BC \in \mathcal{O}^*$, then clearly, $A, C \in \mathscr{E} - \mathcal{N}$ and $A \supset C$. We shall show that $B \mid A$ and $C \mid B$ are both $\succ C \mid A$; this will show that $X \mid X \succ C \mid A$, so $AC \in \mathcal{O}^*$, and also, that $AB, BC \prec^* AC$.

By the strict clause of Axiom 6′, $B \mid B \succ C \mid B$ and $B \mid A \sim B \mid A$ yield $B \mid A \succ C \mid A$. Similarly, $C \mid B \sim C \mid B$ and $B \mid B \succ B \mid A$ yield $C \mid B \succ C \mid A$. (Note that Axioms 1 and 3 were used implicitly in the above proof.)

4.  This follows immediately from Axiom 6′.

5.  Suppose that $AB, CD \in \mathcal{O}^*$, with $AB \succ^* CD$. By Axiom 8′, applied to $D \mid C \succ B \mid A$, we have $C'$ and $D'$, with $A \supset C'$, $D'$ and $C', D' \supset B$, such that

$$ B \mid C' \sim D \mid C \sim D' \mid A. $$

By Axiom 5′ or 8′, $C' \in \mathscr{E} - \mathcal{N}$; by Axiom 5′, $D' \in \mathscr{E} - \mathcal{N}$. Thus, $AD'$, $C'B \in \mathcal{O}^*$, and $AD' \sim^* CD \sim C'B$. To complete the proof we must show that $AC'$, and $D'B \in \mathcal{O}^*$, i.e., that $X \mid X \succ$ both $C' \mid A$ and $B \mid D'$. If $X \mid X \sim C' \mid A$, then apply the strict clause of Axiom 6 to $B \mid C' \succ B \mid A$, $C' \mid A \sim A \mid A$, to obtain $B \mid A \succ B \mid A$, a contradiction. Similarly, $B \mid D' \sim B \mid B$ and $D' \mid A \succ B \mid A$ would yield $B \mid A \succ B \mid A$; hence, $X \mid X \succ B \mid D'$.

6.  $\{A_i\}$ is a standard sequence, in the sense of Definition 4.1, if and only if it is a standard sequence in the sense of Definition 8. (This is immediate from the definitions.) Thus, any standard sequence in $\langle \mathscr{E} - \mathcal{N}, \mathcal{O}^*, \gtrsim^* \rangle$ is finite.[2]

---

[2] The reader may be puzzled by the fact that all standard sequences are finite. This is because all standard sequences are ascending, hence, all are strictly bounded by $XA_1$. If we permitted descending standard sequences, $A_1, A_2, \ldots$, such that $A_i \supset A_{i+1}$ and $A_{i+1} \mid A_i \sim A_i \mid A_{i-1}$, there could very well be infinite standard sequences. But it would still be true that any strictly bounded standard sequence is finite. By solvability, any strictly bounded descending standard sequence could be converted into an ascending standard sequence. This is really the reason why only one direction needs to be considered.

This completes the verification of the axioms for positive differences. By Theorem 4.1, there exists a positive-valued function $\psi$ on $\mathscr{E} - \mathscr{N}$ such that $AB \succsim^* CD$ iff $\psi(AB) \geqslant \psi(CD)$, and for $AB$, $BC \in \mathscr{O}^*$, $\psi(AC) = \psi(AB) + \psi(BC)$. We define a function $Q$ as

$$Q(A) = \begin{cases} 1, & \text{if } X \mid X \sim A \mid X, \\ \exp[-\psi(XA)], & \text{if } XA \in \mathscr{O}^*. \end{cases}$$

We must show that $Q$ is well defined and that its domain is $\mathscr{E} - \mathscr{N}$; that is, the two conditions in its definition are mutually exclusive, and $A \in \mathscr{E} - \mathscr{N}$ iff one of them holds.

By definition of $\mathscr{O}^*$, $XA \in \mathscr{O}^*$ excludes both $X \mid X \sim A \mid X$ and $A \in \mathscr{N}$. Also, $X \mid X \sim A \mid X$ excludes $A \in \mathscr{N}$ (otherwise, $X \mid X \sim A \mid X \sim \varnothing \mid X$ yields $X \in \mathscr{N}$). Conversely, if $A \notin \mathscr{N}$ and $X \mid X \succ A \mid X$, then $XA \in \mathscr{O}^*$. Thus, $Q$ is well defined on $\mathscr{E} - \mathscr{N}$.

Next we note that if $AB \in \mathscr{O}^*$, then

$$Q(B)/Q(A) = \exp[-\psi(AB)].$$

For if $XA \in \mathscr{O}^*$, then $XB \in \mathscr{O}^*$, so

$$\begin{aligned} Q(B)/Q(A) &= \exp[-\psi(XB) + \psi(XA)] \\ &= \exp[-\psi(AB)]. \end{aligned}$$

Or if $XA \notin \mathscr{O}^*$, then $X \mid X \sim A \mid X$, and by Axiom 6', $B \mid X \sim B \mid A$ (otherwise, the strict clause of Axiom 6' yields $B \mid X \succ B \mid X$). Consequently, $XB \in \mathscr{O}^*$ and $\psi(AB) = \psi(XB)$; the equation follows, since $Q(A) = 1$.

We can now show that (ii) of Theorem 8 is satisfied. First, if $AB$, $CD \in \mathscr{O}^*$, then the fact that $\exp[-\psi(AB)]$ is a strictly decreasing function of $\psi(AB)$ gives the required result (since $\psi$ is order preserving on $\mathscr{O}^*$, and the ordering $\succsim^*$ on $\mathscr{O}^*$ is the converse of the corresponding $\succsim$ ordering). This extends easily to $A$, $B$, $C$, $D \in \mathscr{E} - \mathscr{N}$ with $A \supset B$, $C \supset D$, but $AB$ or $CD \notin \mathscr{O}^*$. For if $AB \notin \mathscr{O}^*$, then $X \mid X \sim B \mid A$; but this entails $A \mid X \sim B \mid X$. (If $A \mid X \succ B \mid X$, then $B \mid A \sim B \mid B$ and Axiom 6' yield $B \mid X \succ B \mid X$; etc.) Hence, $Q(A) = Q(B)$. Similarly, if $CD \notin \mathscr{O}^*$, $X \mid X \sim D \mid C$ and $Q(C) = Q(D)$. Thus, if both are not in $\mathscr{O}^*$, $B \mid A \sim D \mid C$ and $Q(B)/Q(A) = 1 = Q(D)/Q(C)$; if only $AB \notin \mathscr{O}^*$, then $B \mid A \succ D \mid C$ and $Q(B)/Q(A) = 1 > \exp[-\psi(CD)] = Q(D)/Q(C)$. Finally, the representation extends to $A$, $B$, $C$, $D$ arbitrary in $\mathscr{E} - \mathscr{N}$ by using the representation for $A \cap B$, $A$, $C \cap D$, $C$, and employing Axiom 4.

To prove uniqueness, note that if $Q'$ is another representation, then $\psi'(AB) = -\log[Q'(B)/Q'(A)]$ gives another representation for $\langle \mathscr{E} - \mathscr{N},$

$\mathcal{A}^*, \gtrsim^*\rangle$, hence, by Theorem 4.1, $\psi' = \alpha\psi$. Thus,

$$Q'(A) = Q'(A)/Q'(X)$$
$$= \exp[-\psi'(XA)]$$
$$= \exp[-\alpha\psi(XA)]$$
$$= [Q(A)]^\alpha,$$

if $X \mid X > A \mid X$. Since $1^\alpha = 1$, we have $Q' = Q^\alpha$ on $\mathcal{E} - \mathcal{N}$.                    $\Diamond$

## 5.8 INDEPENDENT EVENTS

Intimately connected with the concept of conditional probability is that of (statistical) independence. Given a probability measure $P$, events $A$, $B$ are independent if and only if $P(A \cap B) = P(A) P(B)$ [or, for $P(B) \neq 0$, if and only if $P(A \mid B) = P(A)$]. In qualitative terms, using the primitives of the preceding sections, we can define $A$, $B$ to be $\sim$-independent if and only if either $B$ is in $\mathcal{N}$ or $A \mid B \sim A \mid X$. (Or again, using $*$ defined in Section 5.7, $A$, $B$ are $\sim$-independent when $\mathbf{A} * \mathbf{B} = \mathbf{A} \cap \mathbf{B}$.)

An interesting idea, introduced by Domotor (1969), is to accept independence of events as a primitive notion, in which case the primitives for a theory of conditional probability consist of a set $X$, an algebra $\mathcal{E}$ on $X$, an ordering $\gtrsim^*$ on $\mathcal{E}$, and a new binary relation (interpreted as independence) $\perp$ on $\mathcal{E}$. Of course, Theorem 2 is then to be enriched by adding to the constraints on the representation the following one:

$$A \perp B \quad \text{iff} \quad P(A \cap B) = P(A) P(B).$$

It will be noted that this property of the representation together with the axiom $A \perp X$ (see Definition 9 below) implies $P(X) = 1$. Without such an additional constraint, we can obtain an absolute scale of probability only by the fiat of normalization. From the standpoint of unconditional probability, only ratios of probabilities are meaningful. This further constraint, based on the new primitive $\perp$, yields a genuine absolute scale. Another way to understand the absolute scale is from the standpoint of conditional probability: in Section 5.7.3, $\mathbf{X}$ is the natural multiplicative identity with respect to $*$, forcing $P(X) = 1$.

We start by introducing the axioms of a pure independence structure $\langle X, \mathcal{E}, \perp \rangle$.

DEFINITION 9. *Suppose $\mathcal{E}$ is an algebra of sets on $X$ and $\perp$ is a binary* *relation on $\mathscr{E}$. Then $\perp$ is an* independence relation[3] *iff*

   1.  $\perp$ *is symmetric.*

   2.  *For $A \in \mathscr{E}$, the set $\{B \mid B \in \mathscr{E}$ and $A \perp B\}$ is a QM-algebra of sets on $X$ (Definition 5), i.e.,*

      (i)   $A \perp X$;

      (ii)  *if $A \perp B$, then $A \perp -B$; and*

      (iii)  *if $A \perp B$, $A \perp C$, and $B \cap C = \varnothing$, then $A \perp B \cup C$.*

It is easy to demonstrate that these axioms are necessary for the representation, e.g., for 2 (ii), if $P(A \cap B) = P(A) \, P(B)$, then

$$
\begin{aligned}
P(A \cap -B) &= P(A) - P(A \cap B) \\
&= P(A) - P(A) \, P(B) \\
&= P(A)[1 - P(B)] \\
&= P(A) \, P(-B).
\end{aligned}
$$

It is definitely not true that $\{B \mid A \perp B\}$ is an ordinary subalgebra of $\mathscr{E}$; there can be events $A$, $B$, and $C$ for which $A \perp B$ and $A \perp C$ but not $A \perp B \cap C$ and not $A \perp B \cup C$. This corresponds to the fact that pairwise independence of events, in the usual probabilistic sense, does not entail independence of the entire set of three or more events. Only if $A$, $B$, $C$ form an independent family of sets is each event independent of the subalgebra generated by the other events. This motivates the definition of an abstract independence relation for sets of three or more events, based on $\perp$, using subalgebras.

If $\mathscr{S}$ is a subset of $\mathscr{E}$, then the *smallest subalgebra containing* $\mathscr{S}$ is given as the intersection of all subalgebras of $\mathscr{E}$ that contain $\mathscr{S}$. (Note that $\mathscr{E}$ itself is a subalgebra containing $\mathscr{S}$, so this intersection is defined and it contains $\mathscr{S}$; and the intersection of arbitrarily many subalgebras is easily seen to be a subalgebra.) The smallest subalgebra containing $A$ consists of $\{\varnothing, A, -A, X\}$.

Two subalgebras of $\mathscr{E}$, $\mathscr{E}_1$ and $\mathscr{E}_2$ are said to be $\perp$-*independent* if and only if $A_1 \perp A_2$ whenever $A_1$ is in $\mathscr{E}_1$ and $A_2$ is in $\mathscr{E}_2$. Note that if $A \perp B$, then $\{\phi, A, -A, X\}$ and $\{\varnothing, B, -B, X\}$ are $\perp$-independent. This leads to the generalization: $A_1, ..., A_m$ are $\perp$-independent if and only if the smallest subalgebras containing every proper subset of the $A_i$ and its complement are $\perp$-independent. Formally, we have:

DEFINITION 10.  *Let $\mathscr{E}$ be an algebra of sets and $\perp$ an independence*

---

[3] This concept is different from that of a relation being independent implicitly defined in Definition 1.3 of Section 1.3.2.

*relation on* $\mathscr{E}$. *For* $m \geqslant 2$, $A_1,..., A_m \in \mathscr{E}$ *are* $\perp$-*independent iff, for every proper subset* $M$ *of* $\{1,..., m\}$, *every* $B$ *in the smallest subalgebra containing* $\{A_i \mid i \in M\}$, *and every* $C$ *in the smallest subalgebra containing* $\{A_i \mid i \notin M\}$, *we have* $B \perp C$.

As we noted in motivating this definition, if $m = 2$, we have $A$ and $B$ are $\perp$-independent if and only if $A \perp B$.

We can now define what we mean by a structure of qualitative probability with independence. We take a structure of qualitative probability $\langle X, \mathscr{E}, \gtrsim \rangle$; an independence relation $\perp$ on $\mathscr{E}$; and an axiom interlocking $\gtrsim$ and $\perp$. The interlocking axiom is very natural, given the representation: probabilities of independent events multiply, and multiplication is order preserving, therefore, intersections of independent events preserve order. In addition, we include in the following definition a very strong structural condition that uses the defined notion of $\perp$-independence of sets of events.

DEFINITION 11.   *Suppose that* $X$ *is a nonempty set,* $\mathscr{E}$ *is an algebra of sets on* $X$, *and* $\gtrsim$ *and* $\perp$ *are binary relations on* $\mathscr{E}$. *The quadruple* $\langle X, \mathscr{E}, \gtrsim, \perp \rangle$ *is a* structure of qualitative probability with independence *iff the following axioms hold*:

1.   $\langle X, \mathscr{E}, \gtrsim \rangle$ *is a structure of qualitative probability (Definition 4).*

2.   $\perp$ *is an independence relation (Definition 9).*

3.   *Suppose that* $A, B, C, D \in \mathscr{E}$, $A \perp B$, *and* $C \perp D$. *If* $A \gtrsim C$ *and* $B \gtrsim D$, *then* $A \cap B \gtrsim C \cap D$; *moreover, if* $A \succ C$, $B \gtrsim D$, *and* $B \succ \varnothing$, *then* $A \cap B \succ C \cap D$.

*The structure* $\langle X, \mathscr{E}, \gtrsim, \perp \rangle$ *is* complete *iff the following additional axiom holds*:

4.   *For any* $A_1,..., A_m, A \in \mathscr{E}$, *there exists* $A' \in \mathscr{E}$ *with* $A' \sim A$ *and* $A' \perp A_i$, $i = 1,..., m$. *Moreover, if* $A_1,..., A_m$ *are* $\perp$-*independent, then* $A'$ *can be chosen so that* $A_1,..., A_m, A'$ *are also* $\perp$-*independent.*

In a complete structure of qualitative probability with independence, we can introduce a defined relation $\gtrsim'$ on pairs $A \mid B$ and $C \mid D$. Clearly, if $A \subset B$, $C \subset D$, $A \perp D$, and $C \perp B$, we will want to assert that $A \mid B \gtrsim' C \mid D$ if and only if $P(A) P(D) \geqslant P(B) P(C)$, or, in qualitative terms, when $A \cap D \gtrsim C \cap B$. If the requisite independence relations for $A, D$ and $C, B$ do not hold, then using completeness (Axiom 4), we may replace these sets by the equivalent ones for which independence does hold. The theorem we obtain is that, with suitable definitions of $\mathscr{N}$ and $\gtrsim'$, a complete structure of qualitative probability with independence gives rise to a structure of qualitative conditional probability.

DEFINITION 12. *Suppose* $\langle X, \mathscr{E}, \succsim, \perp \rangle$ *is a structure of qualitative probability with independence. Let* $\mathscr{N} = \{A \mid A \in \mathscr{E}, A \sim \varnothing\}$. *If* $A, C \in \mathscr{E}$ *and* $B, D \in \mathscr{E} - \mathscr{N}$, *define*

$$A \mid B \succsim' C \mid D$$

*iff there exist* $A', B', C', D' \in \mathscr{E}$, *with*

$$A' \sim A \cap B, \quad B' \sim B, \quad C' \sim C \cap D, \quad D' \sim D;$$
$$A' \perp D' \quad \text{and} \quad C' \perp B';$$

*and*

$$A' \cap D' \succsim C' \cap B'.$$

THEOREM 10. *Suppose that* $\langle X, \mathscr{E}, \succsim, \perp \rangle$ *is a complete structure of qualitative probability with independence and that* $\mathscr{N}$ *and* $\succsim'$ *are given by Definition 12. Then* $\langle X, \mathscr{E}, \mathscr{N}, \succsim' \rangle$ *is a structure of qualitative conditional probability* (*Definition 8*).

If we now make the additional assumptions that the Archimedean and solvability axioms of Section 5.6 are satisfied by $\langle X, \mathscr{E}, \mathscr{N}, \succsim' \rangle$, then Theorem 10 and these assumptions allow us to deduce the representation $P$ of Theorem 7. This gives the desired representation for $\perp$. For if $A \perp B$, with $B$ not in $\mathscr{N}$, then Definition 12 gives $A \mid B \sim' A \mid X$, hence,

$$P(A \cap B)/P(B) = P(A).$$

(If $A, B$ are both $\sim \varnothing$, then $A \cap B \sim \varnothing$, so the desired result is trivial.)

Of course, the Archimedean and solvability axioms, stated in terms of $\succsim'$, can be translated into terms of $\succsim$ and $\perp$. The resulting axioms are not very natural. A simple axiomatization directly in terms of $\succsim$ and $\perp$ would be preferable.

## 5.9 PROOF OF THEOREM 10

*A complete structure of conditional probability with independence induces, by Definition 12, a structure of qualitative conditional probability.*

*PROOF.* We remark first that by Axiom 3 of Definition 11, the definition of $\succsim'$ is independent of the particular choice of $A', B', C', D'$ in Definition 12. For if $A' \sim A''$, $B' \sim B''$, $C' \sim C''$, and $D' \sim D''$, with $A' \perp D'$, $A'' \perp D''$, $C' \perp B'$, and $C'' \perp B''$, then $A' \cap D' \sim A'' \cap D''$ and $C' \cap B' \sim C'' \cap B''$.

By completeness (Axiom 4), $\succsim'$ is a connected relation; for let $A' = A \cap B$, $C' = C \cap D$, and use the axiom twice with $m = 1$ to obtain $D' \sim D$ with $A' \perp D'$ and $B' \sim B$ with $C' \perp B'$; then compare $A' \cap D'$ with $C' \cap B'$.

To show that $\succsim'$ is transitive, use completeness successively with $m = 1, 2,$
3, 4, and 5 to obtain $A' = A \cap B$, $B' \sim B$, $C' \sim C \cap D$, $D' \sim D$,
$E' \sim E \cap F$, and $F' \sim F$, such that $A', B', C', D', E', F'$ are $\perp$-independent.
By the remark above on the definition of $\succsim'$, we infer, from $A \mid B \succsim' C \mid D$
and $C \mid D \succsim' E \mid F$, that $A' \cap D' \succsim C' \cap B'$ and $C' \cap F' \succsim D' \cap E'$. By the
definition of $\perp$-independence, we have

$$A' \cap D' \perp C' \cap F', \qquad C' \cap B' \perp D' \cap E'.$$

Therefore, Axiom 3 yields

$$A' \cap F' \cap C' \cap D' \succsim E' \cap B' \cap C' \cap D'.$$

If not $A \mid B \succsim' E \mid F$, then $E' \cap B' \succ A' \cap F'$. If $C' \cap D' \succ \varnothing$, then the
strict clause of Axiom 3 yields $E' \cap B' \cap C' \cap D' \succ A' \cap F' \cap C' \cap D'$
(since $\perp$-independence gives $E' \cap B' \perp C' \cap D'$ and $A' \cap F' \perp C' \cap D'$).
This contradiction shows that $A \mid B \succsim E \mid F$ as required. On the other hand,
if $C' \cap D' \sim \varnothing$, then the strict clause of Axiom 3, applied to $C' \perp D'$,
$\varnothing \perp \varnothing$, and $D' \succ \varnothing$ ($D' \in \mathscr{E} - \mathscr{N}$) yields $C' \sim \varnothing$; thence,

$$C' \cap F' \sim \varnothing \succsim D' \cap E';$$

and since $D' \succ \varnothing$, the same arguments give $E' \sim \varnothing$ and $E' \cap B' \sim \varnothing$.
Thus, in this case also, $A' \cap F' \succsim E' \cap B'$, as required.

The second axiom of Definition 8, that $X \in \mathscr{E} - \mathscr{N}$ and $A \in \mathscr{N}$ iff
$A \mid X \sim' \varnothing \mid X$, is immediate; obviously, $A \mid X \sim' B \mid X$ iff $A \sim B$. Similarly,
the third axiom of Definition 8, that $X \mid X \succsim' A \mid B$ and $X \mid X \sim' A \mid A$, is
immediate; and the fourth axiom, $A \mid B \sim' A \cap B \mid B$, follows from the
definition of $\succsim'$. Thus, only the additivity and strong 6-tuple conditions need
to be established.

Suppose that $A \mid C \succsim' D \mid F$ and $B \mid C \succsim' E \mid F$, where $A \cap B = \varnothing = $
$D \cap E$. Without loss of generality, assume $A \cup B \subset C$, $D \cup E \subset F$. By
completeness, choose $C' \sim C$ with $C' \perp D$, $E$ and choose $F' \sim F$ with
$F' \perp A$, $B$. By definition of $\succsim'$ we have

$$A \cap F' \succsim D \cap C'$$
$$B \cap F' \succsim E \cap C'.$$

By Lemma 2, which uses only qualitative probability,

$$(A \cup B) \cap F' \succsim (D \cup E) \cap C',$$

and if either antecedent inequality is strict, so is the conclusion. By the disjoint
union property of $\perp$ [Axiom 2 (iii) of Definition 9] we have $C' \perp D \cup E$
and $F' \perp A \cup B$. Hence, $A \cup B \mid C \succsim' D \cup E \mid F$, with $\succ'$ holding if either
antecedent is strict.

Finally, suppose that $A \supset B \supset C$, $D \supset E \supset F$, $B \mid A \gtrsim F \mid E$ and $C \mid B \gtrsim' E \mid D$. Choose $A'$, $B'$, $C'$, $D'$, $E'$, $F'$, respectively $\sim A, B, C, D, E, F$ and $\perp$-independent (completeness with $m = 1, 2, 3, 4, 5$). We have $B' \cap E' \gtrsim F' \cap A'$ and $C' \cap D' \gtrsim E' \cap B'$ (with strict $\gtrsim'$ going to strict $\gtrsim$). By transitivity of $\gtrsim$, $C' \cap D' \gtrsim F' \cap A'$ (or strict inequality, if there is a strict antecedent), hence $C \mid A \gtrsim$ (or $>$, as required) $F \mid D$.  $\diamondsuit$

## EXERCISES

**1.** For each of the following three families $\mathscr{E}$ of sets, either show that $\mathscr{E}$ is an algebra of sets (Definition 1) or show a violation of one of the axioms.

(i) Let $X$ be a nonempty set and $A$ a nonempty subset of $X$. Then $\mathscr{E}$ consists of all subsets of $X$ that either include $A$ or are disjoint from $A$.

(ii) Let $X$ be a nonempty set and $A$ a nonempty subset of $X$. Then $\mathscr{E}$ consists of $\varnothing$ and all subsets of $X$ that intersect both $A$ and $-A$.

(iii) Let $\mathscr{F}$ and $\mathscr{G}$ be algebras of sets on $X$ and $Y$, respectively. Then $\mathscr{E}$ consists of those subsets of $X \cup Y$ of the form $A \cup B$ where $A$ is in $\mathscr{F}$ and $B$ is in $\mathscr{G}$.    (5.1)

**2.** Suppose that $\langle X, \mathscr{E}, P \rangle$ is a finitely additive probability space (Definition 2). Prove that, for all $A, B \in \mathscr{E}$

$$P(A \cup B) = P(A) + P(B) - P(A \cap B).   (5.1)$$

**3.** Prove the independence of the axioms of qualitative probability (Definition 4).    (5.2.1)

**4.** In a structure of qualitative probability (Definition 4), either prove that the relation $\sim^*$ (defined on p. 206 in connection with the definition of tight) is an equivalence relation or give a counterexample.    (5.2.3)

**5.** Suppose that $\langle X, \mathscr{E}, \gtrsim \rangle$ is a structure of qualitative probability (Definition 4) and that $A, B, C, D \in \mathscr{E}$ are such that $A \cup C = B \cup D = X$. Prove that if $A \gtrsim B$ and $C \gtrsim D$, then $A \cap C \gtrsim B \cap D$.    (5.2.1, 5.3.1)

**6.** Suppose that $\langle X, \mathscr{E}, \gtrsim \rangle$ is a structure of qualitative probability (Definition 4) and that $\mathscr{N} = \{A \mid A \in \mathscr{E} \text{ and } A \sim \phi\}$. Prove that $\mathscr{N}$ has the following properties:

(i) If $A \in \mathscr{N}$, $B \in \mathscr{E}$, and $B \subset A$, then $B \in \mathscr{N}$.

(ii) If $A, B \in \mathscr{N}$, then $A \cup B$, $A \cap B \in \mathscr{N}$.

(iii) If $A \in \mathscr{N}$, then $-A \in \mathscr{E} - \mathscr{N}$.    (5.2.1, 5.3.1)

**7.** Let $X = \{a, b, c, d, e\}$ and $\mathscr{E} = 2^X$. Suppose that $\gtrsim$ is a weak ordering of $\mathscr{E}$ such that:

   (i)  $\{a\} \sim \{b\}$;

  (ii)  $\{c\} \sim \{d\}$;

 (iii)  $\{a, b\} \sim \{c, d, e\}$;

 (iv)  $\{a, e\} \sim \{c, d\}$.

Find a probability representation consistent with $\gtrsim$. Is it unique?
$$(5.4.3, \ 5.5.3)$$

**8.** Does the probability space in Exercise 7 satisfy the hypotheses of Theorem 6? Can you define general circumstances in which a finite structure of qualitative probability has a unique representation?     (5.4.3, 9.2.2).

**9.** Suppose that if there exists an $A$ with $0 < P(A) < 1$, then $\langle X, \mathscr{E}, P \rangle$ is a finitely additive probability space (Definition 2) with the following property: for every $A$ in $\mathscr{E}$ with $0 < P(A) < 1$, there exists a $B$ in $\mathscr{E}$ with $0 < P(B) < 1$ that is independent of $A$ (p. 238). Prove that this space has no atoms (p. 216).     (5.1, 5.4.2, 5.8)

*In the following three exercises we suppose that $\langle X, \mathscr{E}, \mathscr{N}, \gtrsim \rangle$ is a structure of qualitative conditional probability (Definition 8, Section 5.6.1). In the proofs you may use the axioms of Definition 8 and the lemmas of Section 5.7.1, but not Theorem 7.*

**10.** If $A \ \varepsilon \mathscr{E}, B, C \varepsilon \mathscr{E} - \mathscr{N}, A \ \ B = A \ \ C$, and $B \ \ C$, then $A \mid B \gtrsim A \mid C$.
$$(5.6.1, \ 5.7.1)$$

**11.** If $A, A', B, B' \in \mathscr{E}, C, D \in \mathscr{E} - \mathscr{N}, A \mid C \gtrsim A' \mid D$, and $B \mid C \lesssim B' \mid D$, then $(A - B) \mid C \gtrsim (A' - B') \mid D$.     (5.6.1, 5.7.1)

**12.** Suppose that $A, B, C \in \mathscr{E}$ and that $C \cap B, \ C - B \in \mathscr{E} - \mathscr{N}$. Prove that $A \mid C \sim A \mid C \cap B$ iff $A \mid C \sim A \mid (C - B)$. Thus, letting $C = X$, if $A$ is independent of $B$ (see Section 5.8), then $A \mid B \sim A \mid -B$.
$$(5.6.1, \ 5.7.1, \ 5.8)$$

**13.** Prove Exercises 10–12 using the representation of Theorem 7.     (5.6.2)

# Chapter 6   *Additive Conjoint Measurement*

## ☐ 6.1   SEVERAL NOTIONS OF INDEPENDENCE

Not all attributes that we wish to measure have an internal, additive structure of the sort we studied in Chapters 3 and 5 (temperature and density are just two examples), and when no extensive concatenation is apparent we must use some other structure in measuring the attribute. One of the most common alternative structures is for the underlying entities to be composed of two or more components, each of which affects the attribute in question. A simple example is the attribute momentum which is exhibited by physical objects and which is affected both by their mass and by their velocity. Another example, discussed in Section 1.3.2, is the judged comfort of various humidity and temperature combinations.

The objects in these examples cannot readily be concatenated, nevertheless, they can be treated as composite entities. The rest of this volume, except for the last chapter, concerns theories leading to construction of measurement scales for composite objects which preserve their observed order with respect to the relevant attribute (e.g., preference, comfort, momentum) and where the scale value of each object is a function of the scale values of its components. Since such theories lead to simultaneous measurement of the objects and their components, they are called conjoint measurement theories. The present chapter deals with the additive case; more complicated rules of combination are studied in Chapters 7, 8, and 9.

The first part of this chapter (Sections 6.1–6.3) deals with the two-component case. Empirical examples from physics and psychology are presented in Section 6.4, and modifications of the basic theory are introduced in Sections 6.5 and 6.6. Exchange relations between components are often expressed as indifference curves, i.e., sets of objects that are equivalent with respect to the attribute under study. An analysis of additivity from this standpoint is included in Sections 6.7 and 6.8. Sections 6.9 and 6.10 apply conjoint measurement to certain structures with binary operations. Finally, the theory of the $n$-component case, which is surprisingly simpler than that of the two-component case, is investigated in Sections 6.11 and 6.12.

It is basic to the success of any conjoint measurement procedure that the attribute and the underlying entities exhibit two quite distinct forms of independence, which we now discuss.

### 6.1.1  Independent Realization of the Components

The first notion of independence is that the two components shall be independently realizable or, as is often said, that they are independent variables. We avoid the "variable" terminology because, among other reasons, it suggests that numerical representations already exist, and we do not want to assume that they do. Formally, all we are saying is that for any $a$ in $A_1$ and $p$ in $A_2$, $ap$ must be a realizable entity[1]; put another way, the value for each component can be chosen without regard to the value for the other. This postulate can fail for either of two quite different reasons.

1.  Practical considerations may limit us to some (nonetheless rich) subset $B$ of $A_1 \times A_2$. For example, in studies of momentum it may not be feasible to realize very rapidly moving, very massive bodies. Such a limitation is not conceptually inherent in the attribute, and presumably we should be able to construct a theory to encompass it in much the same way as we did for restricted concatenation in extensive measurement; however, as yet, no such theory exists, and so we are forced to confine ourselves to the cases where $B = A_1 \times A_2$. Why this is not an impossibly severe limitation and why, nonetheless, it should be eliminated is explained in Section 6.5.5.

2.  Alternatively, the constraint may be inherent because an empirical law relates the two components. Such laws, which are of great interest, are investigated in their own right (see Chapter 10). Density is a familiar example. Suppose that we have a homogeneous substance and let $A_1$ denote the continuum of possible masses and $A_2$, the continuum of possible volumes. Once the mass has been selected, then under fixed experimental conditions only one volume can occur. That is, mass and volume are not independent components for a given substance since $m = dV$, where $d$ is known as the

---

[1] To simplify the notation in this chapter, we write $ap$ rather than the more usual $(a, p)$ for ordered pairs.

density of the substance. Contrast this with momentum where, in principle, any velocity may be paired with any mass.

Theories of conjoint measurement apply only when the values of the components can be selected independently. Formally, this assumption is incorporated into the theory simply by assuming that $\gtrsim$ is a relation on the whole of $A_1 \times A_2$.

### 6.1.2 Decomposable Structures

A second notion of independence is that the two components contribute their effects independently to the attribute in question. This simply indicates the type of representation theorem we desire to prove. Specifically, we are interested in discovering axioms about the triple $\langle A_1, A_2, \gtrsim \rangle$ that, on the one hand, appear to be empirically valid laws for at least one interpretation of the axiom system and that, on the other hand, permit us to construct real-valued functions $\phi_i$ on $A_i$, $i = 1, 2$, and a function $F$ from Re $\times$ Re into Re, 1:1 in each variable, such that, for all $a, b$ in $A_1$ and $p, q$ in $A_2$,

$$ ap \gtrsim bq \qquad \text{iff} \qquad F[\phi_1(a), \phi_2(p)] \geqslant F[\phi_1(b), \phi_2(q)]. \qquad (1) $$

Put into words, there are numerical scales $\phi_i$ on the two components and a rule $F$ for combining them such that the resultant measure preserves the qualitative ordering of the attribute. When such a representation exists, we say that the structure $\langle A_1, A_2, \gtrsim \rangle$ is *decomposable*. Although Equation (1) is exceedingly general, which means that very many structures $\langle A_1, A_2, \gtrsim \rangle$ are decomposable, it is by no means a trivial restriction. To illustrate this, we provide at the end of the next subsection a class of interesting examples of nondecomposable structures.

### 6.1.3 Additive Independence

In those sciences heavily influenced by inferential statistics, particularly by the analysis of variance, the notion of components contributing non-interactively to a measure has come to have the very special meaning of additive independence, i.e., $F(x, y) = x + y$, and so Equation (1) reduces simply to

$$ ap \gtrsim bq \qquad \text{iff} \qquad \phi_1(a) + \phi_2(p) \geqslant \phi_1(b) + \phi_2(q). \qquad (2) $$

Although our concern in this chapter is exclusively with additive independence, we cannot overly stress the fact that any representation of the form given in Equation (1) establishes a measure of the attribute in which the components contribute to it independently, albeit nonadditively. In Chapters 7 and 8 we study some nonadditive models where $F$ is a polynomial function of its arguments. Other classes of functions are of interest, especially those that have arisen in physics, but as yet they have not been studied within a measurement context. We should perhaps add the informal and imprecise, but nonetheless important, observation that such measurement theories are likely to be taken seriously—or at least remembered—only when $F$ is a comparatively simple function.

Our aim is to axiomatize a relation $\gtrsim$ on $A_1 \times A_2$ so as to justify the representation of Equation (2), and to do this without introducing any other primitives, in particular, without any form of concatenation. This was done for finite equally spaced structures in Section 1.3.2, and the general idea was further discussed in connection with difference structures in Section 4.1.2. There we indicated the use of additivity to generate a standard sequence on one component by using repeated trade-offs with a fixed difference on the second one. This chapter develops the idea fully.

By setting $\psi_i = \exp \phi_i$, any additive representation into the real numbers can be replaced by the following multiplicative representation into the positive real numbers:

$$ap \gtrsim bq \quad \text{iff} \quad \psi_1(a)\,\psi_2(p) \geqslant \psi_1(b)\,\psi_2(q). \tag{3}$$

The general issue of alternative representations, as illustrated in Equation (3), is taken up in Section 6.5.2.

It is now convenient to give the example of nondecomposable structures that was promised earlier. Suppose that $\phi_i$ and $\psi_i$ are nonmonotonic, real-valued functions on $A_i$, $i = 1, 2$, and let the relation $\gtrsim$ be defined on $A_1 \times A_2$ by $ap \gtrsim bq$ iff

$$\phi_1(a) + \phi_2(p) + \psi_1(a)\,\psi_2(p) \geqslant \phi_1(b) + \phi_2(q) + \psi_1(b)\,\psi_2(q).$$

In the language of analysis of variance, there is an (additive) interaction term $\psi_1\psi_2$, which itself exhibits multiplicative independence. Simple though this representation is to state, nothing whatsoever is known about the relations for which it is a representation. Aside from the weak ordering of the relation, no other necessary conditions are known. A measurement theory for this special type of interaction might be of interest. Such structures are not decomposable in general since two distinct functions are required for each component.

### 6.1.4 Independent Relations

Clearly, if Equation (1) holds the relation must be a weak order. Our earlier comments (Sections 1.3 and 1.5) apply here without change, and so no more need be said about this assumption.

We turn next to features of the relation that are consequences of additive independence. By far, the most important of these properties is the one that we arrived at in Section 1.3.2, namely, that if two entities have a common value on one component, then the ordering is unaffected when that value is changed to any other common one. The simple proof (Section 1.4.1) bears repeating.

$$ap \gtrsim bp \quad \text{iff} \quad \phi_1(a) + \phi_2(p) \geqslant \phi_1(b) + \phi_2(p)$$
$$\text{iff} \quad \phi_1(a) \geqslant \phi_1(b) \quad [\text{subtracting} \quad \phi_2(p)]$$
$$\text{iff} \quad \phi_1(a) + \phi_2(q) \geqslant \phi_1(b) + \phi_2(q) \quad [\text{adding} \quad \phi_2(q)]$$
$$\text{iff} \quad aq \gtrsim bq.$$

This property is sufficiently important—nay, crucial—that we give it a formal definition and the name "independence"; some authors call it "monotonicity."

DEFINITION 1.   *A relation $\gtrsim$ on $A_1 \times A_2$ is independent[2] iff, for $a, b \in A_1$, $ap \gtrsim bp$ for some $p \in A_2$ implies that $aq \gtrsim bq$ for every $q \in A_2$; and, for $p, q \in A_2$, $ap \gtrsim aq$ for some $a \in A_1$ implies that $bp \gtrsim bq$ for every $b \in A_1$.*

Perhaps the most important feature of an independent relation is that it induces the following natural ordering on each component.

DEFINITION 2.   *Suppose that $\gtrsim$ is an independent relation on $A_1 \times A_2$.*

(i)   *Define $\gtrsim_1$ on $A_1$: for $a, b \in A_1$, $a \gtrsim_1 b$ iff for some $p \in A_2$, $ap \gtrsim bp$;* and

(ii)   *define $\gtrsim_2$ on $A_2$ similarly.*

The following results are immediate consequences of the definitions; the proofs are left as Exercise 1.

LEMMA 1.   *If $\gtrsim$ is an independent weak ordering of $A_1 \times A_2$, then*

(i)   *$\gtrsim_i$ of Definition 2 is a weak ordering of $A_i$.*

(ii)   *For $a, b \in A_1$ and $p, q \in A_2$, if $a \gtrsim_1 b$ and $p \gtrsim_2 q$, then $ap \gtrsim bq$.*

(iii)   *If either antecedent inequality of (ii) is strict, so is the conclusion.*

(iv)   *For $a, b \in A_1$ and $p, q \in A_2$, if $ap \sim bq$, then $a \gtrsim_1 b$ iff $q \gtrsim_2 p$.*

---

[2] In Definition 1.3 we defined the concept of independent conjoint structures in which $\gtrsim$ is a weak ordering as well as independent. Here we isolate the concept of independence. Note that this concept is unrelated to the independence relation defined on an algebra of sets (Definition 5.9, Section 5.8).

It should be noted that for any decomposable structure, i.e., one with a representation of the form given by Equation (1), if $F$ is strictly monotonic in each variable, then the relation is independent. The proof is left as Exercise 2.

## □ 6.2  ADDITIVE REPRESENTATION OF TWO COMPONENTS

### 6.2.1  Cancellation Axioms

Experimentalists prize the qualitative property of (additive) independence, and when data cross in the sense that $ap > bp$ and $bq > aq$ are both unambiguously true, they correctly conclude that no additive representation can possibly exist. Unfortunately, when independence is confirmed in a two-component experiment, we still cannot be at all confident that additivity holds, because other necessary consequences of additivity are not logical consequences of independence. They are not even consequences of it when certain restrictive structural assumptions are made; however, as we shall see in Sections 6.11 and 6.12, additivity does follow from an obvious generalization of independence to $n \geqslant 3$ components plus the structural conditions.

After independence, the next most simple consequence of an additive representation is derived as follows.

$$ax \gtrsim fq \quad \text{iff} \quad \phi_1(a) + \phi_2(x) \geqslant \phi_1(f) + \phi_2(q),$$
$$fp \gtrsim bx \quad \text{iff} \quad \phi_1(f) + \phi_2(p) \geqslant \phi_1(b) + \phi_2(x).$$

Adding the two numerical inequalities,

$$\phi_1(a) + \phi_2(x) + \phi_1(f) + \phi_2(p) \geqslant \phi_1(f) + \phi_2(q) + \phi_1(b) + \phi_2(x).$$

Since $\phi_1(f) + \phi_2(x)$ is common to both sides, it may be subtracted, leaving

$$\phi_1(a) + \phi_2(p) \geqslant \phi_1(b) + \phi_2(q)$$

which is equivalent to $ap \gtrsim bq$. Summarizing, if an additive representation exists, then the following property must also hold.

DEFINITION 3. *A relation $\gtrsim$ on $A_1 \times A_2$ satisfies* double cancellation *provided that, for every $a, b, f \in A_1$ and $p, q, x \in A_2$, if $ax \gtrsim fq$ and $fp \gtrsim bx$, then $ap \gtrsim bq$. The weaker condition in which $\gtrsim$ is replaced by $\sim$ is the* Thomsen condition.

To illustrate Definition 3, consider indifference curves of the form $\{bq \in A_1 \times A_2 \mid bq \sim ap\}$, displayed in Figure 1. The Thomsen condition, then, asserts that if $ax$ and $fq$ lie on the same indifference curve while the same holds for $fp$ and $bx$, then $ap$ and $bq$ also lie on the same indifference curve. This property has been studied extensively in the theory of webs, see e.g., Aczél, Belousov, and Hosszú (1960). Double cancellation can be interpreted similarly in terms of Figure 1.

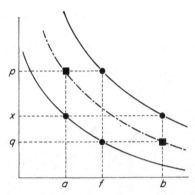

**FIGURE 1.** A graphical illustration of the Thomsen condition (a special case of double cancellation). The two solid indifference curves are those of the two given indifferences $ax \sim fq$ and $fp \sim bx$ (the four points are shown as dots). The conclusion, $ap \sim bq$, is that the two points shown as ■ must lie on a common (broken) indifference curve.

We refer to the property of Definition 3 as a cancellation property because the symbols $f$ and $x$ appear on both sides of the antecedent inequalities and they cancel in the sense that they do not appear in the conclusion. And we refer to it as double cancellation because it has two antecedent inequalities. Below we present a triple cancellation property which has three antecedent inequalities. Observe that were we to follow this terminology rigidly, independence would be called "single cancellation" and transitivity ($ap \gtrsim fx$ and $fx \gtrsim bq$ imply $ap \gtrsim bq$) would be a second form of double cancellation. Of course, other pairs of inequalities must be examined to see if they imply, via the representation, some third inequality. When we do this exhaustively, we discover nothing new: either they imply nothing at all (because the sum of the two numerical inequalities does not reduce to just one term from each component on each side), or they are another version of double cancellation, or they assert the transitivity of $\gtrsim$.

So, to discover further necessary conditions, we must turn to three or more inequalities in the hypothesis. By a completely parallel argument, we can show that the following is a necessary condition.

A relation $\gtrsim$ on $A_1 \times A_2$ satisfies triple cancellation *provided that, for every $a, b, f, g \in A_1$ and $p, q, x, y \in A_2$, if $ax \gtrsim by, fy \gtrsim gx$, and $gp \gtrsim fq$, then $ap \gtrsim bq$.*

This property with $\gtrsim$ replaced by $\sim$ is called the *Reidemeister condition* in the theory of webs. It is illustrated in Figure 2. Here there are three pairs of given, related points on three indifference curves, and the requirement is that two points defined in terms of the six lie on a fourth curve.

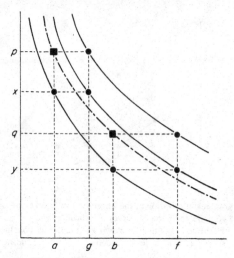

**FIGURE 2.** A graphical illustration of the Reidemeister condition (a special case of triple cancellation). The three given equivalences, $ax \sim by$, $fy \sim gx$, and $gp \sim fq$, are shown as dots on solid indifference curves. The conclusion $ap \sim bq$ is shown as ■ on a broken indifference curve.

An exhaustive examination of triples of inequalities reveals two more, inherently different, forms of triple cancellation. Note that in the definition of triple cancellation $a$ and $b$ both appear in one inequality and $p$ and $q$ both appear in another one. One obviously different case arises when $a$ and $b$ appear in two different inequalities and $p$ and $q$ both appear in the third, and yet another when $p$ and $q$ are in different ones and $a$ and $b$ are both in the third. The case where both $a$ and $b$ appear in two inequalities and $p$ and $q$ also appear in two inequalities does not need to be considered since the desired conclusion derives from independence, double cancellation, and transitivity; see Exercise 4. Restricted versions of these added types of triple cancellation are stated in Lemma 8, Section 6.6.1.

Clearly, we can continue to derive necessary conditions for any number of given inequalities, and it is evident that as the number of inequalities increases, so does the number of inherently different cancellation conditions. Our concern here, however, is to develop an axiom system that is adequate to prove the additive representation without assuming very many or very complex cancellation properties. That they are not all independent of each other seems evident. For example, if a reflexive relation on $A_1 \times A_2$ satisfies triple cancellation, then it is independent. The proof is left as Exercise 5.

Thus far, we have encountered in this volume several types of axioms that are necessary for the construction of a numerical representation of the additive type. In Chapter 3 we have studied the monotonicity axiom (Definition 3.1, Section 3.2.1) of the form $a \succsim b$ implies $a \circ c \succsim b \circ c$. Essentially the same property (see Axiom 3 of Definition 5.4, Section 5.2.1) plays a central role in Chapter 5, while a somewhat different form of the monotonicity axiom is employed in Chapter 4 (see, e.g., Axiom 3 of Definition 4.3, Section 4.4.1). Finally, two additional types of axioms, independence (p. 249) and cancellation (p. 250), are introduced in the present chapter. It is important to note that despite the different names and forms, all these axioms express essentially the same idea, namely the compatibility of the order with the additive operation. The form of each particular axiom depends on the form that the additive operation takes in each structure.

### 6.2.2 Archimedean Axiom

Our last necessary condition is an Archimedean axiom. As in the other theories we have examined, we first formulate a notion of a standard sequence and then the axiom simply asserts that every strictly bounded standard sequence is finite. In this context a standard sequence is a set of elements on one component that are equally spaced. We first formulate the definition and then discuss it.

DEFINITION 4.   *Suppose $\succsim$ is an independent weak ordering of $A_1 \times A_2$. For any set $N$ of consecutive integers (positive or negative, finite or infinite), a set $\{a_i \mid a_i \in A_1, i \in N\}$ is a* standard sequence *(on component 1) iff there exist $p, q \in A_2$ such that not $(p \sim_2 q)$ and for all $i, i + 1 \in N, a_i p \sim a_{i+1} q$. A standard sequence $\{a_i \mid i \in N\}$ is* strictly bounded *iff there exist $b, c \in A_1$ such that for all $i \in N, c \succ_1 a_i \succ_1 b$. A parallel definition holds for the second component.*

By Lemma 1, if $p \succ_2 q$, then $a_{i+1} \succ_1 a_i$ and the standard sequence is increasing, and if $q \succ_2 p$, then $a_i \succ_1 a_{i+1}$ and the sequence is decreasing. Moreover, each interval from $a_i$ to $a_{i+1}$ is equivalent to the interval from $p$ to $q$, and so they are equivalent to each other. Thus, such a sequence of intervals parallels the standard sequences of extensive and difference measurement, so we have used the same term again; and the argument for assuming the Archimedean property is also the same.

It is not especially easy to work with standard sequences in the plots of indifference curves we used in Figures 1 and 2. An alternative graphical device we have found useful is to plot, in the same units, the two coordinates of the additive representation as parallel, rather than perpendicular, straight lines. A standard sequence is a set of equally spaced points on one line. If a dotted line is located midway between and parallel to the two coordinates, then the ordering of the entities is given by the intersection of the straight line connecting its component values with the dotted line (obviously, this yields $[\phi_1(a) + \phi_2(p)]/2$, but the factor of $\frac{1}{2}$ does not affect the ordering; see Figure 3).

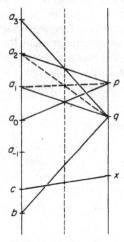

**FIGURE 3.** A second way to plot conjoint orderings. This is especially suited to showing standard sequences and inequalities and equivalences of specific pairs of points. Note that the line connecting $b$ with $q$ cuts the dotted midline above the point where the line connecting $c$ with $x$ cuts it. This corresponds to the algebraic inequality $bq > cx$. When two such lines intersect on the dotted line, as for example $a_0p$ and $a_1q$ do, this corresponds to the equivalence, $a_0p \sim a_1q$. Note that $a_0$, $a_1$, $a_2$, and $a_3$ form a standard sequence relative to $p$ and $q$.

### 6.2.3  Sufficient Conditions

The most important structural condition concerns the solution of equations of the form $ap \sim bq$ when three of the elements are given and the fourth is to be solved for. Conceptually, this property is closely related to the solvability axioms that have arisen in extensive, difference, and probability structures. As in those theories, the assumption that such solutions exist implies something about the two components $A_1$ and $A_2$: either they are dense in the way the real or rational numbers are or they are equally spaced in the way the integers are. In addition, in the presence of the other axioms, simple unrestricted solvability implies that both $A_1$ and $A_2$ are infinite and that the additive representation, assuming one exists, is unbounded. Generally, this is unsatisfactory. For example, suppose that we are measuring momentum, then we cannot help but use a bounded range of masses and of velocities. If $a$ is a mass near the upper bound and $b$ is one near the lower bound, and if $p$ is a velocity near its upper bound, then it is not possible to find a velocity $q$ within the bounded range such that $ap \sim bq$. In such an example, the failure to find a solution is not conceptually inherent to the attribute, but only to the data we propose to collect. In other cases, the failure may be conceptually inherent. For example, the subjective loudness of a tone depends upon its frequency as well as its intensity (witness the loudness control on good amplifiers), both of whose ranges are limited by properties—either auditory or pain—of the auditory system. The forms of this region and of the equal-loudness contours are shown in Figure 4. Some equal-loudness equations can be solved, but others cannot; an example of an impossible one is shown in the figure.

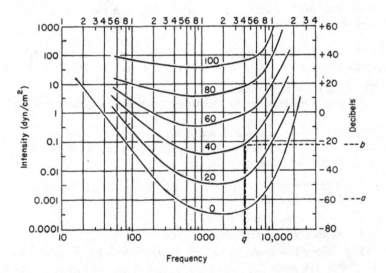

**FIGURE 4.** Equal-loudness contours. Each curve represents the intensity required at each frequency to produce a tone judged equal in loudness to a 1000-cps. tone whose level above threshold appears on the curve. Adapted from *Theory of Hearing* by E. G. Wever (1949, p. 307). The data are due to Fletcher and Munson (1933). Note that the equation $ap \sim bq$ cannot be solved for $p$.

In both examples, the continuity of the components makes clear that $ap \sim bq$ has a solution $b$ if and only if there exist $\bar{b}$, $\underline{b} \in A_1$ such that $\bar{b}q \gtrsim ap \gtrsim \underline{b}q$; a similar statement holds for the second component. With this restriction added, the solvability of equations in conjoint measurement has a similar empirical bite to Axiom 5 of Definition 4.1 (Section 4.2) in difference measurement.

**DEFINITION 5.** *A relation $\gtrsim$ on $A_1 \times A_2$ satisfies* unrestricted solvability *provided that, given three of $a$, $b \in A_1$ and $p$, $q \in A_2$, the fourth exists so that $ap \sim bq$; it satisfies* restricted solvability *provided that:*

(i) *whenever there exist $a$, $\bar{b}$, $\underline{b} \in A_1$ and $p$, $q \in A_2$ for which $\bar{b}q \gtrsim ap \gtrsim \underline{b}q$, then there exists $b \in A_1$ such that $bq \sim ap$;*

(ii) *a similar condition holds on the second component.*

Our second sufficient condition simply requires that each component actively affect the attribute in question. This is hardly a limiting assumption within the context of conjoint measurement.

**DEFINITION 6.** *Suppose that $\gtrsim$ is a relation on $A_1 \times A_2$. Component $A_1$ is* essential *iff there exist $a, b \in A_1$ and $p \in A_2$ such that not $(ap \sim bp)$. A similar definition holds for $A_2$.*

**LEMMA 2.** *Suppose that $\gtrsim$ is an independent relation on $A_1 \times A_2$. Then component $A_1$ is essential iff there exist $a, b \in A_1$ such that $a \succ_1 b$.*

The proof is left as Exercise 7.

The following definition formulates the basic two-component structure:

**DEFINITION 7.** *Suppose $A_1$ and $A_2$ are nonempty sets and $\gtrsim$ is a binary relation on $A_1 \times A_2$. The triple $\langle A_1, A_2, \gtrsim \rangle$ is an* additive conjoint structure *iff $\gtrsim$ satisfies the following six axioms:*

1. *Weak ordering.*

2. *Independence (Definition 1).*

3. *Thomsen condition (Definition 3).*

4. *Restricted solvability (Definition 5).*

5. *Archimedean property, i.e., every strictly bounded standard sequence (Definition 4) is finite.*

6. *Each component is essential (Definition 6).*

*The structure is* symmetric *iff, in addition,*

7. *For $a, b \in A_1$, there exist $p, q \in A_2$ such that $ap \sim bq$, and for $p', q' \in A_2$, there exist $a', b' \in A_1$ such that $a' p' \sim b'q'$.*

Examples in Section 6.3.1 show that neither independence nor the Thomsen condition can be derived from the remaining axioms. As the next theorem shows, we can drop the assumption of independence provided that we are willing to substitute double cancellation for the Thomsen condition and un-restricted for restricted solvability; the latter, however, is empirically unac-ceptable in many cases.

THEOREM 1. *Suppose $\langle A_1, A_2, \gtrsim \rangle$ is a structure for which the weak ordering, double cancellation, unrestricted solvability, and the Archimedean axioms hold. If at least one component is essential, then $\langle A_1, A_2, \gtrsim \rangle$ is a symmetric, additive conjoint structure, i.e., all axioms of Definition 7 hold.*

It may be worth mentioning why we call a structure "symmetric" when it satisfies Axiom 7. Intuitively, a symmetric structure is one in which, for any interval $(a, b)$ on the first component, we can find an interval $(q, p)$ on the second component that is equivalent to it, and vice versa. Assuming the additive representation, consider a plot of the structure in the plane using $\phi_1$ and $\phi_2$ as the coordinate values. The indifference curves, $\phi_1(a) + \phi_2(p) =$ constant, are all straight lines with slope $-1$. In the general case, the outline of the image of the structure is a rectangle, but when Axiom 7 holds it is a square, which is the symmetric rectangle.

### 6.2.4  Representation Theorem and Method of Proof

The basic representation theorem is:

THEOREM 2. *Suppose $\langle A_1, A_2, \gtrsim \rangle$ is an additive conjoint structure. Then there exist functions $\phi_i$ from $A_i$, $i = 1, 2$, into the real numbers such that, for all $a, b \in A_1$ and $p, q \in A_2$,*

$$ap \gtrsim bq \quad iff \quad \phi_1(a) + \phi_2(p) \geq \phi_1(b) + \phi_2(q).$$

*If $\phi_i'$ are two other functions with the same property, then there exist constants $\alpha > 0$, $\beta_1$, and $\beta_2$ such that*

$$\phi_1' = \alpha\phi_1 + \beta_1 \quad and \quad \phi_2' = \alpha\phi_2 + \beta_2.$$

Note that the change in unit, $\alpha$, is the same for both components.

Because the method of proof, due to Holman (1971), is of some interest, we discuss it here although the actual details are not given until Sections 6.3.3–6.3.4. We begin with the special case of a bounded, symmetric structure within which we are able to construct an extensive structure on each com-

ponent. Then the function described in Theorem 3.3 is used to prove
Theorem 2. Specifically, suppose $A_1 \times A_2$ forms a "square" bounded by
points $\bar{a} > \underline{a}$ and $\bar{p} > \underline{p}$ such that

$$\underline{a}\bar{p} \sim \bar{a}\underline{p} .$$

For any $a$ in $A_1$, we can define its image on $A_2$, $\pi(a)$, as the solution to
$\underline{a}\pi(a) \sim a\underline{p}$. In essence, the interval from $\underline{p}$ to $\pi(a)$ on $A_2$ is equivalent to
the interval from $\underline{a}$ to $a$ on $A_1$. Now, with $a, b$ in $A_1$, we can form their
concatenation by finding the image of one on $A_2$ and finding the element
$a \circ b$ on $A_1$ that solves $(a \circ b) \underline{p} \sim a\pi(b)$. The only problem is to restrict our
attention to pairs $a, b$ for which a solution exists.

DEFINITION 8. *Suppose $\langle A_1, A_2, \gtrsim \rangle$ is a symmetric, additive conjoint
structure. It is* bounded *iff there are $\underline{a}, \bar{a} \in A_1$, $\underline{p}, \bar{p} \in A_2$ such that*

$$\underline{a}\bar{p} \sim \bar{a}\underline{p}$$

*and, for $a \in A_1$ and $p \in A_2$,*

$$\bar{a} \gtrsim_1 a \gtrsim_1 \underline{a} \quad \text{and} \quad \bar{p} \gtrsim_2 p \gtrsim_2 \underline{p}.$$

*Moreover, for $a, b \in A$, we define: $\pi(a) \in A_2$ is the (unique up to $\sim_2$) solution to
$\underline{a}\pi(a) \sim a\underline{p}$; $B_1 = \{ab \mid a, b >_1 \underline{a}$ and $\bar{a}\underline{p} \gtrsim a\pi(b)\}$; for $ab \in B_1$, $a \circ b$ is the
(unique up to $\sim_1$) solution to $(a \circ b) \underline{p} \sim a\pi(b)$. Similar definitions hold for $A_2$
with $\alpha(p)$ playing the role of $\pi(a)$.*

These defined notions play a key role in the proof of the following intermediate result (Lemma 5):

*If $\langle A_1, A_2, \gtrsim \rangle$ is a bounded, symmetric, additive conjoint structure, and if
$B_1$ is nonempty, then $\langle A_1, \gtrsim_1, B_1, \circ \rangle$ is an extensive structure with no
essential maximum (Definition 3.3).*

To prove Theorem 2 in the general case of a nonsymmetric structure, we
show that it includes a substructure that is bounded and symmetric (actually,
it usually includes many) and that there is a systematic way to extend the
additive representation of the substructure to the whole of the original
structure.

If triple cancellation (p. 251) is assumed in place of independence and the
Thomsen condition, the conclusions of Theorem 2 still hold, and the proof
is easier and more intuitive, being based directly on positive-difference
structures as discussed in Section 4.1.2. We feel that the present version of
Theorem 2 is more attractive, since independence and the Thomsen condition
are conceptually simpler and are ordinarily easier to test than triple cancellation. But for purposes of understanding the proof and constructing the
representation, triple cancellation is superior.

The proof based on triple cancellation begins with the very simple proof of independence, which allows a weak order $\gtrsim_i$ to be defined on $A_i$. Then the set $A_1^*$ of positive $A_1$ differences is defined (it coincides with the relation $\succ_1$) and is ordered by

$ab \gtrsim_1^* a'b'$ *iff there exist* $x, y \in A_2$ *with* $ax \gtrsim by, b'y \gtrsim a'x$.

Triple cancellation is used to show that $\gtrsim_1^*$ is a weak order; one then easily shows that $\langle A_1, A_1^*, \gtrsim_1^* \rangle$ is a positive-difference structure. The monotonicity axiom, key to positive-difference structures, follows from transitivity of $\gtrsim$, if the symmetry of the structure is exploited to obtain $x, y, z$ with

$$ax \gtrsim by, \qquad b'y \gtrsim a'x,$$
$$by \gtrsim cz, \qquad c'z \gtrsim b'y.$$

The reader will find it instructive to diagram this configuration.

Alternative proofs based on positive-difference structures, using either double or triple cancellation, are developed through Exercises 11–18.

### 6.2.5 Historical Note

Additive conjoint measurement has been of concern for some time, although other names have been commonly used for it. In the 19th century, utility functions were sometimes assumed to be additive over the components of commodity bundles, but interest in that hypothesis faded when it was recognized that it could not be generally valid—e.g., it surely fails when there are complementary commodities such as bread and butter or record players and records—and when Pareto argued that ordinal utility functions were adequate for the economic theory of that time (recall Lange's criticism mentioned in Section 4.1.1). In this century, statisticians and, therefore, the consumers of statistics have frequently postulated additive models (especially, in the analysis of variance) which were, and still are, used to test hypotheses that several variables contribute noninteractively, i.e., additively to an overall measure of some attribute. In both cases, and in others like them, the additive representation was assumed, not justified; only recently has there been progress in finding qualitative conditions that are necessary and/or sufficient for such an additive representation to exist.

Early analyses of additivity were often related to marginal utility ideas. The work of Frisch (1926, 1932, 1937) is noteworthy for its formal axiomatic character. Fisher (1892, 1927), while informal and discursive, included many of the key qualitative ideas: independence (1927, p. 175ff), connectedness (p. 179ff), solvability in the form of highly divisible commodities (p. 181), and something resembling a standard sequence (p. 183). It was not, however, until after World War II that the whole representation-theorem approach was fully developed, in part stimulated by the successful axiomatization of expected-utility theory by von Neumann and Morgenstern (1947). Subsequent developments in the context of expected-utility theory are reviewed in Section 8.6.

Adams and Fagot (1959) discussed necessary cancellation properties, including those that we have defined explicitly (and in a preliminary technical report, 1956, they obtained partial results about conditions sufficient for a representation to exist). In the following year, Debreu published a full-blown representation theorem. It is somewhat like those given here but, as with much of the economic literature, some of the assumptions were cast in topological terms. Nevertheless, in some respects it was the best result available until recently. Debreu's result and an improved version of it are given in Section 6.11. His proof reduced the problem to a theorem of Blaschke (Blaschke & Bol, 1938) that states conditions under which it is possible to map the plane homeomorphically into itself in such a way that three systems of curves map into three systems of parallel straight lines (Theorem 7, Section 6.7.1). This result of web theory was reformulated in algebraic terms by Aczél, Belousov, and Hosszú (1960) and Aczél, Pickert, and Radó (1960), and it is possible to show from their results that a system with unrestricted solvability has an additive representation. Luce and Tukey (1964), who were unaware of these two papers, constructed the representation directly without drawing upon other theorems. An alternative proof, based on Hölder's theorem, plus some other results were given by Krantz (1964). He defined concatenation on $A_1 \times A_2$ as follows: Let $a_0$ in $A_1$ and $p_0$ in $A_2$ be fixed, and for $ap, bq$ in $A_1 \times A_2$, let $c$ solve $cp_0 \sim ap$ and $r$ solve $a_0 r \sim bq$. Define $cr = ap \circ bq$. The proof given here is still different from any of these.

Although restricted solvability has appeared in other contexts, e.g., Debreu (1960) and Pfanzagl (1959a), it was first studied in an algebraic version of conjoint measurement by Luce (1966). [Somewhat parallel papers in web theory are Radó (1965) and Havel (1966).] Luce's system differs from the one given in Definition 7 in that instead of assuming that each component is essential (Axiom 6), a somewhat similar but considerably more complex condition, phrased in the language of standard sequences, was used. His proof of the representation involved imbedding the given system isomorphically into an artificially constructed system with unrestricted solvability. The restriction of the representation of that system to the image of the given system completed the proof. Holman's proof, which we give, constructs extensive structures on each component. It is the most transparent one that we know. An alternative to his, which assumes double cancellation and uses positive-difference structures, is outlined in Exercises 11–18; however, it is longer than Holman's proof.

References to other, related papers which are concerned with necessary and sufficient conditions for an additive representation (Chapter 9), with generalizations to $n$ components (Section 6.11), and with nonadditive representations (Chapters 7 and 8) are cited as these topics are discussed. For a compact summary of the major features of the various theories up to 1966, including a number of theories that combine additivity with the expected utility hypothesis (Chapter 8), see Fishburn (1966, 1967a, 1970).

## 6.3 PROOFS

### 6.3.1 Independence of the Axioms of Definition 7

The routine verification of the following examples is left as Exercise 10.

1.  $A_1 = A_2 = \text{Re}$; $ap \gtrsim a'p'$ iff either

    (i)  $a > 0 > a'$

or

    (ii) $a = a', p \geqslant p'$

This relation satisfies Axioms 2–6 of Definition 7 but not weak ordering; it is transitive and reflexive, but not connected; there are no standard sequences. An example of a relation that satisfies Axioms 2–6 and is connected but not transitive is given by $A_1 = \{a, a'\}$, $A_2 = \{p, p'\}$, $a' >_1 a$, $p' >_2 p$, $a'p \sim ap'$, and $a'p' \sim ap$.

2.  $A_1 = A_2 = \text{Re} - \{0\}$ and $ap \gtrsim bq$ iff $ap \geqslant bq$. Independence fails since $p$ and $q$ may have opposite signs.

3.  $A_1 = A_2 = \text{Re}^+$ and $ap \gtrsim bq$ iff $\phi(a, p) \geqslant \phi(b, q)$, where

$$\phi(a, p) = \begin{cases} ap + a^p, & \text{if } a, p \geqslant 1, \\ a(p + 1), & \text{if } 0 < a \leqslant 1 \leqslant p, \\ 2ap, & \text{if } 0 < p \leqslant 1. \end{cases}$$

The Thomsen condition fails with $a = \frac{1}{2}, b = \frac{2}{3}, f = 1, p = 2, q = \frac{1}{2}, x = 1$.

4.  $A_1 = A_2 = \{2^n \mid n \text{ is positive integer}\}$ and $ap \gtrsim bq$ iff $a + p \geqslant b + q$. Axiom 4 fails because $2^2 + 2^3 > 2^1 + 2^3 > 2^2 + 2^2$.

5.  Let $\mathscr{A} = \langle A_1, A_2, \gtrsim \rangle$ and $\mathscr{B} = \langle B_1, B_2, \gtrsim^* \rangle$ be any two additive conjoint structures for which there is an infinite standard sequence $\{b_i\}$ on component $B_1$. Select any $a_0 \in A_1$ and $p_0 \in A_2$ and form the structure $\langle C_1, C_2, \gtrsim \rangle$, where $C_1 = (A_1 - \mathbf{a_0}) \cup B_1$, $C_2 = (A_2 - \mathbf{p_0}) \cup B_2$, and for $a, b \in C_1$ and $p, q \in C_2$, define $ap \gtrsim bq$ iff either (1) $a, b \in B_1, p, q \in B_2$, and $ap \gtrsim^* bq$ or (2) at least one of the elements $a, b, p, q$ is from $\mathscr{A}$ and that when

those that are from $\mathscr{B}$ are replaced by $a_0$ or $p_0$, as is appropriate, the resulting inequality in $\mathcal{C}$ is a valid one. This structure is not Archimedean because $\{b_i\}$, which by choice is infinite, is obviously bounded, and it is routine to show that the other axioms hold. (Theorem 3 of Section 6.5.1 makes clear that the example must be of this character.)

It is perhaps worth commenting in this connection that in the absence of the Archimedean postulate, triple (and many other forms of) cancellation cannot be proved to hold. For suppose, in the above example, we have $a_1$ in $A_1$ and $p_1$ in $P_1$ such that $a_0 p_1 \sim a_1 p_0$ and a standard sequence $\{q_j\}$ in $B_2$ such that $b_i q_{j-1} \sim b_{i-1} q_j$. Then, $a_1 q_i \sim a_1 q_j$, $b_{i+1} q_j \succ b_j q_i$, and $b_j p_1 \sim b_{i+1} p_1$, but it is not true that $a_1 p_1 \succ a_1 p_1$.

6.   $A_1 = \{1\}$, $A_2 = \mathrm{I}^+$, $ap \succsim bq$ iff $a + p \geqslant b + q$.

### 6.3.2   Theorem 1   (p. 257)

*If a structure satisfies the weak ordering, double cancellation, unrestricted solvability, and Archimedean axioms, and if one component is essential, then it also satisfies independence, the other component is essential, and the structure is symmetric.*

PROOF.   *Independence.*   Suppose $ap \succsim bp$ and $q \in A_2$. By unrestricted solvability, there exists $c \in A_1$ such that $bq \sim cp$. By double cancellation, $aq \succsim cp \sim bq$. The result follows by transitivity. The other case is similar.

*Essentialness.*   Suppose that $A_1$ is essential. Then by Lemma 2 there exist $a, b \in A_1$ such that $a \succ_1 b$. For any $p \in A_2$, unrestricted solvability insures that $q$ exists for which $ap \sim bq$, and by Lemma 1 $q \succ_2 p$. So $A_2$ is also essential.

*Symmetry.*   If $a, b \in A_1$, then for any $p \in A_2$, unrestricted solvability insures that there is a $q \in A_2$ such that $ap \sim bq$, and similarly for $p', q' \in A_2$.   $\diamond$

### 6.3.3   Preliminary Lemmas for Bounded Symmetric Structures

The hypothesis of the following four lemmas is that $\langle A_1, A_2, \succsim \rangle$ is a bounded, symmetric, additive conjoint structure. The symbols $\pi$, $B_1$, and $\circ$ are defined in Definition 8 (Section 6.2.4). Some lemmas are stated only for the first component; a similar statement holds for the second.

LEMMA 3.   *The operation $\circ$ is well defined on $B_1$.*

PROOF.   Since

$$\underline{a}\bar{p} \sim \bar{a}p \succsim ap \succsim \underline{a}p,$$

$\pi(a)$ exists by restricted solvability. By definition of $B_1$,

$$\bar{a}p \gtrsim a\pi(b) \gtrsim ap,$$

so $a \circ b$ exists by restricted solvability. It is unique up to $\sim_1$ by independence. ◇

LEMMA 4. $a\pi(b) \sim b\pi(a)$, $B_1$ is symmetric, and $a \circ b \sim_1 b \circ a$.

*PROOF.* The Thomsen condition yields the first statement, the second is immediate, and the third follows from Lemma 1. ◇

LEMMA 5. *If* $\langle A_1, A_2, \gtrsim \rangle$ *is a bounded, symmetric, additive conjoint structure and if $B_1$ is nonempty, then* $\langle A_1, \gtrsim_1, B_1, \circ \rangle$ *satisfies Definition 3.3 (Section 3.4.3).*

*PROOF.* We prove, out of order, the six axioms of Definition 3.3.

1.  $\gtrsim_1$ is a weak order by Lemma 1.
5.  Since $b \succ_1 a$, Lemma 1 implies $\pi(b) \succ_2 p$, so

$$(a \circ b)\, p \sim a\pi(b) \succ ap,$$

whence $a \circ b \succ_1 a$.

2.  Suppose $ab, (a \circ b)\, c \in B_1$. By what we have shown.

$$\bar{a}p \gtrsim (a \circ b)\, \pi(c) \succ b\pi(c),$$

so $bc \in B_1$. The Thomsen condition on

$$b\pi(a) \sim (a \circ b)\, p \qquad \text{and} \qquad (b \circ c)\, p \sim b\pi(c)$$

yields $(b \circ c)\, \pi(a) \sim (a \circ b)\, \pi(c)$. Thus, by Lemma 4,

$$\bar{a}p \gtrsim (a \circ b)\, \pi(c) \sim (b \circ c)\, \pi(a) \sim a\pi(b \circ c),$$

so $a(b \circ c) \in B_1$. Moreover,

$$[a \circ (b \circ c)]\, p \sim a\pi(b \circ c) \sim (b \circ c)\, \pi(a) \sim (a \circ b)\, \pi(c) \sim [(a \circ b) \circ c]\, p.$$

3.  Suppose $ac \in B_1$ and $a \succ_1 b$. By independence and Lemma 4, $a\pi(c) \gtrsim b\pi(c) \sim c\pi(b)$, so $cb \in B_1$ and $a \circ c \gtrsim c \circ b$.

4.   Suppose $a >_1 b$. Since

$$a\pi(b) > ap \gtrsim bp \sim \underline{a}\pi(b),$$

restricted solvability implies the existence of $c$ such that

$$ap \sim c\pi(b) \sim b\pi(c) \sim (b \circ c) p,$$

whence $b \circ c \sim_1 a$.

6.   Defining $na$ in the usual way, we see that

$$[na]p \sim [(n-1) a \circ a]p \sim [(n-1) a]\pi(a),$$

so $na$ is a conjoint standard sequence relative to $p, \pi(a)$. Since it is bounded, it is finite.                                                                            $\diamondsuit$

LEMMA 6.   *If $ap, bq \lesssim \bar{a}p$, then*

$$ap \gtrsim bq \qquad iff \qquad a \circ \alpha(p) \gtrsim_1 b \circ \alpha(q).$$

*PROOF.*   Observe,

$$\underline{a}p \sim \alpha(p) p \sim \underline{a}\pi[\alpha(p)],$$

so by independence,

$$ap \sim a\pi[\alpha(p)] \sim [a \circ \alpha(p)] p.$$

Thus $ap \gtrsim bq$ iff $[a \circ \alpha(p)]p \gtrsim [b \circ \alpha(q)]p$, and the result follows by independence.                                                                    $\diamondsuit$

### 6.3.4   Theorem 2   (p. 257)

*Suppose that $\langle A_1, A_2, \gtrsim \rangle$ is an additive conjoint structure. Then it has an additive representation that is unique up to positive linear transformations with a common unit.*

*PROOF.*   First assume a bounded symmetric structure. If $B_1 = \varnothing$, there is no $a$ with $\bar{a} >_1 a >_1 \underline{a}$. [Otherwise, solving $ap \sim \bar{a}p$ for $p$ yields $a\alpha(p) \in B_1$.] Hence $\phi_1(\bar{a}) = \phi_2(\bar{p}) = 1$, $\phi_1(\underline{a}) = \phi_2(\underline{p}) = 0$ is an additive representation. If $B_1 \neq \varnothing$, then Lemma 5 and Theorem 3.3 yield a representation $\phi_1$ for $\langle A_1, \gtrsim_1, B_1, \circ \rangle$. For $p \in A_2$, choose $\alpha(p)$ with $\alpha(p) p \sim \underline{a}p$ and define $\phi_2(p) = \phi_1[\alpha(p)]$. By Lemma 6 and additivity of $\phi_1$, we have, for $ap, bq \lesssim \bar{a}p$, $ap \gtrsim bq$ iff $\phi_1(a) + \phi_2(p) \geqslant \phi_1(b) + \phi_2(q)$. To show that additivity holds throughout the structure, it suffices to consider the case where $ap$, $bq \gtrsim \bar{a}p$. Suppose $ap \sim bq$, and, with no loss of generality, $b \gtrsim_1 a$ and $p \gtrsim_2 q$. Since $b\bar{p} \gtrsim \underline{a}\bar{p} \sim \bar{a}p \gtrsim bp$, there exists $r$ such that $br \sim \underline{a}\bar{p}$. Similarly, there exists $c$ such that $cp \sim \underline{a}\bar{p}$. By the Thomsen condition, $ap \sim bq$ and $br \sim cp$ yield $ar \sim cq$. By independence, $\underline{a}\bar{p} \gtrsim ar$, $cq$. So, by what we have already shown,

$$\phi_1(b) + \phi_2(r) = \phi_1(c) + \phi_2(p) \qquad \text{and} \qquad \phi_1(a) + \phi_2(r) = \phi_1(c) + \phi_2(q),$$

whence

$$\phi_1(a) + \phi_2(p) = \phi_1(b) + \phi_2(q).$$

If $ap > bq$, then by restricted solvability, there exists $d$ with $a >_1 d$ and $dp \sim bq$. So

$$\phi_1(a) + \phi_2(p) > \phi_1(d) + \phi_2(p) = \phi_1(b) + \phi_2(q),$$

as required.

We now drop the assumption that $\langle A_1, A_2, \gtrsim \rangle$ is bounded and symmetric. By essentialness and restricted solvability, there exist $\bar{a} >_1 \underline{a}$ and $\bar{p} >_2 \underline{p}$ with $\bar{a}\underline{p} \sim \underline{a}\bar{p}$. It is readily shown that these define the boundaries of a bounded symmetric conjoint structure. It suffices to show that the additive representation on any such bounded symmetric substructure can be extended to $\langle A_1, A_2, \gtrsim \rangle$. We use the following lemma.

LEMMA 7. *Let* $\bar{a}, \underline{a}, \bar{p}, \underline{p}$ *be boundaries of a bounded symmetric substructure of* $\langle A_1, A_2, \gtrsim \rangle$. *If* $a >_1 \bar{a}$, *then there is a decreasing standard sequence, denoted* $a, a - 1,..., a - n$, *relative to* $\bar{p}, \underline{p}$, *with* $\bar{a} \gtrsim_1 a - n >_1 \underline{a}$; $n$ *is* $\geqslant 1$ *and unique. Likewise for* $\underline{a} >_1 a$, *there is an increasing standard sequence* $a, a + 1,..., a + m$, *relative to* $\underline{p}, \bar{p}$, *with* $\bar{a} >_1 a + m \gtrsim_1 \underline{a}$, *with* $m \geqslant 1$ *and unique. Similar statements hold for* $A_2$.

PROOF. Let $\{a_i\}$ be a maximal standard sequence with $a_1 = \bar{a}$, $a_0 = \underline{a}$. For $a >_1 \bar{a}$, let $n \geqslant 1$ be maximal with $a >_1 a_n$. We have $a_n\bar{p} \gtrsim ap$; this follows from $a_{n+1} \gtrsim_1 a$ and $a_{n+1}p \sim a_n\bar{p}$, if $a_{n+1}$ exists; otherwise, it follows from maximality of $\{a_i\}$ since if $ap \gtrsim a_n\bar{p}$, $a_{n+1}$ can be constructed. By restricted solvability we can find $a - 1$ with $(a - 1)\bar{p} \sim ap$; clearly, $a_n \gtrsim_1 a - 1 >_1 a_{n-1}$. The result now follows easily by induction, and the other statements are proved analogously.                                    ◇

Returning to the main proof, we extend an additive representation $\phi_1$, $\phi_2$, on a bounded symmetric substructure, by normalizing $\phi_1(\bar{a}) = 1 = \phi_2(\bar{p})$, $\phi_1(\underline{a}) = 0 = \phi_2(\underline{p})$, and letting $\phi_1(a) = n + \phi_1(a - n)$ or $-m + \phi_1(a + m)$, depending on which construction of Lemma 7 applies. Extend $\phi_2$ similarly. These extensions are clearly necessary if $\phi_1$, $\phi_2$ are to be additive; this proves the uniqueness part of Theorem 2, since the uniqueness on a bounded symmetric substructure follows from Theorem 3.3. We show that the extended functions are an additive representation.

First, it can easily be shown that $\phi_i$ preserves $\succsim_i$, $i = 1, 2$. It therefore suffices to show that $ap \sim bq$ implies $\phi_1(a) + \phi_2(p) = \phi_1(b) + \phi_2(q)$. For then, if $ap \succ bq$, with, say, $a \succ_1 b$, solve $a'p \sim bq$; then $\phi_1(a) + \phi_2(p) > \phi_1(a') + \phi_2(p) = \phi_1(b) + \phi_2(q)$.

Suppose $ap \sim bq$, with $p$, $q$ between $\bar{p}$, $\underline{p}$. Take $a \succ_1 b$ and $a \succ_1 \bar{a}$; the proof is similar in other cases. First, suppose that the integer $n$ of Lemma 7 is the same for $a$, $b$. Solve $a'\bar{p} \sim ap$, $b'\bar{p} \sim bq$. By the Thomsen condition [using $ap \sim (a - 1)\,\bar{p}$, etc.] we have $a'p \sim (a - 1)\,p$ and $b'p \sim (b - 1)\,q$, hence, $(a - 1)\,p \sim (b - 1)\,q$. Inductively, $(a - n)\,p \sim (b - n)\,q$, and $\phi_1(a) + \phi_2(p) = \phi_1(b) + \phi_2(q)$ follows. If the integers of Lemma 7 are different for $a$, $b$, construct a standard sequence $a_0$, $a_1$, ..., $a_n$ with $a_0 = \underline{a}$, $a_1 = \bar{a}$, $a \succ_1 a_n \succsim_1 b$, and $a_n \bar{p} \succsim ap$. Solve $a_n p' \sim ap$. Necessarily $b \succ_1 a_{n-1}$. By the preceding argument we have $(a - n)\,p \sim a_0 p'$ and $(b - n + 1)\,q \sim a_1 p'$, whence the required result follows.

The above argument shows that there is an additive representation for $A_1 \times \{p \mid \bar{p} \succsim_2 p \succsim_2 \underline{p}\}$, for any $\bar{p} \succ_2 \underline{p}$ such that there exist $\bar{a}$, $\underline{a}$ with $\underline{a}p \sim \bar{a}\underline{p}$. A symmetric argument applies for any $\bar{a}$, $\underline{a}$ and $A_2$. Therefore triple cancellation (p. 251) holds in $\langle A_1, A_2, \succsim \rangle$, since it can be deduced from the representation over an appropriate substructure. We now conclude the proof using triple cancellation.

If $a \succ_1 \bar{a}$ and $p \succ_2 \underline{p}$ we use triple cancellation to get $ap \sim (a - 1)(p + 1)$, and by definition,

$$\phi_1(a) + \phi_2(p) = \phi_1(a - 1) + \phi_2(p + 1).$$

This can be continued until either $\bar{a} \succsim_1 (a - n)$ or $(p + m) \succsim_2 \bar{p}$. Thus, if $ap \sim bq$, we can restrict attention to two cases: $a, b \succsim_1 \underline{a}$ and $p, q \succsim_2 \bar{p}$, or else $\bar{a} \succsim_1 a, b$ and $\bar{p} \succsim_2 p, q$. By symmetry we consider only the first. By what has been shown above, it suffices to consider only the case where $a \succ_1 \bar{a}$ and $q \succ_2 \bar{p}$. By triple cancellation, $(a - 1)\,q \sim a(q - 1)$, whence, by the Thomsen condition, $(a - 1)\,p \sim b(q - 1)$. This can be continued until $(a - i)(p - j) \sim (b - m)(q - n)$, where $0 \leqslant i, j, m, n$ and $i + j = m + n$ and at least three of the four elements $a - i, b - m, p - j, q - n$ are between $\bar{a}$, $\underline{a}$ or between $\bar{p}$, $\underline{p}$. By the previously considered case,

$$\phi_1(a - i) + \phi_2(p - j) = \phi_1(b - m) + \phi_2(q - n),$$

and the result follows.

## □ 6.4  EMPIRICAL EXAMPLES

### 6.4.1  Examples from Physics

We cite only two physical examples for which additive conjoint measurement provides a possible method of fundamental measurement; the reader will think of many others. The first is momentum (p. 246). It is trivial to show that if our qualitative ordering is determined by $p = mv$ and if both $m$ and $v$ are onto $\mathrm{Re}^+$, the axioms of Definition 7 must be satisfied. Therefore, by Theorem 2 expressed in multiplicative form [Equation (3), Section 6.1.3], there exists a representation $\psi_1(m)\,\psi_2(v)$, where the $\psi_i$ are both strictly increasing functions. This representation is unique up to power transformations with common exponent, i.e., any other representation has the form $\psi_1' = \beta_1\psi_1{}^\alpha$, $\psi_2' = \beta_2\psi_2{}^\alpha$, where $\alpha$, $\beta_1$, $\beta_2 > 0$.

We seem to have failed on two scores to measure momentum in the same way as the usual $mv$ formula. First, we have determined the measure only up to a free positive exponent. However, this difference is only apparent since we could rewrite all of physics in terms of $p' = p^\alpha = m^\alpha v^\alpha$, where $\alpha > 0$. It is pure convention that we choose $\alpha = 1$. What is not a convention however, is that when mass and velocity are measured in the usual extensive fashion, the ratio of their exponents is 1 rather than some other number. This ratio establishes the trading relation, in this case, $x{:}1/x$ between $m$ and $v$ in their contributions to momentum. For kinetic energy, $\tfrac{1}{2}mv^2$, the trading relation is $x{:}1/x^{1/2}$. Second, and rather more subtle, we have merely established that the mass and velocity contributions to the conjoint measures of momentum are some arbitrary functions of, respectively, the extensive scales of mass and velocity, and not necessarily the same function at that. It is clear that we shall be satisfied with the conjoint measures only if we can show that

$$\psi_1(m) = \beta_1 m^\alpha \qquad \text{and} \qquad \psi_2(v) = \beta_2 v^\alpha.$$

Suffice it to say for now that this can be proved when we take into account the trading relation given by $mv$ and the fact that both $m$ and $v$ are extensive measures. Since this problem is concerned with the form of numerical laws, we postpone its formulation and solution until Chapter 10.

Our second example, which is similar and so can be treated briefly, concerns the measurement of the temperature of an ideal gas. Recall that if $p$ denotes the pressure, $V$ the volume, and $T$ the absolute (Kelvin) temperature, then for an ideal gas there exists a constant $R$ such that $pV = RT$, or rewriting, $p = RT/V$. Thus, if we use pressure to order our observations and manipulate both the volume of and the heat applied to the gas, then a multiplicative representation can be found, one term of which is the temperature contribution to the pressure. It will require a consideration of trading relations, somewhat like those for momentum, to show that the conjoint measures $\psi_1\psi_2$ and $\psi_2$ are power functions of $p$ and $V$, that the ratio of their exponents is $-1$, and that $T$ is defined by $\psi_1$. All of this is explored in great detail in Section 10.7.

It is a curious fact of history that neither physicists nor philosophers of science have undertaken a serious qualitative analysis of the multiplicative structure of physical variables—certainly not to the same extent and depth that they have studied additivity within a single attribute. One might have expected the theories of extensive and multiplicative conjoint measurement to have been developed at about the same time and by the same people; in fact, the latter theory appeared more than half a century after the former and it was created by behavioral scientists and statisticians, not by physicists or philosophers of physics.

### 6.4.2   Examples from the Behavioral Sciences

Many empirical investigations in the social sciences have focused on exchange relations among two or more variables. In economics, the study of substitution rates and indifference curves play a central role in many theoretical developments. In psychology, many experimental attempts have been made to characterize equal-$x$ contours or curves, where $x$ is an attribute such as loudness (see Figure 4) or utility. Such contours are of theoretical as well as of empirical interest because they permit a direct test of the axioms of the additive model. Furthermore, they are often sufficient for the construction of the relevant scales, provided the axioms are satisfied. In this section we mention two examples where the study of indifference curves has led to the finding of additive decomposition. A more detailed discussion of the problems of scaling and testing is deferred to Section 9.4; see also Fishburn (1967a, b).

In a systematic program of study of indifference curves, Campbell and Masterson (1969) placed a rat in a small rectangular cage on one side of which he received one aversive stimulus and on the other side a different one. The stimuli are judged indifferent when he spends half the trials on each side of the cage. In one study, 60Hz electric shocks were used. On one side a reference source of 150 k$\Omega$ was paired with one of five voltages $V_r = 45, 72, 115, 185,$ and $300$ V, which are approximately equally spaced on a log–voltage scale and range from about the aversion threshold to the tetanization threshold. On the other side, one of six sources, $Z = 0, 35, 75, 150,$ 300, and 600 k$\Omega$, was paired with a voltage $V_z$ that was found to establish behavioral indifference. It was noticed that to a good approximation,

$$V_z = \alpha + \beta V_r + \gamma Z + \delta V_r Z.$$

Is this formula consistent with additive independence and, if so, what are the scales? The answer is simple since the equation can be rewritten as

$$V_z = [(\gamma + \delta V_r)(\beta + \delta Z) + \alpha\delta - \beta\gamma]/\delta$$

and hence

$$\log(\delta V_z + \beta\gamma - \alpha\delta) = \log(\gamma + \delta V_r) + \log(\beta + \delta Z),$$

which establishes additive independence. Actually, there are only two free constants if we take into account the fact that when $Z = Z_r$ (which is the reference impedence), then $V_z = V_r$. Substituting, it can be seen that

$$V_z = \gamma(Z - Z_r) + V_r[1 + \delta(Z - Z_r)],$$

and so

$$\log(\delta V_z + \gamma) = \log(\delta V_r + \gamma) + \log[1 + \delta(Z - Z_r)].$$

The quality of the fit to the data is shown in Figure 5 ($\gamma = 0.0872$, $\delta = 0.00530$).

This example raises a problem. Suppose that we are given (or, in practice, find) a function $F$ of two numerical measures, say $x$ and $y$. It is not always apparent whether it can be transformed into an additive representation. We need a systematic procedure to find those functions $f, f_1$, and $f_2$, if they exist, such that

$$f[F(x, y)] = f_1(x) + f_2(y).$$

Such a procedure is provided in Section 6.5.3.

In a study of binaural loudness, Levelt, Riemersma, and Bunt (1972) formed 36 stimuli by presenting all possible combinations of one of six different 1000-Hz signals to each of the ears. The signals were in equal-decibel steps from 20 to 70 dB Sound Pressure Level (SPL). Two subjects ordered, according to loudness, those pairs for which the ordering is not predictable by independence. Using a computer program, a best fitting additive representation was constructed. To an excellent first approximation, it was found to be of the form

$$\alpha_r I_r^{\beta_r} + \alpha_l I_l^{\beta_l} + \gamma,$$

where $I_r, I_l$ are the physical intensities to the left and right ears. Not only does this say that the data are approximately additive, but that each additive scale is a power function of intensity. Moreover, the dependence is different for the two ears. Combining the above data with other related data, the pairs

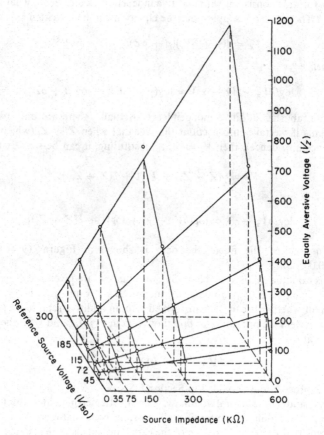

**FIGURE 5.** Representation of contours of equal aversiveness to shock for rats for the equation $V_z = \gamma(Z - Z_r) + V_r[1 + \delta(Z - Z_r)]$. (From Campbell & Masterson, 1969, p. 23.)

of exponents were 0.44, 0.61 and 0.41, 0.47 for these subjects. A power-function representation of loudness has been advocated by Stevens (1957, 1959a,b, 1960, 1966) on the basis of quite different considerations. No attempt has yet been made to check whether the exponents obtained by Stevens' methods agree with those obtained by Levelt's.

Also included in this study was an experimental attempt to construct standard sequences. It was partially successful in that the resulting functions agreed well with those described above; however, severe difficulties were encountered in the tendency for error to accumulate in the construction.

## 6.5 MODIFICATIONS OF THE THEORY

### 6.5.1 Omission of the Archimedean Property

As in extensive measurement, the Archimedean axiom has a character somewhat different from the rest of the axioms; this is true both of its formulation and its testability. This leads us to ask what happens when we omit it. Although we explore the effect of its omission only within the context of additive conjoint measurement, we conjecture that substantially the same conclusions are true for other measurement systems (see below); however, we know of no systematic studies to prove this. A useful research problem would be to formulate the Archimedean property in a highly abstract schema of measurement systems and then to investigate its role there, once and for all. Some aspects of this problem are studied in Chapter 18 of Volume II.

Since the Archimedean property states that a strictly bounded standard sequence is finite, its omission admits the possibility of strictly bounded, infinite standard sequences. We wish to investigate the interrelations of two infinite standard sequences. If $\{a_i \mid N\}$ is a standard sequence, its *convex cover* is

$$\mathscr{C}(a_i \mid N) = \{a \mid a \in A_1 \text{ and } a_i \gtrsim_1 a \gtrsim_1 a_j \text{ for some } i, j \in N\}.$$

Also, define (for an increasing standard sequence)

$$\mathscr{C}(a_{i-k}, a_i) = \{a \mid a \in A_1 \text{ and } a_i \gtrsim_1 a \gtrsim_1 a_{i-k}\}.$$

The following can then be proved:

THEOREM 3. *Suppose that* $\langle A_1, A_2, \gtrsim \rangle$ *is a structure satisfying the axioms of Definition 7 (p. 256), double cancellation, but not the Archimedean axiom. Let* I *denote the set of all integers* (*positive, negative, and zero*). *Suppose that* $\{a_i \mid \mathrm{I}\} \in A_1$ *and* $\{p_i \mid \mathrm{I}\} \in A_2$ *are dual standard sequences in the sense that each is defined relative to the other's 0.1 elements and that* $\{b_i \mid \mathrm{I}\}$, $\{q_i \mid \mathrm{I}\}$ *are also dual standard sequences. Then exactly one of the following holds.*

(i)  $\mathscr{C}(a_i \mid \mathrm{I}) = \mathscr{C}(b_i \mid \mathrm{I})$;

(ii)  $\mathscr{C}(a_i \mid \mathrm{I}) \cap \mathscr{C}(b_i \mid \mathrm{I}) = \varnothing$;

(iii)  *there exists* $k \in \mathrm{I}$ *such that* $\mathscr{C}(a_{k-1}, a_{k+1}) \supset \mathscr{C}(b_i \mid \mathrm{I})$; *or*

(iv)  *there exists* $k \in \mathrm{I}$ *such that* $\mathscr{C}(b_{k-1}, b_{k+1}) \supset \mathscr{C}(a_i \mid \mathrm{I})$.

*Similar conditions obtain for* $\mathscr{C}(p_j \mid \mathrm{I})$ *and* $\mathscr{C}(q_j \mid \mathrm{I})$.

That is, the extent (convex cover) of two infinite standard sequences is either identical, or they do not overlap at all, or one is included within a very small region of the other. The following corollary makes clear just how small that region really is. In this corollary we postulate a real-valued function having some of the properties of an additive representation (Theorem 2).

**COROLLARY.** *Suppose that the hypotheses of the theorem are true and that case* (iii) *holds. Suppose further that* $\phi : \mathscr{C}(a_i \mid \mathrm{I}) \to \mathrm{Re}$ *is such that*

(i)  *if* $a \gtrsim_1 b$, *then* $\phi(a) \geqslant \phi(b)$;

(ii)  *for* $i, j \in \mathrm{I}$, $\phi(b_i) - \phi(b_j) = (i - j)[\phi(b_1) - \phi(b_0)]$.

*Then for all* $i \in \mathrm{I}$, $\phi(b_i) = constant$.

Thus, if the function $\phi$ is capable of spanning the infinite standard sequence $\{a_i \mid \mathrm{I}\}$ in such a way as to preserve the equality of its successive intervals, then it must ignore the ordering within the strictly bounded infinite standard sequence $\{b_i \mid \mathrm{I}\}$ and so map all of these elements into a single number. The $b$ intervals are not commensurate with the $a$ ones.

For additive conjoint structures satisfying unrestricted solvability but not the Archimedean property, Luce and Tukey (1964) showed directly (without reducing it to a known representation theorem) that for each standard sequence $\{a_i \mid \mathrm{I}\}$ in $A_1$ there exists a real-valued function $\phi_1$ on $\mathscr{C}(a_i \mid \mathrm{I})$ such that, for all $a, b, \in \mathscr{C}(a_i \mid \mathrm{I})$,

(i)  $\phi_1(a_i) = i$;

(ii)  $a \gtrsim_1 b$ implies $\phi_1(a) \geqslant \phi_1(b)$.

A similar statement is true for $A_2$. Adding the Archimedean axiom allows one to show that the convex cover of any infinite standard sequence in $A'_1$ equals $A_1$ and that the implication in (ii) can be replaced by an equivalence.

One suspects that similar results hold for most, if not all, of the other measurement structures discussed in this volume, but none exist in the literature. It would be interesting to know what restrictions, if any, are needed to show the following metameasurement-theoretical result. Suppose that a set $\mathcal{A}$ of axioms plus an Archimedean axiom (in the form that every strictly bounded standard sequence is finite) is sufficient to construct a real-valued, interval-scale homomorphism for which standard sequences map into equally spaced points. Then $\mathcal{A}$ without the Archimedean axiom is sufficient to prove for each standard sequence the existence of an appropriate scale satisfying properties (i) and (ii) above. (For a metamathematical analysis of the status of the Archimedean axiom in extensive, difference, and additive conjoint measurement, see Chapter 18 in Volume II.)

## □ 6.5.2 Alternative Numerical Representations

The task of describing all possible numerical representations of an additive conjoint structure parallels closely the same problem for extensive measurement, which was described in detail in Section 3.9. We, therefore, merely outline the results. Suppose that $\langle A_1, A_2, \gtrsim \rangle$ is an additive conjoint structure, that $\phi = \phi_1 + \phi_2$ is a representation (into the additive reals), and that $\psi = \psi_1 \oplus \psi_2$ is a numerical representation involving some binary numerical operation $\oplus$ different from $+$. As in the extensive case, one can show that there exist strictly increasing functions $f, f_1$, and $f_2$ such that

$$\phi = f\psi, \qquad \phi_i = f_i\psi_i, \qquad i = 1, 2,$$
$$x \oplus y = f^{-1}[f_1(x) + f_2(y)].$$

Conversely, any such system of functions provides a representation into the real numbers that is additive in $\oplus$.

The typical transformations that preserve such a representation are $T^{f_i}_{\alpha,\beta_i} = f_i^{-1}(\alpha f_i + \beta_i)$, $i = 0, 1, 2$, where $f_0 = f$, $\beta_0 = \beta_1 + \beta_2$, and $\alpha > 0$. These are all the same two-parameter group in which $T_{1,0}$ is the identity and $T_{1/\alpha, -\beta/\alpha}$ is inverse to $T_{\alpha,\beta}$.

The most commonly used alternative representation is the multiplicative one [Equation (3), Section 6.1.3] that is characterized by

$$f = f_1 = f_2 = \log$$
$$x \oplus y = \exp(\log x + \log y) = xy,$$
$$T_{\alpha,\beta}(\psi) = \exp(\alpha \log \psi + \beta) = \gamma \psi^\alpha,$$

where $\gamma = \exp \beta$. We have met it several times, and it will play an important role in Chapter 10.

### 6.5.3 Transforming a Nonadditive Representation into an Additive One

As we saw in Section 6.4.2, a nonadditive representation can arise when, for example, the two components are physical quantities described in their usual measures, $\psi_1$ and $\psi_2$, and some function $F$ of them is found to reproduce the qualitative ordering of the attribute of interest. Thus, $F$ defines $\oplus$ as

$$\psi(a, p) = \psi_1(a) \oplus \psi_2(p) = F[\psi_1(a), \psi_2(p)].$$

With the data in this form, we can ask whether an additive representation also exists, which amounts to finding functions $f, f_1$ and $f_2$ such that

$$f[F(x, y)] = f_1(x) + f_2(y),$$

where we have substituted $x = \psi_1(a)$ and $y = \psi_2(p)$. If we suppose that this equation holds for open intervals of real numbers and if we impose suitable smoothness conditions on the functions, then the problem has a simple solution (Scheffé, 1959). In the following statement of the result, we denote partial differentiation by subscripts and ordinary differentiation by primes.

THEOREM 4.   *Suppose that $X$, $Y$, and $Z$ are open intervals of real numbers and that $F : X \times Y \xrightarrow{onto} Z$ is such that $F_{xy}$ exists everywhere on $X \times Y$ and that $F_x F_y \neq 0$. There exist $f : Z \to \mathrm{Re}$, $f_1 : X \to \mathrm{Re}$, $f_2 : Y \to \mathrm{Re}$ such that $f''$ exists, $f' \neq 0$, and*

$$f[F(x, y)] = f_1(x) + f_2(y) \tag{4}$$

*iff there is an integrable function $G : Z \to \mathrm{Re}$ such that*

$$F_{xy}/F_x F_y = G(F). \tag{5}$$

*Moreover,*

$$f(F) = C_1 \int \exp\left[-\int G(F)dF\right] dF + C_2 \tag{6}$$

$$f_1(x) = \int F_z(z, y) f'[F(z, y)]\, dz + C_3 \tag{7}$$

$$f_2(y) = f[F(x, y)] - f_1(x). \tag{8}$$

In the example of Section 6.4.2,

$$F(x, y) = \alpha + \beta x + \gamma y + \delta xy,$$

so

$$F_x = \beta + \delta y, \qquad F_y = \gamma + \delta x, \qquad F_{xy} = \delta,$$

hence,

$$\frac{F_{xy}}{F_x F_y} = \frac{\delta}{(\beta + \delta y)(\gamma + \delta x)} = \frac{\delta}{\beta\gamma - \alpha\delta + \delta F} = G(F).$$

## 6.5.4   Subtractive Structures

For some interpretations of $\langle A_1, A_2, \succsim \rangle$, it is more natural to think in terms of numerical differences rather than sums. For example, if $A_1$ and $A_2$ are two different sets of commodities and we ask a subject to judge whether the utility increment in having $a$ rather than $p$ exceeds that of having $b$ rather than $q$, then a representation of the form $\phi_1(a) - \phi_2(p) \geq \phi_1(b) - \phi_2(q)$ is suggested. Since this is equivalent to $\phi_1(a) + \phi_2(q) \geq \phi_1(b) + \phi_2(p)$, we are led to define the dual relations $\succsim$ and $\succsim'$ as follows: for $a, b$ in $A_1$ and $p, q$ in $A_2$,

$$ap \succsim bq \qquad \text{iff} \qquad aq \succsim' bp. \tag{9}$$

The interrelations between these relations are simple; some of them are summarized as follows.

THEOREM 5. *If two relations are dual in the sense of Equation* (9), *then transitivity and double cancellation are dual properties, and independence, restricted and unrestricted solvability, and the Archimedean property are self-dual properties.*

The proof is left as Exercise 20.

It is not difficult to see, then, that Definition 7 or the hypotheses of Theorem 1 provide an alternative axiomatization of algebraic differences (Definition 4.3, Section 4.4.1). Basically independence and double cancellation replace Axioms 2 and 3 of Definition 4.3, and the solvability and Archimedean axioms are slightly modified.

### 6.5.5 Need for Conjoint Measurement on $B \subset A_1 \times A_2$

Two reasons suggest that it would be desirable to search for axiom systems for additive conjoint measurement in which the realizable entities comprise only a subset $B$ of those logically possible. The first is that this is often true in experiments. Examples were mentioned in Section 6.1.1. In practice we can bypass this limitation of the theory, even though theoretically it is most inelegant, by patching together a representation from our present results. Given some subset $B$ that is subject to reasonable restrictions (roughly, that it corresponds to something resembling a convex region of the plane), then we can almost cover it by a system of overlapping rectangles, each of which satisfies Definition 7 (see Figure 6). An additive representation holds

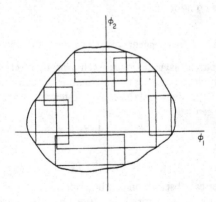

FIGURE 6.  See text for discussion.

over each of the rectangles. If the regions of overlap are sufficiently large so that each contains a rectangular substructure that satisfies Definition 7, then by the uniqueness of the representation we know that the representations over the two rectangles can be chosen so that they coincide over the substructure. In this way, the representation can be extended throughout $B$ provided that certain consistency requirements are fulfilled. Formulating these requirements would amount to the theory we wish.

A second reason for trying to devise such a theory is that it would reveal something more of the interrelations between extensive and additive conjoint measurement. Our proof, which reduced additive conjoint measurement to extensive measurement, superficially suggests that extensive measurement may be the more basic of the two. However, this is probably a misleading conclusion. For example, had we had extensive measurement only for $B = A \times A$ and with *strong solvability*, i.e., if $a \succ b$, there exists $c$ in $A$ such that $a \sim b \circ c$, we probably would have concluded just the opposite. The present proof of the conjoint representation would not have been possible, but a direct one, similar to that of Luce (1966), could have been provided. Then we could have used the following result to arrive at the extensive representation from the conjoint one.

THEOREM 6.  *Suppose $\langle A, A \times A, \succsim, \circ \rangle$ is a nontrivial (i.e., there exist $a, b \in A$ with $a \succ b$) extensive structure with no essential maximum. If strong solvability holds, then $\langle A, A, \succsim' \rangle$, where, for $a, b, p, q \in A$, $ap \succsim' bq$ iff $a \circ p \succsim b \circ q$, is a symmetric, additive conjoint structure.*

This is proved in Section 6.6.4. Although no generalization of Definition 7 has yet been suggested, we suspect that there is one for which Theorem 6 is still true with $B \subset A \times A$ in the extensive structure and with symmetry dropped from the conjoint one.

## 6.5.6   Symmetries of Independent and Dependent Variables

Despite the apparent asymmetry between a product set $A$ and its factors, $A_1$ and $A_2$, additive conjoint measurement is actually symmetric in the independent and dependent variables. This statement is based in part on the trivial fact that when

$$\phi = \phi_1 + \phi_2,$$

we can also write

$$\phi_1 = \phi + (-\phi_2).$$

This simple numerical observation can be extended in a way that sheds additional light on the various factorial structures that arise. If $A = A_1 \times A_2$, it does not follow at all that $A_1 = A \times A_2$. Nevertheless, if we order $A \times A_2$ by the function $\phi(a, p) - \phi_2(q)$, or $\phi_1(a) + \phi_2(p) - \phi_2(q)$, this ordering clearly reduces to the $\gtrsim_1$ ordering on $A_1$, if $a$ in $A_1$ is identified with $((a, p), p)$ in $A \times A_2$. For a geometric formulation of this argument, note that if $\phi_1$ and $\phi_2$ are taken as coordinates, the indifference curves form a set of lines, $\phi_1 + \phi_2 = $ constant, with slope $-1$. The $\phi_1$-coordinate can be regarded as a family of vertical lines, $\phi_1 = $ constant, which express $\sim_1$-equivalences. Similarly, the $\phi_2$ coordinate can be regarded as a set of horizontal lines ($\sim_2$-equivalence). Now, one may regard the horizontal lines and the lines with slope $-1$ as a set of oblique coordinates, and treat the vertical lines as indifference curves.

A third form of the argument, due to Krantz (1964), is suggested by the above geometric argument. If the two factors, $A_1$ and $A_2$, are really equivalence classes, then it is most natural to consider a single set $A$ of objects and three order relations, $\gtrsim_1$, $\gtrsim_2$, $\gtrsim_3$ on $A$. We can take the $\sim_1$-equivalence class to be the first coordinate, the $\sim_2$-equivalence class to be the second coordinate, and let $\gtrsim$ be the ordering induced by $\gtrsim_3$ on $(A/\sim_1) \times (A/\sim_2)$. If this seems strange, it should be realized that it is the usual situation in applications of conjoint measurement to physics. For example, moving objects can be ordered in three different ways (at least): $\gtrsim_m$ (mass ordering), $\gtrsim_v$ (velocity ordering), and $\gtrsim_p$ (momentum ordering). When we write an object with mass $a$ and velocity $q$ as an ordered pair $aq$ and consider the momentum ordering $\gtrsim_p$ on the product set, e.g., $aq \gtrsim_p a'q'$, we are really *constructing* a product set by observing $\sim_m$ and $\sim_v$. There is only one set of objects with no inherent product structure; such a structure is imposed by observing two variables, $m$ and $v$, for each object. Clearly, if there are three orderings, which two we select as independent variables is a matter of convention. The particular convention chosen in the case of momentum results from the fact that $m$ and $v$ are also extensively measurable and that a simple qualitative law exists which links the extensive and conjoint structures (see Section 10.7.2, Laws of Exchange). Thus, $p$ is usually calculated from $m$ and $v$ after the latter have been extensively measured; $p$ is in that sense a derived (exchange-derived or *e*-derived) measure.

Some objects, such as commodity bundles, have a natural product structure, and this determines the convention as to which variable is regarded as dependent. In other applications, there may be no natural choice. For example, suppose that observers are presented objects of various sizes at a fixed distance and are asked to judge

   (i) the apparent physical size, $s$, i.e., their best guess as to the actual size;

  (ii) the perceived size $S$, i.e., how big the object seems;

 (iii) the perceived distance $D$, i.e., how far it seems.

If $d$ denotes apparent physical distance, then it is natural to hypothesize that

$$S/s = D/d.$$

If $d$ is fixed, e.g., apparent physical distance $d$ is always judged correctly in the particular experimental situation, although apparent physical size $s$ is misjudged because of the use of unusual sizes for familiar objects, we have a hypothesized multiplicative conjoint structure with three dependent variables, any two of which can be used as coordinates.

Note that an axiomatization of conjoint measurement based on three orderings, $\gtrsim_1$, $\gtrsim_2$, $\gtrsim_3$, on a set $A$, requires one essentially new axiom.

*If $a \sim_1 b$, then $a \gtrsim_2 b$ iff $a \gtrsim_3 b$; and the two similar statements obtained by permuting the indices.*

This axiom says, in part, that if $a$ and $b$ are equivalent in any two orderings, they are equivalent in the third one also. This is implicit in our usual axiomatizations. When we let $\gtrsim$ be a relation on $A_1 \times A_2$, we automatically build into the formulation the idea that given the two coordinates, the position of the object in the ordering is known. When the coordinates are constructed, however, this assumption must be made explicit: we cannot identify an element of $A$ with an element of $(A/\sim_1) \times (A/\sim_2)$ unless $\gtrsim_3$ induces a well-defined ordering on $(A/\sim_1) \times (A/\sim_2)$.

We conclude that additive conjoint measurement involves measurement of three variables, $\phi_1$, $\phi_2$, and $\phi$, which are linearly related, and there is no real asymmetry; all three measures have essentially the same status and either component of the Cartesian product could itself be identified as a Cartesian product, using the other two variables as coordinates. The choice of independent variables is a convention, from the standpoint of the theory, although there are often good reasons for such conventions, based on other considerations, among them extensive measurement of variables, natural product structures, experimental control of certain variables, or assumptions about error of the type made in linear statistical models.

## 6.5.7   Alternative Factorial Decompositions

In many applications several alternative factorial decompositions of the same set of objects may be considered, and the selection of the appropriate structure poses a nontrivial problem. Rectangles, for example, may be characterized by their height $H$, and width $W$, or by their area $HW$, and shape, $H/W$. If rectangles are ordered by some sort of psychological judgment (e.g., judged area, aesthetic value), it is an empirical question which of the two structures, if either, satisfies the axioms of the theory. The answer is likely to depend on the nature of the ordering, e.g., subjective area or aesthetic value. Similarly, a set of gambles for which one has a fifty-fifty chance to win \$a or to lose \$p can be analyzed in terms of potential profit $a$ and potential loss $p$ or in terms of expected value $(a - p)/2$ and the variance $(a + p)^2/4$. The theory of expected utility (see Chapter 8) implies that the ordering of such gambles with respect to preference satisfies the additive theory relative to the former factorial decomposition. The theory of risk described in Section 3.14.1, on the other hand, implies that the ordering of such gambles with respect to risk satisfies the additive theory relative to the latter factorial decomposition. The acceptance or rejection of additive conjoint measurement for a particular domain, therefore, is always relative to a particular ordering and a particular factorial decomposition. There are no general methods, however, for discovering whether a particular set of objects can be decomposed so as to satisfy additivity.

## 6.6 PROOFS

### 6.6.1 Preliminary Lemmas

The common hypothesis of the following four lemmas is that $\langle A_1, A_2, \gtrsim \rangle$ is an additive conjoint structure satisfying double cancellation but not necessarily satisfying the Archimedean axiom. The first one formulates two special kinds of triple cancellation (p. 251).

LEMMA 8. *If* $ax \gtrsim fy$, $gy \gtrsim bx$, $fp \gtrsim gq$, *and if there exist solutions* $k, l \in A_1$ *to* $kx \sim fp$ *and* $lx \sim gq$, *then* $ap \gtrsim bq$. *If* $fp \gtrsim gx$, $gy \gtrsim fq$, $ax \gtrsim by$, *and if there exist solutions* $u, v \in A_2$ *to* $fu \sim ax$ *and* $fv \sim by$, *then* $ap \gtrsim bq$.

PROOF. By double cancellation (Axiom 3), $fp \sim kx$ and $ax \gtrsim fy$ imply $ap \gtrsim ky$. Since $kx \sim fp \gtrsim gq \sim lx$, independence (Axiom 2) implies that $ky \gtrsim ly$. And by double cancellation, $lx \sim gq$ and $gy \gtrsim bx$ imply that $ly \gtrsim bq$. The conclusion follows from Axiom 1. The proof of the second statement is similar. ◇

LEMMA 9. *If* $\{a_i\}$ *and* $\{p_j\}$ *are doubly infinite standard sequences that are dual, i.e.,* $a_0 p_1 \sim a_1 p_0$, *then unrestricted solvability holds for elements in* $\mathscr{C}(a_i \mid \mathrm{I})$ *and* $\mathscr{C}(p_j \mid \mathrm{I})$.

PROOF. We prove the lemma for one of the two cases; the other is similar. Let $a, b \in \mathscr{C}(a_i \mid \mathrm{I})$ and $p \in \mathscr{C}(p_j \mid \mathrm{I})$. We must show that there is a $q \in A_2$ such that $ap \sim bq$. If the standard sequences are increasing, there are elements such that $p_j \gtrsim_2 p \gtrsim_2 p_{j-1}$, $a_i \gtrsim_1 a \gtrsim_1 a_{i-1}$ and $a_k \gtrsim_1 b \gtrsim_1 a_{k-1}$.

We first assume that $a \gtrsim_1 b$. Then we have at once that

$$a_i p_j > ap \gtrsim bp > bp_{j-1}.$$

One can show by induction (using restricted solvability) that for $l = i + j - k + 1, a_{k-1}p_l \sim a_i p_j$. So $bp_l \gtrsim a_{k-1}p_l$, and by transitivity,

$$bp_l \gtrsim ap \gtrsim bp_{j-1}.$$

Hence, by restricted solvability there is a $q$ such that $ap \sim bq$, as desired. The proof is similar for $b \gtrsim_1 a$ and for decreasing sequences.   ◇

For $a$ in $A_1$, we define $a + 1, a - 1$ in $A_1$ as the solutions, if they exist, to $(a + 1) p_0 \sim a p_1, (a - 1) p_1 \sim a p_0$. For $p$ in $A_2$, $p + 1, p - 1$ in $A_2$ are defined similarly. It is easily verified that if $a >_1 a_1$, then $a - 1$ exists; if $a_0 >_1 a$, then $a + 1$ exists; and similarly for $p$.

LEMMA 10.   *Under the hypotheses of Lemma 9, for all $a, b \in \mathscr{C}(a_i \mid \mathrm{I})$ and $p, q \in \mathscr{C}(p_j \mid \mathrm{I})$, $ap \gtrsim bq$ implies $(a - 1) p \gtrsim (b - 1) q, (a + 1) p \gtrsim (b + 1) q$, $a(p - 1) \gtrsim b(q - 1)$, and $a(p + 1) \gtrsim b(q + 1)$.*

PROOF.   Lemma 9 assures the existence of $a - 1$, $a + 1$, etc., and the existence of the solutions in the hypothesis of Lemma 8. By that lemma $(a - 1) p_1 \sim a p_0, bp_0 \sim (b - 1) p_1$, and $ap \gtrsim bq$ imply $(a - 1) p \gtrsim (b - 1)q$. The other three cases are similar.   ◇

### 6.6.2   Theorem 3   (p. 271)

*Suppose that $\langle A_1, A_2, \gtrsim \rangle$ satisfies double cancellation and the axioms of Definition 7 except the Archimedean one, and that both $\{a_i \mid \mathrm{I}\}, \{p_j \mid \mathrm{I}\}$ and $\{b_i \mid \mathrm{I}\}, \{q_j \mid \mathrm{I}\}$ are pairs of doubly infinite dual standard sequences. Then, either (i) $\mathscr{C}(a_i \mid \mathrm{I}) = \mathscr{C}(b_i \mid \mathrm{I})$; (ii) $\mathscr{C}(a_i \mid \mathrm{I}) \cap \mathscr{C}(b_i \mid \mathrm{I}) = \varnothing$; (iii) there is $k \in \mathrm{I}$ such that $\mathscr{C}(a_{k-1}, a_{k+1}) \supset \mathscr{C}(b_i \mid \mathrm{I})$; or (iv) there is $k \in \mathrm{I}$ such that $\mathscr{C}(b_{k-1}, b_{k+1}) \supset \mathscr{C}(a_i \mid \mathrm{I})$. A similar disjunction holds for $\mathscr{C}(p_j \mid \mathrm{I})$ and $\mathscr{C}(q_j \mid \mathrm{I})$.*

PROOF.   It is sufficient to prove the theorem for the first component. Let us assume that neither $\mathscr{C}(a_i) = \mathscr{C}(b_i)$ nor $\mathscr{C}(a_i) \cap \mathscr{C}(b_i) = \varnothing$ is true. Then at least one of the following holds for some integer $k$ and for all integers $i$: $b_k \gtrsim a_i$, $a_i \gtrsim b_k$, $a_k \gtrsim b_i$, or $b_i \gtrsim a_k$. For suppose that none were true, then by the negation of the first, for any $k$, there exists an $i$ such that $a_i > b_k$ and by the negation of the second there exists a $j$ such that $b_k > a_j$, hence $\mathscr{C}(b_i) \subset \mathscr{C}(a_i)$. Similarly, $\mathscr{C}(a_i) \subset \mathscr{C}(b_i)$, hence they are equal, contrary to assumption.

Let us suppose that the first holds; the arguments for the other cases are similar. If $\{b_i \mid \mathrm{I}\}$ is increasing (decreasing) let $k$ be the smallest (largest) integer such that, for all $i$, $b_{k+1} \gtrsim_1 a_i$. This integer exists since if it did not then either $b_i$ would not be increasing (decreasing) or $\mathscr{C}(a_i) \cap \mathscr{C}(b_i) = \varnothing$, contrary to assumption. With no loss of generality, we assume that $\{b_i\}$ is increasing. So for at least one $j$, $a_j >_1 b_k$. If for all $i$, $a_i \gtrsim_1 b_{k-1}$, we are done; so we assume that for some $m$, $b_{k-1} >_1 a_m$. Since $\{b_i \mid I\}$ and $\{q_j \mid I\}$ are dual,

$$a_j q_0 > b_k q_0 \sim b_{k-1} q_1 > a_m q_1.$$

Using Lemma 10 inductively, $a_j q_0 > a_m q_1$ implies that

$$a_{2j-m} q_0 > a_j q_1 > b_k q_1 \sim b_{k+1} q_0.$$

Therefore, $a_{2j-m} >_1 b_{k+1}$, contrary to the choice of $k$. This proves part (iv); the other possibilities lead to either (iii) or (iv).

We next establish the corollary that *if $\phi$ preserves $\gtrsim_1$ on $A_1$, $\phi(b_i) - \phi(b_j) = (i-j)[\phi(b_1) - \phi(b_0)]$, and case (iii) holds, then $\phi$ is constant over $\mathscr{C}(b_i \mid I)$.* For any positive integer $n$,

$$
\begin{aligned}
n \mid \phi(b_1) - \phi(b_0) \mid &= \mid n[\phi(b_1) - \phi(b_0)] \mid \\
&= \mid \phi(b_n) - \phi(b_0) \mid \\
&\leqslant \mid \phi(a_{k+1}) - \phi(a_{k-1}) \mid.
\end{aligned}
$$

This can hold for all $n$ only if $\phi(b_1) = \phi(b_0)$. Therefore, for all $n$, $\phi(b_n) = \phi(b_0)$. $\qquad\qquad \diamondsuit$

### 6.6.3 Theorem 4 (p. 274)

*Suppose that $X$, $Y$, and $Z$ are open intervals of real numbers and that $F$: $X \times Y \xrightarrow{\text{onto}} Z$ is such that $F_x F_y \neq 0$. It can be transformed into an additive representation by a twice differentiable function $f$ with nonzero slope iff $F_{xy}/F_x F_y$ is a function only of $F$. The transformation is given by Equation (6) and the additive components by Equations (7) and (8).*

*PROOF.* Suppose, first, that $f[F(x, y)] = f_1(x) + f_2(y)$. The partial derivative of this with respect to $x$ yields

$$f'(F) F_x = f_1'(x),$$

and the partial derivative of that with respect to $y$ yields

$$f''(F) F_x F_y + f'(F) F_{xy} = 0.$$

Since $f'(F) \neq 0$ and since $F_x F_y \neq 0$,

$$\frac{F_{xy}}{F_x F_y} = -\frac{f''(F)}{f'(F)} = -\frac{d}{dF}\log f'(F) = G(F),$$

which establishes Equation (5). Equations (6) and (7) follow by integration.

Conversely, assuming (5), define $f, f_1, f_2$ by Equations (6), (7), and (8), respectively. To show $f_1$ is a function of $x$ only, take the partial derivative of $F_x f'(F)$ with respect to $y$:

$$F_{xy} f'(F) + F_x F_y f''(F).$$

This vanishes because $F_{xy}/F_x F_y = G(F) = -f''(F)/f'(F)$ by (5) and (6). Similarly, take the partial derivative of $f_2$ with respect to $x$: $f'(F)F_x - f_1'(x)$. This vanishes by (7).  $\diamondsuit$

### 6.6.4  Theorem 6  (p. 276)

*Suppose that $\langle A, \succsim, A \times A, \circ \rangle$ is a nontrivial extensive structure with no essential maximum for which strong solvability holds. Then $\langle A, A, \succsim' \rangle$, where $ap \succsim' bq$ iff $a \circ p \succsim b \circ q$, is a symmetric, additive conjoint structure.*

*PROOF.*  We verify the seven axioms of Definition 7 (Section 6.2.3).

1.  $\succsim'$ is a weak ordering of $A \times A$ by Axiom 1 of Definition 3.3.

2.  Assume $ap \succsim bp$, and so $a \circ p \succsim b \circ p$. By Exercise 3.15, we have $a \succsim b$, and so by Axiom 3 and Lemma 3.8, $a \circ q \succsim b \circ q$, as desired. The other case of independence is similar.

3.  Suppose that $a \circ x \succsim f \circ q$ and $f \circ p \succsim b \circ x$, then by Lemmas 3.7, 3.8, and 3.9 of Section 3.5.2, $(a \circ p) \circ (f \circ x) \succsim (b \circ q) \circ (f \circ x)$, and so by Exercise 3.15, $a \circ p \succsim b \circ q$.

4.  Suppose that $\underline{b} \circ q \succsim a \circ p \succsim \underline{b} \circ q$. If $a \circ p \sim \underline{b} \circ q$, we are done. If not, then $a \circ p \succ \underline{b} \circ q$, and so by strong solvability there exists $x$ such that $a \circ p \sim \underline{b} \circ q \circ x \sim (\underline{b} \circ x) \circ q$, where we have used Lemmas 3.8 and 3.9. So $b = \underline{b} \circ x$ solves the equation. The other case is similar.

5.  Suppose that $\{a_i \mid i \in N, \text{not}(p \sim q), a_i \circ p \sim a_{i+1} \circ q\}$ is strictly bounded and $N$ is infinite. With no loss of generality, we may assume that $p \succ q, a_{i+1} \succ a_i$, and there exists $b \succ a_i$, $i \in N$. By strong solvability, there exists a $c$ such that $p \sim q \circ c$. By Lemma 3.7,

$$a_{i+1} \circ q \sim a_i \circ p \sim a_i \circ c \circ q,$$

and so by Exercise 3.15, $a_{i+1} \sim a_i \circ c$. By induction, $a_{i+1} \sim a_1 \circ ic$. Since $b > a_1$, by strong solvability, there exists $d$ such that $b \sim a_1 \circ d$. Thus, $a_1 \circ d > a_{i+1} \sim a_1 \circ ic$, and so by Exercise 3.15, $d > ic$ for $i \in N$. The assumption that $N$ is infinite violates Axiom 6.

6.   Since there exist $a > b$, $A_1 = A_2 = A$ is essential.

7.   The structure is symmetric since, by Lemma 3.8, $a \circ b \sim b \circ a$ for all $a, b \in A$.    $\Diamond$


## 6.7   INDIFFERENCE CURVES AND UNIFORM FAMILIES OF FUNCTIONS

As we noted in Section 6.4.2, data are often reported not as an ordering relation, as assumed in Definition 7, but as indifference curves described in terms of some more or less arbitrary measures. Let us assume that these measures lie on real intervals, then the family of indifference curves can be described as $\{F_\lambda \mid \lambda \in \Lambda\}$, where $F_\lambda$ is a function from one interval onto the other and $\lambda$ is simply an index to identify each curve. Thus, $xy \sim x'y'$ if and only if for some $\lambda \in \Lambda$, $y = F_\lambda(x)$ and $y' = F_\lambda(x')$. The problem we consider here is whether there exist transformations of the given measures that permit us to map the indifference curves onto those of an additive structure and, if so, to characterize those transformations. It is evident that the indifference curves of an additive structure,

$$\phi_1(a) + \phi_2(p) = \text{constant},$$

form a family of parallel straight lines with slope $-1$ when plotted in the $\phi_1$, $\phi_2$ coordinates. Equally well, if we use the $\phi_1$, $-\phi_2$ coordinates, the family consists of parallel straight lines with slope 1. So it does not really matter whether we view the problem in terms of increasing or decreasing families of functions. Some of the proofs, however, are simpler in the increasing case because an increasing function of an increasing function is increasing, whereas, a decreasing function of a decreasing function is increasing, not decreasing.

Let us begin by supposing that an additive representation exists and that the transformations from the given measures into the additive ones are $\phi_1(x) + \phi_2(y) = \text{constant} = \lambda$. Furthermore, let us assume that the $\phi_i$ are strictly increasing and 1:1 functions from the reals onto the reals. Observe that they must, therefore, be continuous functions. It will prove convenient to refer to strictly monotonic, 1:1 functions of the real numbers onto the real numbers as *scales*. So, we assume that the $\phi_i$ are increasing scales. We *have* lost generality in this assumption. For example, if the equal-loudness contours of Figure 4 can be mapped into an additive family of indifference curves, the frequency transformation clearly must be nonmonotonic. Returning to our restrictive case, we may rewrite this indifference curve as

$$y = \phi_2^{-1}[\lambda - \phi_1(x)]$$
$$= G_\lambda[\phi_1(x)]$$
$$= F_\lambda(x),$$

where we have defined

$$G_\lambda(z) = \phi_2^{-1}(\lambda - z)$$

and

$$F_\lambda = G_\lambda\phi_1 .$$

There are three observations to be made. First, $G_\lambda$ is a decreasing scale because $\phi_2$ is an increasing one, and $F_\lambda$ is also a decreasing scale because $G_\lambda$ is decreasing and $\phi_1$ is increasing. Second,

$$G_\lambda(z) = \phi_2^{-1}(\lambda - z) = \phi_2^{-1}(\mu + \lambda - \mu - z) = G_\mu[z - (\lambda - \mu)].$$

Thus, the $G$-scales are parallel in the sense that any two can be superimposed by an additive shift of the independent variable. Finally, it is easy to see that no two distinct $G$-scales intersect and that no two distinct $F$-scales intersect.

If we begin with the subtractive representation $\phi_1(x) - \phi_2(y) = \lambda$, then a similar argument shows that increasing scales $G_\lambda$ and $F_\lambda$ can be defined such that the $G$'s are parallel, $G_\lambda = F_\lambda\phi_1^{-1}$, and $F_\lambda$ are indifference curves.

Because, as Levine (1970) has emphasized, such families of scales arise in a surprising number of contexts, including probability theory, it is useful to give them a formal definition.

DEFINITION 9.   *A family* $\{F_\lambda \mid \lambda \in \Lambda\}$ *of scales is* uniform[3] *iff there exists an increasing scale* $\phi$, *called a* parallelizer, *such that the family* $\{F_\lambda\phi^{-1} \mid \lambda \in \Lambda\}$ *is parallel in the sense that, for every* $\lambda, \mu \in \Lambda$, *there exists a real constant* $k_{\lambda\mu}$ *such that, for all real* x,

$$F_\lambda\phi^{-1}(x) = F_\mu\phi^{-1}(x - k_{\lambda\mu}).$$

---

[3] Levine introduced the term "uniform system"; however, as a set of functions is usually called a "family," a "collection," or a "class" rather then a "system," we prefer to define only the adjective and let it modify whatever noun seems convenient to use. He also used the term "solution" where we use "parallelizer" and "crossed" where we use "intersect."

So, we see that if the indifference curves can be put in additive or subtractive form, then the curves form a uniform family. Conversely, suppose
that $\{F_\lambda \mid \lambda \in \Lambda\}$ is a uniform family of increasing scales with a parallel-
izer $\phi_1$. Let $G_\lambda = F_\lambda \phi_1^{-1}$ and, for some $\lambda_0 \in \Lambda$, define $\phi_2(y) = G_{\lambda_0}^{-1}(y)$.
Clearly, $\phi_2$ is an increasing scale because $G_{\lambda_0}$ is. Moreover, if $y = F_\lambda(x)$, then

$$\phi_2(y) = G_{\lambda_0}^{-1} F_\lambda(x)$$

$$= G_{\lambda_0}^{-1} F_\lambda \phi_1^{-1} \phi_1(x)$$

$$= G_{\lambda_0}^{-1} G_\lambda \phi_1(x)$$

$$= G_{\lambda_0}^{-1} G_{\lambda_0}[\phi_1(x) - k_{\lambda\lambda_0}]$$

$$= \phi_1(x) - k_{\lambda\lambda_0}$$

which is a subtractive indifference curve.

So our problem of finding when a given family of indifference curves can
be transformed into those of an additive conjoint structure is equivalent to
discovering conditions when the family is uniform. We deal first, in Section
6.7.1, with the case where each point of the plane lies on some indifference
curve. The problem actually faced by the experimenter is different from that
because he has only a finite set of indifference curves, i.e., $\Lambda$ is finite, and they
are imperfectly known. Levine's results for a finite number of completely
specified indifference curves are described in Section 6.7.2. In Levine(1972),
these results are extended to families of functions that can be rendered
parallel by making both a translation and a change of scale.

### 6.7.1  A Curve Through Every Point

As we noted above, a necessary condition for a family to be uniform is that
any two distinct members of the family not intersect. This is by no means a
sufficient condition. It turns out that when the family completely covers the
plane another necessary condition added to nonintersection is sufficient.
This second property is the Thomsen condition (see Definition 3, p. 250),
which we have illustrated in Figure 1. Note that by Theorem 5, the statement of the result is independent of whether we use increasing or decreasing
scales (subtractive or additive structures) because transitivity of equivalence
and the Thomsen condition are dual properties.

THEOREM 7.   *Suppose that* $\{F_\lambda \mid \lambda \in \Lambda\}$ *is a family of increasing scales such that for every real* $x, y$ *there exists* $\lambda \in \Lambda$ *with* $y = F_\lambda(x)$, *i.e., through every point in the plane some member of the family passes. The family is uniform iff no two distinct members of the family intersect and the Thomsen condition holds. The parallelizer is unique up to a positive linear transformation.*

This theorem, which is due to Blaschke (Blaschke & Bol, 1938), is proved in Section 6.8.1.

The general theory of mapping three families of curves in the plane, each of which is nonintersecting, into three families of parallel straight lines in the plane is known as the *theory of webs.* (Usually, two of the three families are interpreted as coordinates.) It is the topic of Blaschke and Bol's work. In the 1960's various purely algebraic generalizations of webs were studied. They are related to axiomatizations such as Definition 7, and they can be used to prove measurement representation theorems. See Aczél, Belousov, and Hosszú (1960), Aczél, Pickert, and Radó (1960), Havel (1966), and Radó (1960, 1965).

### 6.7.2   A Finite Number of Curves

As we mentioned, the problem in practice is not the one just discussed, but the corresponding one when only a finite number of indifference curves are given (and, imperfectly, at that—but we do not treat the problem of fallible data here). Obviously, there is no problem with one scale. We turn to two. In this case, the necessary condition that they not intersect is also sufficient (Levine, 1970).

THEOREM 8.   *Suppose that* $F$ *and* $G$ *are distinct increasing scales. The pair* $\{F, G\}$ *is uniform iff* $F$ *and* $G$ *do not intersect. Let* $H = F^{-1}G$ *and* $H^n = H^{n-1}H$. *Then*

$$\phi(x) = n + H^{-n}(x)/H(0), \qquad for \quad H^n(0) \leqslant x < H^{n+1}(0),$$

*is a parallelizer of* $\{F, G\}$. *If* $\phi'$ *is another parallelizer, then there is a continuous periodic function* $\pi$ *with period* 1, *i.e.,* $\pi(x + 1) = \pi(x)$, *and real* $\alpha > 0$ *such that* $\phi' = \alpha(\phi + \pi\phi)$. *Moreover, if* $\phi + \pi\phi$ *is strictly increasing, then* $\phi' = \alpha(\phi + \pi\phi)$ *is a parallelizer.*

The uniqueness of the parallelizer is not really satisfactory when there are only two scales in the family because transformations more general than positive linear ones are acceptable. Fortunately, the difficulty evaporates when we go to families of at least three scales, provided that they do not degenerate, in a sense to be specified in Theorem 9, to the case of two scales.

Within families of three or more indifference curves, distinct scales must not intersect, but as we should anticipate from Theorem 7 this is no longer a sufficient condition. In order to make it sufficient we must look not only at the given family of scales but also at those that can be generated from it by certain compositions and certain limits of the scales. If all of these do not intersect, then Levine (1970) showed that the given family is uniform. We make this precise in Theorem 9.

Suppose that $\mathcal{F} = \{F_\lambda \mid \lambda \text{ in } \Lambda\}$ is a nonempty family of increasing scales. Its *associated group* $\mathcal{G}$ is the set of all scales generated by repeated compositions of scales from the set $\{F_{\lambda_0}^{-1}F_\lambda, (F_{\lambda_0}^{-1}F_\lambda)^{-1} \mid \lambda \text{ in } \Lambda\}$, where $\lambda_0$ in $\Lambda$. We have not subscripted $\mathcal{G}$ with $\lambda_0$ because $\mathcal{G}$ is independent of that choice; this follows from the observation that

$$F_{\lambda_1}^{-1}F_\lambda = (F_{\lambda_0}^{-1}F_{\lambda_1})^{-1} F_{\lambda_0}^{-1}F_\lambda .$$

Moreover, it is easy to show that $\mathcal{G}$ is a group and that the identity function $e$ is its identity element.

Observe that when the scales of $\mathcal{G}$ do not intersect, we may order them naturally in the following way: for all $f, g \in \mathcal{G}, f \gtrsim g$ if and only if $f(x) \geqslant g(x)$ for some real $x$ and so, by the nonintersection assumption, for all real $x$.

To establish the uniformity of $\mathcal{F}$, it is not quite sufficient to suppose that the scales of $\mathcal{G}$ do not intersect. Levine (1970) has shown that it suffices to require that certain limits of sequences in $\mathcal{G}$ which are functions also do not intersect. This result is formulated precisely in Theorem 9 below. Some delicacy is required in the exact statement because a pointwise limit of scales from $\mathcal{G}$ may not be a function and, if it is, it may not be increasing (although, as is easily seen, it is nondecreasing).

Let $\mathcal{H}$ be the family of all functions from Re into Re that are pointwise limits of sequences of functions in $\mathcal{G}$. $\mathcal{H}$ is called the *completion* of $\mathcal{G}$ (Levine, 1970). Not every sequence in $\mathcal{G}$ defines an element of $\mathcal{H}$, but if $g_i \gtrsim g_{i+1} \gtrsim g$, $i = 1, 2,...$, then $g_i$ converges to a function $h \gtrsim g$ in $\mathcal{H}$. Clearly, $\mathcal{G} \subset \mathcal{H}$.

THEOREM 9. *A nonempty family $\mathcal{F}$ of increasing scales is uniform iff any two distinct functions in the completion $\mathcal{H}$ of its associated group $\mathcal{G}$ do not intersect.*

*If $\mathcal{G}$ is cyclic the parallelizers of $\mathcal{F}$ are exactly those of the uniform family consisting of the generator and the identity function (uniqueness part of Theorem 8). If $\mathcal{G}$ is noncyclic, $\mathcal{H}$ fulfills the conditions of Theorem 7 and the parallelizer is unique up to positive linear transformations.*

Recall that a group $\mathcal{G}$ is *cyclic* if and only if there is an element $g$ (the generator) such that

$$\mathcal{G} = \{g^n \mid n = 0, \pm 1,...\},$$

where $g^0 = e$.

The proof in Sections 6.8.3 and 6.8.4 is essentially that of Levine. We first show (Lemma 11) that the problem of finding a parallelizer for a family $\mathscr{F}$ is the same as finding one for its associated group $\mathscr{G}$. Next, Lemma 12 shows that if the scales of $\mathscr{G}$ do not intersect, then it is an Archimedean ordered group. The resulting homomorphism of $\mathscr{G}$ into $\langle \text{Re}, \geqslant, + \rangle$ is used in Lemma 13 to show that if $\mathscr{G}$ is noncyclic and distinct functions in $\mathscr{H}$ do not intersect, then for every $x$, $\{g(x) \mid g \in \mathscr{G}\}$ is dense in Re. This shows that $\mathscr{H}$ contains a curve through every point, and allows us to reduce Theorem 9 to Theorem 7 for the noncyclic case. (Levine established the existence of a parallelizer directly, whereas we reduce it via Theorem 7 to additive conjoint measurement.) The cyclic case reduces immediately to Theorem 8.

## 6.8 PROOFS

### 6.8.1 Theorem 7 (p. 286)

*Suppose that a family of increasing scales is such that at least one passes through each point of the plane. Then the family is uniform iff no two distinct scales intersect and the Thomsen condition holds. The parallelizer is an interval scale.*

PROOF. The necessity of the two conditions is obvious.

To show their sufficiency, we show that the given scales define an additive, and so by Theorem 5 a subtractive, conjoint structure satisfying unrestricted solvability; hence by Theorems 1 and 2 there is a subtractive representation. Since the scales are nonintersecting, we may order them by the property $F_\lambda \geqslant F_\mu$ iff for some $x$ and hence, by their continuity and the fact that they do not intersect, for all $x$, $F_\lambda(x) \geqslant F_\mu(x)$. Define $\succsim$ on $\text{Re} \times \text{Re}$ as follows: $xy \succsim x'y'$ if and only if there is $F_\lambda$ through the point $xy$ and $F_\mu$ through $x'y'$ such that $F_\lambda \geqslant F_\mu$. Clearly, $\succsim$ is a well-defined, independent weak order. Moreover, $x \succsim_1 y$ if and only if $x \leqslant y$ and $x \succsim_2 y$ if and only if $x \geqslant y$ follow readily from the definitions.

Unrestricted solvability holds since there is a scale through each point and scales are continuous and onto.

To show double cancellation, suppose that $ax \succsim fq$ and $fp \succsim bx$. By solvability, there exist $a', b'$ with $a' \leqslant a$, $b' \geqslant b$, $a'x \sim fq$, and $fp \sim b'x$. By the Thomsen condition and the definition of $\succsim$, $ap \succsim a'p \sim b'q \succsim bq$.

To show the Archimedean property, suppose that $\{a_i\}$ is a standard sequence, i.e., for some $p, q \in \text{Re}$, $a_{i+1}q \sim a_i p$. Suppose that it is an increasing sequence, then since we are working with a subtractive structure, $p < q$. If there is a strict upper bound of the sequence, then there is a supremum $b$. Let $F_\lambda$ be a curve through $bq$ and solve $cp \sim bq$ for $c$. We have $c < b$, hence for some $i$, $a_i > c$. Choose $F_\mu$ through $a_i p$, then by hypothesis, $F_\mu$ also goes through $a_{i+1}q$. We have $F_\mu(a_{i+1}) = q = F_\lambda(b) > F_\lambda(a_{i+1})$, hence $F_\mu(a_i) > F_\lambda(a_i)$. Thus, $F_\lambda(c) = p = F_\mu(a_i) > F_\lambda(a_i)$, contradicting $a_i > c$. A similar argument on a lower bound shows that a strictly bounded standard sequence is finite.

The uniqueness of the parallelizer follows from the uniqueness of the additive representation (Theorem 2). $\diamond$

### 6.8.2 Theorem 8 (p. 286)

*Suppose that $F$ and $G$ are distinct increasing scales, then $\{F, G\}$ is uniform (Definition 9) iff $F$ and $G$ do not intersect. If $H = F^{-1}G$, then*

$$\phi(x) = n + H^{-n}(x)/H(0), \quad \text{for} \quad H^n(0) \leqslant x < H^{n+1}(0)$$

*is a parallelizer. If $\phi'$ is another parallelizer, then $\phi' = \alpha(\phi + \pi\phi)$, where $\alpha > 0$ and $\pi$ is periodic with period 1.*

*PROOF.* Suppose, first, that $\{F, G\}$ is uniform, i.e., there exists an increasing scale $\phi$ and a constant $k \neq 0$ such that $F\phi^{-1}(x) = G\phi^{-1}(x - k)$. Setting $y = \phi^{-1}(x)$, then $F(y) = G\phi^{-1}[\phi(y) - k]$. If for some $y$, $F(y) = G(y)$, then $y = \phi^{-1}[\phi(y) - k]$, and so $k = 0$, which is contrary to assumption. Therefore, $F$ and $G$ do not intersect.

Conversely, suppose that distinct scales $F$ and $G$ do not intersect. Define $\phi$ as in the statement of the theorem. Observe that $\phi$ is defined for all $x$ since clearly $H(0) \neq 0$ and if $h$ is the supremum of $H^n(0)$, then $h = \lim H^n(0) = H \lim H^{n-1}(0) = H(h)$, which is impossible because $H$ is a strictly increasing function. A similar argument shows that there is no infimum. Next we show that $\phi$ is an increasing scale. It is obviously onto, so it is sufficient to show that it is strictly increasing. Suppose that $x < y$. If both $x$ and $y$ are in the same interval $[H^n(0), H^{n+1}(0))$, then $\phi(x) < \phi(y)$ because $H^{-n}$ is strictly increasing. If they are in different intervals, the same conclusion follows because when $x < H^{n+1}(0)$, the fact that $H^{-n}$ is increasing implies

$$H^{-n}(x)/H(0) < H^{-n}H^{n+1}(0)/H(0) = H(0)/H(0) = 1.$$

Observe that $\phi$ satisfies Abel's functional equation

$$\phi H(x) - \phi(x) = n + 1 + \frac{H^{-n-1}H(x)}{H(0)} - n - \frac{H^{-n}(x)}{H(0)} = 1,$$

whence applying $F$ to $H$,

$$G(x) = F\phi^{-1}[\phi(x) + 1],$$

and so $F\phi^{-1}$ and $G\phi^{-1}$ are parallel.

The uniqueness properties of Abel's equation are well known; the details are left as Exercise 25.                                                    ◇

### 6.8.3 Preliminary Lemmas About Uniform Families

Throughout this and the following subsection, we let $\mathscr{F}$ denote a nonempty family of increasing scales and $\mathscr{G}$ its associated group (see Section 6.7.2 for definitions).

LEMMA 11.  *$\mathscr{F}$ is a uniform family iff $\mathscr{G}$ is a uniform family, and the two sets of parallelizers are identical.*

*PROOF.*  Suppose that $\mathscr{F}$ is a uniform family with parallelizer $\phi$. There exist constants $k_\lambda$, where $\lambda \in \Lambda$, such that

$$F_\lambda(x) = F_{\lambda_0}\phi^{-1}[\phi(x) - k_\lambda].$$

Thus,

$$g_\lambda(x) = \phi^{-1}[\phi(x) - k_\lambda] \quad \text{and} \quad g_\lambda^{-1}(x) = \phi^{-1}[\phi(x) + k_\lambda],$$

where $g_\lambda = F_{\lambda_0}^{-1}F_\lambda$ and $g_\lambda^{-1}$ are the generators of $\mathscr{G}$. From the form of these generators, it follows that any scale in $\mathscr{G}$ is of the form

$$g(x) = \phi^{-1}[\phi(x) + k],$$

for some real $k$. Therefore,

$$g\phi^{-1}(y - k) = \phi^{-1}(y),$$

hence the $g\phi^{-1}$ are parallel, and so $\mathscr{G}$ is a uniform family with the parallelizer $\phi$.

Conversely, suppose that $\mathscr{G}$ is a uniform family with a parallelizer $\phi$. Let $k_\lambda$ be the constants that render the generators of $\mathscr{G}$, $g_\lambda = F_{\lambda_0}^{-1}F_\lambda$, parallel, i.e.,

$$F_{\lambda_0}^{-1}F_\lambda\phi^{-1}(x - k_\lambda) = g_\lambda\phi^{-1}(x - k_\lambda)$$

$$= g_\mu\phi^{-1}(x - k_\mu)$$

$$= F_{\lambda_0}^{-1}F_\mu\phi^{-1}(x - k_\mu).$$

Applying $F_{\lambda_0}$ to this equation yields the conclusion that $\mathscr{F}$ is a uniform family with $\phi$ a parallelizer.                                                                      $\diamondsuit$

In the following two lemmas it is assumed that the functions in $\mathscr{G}$ do not intersect one another. This permits us to define the following ordering relation on $\mathscr{G}$: If $f$, $g$ in $\mathscr{G}$, then $f \gtrsim g$ if and only if for some real $x$, and since the scales of $\mathscr{G}$ do not intersect, for all real $x$, $f(x) \geqslant g(x)$.

LEMMA 12. *If the scales in $\mathscr{G}$ do not intersect, then under the composition of functions $\langle \mathscr{G}, \gtrsim \rangle$ is an Archimedean ordered group.*

PROOF. We noted earlier that $\mathscr{G}$ is a group and that the identity function $e$ is its identity element. The ordering is obviously a simple one. If $f, f', g \in \mathscr{G}$, then $f \gtrsim f'$ is equivalent to $fg \gtrsim f'g$ and $gf \gtrsim gf'$ because the scales do not intersect and they are strictly increasing. Suppose that $\{f^n\}, f \in \mathscr{G}$, is a standard sequence of the group. We deal explictly only with the case where $f > e, n \geqslant 1$, and there is a strict upper bound. The other cases differ only in trivial ways. Since $\{f^n\}$ has a strict upper bound, say $g$, then the sequence of numbers $\{f^n(0)\}$ has the upper bound $g(0)$, and so $\alpha = \sup\{f^n(0)\}$ exists. Since $f > e$ and $f$ is continuous, for any $\epsilon > 0$ there exists a $\delta > 0$ such that for all $x$ with $\alpha - \epsilon \leqslant x \leqslant \alpha, f(x) > x + \delta$. If the sequence is infinite, we may choose $n$ so large that $\alpha - \min(\delta, \epsilon) \leqslant f^n(0) \leqslant \alpha$. Thus, $f^{n+1}(0) = ff^n(0) > f^n(0) + \delta \geqslant \alpha$, which contradicts the choice of $\alpha$. Therefore, the sequence must be finite, i.e., the ordering is Archimedean.   $\diamondsuit$

LEMMA 13. *If $\mathscr{G}$ is noncyclic and any distinct functions that are pointwise limits of sequences in $\mathscr{G}$ do not intersect, then for any $x \in \mathrm{Re}$, $\{g(x) \mid g \in \mathscr{G}\}$ is dense in $\mathrm{Re}$.*

PROOF. By Lemma 12 and Hölder's Theorem (2.5) there exists a homomorphism $\psi$ of $\mathscr{G}$ onto a dense subgroup of $\langle \mathrm{Re}, \geqslant, + \rangle$. Let $g_1, g_2,...$ be a sequence in $\mathscr{G}$ such that $\psi(g_i)$ decreases to zero in $\mathrm{Re}$. Then the sequence $g_i$ decreases to a function $h \in \mathscr{H}$. We show that $h = e$. Clearly $h \gtrsim e$, and since $h$ is nondecreasing, $h^2 \gtrsim h$. On the other hand, if $\psi(g_j) \leqslant \frac{1}{2}\psi(g_i)$, we have $g_j^2 \lesssim g_i$; and letting $i \to \infty$, $h^2 \lesssim h$. Thus $h^2 \sim h$, and since $h(h(x)) = h(x) = e(h(x))$, $h$ and $e$ intersect and by hypothesis, $h = e$.

Since the $g_i$ are continuous and converge monotonically to a continuous limit, convergence is uniform on any compact $K \subset \mathrm{Re}$. Thus for every $\delta > 0$ and compact $K$, there exists $g > e$ with $|g(y) - y| < \delta$ for $y \in K$. In particular,

$$|g^{n+1}(x) - g^n(x)| < \delta \quad \text{for} \quad g^n(x) \in K.$$

For $g > e$, $\{g^n(x)\}$ is unbounded. For suppose $\sup g^n(x) = u < +\infty$.

Then

$$g(u) = g[\lim_{n\to\infty} g^n(x)] = \lim_{n\to\infty} g[g^n(x)] = u,$$

contradicting $g > e$; and likewise, letting $n \to -\infty$, $\inf g^n(x) = -\infty$. Therefore, for any $\delta > 0$ and any finite interval $K$ containing $x$, there exists $g \in \mathscr{G}$ such that $\{g^n(x)\}$ covers all of $K$ at a spacing of less than $\delta$; and the assertion of the lemma follows.                                                                         ◇

### 6.8.4  Theorem 9  (p. 287)

*A nonempty family $\mathscr{F}$ of increasing scales is uniform iff the scales of its associated group $\mathscr{G}$ do not intersect and all pointwise limits of bounded ascending sequences from $\mathscr{G}$ do not intersect.*

PROOF. By Lemma 11, there is no loss of generality in assuming $\mathscr{F} = \mathscr{G}$. Suppose that $\mathscr{G}$ is uniform and $\phi$ is a parallelizer of $\mathscr{G}$. By definition, if $g \in \mathscr{G}$, then $g\phi^{-1}$ is parallel to $e\phi^{-1}$ and so the function $\phi g\phi^{-1}$ is parallel to $e$, i.e., it is a straight line of the form $x + k$. Any bounded ascending sequence of such functions obviously converges to a function of the same form, so the inverse increasing transformations show that these limits are, in fact, increasing scales that do not intersect one another.

Conversely, suppose that $\mathscr{G}$ has the property that its elements do not intersect and that functions which are limits of bounded ascending sequences from $\mathscr{G}$ do not intersect one another. Let $\mathscr{G}^+ = \{g \mid g \in \mathscr{G} \text{ and } g > e\}$. Suppose that $\mathscr{G}^+$ has a least element $g_1$. By the Archimedean property of $\mathscr{G}$ (Lemma 12), for any $f \in \mathscr{G}$ there is an integer $n$ such that $g_1^{n+1} > f \gtrsim g_1^n$. If $f > g_1^n$, then apply $g_1^{-n}$ to this inequality to obtain $g_1 > g_1^{-1}f > e$, which is contrary to the choice of $g_1$. Thus, $f = g_1^n$ and $\mathscr{G}$ is cyclic. By Theorem 8, $g_1$ and $e$ are parallel with some parallelizer $\phi$, i.e., $\phi g_1\phi^{-1}(x) = x + k$. By induction, $\phi f\phi^{-1}(x) = \phi g_1^n\phi^{-1}(x) = x + nk$, hence $\mathscr{G}$ is uniform and has the set of parallelizers given in Theorem 8.

So we assume that $\mathscr{G}^+$ has no least element or, what is the same thing, that $\mathscr{G}$ is noncyclic. For every real $x, y$, Lemma 13 implies that there is a bounded ascending sequence of scales $g_i \in \mathscr{G}$ such that $g_i$ converges to a function $g$ for which $g(x) = y$. To show that $g$ is increasing we first note that it is nondecreasing and that there exists a bounded ascending sequence $f_i \in \mathscr{G}$ such that $f_i$ converges to a function $f$ for which $f(y) = x$. It is easy to see that $f_i g_i$ and $g_i f_i$ are bounded ascending sequences converging to $fg$ and $gf$, respectively, and that $fg(x) = x$ and $gf(y) = y$. Thus, by the nonintersection hypothesis, $fg = gf = e$. So $f = g^{-1}$ and $g$ is a scale. Thus, these limits, which include $\mathscr{G}$, form a family $\mathscr{H}$ of nonintersecting, increasing scales with the property that at least one passes through every point.

We next show that the elements of $\mathcal{H}$ commute. By Lemma 12 and Theorem 2.5, those of $\mathcal{G}$ commute. Suppose there exist real $x$ and $f, g \in \mathcal{H}$ such that $fg(x) > gf(x)$. If $f_i$, $g_i \in \mathcal{G}$ converge to $f, g$, then for some sufficiently large $i$, $f_i g_i(x) > g_i f_i(x)$, which is impossible.

The proof that $\mathcal{G}$ is uniform is concluded if we can show that the Thomsen condition holds in $\mathcal{H}$, since $\mathcal{G} \subset \mathcal{H}$ and Theorem 7 shows that $\mathcal{H}$ is uniform. Suppose that the points $ax$ and $dq$ lie on one curve, $f$, i.e., $x = f(a)$ and $q = f(d)$, and that $dp$ and $bx$ lie on another, $g$, i.e., $p = g(d)$ and $x = g(b)$. By what we have just shown there is a curve $h$ through $ap$, i.e., $p = h(a)$. The Thomsen condition is established[4] by showing that $bq$ lies on $h$. Substituting and using the commutativity of $\mathcal{H}$,

$$q = f(d) = fg^{-1}(p) = fg^{-1}h(a) = fg^{-1}hf^{-1}(x) = fg^{-1}hf^{-1}g(b) = h(b).$$

We next prove the desired uniqueness of the parallelizers. Let $\phi$ and $\psi$ be two parallelizers of a noncyclic $\mathcal{G}$. Thus, for any $g \in \mathcal{G}$ there are constants $k_g$ and $l_g$ such that

$$g(x) = \phi^{-1}[\phi(x) + k_g] = \psi^{-1}[\psi(x) + l_g].$$

If we let $\theta = \psi\phi^{-1}$ and $y = \phi(x)$, then

$$\theta(y + k_g) = \theta(y) + l_g.$$

Setting $y = 0$ yields $\theta(k_g) - \theta(0) = l_g$, so letting $\xi = \theta - \theta(0)$,

$$\xi(x + k_g) = \xi(x) + \xi(k_g).$$

Since $k_g = \phi g(x) - \phi(x)$ and $\phi$ is increasing and continuous, the values of $k_g$ are dense in the real numbers because, as was just shown, $g(x)$ are dense in the reals. Since $\xi$ is continuous and increasing, it is well known (Aczél, 1966, p. 34) that $\xi(x) = \alpha x, \alpha > 0$, and so $\theta(x) = \alpha x + \beta$, from which it follows that $\psi = \alpha\phi + \beta$. The converse is obvious. $\diamondsuit$

## 6.9 BISYMMETRIC STRUCTURES

### 6.9.1 Sufficient Conditions

The next system of measurement looks, at first glance, as though it should have been included in Chapter 3 on extensive measurement because a binary operation, somewhat like concatenation, is postulated. Since, however, this operation is not necessarily assumed to be either associative or commu-

---

[4] This part of the proof is due to A. A. J. Marley (personal communication).

tative and it may be idempotent ($a \circ a \sim a$), it is more general than those studied in Chapter 3; and it turns out to be more natural to subsume the system under additive conjoint measurement.

Pfanzagl (1959 a,b) first introduced bisymmetric structures as a qualitative reformulation of a numerical functional equation, which had been studied by Aczél (1948) and many others (see the references given on p. 278–279 of Aczél, 1966). This equation arose from a consideration of the properties of means of pairs of numbers. Thus, one possible interpretation of $a \circ b$ in this structure is as the element whose value equals the mean of the values of $a$ and $b$. As we shall see from the representation theorem, the structure is actually a good deal more general than this.

DEFINITION 10. *Suppose that $A$ is a nonempty set, that $\gtrsim$ is a binary relation on $A$, and that $\circ$ is a binary operation from $A \times A$ into $A$. The triple $\langle A, \gtrsim, \circ \rangle$ is a* bisymmetric structure *iff, for all $a, b, \bar{b}, \underline{b}, c, d \in A$, the following five axioms hold:*

1. *$\langle A, \gtrsim \rangle$ is a weak order.*

2. *Monotonicity: $a \gtrsim b$ iff $a \circ c \gtrsim b \circ c$ iff $c \circ a \gtrsim c \circ b$.*

3. *Bisymmetry: $(a \circ b) \circ (c \circ d) \sim (a \circ c) \circ (b \circ d)$.*

4. *Restricted solvability: If $\bar{b} \circ c \gtrsim a \gtrsim \underline{b} \circ c$ (or $c \circ \bar{b} \gtrsim a \gtrsim c \circ \underline{b}$), then there exists $b' \in A$ such that $b' \circ c \sim a$ (or $c \circ b' \sim a$).*

5. *Archimedean: Every strictly bounded standard sequence is finite, where $\{a_i \mid a_i \in A, i \in N\}$ is a* standard sequence *iff there exist $p, q \in A$ such that $p \succ q$ and, for all $i, i + 1 \in N$, $a_i \circ p \sim a_{i+1} \circ q$ or, for all $i, i + 1 \in N$, $p \circ a_i \sim q \circ a_{i+1}$.*

The critically new axiom is bisymmetry, which as we said was motivated by the interpretation in terms of means. If $\phi$ is some numerical representation, then the mean of $\phi(a)$ and $\phi(b)$ is $[\phi(a) + \phi(b)]/2$, and so

$$\frac{1}{2}\{[\phi(a) + \phi(b)]/2 + [\phi(c) + \phi(d)]/2\}$$
$$= \frac{1}{4}[\phi(a) + \phi(b) + \phi(c) + \phi(d)]$$
$$= \frac{1}{2}\{[\phi(a) + \phi(c)]/2 + [\phi(b) + \phi(d)]/2\},$$

which corresponds to Axiom 3. For this interpretation, several other similar equations also follow because the operation of means is commutative, but their qualitative analogs are not assumed because we do not necessarily want $\circ$ to be commutative.

The psychophysical method of bisection is a possible example of an empirical bisymmetric operation. Suppose that $a$ and $b$ denote tones of the same frequency, but different intensities, presented for short periods of time in the order given. Let $a \circ b$ denote the intensity that a subject deems to be half-

way between *a* and *b*. Thus, $a \circ b$ psychologically bisects the interval
from *a* to *b*. In general, data make clear that $a \circ b \not\sim b \circ a$. Empirical support
for the bisymmetry axiom, using pure tones varying in intensity as stimuli,
has been reported by Cross (1965). (Figure 7 provides a graphical illustration
of the meaning of bisymmetry.)

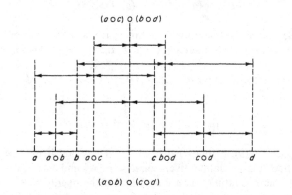

**FIGURE 7.** A graphical illustration of the bisymmetry condition.

The definition of a standard sequence parallels closely one for a conjoint
structure and not that of an extensive one. The temptation to replace it by
a defined notion of *na*, as in extensive measurement, quickly fades when one
recognizes how many different possible definitions of *na* there are. Because
$\circ$ need not be associative, none of the following need be equivalent:
$(a \circ a) \circ (a \circ a)$, $((a \circ a) \circ a) \circ a$, $(a \circ (a \circ a)) \circ a$, $a \circ (a \circ (a \circ a))$, or
$a \circ ((a \circ a) \circ a)$. And as the value of *n* increases, so does the number of non-
equivalent candidates for *na*.

The representation that can be proved is:

THEOREM 10. *Suppose $\langle A, \gtrsim, \circ \rangle$ is a bisymmetric structure, then there
exist real numbers $\mu > 0$, $\nu > 0$, and $\lambda$ and a real-valued function $\phi$ on A
such that, for $a, b \in A$,*

(i)   $a \gtrsim b$ iff $\phi(a) \geqslant \phi(b)$;

(ii)  $\phi(a \circ b) = \mu\phi(a) + \nu\phi(b) + \lambda$;

(iii) *if $\mu'$, $\nu'$, $\lambda'$, $\phi'$ is another representation fulfilling (i) and (ii), then there
exist constants $\alpha > 0$ and $\beta$ such that*

$$\phi' = \alpha\phi + \beta,$$
$$\lambda' = \alpha\lambda + \beta(1 - \mu - \nu),$$
$$\mu' = \mu, \qquad \nu' = \nu.$$

Three simple facts about this representation help to give some insight into the system. First, if the operation is idempotent, i.e., for all $a \in A$, $a \circ a \sim a$, as is true for means and for what are called *intensive quantities*, then $\lambda = 0$ and $\mu + \nu = 1$. This follows immediately from

$$\phi(a) = \phi(a \circ a) = \mu\phi(a) + \nu\phi(a) + \lambda$$

for all $a$. Second, if the operation is commutative, as it is for means and for extensive measures, and if there exist $a > b$, then $\mu = \nu$. This follows from

$$\mu\phi(a) + \nu\phi(b) + \lambda = \phi(a \circ b) = \phi(b \circ a) = \mu\phi(b) + \nu\phi(a) + \lambda,$$

whence

$$(\mu - \nu)[\phi(a) - \phi(b)] = 0.$$

So from $\phi(a) > \phi(b)$, it follows that $\mu = \nu$. Therefore, if the operation is both idempotent and commutative, then $\mu = \nu = \frac{1}{2}$ and $\lambda = 0$, which is the appropriate representation for means. Third, if the operation is associative as well as commutative, as it is in extensive measurement, and if there exist $a > b$, then $\mu = \nu = 1$. By commutativity, $\mu = \nu$, and so

$$\begin{aligned}
\mu\phi(a) + 2\mu^2\phi(b) + (\mu + 1)\lambda &= \phi[a \circ (b \circ b)] \\
&= \phi[(a \circ b) \circ b] \\
&= \mu^2\phi(a) + (\mu^2 + \mu)\,\phi(b) + (\mu + 1)\,\lambda,
\end{aligned}$$

whence

$$\mu(1 - \mu)[\phi(a) - \phi(b)] = 0.$$

Since $\phi(a) > \phi(b)$, $\mu = 0$ or $1$. If $\mu = 0$, then $\phi(a \circ b) = \lambda = $ constant, which violates the order-preserving property, so $\mu = 1$.

A detailed investigation of idempotence and commutativity of a bi-symmetric operation was carried out by Kristof (1968). Using axioms that are somewhat stronger than those of Definition 10, he proved that there is exactly one idempotent element or all elements are idempotent, and that either every pair of elements is commutative or none is.

Our proof of Theorem 10 involves reducing it to Theorem 2 via the relation $\succsim'$ defined on $A \times A$ by

$$ap \succsim' bq \quad \text{iff} \quad a \circ p \succsim b \circ q,$$

and showing that $\langle A, A, \succsim' \rangle$ is an additive conjoint structure. Pfanzagl (1959a,b) gave a direct proof of a slightly less general theorem. He included

three of the necessary conditions of Definition 10—weak ordering, monotonicity, and bisymmetry—but substituted two topological assumptions for
our restricted solvability and Archimedean assumptions (see Kelley, 1955, for
the relevant topological definitions). They are formulated precisely in the
following version of his theorem.

THEOREM 11. *Suppose that $\langle A, \gtrsim, \circ \rangle$ is a structure that satisfies Axioms
1–3 of Definition 10. In the order topology of $A$, if $A$ is connected and the
operation $\circ$ is continuous in both of its arguments, then $\langle A, \gtrsim, \circ \rangle$ satisfies all
of the axioms of Definition 10 and the function $\phi$ of Theorem 10 is continuous.*

### 6.9.2 A Finite, Equally Spaced Case

As usual when we arrive at a finite, equally spaced version of the axioms,
we drop the solvability and Archimedean axioms. But this time a number of
other changes will be made because we shall deal only with the special case
of an idempotent and commutative operation which has the representation
$\phi(a \circ b) = [\phi(a) + \phi(b)]/2$. These two assumptions are explicitly included.
Moreover, in the finite case, the operation $\circ$ cannot possibly be closed, so
we must include axioms that characterize the set $B \subset A \times A$ over which $\circ$
is defined, just as we did in Section 3.4 for extensive measurement. We will
retain the monotonicity property and introduce an equal-spacing assumption.
When we do this, it turns out that we can drop bisymmetry altogether. Since
we are dealing with such a special case, we do not provide a separate definition of the structure, but merely state it as the hypothesis of the representation theorem.

THEOREM 12. *Suppose that $A$ is a finite set, $\gtrsim$ a binary relation on
$A$, $B$ a nonempty subset of $A \times A$, and $\circ$ a function from $B$ into $A$, such
that, for all $a, b, c, d \in A$:*

1. *$\langle A, \gtrsim \rangle$ is a weak order.*
2. *If $ac, bc \in B$, then $a \gtrsim b$ iff $a \circ c \gtrsim b \circ c$.*
3. *If $ab \in B$, then $ba \in B$ and $a \circ b \sim b \circ a$.*
4. *$aa \in B$ and $a \circ a \sim a$.*
5. *If $bc \in B$, $a J b$, and $c J d$, then $ad \in B$ and $a \circ d \sim b \circ c$, where $J$ is
given by Definition 1.5 (Section 1.3.2).*

*Then there exists a real-valued function $\phi$ on $A$ such that, for every $a, b \in A$,*

(i) *$a \gtrsim b$ iff $\phi(a) \geq \phi(b)$;*
(ii) *If $ab \in B$, then $\phi(a \circ b) = [\phi(a) + \phi(b)]/2$.*

*Moreover, any other real-valued function that satisfies* (i) *and* (ii) *is related to* $\phi$ *by a positive linear transformation.*

The proof is left as Exercises 30 and 31.

## 6.10 PROOFS

### 6.10.1 Theorem 10 (p. 295)

*If* $\langle A, \succsim, \circ \rangle$ *is a bisymmetric structure, then there is has a real-valued order-preserving function* $\phi$ *and constants* $\mu > 0$, $\nu > 0$, *and* $\lambda$ *such that* $\phi(a \circ b) = \mu\phi(a) + \nu\phi(b) + \lambda$; $\phi$ *is an interval scale.*

PROOF. We assume the axioms of a bisymmetric structure (Definition 10) and that there exist $a, b \in A$ such that $a > b$, and we prove those of an additive conjoint structure (Definition 7, Section 6.2.3) for $\succsim'$, where $ap \succsim' bq$ iff $a \circ p \succsim b \circ q$.

1.  Axiom 1.

2.  If $ap \succsim' bp$, then $a \circ p \succsim b \circ p$, so by Axiom 2, $a \succsim b$ and, by the same axiom, $a \circ q \succsim b \circ q$, whence by definition $aq \succsim' bq$. The other case of independence is similar.

3.  If $ax \succsim' fq$ and $fp \succsim' bx$, then by Axioms 2 and 3,

$$(a \circ p) \circ (x \circ x) \sim (a \circ x) \circ (p \circ x)$$
$$\succsim (f \circ q) \circ (p \circ x)$$
$$\sim (f \circ p) \circ (q \circ x)$$
$$\succsim (b \circ x) \circ (q \circ x)$$
$$\sim (b \circ q) \circ (x \circ x),$$

and so, by Axiom 2, $ap \succsim' bq$.

4.  Axiom 4.

5.  Axiom 5.

6.  By assumption, there exist $a > b$, so both components are essential.

If $a \sim b$ for all $a, b \in A$, then $\phi(a) =$ constant, $\mu + \nu = 1$, and $\lambda = 0$ clearly fulfills the theorem. So we assume there exist $a > b$, in which case $\langle A, A, \succsim' \rangle$ is an additive conjoint structure. By Theorem 2, it has a real-valued additive representation $\phi_1$, $\phi_2$ which is unique up to linear transformations with a common unit. For any fixed $p \in A$, define

$$\phi_{ilp}(a) = \phi_i(p \circ a), \qquad \phi_{irp}(a) = \phi_i(a \circ p), \qquad i = 1, 2.$$

Observe that by Axioms 2 and 3,

$$\phi_{1lp}(a) + \phi_{2lp}(b) \geqslant \phi_{1lp}(c) + \phi_{2lp}(d)$$
$$\text{iff } \phi_1(p \circ a) + \phi_2(p \circ b) \geqslant \phi_1(p \circ c) + \phi_2(p \circ d)$$
$$\text{iff } (p \circ a)(p \circ b) \gtrsim' (p \circ c)(p \circ d)$$
$$\text{iff } (p \circ a) \circ (p \circ b) \gtrsim (p \circ c) \circ (p \circ d)$$
$$\text{iff } (p \circ p) \circ (a \circ b) \gtrsim (p \circ p) \circ (c \circ d)$$
$$\text{iff } a \circ b \gtrsim c \circ d$$
$$\text{iff } ab \gtrsim' cd.$$

Thus $\phi_{1lp}$, $\phi_{2lp}$ form an additive representation. Similarly, so do $\phi_{1rp}$, $\phi_{2rp}$. By the uniqueness of additive representations,

$$\phi_i(p \circ a) = \alpha_l(p)\,\phi_i(a) + \beta_{il}(p), \qquad i = 1, 2,$$
$$\phi_i(a \circ p) = \alpha_r(p)\,\phi_i(a) + \beta_{ir}(p), \qquad i = 1, 2,$$

where $\alpha_l > 0$ and $\alpha_r > 0$. From Axiom 3.

$$\phi_1(a \circ b) + \phi_2(c \circ d) = \phi_1(a \circ c) + \phi_2(b \circ d),$$

and so,

$$\alpha_l(a)\,\phi_1(b) + \beta_{1l}(a) + \alpha_r(d)\,\phi_2(c) + \beta_{2r}(d)$$
$$= \alpha_l(a)\,\phi_1(c) + \beta_{1l}(a) + \alpha_r(d)\,\phi_2(b) + \beta_{2r}(d),$$

or rewriting,

$$\alpha_l(a)[\phi_1(b) - \phi_1(c)] = \alpha_r(d)[\phi_2(b) - \phi_2(c)].$$

Thus, there are constants $\mu$, $\nu$ such that for all

$$a, d \in A, \, \alpha_l(a) = \nu > 0, \, \alpha_r(d) = \mu > 0.$$

Substitute again in the bisymmetry equation,

$$\mu\phi_1(a) + \beta_{1r}(b) + \mu\phi_2(c) + \beta_{2r}(d) = \mu\phi_1(a) + \beta_{1r}(c) + \mu\phi_2(b) + \beta_{2r}(d),$$

so

$$\beta_{1r}(b) - \beta_{1r}(c) = \mu[\phi_2(b) - \phi_2(c)] = \nu[\phi_1(b) - \phi_1(c)].$$

Thus, setting $\phi = \phi_1$ and choosing some fixed $c \in A$,

$$\phi(a \circ b) = \mu\phi(a) + \beta_{1r}(b)$$
$$= \mu\phi(a) + \nu[\phi(b) - \phi(c)] + \beta_{1r}(c)$$
$$= \mu\phi(a) + \nu\phi(b) + \lambda$$

where $\lambda = \beta_{1r}(c) - \nu\phi_1(c)$. This establishes the decomposition property of $\phi$.

Since $\phi = \phi_1$, it is order preserving by Theorem 2.

To show uniqueness, let us suppose that $\phi', \mu', \nu', \lambda'$ is another representation. Since $\mu\phi, \nu\phi$ and $\mu'\phi', \nu'\phi'$ both form additive representations of $\langle A, A, \gtrsim' \rangle$, we know by Theorem 2 that there are constants $\alpha' > 0, \beta_1, \beta_2$ such that

$$\mu'\phi' = \alpha'\mu\phi + \beta_1, \qquad \nu'\phi' = \alpha'\nu\phi + \beta_2.$$

Eliminating $\phi$ and noting that the resulting equation holds for all $a \in A$, we have $\beta_1/\mu' = \beta_2/\nu' = \beta$ and $\nu/\nu' = \mu/\mu'$. From this it follows that $\phi' = \alpha\phi + \beta$, where $\alpha = \alpha'\mu/\mu'$. Substituting this into the expression for $\phi(a \circ b)$, it is easy to show that $\mu' = \mu, \nu' = \nu$, and $\lambda' = \alpha\lambda + \beta(1 - \mu - \nu)$.

$\diamond$

## 6.10.2 Theorem 11 (p. 297)

*Suppose that $\langle A, \gtrsim, \circ \rangle$ satisfies the weak order, monotonicity, and bisymmetry axioms of Definition 10. If, in addition, $A$ is connected and $\circ$ is continuous in each argument, then $\langle A, \gtrsim, \circ \rangle$ also satisfies the restricted solvability and Archimedean axioms and the function $\phi$ of Theorem 10 is continuous.*

*PROOF.* First, we establish that restricted solvability holds. Suppose that $\bar{b} \circ c \gtrsim a \gtrsim \underline{b} \circ c$, and let $\underline{B} = \{x \mid a \gtrsim x \circ c\}$ and $\bar{B} = \{x \mid x \circ c \gtrsim a\}$. By the continuity of $\circ$, $\underline{B}$ and $\bar{B}$ are closed and, by assumption, they are nonempty; clearly $A = \underline{B} \cup \bar{B}$. So, by connectedness, $\underline{B} \cap \bar{B} \neq \varnothing$, i.e., for some $b \in A$, $b \circ c \sim a$. Right solvability is proved similarly.

Next, we show the Archimedean axiom. Suppose that $p > q$ and that $\{a_i\}$ is such that, for all $i$, $a_i \circ p \sim a_{i+1} \circ q$. By the monotonicity axiom, $a_{i+1} > a_i$. Let $B = \{b \mid \text{for all } a_i, b > a_i\}$. We show that if the sequence is infinite, then $B = \varnothing$. Suppose not, then the sequence is strictly bounded. With no loss of generality, let us suppose it is bounded from above, then it has a supremum $a^*$ to which it converges because the order topology is connected. By continuity $a_i \circ q$ converges to $a^* \circ q$. Thus, by bisymmetry,

$$(a_i \circ q) \circ (p \circ p) \sim (a_i \circ p) \circ (q \circ p)$$
$$\sim (a_{i+1} \circ q) \circ (q \circ p),$$

and taking limits,

$$(a^* \circ q) \circ (p \circ p) \lesssim (a^* \circ q) \circ (q \circ p).$$

Thus $p \lesssim q$, a contradiction. A similar proof holds starting with $p \circ a_i \sim q \circ a_{i+1}$.

Since $\phi(a \circ b) = \mu\phi(a) + \nu\phi(b) + \lambda$, a discontinuity of $\phi$ at $a$ goes into one at $a \circ b$, for all $b$. It follows by connectedness that any discontinuity of $\phi$ leads to uncountably many jump discontinuities, which is impossible, so $\phi$ is continuous. ◇

## □ 6.11 ADDITIVE REPRESENTATION OF $n$ COMPONENTS

### 6.11.1 The General Case

Sometimes three or more components affect the attribute of interest, and so we need to generalize the definitions and theorems to that case. Since all of the definitions we need are natural generalizations of those given for $n = 2$, we introduce them without further comment. Let $N = \{1, 2,..., n\}$.

DEFINITION 11. *A relation $\gtrsim$ on $\mathsf{X}_{i=1}^n A_i$ is independent iff, for every $M \subset N$, the ordering $\gtrsim_M$ induced by $\gtrsim$ on $\mathsf{X}_{i \in M} A_i$ for fixed choices $a_i \in A_i$, $i \in N - M$, is unaffected by those choices.*

It is trivial to show that if $\gtrsim$ is a weak ordering, then so is $\gtrsim_M$.

DEFINITION 12. *A relation $\gtrsim$ on $\mathsf{X}_{i=1}^n A_i$ satisfies restricted solvability iff, for each $i \in N$, whenever*

$$b_1 \cdots \bar{b}_i \cdots b_n \gtrsim a_1 \cdots a_i \cdots a_n \gtrsim b_1 \cdots \underline{b}_i \cdots b_n ,$$

*then there exists $b_i \in A_i$ such that*

$$b_1 \cdots b_i \cdots b_n \sim a_1 \cdots a_i \cdots a_n .$$

Since both the Archimedean property, Definition 7, and the essentialness of components, Definition 6, were stated for individual components, there is no need to define them again.

DEFINITION 13. *Suppose that $A_i$, $i \in N, n \geqslant 3$, are nonempty sets and that $\gtrsim$ is a binary relation on $\mathsf{X}_{i=1}^n A_i$. The structure $\langle A_1,..., A_n, \gtrsim \rangle$ is an $n$-component, additive conjoint structure iff $\gtrsim$ satisfies the following five axioms:*

1. *Weak ordering.*
2. *Independence (Definition 11).*
3. *Restricted solvability (Definition 12).*
4. *Every strictly bounded standard sequence (Definition 4) is finite.*
5. *At least three components are essential (Definition 6).*

THEOREM 13. *If* $\langle A_1 ,..., A_n , \gtrsim \rangle$, $n \geqslant 3$, *is an n-component, additive conjoint structure, then there exist real-valued functions* $\phi_i$ *on* $A_i$, $i \in N$, *such that for all* $a_i , b_i \in A_i$,

$$a_1 \cdots a_n \gtrsim b_1 \cdots b_n \quad \textit{iff} \quad \sum_{i=1}^{n} \phi_i(a_i) \geqslant \sum_{i=1}^{n} \phi_i(b_i).$$

*If* $\{\phi_i{'}\}$ *is another such family of functions, then there exist numbers* $\alpha > 0$ *and* $\beta_i$, $i \in N$, *such that*

$$\phi_i{'} = \alpha \phi_i + \beta_i .$$

This result, which obviously generalizes Theorem 2, is a bit surprising in that no mention is made of a property similar to double cancellation. As a matter of fact, for any two components, it can be derived from the assumptions (Lemma 14 of Section 6.12.1). Independence for three or more components is much stronger than it is for only two. In practice this means that if $n \geqslant 3$ and if we are persuaded of restricted solvability and the Archimedean property, then, unlike the two-component case, all we need to check experimentally are the weak ordering and independence assumptions. Both of these assumptions seem simpler to check than double cancellation.

The first published version of this theorem was proved by Debreu (1960). Some of his assumptions are topological. We obtain a somewhat more general result by replacing them with solvability and Archimedean axioms. For example, his theorem does not apply to rational components since the set of rational numbers is not connected. We first give a topological definition, and then the statement of Debreu's theorem, which we prove in Section 6.12.3. For the other topological definitions see, for example, Kelley (1955).

DEFINITION 14. *Let* $A_1 ,..., A_n$ *be topological spaces. The* product topology *on* $\mathsf{X}_{i=1}^{n} A_i$ *consists of the following collection of open sets*:

1.  *If* $U_i$ *is an open subset of* $A_i$, $i = 1,..., n$, *then* $\mathsf{X}_{i=1}^{n} U_i$ *is open.*

2.  *Any union of arbitrarily many subsets of* $\mathsf{X}_{i=1}^{n} A_i$, *each of which is open according to provision 1, is itself open.*

It can easily be verified that this collection of open sets forms a topology.

THEOREM 14. *Suppose that* $n \geqslant 3$ *and that* $\langle A_1 ,..., A_n , \gtrsim \rangle$ *satisfies Axioms 1, 2, and 5 of Definition 13. If, in addition, each* $A_i$ *is a connected topological space and if, for every* $b \in \mathsf{X}_{i=1}^{n} A_i$, *the sets* $\{a \mid a \in \mathsf{X}_{i=1}^{n} A_i$ *and* $b \gtrsim a\}$ *and* $\{a \mid a \in \mathsf{X}_{i=1}^{n} A_i$ *and* $a \gtrsim b\}$ *are closed in the product topology on* $\mathsf{X}_{i=1}^{n} A_i$, *then the structure also satisfies Axioms 3 (restricted solvability) and 4 (Archimedean), and so, by Theorem 13, there exists an additive, order-*

*preserving representation of $\gtrsim$ that is unique up to positive linear transformations.*

We remark that although Debreu assumed and used separability of the $A_i$ in his proof, we do not in ours.

Before turning to proofs of Theorems 13 and 14, we should mention the possibility of generalizing the above results to a countable number of factors. There is an intermediate result which begins with a partial representation and gives conditions sufficient to complete it, but it is not really a representation theorem in the sense employed in this book (Fishburn, 1966a). We have been informed that Grodal and Mertens (1968) cover not only the countable case, but more general ones.

### 6.11.2   The Case of Identical Components

If $\langle A_1 ,..., A_n , \gtrsim \rangle$ is an additive conjoint structure, and $A_i = A_j$, then it is natural to ask whether $\phi_i = \phi_j$ in the representation or, more generally, whether $\phi_i$ and $\phi_j$ are proportional. (Since the zeros are arbitrary, we can always choose the representation so that $\phi_i$ and $\phi_j$ have the same zero point.)

An example with $n = 2$ is found in the additive representation for an algebraic-difference structure (Section 4.4): if $\gtrsim$ is an ordering of $A \times A$ and $ab \gtrsim cd$ if and only if $\phi(a) - \phi(b) \geqslant \phi(c) - \phi(d)$, define $\gtrsim^*$ on $A \times A$ by $ab \gtrsim^* cd$ if and only if $ad \gtrsim cb$. Clearly, $\langle A \times A, \gtrsim^* \rangle$ has an additive representation $(\phi, \phi)$.

If we take the characteristic sign-reversal axiom of algebraic-difference structures,

$$ad \gtrsim cb \qquad \text{iff} \qquad bc \gtrsim da,$$

this translates in to a $\gtrsim^*$ axiom

$$ab \gtrsim^* cd \qquad \text{iff} \qquad ba \gtrsim^* dc.$$

This amounts to saying that $ab \sim^* ba$, which is obviously a necessary condition for $\phi_1 = \phi_2$. The same condition can easily be shown to be sufficient (given an additive conjoint structure), and moreover, it generalizes to the case $n \geqslant 2$. If some of the $A_i$ are identical in an additive conjoint structure, then the corresponding $\phi_i$ are identical if and only if, for any $a$ in $\mathsf{X}_{i=1}^{n} A_i$ and any $b$ obtained from $a$ by a permutation of the components from identical $A_i$, $a \sim b$ (see Exercise 36).

An interesting application of additive conjoint measurement with $n$ identical components is to the problem of deferred income or consumption. Let $a_1 \cdots a_n$ denote the outcome of receiving $a_1$ in time period 1, $a_2$ in time period 2,..., $a_n$ in time period $n$. If preferences over the basis set $A$ of commodities are stable over time periods, i.e., no matter how long the income is deferred (a condition that is easily violated, e.g., by a 1971 automobile to be received in 1971, 1972,..., 1990), then a reasonable model may be

$$\phi(a_1 \cdots a_n) = \sum_{i=1}^{n} \lambda_i u(a_i),$$

where $u$ is the utility function on $A$ and $\lambda_i$ is a discount factor. If the discount factors are unequal, we no longer obtain equivalent elements by permutation. Most people would rather receive $1000 now and $500 next year than $500 now and $1000 next year.

A necessary and sufficient condition for all the $\phi_i$ to be proportional to $u$ is that standard sequences be invariant across factors. That is, if $a^{(1)}, a^{(2)},..., a^{(j)},...$ is a standard sequence in one factor, then it cannot have unequal spacing in some other factor. Clearly, it suffices to assume this for three-term standard sequences: if $a_i, b_i, c_i$ is a standard sequence, $a_i = a_j, b_i = b_j$, and $c_i = c_j$, then $a_j, b_j, c_j$ cannot be unequally spaced. We formulate this result in Theorem 15 below. A slightly more complex theorem, formulated in terms of a difference structure on $A^n \times A^n$, was obtained by Fishburn (1970, Chapter 7). Still another version of the result interprets the factors $\lambda_i$ as probabilities. Here the problem is complicated by requiring that these factors constitute a probability measure. This version, due to Luce and Krantz (1971), is presented in Chapter 8; the standard-sequence condition is formulated there as Axiom 5 of Definition 1, Section 8.2.6.

A special form of $\lambda_i$ is given by the exponential discount rate, $\lambda_i = \lambda^{i-1}$, where $\lambda > 0$ (and usually, $\lambda < 1$). This is characterized by a qualitative property of stationarity formulated by Koopmans (1960); it was called temporal consistency by Williams and Nassar (1966).

DEFINITION 15.  *A binary relation $\gtrsim$ on $A^n$, $n \geqslant 3$, is* stationary *iff there exists $x \in A$ such that for all $a^{(1)},..., a^{(n-1)}, b^{(1)},..., b^{(n-1)} \in A$,*

$$a^{(1)} \cdots a^{(n-1)}x \gtrsim b^{(1)} \cdots b^{(n-1)}x$$

*iff*

$$xa^{(1)} \cdots a^{(n-1)} \gtrsim xb^{(1)} \cdots b^{(n-1)}.$$

This condition is clearly weaker than permutability, since only certain equivalences are preserved, and only for special permutations. On the other hand, it is easily seen to imply the invariance-of-standard-sequences property.

For example, suppose $a, b, c$ is a standard sequence in the first component, with mesh equal to the $p, q$ interval in the second factor, so that

$$apa^{(3)} \cdots a^{(n-1)}x \sim bqa^{(3)} \cdots a^{(n-1)}x,$$
$$bpa^{(3)} \cdots a^{(n-1)}x \sim cqa^{(3)} \cdots a^{(n-1)}x.$$

Applying stationarity moves $a, b, c$ into the second factor, where they still form a standard sequence, with mesh equal to $p, q$ in the third factor. (This argument is applicable except to a standard sequence in the first or last factor with mesh elements in the last or first factor, respectively; here, a more complicated argument is required.) So stationarity first yields the result that the representation can be put in the form $\sum \lambda_i u(a_i)$, because of the standard-sequence property. Then, the argument that in this case $\lambda_i = \lambda^{i-1}$ is quite easy (Exercise 35).

A somewhat more general theory where the preference is over countably many copies of $A$ (infinite time horizon) can be found in Koopmans (1960) and in Koopmans, Diamond, and Williamson (1964).

We summarize the results outlined in this section by the following theorem:

THEOREM 15. Let $\langle A^n, \gtrsim \rangle$ be an $n$-component additive conjoint structure (Definition 13 if $n \geqslant 3$ or Definition 7 if $n = 2$).

(i)   There exists an interval scale $u$ on $A$ and nonzero numbers $\lambda_1, ..., \lambda_n$ such that $\sum_{i=1}^n \lambda_i u(a_i)$ is order preserving iff for any standard sequence $a_i, b_i, c_i$, any distinct $j, k, 1 \leqslant j, k \leqslant n$, and any $a_j, b_j, c_j, p_k, q_k \in A$, if

$$a_i = a_j, \qquad b_i = b_j, \qquad c_i = c_j,$$

and

$$a_j p_k \sim b_j q_k,$$

then

$$b_j p_k \sim c_j q_k.$$

The $\lambda_i$ are unique up to multiplication by a positive constant.

(ii)   There exists an interval scale $u$ on $A$ and a unique number $\lambda > 0$, such that $\sum_{i=1}^n \lambda^{i-1} u(a_i)$ is order preserving iff stationarity holds (Definition 15).

(iii)   There exists an interval scale $u$ on $A$ such that $\sum_{i=1}^n u(a_i)$ is order preserving iff for all $a_1 \cdots a_n$ and $b_1 \cdots b_n \in A_n$ such that the $b_i$ are a permutation of the $a_i$

$$a_1 \cdots a_n \sim b_1 \cdots b_n.$$

The proof is given in Section 6.12.4 and Exercises 35 and 36.

## 6.12  PROOFS

### 6.12.1  Preliminary Lemma

Throughout the following proofs we focus on two or three components. It will be assumed that the other components have arbitrary, but fixed, values which because of independence are immaterial, and so they will be suppressed from the notation. The induced relation $\gtrsim_{ijk}$ on components $A_i \times A_j \times A_k$ is simply denoted $\gtrsim$, that on $A_i \times A_j$ by $\gtrsim_{ij}$, and that on $A_i$ by $\gtrsim_i$.

**LEMMA 14.**  *Suppose that $n \geqslant 3$ and that $\langle A_1, ..., A_n, \gtrsim \rangle$ is an n-component, additive conjoint structure. For $i, j \in N$, $i \neq j$, the relation $\gtrsim_{ij}$ satisfies double cancellation.*

*PROOF.*  Suppose that for $a, b, f \in A_i$ and $p, q, x \in A_j$, $i \neq j$, $ax \gtrsim_{ij} fq$ and $fp \gtrsim_{ij} bx$, then we show $ap \gtrsim_{ij} bq$. Since the conclusion is obvious from independence if $a \gtrsim_i b$ and $p \gtrsim_j q$, we assume either $a \gtrsim_i b$ and $q >_j p$ or $b >_i a$ and $p \gtrsim_j q$; the proof is given for the former case only; it is similar for the latter.

If for any distinct pair $c, d \in \{a, b, f\}$ there exist $u, v \in A_k$, $k \neq i, j$, such that $cu \sim_{ik} dv$ or if for any distinct pair, $r, s \in \{p, q, x\}$ there exist $u', v' \in A_k$, $k \neq i, j$, such that $ru' \sim_{jk} sv'$, then the conclusion follows. We illustrate the proof for two cases, and remark that the other four are similar. First, suppose that $c = a$ and $d = f$, i.e., $au \sim_{ik} fv$, then by hypothesis and independence,

$$fxv \sim axu \gtrsim fqu$$

which implies $xv \gtrsim_{jk} qu$, from which we obtain

$$apu \sim fpv \gtrsim bxv \gtrsim bqu$$

and so by independence, $ap \gtrsim_{ij} bq$. Second, suppose that $pu' \sim_{jk} qv'$, then

$$axv' \gtrsim fqv' \sim fpu' \gtrsim bxu',$$

which implies $av' \gtrsim_{ik} bu'$, from which we obtain

$$apv' \gtrsim bpu' \sim bqv',$$

and so $ap \gtrsim_{ij} bq$.

So, we suppose that no such elements in $A_k$ exist. By restricted solvability, this means either that, for all $u, v$, $au >_{ik} bv$ or, for all $u, v$, $bv >_{ik} au$. Since we have assumed $a \gtrsim_i b$, $au \gtrsim_{ik} bu$, and so the former must hold. Similarly, for all $u, v$, $qu >_{jk} pv$. Since $A_k$ is essential, there are $u, v$ such that $u >_k v$.

We construct decreasing dual standard sequences in $A_i$, $A_j$, starting at $a$, $q$, with mesh equal to $uv$. In fact, suppose that $a^{(0)},..., a^{(n)}$ and $q^{(0)},..., q^{(n)}$ are decreasing standard sequences relative to $u$, $v$; then for $0 \leqslant l < n$ and $0 < m \leqslant n$, we have $a^{(l)}q^{(m)}u \sim a^{(l)}q^{(m-1)}v \sim a^{(l+1)}q^{(m-1)}u$, so that $\{a^{(l)}\}$, $\{q^{(m)}\}$ are in fact dual. To carry out the construction, let $a^{(0)} = a$, $q^{(0)} = q$, and suppose that $a^{(0)},..., a^{(n-1)}$ and $q^{(0)},..., q^{(n-1)}$ have been constructed, with $n \geqslant 1$. If $a^{(n-1)}v >_{ik} bu$, then by restricted solvability there exists $a^{(n)}$ with $a^{(n)}u \sim_{ik} a^{(n-1)}v$. Now if $pu \gtrsim_{jk} q^{(n-1)}v$ we have

$$apu = a^{(0)}pu \gtrsim a^{(0)}q^{(n-1)}v \sim a^{(n-1)}q^{(0)}v > bq^{(0)}u = bqu,$$

so $ap \gtrsim_{ij} bq$ as desired. Assume $q^{(n-1)}v >_{jk} pu$, then by restricted solvability, $q^{(n)}$ can be constructed.

By the Archimedean axiom, this construction terminates with some $n \geqslant 1$ such that $a^{(n-1)}v >_{ik} bu \gtrsim_{ik} a^{(n)}v$.

Note that if $a^{(l)}x \gtrsim_{ij} fq^{(l)}$, then since $a^{(l+1)}q^{(l)} \sim_{ij} a^{(l)}q^{(l+1)}$ and $a^{(l+1)}u \sim_{ik} a^{(l)}v$, the first part of the proof gives $a^{(l+1)}x \gtrsim_{ij} fq^{(l+1)}$. Since this holds for $l = 0$, by hypothesis it holds up to $a^{(n)}x \gtrsim_{ij} fq^{(n)}$. From this and $fp \gtrsim_{ij} bx$ we can infer $a^{(n)}p \gtrsim_{ij} bq^{(n)}$ by the first part of the proof; because

$$a^{(n)}u >_{ik} bu \gtrsim_{ik} a^{(n)}v,$$

we know there exists $w \in A_k$ with $a^{(n)}w \sim_{ik} bu$, and the first part of the lemma applies. At this point we work backward: $a^{(l+1)}p \gtrsim_{ij} bq^{(l+1)}$ and $a^{(l)}q^{(l+1)} \sim_{ij} a^{(l+1)}q^{(l)}$ yield, by the first part of the lemma, $a^{(l)}p \gtrsim_{ij} bq^{(l)}$; and since we have the result for $l + 1 = n$, we obtain it for $l = 0$, i.e., $ap \gtrsim_{ij} bq$.                                                                 ◇

It is worth mentioning that the proof of this lemma is very much simpler if unrestricted solvability holds.

### 6.12.2 Theorem 13 (p. 302)

*An $n \geqslant 3$ component, additive conjoint structure has an additive order-preserving, interval-scale representation.*

*PROOF.* There is no loss of generality in assuming that all components are essential. We first show that there is a symmetric substructure for which the theorem is true. Let $a_0$, $a_1' \in A_i$ be such that $a_1' >_i a_0$ and $p_0'$, $p_1 \in A_j$ such that $p_1 >_j p_0'$. If $a_0p_1 \sim_{ij} a_1'p_0'$, we accept the given elements. If $a_1'p_0' >_{ij} a_0p_1$, then since $a_0p_1 >_{ij} a_0p_0'$, there exists, according to restricted solvability, an $a_1 \in A$ such that $a_1 >_i a_0$ and $a_1p_0' \sim_{ij} a_0p_1$. If $a_0p_1 >_{ij} a_1'p_0'$, then a similar argument shows that $p_0 \in A_j$ exists such that $p_1 >_j p_0$ and $a_1'p_0 \sim_{ij} a_0p_1$. Dropping all primes, we can also construct $u_1$, $u_0 \in A_k$ such that $p_0u_1 \sim_{jk} p_1u_0$, and by independence, it follows that

$a_0 u_1 \sim_{ik} a_1 u_0$. We can continue inductively, and it is easy to see that each two-component substructure bounded by these elements is symmetric in addition to fulfilling the original conditions.

By Lemma 14, each $\langle A_i, A_j, \gtrsim_{ij} \rangle$ satisfies the conditions of Theorem 2 and so has an additive representation $\phi_i$, $\phi_j$. In fact, we may select $\phi_1, ..., \phi_n$ such that any pair is additive over the symmetric substructure. To show this, it is sufficient to show for any distinct $i$, $j$, $k$ that the group operations (Definition 8, p. 258) induced on $j$ by $i$ and $k$ are identical. By independence, for any $a \in A_j$, $\underline{a}_i a_j \pi_k(a) \sim \underline{a}_i a \underline{a}_k \sim \pi_i(a) \, \underline{a}_j \underline{a}_k$. Using this and independence, for $a, b \in A_j$,

$$\pi_i(b)(a \circ_k b) \, \underline{a}_k \sim \pi_i(b) \, a\pi_k(b) \sim \underline{a}_i(a \circ_i b) \, \pi_k(b) \sim \pi_i(b)(a \circ_i b) \, \underline{a}_k \,,$$

whence $a \circ_k b \sim_j a \circ_i b$. Thus, if $\phi_i$, $\phi_j$ and $\phi_j'$, $\phi_k'$ are additive representations, by the uniqueness of Theorem 3.3 there is a linear transformation of the latter pair into $\phi_j$, $\phi_k$. Finally, it is not difficult to show that $\phi_i$, $\phi_k$ is also additive.

Suppose that $a, b \in A_i$, $p, q \in A_j$, and $u, v \in A_k$ are all within the symmetric substructure and that $apu \gtrsim bqv$, then we show that

$$\phi_1(a) + \phi_2(p) + \phi_3(u) \geqslant \phi_1(b) + \phi_2(q) + \phi_3(v).$$

This is obvious if $a \gtrsim_i b$, $p \gtrsim_j q$, and $u \gtrsim_k v$, so we assume that at least one inequality is reversed. With no loss of generality, we may assume either

(1)  $a \gtrsim_i b, p \gtrsim_j q, v >_k u$     or    (2)  $a \gtrsim_i b, q >_j p, v >_k u$.

(1)  If  $au \gtrsim_{ik} bv$,  then  $\phi_i(a) + \phi_k(u) \geqslant \phi_i(b) + \phi_k(v)$  and  from $p \gtrsim_j q, \phi_j(p) \geqslant \phi_j(q)$, so the result follows by addition of inequalities. So, we assume that $bv >_{ik} au$, which with $av \gtrsim_{ik} bv$ implies, by restricted solvability, that there exists $z \in A_k$ such that $bv \sim_{ik} az$. Thus, $apu \gtrsim bqv \sim aqz$, so $pu \gtrsim_{jk} qz$, whence $\phi_j(p) + \phi_k(u) \geqslant \phi_j(q) + \phi_k(z)$. From

$$az \sim_{ik} bv, \phi_i(a) + \phi_k(z) = \phi_i(b) + \phi_k(v).$$

Adding the two inequalities and subtracting $\phi_k(z)$ yields the result.

(2)  From $q >_j p$ and $v >_k u$,

$$apv > apu \gtrsim bqv > bpv$$

so by restricted solvability, there exists $c \in A_i$ such that $cpv \sim apu \gtrsim bqv$. Thus,  $\phi_i(a) + \phi_k(u) = \phi_i(c) + \phi_k(v)$  and  $\phi_i(c) + \phi_j(p) \geqslant \phi_i(b) + \phi_j(q)$. The result follows by adding the inequalities and subtracting $\phi_j(c)$.

Since strict inequalities under $\gtrsim_{ij}$ go into strict numerical inequalities, the converse follows. A simple induction extends this result to any $n$.

To extend additivity, use triple cancellation as in Theorem 2, reducing $a \sim b$ to the case where $a$ is in the symmetric substructure and at most one $b_i$ is outside it. The argument for this remaining case is exactly like that of Theorem 2.

The uniqueness of the representation follows immediately from independence and Theorem 2. $\qquad\qquad\qquad\qquad\qquad\qquad\qquad\qquad\qquad\qquad\qquad\qquad\diamond$

### 6.12.3 Theorem 14  (p. 302)

*If $\langle A_1 ,..., A_n , \gtrsim\rangle$, $n \geqslant 3$, satisfies the weak ordering, independence, and essentialness axioms of Definition 13, if each $A_i$ is a connected topological space, and if for every $b$ the sets $\{a \mid a \gtrsim b\}$ and $\{a \mid b \gtrsim a\}$ are closed in the product topology on $\mathsf{X}_{i=1}^{n} A_i$, then $\langle A_1 ,..., A_n , \gtrsim\rangle$ is an n-component, additive conjoint structure (Definition 13).*

PROOF. We first show restricted solvability for $A_1$; a parallel proof holds for any other component. Suppose that

$$\bar{b}_1 b_2 \cdots b_n \gtrsim a \gtrsim \underline{b}_1 b_2 \cdots b_n .$$

Let
$$\bar{B} = \{b_1 \mid b_1 b_2 \cdots b_n \gtrsim a\},$$
$$\underline{B} = \{b_1 \mid b_1 b_2 \cdots b_n \lesssim a\}.$$

Then $\bar{B}$ and $\underline{B}$ are nonempty subsets of $A_1$ since $\bar{b}_1 \in \bar{B}$ and $\underline{b}_1 \in \underline{B}$, and $\bar{B} \cup \underline{B} = A_1$. If we can show that $\bar{B}, \underline{B}$ are closed, then since a connected topological space cannot be the disjoint union of two nonempty closed sets, it follows that $\bar{B} \cap \underline{B} \neq \varnothing$. But for $b_1 \in \bar{B} \cap \underline{B}, b_1 b_2 \cdots b_n \sim a$ as required for solvability.

To show that $\bar{B}$ is closed, we show that its complement $-\bar{B}$ is open. If $b_1 \in -\bar{B}$, then $b_1 b_2 \cdots b_n < a$. Let $C = \{b \mid b \in \mathsf{X}_{i=1}^{n} A_i \text{ and } b < a\}$. By hypothesis, $C$ is open, and $b_1 b_2 \cdots b_n \in C$. By the definition of product topology, there is a set of form $\mathsf{X}_{i=1}^{n} U_i$ such that each $U_i$ is open in $A_i$, and $b_1 b_2 \cdots b_n \in \mathsf{X}_{i=1}^{n} U_i \subset C$. In particular, for every $b_1' \in U_1$, $b_1' b_2 \cdots b_n \in C$, so $b_1' b_2 \cdots b_n < a$. Hence, $U_1 \subset -\bar{B}$. Since an arbitrary element $b_1$ of $-\bar{B}$ is contained in an open set $U_1$ included in $-\bar{B}$, it follows that $-\bar{B}$ is the union of open sets of $A_1$, hence it is open, and $\bar{B}$ is closed. The proof that $\underline{B}$ is closed is the same except that the inequalities are reversed.

To show the Archimedean property, let $\{a_{1i} \mid i \in N\}$ be an infinite standard sequence on component $A_1$ (Definition 4) which is increasing and has an index set $N$ which is unbounded from above. (The arguments for other components $A_i$ and for other varieties of infinite standard sequences are similar.)

We show that $\{a_{1i}\}$ has no strict upper bound in $A_1$. For $i \in N$, let $B_{1i} = \{b_1 \mid b_1 \in A_1, b_1 \prec a_{1i}\}$. Let $B_1 = \bigcup_{i \in N} B_{1i}$. By the same argument used to prove restricted solvability, each $B_{1i}$ is open; thus $B_1$ is open, and it is clearly nonempty. The complement $-B_1$ is the set of all upper bounds of $\{a_{1i}\}$. To show that $-B_1$ is empty, it suffices to show that it is open, since $A_1$ cannot be the disjoint union of nonempty open sets.

Suppose that there exists $b_1 \in -B_1$. Since $\{a_{1i}\}$ is an increasing standard sequence, there exist elements $p = a_2 \cdots a_n$ and $q = a_2' \cdots a_n'$ in $\mathsf{X}_{i=2}^n A_i$, with $p \succ q$, such that

$$a_{1,i+1}q \sim a_{1i}p, \qquad \text{all} \quad i \in N.$$

Hence we have

$$b_1p \succ b_1q \succsim a_{1,i+1}q \sim a_{1i}p, \qquad \text{all} \quad i \in N.$$

By restricted solvability, there exists $c_1 \in A_1$ such that $c_1p \sim b_1q$. Clearly, $b_1 \succ c_1$, and $c_1$ is also in $-B_1$, since $c_1p \succsim a_{1i}p$, all $i \in N$. Thus $\{a_1 \mid a_1 \succ c_1\}$ is an open subset of $A_1$ that contains $b_1$ and is included in $-B_1$. It follows that $-B_1$ is open.                                                                  ◇

### 6.12.4  Theorem 15  [p. 305; part (i)]

If $\langle A^n, \succsim \rangle$ is an n-component additive conjoint structure, it has a representation $\sum_{i=1}^n \lambda_i u(a_i)$ iff no standard sequence in one component is unequally spaced in any other component; if this condition holds, u is an interval scale and the $\lambda_i$ are unique up to multiplication by a positive constant.

PROOF. The necessity of the standard-sequence condition and the uniqueness statement are easily shown; we prove only sufficiency. By hypothesis, there is a representation of form $\sum_{i=1}^n \phi_i(a_i)$. Without loss in generality, let $\phi_1$ have a range at least as large as that of any other $\phi_i$; thus, by restricted solvability, for any $a_j$, $b_j$ ($j \neq 1$), there exist $p_1$, $q_1$ with $p_1a_j \sim q_1b_j$ in the induced ordering. So if $a^{(1)}, \cdots, a^{(n)}$ is any finite standard sequence in the first component, we can find $p_1, q_1$ with $p_1a_j^{(i+1)} \sim q_1a_j^{(i)}$ (letting $a_j^{(i)}$ denote $a^{(i)}$, regarded as an element of the $j$th component) and so $a^{(1)},\ldots,a^{(n)}$ is also a standard sequence in the $j$th component.

Let $\phi_j(b^{(0)}) = 0$, all $j$, $\phi_1(b^{(1)}) = 1$, and for $j \neq 1$, $\phi_j(b^{(1)}) = \lambda_j$. If $b \succ_1 b^{(1)}$, construct any standard sequence from $b^{(0)}$ to $b$, in the first component (the "distance" of the last term in the sequence from $b$ being less than the mesh of the sequence). The standard sequence gives us the approximations

$$\phi_1(b) \approx n[\phi_1(a^{(1)}) - \phi_1(a^{(0)})],$$
$$1 = \phi_1(b^{(1)}) \approx m[\phi_1(a^{(1)}) - \phi_1(a^{(0)})],$$

or $\phi_1(b) \approx n/m$. Since the sequence is also a standard sequence in the $j$th component, we obtain the same approximations, with

$$\lambda_j = \phi_j(b^{(1)}) \approx m[\phi_j(a^{(1)}) - \phi_j(a^{(0)})].$$

Thus, we obtain $\phi_j(b)/\lambda_j \approx n/m \approx \phi_1(b)$. Similar constructions hold for other cases, with $b^{(1)} \succ_1 b$. (Compare with Lemma 8.7.)

If $A$ consists of only one standard sequence in the first component, these approximations become exact for that standard sequence; otherwise, they become exact in the limit. So we let $\phi_1 = u$, $\phi_j = \lambda_j u$.                $\diamondsuit$

Parts (ii) and (iii) of the Theorem are left as Exercises 35 and 36.

## 6.13  CONCLUDING REMARKS

At various points in the chapter we have left problems unresolved, some of which require new research and others of which are dealt with later. It seems useful to recapitulate them.

Among the unsolved problems, three stand out. First, the proofs of the major representation theorems (2 and 13) are exceedingly long; can they be shortened significantly? Second, these theorems have been proved only for order relations on the full product set $\mathsf{X}_{i=1}^n A_i$, whereas in practice and for theoretical reasons we would like to establish them for relations on certain subsets $B \subset \mathsf{X}_{i=1}^n A_i$. How should these $B$'s be characterized? And third, can an axiomatization be arrived at that leads to a representation with a multiplicatively independent, additive interaction term, i.e., to a representation of the form $\phi_1 + \phi_2 + \psi_1\psi_2$?

The four most important unresolved problems that will be treated later are the following. First, we have provided only sufficient conditions for an additive representation, and surely it would be nice to know necessary and sufficient conditions. For the finite case, an answer is known (Chapter 9). Second, nothing has yet been said about the practical problems of actually determining additive representations from a finite amount of data. This too is discussed in Chapter 9. Third, additivity may fail to hold; what then? Indeed, in some cases such as preferences among gambles—the simplest of which can be thought of as a triple $a\,E\,b$ where $a$ and $b$ are outcomes and $E$ is a chance event that determines which of $a$ and $b$ is received—additivity is not expected to hold. The gambling problem is handled in Chapter 8, and certain more straightforward modifications of the additive representation are treated in Chapter 7. Fourth, in some (physical) examples, we may have both extensive and conjoint measures of the two (or more) components. Because they are both order preserving, these measures are monotonically related. When can we say more than this; when, for example, can we say that the two measures are substantially the same? This we treat in Chapter 10.

**EXERCISES**

**1.** Prove Lemma 1.    (6.1.4)

**2.** If Equation (1) holds and if $F$ is strictly monotonic in each variable, prove that $\succsim$ is an independent relation.    (6.1.2, 6.1.4)

**3.** Prove that in the presence of independence transitivity and double cancellation exhaust the possible cancellation properties with just two antecedent inequalities.    (6.2.1)

**4.** Suppose that there are three inequalities in $\langle A_1, A_2, \succsim \rangle$ which, through cancellation, might lead to a fourth of the form $ap \succsim bq$, that $a$ and $b$ appear in two distinct antecedent inequalities, and that $p$ and $q$ appear in two distinct antecedent inequalities. Show that the conclusion follows from independence, double cancellation, and transitivity.    (6.2.1)

**5.** If a reflexive relation on $A_1 \times A_2$ satisfies triple cancellation, then it is independent.    (6.2.1)

**6.** Suppose that the plane is ordered as follows:

$$xy \succsim x'y' \quad \text{iff} \quad x + y + 2xy \geqslant x' + y' + 2x'y'.$$

Describe fully all standard sequences on the $x$-coordinate for any pair of fixed values $y_0 < y_1$.    (6.2.2)

**7.** Prove Lemma 2.    (6.2.3)

**8.** The lexicographic ordering of the plane, $\langle \text{Re} \times \text{Re}, \succsim_l \rangle$ is defined as follows: if $x, y, x', y' \in \text{Re}$,

$$xy \succsim_l x'y' \quad \text{if either} \quad x > x' \quad \text{or} \quad x = x' \quad \text{and} \quad y \geqslant y'.$$

Which axioms of Definition 7 are satisfied? In each case, provide a proof or a counterexample.    (6.2.3)

**9.** Suppose that $\langle A_1 \times A_2, \sim \rangle$ is an equivalence relation that satisfies unrestricted solvability and double cancellation. Fix $a_0 \in A_1$ and $p_0 \in A_2$ and let $c$ solve $cp_0 \sim ap$ and $r$ solve $a_0r \sim bq$. Define $\circ$ on the equivalence classes of $A_1 \times A_2$ by: $\mathbf{ap} \circ \mathbf{bq} = \mathbf{cr}$. Prove $\langle A_1 \times A_2/\sim, \circ \rangle$ is a commutative group with identity $\mathbf{a_0p_0}$.    (6.2.3, 6.2.5)

**10.** Verify that the various examples of Section 6.3.1 satisfy the axioms they were said to satisfy.    (6.3.1)

*The following eight exercises provide an alternate proof via positive-difference structures of the existence of an additive representation when $\langle A_1, A_2, \succsim \rangle$ is a bounded, symmetric, additive, conjoint structure for which* *double cancellation holds. The positive-difference structure defined below is quite complicated. As an alternative to these exercises, the reader may develop a simpler proof, using triple cancellation and the positive-difference structure sketched on p. 259.*

*We introduce the following definitions:*

$A_1^* = \{ab \mid a, b \in A_1 \text{ and } a >_1 b\}.$

$A_2^* = \{pq \mid p, q \in A_2 \text{ and } p >_2 q\}.$

*For $ab \in A_1^*$, $\pi(a, b)$ solves $b\pi(a, b) \sim ap$.*

*For $pq \in A_2^*$, $\alpha(p, q)$ solves $\alpha(p, q)q \sim \underline{a}p$.*

*For $ab \in A_1^*$, $\overline{ab} \sim_1 \alpha[\pi(a, b), p]$*

*For $pq \in A_2^*$, $\overline{pq} \sim_2 \pi[\alpha(p, q), \underline{a}]$.*

*For $ab, cd \in A_1^*$, $ab \succsim_1^* cd$ iff $\overline{ab} \succsim_1 \overline{cd}$.*

**11.** $\pi(\overline{ab}, \underline{a}) \sim_2 \pi(a, b)$.

**12.** If $a \succsim_1 b >_1 c$, then $\pi(a, c) \succsim_2 \pi(b, c)$, and if $a_1 >_1 b_1$ then $\pi(a, c) >_2 \pi(b, c)$.

**13.** $a\pi(b, \underline{a}) \sim b\pi(a, \underline{a})$.

**14.** If $a >_1 b >_1 c$, then $\overline{ac}\, p \sim \overline{ab}\, \pi(\overline{bc}, \underline{a})$.

**15.** $\alpha(p, p)\, \pi(a, \underline{a}) \sim ap$.

**16.** $\overline{ab} \succsim_1 \overline{\alpha(q, p)\, \alpha(p, p)}$ iff $ap \succsim bq$.

**17.** $\langle A_1, A_1^*, \succsim_1^* \rangle$ is a positive-difference structure (Definition 4.1, Section 4.2) and $A_1^*$ is connected, i.e., if not ($a \sim_1 b$), then $ab \in A_1^*$ or $ba \in A_1^*$.

**18.** Let $\phi_1$ be the difference representation yielded by Exercise 17 and Theorem 4.1. Define $\phi_2(p) = \phi_1[\alpha(p, p)]$. Prove that

$$ap \succsim bq \quad \text{iff} \quad \phi_1(a) + \phi_2(p) \geq \phi_1(b) + \phi_2(q).$$

**19.** Show that the function $F(x, y) = y/(y + e^{-x})$, $x, y > 0$, can be transformed into an additive form and find the transformations.      (6.5.3)

**20.** Prove Theorem 5.      (6.5.4)

**21.** Develop an explicit axiom system and representation theorem for the algebraic-difference structure (Section 4.4) that arises by letting $A_1 = A_2$ in Definition 7 and Theorem 2, replacing the additive structure by a subtractive one, requiring $\phi_1 = \phi_2$, and using Theorem 5.      (6.5.4)

**22.** Using Definition 7 and the lemmas of Section 6.6.1, but not Theorem 2, prove that if $ap \gtrsim bq$ and if $a - 1$ and $q - 1$ exist, then $(a - 1)p \gtrsim b(q - 1)$.
(6.6.1)

**23.** Prove that the associated group of a nonempty family of increasing scales is in fact a group.    (6.7.2)

**24.** Prove directly, i.e., do not use Lemma 12, that the associated group of a uniform family of scales is commutative.    (6.7.2)

**25.** Prove in detail the uniqueness statement of Theorem 8.    (6.8.2)

**26.** Prove that a noncyclic subgroup of the additive real numbers is dense in the real numbers.    (6.8.3)

**27.** Suppose that the operation ○ of a bisymmetric structure (p. 294) is both associative and idempotent, then show that in the representation of Theorem 10, $\lambda = 0$, $\mu = 0$ or 1, and $\nu = 1 - \mu$.    (6.9.1)

**28.** Which axioms of Definition 10 are satisfied (proof or counterexample) by the structure $\langle \text{Re}^+, \geqslant, ○ \rangle$ where ○ is defined by: if $x, y \in \text{Re}^+$, $x ○ y = \max(x, y)$?    (6.9.1)

**29.** Suppose that $\langle A, \gtrsim, ○ \rangle$ is a positive bisymmetric structure (Definition 10). If ○ is associative, show that any sequence of the form $\{na\}$, where $1a = a$ and $na = (n - 1) a ○ a$, is a standard sequence of the structure. (6.9.1)

**30.** Suppose that $\langle A, \gtrsim, B, ○ \rangle$ satisfies the hypotheses of Theorem 12, then prove that:

1.  If $ab \in B$ and $a > b$, then $a > a ○ b > b$.

2.  If $aJ^n b$ and $bJ^n c$, $n$ a positive integer, then $ac \in B$ and $b \sim a ○ c$.

3.  If $ac \in B$, $a > c$, and $b \sim a ○ c$, then there exists a positive integer $n$ such that $aJ^n b$ and $bJ^n c$.    (6.9.2)

**31.** Using the results of Exercise 30 and following the method of proof in Section 4.9.2, prove Theorem 12.    (6.9.2).

**32.** Using the axioms of a bisymmetric structure (p. 294), but not Theorem 10, prove triple cancellation holds for the relation $\gtrsim'$ defined in Section 6.10.1.    (6.10.1)

**33.** Verify that the product topology (Definition 14) is in fact a topology.
(6.11.1)

**34.** Formulate a two-component analog of Theorem 14 and show that it is a special case of Definition 7. Then prove that a bisymmetric structure satisfying the hypotheses of Theorem 11 is a special case of your structure.

In doing this, accept the algebraic deductions in the proof of Theorem 10 and simply prove that the topological ones follow from those of the bisymmetric structure.      (6.2.3, 6.9, 6.10, 6.11, 6.12.3)

**35.** Prove part (ii) of Theorem 15.      (6.11.2)

**36.** Prove part (iii) of Theorem 15.      (6.11.2)

# Chapter 7 Polynomial Conjoint Measurement

## □ 7.1 INTRODUCTION

Many orderings of objects can be decomposed into the effects of two or more factors but cannot be scaled so as to combine additively, as in the previous chapter. Rather, some other combination rule is appropriate. Two examples serve to illustrate these nonadditive combination rules.

*EXAMPLE* 1. Suppose that ordinal measures of rats' performance are obtained for various levels $d$ of food deprivation, $k$ of food reward, and $h$ of prior learning. The observed performance is denoted $dkh$. Hull (1952) proposed that these three factors combine multiplicatively in the sense that there exist functions $\phi_D, \phi_K, \phi_H$ (measuring drive, incentive, and habit strength, respectively) such that

$$dkh \gtrsim d'k'h' \quad \text{iff} \quad \phi_D(d)\,\phi_K(k)\,\phi_H(h) \geq \phi_D(d')\,\phi_K(k')\,\phi_H(h').$$

Spence (1956) proposed a somewhat different model which in ordinal form is

$$dkh \gtrsim d'k'h' \quad \text{iff} \quad [\phi_D(d) + \phi_K(k)]\,\phi_H(h) \geq [\phi_D(d') + \phi_K(k')]\,\phi_H(h').$$

We show in Section 7.4.3 how to test between these alternatives.

*EXAMPLE* 2. Suppose that a stimulus $b$ is specified by nominal values

316

$b_1, ..., b_n$ on $n$ dimensions. A pair of stimuli $b, b'$ is, thus, specified by the levels of $2n$ factors, $b_1, ..., b_n, b_1', ..., b_n'$. Suppose that an observer orders the pairs of stimuli with respect to dissimilarity. A Euclidean model for ordinal dissimilarity asserts that there are functions $\phi_1, ..., \phi_n, \phi_1', ..., \phi_n'$ (in most current applications, $\phi_i' = \phi_i$) such that

$$bb' \gtrsim cc' \qquad \text{iff} \qquad \sum_{i=1}^{n} [\phi_i(b_i) - \phi_i'(b_i')]^2 \geqslant \sum_{i=1}^{n} [\phi_i(c_i) - \phi_i'(c_i')]^2.$$

In the first example, the two rules of combination for three variables, $x_1, x_2, x_3$, are $F(x_1, x_2, x_3) = x_1 x_2 x_3$ and $F(x_1, x_2, x_3) = (x_1 + x_2) x_3$. In Example 2, the rule of combination for $2n$ variables $x_1, ..., x_n, x_1', ..., x_n'$ is $F(x_1, ..., x_n, x_1', ..., x_n') = \sum_{i=1}^{n} (x_i - x_i')^2$.

Our plan is, first, to consider a rather general assumption about the ordering of the joint effects of $n$ factors, namely, that there exists some function $F$ of $n$ variables and there exist scales $\phi_1, ..., \phi_n$ on the $n$ factors such that $F(\phi_1, ..., \phi_n)$ preserves the empirical ordering (Section 7.2). This is called *decomposability*. (Two-factor decomposability was discussed briefly in Sections 6.1.2 and 6.1.3.) It is rather easy to give necessary and sufficient conditions for two slightly different versions of decomposability (Theorem 1). The remainder of the chapter is devoted to the study of the conditions that must hold if the combination rule $F$ is some specific polynomial in $n$ variables (note that the combination rules in both examples above were polynomials).

## □ 7.2 DECOMPOSABLE STRUCTURES

The concept of decomposability mentioned above has an interesting variant in which the function $F$ is strictly increasing in each variable. The two concepts are made precise in the following definition.

DEFINITION 1. *Suppose that $\gtrsim$ is a binary relation on $A = \mathsf{X}_{i=1}^{n} A_i$. The structure $\langle \mathsf{X}_{i=1}^{n} A_i, \gtrsim \rangle$ is* decomposable *iff there exist real-valued functions $\phi_i$ on $A_i$, $i = 1, ..., n$, and a real-valued function $F$ of $n$ real variables that is one to one in each variable separately such that, for all $a, b \in A$,*

$$a \gtrsim b \qquad \text{iff} \qquad F[\phi_1(a_1), ..., \phi_n(a_n)] \geqslant F[\phi_1(b_1), ..., \phi_n(b_n)].$$

*The structure $\langle \mathsf{X}_{i=1}^{n} A_i, \gtrsim \rangle$ is called* monotonically decomposable *iff it is decomposable with a function $F$ that is strictly increasing in each variable separately.*

## 7.2.1 Necessary and Sufficient Conditions

If $\langle \mathsf{X}_{i=1}^{n} A_i , \gtrsim \rangle$ is decomposable, then $\phi(a) = F[\phi_1(a_1),..., \phi_n(a_n)]$ is a homomorphism of $\langle A, \gtrsim \rangle$ into $\langle \text{Re}, \gtrsim \rangle$. Thus, two necessary conditions follow immediately: $\langle A, \gtrsim \rangle$ is a weak order and $A/\sim$ has a countable order-dense subset (Theorem 2.2).

The fact that $F$ is one to one in each variable means that if

$$F(y_1 ,..., y_{i-1} , x_i , y_{i+1} ,..., y_n) = F(y_1 ,..., y_{i-1} , x_i' , y_{i+1} ,..., y_n),$$

then $x_i = x_i'$. Because $F$ is well defined, the equality will still hold if $z_j$ is substituted for $y_j$, $j = 1,..., i - 1, i + 1,..., n$, on both sides. Thus, a third necessary condition for decomposability is *substitutability*: for all $i$, $1 \leqslant i \leqslant n$, all $a_i , a_i'$ in $A_i$ , and all $b, c$ in $A$,

$$b_1 \cdots b_{i-1}a_ib_{i+1} \cdots b_n \sim b_1 \cdots b_{i-1}a_i'b_{i+1} \cdots b_n$$

if and only if

$$c_1 \cdots c_{i-1}a_ic_{i+1} \cdots c_n \sim c_1 \cdots c_{i-1}a_i'c_{i+1} \cdots c_n .$$

Another way to say this is: for all $b$, the equivalence relation $\sim_i(b)$ induced on $A_i$ by $\sim$ when all $n - 1$ components other than the $i$th are fixed at $b_j$ is independent of the choice of $b$.

If $F$ is strictly increasing in each component, then $\sim$ can be replaced by $\gtrsim$ in the above statement, and so the induced weak order $\gtrsim_i$ is independent of the fixed components in the other $n - 1$ factors. We say in this case that $A_i$ is *independent of* $\mathsf{X}_{j \neq i} A_j$ (see Definition 3, Section 7.4.1). A necessary condition for monotone decomposability is that each $A_i$ be independent of $\mathsf{X}_{j \neq i} A_j$.

The necessary conditions just derived are also sufficient. More precisely, we have the following theorem.

THEOREM 1.   *Suppose that $\gtrsim$ is a binary relation on $A = \mathsf{X}_{i=1}^{n} A_i$ . The structure $\langle \mathsf{X}_{i=1}^{n} A_i , \gtrsim \rangle$ is decomposable iff the following axioms are satisfied:*

1. $\gtrsim$ *is a weak order.*

2. $A/\sim$ *has a countable order-dense subset.*

3. $\sim$ *satisfies substitutability.*

*The structure $\langle \mathsf{X}_{i=1}^{n} A_i , \gtrsim \rangle$ is monotonically decomposable iff Axioms 1 and 2 hold and 3 is replaced by:*

3'.   *Each $A_i$ is independent of $\mathsf{X}_{j \neq i} A_j$ .*

Necessity has already been shown. Sufficiency is very elementary. By

Theorem 2.2, there is a homomorphism $\phi$ of $\langle A/\sim, \gtrsim \rangle$ into $\langle \text{Re}, \geqslant \rangle$. One can now fix any $b \in A$ and define $\phi_i$ on $A_i$ by

$$\phi_i(a_i) = \phi(b_1 \cdots b_{i-1} a_i b_{i+1} \cdots b_n).$$

Then define $F$ by $F[\phi_1(a_1),..., \phi_n(a_n)] = \phi(a)$. Axiom 3 (or 3') is used to prove that $F$ is well defined and is $1:1$ (or strictly increasing) in each variable. A complete proof is given in Section 7.2.2.

We remark that Theorem 1 is not a representation theorem in the usual sense because it does not assert the existence of an isomorphism $(\phi_1 ,..., \phi_n)$ of $\langle \mathsf{X}_{i=1}^n A_i , \gtrsim \rangle$ into some definite numerical relational structure $\langle \text{Re}^n, \gtrsim' \rangle$. Instead, it asserts the existence of a numerical relational structure, $\langle \text{Re}^n, \gtrsim' \rangle$, that is isomorphic to $\langle \mathsf{X}_{i=1}^n A_i , \gtrsim \rangle$, where $\gtrsim'$ has the form

$$(x_1 ,..., x_n) \gtrsim' (y_1 ,..., y_n)$$

if and only if $F(x_1 ,..., x_n) \geqslant F(y_1 ,..., y_n)$. In the rest of this chapter (Section 7.4 and beyond) we consider fixed numerical relational structures, corresponding to definite choices of functions $F$.

Two kinds of questions about uniqueness may be raised. First, for a fixed $F$ (fixed numerical relational structure) defined on a subset of $\text{Re}^n$ that includes the Cartesian product of the ranges of the scales $\phi_i$, what transformations $\phi_i \rightarrow \phi_i'$ preserve the isomorphism? This is the conventional uniqueness question, but it is difficult to give a general answer valid for any $F$. Particular cases will be dealt with in later sections where specific functions $F$ are assumed. Second, what transformations $\phi_i \rightarrow \phi_i', F \rightarrow F'$ are possible? The answer here is rather obvious. Note that the construction of the $\phi_i$ and the $F$, above, started from an arbitrary isomorphism $\phi$ of $\langle A, \gtrsim \rangle$ into $\langle \text{Re}, \geqslant \rangle$, and that $F[\phi_1(a_1),..., \phi_n(a_n)] = \phi(a)$. The functions $\phi_i$ were defined from $\phi$ by picking a fixed $b$ in $A$. Suppose that we start instead with an isomorphism $\phi'(a) = h[\phi(a)]$, where $h$ is a strictly increasing function from Re to Re, and with $b'$ in $A$ fixed. Then

$$\phi_i'(a_i) = \phi'(b_1' \cdots b_{i-1}' a_i b_{i+1}' \cdots b_n')$$

$$= h[\phi(b_1' \cdots b_{i-1}' a_i b_{i+1}' \cdots b_n')]$$

$$= h(F[\phi_1(b_1'),..., \phi_{i-1}(b_{i-1}'), \phi_i(a_i), \phi_{i+1}(b_{i+1}'),..., \phi_n(b_n')])$$

$$= h_i[\phi_i(a_i)].$$

The function $h_i$ is well defined and $1:1$ if decomposability holds, and it is well defined and strictly increasing if monotone decomposability holds. In either case, the functions $h_i^{-1}$ are defined, so

$$
\begin{aligned}
F'[\phi_1'(a_1),..., \phi_n'(a_n)] &= \phi'(a) \\
&= h[\phi(a)] \\
&= h(F[\phi_1(a_1),..., \phi_n(a_n)]) \\
&= h(F[h_1^{-1}\phi_1'(a_1),..., h_n^{-1}\phi_n'(a_n)]).
\end{aligned}
$$

Thus, the transformations $\phi_i'(a_i) = h_i[\phi_i(a_i)]$ and

$$
F'(x_1,..., x_n) = h(F[h_1^{-1}(x_1),..., h_n^{-1}(x_n)])
$$

are admissible, where the functions $h_i$ are 1:1 or strictly increasing depending on whether decomposability or monotone decomposability holds, and $h$ is always strictly increasing. Clearly, any strictly increasing function $h$ on the range of $F$ can occur, but whether the $h_i$ are arbitrary 1:1 or increasing functions depends on the structure of $A$, i.e., what variety of choices $b'$ is possible. It is not difficult to show that any functions $\phi_i'$, $F'$ must be related to $\phi_i$, $F$ by transformations of the above type.

Thus, if the numerical relational structure (defined by $F$) is permitted to vary, then the $\phi_i$ may be regarded only as nominal scales (decomposability) or ordinal scales (monotone decomposability). This uniqueness result is closely related to those results in Section 6.5.2 where the numerical relational structure was allowed to vary for additive conjoint structures. Naturally, the uniqueness results for fixed $F$ are much stronger.

### 7.2.2 Proof of Theorem 1 (p. 318)

$\langle \mathsf{X}_{i=1}^n A_i, \gtrsim \rangle$ is decomposable (alternatively, monotonically decomposable) iff $\gtrsim$ is a weak order, $A/\sim$ has a countable, order-dense subset, and $\sim$ satisfies substitutability (alternatively, each $A_i$ is independent of $\mathsf{X}_{j\neq i}^n A_j$).

PROOF. The proof of necessity was given above. If $\gtrsim$ is a weak order and $A/\sim$ has a countable, order-dense subset, then by Theorem 2.2, there exists a real-valued function $\phi$ on $A$ such that $a \gtrsim b$ iff $\phi(a) \geq \phi(b)$. Let $b \in A$ be fixed, and define

$$
\begin{aligned}
\phi_i(a_i) &= \phi(b_1 \cdots b_{i-1} a_i b_{i+1} \cdots b_n), \\
F[\phi_1(a_1),..., \phi_n(a_n)] &= \phi(a).
\end{aligned}
$$

To show that $F$ is well defined, suppose that $\phi_i(a_i) = \phi_i(a_i')$, $i = 1,..., n$. By definition,

$$
b_1 \cdots b_{i-1} a_i b_{i+1} \cdots b_n \sim b_1 \cdots b_{i-1} a_i' b_{i+1} \cdots b_n,
$$

$i = 1,..., n$. By substitutability (substituting $a_j$ for $b_j$, $1 \leqslant j \leqslant i - 1$, $a_j'$ for $b_j$, $i + 1 \leqslant j \leqslant n$) we have

$$a_1 \cdots a_{i-1}a_i a_{i+1}' \cdots a_n' \sim a_1 \cdots a_{i-1}a_i' a_{i+1}' \cdots a_n',$$

for $i = 1,..., n$. By the transitivity of $\sim$, these last $n$ equivalences yield $a_1 \cdots a_n \sim a_1' \cdots a_n'$, or $\phi(a) = \phi(a')$. This shows that $F$ is well defined if substitutability (or independence, which implies substitutability) holds. To show that substitutability implies that $F$ is 1:1 in each variable note that if

$$F[\phi_1(c_1),..., \phi_{i-1}(c_{i-1}), \phi_i(a_i), \phi_{i+1}(c_{i+1}),..., \phi_n(c_n)]$$
$$= F[\phi_1(c_1),..., \phi_{i-1}(c_{i-1}), \phi_i(a_i'), \phi_{i+1}(c_{i+1}),..., \phi_n(c_n)],$$

then

$$c_1 \cdots c_{i-1}a_i c_{i+1} \cdots c_n \sim c_1 \cdots c_{i-1}a_i' c_{i+1} \cdots c_n.$$

By substitutability

$$b_1 \cdots b_{i-1}a_i b_{i+1} \cdots b_n \sim b_1 \cdots b_{i-1}a_i' b_{i+1} \cdots b_n,$$

hence, $\phi_i(a_i) = \phi_i(a_i')$.

To show that independence implies that $F$ is strictly increasing in each variable, carry out the above proof with $>$ substituted for $=$ and $\succ$ for $\sim$. ◇

## □ 7.3  POLYNOMIAL MODELS

A function $F$ of $n$ real variables $x_1,..., x_n$ is a *polynomial* if it is a linear combination of products of the variables raised to nonnegative, integral powers, i.e.,

$$F(x_1,..., x_n) = \sum_{k=1}^{M} \alpha_k x_1^{\beta_{1k}} x_2^{\beta_{2k}} \cdots x_n^{\beta_{nk}},$$

where $\alpha_k \in \mathrm{Re}$ and $\beta_{ik} \in I^+ \cup \{0\}$. For clarity, multipliers of form $x_i^0 = 1$ are omitted. Thus, typical polynomials in three variables are $1 + x_1 + 2x_2 x_3$ and $x_1^3 x_2 - x_1^3 x_3^2$.

### 7.3.1  Examples

Examples 1 and 2 (Section 7.1) illustrate the use of the polynomials $x_1 x_2 x_3$ and $(x_1 + x_2) x_3$ as suggested rules of combination for the influences of drive, incentive, and habit strength on performance of rats, and of $\sum_{i=1}^n (x_i - x_i')^2$ (writing out the squares as $x_i^2 - 2x_i x_i' + (x_i')^2$ makes this a polynomial) as a model for dissimilarity of stimuli varying on $n$ dimensions. Quite a few other empirical examples can be listed. We present several here to illustrate some of the possible scope of applicability of the methods developed in Sections 7.4–7.6. In most cases, the laws actually proposed are not ordinal ones, but they can all be reformulated ordinally; in some cases, the ordinal reformulation may actually be more reasonable than the original.

*EXAMPLE* 3. Physics is full of polynomial combination rules. Two examples are the ideal gas law,

$$p = kT(1/V),$$

where $k$ is a constant, $p$ is pressure, $T$ is absolute temperature, and $1/V$ is the inverse of volume; and the formula for kinetic energy,

$$E = \tfrac{1}{2}mv^2,$$

where $E$ is energy, $m$ is mass, and $v$ is velocity. An example of a slightly different kind is the law of acceleration of a spherical body in a viscous fluid,

$$ma = F - kr\eta v,$$

where $m$ is the mass of the body, $a$ its acceleration, $k$ is a dimensionless constant, $F$ is driving force, $r$ is the radius of the body, $v$ its velocity, and $\eta$ is the viscosity of the fluid. This is a polynomial combination rule in five variables, where the observed quantity, which might be used to order the quintuples $(F, 1/m, r, \eta, v)$ is the second derivative of position with respect to time. In these physical examples the procedures of conjoint measurement are not applied in practice; most of the variables involved are measured by other means (although the acceleration law could be used to measure viscosity). The relation between conjoint and extensive measures of the same variables is discussed in Chapter 10.

*EXAMPLE* 4. Monotonic relations between a derivative and several causal factors are not limited to physics. In economics, for example, the theory of price dynamics postulates the relation

$$dp/dt = H(D - S),$$

where $dp/dt$ is the derivative of price, $H$ is a strictly increasing function, $D$ is the demand and $S$ is the supply; $D, S$ depend on price, among other factors. (See Samuelson, 1947, p. 263.) This is just an additive combination rule for the effects of demand and supply on the ordering of price movements. The function $H$ makes the formulation an ordinal one.

*EXAMPLE* 5. Richardson (1939) postulated that rate of change of armaments, $dx/dt$, by one of two rival nations can be expressed as

$$dx/dt = -ax + my + g,$$

where $x$ is the amount of armaments already at hand, $y$ is the amount held by the rival, and $a$, $m$, $g$ are constants ($a, m \geqslant 0$). If the constants $a, m, g$ are considered to depend on identifiable factors, this could be reformulated as a polynomial model in five variables.

*EXAMPLE* 6. Suppose that a person chooses among lotteries of the following sort: if event $a_3$ occurs, he receives two prizes, $a_1 \in A_1$ and $a_2 \in A_2$; if $a_3$ does not occur, he receives nothing. If $\phi_1$ and $\phi_2$ are (additive) utilities for commodities in $A_1$ and $A_2$, and $\phi_3$ is subjective probability, then the expected subjective utility of such a lottery is

$$[\phi_1(a_1) + \phi_2(a_2)]\, \phi_3(a_3).$$

Such lotteries were studied, in relation to the above combination rule, by Tversky (1967b); see Section 9.4.2.

Other combination rules arising in expected utility theory are considered in Chapter 8.

If a person receives a sum of money $y > 0$ when a coin falls heads and receives $z < 0$ (a loss) when it falls tails, and if this is independently repeated $c$ times, then the perceived risk of the $c$-fold combination was found by Coombs and Huang (1970) to be well described by a polynomial

$$[\phi_1(a) + \phi_2(b)]\, \phi_3(c),$$

where $a = \frac{1}{2}(y - z)$ is the expected regret, $b = \frac{1}{2}(y + z)$ is the expected value, and $c$ is the number of independent plays. Note that the same polynomial is used twice in this example, but that two different orderings (preference and perceived risk) are involved, with different sets of three factors considered.

*EXAMPLE* 7. Atkinson (1957) postulated that the attractiveness of a risky task is given by the formula

$$I_s P_s M_s - I_f P_f M_f,$$

where $s$ and $f$ denote success and failure, $I$ the incentive values of these two states, $P$ the probabilities of their occurrence, and $M$ the strengths of motives for achieving success and avoiding failure. In the simple case where incentive is inversely related to probability, he postulated $I_s = 1 - P_s = P_f$, and similarly, $I_f = P_s$; the resulting combination is

$$(M_s - M_f)\, Q,$$

(where $Q = I_s P_s = I_f P_f$), which is of the same form as Example 6.

*EXAMPLE* 8.   Still another context in which the polynomial of Examples 1, 6, and 7 arises is the quantification of opponent colors theory by Hurvich and Jameson (1955). They employed a formula that reduces to

$$S/(1 - S) = (\mid r\text{--}g \mid + \mid y\text{--}b \mid)/\mid w\text{--}bk \mid,$$

where $S$ denotes the saturation of a color and $\mid w\text{--}bk \mid$, $\mid r\text{--}g \mid$, and $\mid y\text{--}b \mid$ are the absolute values of the achromatic white–black, the chromatic red–green, and the chromatic yellow–blue responses, respectively. (See Chapter 14.)

*EXAMPLE* 9.   The models of performance postulated by Hull and Spence (Example 1) can be elaborated by subtracting off a term due to inhibition. This leads to the polynomials

$$(x_1 + x_2)\, x_3 + x_4 \qquad \text{or} \qquad x_1 x_2 x_3 + x_4.$$

If $x_2$ is held constant, then either of these reduces essentially to $x_1 x_3 + x_4$.

*EXAMPLE* 10.   There are a number of instances in psychology where a multiplicative rule is dictated by the fact that some factor has a natural zero point, with values on opposite sides of the zero point producing opposite effects on the ordering of another factor. For example, suppose that combinations of adverbs of quantity (*very*, *slightly*, etc.) and adjectives (*evil*, *pleasant*, etc.) are rated in respect to moral connotation (good to bad). One will observe (see Cliff, 1959; Gollob, 1968)

$$slightly\ evil > very\ evil,$$
$$very\ pleasant > slightly\ pleasant.$$

The adjective factor has a natural zero with respect to moral connotation (e.g., *slightly orange* $\sim$ *very orange*) and adjectives like *evil* and *pleasant* produce opposite induced orderings over the adverb factor. This makes the polynomial $x_1 x_2$ a natural candidate, with $\phi_2(evil) < 0$, $\phi_2(orange) = 0$, $\phi_2(pleasant) > 0$, and all values of $\phi_1$ nonnegative.

Another instance is found in the work of Rosenberg (1956), who had subjects rate both certain abstract (moral) values and the perceived instrumentalities of a given statement toward each value. He assumed that a subject's overall attitude toward a given statement depends on the sum

$\sum_{i=1}^{n} V_i P_i$, where $V_i$ is his rating of the $i$th value and $P_i$ is the perceived instrumentality of the statement toward the $i$th value. Thus, positive contributions to the overall attitude are expected either when the rating of a value and the instrumentality toward that value are both positive or when they are both negative.

### 7.3.2 Decomposability and Equivalence of Polynomial Models

A polynomial, as defined above, need not be strictly increasing or even 1:1 in each variable. If each coefficient $\alpha_k$ is nonnegative and if the domain of each variable $x_i$ is confined to Re$^+$, then $F$ is clearly strictly increasing in each variable. Thus, if a structure $\langle \boldsymbol{\mathsf{X}}_{i=1}^{n} A_i , \gtrsim \rangle$ satisfies a polynomial model, with each $\phi_i$ having positive values and with the coefficients $\alpha_k$ of the polynomial nonnegative, then the structure satisfies monotone decomposability.

If $\langle \boldsymbol{\mathsf{X}}_{i=1}^{n} A_i , \gtrsim \rangle$ satisfies two different polynomial models, $F(\phi_1, ..., \phi_n)$ and $F'(\phi_1', ..., \phi_n')$, both defined on (Re$^+$)$^n$ and having nonnegative coefficients, then by the results of Section 7.2, these two polynomial models are connected by the equation

$$F'(x_1, ..., x_n) = h\{F[h_1^{-1}(x_1), ..., h_n^{-1}(x_n)]\},$$

where $h$ and the $h_i$ are strictly increasing functions on Re$^+$. The question of which polynomial models are equivalent therefore leads to a subproblem: Given a polynomial $F$ on (Re$^+$)$^n$ with nonnegative coefficients, when do functions $h, h_1, ..., h_n$ lead, via the above equation, to another polynomial of the same type? Unfortunately, it seems to be difficult to give any sort of general answer. One sufficient condition is that $h$ and each $h_i^{-1}$ be themselves polynomials in one variable with nonnegative coefficients. For example, suppose

$$F(x_1 , x_2) = x_1 x_2$$
$$h_1(x_1) = (1 + x_1)^{1/2} - 1$$
$$h_2(x_2) = (1 + x_2)^{1/3} - 1$$
$$h(y) = y.$$

Then it is easy to show that $h_1^{-1}(x_1) = x_1^2 + 2x_1$, $h_2^{-1}(x_2) = x_2^3 + 3x_2^2 + 3x_2$, and

$$F'(x_1 , x_2) = x_1^2 x_2^3 + 3x_1^2 x_2^2 + 2x_1 x_2^3 + 3x_1^2 x_2 + 6x_1 x_2^2 + 6x_1 x_2 .$$

Thus, the rather complicated polynomial $F'$ is entirely equivalent, on (Re$^+$)$^2$, to the much simpler polynomial $x_1 x_2$.

That the transformations $h$, $h_i^{-1}$ need not be polynomials is shown by the following example:

$$F(x_1, \ldots, x_n) = \sum_{i=1}^{n} x_i$$

$$h_i(x_i) = \exp x_i - 1, \qquad h_i^{-1}(x_i) = \log(1 + x_i)$$

$$h(y) = \exp y$$

$$F'(x_1, \ldots, x_n) = \exp \left[ \sum_{i=1}^{n} \log(1 + x_i) \right]$$

$$= \prod_{i=1}^{n} (1 + x_i).$$

If polynomials are defined for nonpositive values of the variables, the above considerations do not apply directly. Nevertheless, many other polynomials do satisfy conditions very much akin to decomposability. For example, consider the polynomial $x_1 x_2$ defined on $\mathrm{Re}^2$. If we exclude zero—i.e., restrict the function to the domain $(\mathrm{Re} - \{0\}) \times (\mathrm{Re} - \{0\})$—then it is 1:1 in each variable separately, i.e., decomposability holds. Moreover, even though monotone decomposability does not hold, the function is either strictly increasing or strictly decreasing in each variable, according to the sign of the fixed value of the other variable. This property is analyzed further in Section 7.4.1. These two properties—decomposability when zero is excluded and strict monotonicity (increasing or decreasing) in each variable—are typical of all the polynomials considered below.

The problem of equivalences between polynomials is slightly more complicated when nonpositive values of the variables are permitted. For example, the polynomial $\sum_{i=1}^{n} x_i$ can always be replaced by $\prod_{i=1}^{n} (1 + x_i)$, but the reverse is not true if the $x_i$ can take arbitrary negative values because $\log (1 + x_i)$ is undefined for $x_i \leqslant -1$. Thus, in looking for polynomials equivalent to a given polynomial, one must examine carefully the domains to be considered.

A final illustration of the difficulties in specifying polynomial equivalence is afforded by the following example. Let $A_1 = A_2 = A_3 = \mathrm{Re}$, and define $\gtrsim$ on $A = A_1 \times A_2 \times A_3$ by

$$a \gtrsim b \qquad \text{iff} \qquad a_1 a_2 + a_3 \geqslant b_1 b_2 + b_3.$$

Clearly, the structure $\langle A, \gtrsim \rangle$ satisfies the polynomial $x_1 x_2 + x_3$, with $\phi_i$ the identity function. Consider the alternative polynomial, $x_1 x_2 + x_3{}^2$. If $\langle A, \gtrsim \rangle$ satisfies this polynomial combination rule, then there exist functions $\phi_1$, $\phi_2$, $\phi_3$, and $h$ from Re into Re, with $h$ strictly increasing, such that, for all $a \in A$,

$$\phi_1(a_1)\, \phi_2(a_2) + \phi_3(a_3)^2 = h(a_1 a_2 + a_3).$$

It is easy to show that $\phi_1(0) = 0$ (set $a_1 = 0$, vary $a_2$, holding $a_3$ constant). Thus, $\phi_3(a_3)^2 = h(a_3)$. But then setting $a_1 = 1$, $\phi_1(a_1) = C_1$, we have

$$C_1 \phi_2(a_2) + h(a_3) = h(a_2 + a_3).$$

It follows that $h$ is linear (from standard theory of functional equations, or on the basis of Theorem 2.4). Hence, $h$ cannot be everywhere positive, contradicting $h = \phi_3{}^2$. We conclude that $x_1 x_2 + x_3{}^2$ cannot be used as a model for $\langle A, \gtrsim \rangle$ as defined.

On the other hand, given any finite subset $B$ of $A$, let $\beta = \inf\{a_3 \mid a \in B\}$. Let $\phi_1$, $\phi_2$ be the identity functions and let $\phi_3(a_3) = (a_3 - \beta)^{1/2}$. Then for $a \in B$,

$$\phi_1(a_1)\, \phi_2(a_2) + \phi_3(a_3)^2 = a_1 a_2 + a_3 - \beta,$$

which is a strictly increasing function of $a_1 a_2 + a_3$. Thus, any finite subset of $A$ can be represented by $x_1 x_2 + x_3{}^2$. More generally, the two polynomials $x_1 x_2 + x_3$ and $x_1 x_2 + x_3{}^2$ are equivalent for any *finite* empirical relational structure, though they may be nonequivalent for infinite ones.

These examples show that the problem of equivalent polynomials for a given empirical relational structure, or for a finite substructure, can be quite complicated.

### 7.3.3  Simple Polynomials

For a special class of polynomials, the problem of specifying sufficient conditions is particularly tractable. These are the polynomials that can be split into either a sum or a product of two polynomials that have no variables in common and such that both of these smaller polynomials can themselves be split in the same way, and so on until one has single variables.

For example, $(x_1 + x_2)\, x_3 x_4$ splits into the product of $x_1 + x_2$ and $x_3 x_4$, which have no variables in common. The factor $x_1 + x_2$ splits into the sum of $x_1$ and $x_2$, and $x_3 x_4$ splits into the product of $x_3$ and $x_4$. (One could also start with $(x_1 + x_2)\, x_3$ times $x_4$, then split the former factor twice more.) An example of a polynomial that does not fall into this special class is $x_1 x_2 + x_2 x_3 + x_1 x_3$ which cannot be factored into independent components because any decomposition into a sum of two polynomials has a variable common to both.

Formally, we define this class of polynomials as follows.

DEFINITION 2. *Let* $X = \{x_1, ..., x_n\}$. *The set* $S(X)$ *of simple polynomials in* $x_1, ..., x_n$ *is the smallest set of polynomials such that*

(i) $x_i \in S(X)$, $i = 1, ..., n$;

(ii) *if* $Y_1$, $Y_2$ *are nonempty subsets of* $X$ *with* $Y_1 \cap Y_2 = \emptyset$, $F_1 \in S(Y_1)$, *and* $F_2 \in S(Y_2)$, *then* $F_1 + F_2$ *and* $F_1 F_2$ *are in* $S(X)$.

This definition is recursive. If $X$ has one element, then $S(X)$ is defined by (i), since (ii) never applies. If $S(Y)$ has been defined for all $Y$ with fewer than $n$ elements, then (i) and (ii) define $S(X)$ for $X$ with $n$ elements.

There are just four classes of simple polynomials in three variables, $x_1$, $x_2$, $x_3$, yielding a total of eight polynomials:

1. *Additive*: $x_1 + x_2 + x_3$.

2. *Distributive*: $(x_1 + x_2) x_3$ and two others obtained by interchanging $x_1$, $x_3$ or $x_2$, $x_3$.

3. *Dual-distributive*: $x_1 x_2 + x_3$ and two others as in 2.

4. *Multiplicative*: $x_1 x_2 x_3$.

There are ten classes of simple polynomials in four variables: eight formed from a three-variable polynomial by either adding or multiplying by a fourth variable, and two more formed from two-variable polynomials, namely, $(x_1 + x_2)(x_3 + x_4)$ and $x_1 x_2 + x_3 x_4$.

The reason for the relative tractability of simple polynomials is that once the analysis has been carried out for one instance of addition or multiplication, the same principles apply to repeated combinations by addition and multiplication, where each such combination involves Cartesian subproducts with no factors in common.

In Section 7.4, we concentrate on those necessary conditions which seem most useful for diagnosing which, if any, simple polynomial is appropriate for an empirical, conjoint relational structure. Figure 1 (p. 345) is a flowchart for diagnosis of three-variable simple polynomials. In Section 7.5, we present representation and uniqueness theorems for the multiplicative, distributive, and dual-distributive cases of three-variable conjoint measurement. (The additive case was treated in Section 6.11, for $n \geqslant 3$ variables.) The extension of the method to four or more variables is also sketched. The theorems of Section 7.5 are proved in Section 7.6. The present chapter develops the results obtained by Krantz (1968). Further discussion, with a slightly different approach to the diagnosis problem, is given by Krantz and Tversky (1971).

## 7.4 DIAGNOSTIC ORDINAL PROPERTIES

☐ **7.4.1 Sign Dependence**

We have already introduced two notions of ordinal independence. In Section 6.11, a relation $\gtrsim$ on $X_{i=1}^n A_i$ was defined to be *independent* if and only if the induced relation on the product of any subset of factors is independent of the fixed components in the remaining factors. In Section 7.2, we defined $A_i$ to be *independent* of $X_{j \neq i} A_j$ if and only if the induced relation on $A_i$ is independent of the fixed components in $A_j$, for all $j \neq i$. The natural generalization of these concepts is to say that $X_{i \in N_1} A_i$ is *independent of* $X_{j \in N_2} A_j$ if and only if the induced relation over the product of the $A_i$ factors, for $i$ in $N_1$, is independent of the fixed components in the $A_j$ factors, for $j$ in $N_2$, it being understood that the components in the remaining factors are held constant. For example, if a three-factor structure satisfies the distributive rule $(x_1 + x_2) x_3$, then $A_1$ is independent of $A_2$ and vice versa. If $\phi_3$ takes only positive (or only negative) values, then $A_1$ is independent of $A_2 \times A_3$; but even if $\phi_3$ takes values throughout Re, $A_1$ is still independent of $A_2$, since for any fixed $a_3 \in A_3$, the ordering of $A_1$-effects does not depend on variation of $a_2$ in $A_2$.

We also note that, for any simple polynomial, if $A_i$ is not independent of $X_{j \neq i} A_j$, then only three different induced relations over $A_i$ can exist, corresponding to multiplication by positive values, zero, or negative values. That is, the induced weak order $\gtrsim_i$ over $A_i$ can be totally reversed for some changes in the element held fixed in $X_{j \neq i} A_j$, and it can be degenerate for others (only one $\sim_i$-equivalence class), but no other possibility is allowed. We describe this by saying that $A_i$ is *sign dependent on* $X_{j \neq i} A_j$, and we generalize this in the same way as independence, i.e., $X_{i \in N_1} A_i$ is sign dependent on $X_{j \in N_2} A_j$ if there are only three possible induced relations on $X_{i \in N_1} A_i$ for varying choices of $a_j$ in $A_j$, $j$ in $N_2$ (one ordering, its reversal, and the degenerate ordering).

Formal definitions of these concepts are given in Definition 3 below. For clarity, we introduce the following notation. Let $A_1, ..., A_n$ be sets and let $N$ be a subset of the indices $\{1, ..., n\}$. For each $a$ in $X_{i=1}^n A_i$, let $a^{(N)}$ denote the vector of components $a_i$ for each $i$ in $N$, and let $A^{(N)}$ denote the set of all such vectors, i.e., $A^{(N)} = X_{i \in N} A_i$.

DEFINITION 3. *Let $\gtrsim$ be a binary relation on $X_{i=1}^n A_i$. Let $N_1, N_2, N_3$ form a partition of the index set $\{1, ..., n\}$. We say that $A^{(N_1)}$ is* independent *of $A^{(N_2)}$ iff, for any fixed $a^{(N_3)} \in A^{(N_3)}$, the relation induced on $A^{(N_1)}$ by fixing a vector $a^{(N_2)} \in A^{(N_2)}$ (and using the fixed $a^{(N_3)}$) is the same for all $a^{(N_2)}$.*

*We say that $A^{(N_1)}$ is* sign dependent *on $A^{(N_2)}$ iff $A^{(N_2)}$ can be partitioned into* *three sets, denoted $S^+(N_1, N_2), S^0(N_1, N_2)$, and $S^-(N_1, N_2)$, such that, for any fixed $a^{(N_2)}$, the relation induced on $A^{(N_1)}$ by choosing $a^{(N_2)}$ depends only on whether $a^{(N_2)} \in S^+(N_1, N_2), S^0(N_1, N_2)$, or $S^-(N_1, N_2)$; specifically, any relation induced by $a^{(N_2)} \in S^-(N_1, N_2)$ is the converse of any one induced by $a^{(N_2)} \in S^+(N_1, N_2)$, and any relation induced by $a^{(N_2)} \in S^0(N_1, N_2)$ is degenerate (the universal relation on $A^{(N_1)}$).*

*For sign dependence, we always assume that at least one of $S^+$ and $S^-$ is nonempty. If exactly one of them is nonempty, the other one and $S^0$ being empty, then independence holds [$A^{(N_1)}$ independent of $A^{(N_2)}$]. When two or three of the sets $S^+, S^0, S^-$ are nonempty, we speak of* proper sign dependence.

As discussed in Section 7.3.2, multiplicative combinations of variables can occur without necessitating any instances at all of proper sign dependence, e.g., when the domain of the polynomial is $(\text{Re}^+)^n$. The observation that each $A_i$ is independent of $\underset{j \neq i}{\times} A_j$ is, therefore, completely undiagnostic as to which simple polynomial might be appropriate (except that the pure multiplicative polynomial, $x_1 x_2 \cdots x_n$, can be put aside since, if it is appropriate here, the additive polynomial must also be appropriate and is used instead). This may be the most common case in empirical applications: monotone decomposability holds, and other criteria must be invoked to narrow down the choice of the function $F$. Diagnosis of the appropriate polynomial in such a case proceeds by using the concept of joint independence, as described in Section 7.4.3. The remainder of this section is relevant only when instances of proper sign dependence exist for a given relational structure. In such cases, it is obvious that much information is available about the location of multiplicative combinations; the precise manner of using such information is discussed below for three-variable polynomials.

It is worth remarking that complete reversal or degeneracy of induced orderings are striking phenomena, when they occur. Sometimes, as with the adverb–adjective combinations (Example 10), they may be predicted from thought experiments. Moreover, when two variables are (mutually) properly sign dependent, then large positive effects can arise in two distinct ways: from positive values of both variables and from negative values of both variables. "The friend of my friend is my friend," and "the enemy of my enemy is my friend." Thus, proper sign dependence is not likely to be missed, and it will be of great value where it occurs, although its range of applications may not be large. As shown by the above aphorisms on friendship and by the work of Rosenberg (1956) mentioned in Example 10, social attitudes may be one of the more fruitful areas of application.

For two- and three-factor structures, the notation used in Definition 3 can be simplified. If $A_i$ is sign dependent on $A_j \times A_k$, then we denote the partition of $A_j \times A_k$ by $(A_j \times A_k)^+, (A_j \times A_k)^0, (A_j \times A_k)^-$, rather

$S^+(\{i\}, \{j, k\})$, etc. Similarly, if $A_j \times A_k$ is sign dependent on $A_i$, the partition of $A_i$ is denoted $A_i{}^+, A_i{}^0, A_i{}^-$. If $A_j$ alone is properly sign dependent on $A_i$ and $A_k$ is independent of $A_i$, we again denote the partition[1] of $A_i$ by $A_i{}^+, A_i{}^0, A_i{}^-$. The corresponding notational simplifications in the two-factor situation are obvious.

If, in the two-factor situation, $A_1$ and $A_2$ are mutually sign dependent, then an additional necessary condition for the multiplicative model $x_1 x_2$ is that, for proper designation of $+$ and $-$ signs, "positive" combinations are greater than "zero" ones which, in turn, are greater than "negative" combinations. Thus, for $a$ in $(A_1{}^+ \times A_2{}^+) \cup (A_1{}^- \times A_2{}^-)$, $b$ in $(A_1{}^0 \times A_2) \cup (A_1 \times A_2{}^0)$, and $c$ in $(A_1{}^+ \times A_2{}^-) \cup (A_1{}^- \times A_2{}^+)$, we must have $a > b > c$. Roskies (1965) assumed this as an additional axiom in his analysis of the model $x_1 x_2$; however, in the presence of strong solvability conditions, it follows from sign dependence. Specifically, we have the following result.

THEOREM 2. *Suppose that $\gtrsim$ is a weak ordering of $A = A_1 \times A_2$ and further that $A_1$, $A_2$ are properly sign dependent on each other. Suppose that for all $a, b \in A$, with $b_i \in A_i - A_i{}^0, i = 1, 2$, there exists $c \in A$ such that*

$$a \sim b_1 c_2 \sim c_1 b_2 \qquad (solvability).$$

*Then $A_i{}^+, A_i{}^-$, for $i = 1, 2$, can be designated such that if*

$$a \in (A_1{}^+ \times A_2{}^+) \cup (A_1{}^- \times A_2{}^-),$$
$$b \in (A_1{}^0 \times A_2) \ \cup (A_1 \times A_2{}^0),$$
$$c \in (A_1{}^+ \times A_2{}^-) \cup (A_1{}^- \times A_2{}^+),$$

*then*

$$a > b > c.$$

The proof is given in Section 7.4.2.

In three-factor structures, Theorem 2 still applies, guaranteeing that an element of $(A_1 \times A_2)^+ \times A_3{}^+$ is larger than an element of $(A_1 \times A_2)^+ \times A_3{}^-$, etc. However, if the multiplicative model $x_1 x_2 x_3$ is appropriate, we have another necessary consistency requirement: $(A_1 \times A_2)^+$ should be equal to

---

[1] One might fear that this notation would run into trouble in the case that $A_j$ and $A_k$ are each properly sign dependent on $A_i$, but with different partitions of $A_i$ required. This does not arise, however. If $A_j$ and $A_k$ are each properly sign dependent on $A_i$, then the only possible three-variable simple polynomials are $(x_j + x_k)x_i$ and $x_j x_k x_i$, and in either case, $A_j \times A_k$ is sign dependent on $A_i$, and the partition into $A_i{}^+, A_i{}^0, A_i{}^-$ from sign dependence of $A_j \times A_k$ serves also for sign dependence of $A_j$ and of $A_k$ separately. Hence, the meaning of this notation, referring in one case to sign dependence of $A_j \times A_k$ on $A_i$, and in another $(x_j x_i + x_k)$ to sign dependence of $A_j$ on $A_i$, is never ambiguous.

$(A_1{}^+ \times A_2{}^+) \cup (A_1{}^- \times A_2{}^-)$, where $(A_1 \times A_2)^+$ is determined from sign dependence of $A_3$ on $A_1 \times A_2$, and $A_1{}^+, A_2{}^+$ are determined from sign dependence of $A_2 \times A_3$ on $A_1$ and $A_1 \times A_3$ on $A_2$, respectively. Other analogous requirements are also necessary. It turns out that these requirements also follow from sign dependence and unrestricted solvability, as shown by the next theorem.

THEOREM 3.  *Suppose that $\gtrsim$ is a weak ordering of $A = A_1 \times A_2 \times A_3$, where each pair of factors and the third factor are properly sign dependent on one another. Suppose that for any $a, b \in A$, with $b_i \in A_i - A_i{}^0$, $i = 1, 2, 3$, there exists $c \in A$ such that*

$$a \sim b_1 b_2 c_3 \sim b_1 c_2 b_3 \sim c_1 b_2 b_3 \qquad \text{(solvability)}.$$

*Then, for $i, j = 1, 2, 3, i \neq j$, $A_i{}^+, A_i{}^-$ and $(A_i \times A_j)^+, (A_i \times A_j)^-$ can be designated such that*

(i)  $(A_i \times A_j)^+ = (A_i{}^+ \times A_j{}^+) \cup (A_i{}^- \times A_j{}^-)$;

(ii)  $(A_i \times A_j)^- = (A_i{}^+ \times A_j{}^-) \cup (A_i{}^- \times A_j{}^+)$;

(iii)  *if*

$$a \in [(A_1 \times A_2)^+ \times A_3{}^+] \cup [(A_1 \times A_2)^- \times A_3{}^-],$$
$$b \in [(A_1 \times A_2)^0 \times A_3] \cup [(A_1 \times A_2) \times A_3{}^0],$$
$$c \in [(A_1 \times A_2)^+ \times A_3{}^-] \cup [(A_1 \times A_2)^- \times A_3{}^+],$$

*then*

$$a > b > c.^2$$

We see that, in the presence of strong solvability conditions, sign dependence implies the kind of consistency of the assignment of $+$ and $-$ signs that is needed for multiplicative models. In checking sign dependence, however, it is well to check the appropriate properties in Theorems 2 and 3, as well as the defining property of Definition 3.

---

[2] The apparent asymmetry of this last statement is illusory; for, by the first statement, we have

$$A^+ = [(A_1 \times A_2)^+ \times A_3{}^+] \cup [(A_1 \times A_2)^- \times A_3{}^-]$$
$$= \{[(A_1{}^+ \times A_2{}^+) \cup (A_1{}^- \times A_2{}^-)] \times A_3{}^+\} \cup \{[(A_1{}^+ \times A_2{}^-) \cup (A_1{}^- \times A_2{}^+)] \times A_3{}^-\}$$
$$= (A_1{}^+ \times A_2{}^+ \times A_3{}^+) \cup (A_1{}^- \times A_2{}^- \times A_3{}^+) \cup (A_1{}^+ \times A_2{}^- \times A_3{}^-)$$
$$\cup (A_1{}^- \times A_2{}^+ \times A_3{}^-),$$

which is exactly the right (fully symmetric) choice of four out of eight octants to be positive. Then $A^0$, $A^-$ are defined similarly, and the final statement of Theorem 3 is $A^+ > A^0 > A^-$, where $A > B$ means that $a > b$ for all $a \in A$, $b \in B$.

A first step toward the choice of a polynomial is to determine, for each pair of factors $A_i$, $A_j$, whether or not $A_i$ is sign dependent on $A_j$. If some form of sign dependence holds for each pair of variables, the results of the tests can be represented by the off-diagonal elements of an $n \times n$ matrix of 0's and 1's as follows: enter 0 in the $i$th row and $j$th column if $A_i$ is independent of $A_j$, and 1 if $A_i$ is properly sign dependent on $A_j$. With three factors $A_1$, $A_2$, $A_3$ there are six off-diagonal entries in this diagnostic matrix, hence, there are $2^6 = 64$ different matrices. (Since proper sign dependence is an irreflexive relation on the index set $\{1,...,n\}$, there are in general $2^{n^2-n}$ such relations.)

Since the indexing of the factors is arbitrary, two sign-dependence matrices are equivalent if they can be transformed into one another by permuting the indices, i.e., by applying the same permutation to the rows and the columns of the matrix. The 64 matrices fall into 16 equivalence classes. The matrix containing all 0's corresponds to monotone decomposability (Theorem 1), but leaves open the additive, distributive, or dual-distributive polynomials as possibilities. Further diagnosis depends on examination of joint independence (Section 7.4.3). There are seven classes of matrices (containing 35 matrices, all told) that are incompatible with any simple polynomial; an example is the matrix

$$
\begin{array}{c|ccc}
 & 1 & 2 & 3 \\
\hline
1 & - & 1 & 0 \\
2 & 0 & - & 1 \\
3 & 0 & 0 & - 
\end{array}.
$$

Since $A_2$ is properly sign dependent on $A_3$, the variables $x_2$ and $x_3$ must multiply, and the domain of $x_3$ must have a variable sign.[3] Since $A_1$ is properly sign dependent on $A_2$, the variables $x_1$ and $x_2$ multiply and $x_2$ has a variable sign. But since $x_2$ multiplies $x_3$ and $x_2$ has a variable sign, $A_3$ must be properly sign dependent on $A_2$, contrary to the matrix. Hence, no simple polynomial is possible.

An unsolved problem of some interest is to characterize axiomatically those finite, irreflexive, binary relations that correspond to the relation of sign dependence for some simple polynomial. However, in practice, there is little difficulty in carrying out the kind of reasoning just illustrated for three-, four-, or five-factor matrices and in deciding which, if any, simple polynomials are compatible with them. We present the results of the analysis

---

[3] The domain of a variable is said to have a variable sign if it is not included wholly in $Re^+$ or wholly in $Re^-$; thus, domains including both zero and positive numbers have variable sign.

for the 16 categories of three-factor matrices in Table 1; the interested reader can verify the details. (See also Exercise 8.)

**TABLE 1**

Patterns of Proper Sign Dependence for Three Factors

| Category number | Number of proper sign dependencies | Example | Number of equivalent matrices | Simple polynomials compatible with the example |
|---|---|---|---|---|
| 0 | 0 | $-$ 0 0<br>0 $-$ 0<br>0 0 $-$ | 1 | All |
| 1 | 1 | $-$ 0 1<br>0 $-$ 0<br>0 0 $-$ | 6 | $x_1x_3 + x_2$<br>$(x_2 + x_3)x_1$ [a] |
| 2 | 2 | $-$ 0 1<br>0 $-$ 1<br>0 0 $-$ | 3 | $x_1x_2x_3$<br>$(x_1 + x_2)x_3$ |
| 3 | 2 | $-$ 1 1<br>0 $-$ 0<br>0 0 $-$ | 3 | $(x_2 + x_3)x_1$ |
| 4 | 2 | $-$ 1 0<br>1 $-$ 0<br>0 0 $-$ | 3 | $x_1x_2 + x_3$ |
| 5 | 3 | $-$ 0 1<br>0 $-$ 1<br>1 0 $-$ | 6 | $(x_1 + x_2)x_3$ [a] |
| 6 | 4 | $-$ 1 0<br>1 $-$ 0<br>1 1 $-$ | 3 | $x_1x_2x_3$ |
| 7 | 4 | $-$ 0 1<br>0 $-$ 1<br>1 1 $-$ | 3 | $(x_1 + x_2)x_3$ |
| 8 | 6 | $-$ 1 1<br>1 $-$ 1<br>1 1 $-$ | 1 | $x_1x_2x_3$ |
| 9 | 2 | $-$ 1 0<br>0 $-$ 1<br>0 0 $-$ | 6 | None |

*Table continued*

**TABLE 1** *(cont.)*

| Category number | Number of proper sign dependencies | Example | Number of equivalent matrices | Simple polynomials compatible with the example |
|---|---|---|---|---|
| 10 | 3 | $-\ 1\ 1$<br>$0\ -\ 1$<br>$0\ 0\ -$ | 6 | None |
| 11 | 3 | $-\ 1\ 1$<br>$1\ -\ 0$<br>$0\ 0\ -$ | 6 | None |
| 12 | 3 | $-\ 1\ 0$<br>$0\ -\ 1$<br>$1\ 0\ -$ | 2 | None |
| 13 | 4 | $-\ 1\ 1$<br>$1\ -\ 1$<br>$0\ 0\ -$ | 3 | None |
| 14 | 4 | $-\ 0\ 1$<br>$1\ -\ 0$<br>$1\ 1\ -$ | 6 | None |
| 15 | 5 | $-\ 0\ 1$<br>$1\ -\ 1$<br>$1\ 1\ -$ | 6 | None |

ᵃ The occurrence of patterns 1 or 5 with the distributive polynomial is possible, but unlikely. Pattern 5, for example, could occur if the domain of $x_1$ were $(-2, -1) \cup (1, 2)$ and the domain of $x_2$ were $(0, 1)$. So for fixed $x_1$, if $x_1 \in (-2, -1)$, then $x_1 + x_2 < 0$ for all $x_2$, whereas if $x_1 \in (1, 2)$, then $x_1 + x_2 > 0$ for all $x_2$. Thus, $A_3$ is not properly sign dependent on $A_2$. But if the domains of each variable have 0 as a limit point, pattern 5 is impossible. Ordinarily, the occurrence of pattern 1 would be strongly diagnostic for dual-distributive polynomials, and pattern 5 would make it doubtful that any polynomial would work.

### 7.4.2 Proofs of Theorems 2 and 3 (p. 331–332)

*Suppose that $\gtrsim$ is a weak order on $A_1 \times A_2$, that $A_1$, $A_2$ are (mutually) sign dependent, and that unrestricted solvability holds. Partitions $A_i^+, A_i^0, A_i^-, i = 1, 2$, can be designated such that, if*

$$A^+ = (A_1^+ \times A_2^+) \cup (A_1^- \times A_2^-)$$
$$A^0 = (A_1^0 \times A_2) \cup (A_1 \times A_2^0)$$
$$A^- = (A_1^+ \times A_2^-) \cup (A_1^- \times A_2^+),$$

*then*

$$A^+ \succ A^0 \succ A^-.$$

*PROOF.* We first prove the theorem subject to the hypothesis that $A_1^+$, $A_1^-$ are both nonempty. Designate $A_2^+$, $A_2^-$ so that $A_2^+ \neq \varnothing$. Choose $a_1^+ \in A_1^+$, $a_2^+ \in A_2^+$, and $a_1^- \in A_1^-$. If $a_1^+a_2^+ < a_1^-a_2^+$, reverse the sign labeling of $A_1^+$, $A_1^-$, and $a_1^+$, $a_1^-$. Thus, we may assume $a_1^+a_2^+ \gtrsim a_1^-a_2^+$. We show that, with this labeling of signs, the conclusion of Theorem 2 holds. The proof is divided into six steps.

I.   Suppose $a_1^+a_2^+ \sim a_1^-a_2^+$, then by sign dependence, for all $b_2 \in A_2$, $a_1^+b_2 \sim a_1^-b_2$. Since $a_1^+ \notin A_1^0$, there exists $b_2$ such that not $a_1^+b_2 \sim a_1^+a_2^+$. But if $a_1^-b_2 \sim a_1^+b_2 > a_1^+a_2^+ \sim a_1^-a_2^+$, then $a_1^-b_2 > a_1^-a_2^+$; this contradicts sign dependence, which states that the ordering of $b_2$ and $a_2^+$ must be opposite relative to $a_1^+$ and $a_1^-$. A similar contradiction arises in the case where $a_1^+b_2 < a_1^+a_2^+$. Hence, it follows that $a_1^+a_2^+ > a_1^-a_2^+$. By sign dependence, we have $a_1^+b_2 > a_1^-b_2$ for all $b_2 \in A_2^+$, and $a_1^+b_2 < a_1^-b_2$, for all $b_2 \in A_2^-$.

II.   For $b_2 \in A_2^+$, $c_2 \in A_2^+$, assume $a_1^+b_2 \gtrsim a_1^+c_2$. Then by step I and sign dependence,

$$a_1^+b_2 \gtrsim a_1^+c_2 > a_1^-c_2 \gtrsim a_1^-b_2 ,$$

from which follows $\{a_1^+\} \times A_2^+ > \{a_1^-\} \times A_2^+$. In the same way, $\{a_1^+\} \times A_2^- < \{a_1^-\} \times A_2^-$.

III.   By solvability, we choose $a_2 \in A_2$ such that $a_1^+a_2 \sim a_1^-a_2^+$. We show that $a_2 \in A_2^-$. By step II, $a_2 \notin A_2^+$, and by step I, $a_1^+a_2^+ > a_1^+a_2$, whence by sign dependence,

$$a_1^-a_2 > a_1^-a_2^+ \sim a_1^+a_2 .$$

This shows that $a_2 \notin A_2^0$, hence, that $a_2 \in A_2^-$. From now on we denote this element $a_2^-$.

IV.   Note that the arguments of steps I and II can be carried out for any $b_1^+ \in A_1^+$, $b_1^- \in A_1^-$, provided that $b_1^+a_2^+ \gtrsim b_1^-a_2^+$. We first show that for any $b_1^+ \in A_1^+$, we have $b_1^+a_2^+ \gtrsim a_1^-a_2^+$. If $b_1^+a_2^+ \gtrsim a_1^+a_2^+$, this follows by transitivity. In the contrary case, sign dependence yields

$$b_1^+a_2^- > a_1^+a_2^- \sim a_1^-a_2^+ .$$

Moreover, by sign dependence, $b_1^+a_2^+ > b_1^+a_2^-$, so that $b_1^+a_2^+ > a_1^-a_2^+$ follows. We conclude that

$$A_1^+ \times A_2^+ > \{a_1^-\} \times A_2^+ , \qquad A_1^+ \times A_2^- < \{a_1^-\} \times A_2^- .$$

Now this conclusion follows for any $b_1^- \in A_1^-$ (replacing $a_1^-$) provided only that $a_1^+a_2^+ > b_1^-a_2^+$. This can be established for any $b_1^-$ : if $a_1^-a_2^+ \gtrsim b_1^-a_2^+$, it is trivial, while in the contrary case, from the fact that $b_1^- \in A_1^-$,

$$b_1^-a_2^- > b_1^-a_2^+ > a_1^-a_2^+ \sim a_1^+a_2^-,$$

whence by sign dependence, $a_1^+a_2^+ > b_1^-a_2^+$. We conclude from this that

$$A_1^+ \times A_2^+ > A_1^- \times A_2^+, \qquad A_1^+ \times A_2^- < A_1^- \times A_2^-.$$

V.  In steps I–IV, the roles of $A_1$, $A_2$ can be exchanged, using the inequality $a_1^+a_2^+ > a_1^+a_2^-$ established in step III to replace the inequality $a_1^+a_2^+ > a_1^-a_2^+$ used throughout. Thus, we conclude that

$$A_1^+ \times A_2^+ > A_1^+ \times A_2^-, \qquad A_1^- \times A_2^+ < A_1^- \times A_2^-.$$

This proves that $A^+ > A^-$ for the case $A_1^+$ and $A_1^-$ are both nonempty.

VI.  Suppose $a_1a_2 \in A_1^+ \times A_2^+$ and $b_1b_2 \in A_1^0 \times A_2$. We show that $b_1b_2 \gtrsim a_1a_2$ leads to a contradiction. In fact, choose $c_2 \in A_2^-$; then

$$a_1c_2 < a_1a_2 \lesssim b_1b_2 \sim b_1c_2 .$$

By sign dependence,

$$a_1a_2 > b_1a_2 \sim b_1b_2 ,$$

a contradiction. Thus $A_1^+ \times A_2^+ > A_1^0 \times A_2$. The other parts of the conclusion $A^+ > A^0$ and $A^0 > A^-$ are demonstrated by symmetrical arguments.

We consider the case where $A_1^- = \varnothing$. Designate $A_2^+$ and $A_2^-$ so that $A_2^+ \neq \varnothing$. Then, $A_2^- = \varnothing$ because, otherwise, we could carry out the argument of steps I–III, with $A_1$, $A_2$ reversed, finally constructing $a_1^- \in A_1^-$. By solvability, it can be shown that $A_1^0 = \varnothing$ if and only if $A_2^0 = \varnothing$. If both $A_1^0$, $A_2^0 = \varnothing$, Theorem 2 is trivial. Otherwise, choose $a_1a_2 \in A_1^+ \times A_2^+$ and $b_1 \in A_1^0$. Suppose that $a_1a_2 \gtrsim b_1a_2$. Then clearly, $a_1a_2 > b_1a_2$, since otherwise $a_2 \in A_2^0$. Thus, for all $c_2 \in A_2^+$, $a_1c_2 > b_1c_2$. Since by solvability, any $d_1d_2 \in A_1^+ \times A_2^+$ is equivalent to some $a_1c_2$ with $c_2 \in A_2^+$, the conclusion to Theorem 2 follows. On the other hand, if $a_1a_2 < b_1a_2$, we can relabel $A_1^+$ as $A_1^-$, and the same proof yields

$$(A_1^0 \times A_2) \cup (A_1 \times A_2^0) > (A_1^- \times A_2^+),$$

as required by Theorem 2.                                                $\Diamond$

We remark that the use of solvability is essential in the proof of Theorem 2. This is shown in the following example.

Let $A_1 = \{a_1, b_1\}$, $A_2 = \{a_2, b_2, c_2\}$, and let the ordering over $A_1 \times A_2$ be that given by the following matrix of ranks

| $A_1$ \ $A_2$ | $a_2$ | $b_2$ | $c_2$ |
|---|---|---|---|
| $a_1$ | 1 | 4 | 5 |
| $b_1$ | 2 | 3 | 6 . |

It is easy to see that $A_1$ is sign dependent on $A_2$, with $A_2^+ = \{a_2, c_2\}$, $A_2^- = \{b_2\}$, and that $A_2$ is independent of $A_1$. But $a_1 b_2$ is between $a_1 a_2$ and $a_1 c_2$, so that no redesignation of signs of $A_2^+$, $A_2^-$ will satisfy $A_1 \times A_2^+ > A_1 \times A_2^-$. In fact, there can be no multiplicative representation of $\langle A_1 \times A_2, \succsim \rangle$, since any such representation would have to have $\phi_2(b_2)$ opposite in sign from, yet in between, $\phi_2(a_2)$ and $\phi_2(c_2)$. The system $\langle A_1 \times A_2, \succsim \rangle$ cannot be embedded in a system that satisfies solvability while maintaining mutual sign dependence.

We turn next to Theorem 3.

*Suppose that $\succsim$ is a weak order on $A_1 \times A_2 \times A_3$, that each pair of factors and the third factor are sign dependent on one another, and that unrestricted solvability holds. The partitions $A_i^+, A_i^0, A_i^-, i = 1, 2, 3$, and $(A_i \times A_j)^+$, $(A_i \times A_j)^0, (A_i \times A_j)^-, i, j = 1, 2, 3, i \neq j$, can be designated such that*

$$(A_i \times A_j)^+ = (A_i^+ \times A_j^+) \cup (A_i^- \times A_j^-),$$
$$(A_i \times A_j)^0 = (A_i^0 \times A_j) \cup (A_i \times A_j^0),$$
$$(A_i \times A_j)^- = (A_i^+ \times A_j^-) \cup (A_i^- \times A_j^+),$$

*and such that $A^+ > A^0 > A^-$, where*

$$A^+ = (A_i \times A_j)^+ \times A_k^+ \cup (A_i \times A_j)^- \times A_k^-,$$

*etc.*

PROOF. We can designate $(A_1 \times A_2)^+, (A_1 \times A_2)^-, A_3^+, A_3^-$ so that the conclusion of Theorem 2 applies to $(A_1 \times A_2) \times A_3$. This proves the second statement of Theorem 3. Next, choose $a_3^+ \in A_3^+$, and consider the induced ordering over $A_1 \times A_2$, for $a_3^+$ fixed. This ordering satisfies the hypotheses of Theorem 2, so we can find subsets of $A_1$, denoted $B_1^+, B_1^0, B_1^-$, and of $A_2$, denoted $B_2^+, B_2^0, B_2^-$, such that the conclusions of Theorem 2 apply, i.e.,

$$B^+ \times \{a_3^+\} > B^0 \times \{a_3^+\} > B^- \times \{a_3^+\},$$

where

$$B^+ = (B_1^+ \times B_2^+) \cup (B_1^- \times B_2^-),$$

and so on. Since $a_3^+ \in A_3^+$, we have the same conclusion for any $b_3 \in A_3^+$ and the opposite inequalities for $b_3 \in A_3^-$. This proves that

$$(A_1 \times A_2)^+ = B^+ = (B_1^+ \times B_2^+) \cup (B_1^- \times B_2^-),$$
$$(A_1 \times A_2)^- = B^- = (B_1^+ \times B_2^-) \cup (B_1^- \times B_2^+).$$

From the definition of $B_i^+, B_i^0, B_i^-, i = 1, 2$, it can be shown that $A_i^+$ is wholly included either in $B_i^+$ or in $B_i^-$, with $A_i^-$ wholly included in the other. Therefore, $A_i^+, A_i^-$ can be designated so that $A_i^+ \subset B_i^+$ and $A_i^- \subset B_i^-$. Clearly, $A_i^0 \subset B_i^0$; thus, since $(B_i^+, B_i^0, B_i^-)$ is a partition of $A_i$, as is $(A_i^+, A_i^0, A_i^-)$, we have $A_i^+ = B_i^+, A_i^- = B_i^-, A_i^0 = B_i^0$. This establishes the first statement of the theorem. $\diamond$

□ **7.4.3 Joint-Independence Conditions**

In Chapter 6 we showed that in the presence of the weak ordering, solvability, and Archimedean axioms, a necessary and sufficient condition for additivity is that each pair of factors be independent of the third (Theorem 6.13 specialized for $n = 3$). The independence of the ordering of the *joint effect* of two or more factors from fixed levels of the other factors is, thus, a much more powerful condition than independence of the ordering of effects of each factor alone. For a multiplicative combination rule in three variables, joint sign dependence, rather than joint independence, is the obvious necessary condition; we shall see (Theorem 4) that, in analogy to Theorem 6.13, it is also the key sufficient condition.

Single-factor independence for all factors does not, of course, imply joint independence. In the distributive and dual-distributive models, with domains in $(Re^+)^3$, exactly one pair of factors is independent of the third. For suppose that

$$(x_1 + x_2) x_3 \geqslant (x_1' + x_2) x_3',$$

i.e., with $x_2$ fixed, $(x_1, x_3) \gtrsim (x_1', x_3')$. Suppose that $x_3 > x_3'$. Then we easily derive

$$x_2 \geqslant (x_1'x_3' - x_1x_3)/(x_3 - x_3').$$

Hence, if $x_2'$ is chosen below the cutoff $(x_1'x_3' - x_1x_3)/(x_3 - x_3')$, we will have the opposite inequality,

$$(x_1 + x_2') x_3 < (x_1' + x_2') x_3'.$$

To put matters differently, consider the distributive combination of drive $D$, incentive $K$, and habit $H$ (Example 1, Section 7.1). Suppose that we choose levels $h, h'$ of habit, with $h > h'$, and levels $k, k'$ of incentive, with $k$ much smaller than $k'$ [so much smaller that $\phi_K(k)\,\phi_H(h) < \phi_K(k')\,\phi_H(h')$, despite the fact that $\phi_H(h) > \phi_H(h')$]. For a very low level $d$ of $D$, the $D \times H$ contribution will be negligible, and performance will be dominated by the $K \times H$ contribution, hence,

$$dkh < dk'h'.$$

For a very high level $d'$ of $D$, the $D \times H$ contribution will dominate the $K \times H$ contribution, and since $h > h'$, multiplication by $d'$ will expand this difference, resulting in

$$d'kh > d'k'h'.$$

The crossover is exactly given by the formula used above: $dkh \sim dk'h'$ if

$$\phi_D(d) = [\phi_K(k')\,\phi_H(h') - \phi_K(k)\,\phi_H(h)]/[\phi_H(h) - \phi_H(h')].$$

We note that the ordinal versions of both the Hull (multiplicative) and Spence (distributive) models satisfy monotone decomposability. Moreover, any two factors, with the third held constant, satisfy all of the properties of additive conjoint measurement. Thus, two-variable experiments cannot possibly distinguish these models. Experiments like those of Seward, Shea, and Davenport (1960) and Pavlik and Reynolds (1963) can only discriminate between the (less reasonable) versions of these models that are formulated in terms of a specific "interval scale" for performance. To distinguish the ordinal laws, one must perform precisely the test sketched above: observe performance as a function of $D$, for two different $K$, $H$ levels, with $h > h'$, $k \ll k'$ (or interchange the roles of $D$ and $K$ in this test).

Diagnosis of the appropriate three-variable polynomial on the basis of joint independence is simply a matter of testing each pair of variables for independence of the third. If each pair is independent of (alternatively sign dependent on) the third, then the additive (alternatively multiplicative) model is appropriate. If exactly one pair is sign dependent on the third, then only the distributive and dual-distributive models are appropriate; if the sign dependence is proper, then, of course, only the distributive model can hold. If none or two pairs are sign dependent, then no simple polynomial can hold.

☐ **7.4.4 Cancellation Conditions**

For the additive and multiplicative models, none of the (logically necessary) cancellation conditions need be assumed because they follow from three-way joint independence and solvability (Lemma 6.14 and Theorem 6.13 for the additive case; Theorem 4 below for the multiplicative version).

For the distributive case, $(x_1 + x_2) x_3$, it is clear, from two-factor additive conjoint measurement that the standard Thomsen condition must hold for the induced order on $A_1 \times A_2$ for any fixed $a_3$. By the same token, regarding $A_1 \times A_2$ as a single factor combining with $A_3$, a multiplicative version of the Thomsen condition (adapted for sign changes, if any) must hold for $(A_1 \times A_2) \times A_3$. Similarly, in the dual-distributive case, $x_1 x_2 + x_3$, the induced order on $A_1 \times A_2$ must satisfy multiplicative versions of the two-factor additive conjoint measurement conditions, whereas $(A_1 \times A_2) \times A_3$ must satisfy the additive versions of those same conditions.

The multiplicative version of the Thomsen condition is given in the next definition.

DEFINITION 4. *Suppose that $\succsim$ is a weak order on $A = A_1 \times A_2$ and that $A_1$, $A_2$ are sign dependent on one another. We say that $\langle A_1, A_2, \sim \rangle$ satisfies the* Thomsen condition, *iff for all $a$, $b$, $c \in A$, whenever*

$$a_1 b_2 \sim b_1 c_2 ,$$
$$b_1 a_2 \sim c_1 b_2 ,$$

*and either $b_1 \notin A_1{}^0$ or $b_2 \notin A_2{}^0$, then*

$$a \sim c.$$

The provision that either $b_1 \notin A_1{}^0$ or $b_2 \notin A_2{}^0$, the special feature introduced for multiplication, avoids "division by zero." This axiom was introduced by Roskies (1965).

If monotone decomposability holds, the multiplicative and additive versions of cancellation conditions are identical, so none of these necessary cancellation conditions can be used to discriminate between the distributive and dual-distributive cases. To distinguish these two polynomials, a new type of cancellation condition is needed. In Section 7.5.2, we shall explain how this new type of cancellation was discovered and how similar conditions can be discovered for simple polynomials in four or more variables. Here, we present the conditions, prove that they are logically necessary, and discuss them. It turns out that in both the distributive and dual-distributive cases the special cancellation implies (with solvability, independence, etc.) appropriate cancellation conditions for the induced orders on $A_1 \times A_2$. Thus, our list of axioms, for each model, consists of independence axioms, solvability and Archimedean axioms, a special cancellation condition, and the Thomsen condition for $(A_1 \times A_2) \times A_3$.

DEFINITION 5. *Suppose that $\succsim$ is a binary relation on $A = A_1 \times A_2 \times A_3$.*

*We say that* distributive cancellation *holds iff, for all a, b, b', c in A, whenever*

$$a_1 b_2 a_3 \gtrsim b_1' c_2 c_3 \,,$$
$$b_1 a_2 a_3 \gtrsim c_1 b_2' c_3 \,,$$
$$b_1' b_2' c_3 \gtrsim b_1 b_2 a_3 \,,$$

*then*

$$a \gtrsim c.$$

The distributive cancellation condition can be elucidated by the following diagram (numbers in the matrices are labels, not ranks).

| | $a_3$ | | | | $c_3$ | |
|---|---|---|---|---|---|---|
| | $a_2$ | $b_2$ | | | $c_2$ | $b_2'$ |
| $a_1$ | 1 | 2 | | $c_1$ | 1' | 3' |
| $b_1$ | 3 | 4 | , | $b_1'$ | 2' | 4' | , |

If $2 \gtrsim 2'$ and $3 \gtrsim 3'$, then $1' > 1$ and $4' \gtrsim 4$ cannot both hold. The reason is that a diagonal (say, 2, 3) of a $2 \times 2$ matrix (the $a_3$ matrix) includes the effects of all four levels of the $2 \times 2$ matrix, multiplied by the $a_3$-effect. Thus, if one diagonal (say, 2, 3) dominates one diagonal of another $2 \times 2$ matrix (say, 2', 3'), then the reverse domination cannot hold for the other pair of diagonals.

Algebraically, the first two inequalities of the distributive cancellation hypothesis imply, given a distributive representation, $\phi_1 , \phi_2 , \phi_3 ,$

$$[\phi_1(a_1) + \phi_2(b_2)] \, \phi_3(a_3) \geqslant [\phi_1(b_1') + \phi_2(c_2)] \, \phi_3(c_3),$$
$$[\phi_1(b_1) + \phi_2(a_2)] \, \phi_3(a_3) \geqslant [\phi_1(c_1) + \phi_2(b_2')] \, \phi_3(c_3).$$

Adding these inequalities gives

$$[\phi_1(a_1) + \phi_2(a_2) + \phi_1(b_1) + \phi_2(b_2)] \, \phi_3(a_3)$$
$$\geqslant [\phi_1(c_1) + \phi_2(c_2) + \phi_1(b_1') + \phi_2(b_2')] \, \phi_3(c_3).$$

However, the third inequality of the hypothesis together with the negation of the conclusion imply the contradictory inequality since, again, the scale value of each row and column component appears exactly once, multiplied by $\phi_3(a_3)$ or $\phi_3(c_3)$ as appropriate.

We have proved that the distributive cancellation condition is necessary for a distributive representation. The same algebraic argument obviously applies to the other conditions, summarized by the qualitative statement that if a diagonal of a $2 \times 2$ matrix dominates a diagonal of another $2 \times 2$ matrix, then the reverse domination cannot hold for the other pair of diagonals.

Note that the proof makes essential use of the distributivity of ordinary multiplication over addition in Re.

Distributive cancellation is also a necessary condition for $x_1 + x_2 + x_3$, and therefore cannot be used to reject additivity. However, it is not a necessary condition for $x_1 x_2 + x_3$, as shown by the following counterexample.

$$x_3 = 0 \qquad\qquad x_3 = 1$$

| $x_1$ \ $x_2$ | 3 | 2 |
|---|---|---|
| 2 | 6 | 4 |
| 1 | 3 | 2 |

| $x_1$ \ $x_2$ | 3 | 1 |
|---|---|---|
| 2 | 7 | 3 |
| 1 | 4 | 2 |

The entries in the $2 \times 2$ matrices are calculated dual distributively, e.g., $7 = 2 \cdot 3 + 1$. Since the entries in the secondary diagonals are matched, and one pair of entries in the main diagonal is matched, distributive cancellation would require that the other pair of entries in the primary diagonal be matched, contrary to fact.

If we put $a_3 = c_3$, $b_i = b_i'$, $i = 1, 2$, then the third hypothesis of distributive cancellation becomes trivial— $b_1 b_2 a_3 \gtrsim b_1 b_2 a_3$ —and the condition reduces to double cancellation for the induced ordering on $A_1 \times A_2$. Thus, we need not assume this double cancellation explicitly in axiomatizing the distributive model.

The intuitive meaning of distributive cancellation is fairly clear: The sum of two row and two column contributions, $\phi_1(a_1) + \phi_2(a_2) + \phi_1(b_1) + \phi_2(b_2)$, can be grouped in either of two ways, as $a_1 a_2$, $b_1 b_2$ or as $a_1 b_2$, $b_1 a_2$ (i.e., along either diagonal of the $2 \times 2$ matrix); there is no interaction, in the sense that either grouping is equivalent. If the dual-distributive rule held, the sum of $\phi_1(a_1) \phi_2(a_2) + \phi_1(b_1) \phi_2(b_2)$ could be very different from the sum yielded by the other grouping, $\phi_1(a_1) \phi_2(b_2) + \phi_1(b_1) \phi_2(a_2)$. The sense in which multiplication produces interaction is clear.

The dual-distributive cancellation condition, to which we next turn, is much more complicated.

DEFINITION 6. *Suppose that $\gtrsim$ is a weak order on $A = A_1 \times A_2 \times A_3$ and that $A_1$ and $A_2$ are sign dependent (not necessarily properly) on one another. We say that* dual-distributive cancellation *holds iff, for all $a, b, c, d,$ and $e \in A$, with $e_2 \notin A_2^0$, whenever*

$$a_1 c_2 c_3 \sim c_1 d_2 b_3,$$
$$d_1 a_2 a_3 \sim b_1 e_2 d_3,$$
$$d_1 c_2 d_3 \sim e_1 d_2 a_3,$$
$$c_1 e_2 e_3 \sim e_1 b_2 e_3,$$

*and*

$$a_1 e_2 e_3 \sim d_1 b_2 e_3 ,$$

*then*

$$a_1 a_2 b_3 \sim b_1 b_2 c_3 .$$

We remark that this complicated property is necessary for $x_1 + x_2 + x_3$ as well as for $x_1 x_2 + x_3$, but is not necessary for $(x_1 + x_2) x_3$. To show the necessity for $x_1 x_2 + x_3$, we translate the above equivalences into equations in the $\phi_i$. For clarity, we drop the notation $\phi_i(\ )$ from the following equations, writing them as though $a_1$, etc. were already real numbers in the dual-distributive representation.

The second and third equivalences yield the equations

$$d_1 a_2 + a_3 = b_1 e_2 + d_3 ,$$
$$d_1 c_2 + d_3 = e_1 d_2 + a_3 .$$

Adding these and canceling $a_3 + d_3$ we have

$$d_1(a_2 + c_2) = b_1 e_2 + e_1 d_2 .$$

Multiplying by $b_2$ gives

$$(d_1 b_2)(a_2 + c_2) = (b_1 b_2) e_2 + (e_1 b_2) d_2 .$$

From the fourth and fifth equivalences, we have

$$c_1 e_2 = e_1 b_2 ,$$
$$d_1 b_2 = a_1 e_2 .$$

Substituting these two in the previous equation gives

$$(a_1 e_2)(a_2 + c_2) = (b_1 b_2) e_2 + (c_1 e_2) d_2 .$$

Since $e_2 \neq 0$ ($e_2$ is not in $A_2{}^0$), we cancel $e_2$, obtaining

$$a_1(a_2 + c_2) = b_1 b_2 + c_1 d_2 .$$

From the first equivalence, we see that

$$a_1 c_2 = c_1 d_2 + b_3 - c_3 .$$

Substituting this for $a_1 c_2$ in the previous equation and canceling $c_1 d_2$ from both sides yields

$$a_1 a_2 + b_3 = b_1 b_2 + c_3 ,$$

proving $a_1 a_2 b_3 \sim b_1 b_2 c_3$ as required.

This condition does not have much intuitive appeal, no matter how long one contemplates it. Its derivation and its role are clarified in Section 7.5.2.

### □ 7.4.5 Diagnosis for Simple Polynomials in Three Variables

The application of the previous three sections to three-factor ordering can be summarized by a flowsheet for diagnostic tests. This is given in Figure 1. One first tests sign dependence for each of the three pairs. If that fails, no simple polynomial is feasible. Otherwise, there are two cases, represented by the left and right branches of the flowsheet. On the left, if independence holds for all pairs, one passes to tests of joint independence. If each of the three pairs of factors is independent of the third factor, additivity holds. If only one pair is independent of the third factor, the distributive cancellation axiom

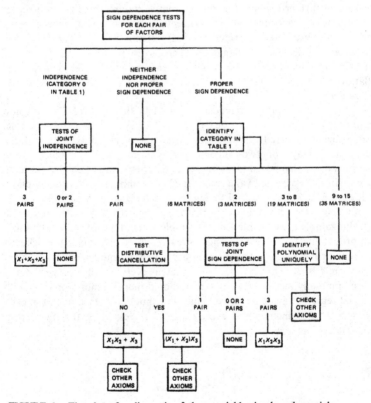

**FIGURE 1.** Flowsheet for diagnosis of three-variable simple polynomials.

must be tested to separate the distributive and dual-distributive models. None of the three-variable simple polynomials implies two independent pairs of factors in general; however, this case occurs with the distributive or dual-distributive representation if the range of one of the variables is highly restricted.

The right branch covers the case of proper sign dependence. Categories 3–15 of Table 1 yield either a unique polynomial (for which other axioms must then be verified) or no compatible polynomial. Category 2, which is compatible with the multiplicative and distributive rules, necessitates joint sign-dependence tests to distinguish them. Finally, category 1, compatible with the dual-distributive or distributive rules (the latter is unlikely—see footnote, Table 1), requires testing of distributive cancellation, as in the case of one jointly independent pair from the left branch of the flowsheet. Any sequence leading to the distributive or dual-distributive models requires checking of further axioms (for the distributive model, there are two routes in the flowsheet, and different axioms remain to be checked, depending on which route was taken), including dual-distributive cancellation if one reaches a dual-distributive diagnosis. There is no purely logical reason for not using the dual-distributive condition, rather than the distributive cancellation one, in the flowsheet for diagnosis, but testing dual distributivity is not very appealing.

A practical example of diagnosis is found in the work of Coombs and Huang (1970), cited in Example 6 (Section 7.3). As noted in the example, their subjects ordered the riskiness (not the desirability) of multiple independent plays of two-outcome even-chance gambles. The ordered objects can be characterized as triples $abc$, where $a$ in $A$ is the expected regret, $b$ in $B$ the expected value, and $c$ in $C$ the number of independent plays. They studied a $3 \times 3$ factorial design in $A \times B$, at each of two levels in $C$, namely $c = 1$ (one play) and $c = 5$ (five independent plays). If $a = b = 0$, then the gamble must be the "zero" gamble where the subject would win 0 with probability $\frac{1}{2}$, otherwise losing 0. Clearly, this should make the riskiness invariant as $c$ varies. That is, $C$ is properly sign dependent on $A$ and on $B$ (hence, on $A \times B$). On the other hand, there is every reason to expect $A$ and $B$ to be mutually independent; riskiness should increase monotonically with expected regret, at every expected value, and should decrease monotonically as expected value increases, at every expected regret. Thus, the sign-dependence matrix has form

|   | $A$ | $B$ | $C$ |
|---|-----|-----|-----|
| $A$ | — | 0 | ? |
| $B$ | 0 | — | ? |
| $C$ | 1 | 1 | — . |

We leave the sign dependence of $A$ and $B$ on $C$ open, since the idea of zero independent plays (which is probably the only source of possible violation of independence) is of doubtful interest. At any rate, the rest of the matrix already narrows down the category to either numbers 3, 7, or 11 of Table 1; thus, the distributive polynomial is the only likely one, *a priori*.

The $A \times B$ matrix tested by Coombs and Huang did not, in fact, include either $a = 0$ or $b = 0$, so no empirical evidence of proper sign dependence was obtained. Rather, pairwise independence was strongly supported. The 16 subjects (out of 28 tested) who were most consistent on replications and who showed little or no violation of transitivity in ordering riskiness also produced no more than one apparent ordering reversal over all six tests of independence ($A$ of $B$, etc.). Following the left side of the flowsheet, joint independence was next tested. Fourteen of the 16 subjects just mentioned showed no more than one reversal in the test that $A \times B$ is independent of $C$ (i.e., not more than one reversal occurred between two orderings of the nine elements in the $A \times B$ matrix—Kendall $\tau \geqslant .94$). The numbers of subjects meeting this same criterion for the other two independence tests ($A \times C$ of $B$ and $B \times C$ of $A$) were seven and 12, respectively. Thus, there was a clear indication that the joint-independence pattern favored the distributive rule. Finally, they also checked distributive cancellation, which was practically never violated by the 16 subjects just mentioned. Thus, the *a priori* diagnosis on the basis of sign dependence was quite well supported.

## 7.5  SUFFICIENT CONDITIONS FOR THREE-VARIABLE SIMPLE POLYNOMIALS

The following theorems depend on the assumption of unrestricted solvability, except where a solution involves "division by zero." We have already used such conditions in the hypotheses of Theorems 2 and 3. A general definition for three factors, suitable for all the simple polynomials, runs as follows.

DEFINITION 7.   *Suppose that $\gtrsim$ is a weak order on $A = A_1 \times A_2 \times A_3$ and that each factor is sign dependent on the other two. We say that* unrestricted solvability *holds iff, for all $a, b \in A$ for which*

$$(b_i, b_j) \notin (A_i \times A_j)^0, \quad i, j = 1, 2, 3, \ i \neq j,$$

*there exists $c \in A$ such that*

$$a \sim b_1 b_2 c_3 \sim b_1 c_2 b_3 \sim c_1 b_2 b_3 .$$

Of course, unrestricted solvability is a very strong condition. It would be desirable to replace it by a restricted solvability condition similar to that used in Chapter 6. We are rather sure that this can be done, but the problems involved seem quite difficult. Undoubtedly, further necessary conditions would have to be added to the list of axioms, just as independence was added in the two-factor theory of Chapter 6 when unrestricted solvability gave way to the restricted version.

### 7.5.1 Representation and Uniqueness Theorems

We have now introduced and discussed all the conditions needed in order to formulate representation and uniqueness theorems for the multiplicative, distributive, and dual-distributive representations. (The ever-present Archimedean condition is simply taken over from Chapter 6.) These theorems follow. They are proved in Section 7.6.

THEOREM 4. *Suppose that $\succsim$ is a binary relation on $A = A_1 \times A_2 \times A_3$ for which the following axioms are satisfied:*

1. *$\succsim$ is a weak order.*
2. *Each pair of factors is sign dependent on the third factor (Definition 3).*
3. *Every strictly bounded standard sequence in one factor (Definition 6.4; the other pair of factors acts as the second component in that definition) is finite.*
4. *Unrestricted solvability holds (Definition 7).*

*Then there exist real-valued functions $\phi_i$ on $A_i$, $i = 1, 2, 3$ such that, for all $a, b \in A$,*

$$a \succsim b \quad iff \quad \phi_1(a_1)\,\phi_2(a_2)\,\phi_3(a_3) \geq \phi_1(b_1)\,\phi_2(b_2)\,\phi_3(b_3).$$

*Moreover, real-valued functions satisfying this property are unique up to the transformations*

$$\phi_i(a_i) \to \begin{cases} \alpha_i[\phi_i(a_i)]^\beta & if \quad \phi_i(a_i) \geq 0, \\ -\alpha_i[-\phi_i(a_i)]^\beta & if \quad \phi_i(a_i) \leq 0, \end{cases}$$

*where the $\alpha_i$ and $\beta$ are real numbers, $\beta > 0$, and $\alpha_1\alpha_2\alpha_3 > 0$.*

THEOREM 5. *Suppose that $\succsim$ is a binary relation on $A = A_1 \times A_2 \times A_3$ for which the following axioms are satisfied:*

1. *$\succsim$ is a weak order.*
2. *$A_1 \times A_2$ and $A_3$ are mutually sign dependent (Definition 3).*

3.  $\langle A_1 \times A_2, A_3, \sim \rangle$ *satisfies the Thomsen condition of Definition* 4.

4.  *Distributive cancellation holds* (*Definition* 5).

5.  *For any induced ordering on* $A_1 \times A_2$, *every strictly bounded standard sequence in one component* (*Definition* 6.4) *is finite.*

6.  *Unrestricted solvability holds* (*Definition* 7).

7.  $(A_1 \times A_2)^0$ *and* $A_3{}^0$ *are nonempty* (*see Definition* 3 *and the notational convention following it.*)

*Then there exist real-valued functions* $\phi_i$ *on* $A_i$, $i = 1, 2, 3$, *such that, for all* $a, b \in A$,

$$a \gtrsim b \qquad iff \qquad [\phi_1(a_1) + \phi_2(a_2)]\, \phi_3(a_3) \geq [\phi_1(b_1) + \phi_2(b_2)]\, \phi_3(b_3).$$

*Moreover, real-valued functions satisfying this property are unique up to the transformations*

$$\phi_1 \rightarrow \alpha\phi_1 + \beta,$$
$$\phi_2 \rightarrow \alpha\phi_2 - \beta,$$
$$\phi_3 \rightarrow \gamma\phi_3,$$

*where* $\alpha, \beta, \gamma$ *are real numbers and* $\alpha\gamma > 0$.

**THEOREM 6**  *Suppose that* $\gtrsim$ *is a binary relation on* $A = A_1 \times A_2 \times A_3$ *for which the following axioms are satisfied:*

1.  $\gtrsim$ *is a weak order.*

2.  $A_1 \times A_2$ *and* $A_3$ *are mutually independent and* $A_1, A_2$ *are mutually sign dependent* (*Definition* 3).

3.  $\langle A_1 \times A_2, A_3, \sim \rangle$ *satisfies the Thomsen condition of Definition* 6.3.

4.  *Dual-distributive cancellation holds* (*Definition* 6).

5.  *Regarding* $\langle A_1 \times A_2, A_3, \gtrsim \rangle$ *as a two-component structure, each component has the property that every strictly bounded standard sequence is finite* (*Definition* 6.4).

6.  *Unrestricted solvability holds* (*Definition* 7).

7.  $A_1{}^0$ *and* $A_2{}^0$ *are nonempty* (*see Definition* 3 *and the notational convention following it*).

*Then there exist real-valued functions* $\phi_i$ *on* $A_i$, $i = 1, 2, 3$, *such that, for all* $a, b \in A$,

$$a \gtrsim b \qquad iff \qquad \phi_1(a_1)\, \phi_2(a_2) + \phi_3(a_3) \geq \phi_1(b_1)\, \phi_2(b_2) + \phi_3(b_3).$$

*Moreover, real-valued functions satisfying this property are unique up to the transformations*

$$\phi_1 \rightarrow \alpha_1 \phi_1 ,$$
$$\phi_2 \rightarrow \alpha_2 \phi_2 ,$$
$$\phi_3 \rightarrow (\alpha_1 \alpha_2) \phi_3 + \beta,$$

*where $\alpha_1$ , $\alpha_2$ , $\beta$ are real numbers and $\alpha_1 \alpha_2 > 0$.*

Note that the statements of Theorems 5 and 6 are closely parallel; we shall see that their proofs are, also.

The purely additive and multiplicative combination rules lead to interval-scale or power transformations for uniqueness, with a common unit or exponent, respectively. When addition and multiplication are combined, as in Theorems 5 and 6, the multiplying variables are true ratio scales, with independent units. Thus, $\phi_1 + \phi_2$ and $\phi_3$ are ratio scales in Theorem 5, whereas $\phi_1$ and $\phi_2$ are ratio scales in Theorem 6.

### 7.5.2 Heuristic Proofs

In proving Theorem 4, we rely on the closely related additive representation, $x_1 + x_2 + x_3$, treated in Theorem 6.13. The first problem is to designate $A_1^+, A_2^+, A_3^+$, etc. so that the sign properties of Theorem 3 can be demonstrated. Next, from the hypotheses of Theorem 4, one shows that $\langle A_1^+ \times A_2^+ \times A_3^+, \gtrsim \rangle$ is a three-component system of additive conjoint measurement, so that Theorem 6.13 yields scales $\psi_i$ on $A_i^+$. Defining $\phi_i = \exp \psi_i$ yields a multiplicative representation, which we extend over other octants such as $A_1^+ \times A_2^+ \times A_3^-$. Some technical difficulties are involved, but there are no really new ideas. Actual construction of scales $\phi_i$ on $A_i$ involves constructing standard sequences in $A_i^+$ and $A_i^-$, exactly as in the case of additive measurement. The generalization to four-variable models of form $x_1 x_2 x_3 x_4$ is obvious; also, the three multiplications in $(x_1 + x_2) x_3 x_4$ would be handled similarly.

The methods used for Theorems 5 and 6 are quite different. The solutions for the additive and multiplicative cases can be used to construct some scales on the factors, but this fails to solve the problem. For example, the two-variable multiplicative case can be applied to $A_1 \times A_2$ and $A_3$ in the case of $(x_1 + x_2) x_3$ ; in fact, Theorem 2, which yields the needed sign properties in two variables, is important for Theorem 5. Similarly, the two-variable additive case applies to $A_1$ and $A_2$ in Theorem 5; and for the polynomial $x_1 x_2 + x_3$, the two-variable additive and multiplicative solutions also enter. However, construction of scales by use of the two-variable partial decompositions does not lead directly to a three-variable solution. For example, the multiplicative scales over $A_1 \times A_2$ and $A_3$, obtained under the hypotheses of Theorem 5, are unique only up to power transforms; but there is at most one "correct" exponent, for which the scale over $A_1 \times A_2$ decomposes additively. Similarly, the additive scales over $A_1 \times A_2$ and $A_3$, obtained under the hypotheses of Theorem 6, are unique only up to linear transformations, but there can be only one "correct" zero for which the scale over $A_1 \times A_2$ decomposes multiplicatively.

To understand better the difficulty of using scales constructed by additivity or by multiplicativity alone, recall that there are many definitions of "addition" of real numbers that exhibit the properties of the usual definition; e.g., if we define, for real numbers $x, y$,

$$x \oplus y = x + y - \beta,$$

where $\beta$ is constant, then this has all the usual properties of addition, provided that we let $\beta$ play the role of 0; i.e., $x \oplus \beta = x + \beta - \beta = x$.

Similarly, multiplication can be defined alternatively for nonnegative numbers by

$$x * y = (xy)^\alpha, \qquad \text{where} \quad \alpha > 0.$$

Here, nothing plays the role of 1, but the other properties of $*$ are similar to those of multiplication. However, these definitions do not permit addition and multiplication to interact properly. In fact, we have

$$(x \oplus y) * z = (x + y - \beta)^\alpha z^\alpha,$$
$$(x * z) \oplus (y * z) = x^\alpha z^\alpha + y^\alpha z^\alpha - \beta,$$

which are not equal for all $x, y, z \geqslant 0$ unless $\beta = 0$ and $\alpha = 1$. Thus, the usual definitions are the only acceptable ones when addition and multiplication are considered together.

The method of treating addition and multiplication together is based on an idea introduced by Krantz (1964), who showed that a concatenation operation can be introduced directly into a system of additive conjoint measurement with unrestricted solvability (Definition 6.5) as indicated in Section 6.2.5. For the case of $(x_1 + x_2) x_3$, we define a concatenation operation $\oplus$ in the system $A_1 \times A_2 \times \{a_3\}$, for some suitably fixed $a_3$. We take care, in so doing, that the zero for $\oplus$ is in $(A_1 \times A_2)^0$, so it will behave as a zero with respect to multiplication as well. Similarly, we define $*$ in $(A_1 \times A_2) \times A_3$, where $A_1 \times A_2$ is considered as one factor. Since we have the correct definition of $\oplus$, using the multiplicative zero, we can hope that the resulting system $\langle A, \oplus, * \rangle$ behaves like a system of addition and multiplication of real numbers; in particular, that the distributive law

$$(x \oplus y) * z = (x * z) \oplus (y * z)$$

is satisfied. The condition needed to prove that the distributive law holds is precisely the distributive cancellation property (Definition 5).

What we actually show is that, under the hypotheses of Theorem 5, if $(a_1^0, a_2^0)$ is a pair chosen in $(A_1 \times A_2)^0$, $a^1$ is an element of $A$ for which $(a_1^1, a_2^1) \notin (A_1 \times A_2)^0$, and $a_3^1 \notin A_3^0$, then we can define $\oplus$ and $*$ in $A/\sim$ such that $\langle A/\sim, \gtrsim, \oplus, * \rangle$ is an Archimedean ordered ring and, for all $\mathbf{b} \in A/\sim$,

$$\mathbf{b} = (\mathbf{b}_1\mathbf{a}_2^0\mathbf{a}_3^1 \oplus \mathbf{a}_1^0\mathbf{b}_2\mathbf{a}_3^1) * (\mathbf{a}_1^1\mathbf{a}_2^1\mathbf{b}_3). \tag{1}$$

Now a ring isomorphism $\phi$ into $\langle \text{Re}, \geqslant, +, \cdot \rangle$ (Theorem 2.6) yields

$$\phi(\mathbf{b}) = [\phi(\mathbf{b}_1\mathbf{a}_2^0\mathbf{a}_3^1) + \phi(\mathbf{a}_1^0\mathbf{b}_2\mathbf{a}_3^1)] \cdot \phi(\mathbf{a}_1^1\mathbf{a}_2^1\mathbf{b}_3).$$

Letting

$$\phi_1(b_1) = \phi(\mathbf{b}_1\mathbf{a}_2^0\mathbf{a}_3^1),$$
$$\phi_2(b_2) = \phi(\mathbf{a}_1^0\mathbf{b}_2\mathbf{a}_3^1),$$
$$\phi_3(b_3) = \phi(\mathbf{a}_1^1\mathbf{a}_2^1\mathbf{b}_3),$$

we obtain that $[\phi_1(b_1) + \phi_2(b_2)] \phi_3(b_3) = \phi(\mathbf{b})$ is order preserving on $A/\sim$, as required for Theorem 5. The permissible transformations come not from transformations of $\phi$ (which by Theorem 2.6 is unique) but from the arbitrary choices of $a_1^0$, $a_2^0$, and $a^1$.

Similarly, under the hypotheses of Theorem 6, we define different operations $\oplus$, $*$ on $A/\sim$, satisfying

$$\mathbf{b} = [(\mathbf{b}_1\mathbf{a}_2^1\mathbf{a}_3^0) * (\mathbf{a}_1^1\mathbf{b}_2\mathbf{a}_3^0)] \oplus \mathbf{a}_1^0\mathbf{a}_2^0\mathbf{b}_3 , \tag{2}$$

where $a^0$, $a_1^1$, $a_2^1$ are suitably chosen, and the ring isomorphism now gives the dual-distributive representation.

In each case, the pivotal step is the proof that the system $\langle A/\sim, \oplus, * \rangle$ satisfies distributive laws and, hence, is a ring. This is where the distributive and dual-distributive cancellation laws are used. We can define $\oplus$ and $*$ appropriately in terms of the desired decomposition, and then we derive the needed cancellation law from the requirement that $*$ is distributive over $\oplus$. We illustrate this by deriving the distributive cancellation law (Definition 5).

The distributive law for $\oplus$, $*$, in the case of the definitions of $\oplus$, $*$ for Theorem 5, can be written

$$(\mathbf{b}_1\mathbf{a}_2^0\mathbf{a}_3^1 \oplus \mathbf{a}_1^0\mathbf{b}_2\mathbf{a}_3^1) * (\mathbf{a}_1^1\mathbf{a}_2^1\mathbf{b}_3)$$
$$= [(\mathbf{b}_1\mathbf{a}_2^0\mathbf{a}_3^1) * (\mathbf{a}_1^1\mathbf{a}_2^1\mathbf{b}_3)] \oplus [(\mathbf{a}_1^0\mathbf{b}_2\mathbf{a}_3^1) * (\mathbf{a}_1^1\mathbf{a}_2^1\mathbf{b}_3)].$$

By defining $*$, as $(\mathbf{b}_1\mathbf{b}_2\mathbf{a}_3^1) * (\mathbf{a}_1^1\mathbf{a}_2^1\mathbf{b}_3) = \mathbf{b}_1\mathbf{b}_2\mathbf{b}_3$ , the right side reduces to

$$\mathbf{b}_1\mathbf{a}_2^0\mathbf{b}_3 \oplus \mathbf{a}_1^0\mathbf{b}_2\mathbf{b}_3 , \tag{3}$$

whereas the left side, by Equation (1), is $\mathbf{b_1 b_2 b_3}$. To calculate expression (3), we use solvability to obtain $c_1$, $c_2$ such that

$$b_1 a_2{}^0 b_3 \sim a_1{}^0 c_2 a_3{}^1,$$

(4)

$$a_1{}^0 b_2 b_3 \sim c_1 a_2{}^0 a_3{}^1.$$

By defining $\oplus$ as $\mathbf{c_1 a_2{}^0 a_3{}^1} \oplus \mathbf{a_1{}^0 c_2 a_3{}^1} = \mathbf{c_1 c_2 a_3{}^1}$, expression (3) reduces to $\mathbf{c_1 c_2 a_3{}^1}$, i.e., the distributivity of $*$ over $\oplus$ holds if and only if, given the hypotheses (4), the desired conclusion

$$b_1 b_2 b_3 \sim c_1 c_2 a_3{}^1$$

(5)

obtains. Thus, we could have taken as distributive cancellation the condition that Equation (4) implies Equation (5), with the side condition that $(a_1{}^0, a_2{}^0) \in (A_1 \times A_2)^0$. The actual condition of Definition 5 is somewhat stronger than the simple statement that Equation (4) implies Equation (5), from which it evolved as we tried to see what was the essential qualitative law involved. The dual-distributive cancellation condition was derived by an analogous process.

This completes the heuristic proofs of Theorems 5 and 6. Our final remark is that these proofs are based on the Corollary of Theorem 2.6, i.e., that an Archimedean ordered ring is isomorphic to a unique subring of real numbers. The construction of the asserted isomorphism is based on construction of multiples, using the addition operation just as in extensive measurement. Thus, the construction of measurement scales $\phi_i$ on $A_i$, for the simple polynomials discussed here, is based on the construction of additive standard sequences, much as in the case of additive conjoint measurement. The main difference is that the zero points of such sequences are not arbitrary here, but depend on the multiplicative 0 intrinsically defined by sign dependence.

The independence, sign dependence, and cancellation conditions derived for a given polynomial model are all necessary, and they can be tested even when the factors do not include natural multiplicative zeros. However, the problems of finding sufficient conditions and methods for constructing scales in these cases remain to be solved.

### 7.5.3 Generalization to Four or More Variables

We can now outline a general schema for developing a sufficient set of conditions for any given simple polynomial. We assume all the obvious independence, sign dependence, solvability, and Archimedean conditions, and in addition, we introduce three classes of cancellation conditions:

(i) Thomsen conditions which are needed to permit construction of a binary operation corresponding to each addition or multiplication that enters into the polynomial;

(ii) Cancellation conditions which guarantee that two different concatenations, both corresponding to addition in the polynomial or both corresponding to multiplication, do in fact coincide;

(iii) A cancellation condition which guarantees that the distributive law holds for multiplication over addition.

It is this schema for generating necessary cancellation conditions which, together with solvability, etc., yield a set of sufficient conditions, which makes the analysis of simple polynomials so tractable.

In the case of three-variable polynomials, we have only one addition and one multiplication in the definition of our ring, so cancellation laws of type (ii) are not needed. Also, of course, some of the cancellation laws may follow from others or from laws of sign dependence; thus, the system of axioms generated by the above schema may be far from independent, much less optimal in simplicity. For Theorems 5 and 6, we have simplified the sufficient systems as much as we could after deriving them.

To illustrate these principles further, we sketch the axioms needed for the four-variable polynomial, $x_1 x_2 + x_3 x_4$.

Clearly, we want to assume that the pairs $A_1$, $A_2$ and $A_3$, $A_4$ are both mutually sign dependent and that $A_1 \times A_2$ and $A_3 \times A_4$ are mutually independent. Also, Thomsen conditions are needed for

$$\langle A_1 \times A_2, A_3 \times A_4, \gtrsim \rangle,$$

and for the induced orders on $A_1 \times A_2$ and $A_3 \times A_4$.

For the type (ii) cancellation law, we remark that multiplication can be defined on $A/\sim$ in two different ways ($a_i^0 \in A_i^0$, $a_i^1 \notin A_i^0$, $i = 1, 2, 3, 4$)

$$b_1 a_2^1 a_3^0 a_4^0 *_{12} a_1^1 b_2 a_3^0 a_4^0 = b_1 b_2 a_3^0 a_4^0,$$

or

$$a_1^0 a_2^0 b_3 a_4^1 *_{34} a_1^0 a_2^0 a_3^1 b_4 = a_1^0 a_2^0 b_3 b_4.$$

For the two multiplications to be the same, the units must be the same, i.e.,

$$a_1^1 a_2^1 a_3^0 a_4^0 \sim a_1^0 a_2^0 a_3^1 a_4^1.$$

This leads to the cancellation-type condition:

*For all* $a_i{}^0 \in A_i{}^0$, $a_i{}^1 \in A_i - A_i{}^0$, $i = 1, 2, 3, 4$ *if*

$$b_1 a_2{}^1 a_3{}^0 a_4{}^0 \sim a_1{}^0 a_2{}^0 b_3 a_4{}^1,$$
$$a_1{}^1 b_2 a_3{}^0 a_4{}^0 \sim a_1{}^0 a_2{}^0 a_3{}^1 b_4,$$
$$a_1{}^1 a_2{}^1 a_3{}^0 a_4{}^0 \sim a_1{}^0 a_2{}^0 a_3{}^1 a_4{}^1,$$

*then*

$$b_1 b_2 a_3{}^0 a_4{}^0 \sim a_1{}^0 a_2{}^0 b_3 b_4.$$

This condition is clearly necessary for a representation of form $x_1 x_2 + x_3 x_4$, and it is a key element in a set of sufficient conditions since it is used to show $*_{12} = *_{34}$.

A more general sort of cancellation condition, which is also necessary and which accomplishes the same purpose, can be found by replacing the elements $a_i{}^0 \in A_i{}^0$ by more general elements $c_i \in A_i$, which satisfy the requirements

$$c_1 c_2 \sim c_1 b_2 \sim c_1 a_2{}^1 \quad \text{and} \quad c_3 c_4 \sim b_3 c_4 \sim a_3{}^1 c_4,$$

in any induced orderings on $A_1 \times A_2$, $A_3 \times A_4$. Thus, we obtain the condition:

*For all* $a_2{}^1 \in A_2 - A_2{}^0$, $a_3{}^1 \in A_3 - A_3{}^0$, *if*

$$b_1 a_2{}^1 c_3 c_4 \sim c_1 c_2 b_3 a_4{}^1,$$
$$a_1{}^1 b_2 c_3 c_4 \sim c_1 c_2 a_3{}^1 b_4,$$
$$a_1{}^1 a_2{}^1 c_3 c_4 \sim c_1 c_2 a_3{}^1 a_4{}^1,$$
$$c_1 c_2 \sim c_1 b_2 \sim c_1 a_2{}^1,$$
$$c_3 c_4 \sim b_3 c_4 \sim a_3{}^1 c_4,$$

*then*

$$b_1 b_2 c_3 c_4 \sim c_1 c_2 b_3 b_4.$$

To show that this condition is necessary, we drop the symbols $\phi_i(\ )$ and treat $a_i$, etc., as numbers.

The last two equivalences imply that

$$c_1 = 0 \quad \text{or} \quad c_2 = b_2 = a_2{}^1;$$
$$c_4 = 0 \quad \text{or} \quad c_3 = b_3 = a_3{}^1.$$

If $c_1 = 0 = c_4$, the numerical forms of the first two equivalences can be multiplied together, and $a_1{}^1 a_2{}^1 = a_3{}^1 a_4{}^1$ can be canceled, giving the desired result. If $c_1$ and $c_4$ are both not equal to 0, the numerical forms of the first and third equivalences can be subtracted, yielding

$$(b_1 - a_1{}^1) a_2{}^1 = (b_3 - a_3{}^1) a_4{}^1 = 0,$$

so $b_1 = a_1{}^1$. Similarly, $b_4 = a_4{}^1$ and the desired conclusion is the same as the third hypothesized equivalence. Similar proofs operate where $c_1 = 0$, $c_4 \neq 0$ or vice versa.

Finally, one may obtain a type (iii) cancellation law for $x_1 x_2 + x_3 x_4$, which is used to prove distributivity of $*$ over $\oplus$; this is similar to the cancellation condition for $x_1 x_2 + x_3$ (Definition 6). See Krantz (1968) for a derivation of the distributive cancellation for $x_1 x_2 + x_3 x_4$, and for the original publication of this schema.

## 7.6 PROOFS

### 7.6.1 A Preliminary Result

LEMMA 1. *Suppose that* $A_1, ..., A_n$ *are nonempty sets and that* $\sim$ *is an equivalence relation on* $A = X_{i=1}^{n} A_i$ *satisfying unrestricted solvability.*

(i) *If* $n \geqslant 3$, *then independence of* $\sim$ (*Definition 3 applied to* $\sim$) *implies the Thomsen condition* (*Definition 6.3, Section 6.2.1*).
*By unrestricted solvability define* $\circ$ *on all of* $A/\sim$ *as follows: let* $a \in A$ *be fixed, then*

$$\mathbf{b_1 a_2 a_3} \cdots \mathbf{a_n} \circ \mathbf{a_1 b_2 a_3} \cdots \mathbf{a_n} = \mathbf{b_1 b_2 a_3} \cdots \mathbf{a_n} .$$

(ii) *If* $n \geqslant 2$ *and the Thomsen condition holds, then* $\langle A/\sim, \circ \rangle$ *is a commutative group.*

(iii) *If* $n \geqslant 3$ *and independence holds, then* $\circ$ *satisfies*

$$\mathbf{b_1} \cdots \mathbf{b_n} = \mathbf{b_1 a_2} \cdots \mathbf{a_n} \circ \mathbf{a_1 b_2 a_3} \cdots \mathbf{a_n} \circ \cdots \circ \mathbf{a_1} \cdots \mathbf{a_{n-1} b_n} .$$

*PROOF.* (i) In showing the Thomsen condition, there is no loss of generality in working with the first two components. Suppose $a_1 b_2 \sim b_1 c_2$ and $b_1 a_2 \sim c_1 b_2$. Choose any $b_3 \in A_3$, and let $b_3{}' \in A_3$ satisfy $a_1 b_2 b_3 \sim b_1 b_2 b_3{}' \sim b_1 c_2 b_3$. By independence, we have

$$a_1 a_2 b_3 \sim b_1 a_2 b_3{}',$$
$$b_1 a_2 b_3{}' \sim c_1 b_2 b_3{}',$$
$$c_1 b_2 b_3{}' \sim c_1 c_2 b_3 .$$

By transitivity, $a_1 a_2 \sim c_1 c_2$ as required. (Note: this proof is much simpler than that of Lemma 6.14 because we are assuming unrestricted solvability.)

(ii) Since $a_3 \cdots a_n$ remain fixed throughout and play no role for $n \geqslant 3$, we restrict ourselves to $n = 2$. Suppose that $b_1 a_2 \sim a_1 b_2{}'$, $a_1 b_2 \sim b_1{}' a_2$. By hypothesis, $b_1 b_2 \sim b_1{}' b_2{}'$. But this shows that $\mathbf{b_1 a_2} \circ \mathbf{a_1 b_2} = \mathbf{b_1{}' a_2} \circ \mathbf{a_1 b_2{}'}$

whenever $b_1 a_2 = a_1 b_2'$ and $a_1 b_2 = b_1' a_2$. Therefore $\circ$ is well defined and commutative.

Suppose that $b_1 b_2' \sim a_1 c_2$ and $c_1 b_2' \sim a_1 b_2$. By hypothesis, $b_1 b_2 \sim c_1 c_2$. But

$$
\begin{aligned}
b_1 b_2 &= b_1 a_2 \circ a_1 b_2 \\
&= b_1 a_2 \circ c_1 b_2' \\
&= b_1 a_2 \circ [a_1 b_2' \circ c_1 a_2], \\
c_1 c_2 &= a_1 c_2 \circ c_1 a_2 \\
&= b_1 b_2' \circ c_1 a_2 \\
&= [b_1 a_2 \circ a_1 b_2'] \circ c_1 a_2 .
\end{aligned}
$$

This shows that $\circ$ is associative.

Obviously, $a_1 a_2$ is an identity for $\circ$; and for any $b_1$, there exists $b_2$ such that $b_1 b_2 \sim a_1 a_2$, and so

$$b_1 a_2 \circ a_1 b_2 = a_1 a_2 .$$

Thus, every element has an inverse. This shows that $\circ$ is a commutative group operation.

(iii) For simplicity, we show this for $n = 3$; the general case proceeds along the same principles, using mathematical induction. For any $b \in A$ find $c_1$, $c_2$ such that

$$b_1 b_2 a_3 \sim c_1 a_2 a_3 \qquad \text{and} \qquad a_1 a_2 b_3 \sim a_1 c_2 a_3 .$$

By independence,

$$b_1 b_2 b_3 \sim c_1 a_2 b_3 \qquad \text{and} \qquad c_1 a_2 b_3 \sim c_1 c_2 a_3 .$$

Thus, by transitivity

$$
\begin{aligned}
b_1 b_2 b_3 &= c_1 c_2 a_3 \\
&= c_1 a_2 a_3 \circ a_1 c_2 a_3 \\
&= b_1 b_2 a_3 \circ a_1 a_2 b_3 \\
&= b_1 a_2 a_3 \circ a_1 b_2 a_3 \circ a_1 a_2 b_3
\end{aligned}
$$

as required.                                                                  $\Diamond$

## 7.6.2  Theorem 4  (p. 348)

*Suppose that $\succsim$ is a weak order on $A_1 \times A_2 \times A_3$ for which each pair of factors is sign dependent on the third factor and the Archimedean and unrestricted solvability conditions hold. Then $\langle A_1 \times A_2 \times A_3, \succsim \rangle$ has a multi-*

*plicative representation, which is unique up to power transformations with
common exponent.*

*PROOF.* As a first consequence of Theorem 3, we note that any nonzero
octant $(A_1^+ \times A_2^+ \times A_3^+, A_1^+ \times A_2^+ \times A_3^-,$ etc.) of $A - A^0$ satisfies
unrestricted solvability, in the sense of Definition 6.5. For example, if
$a \in A_1^+ \times A_2^+ \times A_3^+$, and $b_1 b_2 \in A_1^+ \times A_2^+$, then there exists $b_3 \in A_3$ such
that $a \sim b$; moreover, $b_3 \in A_3^+$, since otherwise $a \succ b$.

It follows from this that the conclusion of Theorem 6.13 applies to any
nonempty octant. If the octant contains only one equivalence class,
Theorem 6.13 is trivial. Otherwise, by unrestricted solvability, all three
components are essential (Definition 6.6), and by sign dependence, independence holds within the octant; hence, the hypotheses of Theorem 6.13 hold,
and its conclusion follows.

From the proof of Theorem 2, we see that either there is exactly one
nonempty octant, or all octants are nonempty. In the former case, we can
designate the nonempty octant as $A_1^+ \times A_2^+ \times A_3^+$ or as $A_1^- \times A_2^+ \times A_3^+$,
according to whether its elements are greater or less than those of $A^0$. Let
$\psi_1, \psi_2, \psi_3$ be an additive representation for the octant, and define $\phi_1, \phi_2, \phi_3$
by

$$\phi_i(a_i) = \begin{cases} \delta \exp[\delta \psi_i(a_i)], & \text{if } a_i \notin A_i^0, \\ 0, & \text{if } a_i \in A_i^0, \end{cases}$$

$i = 1, 2, 3$, where $\delta$ is defined by

$$\delta = \begin{cases} +1, & \text{if } A_1^+ \neq \varnothing, \\ -1, & \text{if } A_1^- \neq \varnothing. \end{cases}$$

Clearly, $\phi_1, \phi_2, \phi_3$ provide a multiplicative representation for $A_1 \times A_2 \times A_3$.
The uniqueness part of Theorem 4 follows from the uniqueness part
of Theorem 6.13 plus the fact that an additive decomposition $\psi_i$ is
uniquely recoverable from any multiplicative decomposition over a single
octant.

It remains to demonstrate Theorem 4 in the case where all octants are
nonempty. Consider $\sim$ on $A - A^0$. It satisfies unrestricted solvability, and
by sign dependence of $\succsim$, independence of $\sim$ holds, so Lemma 1 applies. Let
$a_i^+$ be chosen arbitrarily in $A_i^+, i = 1, 2, 3$. Choose $a_1^-$ arbitrarily in $A_1^-$,
and choose $a_2^- \in A_2^-, a_3^- \in A_3^-$ such that

$$a_1^+ a_2^- a_3^+ \sim a_1^- a_2^+ a_3^+,$$
$$a_1^- a_2^+ a_3^+ \sim a_1^+ a_2^+ a_3^-.$$

By transitivity, we have $a_1^+a_2^-a_3^+ \sim a_1^+a_2^+a_3^-$. Thus, in brief, we have

$$a_1^+a_2^- \sim a_1^-a_2^+,$$
$$a_2^+a_3^- \sim a_2^-a_3^+, \qquad (6)$$
$$a_1^+a_3^- \sim a_1^-a_3^+.$$

Let $\circ$ be the concatenation operation defined as in Lemma 1 with $\mathbf{a_1^+a_2^+a_3^+}$ as identity. Concatenating the above equivalences, we obtain

$$a_1^-a_2^-a_3^+ \sim a_1^-a_2^+a_3^- \sim a_1^+a_2^-a_3^-.$$

By solvability and Theorem 3 we can choose $b^+ \in A_1^+ \times A_2^+ \times A_3^+$ such that

$$a_1^-a_2^-a_3^+ \sim a_1^+a_2^+b_3^+ \sim a_1^+b_2^+a_3^+ \sim b_1^+a_2^+a_3^+. \qquad (7)$$

We can now define $\phi_1$, $\phi_2$, $\phi_3$ on $A_1$, $A_2$, and $A_3$. Suppose that $\psi_1, \psi_2, \psi_3$ give an additive representation for $\langle A_1^+ \times A_2^+ \times A_3^+, \gtrsim \rangle$ (Theorem 6.13), and define

$$\phi_i(a_i) = \begin{cases} \exp \psi_i(a_i), & a_i \in A_i^+, \\ 0, & a_i \in A_i^0. \end{cases} \qquad (8)$$

For any $a_1 \in A_1^-$, $a_2 \in A_2^-$, or $a_3 \in A_3^-$, we can find $b_1 \in A_1^+$, $b_2 \in A_2^+$, or $b_3 \in A_3^+$, respectively, such that

$$a_1a_2^+a_3^+ \sim b_1a_2^+a_3^-,$$
$$a_1^+a_2a_3^+ \sim a_1^-b_2a_3^+, \qquad (9)$$
$$\text{or} \qquad a_1^+a_2^+a_3 \sim a_1^+a_2^-b_3,$$

respectively. Clearly we must define $\phi_i(a_i)$, $a_i \in A_i^-$, so that

$$\phi_1(a_1)\,\phi_3(a_3^+) = \phi_1(b_1)\,\phi_3(a_3^-), \text{ etc.}$$

In particular, Equation (6) is a special case of Equation (9), with $a_i = a_i^-$, so that we must have

$$\phi_1(a_1^-)/\phi_1(a_1^+) = \phi_2(a_2^-)/\phi_2(a_2^+) = \phi_3(a_3^-)/\phi_3(a_3^+).$$

By Equation (7), the product of any two of these ratios is equal to $\phi_i(b_i^+)/\phi_i(a_i^+)$, which is defined by Equation (8) above. We denote this constant by $K^2$, so that we are forced to define, for $i = 1, 2, 3$,

$$\phi_i(a_i^-)/\phi_i(a_i^+) = -K, \qquad \text{where } K > 0.$$

Moreover, from Equation (9), we define, for $a_i \in A_i^-$,

$$\phi_i(a_i) = -K\phi_i(b_i), \qquad i = 1, 2, 3. \tag{10}$$

Equations (8) and (10) complete the definition of $\phi_1$, $\phi_2$, $\phi_3$, where in Equation (10) it is understood that $b_i$ satisfies Equation (9). They also prove the uniqueness part of Theorem 4, since the functions $\phi_i$ on $A_i^+$ are unique up to power transformations with common exponent (any set of $\phi_i$ on $A_i^+$ yields a unique set of $\psi_i$ giving an additive representation), while the extensions of $\phi_i$ to $A_i^0$, $A_i^-$ are the only possible ones yielding a multiplicative decomposition.

It remains to show, for any $a, a' \in A$, that $a \gtrsim a'$ iff

$$\phi_1(a_1)\,\phi_2(a_2)\,\phi_3(a_3) \geqslant \phi_1(a_1')\,\phi_2(a_2')\,\phi_3(a_3').$$

By Theorem 3, it suffices to show this for $a, a'$ both in $A^+$ or both in $A^-$. First consider $A^+$. For $a \in A^+$, there exists $a' \in A_1^+ \times A_2^+ \times A_3^+$ with $a \sim a'$. Since the $\phi_i$ have the desired properties on the $(+++)$ octant, it suffices to show that $\phi_1(a_1)\,\phi_2(a_2)\,\phi_3(a_3) = \phi_1(a_1')\,\phi_2(a_2')\,\phi_3(a_3')$. Take $a \in A_1^+ \times A_2^- \times A_3^-$; the proof for the other octants is similar. Choose $b_2b_3$ and $c_2c_3 \in A_2^+ \times A_3^+$ such that $a_1^+a_2a_3^+ \sim a_1^-b_2a_3^+$, $a_1^+a_2a_3 \sim a_1^+a_2^-b_3$ and $a_1b_2b_3 \sim a_1^+c_2c_3$. By definition $\phi_i(a_i) = -K\phi_i(b_i)$ $i = 2, 3$, so

$$\begin{aligned}
\phi_1(a_1)\,\phi_2(a_2)\,\phi_3(a_3) &= K^2\phi_1(a_1)\,\phi_2(b_2)\,\phi_3(b_3) \\
&= K^2\phi_1(a_1^+)\,\phi_2(c_2)\,\phi_3(c_3) \\
&= \phi_1(b_1^+)\,\phi_2(c_2)\,\phi_3(c_3), \tag{11}
\end{aligned}$$

since $K^2 = \phi_1(b_1^+)/\phi_1(a_1^+)$. Therefore it suffices to prove that $a \sim b_1^+c_2c_3$. By repeated use of Lemma 1, and substituting from the above equivalences and those of Equation (7), we have

$$\begin{aligned}
\mathbf{a} &= \mathbf{a_1a_2^+a_3^+} \circ \mathbf{a_1^+a_2a_3^+} \circ \mathbf{a_1^+a_2^+a_3} \\
&= \mathbf{a_1a_2^+a_3^+} \circ \mathbf{a_1^-b_2a_3^+} \circ \mathbf{a_1^+a_2^-b_3} \\
&= \mathbf{a_1a_2^+a_3^+} \circ [\mathbf{a_1^-a_2^+a_3^+} \circ \mathbf{a_1^+b_2a_3^+}] \circ [\mathbf{a_1^+a_2^-a_3^+} \circ \mathbf{a_1^+a_2^+b_3}] \\
&= \mathbf{a_1b_2b_3} \circ \mathbf{a_1^-a_2^-a_3^+} \\
&= \mathbf{a_1^+c_2c_3} \circ \mathbf{b_1^+a_2^+a_3^+} \\
&= \mathbf{b_1^+c_2c_3}, \qquad \text{as required.}
\end{aligned}$$

For $A^-$, it suffices to show that $\phi_1$, $\phi_2$, $\phi_3$ give a multiplicative representation for the $(---)$ octant, since then an argument like the one just provided reduces the other negative octants to that one. For $a \in A_1^- \times A_2^- \times A_3^-$, choose $b \in A_1^+ \times A_2^+ \times A_3^+$ with all three equivalences of Equation (9) holding. By definition,

$$\phi_1(a_1) \; \phi_2(a_2) \; \phi_3(a_3) = (-K)^3 \, \phi_1(b_1) \, \phi_2(b_2) \, \phi_3(b_3). \tag{12}$$

Therefore it suffices to show that $a \gtrsim a'$ iff $b' \gtrsim b$, for $a$, $a'$ in the $(---)$ octant and $b$, $b'$ defined correspondingly above. But an argument using concatenation and Lemma 1, like the one preceding, shows that

$$\mathbf{a} = \mathbf{b} \circ (\mathbf{a_1^- a_2^- a_3^-}).$$

From sign dependence, it follows that concatenation by an element of $A^-$ is order reversing, as required.                                                    $\diamondsuit$

### 7.6.3  Theorem 5  (p. 348)

*Suppose that $\gtrsim$ is a weak order on $A_1 \times A_2 \times A_3$ for which $A_1 \times A_2$ and $A_3$ are sign dependent on one another, $\langle A_1 \times A_2 , A_3 , \sim \rangle$ satisfies the Thomsen condition, the Archimedean and distributive cancellation (Definition 5) axioms hold, unrestricted solvability holds, and $(A_1 \times A_2)^0$, $A_3^0$ are nonempty. Then there exists a distributive representation for $\langle A_1 \times A_2 \times A_3 , \gtrsim \rangle$, which is unique up to transformations $\phi_1 \to \alpha\phi_1 + \beta$, $\phi_2 \to \alpha\phi_2 - \beta$, $\phi_3 \to \gamma\phi_3$, $\alpha\gamma > 0$.*

*PROOF.*  We proceed, following the outline in Section 7.5.2, by introducing operations $\oplus$, $*$ in $A/\sim$ and proving that we obtain an Archimedean ordered ring.

Specifically, choose $a_1^0 a_2^0 \in (A_1 \times A_2)^0$ and $a^1 \notin A^0$ ($a^1$ exists because sign dependence means that either $A^+$ or $A^-$ is nonempty).

Define $\oplus$ on $A/\sim$ by

$$\mathbf{b_1 a_2^0 a_3^1} \oplus \mathbf{a_1^0 b_2 a_3^1} = \mathbf{b_1 b_2 a_3^1}.$$

And define $*$ on $A/\sim$ by

$$\mathbf{b_1 b_2 a_3^1} * \mathbf{a_1^1 a_2^1 b_3} = \mathbf{b_1 b_2 b_3} = \mathbf{b}.$$

The induced equivalence relation on $A_1 \times A_2$ (fixing $a_3^1 \in A_3 - A_3^0$) satisfies solvability because unrestricted solvability holds in the original system. It also satisfies the Thomsen condition, as we shall now show using distributive cancellation. In fact, suppose that

$$b_1 c_2 a_3^1 \sim c_1 d_2 a_3^1,$$
$$c_1 b_2 a_3^1 \sim d_1 c_2 a_3^1.$$

Then since $c_1 c_2 a_3{}^1 \sim c_1 c_2 a_3{}^1$, the hypotheses of distributive cancellation are satisfied, and the conclusion is $b_1 b_2 a_3{}^1 \sim d_1 d_2 a_3{}^1$, as required for the Thomsen condition. Since the definition of $\oplus$ on $A/\!\!\sim$ is the same as that of $\circ$ (from Lemma 1) on the $a_3{}^1$-induced equivalence classes of $A_1 \times A_2$, it follows that $\oplus$ is a well-defined commutative group operation.

By unrestricted solvability, the system $\langle (A_1 \times A_2) - (A_1 \times A_2)^0, A_3 - A_3{}^0, \sim \rangle$ is also solvable. By the hypothesis of Theorem 5, this same system satisfies the Thomsen condition. Hence, $*$ is a well-defined commutative group operation[4] on $(A/\!\!\sim) - \{A^0\}$. Also, it is obvious from the definition of $*$ that if $b$ is arbitrary, then $A^0 * \mathbf{b}$ is defined and is equal to $A^0$. Thus, to show that the system $\langle A/\!\!\sim, \oplus, * \rangle$ is a ring, only the distributive law need be demonstrated.[5] Of course, the distributive cancellation condition was derived to serve just this purpose. Note that

$$\mathbf{b_1 a_2{}^0 a_3{}^1 * a_1{}^1 a_2{}^1 b_3 = b_1 a_2{}^0 b_3 ,}$$
$$\mathbf{a_1{}^0 b_2 a_3{}^1 * a_1{}^1 a_2{}^1 b_3 = a_1{}^0 b_2 b_3 .}$$

Choose $c_1$, $c_2$ such that

$$b_1 a_2{}^0 b_3 \sim a_1{}^0 c_2 a_3{}^1,$$
$$a_1{}^0 b_2 b_3 \sim c_1 a_2{}^0 a_3{}^1.$$

Since $a_1{}^0 a_2{}^0 \in (A_1 \times A_2)^0$ we have

$$a_1{}^0 a_2{}^0 b_3 \sim a_1{}^0 a_2{}^0 a_3{}^1.$$

Thus by distributive cancellation,

$$b = b_1 b_2 b_3 \sim c_1 c_2 a_3{}^1.$$

But

$$\mathbf{b = [b_1 a_2{}^0 a_3{}^1 \oplus a_1{}^0 b_2 a_3{}^1] * a_1{}^1 a_2{}^1 b_3 ,}$$

and

$$\begin{aligned}
\mathbf{c_1 c_2 a_3{}^1} &= \mathbf{a_1{}^0 c_2 a_3{}^1 \oplus c_1 a_2{}^0 a_3{}^1} \\
&= \mathbf{[b_1 a_2{}^0 b_3] \oplus [a_1{}^0 b_2 b_3]} \\
&= \mathbf{[b_1 a_2{}^0 a_3{}^1 * a_1{}^1 a_2{}^1 b_3] \oplus [a_1{}^0 b_2 a_3{}^1 * a_1{}^1 a_2{}^1 b_3].}
\end{aligned}$$

This proves the distributive law.

We next must show that (for suitable choice of $a^1$) the field $\langle A/\!\!\sim, \oplus, * \rangle$ is an ordered field relative to $\gtrsim$. The key step is to show that $A_1$ and $A_2$

---

[4] Note that $A^0$ is an equivalence class with respect to $\sim$.

[5] In fact, since $*$ is a commutative group operation on $(A/\!\!\sim) - \{A^0\}$, the system is not merely a ring, it is a field.

are mutually independent. To this end, suppose that $c_1 a_2 a_3 \gtrsim d_1 a_2 a_3$. Let $b_2 \in A_2$. If $a_3 \in A_3{}^0$, then $c_1 b_2 a_3 \gtrsim d_1 b_2 a_3$ is trivial. Otherwise, by solvability, there exists $e_1 \in A_1$ such that $d_1 a_2 a_3 \sim e_1 b_2 a_3$. We have

$$c_1 a_2 a_3 \gtrsim e_1 b_2 a_3 ,$$
$$e_1 b_2 a_3 \sim d_1 a_2 a_3 ,$$
$$e_1 a_2 a_3 \sim e_1 a_2 a_3 ,$$

so by distributive cancellation, $c_1 b_2 a_3 \gtrsim d_1 b_2 a_3$. Similarly, $A_2$ is independent of $A_1$.

From mutual independence, it follows easily that addition $\oplus$ preserves inequalities. Also, $a^1$ can be chosen $a^1 \succ a_1{}^0 a_2{}^0 a_3{}^1$. To do this, choose any $a_1 \in A_1 - A_1{}^0$, $a_3{}^1 \in A_3 - A_3{}^0$. If $a_1 a_2{}^0 a_3{}^1 \succ a_1{}^0 a_2{}^0 a_3{}^1$, then we are done, with $a^1 = a_1 a_2{}^0 a_3{}^1$. If $a_1{}^0 a_2{}^0 a_3{}^1 \succ a_1 a_2{}^0 a_3{}^1$, then choose $a_2 \in A_2$ such that $a_1 a_2 a_3{}^1 \sim a_1{}^0 a_2{}^0 a_3{}^1$. Since $A_1$ is independent of $A_2$, we have $a_1{}^0 a_2 a_3{}^1 \succ a_1{}^0 a_2{}^0 a_3{}^1$, so let $a^1 = a_1{}^0 a_2 a_3{}^1$.

With the above choice of $a^1$, multiplication $*$ by elements $\succ A^0$ preserves inequalities, since such elements, by Theorem 2 are in

$$A^+ = [(A_1 \times A_2)^+ \times A_3{}^+] \cup [(A_1 \times A_2)^- \times A_3{}^-],$$

as is $a^1$, and sign dependence does the rest. Similarly, multiplication by elements $\prec A^0$ reverses inequalities. This shows that $\langle A/\sim, \oplus, *, \gtrsim \rangle$ is an ordered field. (See Definition 2.5 and footnote 5.)

Finally, the field is Archimedean; for suppose that there were a positive element whose integral multiples were bounded; this would give rise to a bounded standard sequence in $A_1$ or $A_2$, which is impossible.

From the corollary to Theorem 2.6, we know that there is a unique isomorphism $\phi$ of $\langle A/\sim, \oplus, *, \gtrsim \rangle$ onto a subfield of real numbers. As shown in Section 7.5.2, this yields a distributive representation of $\langle A, \gtrsim \rangle$, with the $\phi_i$ defined by

$$\phi_1(b_1) = \phi(b_1 a_2{}^0 a_3{}^1),$$
$$\phi_2(b_2) = \phi(a_1{}^0 b_2 a_3{}^1), \tag{13}$$
$$\phi_3(b_3) = \phi(a_1{}^1 a_2{}^1 b_3).$$

To establish the uniqueness part of Theorem 5, suppose that $\eta_1, \eta_2, \eta_3$ are any other functions yielding a distributive representation of $\langle A, \gtrsim \rangle$. Define a function $\eta$ from $A/\sim$ to Re by

$$\eta(\mathbf{b}) = \frac{[\eta_1(b_1) + \eta_2(b_2)] \, \eta_3(b_3)}{[\eta_1(a_1{}^1) + \eta_2(a_2{}^1)] \, \eta_3(a_3{}^1)}. \tag{14}$$

This is well defined since $[\eta_1(a_1{}^1) + \eta_2(a_2{}^1)]\,\eta_3(a_3{}^1) > 0$ and because $b \sim c$ if and only if $[\eta_1(b_1) + \eta_2(b_2)]\,\eta_3(b_3) = [\eta_1(c_1) + \eta_2(c_2)]\,\eta_3(c_3)$. Also, one must have $\eta_1(a_1{}^0) + \eta_2(a_2{}^0) = 0$, and this makes it easy to verify that $\eta$ is an isomorphism. Since $\phi$ is unique, $\eta = \phi$. Thus, by Equations (13) and (14),

$$\phi_1(b_1) = \frac{\eta_1(b_1) + \eta_2(a_2{}^0)}{\eta_1(a_1{}^1) + \eta_2(a_2{}^1)},$$

$$\phi_2(b_2) = \frac{\eta_2(b_2) + \eta_1(a_1{}^0)}{\eta_1(a_1{}^1) + \eta_2(a_2{}^1)},$$

$$\phi_3(b_3) = \frac{\eta_3(b_3)}{\eta_3(a_3{}^1)},$$

i.e.,

$$\eta_1 = \alpha\phi_1 + \beta,$$
$$\eta_2 = \alpha\phi_2 - \beta,$$
$$\eta_3 = \gamma\phi_3$$

where

$$\alpha = \eta_1(a_1{}^1) + \eta_2(a_2{}^1),$$
$$\gamma = \eta_3(a_3{}^1),$$
$$\beta = \eta_1(a_1{}^0) = -\eta_2(a_2{}^0),$$

and

$$\alpha\gamma > 0. \qquad\qquad\qquad\qquad \diamondsuit$$

### 7.6.4  Theorem 6  (p. 349)

*Suppose that $\gtrsim$ is a weak order on $A_1 \times A_2 \times A_3$ for which $A_1 \times A_2$ and $A_3$ are mutually independent, $A_1$ and $A_2$ are sign dependent on one another, $\langle A_1 \times A_2, A_3, \gtrsim \rangle$ satisfies the Thomsen condition, the Archimedean and dual-distributive cancellation (Definition 6) axioms hold, unrestricted solvability holds, and $A_1{}^0, A_2{}^0$ are nonempty. Then there exists a dual-distributive representation for $\langle A_1 \times A_2 \times A_3, \gtrsim \rangle$, which is unique up to transformations*

$$\phi_1 \to \alpha_1\phi_1,\ \phi_2 \to \alpha_2\phi_2,\ \phi_3 \to (\alpha_1\alpha_2)\,\phi_3 + \beta,\ \alpha_1\alpha_2 > 0.$$

*PROOF.* In outline, the proof is exactly like that of Theorem 5 in the previous section. In particular, choose $a_i{}^0 \in A_i{}^0$, $a_i{}^1 \in A_i{}^+$, $i = 1, 2$ and $a_3{}^0 \in A_3$. Define $\oplus$ on $A/\sim$ by

$$\mathbf{b_1 b_2 a_3{}^0} \oplus \mathbf{a_1{}^0 a_2{}^0 b_3} = \mathbf{b_1 b_2 b_3} = \mathbf{b}.$$

Define $*$ on $A/\sim$ by

$$\mathbf{b_1a_2^1a_3^0} * \mathbf{a_1^1b_2a_3^0} = \mathbf{b_1b_2a_3^0}.$$

Note that we can assume $A_i^+$ or $A_i^-$ is nonempty, $i = 1, 2$, by mutual sign dependence (see Definition 3); and solvability guarantees that both $A_i^+$ and $A_i^-$ are nonempty, so $a_i^1$ can be chosen in $A_i^+$, $i = 1, 2$.[6]

Since $\langle A_1 \times A_2, A_3, \gtrsim \rangle$ satisfies unrestricted solvability and the Thomsen condition, $\langle A_1 \times A_2, A_3, \sim \rangle$ satisfies the hypotheses of Lemma 1, statement (ii). We conclude that $\oplus$ is a well-defined commutative group operation on $A/\sim$.

The induced equivalence relation on $(A_1 - A_1^0) \times (A_2 - A_2^0)$, for $a_3^0$ fixed, also satisfies solvability. We show that it satisfies the Thomsen condition. Suppose that

$$b_1c_2a_3^0 \sim c_1d_2a_3^0,$$
$$c_1b_2a_3^0 \sim d_1c_2a_3^0.$$

We then have the following five conditions holding:

$$b_1c_2a_3^0 \sim b_1c_2a_3^0,$$
$$c_1b_2a_3^0 \sim d_1c_2a_3^0,$$
$$c_1c_2a_3^0 \sim c_1c_2a_3^0,$$
$$b_1c_2a_3^0 \sim c_1d_2a_3^0,$$
$$b_1c_2a_3^0 \sim c_1d_2a_3^0.$$

These conditions satisfy the hypothesis of dual-distributive cancellation (Definition 6), and the conclusion is

$$b_1b_2a_3^0 \sim d_1d_2a_3^0.$$

This proves that $\sim$ satisfies the Thomsen condition and hence, by Lemma 1, that $*$ is a well-defined commutative group operation on

$$(A_1 - A_1^0) \times (A_2 - A_2^0) \times \{a_3^0\}/\sim.$$

This set of equivalence classes includes all of $A/\sim$ except $\mathbf{a^0}$. Clearly, multiplication $*$ by $\mathbf{a^0}$ is also well defined and yields $\mathbf{a^0}$ as the result. Thus, to prove that $\langle A/\sim, \oplus, * \rangle$ is a field, it suffices to demonstrate the distributive law.

Given $b \in A$, we can choose $c_2, d_2, e_2, \in A_2, c_3 \in A_3$, such that

$$a_1^1b_2b_3 \sim a_1^1c_2a_3^0 \sim b_1e_2a_3^0,$$
$$a_1^0a_2^0b_3 \sim a_1^1d_2a_3^0,$$
$$b_1d_2a_3^0 \sim a_1^0a_2^0c_3.$$

---

[6] This is similar to the choice of $a^1$ in $A^+$ in the proof of Theorem 5.

Then

$$[a_1{}^1b_2a_3{}^0 \oplus a_1{}^0a_2{}^0b_3] * b_1a_2{}^1a_3{}^0 = b_1c_2a_3{}^0,$$
$$[a_1{}^1b_2a_3{}^0 * b_1a_2{}^1a_3{}^0] \oplus [a_1{}^0a_2{}^0b_3 * b_1a_2{}^1a_3{}^0] = b_1b_2c_3.$$

From the above choice of $c_2, d_2, e_2, c_3$, and the fact that $a_1{}^0 \in A_1{}^0$, we obtain the following five equivalences, which are hypotheses for dual-distributive cancellation:

$$b_1d_2a_3{}^0 \sim a_1{}^0a_2{}^0c_3,$$
$$a_1{}^1b_2b_3 \sim b_1e_2a_3{}^0,$$
$$a_1{}^1d_2a_3{}^0 \sim a_1{}^0a_2{}^0b_3,$$
$$a_1{}^0e_2a_3{}^0 \sim a_1{}^0c_2a_3{}^0,$$
$$b_1e_2a_3{}^0 \sim a_1{}^1c_2a_3{}^0.$$

This yields the conclusion $b_1b_2c_3 \sim b_1c_2a_3{}^0$ as required.

The proof that $\langle A/\sim, \oplus, * \rangle$ is an Archimedean ordered field, relative to $\gtrsim$, is straightforward. The inequality-preserving or reversing properties of concatenation by $\oplus$ and $*$ arise from the independence and sign-dependence conditions, together with the fact that $a_1{}^1a_2{}^1a_3{}^0$ has been chosen (by solvability) to be $> a^0$.

The rest of Theorem 6 follows from the ring isomorphism (corollary to Theorem 2.6) in close analogy to the proof of Theorem 5.                    $\Diamond$


## EXERCISES

**1.** Let a rectangle be represented by the symbol *as*, where *a* is its area and *s* is its shape ratio (width divided by height). Suppose that the dissimilarity between rectangles is given by a formula of the following form

$$d(as, a's') = f[\phi(a, a'), \psi(s, s')].$$

Show that this formula can be analyzed as an instance of decomposability.
(7.1, 7.2)

**2.** Suppose that four rectangles $r_1, r_2, r_3, r_4$ are given by the relative height (H) and width (W) values

| W / H | 2.86 | 3.72 |
|---|---|---|
| 1.30 | $r_2$ | $r_4$ |
| 1.00 | $r_1$ | $r_3$ |

Suppose that the mean dissimilarity judgments are as follows. (These are actual data for one subject from an experiment conducted by two of the authors.)

$$d(r_1, r_2) = 11.44,$$
$$d(r_2, r_4) = 9.28,$$
$$d(r_1, r_3) = 9.25,$$
$$d(r_3, r_4) = 12.50.$$

Show that these data violate the formula given in Exercise 1 if $\psi$ is assumed to be symmetric. Are the data consistent with a height $\times$ width decomposability hypothesis? (7.1, 7.2)

**3.** Let $P(a, b)$ denote the probability that a subject chooses $a$, rather than $b$, when forced to choose one or the other. Pairwise choice probabilities, for objects $a, b,...$, in a set $A$, are said to satisfy *simple scalability* iff there is a real-valued function $u$ on $A$ and a function $F$ of two variables, strictly increasing in the first variable and strictly decreasing in the second, such that

$$P(a, b) = F[u(a), u(b)].$$

How is simple scalability related to monotone decomposability? What conditions on pairwise choice probabilities are necessary and/or sufficient for simple scalability (Tversky & Russo, 1969)? (7.2)

**4.** Suppose that $\langle A_1 \times A_2 \times A_3, \gtrsim \rangle$ is decomposable relative to both $x_1 + x_2 + x_3$ and $(x_1 + x_2) x_3$. Write down a functional equation that must hold, and show that this equation has no solution when each of the variables that enter it vary within an open interval. (7.3.2)

**5.** Define a set $S'(X)$ of polynomials in $X = \{x_1,..., x_n\}$ by letting $S'(X)$ be the smallest set of polynomials such that

(i) $x_i x_j \in S'(X)$ if $1 \leqslant i < j \leqslant n$;

(ii) if $Y_1$ and $Y_2$ are subsets of $X$ with at least two elements, such that $Y_1 \cap Y_2 = \varnothing$, $F_1 \in S'(Y_1)$ and $F_2 \in S'(Y_2)$, then $F_1 + F_2$ and $F_1 F_2$ are in $S'(X)$.

Show that $S'(\{x_1, x_2, x_3, x_4\})$ contains only two elements, up to permutations of variables. (7.3.3)

**6.** Let a rank order on $A_1 \times A_2 = \{a_1, b_1, c_1\} \times \{a_2, b_2, c_2\}$ be represented by the following matrix of ranks (1 is lowest, 9 highest).

| $A_1$ \ $A_2$ | $a_2$ | $b_2$ | $c_2$ |
|---|---|---|---|
| $a_1$ | 7 | 2 | 1 |
| $b_1$ | 6 | 5 | 4 |
| $c_1$ | 3 | 8 | 9 |

Show that $A_1$ and $A_2$ are mutually sign dependent; give the signs; and show that the conclusion of Theorem 2 is valid.     (7.4.1)

7.  Find a multiplicative representation for the ordering of Exercise 6. Show that a representation can be found in which $\phi_1(c_1) = \phi_2(c_2) = 1$, $\phi_2(b_2) = \frac{1}{2}$, and $|\phi_1(a_1)|$ is arbitrarily large.     (7.5.1)

8.  Find all the four-variable simple polynomials compatible with each of the following sign-dependence patterns.

(a)     — 0 0 0
        1 — 0 0
        0 0 — 0
        0 0 0 — ,

(b)     — 0 1 0
        0 — 1 0
        1 1 — 0
        0 0 0 — ,

(c)     — 0 1 1
        0 — 1 1
        1 1 — 0
        1 1 0 —
                                                                    (7.4.1)

9.  Let a rank order of $A_1 \times A_2 \times A_3 = \{a_1, b_1\} \times \{a_2, b_2\} \times \{a_3, b_3\}$ be given by the matrices

|       | $a_3$ | | | $b_3$ | |
|-------|-------|-------|---|-------|-------|
|       | $a_2$ | $b_2$ |   | $a_2$ | $b_2$ |
| $a_1$ | 2     | 5     |   | 1     | 6     |
| $b_1$ | 4     | 7     | , | 3     | 8     | .

Find a representation by the appropriate three-variable simple polynomial.
                                                              (7.4.1, 7.5.1)

10.  State a set of sufficient conditions for the polynomial $(x_1 + x_2)(x_3 + x_4)$.
                                                                    (7.5.3)

# *Chapter 8   Conditional Expected Utility*

---

## □ 8.1   INTRODUCTION

Unlike most theories of measurement, which may have both physical and behavioral interpretations, the theory of expected utility is devoted explicitly to the problem of making decisions when their consequences are uncertain. It is probably the most familiar example of a theory of measurement in the social sciences. In this chapter we first sketch the ideas underlying the theory and then formulate it axiomatically.

The basic entities of the theory are a set of uncertain alternatives (in common terms, gambles), and an individual's ordering of them according to his personal preferences. Two examples of gambles are:

G1.   A die is rolled once; if either the 1 or the 2 occurs, then you receive a particular book; otherwise, i.e., if either the 3, 4, 5, or 6 occurs, you lose $1.

G2.   A coin is tossed once; if head occurs you lose $3; if tail occurs you receive a particular phonograph record.

We assume that one either prefers G1 to G2, is indifferent between them, or prefers G2 to G1. Except in certain special cases (see below), we do not assume that different people necessarily agree about which is preferable; however, we do assume and attempt to formulate some form of internal consistency in each person's preferences.

Several features of these and similar gambles should be noted. First, the consequences—be they books, records, money, or whatever—are in no way inherently related to the chance events—dice, coins, or whatever—that determine which consequence is received. Each specific gamble prescribes a particular contingency between events and their consequences, but numerous other gambles can be constructed from the same chance events and consequences. Second, the set of possible consequences can include many different types of things: the gain or loss of money, the receipt of commodities or commodity bundles, the presentation of emotional stimuli of various sorts, etc. Third, the decision maker may have any information whatsoever about the chance events: he may believe or be told that a die is fair, in which case he presumably accepts that each face has a probability of $\frac{1}{6}$ of occurring; or he may be allowed to experience a few, or many, rolls of the die; or he may know nothing at all about the event, or about the mechanism that generates it. In general, therefore, we do not assume that he has access to objective probabilities, even when they can be properly defined.

If we form a number of such gambles and ask a person to choose between pairs of them, with his payoffs being determined both by his choices and by the chance events underlying the gambles, we find that most people exhibit some exchange between the desirability—the utility—of the consequences, and the chance—the probability—of receiving them. In certain especially simple choices, great consistency is then exhibited. Consider the choice between G1 and the gamble G3:

G3.   Same as G1, except that you receive the book rather than lose $1 when 3 occurs.

It is surely plausible that sane people should prefer G3 to G1. The argument is simply that no matter which event (face of the die) occurs, you are at least as well off with G3 as with G1; and for one event, namely 3, you are strictly better off (on the assumption that you prefer winning the book to losing a dollar). This general type of argument is known as the *dominance principle*, and when applicable its logic seems overriding. However, it is rarely applicable. Consider, for example, the gamble

G4.   Same as G1 except that the losses are reduced from $1 to $.50.

Again, the choice between G1 and G4 is resolved in favor of G4 by the dominance principle. But the choice between G3 and G4 cannot be argued so simply if, indeed, it can be argued at all; it seems to rest on a personal judgment that cannot be reduced to a priori principles of rationality. Although in G3 the book is more probable and the loss less probable than in G4, the magnitude of the loss in G3, if it occurs, is greater than in G4. Which way your preference tips depends in some fashion both on your evaluation of the magnitudes of the gain and the losses and on your estimate of the relative chances of gaining and losing.

A possible rule of choice for cases in which the dominance principle does not apply is to select the gamble with the higher expected value. For example, if the value of the book in G3 is $2 and the probability of each face of the die is $\frac{1}{6}$, then the expected return is $\frac{1}{2}(\$2) + \frac{1}{2}(-\$1) = \$.50$; whereas, the expected return of G4 is $\frac{1}{3}(\$2) + \frac{2}{3}(-\$.50) = \$.33$. Thus, by this rule one should choose G3 over G4. There are, however, flaws in this proposal. For one, many consequences cannot easily be assigned money values: even with goods and services, price and value are quite distinct, and health and love are literally priceless. Second, even for money gambles with a simple structure, strict adherence to a rule of maximizing expected return leads to decisions which most people regard as unreasonable. For example, tossing a coin for any stake, be it $10,000, $10, or $1, has an expected return of zero, yet most people distinctly prefer the less risky gambles of $1 and $10, and they regard their preference as eminently reasonable. They cannot afford to lose $10,000. The disutility of losing $10,000 is vastly greater than the utility of winning the same amount.

If we propose instead the maximization of subjective expected utility (SEU) as a rule of choice, these objections are greatly weakened. In such a model, each consequence, monetary or not, is assigned a numerical utility that represents its value to the decision maker, and each event is assigned a personal probability representing the decision maker's opinion about the likelihood of that event. A gamble with $m$ mutually exclusive and exhaustive events, having (nonnegative) subjective probabilities $P_1, ..., P_m$, where $\sum_{i=1}^{m} P_i = 1$, and with associated consequences, having utilities $u_1, ..., u_m$, respectively, has $SEU = \sum_{i=1}^{m} u_i P_i$. The decision maker following this rule chooses, among several gambles offered, the one that maximizes SEU.

A still more general and plausible theory for choices made under conditions of uncertainty is that individuals behave *as if* they maximize SEU. That is, a person's choices may exhibit consistencies, i.e., satisfy axioms, such that we can construct personal utility and probability functions from which his choices can be predicted by calculating SEU. These functions are inferred from preferences rather than reported directly by the decision maker. Hence such a theory does not require that individuals consciously assign numerical utilities and probabilities and calculate SEU. The question becomes: What qualitative properties are needed to establish the existence of a numerical SEU calculus? Once such axioms are found, the construction of the utilities and probabilities constitutes a conjoint measurement procedure.

Observe that the representation associated with a gamble, $\sum u_i P_i$, has a polynomial form, although it differs from the sort discussed in the last chapter in several ways. First, although the number of additive terms is finite, it is not fixed since gambles may differ in the number of their consequences. Second, although each gamble has several terms or components, only two functions are involved: $u$ over consequences and $P$ over events. Third, with respect to the components, the function is bilinear, i.e., it is linear both in $u$ and in $P$. Fourth, as in all conjoint measurement models, the representation admits an exchange between the two measures, $u$ and $P$; however, it is not a totally free exchange. Specifically, the probabilities exhibit the usual constraints: they are numbers that lie between 0 and 1, and the probabilities of events that form a partition of the universal event sum to 1. Therefore, when we alter $u_1$ to $u_1'$ by changing the consequence associated with the first event, we cannot always counteract the resulting change in SEU by finding an event whose probability $P_1'$ satisfies the equation $u_1 P_1 = u_1' P_1'$. For one thing, the solution to this equation need not lie between 0 and 1. Moreover, when $P_1$ is changed, at least one of the other probabilities must also change if the sum of all the probabilities is to remain 1. These remarks make clear that the expected utility problem, although leading to a polynomial representation, is not a special case of the class of polynomial models that were studied in the previous chapter. It is a special problem which must be explored with special techniques.

## ☐ 8.2  A FORMULATION OF THE PROBLEM

In contrast to the other theories that we have studied, opinions differ considerably about the entities that should be taken as primitive. As a result, many axiom systems, varying in form and generality, were developed and shown to yield more or less the same representation theorem. In this chapter we present what appears to us as the most satisfactory system, developed by Luce and Krantz (1971). A brief outline of the other systems and their relation to ours is given in Section 8.6.

### 8.2.1  The Primitive Notions

As in the axiomatizations of qualitative probability, we begin with an algebra $\mathscr{E}$ of subsets of a given nonempty set $X$ (see Definition 5.1). The elements of $\mathscr{E}$ represent chance events to which probabilities ultimately will be assigned. A related primitive is a subfamily $\mathscr{N}$ of (null) events; these events will have probability 0 in the representation, and they are characterized by the axioms (compare with Definition 5.8, Section 5.6.1 and Lemma 5.8,

Section 5.7.1). The third primitive is a nonempty (abstract) set $\mathscr{C}$ which is intended to represent the set of possible consequences that can be associated with events. (Some authors use "outcome" or "payoff" for our "consequence.") From the sets $\mathscr{E} - \mathscr{N}$ and $\mathscr{C}$ we construct a set of entities—actually, functions—called *conditional decisions* (and sometimes, for brevity, simply *decisions*). They generalize what has been called in the literature "gambles," "acts," or "courses of action." The decisions are "conditional" in the sense that the occurrence of a fixed nonnull event is assumed. When the decision is considered, this event becomes the universe of possibilities, and each conditional decision specifies which consequences are associated with each of the possible events in this restricted universe. More formally: for $A$ in $\mathscr{E} - \mathscr{N}$, any function $f_A$ from $A$ into $\mathscr{C}$ is a *decision conditional on* $A$. To every $x$ in $A$ the decision $f_A$ assigns a consequence, denoted $f_A(x)$, from $\mathscr{C}$. For $x$ in $X - A$, $f_A$ is not defined. To avoid ambiguity we make the conditioning event, which is the domain of the decision, explicit in the notation.

It is important to understand what is meant by a conditional decision and why it is introduced in just this way, for this concept constitutes a major point of difference between the present theory and many others. Suppose, for example, that you have decided for some reason to travel from New York to Boston on a particular day, and that you view commercial airline, train, or passenger automobile as the possible ways of travel. Each such possibility involves certain risks of certain consequences, which may vary, at least in part, from one trip to the next. One can readily formulate typical possibilities. The important point is that each choice delimits the role of chance to those events peculiar to that mode of travel, and in evaluating each conditional decision one need only take those risks into account. When thinking about the airplane trip, one considers events such as: arrival on time; a delayed arrival; cancellation of the flight; a crash, etc. One may also attempt to evaluate the probabilities of these events, conditional on choosing air transport, and the utilities of the several consequences associated with these events. Some of these consequences may be unique to the trip in question, whereas others may have to do with the nature of the event. Similarly, when evaluating each of the other modes of travel, one restricts one's attention to the conditioning event, namely, that mode of travel. Observe that if one elects one decision, certain events simply may not occur: flying precludes driving.

A second example, of a type that could be run as a laboratory experiment, is based on a set $X$ and an algebra $\mathscr{E}$ consisting of experimentally realizable events and their unions. For example, suppose that $X$ includes, among its elements, the six possibilities resulting from rolling a particular die. Denote this subset by $A = \{1,2,3,4,5,6\}$, and suppose $A$ is in $\mathscr{E}$. Let $\mathscr{C}$ be any set of consequences that has, among its elements, a particular book $b$ and the loss of \$1, denoted by $-\$1$. Then the following function is a decision conditional on $A$.

$$f_A(x) = \begin{cases} b, & \text{if } x = 1, 2, \\ -\$1, & \text{if } x = 3, 4, 5, 6. \end{cases} \tag{1}$$

Obviously, $f_A$ is the gamble G1 described earlier. Similarly, if the event of tossing a given coin, $B = \{H, T\}$, is in $\mathscr{E}$, and if the phonograph record $r$ is also in $\mathscr{C}$, then the gamble G2 is given by

$$g_B(x) = \begin{cases} -\$3, & \text{if } x = H, \\ r, & \text{if } x = T. \end{cases} \tag{2}$$

A typical experimental trial then involves offering the subject two conditional decisions, such as G1 and G2 [Equations (1) and (2)], and asking the subject which he prefers. When he has decided, the die is rolled or the coin tossed—in accord with his choice—and the event that occurs determines which consequence is received, as prescribed by the conditional decision.

If $\mathscr{D}$ denotes the set of conditional decisions used in the experiment and if all pairs of them are presented, then the subject's choices induce a relation $\gtrsim$ on $\mathscr{D}$. We want to axiomatize this relation and to represent it numerically.

From gambles based upon coins, dice, and the like, others of a more complex nature can be composed. For example, from the two gambles just described we may construct

$$h_{A \cup B}(x) = \begin{cases} b, & \text{if } x = 1, 2, \\ -\$1, & \text{if } x = 3, 4, 5 \text{ or } 6, \\ -\$3, & \text{if } x = H, \\ r, & \text{if } x = T, \end{cases} \tag{3}$$

where $A \cup B$ denotes the event: either the die $A$ is rolled or the coin $B$ is tossed, but not both. The experimenter decides which is to be done. How he decides this is immaterial as far as the theory is concerned, but in practice some scheme such as the following probably would be adopted. The experimenter picks a probability distribution over the set of all events $\mathscr{E}$ such that whenever a decision conditional on $A$ is selected, the probabilities of the subevents of $A$ are determined by the chosen conditional distribution. In particular, when a decision conditional on $A \cup B$, where $A \cap B = \varnothing$, is selected, then the probabilities of $A$ and $B$ occurring are determined by whatever chance mechanisms were chosen in advance by the experimenter.

These mechanisms may or may not be familiar to the subject. Sometimes the subject is fully informed of the details; sometimes he knows nothing about the distributions governing the event; and in yet another design, we might assure him that the die $A$ and coin $B$ are fair, but tell him nothing about the conditional probability mechanism. When $A$ and $B$ are events, such as successful arrival of a flight on time or delayed arrival, then there may be no clearly specified chance mechanism. Whether $A$ or $B$ occurs depends on natural chance mechanisms, not ones created for experimental purposes. The representation theorem stated below asserts the existence of a probability $P(A \mid A \cup B)$ which represents the subject's evaluation of the relative chances for $A$ and for $B$, given $A \cup B$. Even when the chances are controlled by the experimenter, they bear no necessary relation to the subjective probability $P$.

Generalizing the construction of Equation (3) from Equations (1) and (2), if $A$ and $B$ are disjoint events and if $f_A$ and $g_B$ are two conditional decisions, then $f_A \cup g_B$ is the decision conditional on $A \cup B$ defined by

$$(f_A \cup g_B)(x) = \begin{cases} f_A(x), & \text{if } x \in A, \\ g_B(x), & \text{if } x \in B. \end{cases} \tag{4}$$

We have used the union symbol because the defined function is the set theoretic union of the given functions when they are interpreted as subsets of $X \times \mathscr{C}$.

So our ingredients are an algebra of events $\mathscr{E}$, a subfamily $\mathscr{N}$ of null events, a set of consequences $\mathscr{C}$, a set $\mathscr{D}$ of conditional decisions (i.e., functions $f_A$ on events $A$ of $\mathscr{E} - \mathscr{N}$ into $\mathscr{C}$), and a binary preference relation $\succsim$ defined over $\mathscr{D}$. We also have a defined notion of union for some pairs of decision in $\mathscr{D}$.

### 8.2.2  A Restriction on $\mathscr{D}$

So far we have left it ambiguous whether or not $\mathscr{D}$ is the set of all logically possible conditional decisions. If either $\mathscr{E} - \mathscr{N}$ or $\mathscr{C}$ or both are very large sets, then the number of possible decisions is overwhelming. Many of these we might never want to realize, and some we probably could not. For example, if $\mathscr{C}$ includes, as one of its elements, the death of the decision maker, then a decision conditional on a particular automobile trip surely has his death in its range. This is a realistic decision and we can compare it with other decisions—we do so every day. However, few of us would seriously consider decisions that have the death of the decision maker contingent on the roll of a die. It seems sensible, therefore, to limit ourselves to a proper subset of all logically conceivable conditional decisions.

An arbitrary subset $\mathscr{D}$ will not do; it must have some degree of internal structure if we are to obtain the desired representation. We require that $\mathscr{D}$ be closed in two senses: first, if $f_A$ and $g_B$ are in $\mathscr{D}$ and if $A \cap B = \varnothing$, then $f_A \cup g_B$ shall be in $\mathscr{D}$; and second, if $A$ and $B$ are nonnull events, $B \subset A$, and $f_A$ is in $\mathscr{D}$, then the restriction of $f_A$ to $B$ shall also be in $\mathscr{D}$. Note that the first closure condition implies that the union of disjoint nonnull events is again nonnull (otherwise $f_A \cup g_B \in \mathscr{D}$ would be impossible). These closure requirements are formulated below as Axiom 1 of Definition 1.

### 8.2.3 The Desired Representation Theorem

We are now ready to sketch the representation that we want to prove. We wish to develop a list of axioms for $\langle X, \mathscr{E}, \mathscr{N}, \mathscr{C}, \mathscr{D}, \gtrsim \rangle$ that are sufficient to establish that a finitely additive probability measure $P$ exists on $\mathscr{E}$ and that a real-valued function $u$ exists on $\mathscr{D}$ such that for all $f_A$, $g_B \in \mathscr{D}$:

(i)   $R \in \mathscr{N}$ iff $P(R) = 0$;

(ii)  $f_A \gtrsim g_B$ iff $u(f_A) \geqslant u(g_B)$;

(iii) if $A \cap B = \varnothing$, then

$$u(f_A \cup g_B) = u(f_A)\, P(A \mid A \cup B) + u(g_B)\, P(B \mid A \cup B).$$

As stated, this representation is different from the one that we suggested earlier when discussing gambles. First, the utilities are assigned to decisions, not to consequences, i.e., $u$ is a function with domain $\mathscr{D}$, not $\mathscr{C}$. The reason for this is that conditional decisions, unlike gambles, do not necessarily have finite ranges; therefore, we cannot expect to establish a general representation in which the utility of a decision is expressed as a finite weighted sum of utilities over consequences. Conditions under which such a representation is possible are discussed in Section 8.4.1. Second, to the extent that property (iii) embodies the expected-utility principle, it does so in a conditional, rather than an unconditional, form. The reason that conditional probabilities arise is simply the fact that decisions are themselves conditional; it would be unreasonable to expect unconditional probabilities in the theory unless all decisions were defined over the whole of the universal set $X$.

To obtain the above representation, we search for plausible axioms that are sufficient to establish it. We divide the discussion into two phases, first looking for properties that are necessary when the representation is true, and then looking for plausible nonnecessary (structural) ones that provide the added structure which is required for the proof.

### 8.2.4 Necessary Conditions

As always, the representation forces $\gtrsim$ to be a weak ordering. This is Axiom 2 of Definition 1.

Next, let the events $A$ and $B$ be disjoint. If $f_A$ and $g_B$ are indifferent decisions, we show that their union is also indifferent to them, i.e., a mix of two equivalent decisions is equivalent to each of them. By property (ii), $u(f_A) = u(g_B)$, and so by property (iii) and the additivity of $P$,

$$u(f_A \cup g_B) = u(f_A) P(A \mid A \cup B) + u(g_B) P(B \mid A \cup B)$$
$$= u(f_A)[P(A \mid A \cup B) + P(B \mid A \cup B)]$$
$$= u(f_A).$$

Thus, by (ii), $f_A \cup g_B \sim f_A \sim g_B$. This property is formulated as Axiom 3 of Definition 1.

The next property concerns two decisions conditional on the same event. It says, in essence, that if we combine each of these decisions with the same disjoint decision or if we remove from each the same subdecision, we do not alter the direction of the preference. This generalizes to decisions the dominance principle mentioned earlier; it is formally the same as independence in Chapter 6. It is called the extended sure-thing principle. Specifically, suppose that $A$ and $B$ are disjoint and that $f_A^{(1)}$, $f_A^{(2)}$, and $g_B$ are decisions in $\mathscr{D}$. Observe that by part (iii),

$$u(f_A^{(1)} \cup g_B) - u(f_A^{(2)} \cup g_B) = u(f_A^{(1)}) P(A \mid A \cup B) + u(g_B) P(B \mid A \cup B)$$

$$- u(f_A^{(2)}) P(A \mid A \cup B) - u(g_B) P(B \mid A \cup B)$$

$$= [u(f_A^{(1)}) - u(f_A^{(2)})] P(A \mid A \cup B).$$

Since $P(A \mid A \cup B) > 0$, it follows from part (ii) that $f_A^{(1)} \gtrsim f_A^{(2)}$ if and only if $f_A^{(1)} \cup g_B \gtrsim f_A^{(2)} \cup g_B$. This we include as Axiom 4 of Definition 1.

The next two necessary properties appear more complicated. Utility intervals between decisions, both of which are conditional on a fixed event $A$, can be partially ordered in two apparently different ways, but the expected utility property forces the orderings to coincide. One way is by comparisons with equivalent decisions on another event $B$. For example, if $f_A^{(1)} \sim g_B^{(1)}$, $f_A^{(2)} \sim g_B^{(2)}$, $f_A^{(3)} \sim g_B^{(3)}$, and $f_A^{(4)} \sim g_B^{(4)}$, then the interval between $f_A^{(1)}$ and $f_A^{(2)}$ is larger than that between $f_A^{(3)}$ and $f_A^{(4)}$ if and only if the corresponding interval between $g_B^{(1)}$ and $g_B^{(2)}$ is larger than that between $g_B^{(3)}$ and $g_B^{(4)}$. Hence, an ordering of the $g_B$ intervals induces an ordering on the matching $f_A$ intervals. The second way that intervals can be ordered is by using the additive decomposition of property (iii). Suppose, for example, that $A \cap B = \varnothing$ and that

$$f_A^{(1)} \cup g_B^{(2)} \gtrsim f_A^{(2)} \cup g_B^{(1)},$$

$$f_A^{(4)} \cup g_B^{(1)} \gtrsim f_A^{(3)} \cup g_B^{(2)}.$$

Then by (iii),

$$[u(f_A^{(1)}) - u(f_A^{(2)})] \, P(A \mid A \cup B) \geqslant [u(g_B^{(1)}) - u(g_B^{(2)})] \, P(B \mid A \cup B)$$

and

$$[u(g_B^{(1)}) - u(g_B^{(2)})] \, P(B \mid A \cup B) \geqslant [u(f_A^{(3)}) - u(f_A^{(4)})] \, P(A \mid A \cup B).$$

It follows that

$$u(f_A^{(1)}) - u(f_A^{(2)}) \geqslant u(f_A^{(3)}) - u(f_A^{(4)}).$$

Note that this ordering is essentially the same as that defined for differences on one factor in additive conjoint measurement (see p. 259). We require that these two ways of ordering utility intervals be compatible (Axiom 5 of Definition 1), as we did in Section 6.11.2 for identical components.

The property just introduced adumbrates the use of additive conjoint measurement to obtain utility functions, and so leads us to expect an Archimedean property as another necessary axiom. This is indeed required (Axiom 6 of Definition 1).

Our plan, then, is to treat decisions conditional on $A$ and $B$, where $A \cap B = \varnothing$, as additive factors for decisions conditional on $A \cup B$. From this standpoint, Axiom 4 is merely the usual independence condition of additive conjoint measurement. The main complication is that identical components arise because the factor sets are directly comparable, e.g., $f_A \gtrsim g_B$. This necessitates a compatibility axiom (Axiom 5). The treatment is further complicated by the presence of nonessential dimensions (null events) and by the existence of multiple factorial structures resulting from different partitions of the same event.

The next condition, Axiom 7, we list as necessary although, as we shall see, its second part is not quite necessary. First, suppose that event $R$ is null and that $S \subset R$, then by part (i) of the desired representation and by the additivity of probability, $0 \leqslant P(S) \leqslant P(R) = 0$, and so $S$ is also null. It follows that if $A$ is not null, then for any $R$, $A \cup R$ is not null.

Part (ii) of the condition considers a nonnull event $A$, a decision $f_{A \cup R}$, and its restriction to $A$, $f_A$. We shall postulate that $R$ is null if and only if for all nonnull $A$, such that $A \cap R = \varnothing$, $f_{A \cup R} \sim f_A$. In other words, if for all nonnull $A$ the assignment of consequences to $R$ does not affect the utility of a decision on $A \cup R$, then $R$ is viewed as having no chance of occurring, and vice-versa. Arguing from the representation, consider those $A$ for which $-(A \cup R) = B$ is nonnull. Choose $g_B$ in $\mathscr{D}$. Then by parts (i) and (ii) of the representation,

$$
\begin{aligned}
u(f_{A \cup R} \cup g_B) &= u(f_{A \cup R}) \, P(A \cup R \mid X) + u(g_B) \, P(B \mid X) \\
&= u(f_{A \cup R}) \, P(A \mid A \cup B) + u(g_B) \, P(B \mid A \cup B),
\end{aligned}
$$

where we have used

$$P(A \cup R \mid X) = P(A \cup R \mid A \cup B \cup R)$$
$$= [P(A) + 0]/[P(A \cup B) + 0]$$
$$= P(A \mid A \cup B).$$

Similarly,

$$u(f_{A \cup R} \cup g_B) = u(f_R \cup f_A \cup g_B)$$
$$= u(f_A) P(A \mid X) + u(f_R \cup g_B) P(R \cup B \mid X)$$
$$= u(f_A) P(A \mid A \cup B) + u(f_R \cup g_B) P(B \mid A \cup B).$$

Subtracting,

$$[u(f_{A \cup R}) - u(f_A)] P(A \mid A \cup B) = [u(f_R \cup g_B) - u(g_B)] P(B \mid A \cup B).$$

Since the left side is independent of the choice of $g_B$, if there are several such choices possible (this will be assured by later axioms), then $u(f_{A \cup R}) = u(f_A)$. So, by part (ii) of the representation, $f_{A \cup R} \sim f_A$. The argument is obviously not valid when $B = -(A \cup R)$ is null.

### 8.2.5 Nonnecessary Conditions

The necessary conditions just deduced are by no means sufficient to establish the desired representation. The main difficulty is that, despite the closure properties imposed, $\mathscr{D}$ need not be a very rich set of conditional decisions. We must, therefore, impose further requirements that are designed to insure that $\mathscr{D}$ is adequately dense in order to carry out certain constructions.

The first structural condition merely guarantees that there are at least three pairwise disjoint nonnull events and at least two nonequivalent decisions [Axiom 8, (i) and (ii)]. This is hardly restrictive. The requirement of three events permits the use of three-factor additivity, so that all the needed cancellation conditions are deducible from Axiom 4 (see Sections 6.11, 6.12).

The second condition formulates two distinct notions of solvability. Part (i) of Axiom 9 says that if $g_B$ is in $\mathscr{D}$ and if $A$ is a nonnull event, then we can find a decision $f_A$ in $\mathscr{D}$ that is indifferent to $g_B$. This means that the range of possible utility values of conditional decisions is independent of the events on which they are conditional: any utility that can be generated on one nonnull event can be generated on any other nonnull event. Part (ii) is analogous to restricted solvability in additive conjoint measurement. Suppose that $A \cap B = \varnothing$ and that $A$ and $B$ are not null. If

$$h_A^{(1)} \cup g_B \gtrsim f_{A \cup B} \gtrsim h_A^{(2)} \cup g_B,$$

then we assume that there exists $h_A$ in $\mathscr{D}$ such that

$$h_A \cup g_B \sim f_{A \cup B} \,.$$

## 8.2.6 The Axiom System and Representation Theorem

DEFINITION 1. *Suppose that $\mathscr{E}$ is an algebra of subsets of a set $X$ (Definition 5.1), $\mathscr{N}$ is a subset of $\mathscr{E}$, $\mathscr{C}$ is a set, $\mathscr{D}$ is a set of functions whose domains are elements of $\mathscr{E} - \mathscr{N}$ and whose images are subsets of $\mathscr{C}$, and $\gtrsim$ is a binary relation on $\mathscr{D}$. Then $\langle X, \mathscr{E}, \mathscr{N}, \mathscr{C}, \mathscr{D}, \gtrsim \rangle$ is a* conditional decision structure *iff for all $A, B \in \mathscr{E} - \mathscr{N}$, $R, S \in \mathscr{E}$, and all $f_A, f_A^{(i)}, f_{A \cup B}, g_B, g_B^{(i)}, h_A^{(i)}, k_B^{(i)} \in \mathscr{D}$, where $i \in I^+$, the following nine axioms are satisfied:*

1. *Closure:* (i) *If $A \cap B = \varnothing$, then $f_A \cup g_B \in \mathscr{D}$;*
   (ii) *If $B \subset A$, then the restriction of $f_A$ to $B$ is in $\mathscr{D}$.*
2. *Weak order:* $\gtrsim$ *is a weak ordering of $\mathscr{D}$.*
3. *Union indifference:* *If $A \cap B = \varnothing$ and $f_A \sim g_B$, then $f_A \cup g_B \sim f_A$.*
4. *Independence:* *If $A \cap B = \varnothing$, then $f_A^{(1)} \gtrsim f_A^{(2)}$ iff*

$$f_A^{(1)} \cup g_B \gtrsim f_A^{(2)} \cup g_B \,.$$

5. *Compatibility:* *If $A \cap B = \varnothing$, $f_A^{(i)} \sim g_B^{(i)}$, $i = 1, 2, 3, 4$, $f_A^{(1)} \cup k_B^{(1)} \sim f_A^{(2)} \cup k_B^{(2)}$, and $h_A^{(1)} \cup g_B^{(1)} \sim h_A^{(2)} \cup g_B^{(2)}$, then*

$$f_A^{(3)} \cup k_B^{(1)} \gtrsim f_A^{(4)} \cup k_B^{(2)} \quad \textit{iff} \quad h_A^{(1)} \cup g_B^{(3)} \gtrsim h_A^{(2)} \cup g_B^{(4)}.$$

6. *Archimedean:* *If $A \cap B = \varnothing$, $N$ is a sequence of consecutive integers, not $g_B^{(0)} \sim g_B^{(1)}$, and $f_A^{(i)} \cup g_B^{(1)} \sim f_A^{(i+1)} \cup g_B^{(0)}$ for all $i, i + 1 \in N$, then either $N$ is finite or $\{f_A^{(i)} \mid i \in N\}$ is unbounded.*
7. *Nullity:* (i) *If $R \in \mathscr{N}$ and $S \subset R$, then $S \in \mathscr{N}$;*
   (ii) *$R \in \mathscr{N}$ iff, for all $f_{A \cup R} \in \mathscr{D}$ with $A \cap R = \varnothing$, $f_{A \cup R} \sim f_A$, where $f_A$ is the restriction of $f_{A \cup R}$ to $A$.*
8. *Nontriviality:* (i) *$\mathscr{E} - \mathscr{N}$ has at least three pairwise disjoint elements;*
   (ii) *$\mathscr{D}/{\sim}$ has at least two distinct equivalence classes.*
9. *Restricted solvability:* (i) *If $A$ and $g_B$ are given, then there exists $h_A \in \mathscr{D}$ for which $h_A \sim g_B$.*
   (ii) *If $A \cap B = \varnothing$ and $h_A^{(1)} \cup g_B \gtrsim f_{A \cup B} \gtrsim h_A^{(2)} \cup g_B$, then there exists $h_A \in \mathscr{D}$ such that $h_A \cup g_B \sim f_{A \cup B}$.*

Before stating the principal theorem, let us reformulate Axioms 5 and 6 by again introducing the notion of a standard sequence. Take $A$ in $\mathscr{E} - \mathscr{N}$ and let $N$ be a sequence of consecutive integers (finite or infinite, positive or negative or both). A set of decisions $\{f_A^{(i)} \mid f_A^{(i)} \in \mathscr{D}, i \in N\}$ is a *standard sequence* if for some $B$ in $\mathscr{E} - \mathscr{N}$, $A \cap B = \varnothing$, and $g_B^{(0)}, g_B^{(1)}$ in $\mathscr{D}$ with $g_B^{(0)} \not\sim g_B^{(1)}$, then for all $i$, $i + 1$ in $N$,

$$f_A^{(i)} \cup g_B^{(1)} \sim f_A^{(i+1)} \cup g_B^{(0)}.$$

The substitute axioms are:

5′. *If $\{f_A^{(i)} \mid i \in N\}$ and $\{h_B^{(i)} \mid i \in N\}$ are any two standard sequences such that, for some $j, j + 1 \in N, f_A^{(j)} \sim h_B^{(j)}$ and $f_A^{(j+1)} \sim h_B^{(j+1)}$, then for all $i \in N$, $f_A^{(i)} \sim h_B^{(i)}$.*

6′. *Any strictly bounded standard sequence is finite.*

Axiom 6′ is simply a restatement of Axiom 6. Axiom 5′ is the same as the statement of Lemma 5 (Section 8.3.1), which is the only place in the proof of the representation theorem in which Axiom 5 is invoked. Thus, Axioms 5 and 6 may be replaced by 5′ and 6′, respectively. These alternative axioms are, perhaps, more transparent than the original ones.

THEOREM 1. *Suppose that $\langle X, \mathscr{E}, \mathscr{N}, \mathscr{C}, \mathscr{D}, \gtrsim \rangle$ is a conditional decision structure. Then there exist real-valued functions $u$ on $\mathscr{D}$ and $P$ on $\mathscr{E}$ such that $\langle X, \mathscr{E}, P \rangle$ is a finitely additive probability space (Definition 5.2), and for all $A, B \in \mathscr{E} - \mathscr{N}$, $R \in \mathscr{E}, f_A, g_B \in \mathscr{D}$,*

(i) $R \in \mathscr{N}$ *iff* $P(R) = 0$;

(ii) $f_A \gtrsim g_B$ *iff* $u(f_A) \geqslant u(g_B)$;

(iii) *if $A \cap B = \varnothing$, then*

$$u(f_A \cup g_B) = u(f_A)\, P(A \mid A \cup B) + u(g_B)\, P(B \mid A \cup B).$$

*Moreover, $P$ is unique and $u$ is unique up to a positive linear transformation.*

Our proof of Theorem 1, which is given in the next section, involves using Axioms 2, 4, 6, 8, and 9 to construct additive representations of the ordering over those decisions that are conditional on a disjoint union of at least two nonnull events. Axiom 7 then serves to eliminate null events, and Axiom 5, to show that the resulting utility function preserves the order of decisions that are conditional on distinct events.

The notion of a conditional decision structure and the representation theorem given here are due to Luce and Krantz (1971). Earlier, Fishburn (1964) proposed the general idea of formulating decision problems in terms of conditional decisions, compared it with the more familiar statistical model of Savage (1954) (see Section 8.5.4), and suggested the type of representation theorem that one would want to prove. He did not, however, arrive at axioms sufficient to prove such a theorem. Toda and Shuford (1965) discussed many examples of conditional decisions and questioned the usual utility formulation. More recently, Pfanzagl (1967a, 1968) formulated a conditional theory for gambles with two consequences. The present theorem extends Pfanzagl's result to more general conditional decisions while weakening his restrictive structural assumption.

## 8.3  PROOFS

### 8.3.1  Preliminary Lemmas

The common hypothesis of the following five lemmas is that $\langle X, \mathscr{E}, \mathscr{N}, \mathscr{C}, \mathscr{D}, \gtrsim \rangle$ is a conditional decision structure (Definition 1). Note that Axiom 5 is used only once, in Lemma 5.

LEMMA 1.  *If $R, S \in \mathscr{N}$, then $R \cup S \in \mathscr{N}$.*

PROOF.  By Axiom 7 if $A \in \mathscr{E} - \mathscr{N}$, then $A \cup R$ and $A \cup R \cup S \in \mathscr{E} - \mathscr{N}$. For any $f_{A \cup R \cup S} \in \mathscr{D}$, let $f_{A \cup R}$ and $f_A$ be the restrictions of $f_{A \cup R \cup S}$ to $A \cup R$ and $A$, respectively. By Axiom 7, $f_{A \cup R \cup S} \sim f_{A \cup R} \sim f_A$, so by Axiom 2, $f_{A \cup R \cup S} \sim f_A$. Since $f_{A \cup R \cup S}$ was arbitrary, it follows from Axiom 7(ii) that $R \cup S \in \mathscr{N}$.  ◇

LEMMA 2.  *If $A \cap B = \varnothing$, $f_A, f_A{}', g_B, g_B{}' \in \mathscr{D}$, and $g_B \sim g_B{}'$, then $f_A \cup g_B \gtrsim f_A{}' \cup g_B{}'$ iff $f_A \gtrsim f_A{}'$.*

PROOF.  By Axiom 4, $f_A \cup g_B \gtrsim f_A{}' \cup g_B$ iff $f_A \gtrsim f_A{}'$; also $g_B \sim g_B{}'$ iff $f_A{}' \cup g_B \sim f_A{}' \cup g_B{}'$; so by Axiom 2, $f_A \cup g_B \gtrsim f_A{}' \cup g_B{}'$ iff $f_A \gtrsim f_A{}'$.  ◇

For $A \in \mathscr{E} - \mathscr{N}$, let $\mathscr{D}_A = \{f_A \mid f_A \in \mathscr{D}\}$. By Axioms 9(i) and 8(ii), each $\mathscr{D}_A$ contains some $f_A, f_A{}'$ with $f_A > f_A{}'$.

LEMMA 3.  *For $n \geqslant 2$, let $A_1, ..., A_n$ be pairwise disjoint elements of $\mathscr{E} - \mathscr{N}$. Then, for $i = 1, ..., n$, there exist real-valued functions $\phi_i$ on $\mathscr{D}_{A_i}$, such that, for any $f_i, g_i \in \mathscr{D}_{A_i}$,*

$$f_1 \cup \cdots \cup f_n \gtrsim g_1 \cup \cdots \cup g_n$$

*iff*

$$\sum_{i=1}^{n} \phi_i(f_i) \geqslant \sum_{i=1}^{n} \phi_i(g_i).$$

*The $\phi_i$ are unique up to positive linear transformations $\phi_i{}' = \alpha \phi_i + \beta_i$, $\alpha > 0$.*

*PROOF.* First, suppose that $n \geqslant 3$. Define an ordering $\gtrsim'$ on $X_{i=1}^{n} \mathscr{D}_{A_i}$ by the ordering of unions of components. Then $\langle \mathscr{D}_{A_1}, ..., \mathscr{D}_{A_n}, \gtrsim' \rangle$ is an $n$-component *system of additive conjoint measurement (Definition 6.13, Section 6.11.1). Weak ordering follows from Axiom 2, independence from Axiom 4, restricted solvability from Axiom 9(ii), the Archimedean axiom from Axiom 6, and that each $\mathscr{D}_{A_i}$ is essential from Axioms 8 (ii) and 9 (i). The existence and uniqueness of $\phi_1, ..., \phi_n$ satisfying the stated representation follow from Theorem 6.13 (Section 6.11.1).

For $n = 2$, let $A_3 = -(A_1 \cup A_2)$. If $A_3 \in \mathscr{E} - \mathscr{N}$, then by the previous result an additive representation exists for $\langle \mathscr{D}_{A_1}, \mathscr{D}_{A_2}, \mathscr{D}_{A_3}, \gtrsim' \rangle$, which yields a representation for $\langle \mathscr{D}_{A_1}, \mathscr{D}_{A_2}, \gtrsim' \rangle$ as required. Uniqueness follows since the latter is a two-component system of additive conjoint measurement (Definition 6.7, Section 6.2). If $A_3 \in \mathscr{N}$, let $B_1, B_2, B_3'$ be three pairwise disjoint elements of $\mathscr{E} - \mathscr{N}$ [these exist by Axiom 8 (i)] and let $B_3 = -(B_1 \cup B_2)$. Since $B_3' \subset B_3$, by Axiom 7 (i), $B_3 \in \mathscr{E} - \mathscr{N}$. Let $A_i^{(j)} = A_i \cap B_j, i, j = 1, 2, 3$. Then $B_j = A_1^{(j)} \cup A_2^{(j)} \cup A_3^{(j)}$, and since $A_3^{(j)}$ is null and $B_j$ is not, by Lemma 1, either $A_1^{(j)}$ or $A_2^{(j)}$ is nonnull. Therefore, the nonnull $A_i^{(j)}$, $i = 1, 2$, $j = 1, 2, 3$, yield a system satisfying the hypothesis of the present lemma with $n \geqslant 3$, and hence, yield a corresponding additive representation, with functions $\phi_i^{(j)}$ on the nonempty $\mathscr{D}_{A_i^{(j)}}$. Let $\phi_i = \sum_j \phi_i^{(j)}$, where the summation is taken over $j$ such that $A_i^{(j)} \in \mathscr{E} - \mathscr{N}$. For $i = 1, 2$, $A_i$ is nonnull, so for each $i$ there exists $j$ such that $A_i^{(j)}$ is nonnull, consequently $\phi_i$ is defined. By Axiom 7 (ii), the restriction of any $f_{A_i}$ to the union of the nonnull $A_i^{(j)}$ is equivalent to $f_{A_i}$; hence, by Lemma 2, $\phi_1, \phi_2$ define an additive representation for $\langle \mathscr{D}_{A_1}, \mathscr{D}_{A_2}, \gtrsim' \rangle$. Uniqueness again follows because $\langle \mathscr{D}_{A_1}, \mathscr{D}_{A_2}, \gtrsim' \rangle$ is a two-component system of additive conjoint measurement.                                    $\Diamond$

For any $A \in \mathscr{E} - \mathscr{N}$, let $\Phi_A$ denote the family of real-valued functions over $\mathscr{D}_A$ defined as follows: $\phi \in \Phi_A$ iff there exist pairwise disjoint sets $A_1, ..., A_n \in \mathscr{E} - \mathscr{N}$, $n \geqslant 2$, and an additive representation $\phi_1, ..., \phi_n$ of $\langle \mathscr{D}_{A_1}, ..., \mathscr{D}_{A_n}, \gtrsim' \rangle$ such that, for some $r$, $1 \leqslant r \leqslant n$, $A = \bigcup_{i=1}^{r} A_i$ and $\phi = \sum_{i=1}^{r} \phi_i$. [That is, if $f_{A_i}$ is the restriction of $f_A$ to $A_i$, $i = 1, ..., r$, then $\phi(f_A) = \sum_{i=1}^{r} \phi_i(f_A)$.] In other words, $\Phi_A$ is the set of all functions on $\mathscr{D}_A$ that are involved in at least one of the additive representations that have two or more essential components.

**LEMMA 4.** *If $A \in \mathscr{E} - \mathscr{N}$, then*

(i)  *$\Phi_A$ is nonempty.*

(ii) *If $\phi, \psi \in \Phi_A$, then there exist real $\alpha, \beta$ with $\alpha > 0$ such that $\psi = \alpha\phi + \beta$.*

*PROOF.* If $-A \in \mathcal{E} - \mathcal{N}$, then $A, -A$ satisfy the hypothesis of Lemma 3, and the corresponding additive representation yields an element of $\Phi_A$. If $-A \in \mathcal{N}$, let $B_1, B_2'$ be disjoint elements of $\mathcal{E} - \mathcal{N}$, let $B_2 = -B_1$, and let $A_i = A \cap B_i$, $i = 1, 2$; then $A_i \in \mathcal{E} - \mathcal{N}$, for otherwise $B_i = A_i \cup (-A \cap B_i)$ would be null. Let $\phi_1, \phi_2$ be an additive representation of $\langle \mathcal{D}_{A_1}, \mathcal{D}_{A_2}, \succsim' \rangle$, then $\phi = \phi_1 + \phi_2$ is in $\Phi_A$.

To prove (ii), let $A_1,..., A_m$, $B_1,..., B_n$ be in $\mathcal{E} - \mathcal{N}$, with $m, n \geqslant 2$, and $A_i \cap A_j = \varnothing = B_k \cap B_l$, for all $1 \leqslant i < j \leqslant m, 1 \leqslant k < l \leqslant n$. Suppose that for some $1 \leqslant r \leqslant m, 1 \leqslant s \leqslant n$, $A = \bigcup_{i=1}^{r} A_i = \bigcup_{j=1}^{s} B_j$. Let $\phi_1,..., \phi_m$ and $\psi_1,..., \psi_n$ be the corresponding additive representations. Define $\phi = \sum_{i=1}^{r} \phi_i$, $\psi = \sum_{j=1}^{s} \psi_j$. For $i = 1,..., m$, $j = 1,..., n$, let $C_{ij} = A_i \cap B_j$, $C_{i,n+1} = A_i - (\bigcup_{j=1}^{n} B_j)$, $C_{m+1,j} = B_j - (\bigcup_{i=1}^{m} A_i)$. Clearly, the nonnull $C_{ij}$ satisfy the hypotheses of Lemma 3, and yield a corresponding additive representation $\{\theta_{ij}\}$. Let

$$\phi_i' = \sum_j \theta_{ij}, \qquad \psi_j' = \sum_i \theta_{ij},$$

where in these, and in other sums noted in this way, the summation is over those $i$ or $j$ with $C_{ij}$ nonnull. Clearly, at least one $C_{ij}$ is nonnull for each $i$ and each $j$, so $\phi_i', \psi_j'$ are all defined. By Axiom 7 (ii), the restriction of any $f_{A_i}$ (or $f_{B_j}$) to the union of the nonnull $C_{ij}$ in $A_i$ (or $B_j$) is equivalent to $f_{A_i}$ (or $f_{B_j}$). By Lemma 2, $\phi_1',..., \phi_m'$ defines an additive representation for $\langle \mathcal{D}_{A_1},..., \mathcal{D}_{A_m}, \succsim' \rangle$ and $\psi_1',..., \psi_n'$ defines an additive representation for $\langle \mathcal{D}_{B_1},..., \mathcal{D}_{B_n}, \succsim' \rangle$. Hence, for some real $\alpha, \gamma > 0$, and some $\beta_i, \delta_j$, $i = 1,..., m, j = 1,..., n$, $\phi_i = \alpha \phi_i' + \beta_i$ and $\psi_j = \gamma \psi_j' + \delta_j$. Thus,

$$\phi = \sum_{i=1}^{r} \phi_i = \alpha \sum_{i=1}^{r} \phi_i' + \sum_{i=1}^{r} \beta_i = \alpha \sum_{i=1}^{r} \sum_j \theta_{ij} + \beta, \qquad \text{where} \quad \beta = \sum_{i=1}^{r} \beta_i,$$

$$\psi = \sum_{j=1}^{s} \psi_j = \gamma \sum_{j=1}^{s} \psi_j' + \sum_{j=1}^{s} \delta_j = \gamma \sum_{j=1}^{s} \sum_i \theta_{ij} + \delta, \qquad \text{where} \quad \delta = \sum_{j=1}^{s} \delta_j.$$

Since $\bigcup_{i=1}^{r} \bigcup_{j=1}^{n} C_{ij} = A = \bigcup_{j=1}^{s} \bigcup_{i=1}^{m} C_{ij}$, exactly the same $C_{ij}$ must be nonnull for $i = 1,..., r, j = 1,..., n$ as for $j = 1,..., s, i = 1,..., m$; hence, exactly the same $\theta_{ij}$ are included in the sums for $\phi$ and $\psi$. That is, $\phi = \alpha\theta + \beta$, $\psi = \gamma\theta + \delta$, where $\theta = \sum_{i=1}^{r} \sum_j \theta_{ij} = \sum_{j=1}^{s} \sum_i \theta_{ij}$. Hence,

$$\psi = (\gamma/\alpha)\, \phi + (\delta - \beta\gamma/\alpha)$$

as required.                                                                              $\diamond$

LEMMA 5.  *Let $A, B \in \mathcal{E} - \mathcal{N}$ with $A \cap B = \varnothing$ and let $N$ be a sequence of consecutive integers. Suppose that $\{f_i \mid i \in N\}$ and $\{g_i \mid i \in N\}$ are standard* sequences in $\mathscr{D}_A$ and $\mathscr{D}_B$, respectively. If $f_j \sim g_j$, $f_{j+1} \sim g_{j+1}$ for some $j, j+1 \in N$, then $f_i \sim g_i$, for all $i \in N$.

PROOF. By definition of a standard sequence, Axiom 9(i), and Lemma 3, there is no loss in generality in supposing $f_0', f_1' \in \mathscr{D}_A$, $g_0', g_1' \in \mathscr{D}_B$ such that for $i, i+1 \in N$,

$$f_i \cup g_1' \sim f_{i+1} \cup g_0' \quad \text{and} \quad f_1' \cup g_i \sim f_0' \cup g_{i+1}.$$

We proceed by induction. Assume that $f_i \sim g_i$ and that $i, i+1 \in N$. Choose $h \in \mathscr{D}_B$ such that $f_{i+1} \sim h$. Applying Axiom 5 to $f_j \sim g_j$, $f_{j+1} \sim g_{j+1}$, $f_i \sim g_i$, $f_{i+1} \sim h$, $f_j \cup g_1' \sim f_{j+1} \cup g_0'$, $f_1' \cup g_j \sim f_0' \cup g_{j+1}$, we have $f_i \cup g_1' \gtrsim f_{i+1} \cup g_0'$ iff $f_1' \cup g_i \gtrsim f_0' \cup h$. Hence, $f_1' \cup g_i \sim f_0' \cup g_{i+1} \sim f_0' \cup h$ and so $f_{i+1} \sim h \sim g_{i+1}$. Similarly, if $i-1 \in N$, then $f_{i-1} \sim g_{i-1}$. $\diamond$

### 8.3.2 Theorem 1 (p. 381)

If $\langle X, \mathscr{E}, \mathscr{N}, \mathscr{C}, \mathscr{D}, \gtrsim \rangle$ is a conditional decision structure, then there exist $u : \mathscr{D} \to \text{Re}$ and $P : \mathscr{E} \to \text{Re}$ such that $\langle X, \mathscr{E}, P \rangle$ is a finitely additive probability space with $\mathscr{N}$ the set of null events; $u$ is order preserving; and $u$ and $P$ exhibit the conditional expected utility property. $P$ is unique and $u$ is an interval scale.

We break this proof into six parts.

1. *Construction of utility and probability functions.* Choose $f_0, f_1 \in \mathscr{D}$ with $f_1 > f_0$. For $A \in \mathscr{E} - \mathscr{N}$, choose elements of $\mathscr{D}_A$, denoted $\eta_0(A)$ and $\eta_1(A)$ such that $\eta_0(A) \sim f_0$, $\eta_1(A) \sim f_1$. Choose $u_A \in \Phi_A$ normalized so that $u_A[\eta_0(A)] = 0$, $u_A[\eta_1(A)] = 1$. This determines a unique function $u$ on $\mathscr{D}$ that coincides with $u_A$ on $\mathscr{D}_A$ for each $A \in \mathscr{E} - \mathscr{N}$.

For $A, B \in \mathscr{E} - \mathscr{N}$, $A \cap B = \varnothing$, choose an additive representation $\phi_{A,B}$, $\phi_{B,A}$ for $\langle \mathscr{D}_A, \mathscr{D}_B, \gtrsim' \rangle$, normalized so that $\phi_{A,B} + \phi_{B,A} = u_{A \cup B}$. (By Lemma 4, $\phi_{A,B} + \phi_{B,A}$ is linearly related to $u_{A \cup B}$ so this normalization is possible.) Since $\phi_{A,B}$, $\phi_{B,A}$ are, respectively, linearly related to $u_A$, $u_B$, there exist positive constants $P(A \mid A \cup B)$, $P(B \mid A \cup B)$, and real numbers $\beta_{A,B}$, $\beta_{B,A}$, such that

$$\phi_{A,B} = P(A \mid A \cup B) u_A + \beta_{A,B},$$
$$\phi_{B,A} = P(B \mid A \cup B) u_B + \beta_{B,A}.$$

We define $P$ on $\mathscr{E}$ by

$$P(A) = \begin{cases} 0, & \text{if } A \in \mathscr{N}, \\ 1, & \text{if } -A \in \mathscr{N}, \\ P(A \mid A \cup -A) & \text{if } A, -A \in \mathscr{E} - \mathscr{N}. \end{cases}$$

2. *Expected utility decomposition.* By transitivity and the choice of $\eta_0$, $\eta_1$,

$$\eta_0(A) \sim \eta_0(B) \sim \eta_0(A \cup B) \qquad \text{and} \qquad \eta_1(A) \sim \eta_1(B) \sim \eta_1(A \cup B),$$

for $A, B \in \mathscr{E} - \mathscr{N}$, $A \cap B = \varnothing$. By the above normalizations,

$$
\begin{aligned}
0 &= u_{A \cup B}[\eta_0(A \cup B)] \\
&= \phi_{A,B}[\eta_0(A \cup B)] + \phi_{B,A}[\eta_0(A \cup B)] \\
&= P(A \mid A \cup B)\, u_A[\eta_0(A)] + P(B \mid A \cup B)\, u_B[\eta_0(B)] + \beta_{A,B} + \beta_{B,A} \\
&= \beta_{A,B} + \beta_{B,A}\,; \\
1 &= u_{A \cup B}[\eta_1(A \cup B)] \\
&= \phi_{A,B}[\eta_1(A \cup B)] + \phi_{B,A}[\eta_1(A \cup B)] \\
&= P(A \mid A \cup B)\, u_A[\eta_1(A)] + P(B \mid A \cup B)\, u_B[\eta_1(B)] + \beta_{A,B} + \beta_{B,A} \\
&= P(A \mid A \cup B) + P(B \mid A \cup B).
\end{aligned}
$$

Thus, for $A, B$ nonnull and disjoint we have the decomposition,

$$u(f_A \cup g_B) = P(A \mid A \cup B)\, u(f_A) + P(B \mid A \cup B)\, u(g_B),$$

where $P(A \mid A \cup B)$ and $P(B \mid A \cup B)$ are positive numbers that sum to 1. In particular, we have

$$u[\eta_1(A) \cup \eta_0(B)] = P(A \mid A \cup B). \tag{5}$$

3. *Finite additivity for null events.* In this part of the proof, suppose that $A \in \mathscr{E}$, $R \in \mathscr{N}$, and $A \cap R = \varnothing$. If either $A \in \mathscr{N}$ or $-A \in \mathscr{N}$, then by definition of $P$, $P(A \cup R) = P(A) = P(A) + P(R)$. So we may suppose that $A, -A \in \mathscr{E} - \mathscr{N}$. Let $\eta_1(A \cup N)_A$ denote the restriction of $\eta_1(A \cup N)$ to $A$. By Axiom 7 (ii),

$$
\begin{aligned}
\eta_1(A) &\sim \eta_1(A \cup R) \sim \eta_1(A \cup R)_A\,, \\
\eta_0[-(A \cup R)] &\sim \eta_0(-A) \sim \eta_0(-A)_{-(A \cup R)}\,.
\end{aligned}
$$

By Lemma 2 and Axiom 7(ii) in the first and third steps,

$$
\begin{aligned}
\eta_1(A \cup R) \cup \eta_0[-(A \cup R)] &\sim \eta_1(A \cup R)_A \cup \eta_0[-(A \cup R)] \\
&\sim \eta_1(A) \cup \eta_0(-A)_{-(A \cup R)} \\
&\sim \eta_1(A) \cup \eta_0(-A).
\end{aligned}
$$

Since $u$ is order preserving on $\mathscr{D}_X$, by Equation (5),

$$P[A \cup R \mid (A \cup R) \cup -(A \cup R)] = P(A \mid A \cup -A),$$

whence by definition,

$$P(A \cup R) = P(A) = P(A) + P(R).$$

As a corollary we have the following lemma:

LEMMA 6. *Suppose that* $A, -A \in \mathscr{E} - \mathscr{N}, R \in \mathscr{N}, R \subset A$, *and let* $f_{A-R}$ *be the restriction of* $f_A$ *to* $A - R$. *Then*

$$u_A(f_A) = u_{A-R}(f_{A-R}).$$

*PROOF.* By the same arguments as above,

$$f_A \cup \eta_0(-A) \sim f_{A-R} \cup \eta_0(R \cup -A).$$

By the decomposition equation,

$$u_A(f_A)\,P(A) = u_{A-R}(f_{A-R})\,P(A - R).$$

Since $P(A) = P(A - R) + P(R) = P(A - R) \neq 0$, the conclusion follows. $\diamond$

4. *Conditional probability and finite additivity.* Suppose that $A, B$ are disjoint and nonnull. We first show that $P(A \mid A \cup B) = P(A)/P(A \cup B)$. Let $C = -(A \cup B)$. If $C \in \mathscr{E} - \mathscr{N}$, then by Equation (5) and the decomposition property,

$$\begin{aligned}
P(A) &= u[\eta_1(A) \cup \eta_0(B \cup C)] \\
&= u[\eta_1(A) \cup \eta_0(B) \cup \eta_0(C)] \\
&= u_{A \cup B}[\eta_1(A) \cup \eta_0(B)]\,P(A \cup B) + u_C[\eta_0(C)]\,P(C) \\
&= P(A \mid A \cup B)\,P(A \cup B).
\end{aligned}$$

If $C \in \mathscr{N}$, then by Equation (5) and Lemma 6,

$$\begin{aligned}
P(A) = P(A \cup C) &= u_X[\eta_1(A \cup C) \cup \eta_0(B)] \\
&= u_{X-C}[\eta_1(A) \cup \eta_0(B)] \\
&= P(A \mid A \cup B).
\end{aligned}$$

Since $P(A \cup B) = 1$ when $C \in \mathscr{N}$, the desired formula follows.

We can now prove finite additivity: for $A, B$ disjoint and nonnull,

$$\begin{aligned}
&P(A) + P(B) \\
&= P(A \mid A \cup B)\,P(A \cup B) + P(B \mid A \cup B)\,P(A \cup B) = P(A \cup B).
\end{aligned}$$

Since additivity has been already established for the case where either $A$ or $B$ is null, it holds in general.

388

388388 8. CONDITIONAL EXPECTED UTILITY

5. *Preservation of order on $\mathcal{D}$.* Let $A$ and $B$ be nonnull. We wish to show $f_A \gtrsim g_B$ iff $u(f_A) \geq u(g_B)$. Since $u_A$, $u_B$ preserve order, and $\mathcal{D}_A$ can be mapped onto equivalent elements of $\mathcal{D}_B$, it suffices to show $f_A \sim g_B$ implies $u(f_A) = u(g_B)$.

Assume that this holds for any disjoint sets. Let $A$, $B$ be any two sets and suppose that $f_A \sim g_B$. Find $h_{A-B}$, $h'_{A \cap B}$, $h''_{B-A}$, all equivalent to $f_A$ and to $g_B$ (except when some of $A - B$, $A \cap B$, $B - A$ are null in which case the corresponding functions are omitted). From the assumed order-preserving result for disjoint domains,

$$u(h_{A-B}) = u(h'_{A \cap B}) = u(h''_{B \cap A}) = u_1 .$$

From the expected utility decomposition (if $A - B, A \cap B \in \mathcal{E} - \mathcal{N}$), $u(f_A) = u_1$. The same result follows by Lemma 6 if $A - B$ or $A \cap B$ is null. A similar argument holds for $u(g_B)$, so $u(f_A) = u(g_B)$. This reduces the proof of the order preserving property to the case of disjoint sets. For this, we require the following lemma:

LEMMA 7. *Let $\langle F, G, \gtrsim \rangle$ be a two-component additive conjoint structure (Definition 6.7). If $f_0, f_1 \in F$, $g_0, g_1 \in G$, and $(f_1, g_0) \lesssim (f_0, g_1) < (f_1, g_1)$, then there exist a sequence of consecutive integers $N$ including $\{0,1\}$, $\{f^{(i)} \mid i \in N, f^{(i)} \in F\}$, and $g_0', g_1' \in G$ such that:*

(i) $f^{(0)} = f_0, f^{(1)} = f_1$;

(ii) $(f^{(i+1)}, g_0') \sim (f^{(i)}, g_1')$ *for all* $i, i + 1 \in N$;

(iii) *for $f \in F$, there exists $i \in N$ such that either*

$$f^{(i)} \gtrsim f \quad and \quad (f^{(i)}, g_0') \lesssim (f, g_1')$$

*or*

$$f \gtrsim f^{(i)} \quad and \quad (f, g_0') \lesssim (f^{(i)}, g_1').$$

PROOF. Let $f^{(i)} = f_i$, $i = 0, 1$, $g_1' = g_1$, and $g_0'$ be a solution to $(f_1, g_0') \sim (f_0, g_1)$. We proceed to construct $N$ inductively. Suppose that for some finite $N$ including 0 and 1, $\{f^{(i)} \mid i \in N\}$ satisfies (i) and (ii). If $N$ does not satisfy (iii) it can be enlarged in the following fashion: Let $f \in F$ be an element for which the conclusion of (iii) fails. Either $f$ is an upper or lower bound for $\{f^{(i)} \mid i \in N\}$; otherwise, $f^{(i)} \lesssim f \lesssim f^{(i+1)}$ for some $i \in N$, and $(f, g_0') \lesssim (f^{(i+1)}, g_0') \sim (f^{(i)}, g_1')$, contrary to the assumption that (iii) does not hold for $f$. First, suppose that $f$ is an upper bound. By the Archimedean axiom, there is a largest element $f^{(n)}$ in $\{f^{(i)} \mid i \in N\}$. We have $f^{(n)} \lesssim f$, and

$$(f, g_0') > (f^{(n)}, g_1') > (f^{(n)}, g_0').$$

Thus, by restricted solvability, for some $f' \in F, (f', g_0') \sim (f^{(n)}, g_1')$. Enlarge $N$ by adding $n + 1$ to it, and define $f^{(n+1)} = f'$. Similarly, if $f$ is a lower bound, enlarge $N$ from below. By construction, enlarging $N$ in this fashion preserves (i) and (ii). The process either terminates after a finite number of steps, namely, after a finite number of enlargements, $N$ satisfies (iii); otherwise, it ultimately defines an infinite standard sequence in which $N$ includes all positive integers, all negative integers, or all integers. By the earlier argument, $N$ cannot be further enlarged if and only if $N$ satisfies (iii).    ◇

Suppose that $A \cap B = \varnothing$ and $f_A \sim g_B$. We must show $u(f_A) = u(g_B)$. By symmetry, it suffices to consider the case where $f_A \sim g_B > \eta_0(A) \sim \eta_0(B)$ and $f_A \cup \eta_0(B) \lesssim \eta_0(A) \cup g_B$. By Axiom 9 (ii) choose $g_B'$ such that $f_A \cup \eta_0(B) \sim \eta_0(A) \cup g_B'$. For any $h_A, k_B$ such that

$$\eta_0(A) \sim \eta_0(B) < h_A \sim k_B \lesssim g_B' \lesssim g_B \sim f_A,$$

we can apply Lemma 7 to construct standard sequences $\{f_A^{(i)} \mid i \in M\}$, $\{g_B^{(i)} \mid i \in N\}$, such that

$$f_A^{(0)} \sim \eta_0(A) \sim \eta_0(B) \sim g_B^{(0)},$$

$$f_A^{(1)} \sim h_A \sim k_B \sim g_B^{(1)},$$

and for some $m, m' \in M, n, n' \in N,$

$$\eta_1(A) \quad \text{is within one step of} \quad f_A^{(m)},$$

$$f_A \quad \text{is within one step of} \quad f_A^{(m')},$$

$$\eta_1(B) \quad \text{is within one step of} \quad g_B^{(n)},$$

$$g_B \quad \text{is within one step of} \quad g_B^{(n')},$$

in the sense of part (iii) of Lemma 7. By Lemma 5, $f_A^{(i)} \sim g_B^{(i)}$ for all $i$; hence, we can take $m = n$ and $m' = n'$. Furthermore, by the properties of standard sequences, and by the expected utility decomposition,

$$u(f_A^{(i)}) = iu(h_A) \qquad \text{and} \qquad u(g_B^{(i)}) = iu(k_B).$$

We consider two cases. If there exist minimal $h_A, k_B$ such that $h_A \sim k_B > \eta_0(A) \sim \eta_0(B)$, then construct $f_A^{(i)}, g_B^{(i)}$ as above. By minimality, for some $m, m' \in M,$

$$\eta_1(A) \sim f_A^{(m)} \sim g_B^{(m)} \sim \eta_1(B)$$

and

$$f_A \sim f_A^{(m')} \sim g_B^{(m')} \sim g_B \, .$$

Hence,

$$u(g_B) = u(f_A) = m'u(h_A) = m'/m,$$

as required. Otherwise, as $h_A \sim k_B$ is taken arbitrarily close to $\eta_0(A) \sim \eta_0(B)$, then $m, m' \to \infty$. Since $(1/m) \leqslant u(h_A), u(k_B) \leqslant (1/(m-1))$,

$$| u(h_A) - u(k_B)| \leqslant \frac{1}{m-1} - \frac{1}{m} = \frac{1}{m(m-1)}$$

and

$$| u(f_A) - u(g_B)| \leqslant | u(f_A) - u(f_A^{(m')})| + | u(f_A^{(m')}) - u(g_B^{(m')})|$$

$$+ | u(g_B^{(m')}) - u(g_B)|$$

$$\leqslant u(h_A) + m' | u(h_A) - u(k_B)| + u(k_B)$$

$$\leqslant \frac{1}{m-1} + \frac{m'}{m(m-1)} + \frac{1}{m-1} \, .$$

By the same argument as in Theorem 2.4 (construction of a real-valued isomorphism of an Archimedean ordered local semigroup), $m'/m$ stays bounded as $m \to \infty$. From this we conclude that $u(f_A) = u(g_B)$.

This concludes the construction of utility and subjective probability and the proof of their properties.

6.  *Uniqueness.*  Let $u'$, $P'$ be other functions with the properties specified in Theorem 1. For any $A \in \mathscr{E} - \mathscr{N}$, since $u_A' \in \Phi_A$, by Lemma 4, there exist real $\alpha_A > 0$ and $\gamma_A$ such that $u_A' = \alpha_A u_A + \gamma_A$. Calculation shows that $u_A'[\eta_0(A)] = \gamma_A$ and $u_A'[\eta_1(A)] = \alpha_A + \gamma_A$, which by the order-preserving property, are independent of $A$. Thus, $u' = \alpha u + \gamma$, where $\alpha > 0$. Clearly, if $u'$ and $P'$ satisfy Theorem 1, so do $\alpha u' + \gamma$ and $P'$. Hence, for $A$, $B$ disjoint and nonnull

$$P(A \mid A \cup B) = u[\eta_1(A) \cup \eta_0(B)]$$

$$= u'[\eta_1(A)] \, P'(A \mid A \cup B) + u'[\eta_0(B)] \, P'(B \mid A \cup B)$$

$$= P'(A \mid A \cup B).$$

Thus, $P = P'$ for $A \in \mathscr{N}$ or $-A \in \mathscr{N}$, so by the properties of probability functions we have $P = P'$ everywhere. Hence, $u$ is unique up to positive linear transformations, and $P$ is unique. This completes the proof of Theorem 1.                                                              ◇

## 8.4 TOPICS IN UTILITY AND SUBJECTIVE PROBABILITY

### 8.4.1 Utility of Consequences

So far, we have not associated numerical utilities with consequences; Theorem 1 assigns utilities only to decisions. Ultimately, we will view this failure as a virtue. Meanwhile, let us examine the problem involved in assigning utility to consequences.

If $c \in \mathscr{C}$ and $A \in \mathscr{E} - \mathscr{N}$, then the function $c_A$ defined by

$$c_A(x) = c \qquad \text{for any} \quad x \in A,$$

is called a *constant decision*. In other theories of utility, the constant decisions play a crucial role in constructing a utility function over the set of consequences. In such theories, if $c \in \mathscr{C}$ and $u$ is a utility function over $\mathscr{D}$, then the value of $c$ defined by $v(c) = u(c_A)$ seems a sensible assignment, and in those theories it is. There are two reasons why this cannot always be done in our conditional theory. First, we cannot be certain that $\mathscr{D}$ includes any constant decision with $c$ as its consequence. An example of a conditional decision structure that includes no constant decisions is given in Luce and Krantz (1971). Second, even if there are constant decisions in $\mathscr{D}$, we still cannot be sure that the above definition will work, because in the general theory it may happen that $c_A \succ c_B$, $u(c_A) > u(c_B)$, and so $v(c)$ cannot be well defined. If, however, we assume that constant decisions exist and that $c_A \sim c_B$, then the above definition of $v$ on $\mathscr{C}$ has the desired properties, as shown in the next theorem.

We call a conditional decision $f_A \in \mathscr{D}$ a *gamble* if the image of $f_A$ is finite and if, for every $c$ in the image of $f_A$, the set of elements mapped into $c$,

$$f_A^{-1}(c) = \{x \mid x \in A \quad \text{and} \quad f_A(x) = c\},$$

is an event in $\mathscr{E} - \mathscr{N}$. It is not difficult to see that when $f_A$ is a gamble, the sets $f_A^{-1}$ form a partition of $A$ and that $f_A$ is the finite union of constant decisions. This definition of a gamble generalizes our earlier examples, and it accords with common usage.

**THEOREM 2.** *Let* $\langle X, \mathscr{E}, \mathscr{N}, \mathscr{C}, \mathscr{D}, \succsim \rangle$ *be a conditional decision structure and suppose that for every $c \in \mathscr{C}$:*

(i) *There is some $A \in \mathscr{E} - \mathscr{N}$ such that $c_A \in \mathscr{D}$.*

(ii) *If $A, B \in \mathscr{E} - \mathscr{N}$, and $c_A, c_B \in \mathscr{D}$, then $c_A \sim c_B$.*

*Let u and P be the functions constructed in Theorem 1. For $c \in \mathscr{C}$ and $c_A \in \mathscr{D}$, define $v(c) = u(c_A)$. Then for every gamble $f_A \in \mathscr{D}$,*

$$u(f_A) = E[v(f_A)|\, A],$$

*where E is the conditional expectation with respect to P.*

The proof of this theorem is left as Exercise 5.

Observe that since a gamble is of the form

$$f_A = \bigcup_{i=1}^{n} c_{A_i}^{(i)},$$

where $\{A_i\}$ is a finite partition of $A$, part (iii) of Theorem 1 implies that

$$u(f_A) = \sum_{i=1}^{n} v(c^{(i)}) P(A_i \mid A),$$

which is the familiar form of the expected utility property.

Theorem 2 could surely be improved by weakening its assumptions. One possibility is to retain assumption (ii) and weaken assumption (i) in such a way that we could still define a function $v$ over $\mathscr{C}$ such that for some decisions (more general than gambles) $u(f_A) = E[v(f_A)|\, A]$, where the expectation operator is an integral of some sort. In this connection, Pfanzagl (1967b) is of interest.

A second line of development is just the reverse. Retain assumption (i) and relax (ii) so that $c_A$ may not be judged indifferent to $c_B$. For example, consider a set $\mathscr{D}$ of conditional decisions, a real-valued function $v$ on $\mathscr{C}$, a probability measure $P$ on $\mathscr{E}$, and a real-valued function $w$ on $\mathscr{E} - \mathscr{N}$ with the property[1] that for $A, B$ in $\mathscr{E} - \mathscr{N}$ and $A \cap B = \varnothing$,

$$w(A \cup B) = w(A) P(A \mid A \cup B) + w(B) P(B \mid A \cup B). \qquad (6)$$

For those decisions for which the expectation exists, define

$$\begin{aligned} u(f_A) &= E\{[v(f_A) + w(A)] \mid A\} \\ &= E[v(f_A) \mid A] + w(A). \end{aligned} \qquad (7)$$

In particular, if $f_A$ is a gamble, its utility should be of this form. It is not

---

[1] If $\phi$ is any finitely additive measure on $\mathscr{E} - \mathscr{N}$, then it is easy to see that for any real number $\alpha$, and all $A$ in $\mathscr{E} - \mathscr{N}$,

$$w(A) = [\phi(A)/P(A)] + \alpha$$

has the property given in Equation (6), and that every such $w$ is of this form.

difficult to see that the ordering induced on $\mathscr{D}$ by $u$ is consistent with the necessary axioms of Definition 1. The question, then, is what additional assumptions are needed for such a representation to exist. Equation (7) is of particular interest because it admits the possibility that an event may have a utility independent of that generated by the consequences associated with it. In the most general form of the present theory, it is possible to encompass the somewhat elusive concept of utility of gambling. The representation just discussed, in which $v$ and $w$ combine additively is, perhaps, the simplest one in which $u(c_A) \neq u(c_B)$, but it is probably not the only one. No thorough study of the possibilities has yet been conducted.

It should be mentioned that Jeffrey's (1965) book is based on an idea quite similar to that expressed in Equation (6), and that special cases of Equation (7) have been suggested by Siegel (1959), Siegel, Siegel, and Andrews (1964), and Toda and Shuford (1965). In the latter three studies it was assumed that $A = \{x, y\}$ and that $w(A) = f[P(x \mid A)]$ for some function $f$. Siegel took $f$ to be $f(p) = \alpha p(1 - p)$ and interpreted the term $w(A)$ to represent a utility of variability.

### 8.4.2 Relations Between Additive and Expected Utility

Since in many economic contexts, consequences are commodity bundles, i.e., elements in $\mathscr{C} \subset \mathscr{C}_1 \times \mathscr{C}_2 \times \cdots \times \mathscr{C}_n$, we are faced with two alternative ways of measuring utility. On the one hand, we can form gambles on commodity bundles and use expected utility theory, and on the other hand we can use riskless orderings of $\mathscr{C}$ and some form of conjoint measurement theory. In a series of papers, Fishburn (1965 a, b, 1967 c, d, 1969 a) has shown for a whole variety of different expected-utility models (see Section 8.6 for a discussion of some of the alternatives) a condition that is necessary and sufficient for the resulting utility function to be additively decomposable over $\mathsf{X}_{i=1}^{n} \mathscr{C}_i$. We formulate and prove the result for a conditional decision structure in which the utility of a constant decision is of the form $u(c_A) = v(c) + w(A)$.

Suppose that $\langle X, \mathscr{E}, \mathscr{N}, \mathsf{X}_{i=1}^{n} \mathscr{C}_i, \mathscr{D}, \gtrsim \rangle$ is a conditional decision structure and that $u$ and $P$ are the two functions described in Theorem 1. We say that two gambles $f_A$ and $g_A$ on $A$ in $\mathscr{E} - \mathscr{N}$ are *P-equivalent*, denoted $f_A \sim_P g_A$, provided that each $x$ in $\mathscr{C}_i$ has the same probability of occuring in each gamble. More formally, let $\mathscr{C}(i, x)$ denote the set of elements in $\mathscr{C} = \mathsf{X}_{i=1}^{n} \mathscr{C}_i$ which have $x$ as the value of the $i$th component, i.e.,

$$\mathscr{C}(i, x) = \{c \mid c = c_1 \cdots c_i \cdots c_n \in \mathscr{C} \quad \text{and} \quad c_i = x\}.$$

Then, $\sum_{c \in \mathscr{C}(i, x)} P[f_A^{-1}(c)]$ is the (subjective) probability of receiving $x$ on the

$i$th component from gamble $f_A$. And so, $f_A \sim_P g_A$ if and only if, for all $i = 1, 2,..., n$ and all $x$ in $\mathscr{C}_i$,

$$\sum_{c \in \mathscr{C}(i,x)} P[f_A^{-1}(c)] = \sum_{c \in \mathscr{C}(i,x)} P[g_A^{-1}(c)].$$

The critical new assumption is that when $f_A \sim_P g_B$, then $u(f_A) = u(g_B)$. Put in words, if relative to $P$ of the expected utility representation, each component consequence has the same probability of arising, then the two gambles are in a sense not different and so they are judged equivalent in preference, and hence have the same utility.

To prove that this assumption implies an additive decomposition of the utility over $\mathscr{C}$, we must, first, assure ourselves that there is a utility function over $\mathscr{C}$. In addition, we need to make sure that there is some event $E$ in $\mathscr{E} - \mathscr{N}$ that can be partitioned into $n$ equiprobable subevents where $n$ is the number of components of $\mathscr{C}$. Note that this requirement is considerably weaker than Axiom 5' of Section 5.2.3. In fact, it says no more than that $\mathscr{E}$ includes one event such as the throw of a subjectively fair $n$-sided die. Finally, for this special event, we assume that all constant decisions $c_E$ are in $\mathscr{D}$.

**THEOREM 3.** *Suppose that* $\langle X, \mathscr{E}, \mathscr{N}, \mathscr{C}, \mathscr{D}, \gtrsim \rangle$ *is a conditional decision structure (Definition 1, Section 8.2.6), that* $\mathscr{C} = \mathsf{X}_{i=1}^{n} \mathscr{C}_i$, *and that $u$ and $P$ are the functions described in Theorem 1 (Section 8.2.6). Assume that,*

(i) *There exist* $v : \mathscr{C} \to \mathrm{Re}$ *and* $w : \mathscr{E} - \mathscr{N} \to \mathrm{Re}$ *such that for all gambles* $f_A \in \mathscr{D}$, $u(f_A) = E[v(f_A)|A] + w(A)$.

(ii) *There exists* $E \in \mathscr{E} - \mathscr{N}$ *and a partition* $\{E_1 ,..., E_n\}$ *of $E$ such that* $P(E_i | E) = 1/n$ *for* $i = 1, 2,..., n$.

(iii) *For all* $c \in \mathscr{C}$, $c_E \in \mathscr{D}$.

*Then, there exist functions* $v_i : \mathscr{C}_i \to \mathrm{Re}$ *such that, for all* $c = c_1 \cdots c_n \in \mathscr{C}$,

$$v(c) = \sum_{i=1}^{n} v_i(c_i)$$

*iff, for all gambles* $f_A , g_A \in \mathscr{D}$, $f_A \sim_P g_A$ *implies* $u(f_A) = u(g_A)$.

In Fishburn's papers, many variants of this theorem can be found that differ in the exact structure presupposed. In several cases, $\mathscr{C} \subsetneqq \mathsf{X}_{i=1}^{n} \mathscr{C}_i$.

From the point of view of the foundations of measurement, these results, including the one just stated, are not entirely satisfactory. We would like to have an axiomatization of $\langle X, \mathscr{E}, \mathscr{N}, \mathsf{X}_{i=1}^{n} \mathscr{C}_i , \mathscr{D}, \gtrsim \rangle$ that leads directly to a representation theorem in which the utility function over $\mathsf{X}_{i=1}^{n} \mathscr{C}_i$ is additive.

Put another way, we would like to embed the condition $f_A \sim_P g_A$ implies $u(f_A) = u(g_A)$ in axioms that are formulated in terms of the primitives only. Such a result has not been obtained yet.

In an empirical study, Tversky (1967b) investigated the relations between additive and expected utility by comparing risky and riskless preferences for the same consequences. Although both expected utility theory and the additive theory for decomposition of consequences were strongly supported by the data, the resulting utility scales failed to coincide, which suggests the existence of a particular form of utility for gambling. This study is discussed further in Section 9.4.2.

### 8.4.3  The Consistency Principle for the Utility of Money

The expected-utility property does not, by itself, specify any necessary relation between money and the utility of money, but as Pfanzagl (1959) has proved, the relation is closely prescribed if certain other, apparently plausible conditions are added. Specifically, consider simple money gambles, and suppose that the utility function is a continuous and strictly increasing function of money. Moreover, if $a_A \cup b_B$, for $a$, $b$ in Re and $A \cap B = \varnothing$, is a given money gamble, and if $C$ is any nonnull event, we suppose—as is plausible—that there exists a sum of money $c \in$ Re such that $a_A \cup b_B \sim c_C$. Pfanzagl assumed that the indifference is preserved when we augment the stakes by an amount $d$ in Re. More formally, Pfanzagl suggested the following *consistency assumption*,

$$(a + d)_A \cup (b + d)_B \sim (c + d)_C,$$

which appears plausible. The only trouble is that with the usual version of expected utility, the utility functions are highly limited, as shown in the following.

THEOREM 4.  *Suppose that $\mathscr{E}$ is an algebra of events, Re is the real continuum, and for some $A, B, C \in \mathscr{E}, A \cap B = \varnothing$, and all $a, b, c \in$ Re, $\mathscr{D}$ includes all gambles of the form $a_A \cup b_B$ and all constant decisions $c_C$. Suppose that there exists a finitely additive probability measure $P$ on $\mathscr{E}$ and three real-valued functions $u$, $v$, and $w$ on $\mathscr{D}$, Re, and $\mathscr{E}$, respectively, such that:*

(i)  $P(A), P(B), P(C) > 0$;

(ii)  *$v$ is continuous and strictly increasing*;

(iii)  *for every $a, b \in$ Re, there exists $c \in$ Re such that*

$$u(c_C) = u(a_A \cup b_B);$$

(iv)  *for* $a, b \in \mathrm{Re}$,

$$u(a_A \cup b_B) = v(a) P(A \mid A \cup B) + v(b) P(B \mid A \cup B) + w(A \cup B);$$

(v)  *for* $a, b, c \in \mathrm{Re}$, *if* $u(c_C) = u(a_A \cup b_B)$, *then for every* $d \in \mathrm{Re}$,

$$u[(c + d)_C] = u[(a + d)_A \cup (b + d)_B].$$

*Then for all* $a \in \mathrm{Re}$, *either*

$$v(a) = \alpha a + \beta,$$

*where* $\alpha > 0$, *or*

$$v(a) = \alpha \lambda^a + \beta,$$

*where either* $\alpha > 0$ *and* $\lambda > 1$ *or* $\alpha < 0$ *and* $0 < \lambda < 1$.

This formulation of the theorem is somewhat more general than Pfanzagl's, although the conclusion is the same. [He took $B = -A$ and the event $C$ was not mentioned at all, thereby making it possible to state the assumptions just in terms of $v$ with no need to mention $u$. This caused no confusion since he did not have the $w$ term in hypothesis (iv), because it did not arise in his formulation of expected utility theory.] The present proof is based on Aczél's (personal communication) modification of Pfanzagl's proof.

The linear utility function is clearly inadequate, and a careful analysis of the exponential utility function shows that it is also unsatisfactory. In particular, it approaches a finite asymptote in one direction and goes off to infinity rapidly in the other direction. Hence, when the asymptote is nearly reached, an unreasonably large increase in gain (or loss) is required to compensate for even a negligible increase in loss (or gain).

Krantz and Tversky (1965) also investigated the consistency principle. They assumed that the set of simple gambles can be partitioned into those that are acceptable $\mathcal{O}$, those that are unacceptable $\mathcal{U}$, and those that are indifferent to the status quo $\mathcal{I}$. They introduced several rather weak assumptions from which it follows that for any gamble of the form $a_A \cup b_B$ there exists a number $c$ such that for all $d$ in $\mathrm{Re}$, if $d < c$, then $(a - d)_A \cup (b - d)_B$ is in $\mathcal{O}$, and if $d > c$, then $(a - d)_A \cup (b - d)_B$ is in $\mathcal{U}$. This boundary number $c$ is called the maximum buying price of $a_A \cup b_B$, denoted $M(a_A \cup b_B)$. We now show that the existence of a maximum buying price implies that

$$M(a_A \cup b_B) + d = M[(a + d)_A \cup (b + d)_B].$$

Let $c = M(a_A \cup b_B)$, and note the following identity for every $e \in \mathrm{Re}$:

$$(a - e)_A \cup (b - e)_B \equiv [(a + d) - (e + d)]_A \cup [(b + d) - (e + d)]_B.$$

By the property just shown, this gamble is in $\mathcal{O}$ if $e < c$, which is equivalent to $e + d < c + d$, and it is in $\mathcal{U}$ if $e > c$, which is equivalent to $e + d > c + d$. Therefore,

$$M[(a + d)_A \cup (b + d)_B] = c + d,$$

as was to be shown.

If $c = M(a_A \cup b_B)$ and if we assume that $u(a_A \cup b_B) = u(c_C)$, i.e., the preference order of the gambles coincides with that of their maximum buying prices, then hypothesis (v) of Theorem 4 follows, since as we have just shown $M[(a + d)_A \cup (b + d)_B] = c + d$, so

$$u[(c + d)_C] = u[(a + d)_A \cup (b + d)_B].$$

Since the conclusion of Theorem 4 is too restrictive, while the existence of a maximum buying price is unobjectionable, we must reject the equality between the cash equivalent of a gamble [defined by hypothesis (iii) of Theorem 4] and its maximum buying price. Put differently, we must have some pairs of gambles such that the maximum buying price of the preferred gamble does not exceed that of the unpreferred gamble. Although this state of affairs is surely compatible with expected utility theory (in fact, it is a necessary consequence of any utility function that is neither linear nor exponential), it makes the application of the theory exceedingly difficult. Since the preference order now depends on one's financial position, the utility function is defined only for states of total wealth. Consequently, one can no longer speak of the utility of a particular gamble independent of the financial position of the decision maker, which may be very difficult to characterize. The heart of the problem lies in the fact that, even though a person may be indifferent between $a_A \cup b_B$ and $c_C$, possession of the former induces a preference order which does not necessarily coincide with the order induced by possession of the latter. Consequently, financial positions (which actually include many interdependent gambles), cannot be properly summarized by a single monetary index.

As was pointed out by Pfanzagl (1959), most experimental applications of the theory have implicitly accepted the consistency principle in the analysis and the interpretation of the results. Indeed, how to interpret utility functions derived from empirical data is very problematic if the consistency principle does not hold. Thus, the above analysis of the consistency principle poses a serious problem for expected-utility theory: if the principle is accepted, the theory becomes too restrictive; if it is rejected, the applicability of the theory is severely hindered.

### 8.4.4 Expected Utility and Risk

In the absence of a serious contender, expected-utility theory has reigned as the major theory of individual decision making under uncertainty. Nevertheless, some alternative formulations of the problem, based on the notion of risk, have been proposed for both descriptive and prescriptive purposes. The idea underlying these developments is that gambles are perceived as having differing degrees of risk, and that preferences between gambles are determined by their respective risks. Such an approach may, but need not be, compatible with the theory of expected utility.

Allais (1953) argued on the basis of risk considerations that the expectation of the utility distribution alone is inadequate to account for preferences and that the higher moments of the utility distribution (e.g., variance, skewness) should also be taken into account, although he did not show how to do it. Other economists (e.g., Tobin, 1958; Markowitz, 1959) proposed that the risk of a gamble with monetary consequences is expressible in terms of the expectation and the variance of the money distribution and that the decision maker chooses among gambles so as to minimize risk, which decreases with expectation and increases with variance. The main advantages of this approach are its ease of applicability and its compatibility with expected utility theory in the case where utility is a quadratic function of money.

In contrast to the idea of minimizing risk, which was justified by economists on normative grounds, descriptive considerations have led psychologists to propose a more complicated relation between risk and preference. In particular, Coombs (1964) argued that people have an ideal risk level and that, in choosing between gambles, one selects the gamble whose risk is closer to one's ideal risk level.

Following the theory of risk developed in Section 3.14.1, we assume that the risk of a gamble with monetary consequences is expressible in terms of its expectation and variance, and that preference between gambles depends solely on their risks. Furthermore, we also assume that preferences obey expected-utility theory, and we study the relationships between utility and risk. The discussion follows the formulation of Pollatsek and Tversky (1970).

Let $\mathcal{D}$ be a set of gambles with monetary consequences, where $E(f_A)$ and $V(f_A)$ denote, respectively, the conditional expectation and variance of the gamble $f_A$. A preference order is said to depend only on variance and expectation, or to be *VE-dependent*, whenever for all $f_A$, $g_B$ in $\mathcal{D}$,

$$V(f_A) = V(g_B) \quad \text{and} \quad E(f_A) = E(f_B) \quad \text{imply} \quad f_A \sim g_B.$$

Similarly, a preference order is said to depend only on the risk measure, or to be *R-dependent*, whenever for all $f_A$, $g_B$ in $\mathcal{D}$,

$$R(f_A) = R(g_B) \quad \text{implies} \quad f_A \sim g_B,$$

where

$$R(f_A) = \theta V(f_A) - (1 - \theta) E(f_A), \qquad 0 < \theta \leqslant 1.$$

(See Theorem 3.13.) Clearly, any preference order that is $R$-dependent is also $VE$-dependent, but not conversely. To investigate the relationships between expected utility theory, $VE$-dependency, and $R$-dependency, we make the following assumptions. First, we assume that $\mathscr{D}$ contains all gambles with consequences $c^{(1)},..., c^{(n)} \in \mathrm{Re}$ obtained with probabilities $p_1,..., p_n$, respectively. Second, we assume that the utility of consequences (interpreted monetarily) is well defined. And third, we assume that utility of money is expressible as a power series in money. The conclusions are formulated in the following.

THEOREM 5. *Let* $\langle X, \mathscr{E}, \mathscr{N}, \mathscr{C}, \mathscr{D}, \succsim \rangle$ *be a conditional decision structure with* $\mathscr{C} = \mathrm{Re}$ *and suppose that:*

(i) *For all* $n \in I^+$, $c^{(1)},..., c^{(n)} \in \mathrm{Re}$, $p_1,..., p_n \in \mathrm{Re}^+$ *with* $\sum_{i=1}^{n} p_i = 1$, *there exist pairwise disjoint events* $A_1,..., A_n \in \mathscr{E} - \mathscr{N}$ *with* $P(A_i \mid \bigcup_{i=1}^{n} A_i) = p_i$, *such that* $c_{A_i}^{(i)} \in \mathscr{D}$, $i = 1,..., n$;

(ii) *For any* $c \in \mathscr{C}$, $A, B \in \mathscr{E} - \mathscr{N}$, *if* $c_A$, $c_B \in \mathscr{D}$, *then* $c_A \sim c_B$;

(iii) $v(c) = \sum_{m=0}^{\infty} \alpha_m c^m$ *for all* $c \in \mathrm{Re}$, *where* $v(c) = u(c_A)$.

*Then* (1) *a preference order is* $VE$-*dependent if and only if* $v(c) = \alpha_0 + \alpha_1 c + \alpha_2 c^2$, *and* (2) *there is no preference order that is* $R$-*dependent.*

Under the assumptions of Theorem 5, therefore, there is no utility function which is compatible with a preference order that depends only on the risk measure, and the quadratic utility function is the only one that is compatible with a preference order that depends only on variance and expectation. This utility function, however, is not very satisfactory. In the first place, its domain must be bounded by $-\alpha_1/2\alpha_2$ from above or below, if the utility is to be an increasing function of money. In the second place, it is inevitable that the utility of money be concave somewhere, whence $\alpha_2 < 0$. In this case, however, it can be shown that the cash equivalent of any gamble decreases with an increase in wealth—a conclusion that appears unacceptable on both descriptive and prescriptive grounds. [For further details see Pratt (1964), Pollatsek & Tversky (1970).]

The negative nature of the results established in Theorem 5, therefore, can be taken as evidence either against expected-utility theory or against the proposed theory of risk. Alternatively, one may accept both and reject the idea that preferences depend on risk alone. This dependency is, indeed, questionable. It is surely conceivable that an individual may have a preference between two gambles that appear (to him) equally risky. If this is the case, then a more elaborate theory relating risk and preference is called for.

### 8.4.5   Relations Between Subjective and Objective Probability

Our initial discussion of conditional decision theory suggests the following problem. Suppose that an experimenter uses a probability measure $Q$ on the algebra of events $\mathscr{E}$ to induce a conditional distribution $Q(\cdot \mid A)$, which controls the consequence that is received from a decision conditional on a nonnull $A$. Suppose, further, that the decision maker has a subjective probability measure $P$ that reflects his belief concerning the likelihood of the events. How are $P$ and $Q$ related? Existing results provide conditions for which $P = Q$. Since, in general, we do not expect this equality to hold, these results are negative in the sense that they tell us certain hypotheses cannot simultaneously be true. Suppose $P$ and $Q$ are monotonically related. Then it follows readily from Theorem 5.2 that if the ordering of $\mathscr{E}$ induced by these measures satisfies Axiom 5 (Section 5.2.3), then the two measures must coincide because the representation established in Theorem 5.2 is unique. Other assumptions leading to the same conclusion have been formulated in terms of $P$ and $Q$ rather than in terms of the ordering they induce. We prove one, particularly simple, theorem of this type; a closely related result was obtained by Pfanzagl (1968). More elaborate theorems of this kind can be found in Edwards (1962).

THEOREM 6.   *Let $\mathscr{E}$ be an algebra of sets, and let $P$ and $Q$ be finitely additive probability measures on $\mathscr{E}$ that are strictly increasing functions of each other. If for every pair of positive rational numbers $r, s$, satisfying $r + s \leqslant 1$, there exists some $R, S \in \mathscr{E}$ such that $P(R) = r$, $P(S) = s$, and $R \cap S = \varnothing$, then $P = Q$.*

The import of this result is that if $\mathscr{E}$ is sufficiently rich and $P \neq Q$, then the subjective probability must scramble the ordering of the events that is given by the objective probability. This is likely to be the case in many empirical situations.

### 8.4.6   A Method for Estimating Subjective Probabilities

Toda (1963) has observed that a method which elicits a person's true estimates of the probabilities of events must insure that it is inherently disadvantageous for him to misrepresent his judgments. Specifically, suppose that $\{A_1, ..., A_n\}$ is a partition of events whose true subjective probabilities are $P_1, ..., P_n$ and whose reported estimates are $Q_1, ..., Q_n$, where $\sum_{i=1}^{n} Q_i = \sum_{i=1}^{n} P_i = 1$. Our aim is to devise a payoff scheme under which the best consequence is obtained when $Q_i = P_i$, $i = 1, ..., n$. Let the payoff assign the amount of money $F(Q_i)$ to event $A_i$ when $Q_i$ is the reported probability. Then with the subjective estimates $P_i$, the subjective expected value (not utility, unfortunately) is $\sum_{i=1}^{n} P_i F(Q_i)$. Assuming that the person attempts to maximize this quantity, the question is whether there exists a function $F$ that insures the maximum occurs with $Q_i = P_i$. Aczél and Pfanzagl (1966) showed that only a logarithmic payoff function has the desired property.

THEOREM 7. *Suppose that $n > 2$, that $F$ is a once differentiable, real-valued function on $(0,1)$, and that for $i = 1,..., n$, $P_i$, $Q_i \in (0,1)$, $\sum_{i=1}^{n} P_i = \sum_{i=1}^{n} Q_i = 1$. If for all such $P_i$, the function $\sum_{i=1}^{n} P_i F(Q_i)$ achieves its maximum with $Q_i = P_i$, $i = 1,..., n$, then $F(P) = \alpha \log P + \beta$, $\alpha \geqslant 0$.*

The significance of this result is that if we are willing to assume that the utility of money is nearly linear with money in the ranges involved and that people behave as if they maximize subjective expected utility, then it is to their advantage to report subjective probabilities accurately when, and only when, the payoff is linear with the logarithm of the reported probabilities.

Note that the resulting maximum value is, formally, linear with the entropy (or measure of information) of the subjective probability distribution.

A number of other authors have considered this and closely related problems—especially the case where the function $F$ depends on more than just the probability of the component in question. For example, the quadratic expected payoff

$$\sum_{k=1}^{n} P_k \left( 2Q_k - \sum_{i=1}^{n} Q_i{}^2 \right)$$

has the property that it achieves its maximum if and only if $Q_k = P_k$, but $2Q_k - \sum_{i=1}^{n} Q_i{}^2$ depends on more than $Q_k$. Among the relevant papers are Shuford, Albert, and Massengill (1966), Winkler (1969), and Winkler and Murphy (1968).

## 8.5  PROOFS

### 8.5.1  Theorem 3   (p. 394)

*Assume a conditional decision structure with $\mathscr{C} = \bigtimes_{i=1}^{n} \mathscr{C}_i$; the expected utility property holds for gambles; an event $E$ exists that has a partition into $n$ equally probable subevents; and all constant decisions $c_E$ are in $\mathscr{D}$. Then the utility on $\mathscr{C}$ is additive over the components iff P-equivalent gambles have the same utility.*

*PROOF.* Suppose that $v$ on $\mathscr{C}$ is additive and that $f_A \sim_P g_A$, then

$$u(f_A) = \sum_{c \in \mathscr{C}} \{v(c) + w[f_A^{-1}(c)]\} \, P[f_A^{-1}(c) \mid A]$$

$$= \sum_{i=1}^{n} \sum_{c_i \in \mathscr{C}_i} v_i(c_i) \sum_{c \in \mathscr{C}(i, c_i)} P[f_A^{-1}(c) \mid A] + w(A)$$

$$= \sum_{i=1}^{n} \sum_{c_i \in \mathscr{C}_i} v_i(c_i) \sum_{c \in \mathscr{C}(i, c_i)} P[g_A^{-1}(c) \mid A] + w(A)$$

$$= \sum_{c \in \mathscr{C}} \{v(c) + w[g_A^{-1}(c)]\} \, P[g_A^{-1}(c) \mid A]$$

$$= u(g_A).$$

Conversely, for some $d = d_1 \cdots d_n \in \mathscr{C}$, choose numbers $v_i(d_i)$ so that $v(d) = \sum_{i=1}^{n} v_i(d_i)$. Now, consider any $c = c_1 \cdots c_n \in \mathscr{C}$, and let $c^{(i)} = d_1 \cdots d_{i-1} c_i d_{i+1} \cdots d_n$. Define

$$v_i(c_i) = v(c^{(i)}) - \sum_{j \neq i} v_j(d_j).$$

Observe that

$$\sum_{i=1}^{n} v_i(c_i) = \sum_{i=1}^{n} v(c^{(i)}) - (n - 1) \, v(d). \tag{8}$$

Since $c_E$, $c_E^{(i)}$, $d_E \in \mathscr{D}$ and $E_i \in \mathscr{E} - \mathscr{N}$, by Axiom 1 (ii) of Definition 1, $c_{E_i}$, $c_{E_i}^{(i)}$, $d_{E_i} \in \mathscr{D}$. It is not difficult to see that in both gambles $c_{E_i} \cup d_{E-E_i}$ and $\bigcup_{i=1}^{n} c_{E_i}^{(i)}$ only the component consequences $c_i$ and $d_i$ arise, with probabilities $1/n$ and $(n - 1)/n$, respectively. Thus, they are $P$-equivalent and so by hypothesis and Theorem 1,

$$u(c_{E_1}) \, P(E_1 \mid E) + u(d_{E-E_1}) \, P(E - E_1 \mid E) = \sum_{i=1}^{n} u(c_{E_i}^{(i)}) \, P(E_i \mid E).$$

Using the fact that $u(c_A) = v(c) + w(A)$, the expectation property for $w$, and the fact that $E_i$ are equiprobable,

$$v(c)/n + v(d)(n - 1)/n = (1/n) \sum_{i=1}^{n} v(c^{(i)}).$$

Substituting this in Equation (8),

$$\sum_{i=1}^{n} v_i(c_i) = v(c). \qquad \Diamond$$

### 8.5.2 Theorem 4 (p. 395)

*Consider a conditional decision structure with $\mathscr{C} = \mathrm{Re}$; a probability function P on events; and utility functions u, v, and w on gambles, consequences, and events, respectively. If v is continuous and strictly increasing, a form of consequence solvability holds, the usual conditional expectation property is satisfied, and the consistency condition obtains (see p. 395 for an exact statement of the hypotheses), then v is either linear or exponential.*

*PROOF.* For any $a, b, d \in \mathscr{C}$, hypotheses (iii), (iv), and (v) yield

$$v(c) + w(C) = u(c_C)$$
$$= u(a_A \cup b_B)$$
$$= v(a)\, P(A \mid A \cup B) + v(b)\, P(B \mid A \cup B) + w(A \cup B)$$

and

$$v(c + d) + w(C)$$
$$= v(a + d)\, P(A \mid A \cup B) + v(b + d)\, P(B \mid A \cup B) + w(A \cup B).$$

Setting $p = P(A \mid A \cup B)$ and $k = w(A \cup B) - w(C)$, then by hypothesis (ii)

$$pv(a + d) + (1 - p)\, v(b + d) + k = v(c + d) \qquad (9)$$
$$= v\{v^{-1}[pv(a) + (1 - p)\, v(b) + k] + d\}.$$

Treating $d$ as a parameter and setting $V_d(a) = v(a + d)$ in Equation (9),

$$V_d^{-1}[pV_d(a) + (1 - p)\, V_d(b) + k] = v^{-1}[pv(a) + (1 - p)\, v(b) + k].$$

Setting

$$u_1 = V_d, \qquad u_2 = pV_d, \qquad u_3 = (1 - p)\, V_d,$$
$$v_1 = v, \qquad v_2 = pv, \qquad v_3 = (1 - p)\, v,$$

we obtain

$$u_1^{-1}[u_2(a) + u_3(b) + k] = v_1^{-1}[v_2(a) + v_3(b) + k].$$

Now let

$$x = u_2(a), \qquad y = u_3(b) + k, \qquad u_4(b) = u_3(b) + k, \qquad v_4(b) = v_3(b) + k.$$

Then

$$v_1 u_1^{-1}(x + y) = v_2 u_2^{-1}(x) + v_4 u_4^{-1}(y).$$

Since these functions are monotonic and continuous, according to the Corollary of Theorem 1, p. 142 of Aczél (1966) there are constants (depending on $d$) $\gamma(d)$ and $\eta(d)$ such that

$$u_1(a) = \gamma(d)\, v_1(a) + \eta(d),$$

so

$$v(a + d) = \gamma(d)\, v(d) + \eta(d). \tag{10}$$

The conclusion follows by applying Corollary 1 of Theorem 1 on p. 150 of Aczél to Equation (10).                                                    ◇

### 8.5.3  Theorem 5  (p. 399)

*Consider a conditional decision structure with $\mathscr{C} = \mathrm{Re}$. Suppose that the utility of consequences is well defined and is expressible as $v(c) = \sum_{m=0}^{\infty} \alpha_m c^m$, and for $c^{(1)}, ..., c^{(n)} \in \mathrm{Re}$, $p_1, ..., p_n \in (0,1]$, with $\sum p_i = 1$ there are disjoint events $A_1, ..., A_n \in \mathscr{E} - \mathscr{N}$ with $P(A_i \mid \bigcup_{i=1}^{n} A_i) = p_i$ such that $c_{A_i}^{(i)} \in \mathscr{D}$, $i = 1, ..., n$. Then*

1. *A preference order is VE-dependent iff $v(c) = \alpha_0 + \alpha_1 c + \alpha_2 c^2$.*
2. *There is no preference order that is R-dependent.*

PROOF. (1)  Suppose $v(c) = \alpha_0 + \alpha_1 c + \alpha_2 c^2$. Consider $f_A = \bigcup_{i=1}^{n} c_{A_i}^{(i)} \in \mathscr{D}$, where $\{A_1, ..., A_n\}$ is some partition of $A$. For notational convenience, denote $p_i = P(A_i \mid A)$, $c_i = c^{(i)}$. By Theorem 2,

$$u(f_A) = \sum_{i=1}^{n} v(c_i)\, p_i$$

$$= \sum_{i=1}^{n} (\alpha_0 + \alpha_1 c_i + \alpha_2 c_i^2)\, p_i$$

$$= \alpha_0 + \alpha_1 \sum_{i=1}^{n} c_i p_i + \alpha_2 \sum_{i=1}^{n} c_i^2 p_i$$

$$= \alpha_0 + \alpha_1 E(f_A) + \alpha_2 [E^2(f_A) + V(f_A)],$$

and the preference order is, therefore, $VE$-dependent.

Conversely, note that since $v(c) = \sum_{m=0}^{\infty} \alpha_m c^m$, then

$$u(f_A) = \sum_{i=1}^{n} \left( \sum_{m=0}^{\infty} \alpha_m c_i^m \right) p_i = \sum_{m=0}^{\infty} \alpha_m \left( \sum_{i=1}^{n} c_i^m p_i \right).$$

Assuming the preference order is $VE$-dependent, and recalling that $\sum_{i=1}^{n} c_i^2 p_i = E^2(f_A) + V(f_A)$, we obtain $\alpha_m = 0$ for $m > 2$; otherwise, there exist some $g_B, f_A \in \mathcal{D}$ such that $E(f_A) = E(g_B)$, $V(f_A) = V(g_B)$, but $u(f_A) \neq u(g_B)$, contrary to the assumption of $VE$-dependency. Consequently, $u(f_A) = \sum_{i=1}^{n} v(c_i) p_i = \alpha_0 + \alpha_1 \sum_{i=1}^{n} c_i p_i + \alpha_2 \sum_{i=1}^{n} c_i^2 p_i$, and so $v(c) = \alpha_0 + \alpha_1 c + \alpha_2 c^2$, as required.

(2) Suppose there exists a preference order that is $R$-dependent. Hence, there exists some $\phi$ such that for any $f_A \in \mathcal{D}$,

$$u(f_A) = \phi[R(f_A)] = \phi[\theta V(f_A) - (1 - \theta) E(f_A)], \qquad \text{where} \quad 0 < \theta \leqslant 1.$$

Since $R$-dependency implies $VE$-dependency, we also have, by part 1,

$$u(f_A) = \alpha_0 + \alpha_1 \sum_{i=1}^{n} c_i p_i + \alpha_2 \sum_{i=1}^{n} c_i^2 p_i,$$

where $c_i, p_i$ are as in part 1, $i = 1,..., n$. Letting $E(f_A) = 0$ yields $\phi[\theta V(f_A)] = \alpha_0 + \alpha_2 V(f_A)$ for all values of $V$, hence $\phi$ is linear. Letting $V(f_A) = 0$ yields $\phi[(\theta - 1) E(f_A)] = \alpha_0 + \alpha_1 E(f_A) + \alpha_2[V(f_A) + E^2(f_A)]$. But $\phi$ is linear, so $\alpha_2 = 0$, hence $\phi$ does not depend on $V$, i.e., $\theta = 0$, contrary to assumption.                                                                  $\Diamond$

### 8.5.4 Theorem 6 (p. 400)

*If two finitely additive probability measures $P$ and $Q$, on the same algebra $\mathcal{E}$, are strictly increasing functions of each other, and if for any positive rationals satisfying $r + s \leqslant 1$ there exist some disjoint $R, S \in \mathcal{E}$ such that $P(R) = r$ and $P(S) = s$, then $P = Q$.*

*PROOF.* By assumption, there exists a strictly increasing function $f$ such that $Q = f(P)$. Hence, for any positive rationals $r, s$ with $r + s \leqslant 1$,

$$\begin{aligned}
f(r + s) &= f[P(R \cup S)] \\
&= Q(R \cup S) \\
&= Q(R) + Q(S) \\
&= f[P(R)] + f[P(S)] \\
&= f(r) + f(s).
\end{aligned}$$

Furthermore, since $P(\varnothing) = Q(\varnothing) = 0$, and $P(X) = Q(X) = 1$, we obtain $f(0) = 0$ and $f(1) = 1$. By the uniqueness clause of Theorem 2.4, therefore, $f(r) = r$ for any $r \in [0,1]$, and hence $P = Q$.                                          $\diamondsuit$

### 8.5.5   Theorem 7   (p. 401)

*If $F$ is differentiable, $n > 2$, $P$ and $Q$ are $n$-component probability vectors, and $\sum_{i=1}^{n} P_i F(Q_i)$ achieves its maximum with $Q = P$, then $F(t) = \alpha \log t + \beta$, $\alpha \geqslant 0$, for $0 < t \leqslant 1$.*

*PROOF.* Since only $n - 1$ of the $Q_i$ are independent, the function to be maximized may be written as

$$G(Q_1, ..., Q_{n-1}) = \sum_{i=1}^{n-1} P_i F(Q_i) + \left(1 - \sum_{i=1}^{n-1} P_i\right) F\left(1 - \sum_{i=1}^{n-1} Q_i\right).$$

Since $F$ is differentiable, the basic hypothesis that $G$ achieves its maximum with $Q_i = P_i$ implies

$$0 = \partial G(P_1, ..., P_{n-1})/\partial Q_1$$

$$= P_1 F'(P_1) - \left(1 - \sum_{i=1}^{n-1} P_i\right) F'\left(1 - \sum_{i=1}^{n-1} P_i\right),$$

where $F'$ denotes the derivative of $F$. Setting $Z = 1 - \sum_{i=2}^{n-1} P_i$, which is possible since $n > 2$, and $H(P) = PF'(P)$, the last equation may be written as $H(P_1) = H(Z - P_1)$. Since $P_1$ may assume any value in $(0,1)$, $Z$ may assume any value in $(P_1, 1)$. For $P_1 \leqslant \frac{1}{2}$, select $Z = \frac{1}{2} + P_1$, and so $H(P_1) = H(\frac{1}{2}) = \alpha$. If $P_1 > \frac{1}{2}$, then $Z - P_1 < \frac{1}{2}$, and so $H(P) = H(Z - P_1) = \alpha$. Thus, $F'(P) = \alpha/P$, then $F(P) = \alpha \log P + \beta$ by integration. Obviously, the maximum occurs only if $\alpha \geqslant 0$.                                          $\diamondsuit$

## □ 8.6   OTHER FORMULATIONS OF RISKY AND UNCERTAIN DECISIONS

As we noted at the beginning of Section 2, our formulation of the decision problem is by no means the only one that gives rise to an expected-utility representation and, in fact, it is somewhat different from the standard formulation of statistical decision theory. It seems desirable, therefore, to sketch briefly the other approaches, to compare them qualitatively with the one given here, and to cite some of the major references. A complete bibliography can be found in Fishburn (1968).

### 8.6.1 Mixture Sets and Gambles

Although the idea of maximizing expected utility is an old one, the main stimulus for most of the current research was the axiomatization included in the second and third editions of von Neumann and Morgenstern's classic book *Theory of Games and Economic Behavior* (1947, 1953). Strictly speaking their theory is not an example of fundamental measurement because numbers—probabilities—occur in the axioms; however, it is essential background for the understanding of the more recent developments.

Their primitive concepts (in our notation) were a simple order, $\langle \mathscr{U}, \succsim \rangle$, and for each real number $\alpha$ in $(0, 1)$, a closed binary operation on $\mathscr{U}$, mapping $(a, b)$ in $\mathscr{U} \times \mathscr{U}$ into an element of $\mathscr{U}$ that we denote $a\alpha b$. The set $\mathscr{U}$ was interpreted as a set of *abstract utilities*, the ordering relation $\succsim$ was interpreted as preferred or equal, and $a\alpha b$ was interpreted as the abstract utility of an option or strategy that yields abstract utility $a$ with probability $\alpha$ and abstract utility $b$ with probability $1 - \alpha$. Their axioms (apart from simple ordering and closure of the binary operations) were:

1. *If* $a \succ b$, *then* $a \succ a\alpha b \succ b$.
2.   (i)   $a\alpha b = b(1 - \alpha)\, a$.
     (ii)  $(a\alpha b)\, \beta b = a(\alpha\beta)\, b$.
3. *If* $a \succ b \succ c$, *then there exist* $\alpha, \beta \in (0, 1)$ *such that* $a\alpha c \succ b \succ a\beta c$.
4. $a \succsim b$ *iff* $a\alpha c \succsim b\alpha c$.

From these axioms they proved the existence of a real-valued function $u$ on $\mathscr{U}$ that preserves order and satisfies $u(a\alpha b) = \alpha u(a) + (1 - \alpha)\, u(b)$; the function $u$ is an interval scale.

Structures involving an ordering, and binary operations $(a, b) \to a\alpha b$, or $n$-ary operations $(a_1, \ldots, a_n) \to (\alpha_1 a_1, \ldots, \alpha_n a_n)$, where $\alpha_i \geqslant 0$ and $\sum_i \alpha_i = 1$, are called *mixture sets*. Other axiomatic treatments based on mixture sets were given by Herstein and Milnor (1953) and Marschak (1950). Aumann (1962, 1964a) and Kannai (1963) generalized the theory to the case where $\succsim$ is not connected, while Hausner (1954) obtained a multidimensional lexicographic representation by weakening Axiom 3 above.

Some related axiomatizations of utility measurement take as primitive a set of probability distributions over consequences. In these, the mixture operations are not primitive but are defined: if $P$, $Q$ are probability distributions, then convex combinations $\alpha P + (1 - \alpha)\, Q$ are also probability distributions. Such treatments were given by Blackwell and Girshick (1954) and

Fishburn (1970). For closely related developments, see Luce and Raiffa (1957) and Samuelson (1952).

Criticism of the above approach focusses on the use of numerical probability in the axioms. In our view the use of numerical probability in the axiomatization of utility is defensible, although not entirely satisfactory, for the foundations of classical game theory, but it is not adequate for the analysis of decisions under risk or uncertainty, i.e., games against nature.

Consider a two-person game in normal form, where the selection of pure strategies $a_i$ and $b_j$ by players $A$ and $B$ leads to the consequences $A_{ij}$ and $B_{ij}$ for players $A$ and $B$ respectively. A mixed strategy for player $A$, say, involves the selection of each of his pure strategies $a_i$ with probability $\alpha_i \geqslant 0$ where $\sum_i \alpha_i = 1$. Thus, for any pure strategy of $B$, every mixed strategy of $A$ yields a probability distribution over the set of consequences. The selection of a mixed strategy by $B$ now yields a convex combination of the above probability distributions.

If player $A$ wishes to choose his minimax-loss mixed strategy, then he *assumes* a definite mixed strategy of his own (i.e., with assumed numerical probabilities) and considers *all possible* mixed strategies of player $B$. Each of the latter generates a probability distribution over consequences. He orders these distributions (by any means, subject to satisfying the axioms, e.g., those of Blackwell and Girshick) and notes the least preferred one (maximum loss). He repeats this for each of his assumed mixed strategies, obtaining the family of all possible maximum-loss distributions. These he orders and chooses the most preferred; the corresponding mixed strategy is his minimax-loss strategy.

The numerical probabilities that enter above are not probabilities of particular events; they are assumed probabilities, for a possible mixed strategy of player $A$, and arbitrary probabilities (all possible ones must be considered), corresponding to arbitrary mixed strategies of player $B$. Consequently, a preference ordering over the appropriate set of probability distributions allows computation of the minimax strategy.

However, we may ask what is the meaning of preference order over a set of probability distributions? Moreover, how can one implement a calculated mixed strategy? For the preference relation to make sense, the ordered elements must be realizable entities. Hence, the probability distributions must be interpretable, by the player, as lotteries, assigning a consequence to each joint event corresponding to selection of a pair of pure strategies. This is not too unreasonable in game theory, because the player is free to construct the lotteries by using appropriate random mechanisms. Nevertheless, the application of the theory presupposes a correspondence between numerical probability and some family of chance events, and the nature of this correspondence is not analyzed.

This difficulty is somewhat alleviated by using events, rather than probabilities, to generate mixture operations. This was proposed by Pfanzagl (1959, 1967a, 1968), and later generalized from the simple gambles considered in his papers to arbitrary conditional decisions by Luce and Krantz (1971). Instead of probability distributions, the player orders conditional decisions $f_A$, etc. The set $\mathscr{U}$ is defined as $\mathscr{D}/\sim$, i.e., abstract utilities are equivalence classes of conditional decisions. For any disjoint nonnull events $A$, $B$ we have a binary operation $(\mathbf{f}, \mathbf{g}) \to \mathbf{f}(A, B)\,\mathbf{g}$ on $\mathscr{U}$, defined as follows: choose $f_A$ in $\mathbf{f}$, $g_B$ in $\mathbf{g}$, and let the operation $(A, B)$ map the pair $(\mathbf{f}, \mathbf{g})$ into the equivalence class containing $f_A \cup g_B$. Under the axioms of a conditional decision structure, we have a representation $P$, $u$, and we have

$$u[\mathbf{f}(A, B)\,\mathbf{g}] = u(f_A \cup g_B) = u(f_A)\,P(A \mid A \cup B) + u(g_B)\,P(B \mid A \cup B)$$
$$= \alpha u(\mathbf{f}) + (1 - \alpha)\,u(\mathbf{g}),$$

where $\alpha = P(A \mid A \cup B)$.

The above shows one direction leading to conditional decision structures: from mixture spaces with numerical probabilities, via Pfanzagl's event mixtures, to the mixture as the union of conditional decisions. There is another route, via consideration of individual decision making, or games against nature. Although the ordering of numerical probability distributions is a defensible primitive in the context of game theory, it is untenable in the context of individual decision making. That is because the individual no longer considers all possible probability distributions over states of nature in order to calculate a minimax strategy; rather, he must pick his most preferred pure strategy, and his choice must reflect his opinions about the likelihoods of various states of nature. An adequate theory must show in detail how numerical probability is related to preferences among decisions.

The first satisfactory theory for this situation was sketched by Ramsey (1931); it had little influence until attempts were made in the 1950's to improve the von Neumann–Morgenstern system. His central idea was to compare simple gambles based on a partition composed of two equiprobable events, $A$ and $-A$. If $P(A) = P(-A) = \frac{1}{2}$, then under expected-utility theory

$$a_A \cup b_{-A} \gtrsim c_A \cup d_{-A} \quad \text{iff} \quad u(a) + u(b) \geqslant u(c) + u(d). \tag{11}$$

The event $A$ can be identified via the criterion $a_A \cup b_{-A} \sim b_A \cup a_{-A}$ for all $a$, $b$ (see Section 5.2.4). The problem, then, is to provide a suitable set of axioms, stated in terms of gambles of this type, that permit us to construct a utility function with the property given in Equation (11). Ramsey sketched such a theory, and it was worked out in detail for finite sets of consequences by Davidson and Suppes (1956) and for infinite ones by Debreu (1959) and

Suppes and Winet (1955). Several empirical studies based on this approach were performed by Davidson, Suppes, and Siegel (1957); they are discussed in Section 9.4.2.

The reader will recognize, of course, that Equation (11) is just a case of additive conjoint measurement with two identical components; axiomatizations can be obtained based on Chapter 6, or based on the difference-measurement methods of Chapter 4.

Once $u$ is constructed, the subjective-probability function can easily be obtained. For any event $B$, find consequences $a$, $b$, $c$, $d$ such that $a_B \cup b_{-B} \sim c_B \cup d_{-B}$, with $u(a) \neq u(c)$. The expected-utility equation can be solved for $P(B)$ in terms of the $u$ values. One must state further axioms, however, in order to guarantee that, if an algebra $\langle X, \mathscr{E} \rangle$ of events is given, the function $P$ calculated in this way will be a finitely additive probability measure. Prior to Pfanzagl's work, the only satisfactory axiom system for measurement of both $P$ and $u$ was Savage's (which we discuss further in Section 8.6.3). His construction actually starts by obtaining a probability measure, using preferences to define the ordering of events (just as $a_A \cup b_{-A} \sim b_A \cup a_{-A}$ for all $a$, $b$ defines $A \sim -A$, above; see also Section 5.2.4 and Exercise 3 in this chapter). Once the probabilities are known, utilities can be constructed using any event $B$ such that $B$ and $-B$ are both nonnull; it is a question of analyzing the additive representation $u(a_B \cup b_{-B}) = u(a) P(B) + u(b) P(-B)$, where the weighting coefficients $P(B)$, $P(-B)$ are known.

In Pfanzagl's work and in the method of this chapter (see especially Section 8.3.2) the utility and probability functions are constructed simultaneously. This is not at all a peculiarity of the conditional-decision approach, however. In working with simple gambles of form $a_A \cup b_{-A}$, $c_B \cup b_{-B}$, for arbitrary $A$, $B$ one can obtain additive representations $(\varphi_A, \varphi_{-A})$, $(\varphi_B, \varphi_{-B})$ by conjoint-measurement methods; i.e.,

$$a_A \cup b_{-A} \succsim c_A \cup d_{-A} \quad \text{iff} \quad \varphi_A(a) + \varphi_{-A}(b) \geqslant \varphi_A(c) + \varphi_{-A}(d),$$

in analogy with Equation (11), and a similar equation holds for $B$, $-B$. By uniqueness, the $\varphi$'s must be linearly related to the functions entering an expected-utility decomposition, i.e., $u(a) P(A) = \alpha(A) \varphi_A(a) + \beta_1(A)$ and $u(a) P(-A) = \alpha(A) \varphi_{-A}(a) + \beta_2(A)$, where the constants, $\alpha$, $\beta_1$, $\beta_2$ depend on $A$. Similar equations hold for $B$. From these equations we have

$$\frac{u(a) - u(b)}{u(c) - u(d)} = \frac{\varphi_A(a) - \varphi_A(b)}{\varphi_A(c) - \varphi_A(d)} = \frac{\varphi_B(a) - \varphi_B(b)}{\varphi_B(c) - \varphi_B(d)},$$

$$\frac{P(A)}{P(-A)} = \frac{\varphi_A(a) - \varphi_A(b)}{\varphi_{-A}(a) - \varphi_{-A}(b)}.$$

Conditional decision structures can be regarded as generalizing simple gambles as well as mixture sets. This chapter emphasized the connection with simple gambles by using conjoint measurement on $\mathcal{D}_{A_1} \times \cdots \times \mathcal{D}_{A_n}$. But one can also emphasize the connection with mixture sets, regarding $f_{A_1} \cup \cdots \cup f_{A_n}$ not as an $n$-tuple but as the result of an operation on $\mathbf{f}_1, ..., \mathbf{f}_n$, forming the convex combination with nonnumerical coefficients $A_1 \mid \bigcup_{i=1}^{n} A_i, ..., A_n \mid \bigcup_{i=1}^{n} A_i$. This sort of development is found in Luce and Krantz (1971).

Reviews and analyses of expected-utility theory, including different formulations, critical discussions, and historical comments can be found in Adams (1960), Allais (1953), Arrow (1951, 1963), Baumol (1951), Churchman (1961), Churchman, Ackoff and Arnoff (1957), Fishburn (1964, 1968), Guilbaud (1953), Luce and Suppes (1965), and Raiffa (1968).

### 8.6.2 Propositions as Primitives

In general, a rough sort of consensus exists about the primitive terms to be employed in the formulation of the problem of decision making under risk or uncertainty. Nearly everyone seems to agree that there are chance events to which probabilities adhere, consequences which exhibit utilities, and decisions that are more or less arbitrary associations of consequences to events. Jeffrey (1965) took the position that this conceptualization is basically wrong, arguing that both probability and utility (he called it desirability) adhere to the same entities, namely, propositions. For example, the proposition that 10 or more inches of snow will fall in New York next February has both a probability and, if you live or work in New York, a (dis)utility. The fact that you can construct a money gamble based on this event does not alter its inherent utility. It is not a valueless proposition as is, presumably, the proposition that a head will come up when a coin is tossed. This valid point was mostly ignored in earlier discussions of utility theory, except for the frequent, uncomfortable asides about the possibility of a utility of gambling. Notable exceptions are the papers of Siegel (1959), Siegel *et al.* (1964), and Toda and Shuford (1965).

Jeffrey has, however, taken the extreme position, that it is wholly impossible to dissociate utility from probability. To be specific, if $A$ is a proposition, $P(A)$ is its (subjective) probability and $u(A)$ its utility, and if $\land$ and $\lor$ denote "and" and "or," then the basic feature of his theory is the following form of the expected-utility hypothesis: if $P(A \land B) = 0$ and $P(A \lor B) > 0$, then

$$u(A \lor B) = u(A) P(A \mid A \lor B) + u(B) P(B \mid A \lor B).$$

Such a relationship does not seem to provide much room for the construction of gambles. How does one relate the event of snow to the money gambles constructed upon it? Jeffrey's attempt to encompass gambles in his system was extensively criticized by Sneed (1966).

The observant reader has probably noted that if we make the usual mapping from ⟨propositions, ∧, ∨⟩ into ⟨sets, ∩, ∪⟩, then Jeffrey's basic equation translates precisely into Equation (6) of Section 8.4.1 which, as was pointed out, assigns utility to events in a way that is compatible with our general theory of conditional decisions. The advantage of the conditional theory over either Jeffrey's proposal or the classical theory for gambles is that it includes each as a special case, as seen in Equation (7).

Bolker (1966, 1967) provided a possible axiomatization of Jeffrey's system. It is of interest that $u$ is unique not up to linear transformations. Rather, $u$ and $P$ transform by

$$u' = (\alpha u + \beta)/(\gamma u + \delta), \qquad P'/P = \gamma u + \delta.$$

A number of other authors have attempted to axiomatize preference ordering relations over propositions which can be combined by the usual connectives of "and," "or," and "not." Among the relevant papers are Chisholm and Sosa (1966), Hallden (1957), Hansson (1968a), Rescher (1967), and von Wright (1963). Hansson cites serious difficulties with almost all of the axioms that have been proposed and, apparently, numerical representation theorems are not a major goal of this work.

### 8.6.3  Statistical Decision Theory

Some of the theories mentioned in Section 8.6.1 avoided the use of numerical probabilities as a primitive, but they were restricted to finite gambles. Aside from the theory discussed in this chapter, the only other formulation in which both limitations are dropped was first presented by Savage (1954) in his important book *The Foundations of Statistics*. We outline some of its main features.

Suppose that $\mathscr{S}$ is the universe of possible states of nature which are of pertinence to the decision maker. These states are such that exactly one of them obtains, i.e., they are mutually exclusive and exhaustive, and they are so finely subdivided that when a decision (Savage used the term "act") is contemplated, each state results in one and only one consequence from a set $\mathscr{C}$ of possible consequences. That is to say, a decision in this theory is taken to be a function from $\mathscr{S}$ into $\mathscr{C}$.

The basic structure is most easily formulated as a matrix

$$\begin{array}{c} \cdots\; S_j\; \cdots \\[2pt] f_i \begin{bmatrix} \vdots \\ \cdots\; c_{ij}\; \cdots \\ \vdots \end{bmatrix}, \end{array}$$

where $c_{ij} \in \mathscr{C}$ is the consequence of decision $f_i$ when $S_j \in \mathscr{S}$ is the true state of the world, i.e., $f_i(S_j) = c_{ij}$ .

The main additional constraint on $\mathscr{S}$ is imposed by the idea that there exists a probability distribution $P$ over $\mathscr{S}$, which summarizes the decision maker's knowledge and beliefs about the states of nature, and that this distribution is not altered in any way by his selection of a decision. The desired representation theorem asserts that such a probability distribution exists, that a utility function $u$ over $\mathscr{C}$ exists, and that decisions are evaluated according to their expected utility,

$$u(f_i) = \sum_j u(c_{ij}) \, P(S_j).$$

The problem, posed and solved by Savage, is: given an ordering on an (adequately large) set of decisions, formulate properties (axioms) about the ordering that are sufficient to establish the existence of utility and probability functions such that the ordering of the acts coincides with the ordering of their expected utilities.

Savage's axiomatization and proof follow these general lines. First, several axioms are introduced that permit one to define a relation of "more probable than" on subsets of $\mathscr{S}$, and this is shown to satisfy the axioms of qualitative probability (Definition 5.4). Next, an axiom is stated in terms of the qualitative probability (not directly in terms of the given ordering on acts) that allows one to construct an order-preserving probability measure on $\mathscr{S}$ (see the discussion in Section 5.2 of this and related axioms). Several additional axioms are then stated which make it possible to construct (along the lines of von Neumann and Morgenstern's proof) a utility function in terms of the probability function. For explicit statements of the axioms, see Savage (1954) or the summaries given in Fishburn (1970) and in Luce and Raiffa (1957).

For nearly 15 years Savage's system has stood as the most general formulation of the expected-utility hypothesis, but it has several structural features which warrant improvement. We list these limitations and indicate how they are coped with in the conditional theory.

First, the partitioning postulate just mentioned forces $\mathscr{S}$ to be an infinite set. Although a really satisfactory system of sufficient conditions is unlikely when the set of decisions is finite, it seems better not to commit ourselves in advance to locating the infiniteness in the set of events. For some situations, the idealization may be more satisfactory with an infinite set of consequences and a finite set of events. The conditional theory does not require $\mathscr{C}$ to be infinite, although in general $\mathscr{D}$ is.

Second, Savage assumed that any logically possible decision was available;

in particular, he made considerable use of decisions constant over the whole of $\mathscr{S}$. However, many of these decisions may be difficult to realize. In contrast, the conditional decisions of Definition 1 need not include all logical possibilities; in particular, no constant decision need be included. Moreover, should some constant decision be needed, as is convenient when defining utilities over consequences, it seems far more realistic to assume that they are defined only for some events.

Third, one feature of Savage's representation is the fact that the choice of a decision leaves the probability distribution over the states of nature unchanged. This hardly accords with our normal intuitions that certain chance occurrences are modified when we make a decision, sometimes modified to the extent of becoming impossible—we simply cannot have an en route automobile accident when we fly to our destination. The conditional theory obviously differs in this respect, and it may, therefore, provide a more natural formulation of many decision problems. Nonetheless, as we show in the next section, in the finite case neither formulation has a theoretical advantage: each may be transformed into the other. The conditional model does, however, retain a considerable practical advantage since, in general, an equivalent statistical formulation requires a hopelessly large number of states, each specifying an outcome for every possible decision.

### 8.6.4 Comparison of Statistical and Conditional Decision Theories in the Finite Case

Suppose that there are only finitely many different consequences and only finitely many states of nature (or events). Then it is always possible to transform the statistical decision representation into a conditional one, and vice versa. This result was suggested by Fishburn's (1964) analysis of special cases and was established generally by Luce and Krantz (1971).

It seems plausible that the set $\mathscr{C} = \{c_1, ..., c_i, ... c_n\}$ of consequences and the utility function $v$ over it will be the same in the two representations, and so we need only one notation for them. What differ are the events, probability distributions, and decisions. In the statistical theory, let $\mathscr{S} = \{S_1, ..., S_j, ..., S_m\}$ denote the set of states of nature, let $Q : \mathscr{S} \to [0,1]$ be the probability distribution over $\mathscr{S}$, and let a typical decision be $f : \mathscr{S} \to \mathscr{C}$. Decisions are ordered according to

$$u(f) = \sum_{j=1}^{m} v[f(S_j)]\, Q(S_j).$$

In the conditional theory, let $\mathscr{E} = \{\phi, A_1, ..., A_i, ..., A_k\}$ with $A_k = X$ be an algebra of events, let $P$ be a probability measure over $X$, and let a typical decision be $f_A : A \to \mathscr{C}$, where $A$ is in $\mathscr{E}$ and $P(A) > 0$. Decisions are ordered according to

$$u(f_A) = \sum_{i=1}^{n} v(c_i) \, P[f_A^{-1}(c_i)]/P(A).$$

The corresponding matrix displays are

$$\begin{array}{cc}
\text{Statistical} & \text{Conditional} \\
\cdots \, Q(S_j) \, \cdots & \cdots \, v(c_k) \, \cdots \\
f \begin{bmatrix} \vdots & \vdots \\ \cdots & v[f(S_j)] \, \cdots \\ \vdots & \vdots \end{bmatrix}, & f_A \begin{bmatrix} \vdots & \vdots \\ \cdots & P[f_A^{-1}(c_k)]/P(A) \, \cdots \\ \vdots & \vdots \end{bmatrix}.
\end{array}$$

Note that in the statistical display, a consequence $f(S_j)$ is associated with each decision $f$ and state of nature $S_j$; whereas, in the conditional display, an event $f_A^{-1}(c_i)$ is associated with each decision $f_A$ and consequence $c_i$—it is the event that leads to that consequence when that decision is chosen.

Our question, then, is: can we translate one representation into the other? First, suppose that we have the statistical representation. Define: $X = \mathscr{S}$, $\mathscr{E}$ is the set of all subsets of $X$, and $P = Q$, and for every decision $f$, let $f_X = f$. Observe that

$$u(f_X) = \sum_{i=1}^{n} v(c_i) \, P[f_X^{-1}(c_i)]/P(X)$$

$$= \sum_{i=1}^{n} v(c_i) \sum_{S_j \in f_X^{-1}(c_i)} Q(S_j)$$

$$= \sum_{j=1}^{m} v[f(S_j)] \, Q(S_j)$$

$$= u(f).$$

Thus, the translation from the statistical to the conditional model is trivial, with every decision conditional on $\mathscr{S}$. Translation in the other direction is more interesting.

Suppose that we have a conditional representation for which $P(A) > 0$ for $A$ in $\mathscr{E}$ and $A \neq \varnothing$. (This assumption involves no loss in generality since it is obtained from the general finite case by dropping all null atoms.) Define

$$\mathscr{S} = \underset{l=1}{\overset{k}{\times}} A_l \, ,$$

$$Q(S_j) = \prod_{l=1}^{k} P(S_j{}^l)/P(A_l),$$

where $S_j = (S_j{}^1,\ldots, S_j{}^l,\ldots, S_j{}^k) \in \mathscr{S}$, and for every $f_A \in \mathscr{D}$,

$$f(S_j) = f_A(S_j{}^l),$$

where $l$ is the index for which $A_l = A$. Then,

$$u(f) = \sum_{j=1}^{m} v[f(S_j)]\, Q(S_j) = \sum_{i=1}^{n} \sum_{S_j{}^l \in f_A^{-1}(c_i)} v(c_i)\, Q(S_j)$$

$$= \sum_{i=1}^{n} v(c_i) \sum_{S_j{}^l \in f_A^{-1}(c_i)} \prod_{r=1}^{k} P(S_j{}^r)/P(A_r)$$

$$= \sum_{i=1}^{n} v(c_i)\, P[f_A^{-1}(c_i)]/P(A_l)$$

$$= u(f_A).$$

In this reformulation of a conditional problem, we took $\mathscr{S}$ to be the product of all nonnull events in $\mathscr{E}$. Actually, it is not difficult to see that it is sufficient in any particular problem for the product to consist only of those events on which at least one decision is conditional; it is also not difficult to see that no smaller product will do in general (Luce & Krantz, 1971). Thus, some economies of representation may be possible in special cases, but in spite of this, the size of $\mathscr{S}$ for any realistic conditional formulation is overwhelming. For example, if $X$ has $n$ elements and if there are decisions conditional on every subset of two or more elements, then $\mathscr{S}$ has $\prod_{i=1}^{n} i^{\binom{n}{i}}$ elements, where $\binom{n}{i}$ is the binomial coefficient. Now, if $n = 5$, then $\mathscr{S}$ has $5 \times 12^{10}$ elements! The general problem of reducing the statistical system as much as possible is discussed in Marschak (1963).

The above analysis highlights three points. First, the reason for postulating infinitely many states of nature in the statistical theory is apparent. Second, when a problem is simple in statistical decision terms, it is equally simple in its conditional decision reformulation; however, when it is simple in conditional decision terms, it may become impossibly complex in its statistical decision reformulation. (We say "may" because $\mathscr{S}$ can remain reasonably small when only a few conditioning events are under consideration.) Third, the formulation of many decision problems leads naturally to a formalization in conditional decision terms, rather than in statistical decision terms (see Section 8.1).

## 8.7 CONCLUDING REMARKS

We end the chapter with remarks of two quite different sorts. First, we enter briefly into the discussion, which it seems must accompany any presentation of expected-utility theory, about whether it is prescriptive or descriptive. And second, we list three unresolved problems about conditional decision structures.

### 8.7.1 Prescriptive Versus Descriptive Interpretations

The primary intellectual defense invoked for a model of decision making that leads to an expected-utility representation is that the axioms formulate a concept of rational behavior—at least the necessary conditions do, and the structural ones only guarantee a sufficiently rich context. About all that one can do is to invite the reader to examine the axioms—in our case, Axioms 2–7 of Definition 1—to decide for himself whether he feels that it would be rational to abide by them. If so, and if the structural axioms are acceptable, then the representation simply says that such a rational being behaves as if he has a numerical scale of utility over the decisions and a probability function over events such that the utility of the union of a finite set of mutually disjoint decisions equals their conditional expected utility.

Whether the axioms may also be descriptive is a much more complex matter. First, it should be emphasized that all of the published discussion has focused on expectations of a utility function defined over consequences, as in Theorem 2, and not on the more general formulation of Theorem 1.

In general, the criticism against expected-utility theory has taken the form of incompatible experimental findings and of specific examples illustrating decision problems where one's intuitive judgment violates the axioms. Allais (1953) and Ellsberg (1961) have produced several similar examples of this type, one of which we described in detail in Section 5.2.4. We repeat it in slightly modified form.

| Gamble \ Event | $A_1$ red ($\frac{1}{3}$) | $A_2$ black ($\frac{2}{3}$) | $A_3$ white ($\frac{2}{3}$) |
|---|---|---|---|
| I | 100 | 0 | 0 |
| II | 0 | 100 | 0 |
| III | 100 | 0 | 100 |
| IV | 0 | 100 | 100 |

Consider an urn containing 20 black and white balls, in unknown proportion, and 10 red balls. Let each row above represent a gamble, whose consequences are sums of money. Thus, the first event $A_1$ occurs with probability $P_1 = \frac{1}{3}$, whereas, the probabilities $P_2$ and $P_3$ of $A_2$ and $A_3$, respectively, are unknown except for the fact that $P_2 + P_3 = \frac{2}{3}$. Observe that gambles I and II have the same consequence under $A_3$, as do III and IV. Thus, if the expected utility model is correct (in particular, Axiom 4), the choice between each pair depends entirely on the first two events. But since on the first two columns I coincides with III, and II with IV, the theory requires that I is preferred to II if and only if III is preferred to IV. Often, however, people simultaneously prefer I to II and IV to III. See the reasoning given in Section 5.2.4.

Both Ellsberg (1961) and Raiffa (1961) reported that, on the basis of informal questioning, many sophisticated people made these choices, and, what is worse, some were reluctant to modify their choices even when shown that they had violated what is called the extended sure-thing principle (Axiom 4). Recently, MacCrimmon (1968) reported somewhat more formal data on the sure-thing principle and certain other basic assumptions of the theory. His subjects, 38 business executives, were posed a number of choices between gambles, which were formulated in business terms; later they were interviewed about the reasons for their choices, and attempts were made to show them how they had violated various postulates. As anticipated by Allais and Ellsberg, the sure-thing principle appears to have been the greatest source of trouble. Not only did 40 % of the executives violate it at least once, but some (14 %) persisted in their choices when shown explicitly how they violated the principle. In contrast, violations of transitivity were referred to as mistakes by the subjects.

Ellsberg concluded that the failure of people to accept the sure-thing principle reflects a basic conceptual inadequacy of current theories that attempt to transform uncertainty into risk. However, similar violations of the sure-thing principle have been obtained in risky situations (Allais, 1953). Raiffa argued that the principle is normatively compelling and that people who violate it can be persuaded of the inconsistency of their choices by appropriate argument. He suggested that it can usually be done as follows: From a fair coin, form the options $\alpha$ and $\beta$ as

$$\begin{array}{c} & H \quad T \\ \alpha \\ \beta \end{array} \begin{bmatrix} I & IV \\ II & III \end{bmatrix}.$$

Thus, if option $\alpha$ is selected, the coin decides whether gamble I or IV is run, and similarly for option $\beta$. Now, if you really do prefer I to II and IV to III, then there can be no doubt that option $\alpha$ is better than $\beta$. Observe, however, that in both options, $\alpha$ and $\beta$, you obtain 100 with probability $\frac{1}{2}$ and 0 with probability $\frac{1}{2}$ for each of the three states $A_1$, $A_2$, and $A_3$. Thus, $\alpha$ and $\beta$ really are the same option, and so you should be indifferent between them.

Other critical discussions of the sure-thing principle and related matters can be found in Ellsberg (1954), Malinvaud (1952), Manne (1952), Samuelson (1952), Savage (1954), and Wold (1952).

We do not attempt to review here the many experiments conducted to test expected-utility theory. [For summaries and discussions of the pertinent literature, see Becker & McClintock (1967), Coombs, Dawes, & Tversky (1970, Chapter 5), Edwards (1954d, 1961), and Luce & Suppes (1965).] Indeed, it is very difficult to evaluate properly the empirical adequacy of expected-utility theory. This is due, in part, to the problems involved in the empirical interpretation of the theory, in part to the difficulties encountered in experimental tests of the theory, and in part to the inconclusive nature of much of the work. Instead of providing a general evaluation, therefore, we simply mention some of the more salient findings that have emerged from the experimental investigations.

To begin with, it should be noted that in some experimental contexts the theory provides a reasonable account of subjects' preferences, within the limits of their consistency. The question as to whether the studies provide a sufficiently powerful test of the theory, however, is typically left unanswered. At the same time it has been found in some specially devised choice problems that the sure-thing principle (i.e., Axiom 4) is consistently violated (MacCrimmon, 1968). Moreover, there are considerable data showing particular preference patterns that cannot be easily accommodated by expected-utility theory. Examples of such patterns include: preferences for particular probability values (Edwards, 1954a,b); preferences for particular variance and skewness combinations (Edwards, 1954c; Coombs and Pruitt 1960); and preferences reflecting utility for gambling (Tversky, 1967b). Finally, there is a strong indication that in the face of difficult decisions, people try to simplify the problem by ignoring (what appears to them inessential) information and concentrating only on a few major aspects. Such a simplification process may lead to systematic violations not only of the expected-utility principle, but of transitivity as well (Tversky, 1969).

In summary, it appears that, from a descriptive viewpoint, expected-utility theory is neither entirely adequate nor entirely inadequate. On the one hand, it does not provide a complete account of individual decision making under uncertainty. On the other hand, it cannot be discounted as a descriptive model, if only for the fact that, in some situations, people regard the axioms as principles of rational behavior, and modify their preferences so as to satisfy them. The conscious attempt to behave rationally is, in itself, a significant fact about human beings.

## 8.7.2 Open Problems

First, as was mentioned in Section 8.4.1, the problem of how the utility function defined over conditional decisions relates to utility measures on consequences and events, needs further investigation.

Second, within the context of statistical decision theory, an extensive body of work exists on the problem of how a decision maker should use new information (resulting, perhaps, from experiments that he performs) to modify his probability function over the states of nature (Raiffa & Schlaifer, 1961). Since, many problems are much simpler when formulated in terms of conditional decisions, it is interesting to know how to incorporate experimental information into that context. One suspects that the structuring imposed by the decisions may affect the types of experiments that should be performed, even when the Bayesian rule for modifying the probability function continues to be appropriate.

Third, as various studies suggest, the all important sure-thing principle is suspect, at least descriptively, and it would be interesting to know what sorts of representations can be established when it is weakened or eliminated.

## EXERCISES

**1.** Suppose that $P$ is the usual Lebesgue probability measure defined on the unit interval, $\mathscr{E}$ is the family of Lebesgue-measurable sets, $\mathscr{N}$ is the sets of measure 0, $\mathscr{D}_A$ the family of bounded measurable functions on $A \in \mathscr{E} - \mathscr{N}$, $\mathscr{D} = \bigcup_{A \in \mathscr{E}} \mathscr{D}_A$. For $f_A, g_B \in \mathscr{D}$, define $f_A \succsim g_B$ iff $(\int_A f_A \, dP)/P(A) \geqslant (\int_B g_B \, dP)/P(B)$. Prove that the structure $\langle [0, 1], \mathscr{E}, \mathscr{N}, \text{Re}, \mathscr{D}, \succsim \rangle$ satisfies Definition 1.     (8.2.6)

**2.** Keep everything as in Exercise 1 except replace $\mathscr{D}$ by $\mathscr{D}^*$ which is formed from countable (disjoint) unions of functions of the form $f_A(x) = \alpha + x$, $A \in \mathscr{E} - \mathscr{N}$, $\alpha \in \text{Re}$, $x \in A$. Prove that $\langle [0, 1], \mathscr{E}, \mathscr{N}, \text{Re}, \mathscr{D}^*, \succsim \rangle$ satisfies Definition 1.     (8.2.6)

**3.** Suppose that $\langle X, \mathscr{E}, \mathscr{N}, \mathscr{C}, \mathscr{D}, \succsim \rangle$ is a conditional decision structure (Definition 1). Define $\succsim'$ on $\mathscr{E} \times (\mathscr{E} - \mathscr{N})$ as follows: if $A, C \in \mathscr{E}$ and $B, D \in \mathscr{E} - \mathscr{N}$, then $A \mid B \succsim' C \mid D$ iff either

(i) $C \cap D \in \mathscr{N}$,

(ii) $-A \cap B \in \mathscr{N}$,

(iii) $A \cap B$, $-A \cap B$, $C \cap D$, $-C \cap D \in \mathscr{E} - \mathscr{N}$ and there exist $f_{A \cap B}, g_{-A \cap B}, h_{C \cap D}, k_{-C \cap D} \in \mathscr{D}$ such that

$$f_{A \cap B} \sim h_{C \cap D} \succ g_{-A \cap B} \sim k_{-C \cap D}$$

and

$$f_{A \cap B} \cup g_{-A \cap B} \gtrsim h_{C \cap D} \cup k_{-C \cap D}.$$

Which axioms of Definition 5.8 (Section 5.6.1) can you prove? What added conditions enable you to deduce all of them?     (8.2.6, 8.3.1)

**4.** If $f_A$ is a gamble on $A$ (p. 391), prove that $f_A^{-1}$ yields a partition of $A$ and that

$$f_A = \bigcup_{c \in \mathrm{Im} f_A} c_{f_A^{-1}(c)},$$

where $\mathrm{Im} f_A = \{c \mid c \in \mathscr{C} \text{ and for some } x \in A, \ f_A(x) = c\}$.     (8.4.1)

**5.** Prove Theorem 2.     (8.4.1)

*In Exercises 6 and 7, $\langle X, \mathscr{E}, \mathscr{N}, \mathscr{C}, \mathscr{D}, \gtrsim \rangle$ is a conditional decision structure and P and u are the functions described in Theorem 1 (Section 8.2.6). Let $v: \mathscr{C} \to \mathrm{Re}$ and $w: (\mathscr{E} - \mathscr{N}) \to \mathrm{Re}$.*

**6.** Suppose that for all gambles $f_A$,

$$u(f_A) = E[v(f_A) \mid A] + w(A).$$

Assuming that $A, -A \in \mathscr{E} - \mathscr{N}$, that $f_X$ is a gamble, and that

$$w(A) = \alpha[1 - P(A)] + \beta,$$
$$w(-A) = \alpha[1 - P(-A)] + \beta,$$

find the value of $P(A)$ that maximizes $u(f_X)$. Discuss possible experimental tests that use this predicted value (see Siegel *et al*, 1964).     (8.4.1)

**7.** For all $c \in \mathscr{C}$ and $A \in \mathscr{E} - \mathscr{N}$, suppose that $c_A \in \mathscr{D}$ and that $u(c_A) = v(c) + w(A)$. Prove that for $c, d \in \mathscr{C}$ and $A, B \in \mathscr{E} - \mathscr{N}$,

$$v(d) = \frac{u(d_A) u(c_X) - u(c_A) u(d_X)}{u(c_X) - u(c_A)} + v(c) \left[ \frac{u(d_X) - u(d_A)}{u(c_X) - u(c_A)} \right],$$

$$w(B) = \frac{u(c_A) u(d_B) - u(c_B) u(d_A)}{u(c_A) - u(d_A)} + w(A) \left[ \frac{u(c_B) - u(d_B)}{u(c_A) - u(d_A)} \right].$$

Show how you would test the hypothesis that the functions $v$ and $w$ exist, and calculate them.     (8.4.1)

**8.** Prove the assertion made in Footnote 1 (p. 392)     (8.4.1)

**9.** Suppose the hypotheses of Theorem 2 are satisfied, $P(A) = P(-A) = \frac{1}{2}$.

and

$$v(c) = \begin{cases} c^{1/2}, & \text{if } c \geqslant 0, \\ -(-c)^{1/2}, & \text{if } c < 0. \end{cases}$$

Select three monetary consequences $a, b, c \in \mathscr{C}$ such that $c_C \gtrsim a_A \cup b_{-A}$, but $M(a_A \cup b_{-A}) > M(c_C) = c$.     (8.4.3)

**10.** Suppose the hypotheses of Theorem 5 are satisfied, and that the preference order is $VE$-dependent. Show that for all gambles $f_A, g_B \in \mathscr{D}$ with $E(f_A) = E(g_B)$,

$$f_A \gtrsim g_B \quad \text{iff} \quad \delta V(f_A) \geqslant \delta V(g_B), \quad \text{where} \quad \delta \in \{-1, 1\}. \quad (8.4.4)$$

**11.** Let $\mathscr{C} = \{w, x, y, z\}$, $A = \{1, 2, 3\}$, $B = \{3, 4, 5, 6, 7\}$, and let the conditional decisions $f_A, g_A$, and $h_B$ be given by

$$\begin{array}{cccc} & w & x & y & z \\ f_A & \{1\} & \{2\} & \varnothing & \{3\} \\ g_A & \{3\} & \{1, 2\} & \varnothing & \varnothing \\ h_B & \{3, 4\} & \varnothing & \{5, 7\} & \{6\} \end{array}.$$

Reformulate these three decisions as a statistical decision problem.     (8.6.4)

# Chapter 9    Measurement Inequalities

## □ 9.1  INTRODUCTION

The measurement procedures discussed to this point typically have been based on a solvability assumption of one kind or another. In one theory for the (extensive) measurement of length, for instance, we have assumed (Definition 3.3) that whenever $a$ is longer than $b$, there exists $c$ in our object set such that $a$ is no shorter than the combined length of $b$ and $c$. Similarly, in additive conjoint measurement (Definition 6.7) we have required that for any $ap$ in $A_1 \times A_2$ and for certain $b$ in $A_1$, there exists $q$ in $A_2$ such that $ap \sim bq$. In general, solvability assumptions are introduced either to guarantee the existence of elements with certain specified properties or to ensure that the object set is sufficiently rich so that certain constructions can be carried out.

In one sense, therefore, solvability axioms are quite innocuous, since all they assert is that certain equations can be solved without imposing any real constraints on the solutions. Nevertheless, the inclusion of solvability assumptions raises serious problems of both a theoretical and a practical nature. We shall discuss these problems in this section and thereby motivate the topic of this chapter—the study of finite measurement structures that are free of solvability axioms.

As was already mentioned, solvability is an existential axiom—it postulates the existence of certain elements. Consequently, it is typically not a necessary condition for the intended numerical representation. It is certainly possible to construct an additive, order-preserving measurement scale of length, for example, even in situations where for some objects $a, b$ with $a > b$ there is no $c$ such that $a \gtrsim b \circ c$ and the set of objects is not closed under the concatenation operation $\circ$. (It should be noted, however, that in many cases solvability axioms can be made necessary by imposing additional constraints, e.g., continuity, on the desired numerical representation.) It is natural to investigate, therefore, how solvability can be eliminated. The interest in such a result stems mostly from the existence of data that satisfy, for example, the additive conjoint measurement model but fail to satisfy solvability. This occurs when one of the factors is discrete and the entire structure is not equally spaced. The theory developed in Chapter 6 cannot be applied to such data, and it should be supplemented by a theory that contains no solvability assumption and provides necessary and sufficient conditions for the desired representation. (Note that although Theorem 3.1 gives such conditions for extensive measurement when $\circ$ is closed, closure is, in fact, a solvability assumption.)

Aside from the theoretical limitations imposed by the solvability axiom, it presents several practical difficulties as well. Even in contexts where solvability can safely be assumed, the form of the data may render the process of solving the equations practically impossible. This is likely in data where the levels of the various factors have been selected in advance, rather than adjusted to form an equally spaced structure. This difference leads to a distinction between two types of experimental designs: constructive and factorial.

In a constructive design one constructs standard sequences. In a two-dimensional additive situation, for example, we can start by selecting some origin $a_0 p_0$ and some unit $a_1 p_0$. Then we search for the value $p_1$ which solves the equation $a_1 p_0 \sim a_0 p_1$. Having found $p_1$, whose existence is ensured by the solvability axiom, we can now search for $p_2$ that satisfies $a_1 p_1 \sim a_0 p_2$, and for $a_2$ that satisfies $a_1 p_1 \sim a_2 p_0$, etc. Note that $a_2$, for example, can be obtained as the solution of $a_2 p_0 \sim a_1 p_1$, but also as the solution of $a_2 p_0 \sim a_0 p_2$ and the two solutions must, therefore, coincide. Consequently, the uniqueness of the solutions obtained from different equations provides a test of the validity of the model. If the model is valid, the scale values are obtained by counting the number of equally spaced levels along each one of the factors.

Constructive designs are quite common in psychophysics (e.g., in the determination of equal-loudness contours), and they are also found in those studies of decision making that attempt to construct indifference curves, but for several reasons they are quite rare in other areas of the social sciences. One reason is that such designs are difficult to execute because a different construction is required for every subject. Another is that the construction requires a great deal of care because random error is magnified at each successive stage; moreover, time and order biases may introduce systematic error as well.

For these reasons, factorial designs are much more common in the social sciences, and most methods of data analysis (e.g., analysis of variance) have been developed for factorial designs in which a fixed finite set of levels of each factor is selected for study. One then investigates the possibility of an additive (or some other appropriate) representation for the data actually collected. If such a representation exists, one then searches for a method to obtain the scale values and to characterize their degree of uniqueness.

The necessary axioms (e.g., cancellation, independence) can be tested in factorial designs, and the failure of any of these axioms excludes the corresponding models. The nonnecessary axiom (solvability) cannot be tested in a factorial design, but it might be accepted on the basis of other considerations. *Nevertheless, even if the nontested solvability condition is true in the underlying data-generating process and if the tested necessary conditions are true in the obtained factorial data, it does not follow that the obtained data possess a representation of the kind in question.* A simple example is the following hypothetical rank order from a $3 \times 3 \times 2$ factorial design.

|  | $a_3$ | | | |  | $b_3$ | | |
|---|---|---|---|---|---|---|---|---|
|  | $a_1$ | $b_1$ | $c_1$ | |  | $a_1$ | $b_1$ | $c_1$ |
| $a_2$ | 12 | 14 | 18 | | $a_2$ | 11 | 13 | 17 |
| $b_2$ | 4 | 10 | 16 | | $b_2$ | 3 | 9 | 15 |
| $c_2$ | 2 | 6 | 8 , | | $c_2$ | 1 | 5 | 7 . |

It is easy to verify that this ranking does not violate independence. For example, the induced order in $A_2 \times A_3$ is

|  | $a_3$ | $b_3$ |
|---|---|---|
| $a_2$ | 6 | 5 |
| $b_2$ | 4 | 3 |
| $c_2$ | 2 | 1 |

regardless of the value fixed in $A_1$. Nevertheless, this ranking does not admit an additive decomposition, as is shown by the fact that the induced order over $A_1 \times A_2$ violates double cancellation: $a_1 b_2 \lesssim b_1 c_2$, $b_1 a_2 \lesssim c_1 b_2$, yet $a_1 a_2 > c_1 c_2$. In other words, the corresponding set of simultaneous linear inequalities is inconsistent because there are no scales $\phi_1$, $\phi_2$, $\phi_3$ that satisfy

$$\phi_1(a_1) + \phi_2(b_2) + \phi_3(a_3) < \phi_1(b_1) + \phi_2(c_2) + \phi_3(a_3),$$
$$\phi_1(b_1) + \phi_2(a_2) + \phi_3(a_3) < \phi_1(c_1) + \phi_2(b_2) + \phi_3(a_3),$$
$$\phi_1(c_1) + \phi_2(c_2) + \phi_3(a_3) < \phi_1(a_1) + \phi_2(a_2) + \phi_3(a_3).$$

As this illustration shows, the sets of sufficient conditions developed for simple polynomials are such that, even if some conditions are true for the underlying data-generating process (e.g., weak ordering, solvability, Archimedean condition) and the others are true for a particular set of data (e.g., independence), this does not guarantee the existence of an appropriate numerical representation for those data. The existence of such a representation depends on the existence of solutions to a set of simultaneous linear or polynomial inequalities.

To put matters still another way, the data in the above example cannot be embedded in a larger set of ordered data in which solvability is satisfied and independence is maintained, since solvability and independence (on three or more factors) imply double cancellation. Hence, if one believes that solvability holds, then insofar as the data are free from error, they provide indirect evidence against independence; it must be violated elsewhere in the structure. Hence, in order to conclude that a rank order is compatible with some specified numerical representation, it is not enough to test the necessary axioms, even in those contexts where the other axioms are assumed valid. Rather, one must demonstrate that the system obtained by translating the observed order relations into equations and inequalities, according to the appropriate measurement model, has a simultaneous solution. The present chapter investigates necessary and sufficient conditions for the solvability of such systems of equations and inequalities.

The problem of formulating necessary and sufficient conditions for the existence of a linear representation (i.e., additive or subtractive) for finite data structures has been investigated since the late 1950's. Adams and Fagot (1956, 1959) discussed the problem and presented some preliminary results. Scott and Suppes (1958) showed, however, that a finite linear model can not be axiomatized by a first order universal sentence. Scott (1964) was the first to present a characterization of finite linear measurement models based on the theory of linear inequalities. Closely related results were also obtained by Aumann (1964b), Adams (1965), and Tversky (1964, 1967b). We follow the latter development because it is slightly more general (the observed relation is not assumed to be connected) and because it is easier to generalize to nonlinear models. The finite linear case and its applications are discussed in the next three sections. In the last two sections, the results are extended to the infinite polynomial case along the lines developed in Tversky (1967a).


## 9.2  FINITE LINEAR STRUCTURES

We start with the simple case of the two-dimensional additive model discussed in Chapter 6. Let $A_1$ and $A_2$ be finite sets and let $\gtrsim$ be a weak ordering of $A = A_1 \times A_2$. Our purpose is to discover necessary and sufficient conditions for the existence of scales $\phi_1$ and $\phi_2$ such that for all $a, b$ in $A_1$ and $p, q$ in $A_2$

$$ap \gtrsim bq \quad \text{iff} \quad \phi_1(a) + \phi_2(p) \geqslant \phi_1(b) + \phi_2(q). \tag{1}$$

Several consequences of this representation have been discussed in Chapter 6; namely, independence (Definition 6.1), double cancellation (Definition 6.3), and triple cancellation (p. 251). All these conditions have a similar structure. Both transitivity and double cancellation, for instance, can be derived from the additive model by noting that two inequalities (in the data) imply a third one, provided that the antecedent inequalities are selected so that all save two terms on each side of the inequalities can be canceled. Similarly, triple cancellation is a case where three inequalities imply a fourth one. Following the same technique, we can formulate $n$th-order cancellation axioms in which $n$ inequalities imply an additional one. In Chapter 6 we showed that, in the presence of restricted solvability and the Archimedean axiom, independence, and double cancellation are sufficient to establish additivity. All higher-order cancellation axioms, therefore, follow from weak ordering, independence and double cancellation provided the Archimedean property and restricted solvability are satisfied. Once solvability is denied, however, this is no longer true; higher-order cancellation conditions cannot be derived from lower order cancellation conditions. This follows from the observation that a system of $n + 1$ homogeneous linear equations, of the type defined by Equation (1), may fail to have a simultaneous solution, although any subset of $n$ inequalities may have a simultaneous solution.

So, in general, it is impossible to derive an $(n + 1)$st-order cancellation axiom from the $n$th-order cancellation axioms. Moreover, it was shown by Scott and Suppes (1958) that no finite number of cancellation axioms can be sufficient for additivity for all finite sets. (A detailed analysis of this problem is given in Chapter 18, Volume II.) This, of course, does not exclude the possibility of formulating sufficient sets of cancellation axioms, where the size of the set depends on the size and structure of $A$. For example, if $A = \{a_0, ..., a_n\}$ and $B = \{b_0, b_1\}$, then the assumption that $\gtrsim$ is an independent weak order (Definition 6.1) is both necessary and sufficient for the additivity of $A \times B$. The proof of this assertion is left as Exercise 2. However, no general results of this type are available.

We turn now to a theory, based on an irreflexivity condition, which provides necessary and sufficient conditions for additivity. This theory applies to any finite product set, irrespective of the number of factors, and it does not require that the observed relation, denoted $\gtrsim$, be connected. The irreflexivity condition we shall formulate generalizes all types of cancellation axioms, including independence and transitivity. It is an axiom schema that generates infinitely many cancellation axioms. Because of its abstract structure, however, this property is not readily interpretable. The basic theory is developed in terms of the additive conjoint model and then applied to probability structures.

## □ 9.2.1  Additivity

Let $A = A_1 \times \cdots \times A_n$ be a finite set with $A_i$ and $A_j$ pairwise disjoint for $i \neq j$, and let $\gtrsim$ denote a reflexive binary relation on $A$. If neither $a \gtrsim b$ nor $b \gtrsim a$, then $a$ and $b$ are *incomparable*. ($\gtrsim$ is connected if and only if it contains no incomparable pairs). Suppose the sets $A_1, ..., A_n$ contain $k_1, ..., k_n$ elements, respectively, and let $k = \sum_{i=1}^{n} k_i$. Thus, $Y = \bigcup_{i=1}^{n} A_i$ contains $k$ elements, $y_1, ..., y_k$. A 1:1 mapping $v$ of $A$ into $\mathrm{Re}^k$ is defined by $v(a) = \bar{a} = (\bar{a}_1, ..., \bar{a}_k)$ where for any $l = 1, ..., k$,

$$\bar{a}_l = \begin{cases} 1, & \text{if } y_l \text{ is a component of } a, \\ 0, & \text{otherwise.} \end{cases}$$

In this way, each $a$ in $A$ corresponds to a vector $\bar{a}$ in $\mathrm{Re}^k$. Let $\bar{A} = \{\bar{a} \mid a \in A\}$, and let $A^+$ be the closure of $\bar{A}$ with respect to (componentwise) vector addition (i.e., $A^+$ is the smallest semigroup of vectors in $\mathrm{Re}^k$ that contains $\bar{A}$; the elements of $A^+$ all have nonnegative integer components). Formally,

$$A^+ = \{x \in \mathrm{Re}^k \mid x = \bar{a}^{(1)} + \cdots + \bar{a}^{(m)},$$
$$\bar{a}^{(j)} \in \bar{A}, \quad j = 1, ..., m, \quad \text{for} \quad m = 1, 2, ...\}.$$

We define two binary relations on $A^+$ as follows.

DEFINITION 1.  *For $x, y \in A^+$ let $x \sim_{\mathrm{I}} y$ iff there exist $\bar{a}^{(1)}, ..., \bar{a}^{(m)}$, $\bar{b}^{(1)}, ..., \bar{b}^{(m)} \in \bar{A}$ such that $x = \bar{a}^{(1)} + \cdots + \bar{a}^{(m)}$, $y = \bar{b}^{(1)} + \cdots + \bar{b}^{(m)}$, and for $j = 1, ..., m, a^{(j)} \sim b^{(j)}$.*

*Also $x \succ_{\mathrm{I}} y$ iff there exist $\bar{c}^{(1)}, ..., \bar{c}^{(m)}, \bar{d}^{(1)}, ..., \bar{d}^{(m)} \in \bar{A}$ such that $x = \bar{c}^{(1)} + \cdots + \bar{c}^{(m)}, y = \bar{d}^{(1)} + \cdots + \bar{d}^{(m)}, c^{(j)} \gtrsim d^{(j)}, j = 1, ..., m, and for some $j$, not $d^{(j)} \gtrsim c^{(j)}$.*

We let $\gtrsim_1$ be the union of these relations, i.e., $x \gtrsim_1 y$ if and only if $x \sim_1 y$ or $x >_1 y$. Note that since $a \sim a$ for all $a$ in $A$, $\sim_1$ is reflexive and symmetric. But $>_1$ is not necessarily irreflexive nor asymmetric; since different sets of elements of $\bar{A}$ can have the same vector sum, the sum does not determine its summands uniquely. Thus, $>_1$ and $\sim_1$ need not be the asymmetric and symmetric parts of $\gtrsim_1$.

For example, suppose that independence (Definition 6.11) is violated by an ordering $\gtrsim$ on $A_1 \times \cdots \times A_n$. Then there exist $a, a', b, b'$ in $A$ such that

$$a = a_1 \cdots a_i \cdots a_n > b_1 \cdots a_i \cdots b_n = b',$$
$$b = b_1 \cdots b_i \cdots b_n \gtrsim a_1 \cdots b_i \cdots a_n = a'.$$

By Definition 1, $\bar{a} + \bar{b} >_1 \bar{a}' + \bar{b}'$. But for $l = 1,...,k$, $(\bar{a} + \bar{b})_l = 1$ if $y_l$ is one of the $a_i$ or the $b_i$, and $= 0$ otherwise. The same holds for $(\bar{a}' + \bar{b}')_l$, so $\bar{a} + \bar{b}$ equals $\bar{a}' + \bar{b}'$.

This shows that irreflexivity of $>_1$ implies independence of $\gtrsim$. It can be shown similarly that irreflexivity of $>_1$ implies every $n$th-order cancellation axiom, for every $n$ (see the discussion of Scott, 1964, at the end of this subsection).

We leave to the reader the proof that $>_1$ is irreflexive if and only if $>_1$, $\sim_1$ are the asymmetric and symmetric parts of $\gtrsim_1$.

Irreflexivity of $\gtrsim_1$ also implies that $\gtrsim_1$ has no intransitive cycles of the form $x \gtrsim_1 y$, $y \gtrsim_1 z$, $z >_1 x$. For if such a cycle existed, Definition 1 would yield $x + y + z >_1 y + z + x$, contrary to irreflexivity. But $\gtrsim_1$ need not be transitive, since we can have $x \gtrsim_1 y$, $y \gtrsim_1 z$, but $x$ and $z$ are incomparable. (Construct an example.)

The main result is that irreflexivity of $>_1$ is a necessary and sufficient condition for the existence of an additive representation. The proof of necessity is really essential to the understanding of this subsection, so we give it immediately.

Suppose there exists an additive representation $\phi$ for $A$. Let

$$x = \bar{a}^{(1)} + \cdots + \bar{a}^{(m)} = \bar{b}^{(1)} + \cdots + \bar{b}^{(m)}.$$

It follows that

$$\sum_{j=1}^{m} \phi(a^{(j)}) = \sum_{j=1}^{m} \phi(b^{(j)}).$$

To prove this, note that each term of these two sums decomposes into a sum of scale values of the $y_l$ components. Since the vector sums are equal (to $x$), any $y_l$ component occurs equally often in each decomposition of $x$. Therefore, the corresponding numerical sums are equal.

Thus, $x >_1 x$ is impossible, for this would imply the existence of two decompositions such that

$$\phi(a^{(j)}) \geqslant \phi(b^{(j)}), \qquad j = 1, ..., m,$$

with $>$ holding for at least one $j$. Thus, summing these $m$ inequalities,

$$\sum_{j=1}^m \phi(a^{(j)}) > \sum_{j=1}^m \phi(b^{(j)}),$$

contrary to the equality just established through summing by $y_l$ components instead of term by term.

Our first theorem asserts that irreflexivity is not only necessary but also sufficient for the desired representation.

THEOREM 1.   *The relation $>_1$ (Definition 1) is irreflexive iff there exist real-valued functions $\phi$ on $Y$ and $\psi$ on $A$ such that, for all $a, b \in A$:*

(i)   $\psi(a) = \psi(a_1 \cdots a_n) = \sum_{i=1}^n \phi(a_i);$

(ii)   $a \sim b$ *implies* $\psi(a) = \psi(b);$

*and*

(iii)   $a > b$ *implies* $\psi(a) > \psi(b).$

Note that since the $A_i$ are disjoint, we can consider only one function, $\phi$, on $Y = \bigcup_{i=1}^n A_i$, rather than separate functions $\phi_i$ on $A_i$. This is convenient because the desired function $\phi$ is obtained as a vector in $\mathrm{Re}^k$. The restriction of $\phi$ to $A_i$ defines $\phi_i$, but this plays no role.

The necessity of irreflexivity was established above. The proof of sufficiency consists of demonstrating the existence of a vector $z$ in $\mathrm{Re}^k$ such that for all $a, b$ in $A$,

$$a \sim b \text{ implies } z \cdot \bar{a} = z \cdot \bar{b} \quad \text{and} \quad a > b \text{ implies } z \cdot \bar{a} > z \cdot \bar{b}, \quad (2)$$

where $z \cdot \bar{a}$ is the scalar product $\sum_{l=1}^k z_l \bar{a}_l$. Once we have such a vector $z$, define $\phi(y_l) = z_l$. By definition of $\bar{a}$, for $a = a_1 \cdots a_n$

$$\sum_{i=1}^n \phi(a_i) = \psi(a) = z \cdot \bar{a};$$

therefore, $\psi$ is order preserving as required by (ii) and (iii).

The proof that under irreflexivity the desired vector $z$ exists is based on Theorem 2.7, which provides a necessary and sufficient condition for the solvability of a system of homogeneous linear inequalities. The algebraic proof of Theorem 2.7 provides a method for actual construction of $\phi$. (In fact, the problem of finding a numerical solution is reducible to an integer-programming problem; see Gale, 1960.)

Unlike previous theories in which the measurement scales can be described simply as ordinal, interval, or ratio scales, the characterization of the scale type (i.e., the formulation of a uniqueness theorem) in the present case is more complicated. Clearly, not every monotonic transformation can be applied to the scales $\phi_1, ..., \phi_n$ of Theorem 1, and yet they are not quite interval scales. Scales of this type have been called ordered-metric scales, although the term "metric" carries undesirable connotations. To state its uniqueness properties, note that the set of all additive representations of $A$ is isomorphic to the set $X$ of all vectors satisfying Equation (2). Since $X$ is defined as the solution space of a system of (homogeneous) linear inequalities, it is a polyhedral convex cone in $\mathrm{Re}^k$. Using the standard characterization of the solution space, the uniqueness of the scales is summarized in the following:

THEOREM 2. *Suppose that $A$ has an order-preserving additive representation. Then there exist vectors $z^{(1)}, ..., z^{(m)}$ in $\mathrm{Re}^k$, and an integer $j$, with $0 \leqslant j \leqslant m$, such that $z$ is an additive representation of $A$ [i.e., it satisfies Equation (2)] iff*

$$z = \sum_{i=1}^{m} \alpha_i z^{(i)} + c, \tag{3}$$

*where $c$ is any vector whose components are all equal, $\alpha_i \geqslant 0$ for $i \leqslant j$, and $\alpha_i > 0$ for $i > j$.*

The representation is an interval scale if and only if $m = 1$, in which case the solution space is unidimensional and all the solutions are linearly related. The proof of Theorem 2 is left to the reader.

Before turning to other (finite) linear models, we discuss briefly another axiomatization of the additive model. In an investigation of the relation between measurement structures and linear inequalities, Scott (1964) established the following result: Let $A_1$ and $A_2$ be finite sets, and let $\gtrsim$ denote a binary relation on $A_1 \times A_2$. For the structure $\langle A_1 \times A_2, \gtrsim \rangle$ to have an order-preserving additive representation [in the sense of Equation (1)], it is necessary and sufficient that:

(i) $\gtrsim$ is connected.

(ii) For all sequences $a_0, ..., a_n$ in $A_1$, $p_0, ..., p_n$ in $A_2$, and all permutations $\pi, \sigma$ of $\{0, ..., n\}$, if $a_i p_i \gtrsim a_{\pi(i)} p_{\sigma(i)}$, for $i = 0, ..., n - 1$, then $a_{\pi(n)} p_{\sigma(n)} \gtrsim a_n p_n$.

Since the representation [formulated in Equation (1)] is defined in terms of "iff," it is evident that $\gtrsim$ must be connected. In this case, however, it follows that condition (ii) above is equivalent to the irreflexivity of $>_I$. The direct proof of this equivalence, not employing the established representation theorem, is given as Exercise 5. It should be noted that although the various characterizations of the finite additive model (Adams, 1965; Scott, 1964; Tversky, 1964, 1967b) appear different, they can all be regarded as generalizations of the cancellation axioms. Moreover, the various proofs of the representation theorems are all based on the criterion for the solvability of a finite set of homogeneous linear inequalities formulated in Theorem 2.7.

### 9.2.2 Probability

The problem of finding necessary and sufficient conditions for the existence of an order-preserving probability measure on a finite algebra of sets (Definition 5.2) was first stated and solved by Kraft, Pratt, and Seidenberg (1959) and later was reformulated by Scott (1964). Also see Fishburn (1969b). Instead of presenting their results, we show how the solution can be derived using the present methods.

Let $X = \{x_1, ..., x_n\}$ be a finite nonempty set, let $\mathscr{E}$ be an algebra of sets on $X$ (i.e., a collection of subsets of $X$ that is closed under union and complementation), and let $\gtrsim$ be a reflexive binary relation on $\mathscr{E}$. As in Chapter 5, the elements of $\mathscr{E}$ are called events and the relation $\gtrsim$ is interpreted to mean "at least as probable as." Our task is to find conditions equivalent to the existence of a real-valued function $P$ from $\mathscr{E}$ into the unit interval $[0, 1]$, with the following properties. For all $A, B$ in $\mathscr{E}$,

(i)   $A > B$ implies $P(A) > P(B)$;

(ii)  $A \sim B$ implies $P(A) = P(B)$;

(iii) $P(A) \geqslant 0$;                                                      (4)

(iv)  $P(X) = 1$;

(v)   if $A \cap B = \varnothing$, then $P(A \cup B) = P(A) + P(B)$.

We define the vectorial characteristic function $v$ for $\mathscr{E}$ as the 1:1 mapping of $\mathscr{E}$ into $\mathrm{Re}^n$ defined by $v(A) = \bar{A} = (\bar{A}_1, ..., \bar{A}_n)$, where for any $i = 1, ..., n$

$$\bar{A}_i = \begin{cases} 1, & \text{if } x_i \in A, \\ 0, & \text{if } x_i \notin A. \end{cases}$$

Thus, each $A$ in $\mathscr{E}$ is represented by a unique vector of zeros and ones in $\mathrm{Re}^n$. Let $\bar{\mathscr{E}} = \{\bar{A} \mid A \in \mathscr{E}\}$ and $\mathscr{E}^+$ be its additive closure with respect to (componentwise) vectorial addition. Following the earlier development, define $>_1$ and $\sim_1$ by Definition 1, where $\mathscr{E}^+$ and $\mathscr{E}$ take the place of $A^+$ and $A$, respectively. It is readily seen that the irreflexivity of $>_1$ is necessary for the desired representation. Moreover, by Theorem 1, there exists a vector $z$ in $\mathrm{Re}^n$ such that $A \sim B$ implies $z \cdot \overline{A} = z \cdot \overline{B}$, while $A > B$ implies $z \cdot \overline{A} > z \cdot \overline{B}$. Define $P(A) = (z \cdot \overline{A})/(z \cdot \overline{X})$. Clearly, $P$ is order preserving, and $P(X) = 1$. Moreover, if $A \gtrsim \varnothing$, for any $A$ in $X$, while $z \cdot \overline{\varnothing} = z \cdot 0 = 0$, $P(A) \geqslant 0$. Finally if $A \cap B = \varnothing$, then

$$P(A \cup B) = \frac{z \cdot \overline{(A \cup B)}}{z \cdot \overline{X}} = \frac{z \cdot (\overline{A} + \overline{B})}{z \cdot \overline{X}} = \frac{(z \cdot \overline{A}) + (z \cdot \overline{B})}{z \cdot \overline{X}} = P(A) + P(B).$$

Note that if $A \cap B \neq \varnothing$, then $\overline{A \cup B} \neq \overline{A} + \overline{B}$; the latter does not correspond to an event in $\mathscr{E}$. The discussion is summarized by:

THEOREM 3. *Let $X$ be a finite nonempty set, $\mathscr{E}$ an algebra of sets on $X$, and $\gtrsim$ a reflexive binary relation on $\mathscr{E}$ satisfying $X > \varnothing$ and $A \gtrsim \varnothing$ for all $A \in \mathscr{E}$. There exists $P$ satisfying conditions (4) iff $>_1$ is irreflexive.*

For more recent developments along these lines, see Domotor (1969).

## 9.3 PROOF OF THEOREM 1

*The relation $>_1$ is irreflexive iff there exist real-valued functions $\phi$ on $Y$ and $\psi$ on $A$ such that for all $a, b \in A$:*

(i) $\psi(a) = \sum_{i=1}^n \phi(a_i)$;

(ii) *if $a \sim b$, then $\psi(a) = \psi(b)$;*

*and*

(iii) *if $a > b$, then $\psi(a) > \psi(b)$.*

PROOF. By finiteness, we can list the pairs in the relations $>$ and $\sim$:

$$a^{(1)} > b^{(1)}, ..., a^{(j)} > b^{(j)}; \qquad a^{(j+1)} \sim b^{(j+1)}, ..., a^{(m)} \sim b^{(m)},$$

where $a^{(i)}, b^{(i)} \in A$, $i = 1, ..., m$. If $j = 0$, $>$ is empty and we can let $\phi$ be identically zero. Suppose $j \geqslant 1$. Let $r^{(i)} = \overline{a}^{(i)} - \overline{b}^{(i)}$, $i = 1, ..., m$. We need to find a vector $z$ such that $z \cdot r^{(i)} > 0$, $i = 1, ..., j$ and $z \cdot r^{(i)} = 0$, $i = j + 1, ..., m$. As was shown above, $\phi(y_i) = z_i$ then gives the additive representation. If no such vector $z$ exists, then we show that $>_1$ is not irreflexive, by Theorem 2.7. For by that theorem, if there is no solution to

$z \cdot r^{(i)} > 0$, $i = 1,..., j$, $z \cdot r^{(i)} = 0$, $i = j + 1,..., m$, then there exist rationals $\rho_i$, with $\rho_i \geqslant 0$, $i = 1,..., j$ and $\rho_i > 0$ for some $i$, $1 \leqslant i \leqslant j$, satisfying

$$\sum_{i=1}^{m} \rho_i r^{(i)} = 0.$$

Interchanging $a^{(i)}$ and $b^{(i)}$, if necessary, for $i > j$, we can assume that all the $\rho_i$ are nonnegative; in addition, we can multiply them all by a common denominator, obtaining the above vector equation with nonnegative integer coefficients. The equation then reduces to

$$\sum_{i=1}^{m} \rho_i \bar{a}^{(i)} = \sum_{i=1}^{m} \rho_i \bar{b}^{(i)} = x,$$

where $x$ is an element of $A^+$. On the other hand, since $\rho_i > 0$ for some $i \leqslant j$, Definition 1 yields

$$\sum_{i=1}^{m} \rho_i \bar{a}^{(i)} >_I \sum_{i=1}^{m} \rho_i \bar{b}^{(i)},$$

or $x >_I x$. Thus, $>_I$ is not irreflexive.                                          ◇

## 9.4  APPLICATIONS

### 9.4.1  Scaling Considerations

An attempt to apply a finite linear measurement model to empirical data is all too often confronted with the unfortunate situation where the system of equations and inequalities derived from the data (via the measurement model) is inconsistent. In part, at least, this may result from sampling errors that can render the system unsolvable, even if the underlying model is basically valid. To overcome this difficulty, two approaches are available. One is to reduce sampling errors by controlling those factors that contribute to the variability of the data and by increasing the number of replications. Reducing error variance is clearly desirable; nonetheless, the model can be so sensitive even to relatively small sampling errors that this approach can turn out to be too costly.

A second approach to the problem recognizes the fallibility of the data and searches for some "best possible" representation in the presence of measurement errors. To justify such an approach we need a method to decide whether the observed inconsistencies are plausibly attributed to sampling errors. The development of adequate testing procedures, however, presents serious difficulties since it requires embedding the axioms of measurement theory in a statistical model. This problem is investigated in Chapter 17, Volume II, which is devoted to statistical procedures.

The problem is somewhat easier in a factorial experiment with several numerical observations in each cell, i.e., in each treatment combination. In this case we can search for a monotonic transformation of the dependent variable such that each cell mean of the transformed variable can be expressed (as nearly as possible) as an additive combination of its components. Using the terminology of the analysis of variance, we look for an order-preserving transformation that minimizes all interaction effects. The standard test for the significance of the interaction may, then, be used to test for the additivity of the data. A computer program that obtains the desired transformation has been developed by Kruskal (1965) and applied to various sets of real and imaginary data. A related method has been proposed by Box and Cox (1964), who have also developed a special statistical test for additivity. The main difference between the two methods is that in the latter one searches for the best monotone transformation within a particular one-parameter family of transformations, whereas in the former no such constraint is imposed. For a comparison of the methods, see Kruskal (1965).

If the actual data are given in an ordinal rather than in a numerical form and if the inequalities are inconsistent, then one may still want to find an additive representation that satisfies as many order statements as possible. More specifically, let $\phi$ be an additive function on $A$ which is not necessarily order preserving; then $a$, $b$ in $A$ are said to form an *inversion* with respect to $\phi$ if $a \gtrsim b$ but $\phi(a) < \phi(b)$. Thus, one may look for a solution that minimizes the number of inversions or, equivalently, that maximizes the rank order correlation (Kendall's $\tau$) with the observed relation. Such a solution can be found by systematic elimination of inequalities from the system till a (maximal) solvable subset is obtained. Note, however, that a maximal solvable subset is not necessarily unique. A computer program, based on a standard linear programming method, for finding a solution which minimizes the number of inversions is described in Tversky and Zivian (1966).

Another approach to the problem of inconsistent inequalities was pursued by Davidson *et al.* (1957). To describe their procedure, suppose that $\gtrsim$ is connected, and let $A = A_1 \times A_2$. Instead of the usual additive representation they proposed to find scales $\phi_1$, $\phi_2$ and a nonnegative $\theta$ such that for all $a, b$ in $A_1$, $p, q$ in $A_2$

$$ap \gtrsim bq \quad \text{iff} \quad \phi_1(a) + \phi_2(p) + \theta \geqslant \phi_1(b) + \phi_2(q).$$

Clearly, by selecting $\theta$ to be sufficiently large, these equations can always be satisfied. More interestingly, linear programming methods can be used to solve the above equations while minimizing the value of $\theta$, subject to the constraint that the smallest interval between scale values be of fixed length. Intuitively, $\theta$ may be thought of as a threshold of preference in the sense that if for some $x, y$ in $A$, $| \phi(x) - \phi(y)| \leq \theta$, then both $x \gtrsim y$ and $y \succ x$ may be observed. Consequently, $\gtrsim$ is not transitive for alternatives whose scale difference is less than $\theta$. (A similar idea serves as a basis for an algebraic error theory described in Chapter 15.) Unfortunately, the linear programming procedure for minimizing $\theta$ is not very satisfactory because all too often it results in scales where the differences between adjacent scale values are all equal to the minimum length. For further details see Davidson et al. (1957, pp. 82–103).

Once a consistent set of inequalities has been obtained, its solution is still not unique, as shown in Theorem 2. Hence, in order to obtain a particular numerical assignment, additional criteria have been introduced. Perhaps the most satisfactory solution is that based on the maximin $r$ criterion proposed by Abelson and Tukey (1959, 1963). This criterion selects the solution whose minimum product-moment correlation with any other solution is maximal. That is, let $X$ denote the solution space, then $z$ in $X$ is a *maximin r solution* if it maximizes $\min_{y \in X} r_{zy}$. Other proposed criteria have been justified in terms of specific assumptions or computational convenience. Abelson and Tukey (1959, 1963) have also shown how the value of the maximin $r$ increases with the number of order statements, i.e., with the cardinality of $\gtrsim$. To illustrate, consider the finite subtractive model (i.e., the algebraic difference model, see Section 4.4) and let $A = \{a, b, c, d\}$. If only the chain $ad \succ ac \succ ab \succ aa$ is given, then maximin $r = 0.80$. If, in addition, the set of adjacent elements $\{ab, bc, cd\}$ is weakly ordered, then maximin $r$ ranges between 0.82 and 0.97, depending on the particular ordering. Finally, if $A \times A$ is weakly ordered, the range of maximin $r$ is between 0.95 and 0.98. These results suggest that when the observed relation is connected (or nearly so) then, despite the lack of uniqueness, the proposed solution is quite "tight" in the sense that its departures from any other solution are negligible.

## 9.4.2 Empirical Examples

In this section we describe several applications of finite linear measurement models to psychological data. In each case, an attempt has been made to test the validity of the model, as well as to construct measurement scales by solving systems of homogeneous linear inequalities. In discussing the applications, we emphasize measurement models and scaling methods rather than experimental procedures and substantive conclusions.

a.  Our first example is the study of Davidson *et al.* (1957), designed to measure utility and subjective probability (see Section 8.6). In their experiment, subjects chose between two risky options (denoted I and II) of the general form

$$a_A \cup b_{-A} \quad \text{vs} \quad c_A \cup d_{-A},$$

where $a_A \cup b_{-A}$ means the option in which $a$ is the consequence if $A$ occurs and $b$ is the consequence if $-A$ occurs. In this special case it is convenient to display the option as

$$
\begin{matrix}
& \text{I} & \text{II} \\
A & \begin{bmatrix} a & c \\ b & d \end{bmatrix}, \\
-A &
\end{matrix}
$$

where $A$ represents a chance event and $a, b, c, d$, stand for some well-defined consequences such as winning or losing various amounts of money. Suppose the subject prefers I to II, i.e., $I \succ II$, then according to expected-utility theory, there exist a utility function $u$ and a subjective probability function $P$, such that

$$P(A)\,u(a) + P(-A)\,u(b) > P(A)\,u(c) + P(-A)\,u(d). \tag{5}$$

The method for constructing the utility function, which is based on an idea of Ramsey (1931), consists of first finding an event whose subjective probability equals $\frac{1}{2}$. To find such an event, it is both necessary and sufficient that the subject be indifferent between I and II whenever $a = d$ and $b = c$. For, suppose $a$ is preferred to $b$, then if $A$ is considered more likely than $-A$, $a_A \cup b_{-A}$ should be preferred to $b_A \cup a_{-A}$ whereas if $A$ is considered less likely than $-A$, the preference should be inverted. Under expected-utility theory, the indifference between I and II implies $P(A) = P(-A) = \frac{1}{2}$. Once an event (denoted $A^*$) that satisfies the above condition has been identified, Equation (5) reduces to $u(a) + u(b) > u(c) + u(d)$ or equivalently $u(a) - u(d) > u(c) - u(b)$.

Hence, preferences between options can be translated (under the theory) into linear inequalities of utility intervals. Davidson *et al.* (1957) presented an axiomatic analysis of the finite subtractive model, based on an equal spacing assumption, which results in an interval-scale measurement of utility. In order to apply the model to an experiment where only strict preferences are observed, an approximation procedure which yields bounds for points on the utility function was developed. It should be noted that the theory presented in Section 9.2.1 is directly applicable to the present experiment; yet it too fails to yield interval-scale measurement.

To illustrate the approximation procedure, define $(a, b) \gtrsim (c, d)$ iff $I \gtrsim II$ in the following choice situation:

$$
\begin{array}{cc}
 & \text{I} \quad \text{II} \\
A^* & \begin{bmatrix} a & c \\ b & d \end{bmatrix}.
\\ -A^* &
\end{array}
$$

Recall that $I \gtrsim II$ iff $u(a) + u(b) \geqslant u(c) + u(d)$. Now suppose $(b, b) \gtrsim (c, d)$ and $(c, d + 1) \gtrsim (b, b)$. Letting $u(b) = -1$, and $u(c) = 1$ yields $u(d + 1) \geqslant -3 \geqslant u(d)$. Hence, if $u(a) = -3$, then $a$ must be bounded between $d$ and $d + 1$. Next, we find some $e$ such that $(e, d) \gtrsim (b, c)$ and $(e - 1, d) \lesssim (b, c)$. Similarly, we find some $f$ such that $(b, c) \gtrsim (f, d + 1)$ and $(f + 1, d + 1) \gtrsim (b, c)$. Hence, $u(e) + u(d) \geqslant 0 \geqslant u(e - 1) + u(d)$ and $u(f + 1) + u(d + 1) \geqslant 0 \geqslant u(f) + u(d + 1)$, but since $u(d + 1) \geqslant -3 \geqslant u(d)$, $u(e) \geqslant 3 \geqslant u(f)$, and 3 is, therefore, bounded between $u(e)$ and $u(f)$. By continuing the process new monetary values can be bounded by the utility scale. This method, however, has a drawback: as one proceeds in constructing new bounds, errors of measurement accumulate and lengths of bounded intervals increase.

A summary of the data showing the bounds (in cents) for four fixed points on the utility scale is shown in Table 1 for 15 of the students that participated in the experiment. (The data of 4 additional subjects are excluded because the data failed to satisfy some tests required by the theory.) A closer examination of the data revealed that of the 60 pairs of bounds, 19 of the intervals were entirely above the linear money value, 21 were entirely below it, and 20 included it.

An attempt to remeasure the utility scales of 10 subjects some time after the initial session yielded results that were quite similar to those originally obtained, which indicates that the scale values were relatively stable. For further details, including a related procedure for measuring subjective probability and a somewhat different method for measuring utility, see Davidson et al. (1957).

b. Next, we discuss a study of grade expectation, conducted by Coombs (1964, pp. 92–102), to which data a subtractive model was applied. More specifically, the unidimensional folding model (see Section 4.12) was used to scale students' expectations of their course grades.

Sixty-two graduate students in a particular course were presented with all pairs of seven letter grades (from A+ to C+) and were asked to indicate, for each pair, which grade was closer to the grade they expected to receive in that particular course. Let $G$ be the set of grades and $S$ be the set of subjects, and let $x \succ_s y$ denote the observation that subject $s$ judges $x$ to be closer than $y$ to the grade he expects in the course, denoted $I(s)$. According to

**TABLE 1**

Summary of Data Determining Bounds in Cents for Fixed Points on the Utility Scale with $u(-4) = -1$ and $u(6) = 1$.[a]

| Subject | Bounds for $f$ where $u(f) = -5$ (cents) | Bounds for $c$ where $u(c) = -3$ (cents) | Bounds for $d$ where $u(d) = 3$ (cents) | Bounds for $g$ where $u(g) = 5$ (cents) |
|---|---|---|---|---|
| 1 | $-18$ to $-15$ | $-11$ to $-10$ | 11 to 12 | 14 to 18 |
| 2 | $-34$ to $-30$ | $-12$ to $-11$ | 12 to 18 | 31 to 36 |
| 3 | $-18$ to $-11$ | $-8$ to $-7$ | 10 to 13 | 14 to 22 |
| 4 | $-29$ to $-24$ | $-15$ to $-14$ | 14 to 17 | 25 to 31 |
| 5 | $-21$ to $-14$ | $-10$ to $-9$ | 10 to 12 | 16 to 24 |
| 6 | $-25$ to $-21$ | $-14$ to $-13$ | 13 to 15 | 19 to 23 |
| 7 | $-18$ to $-7$ | $-7$ to $-6$ | 7 to 14 | 10 to 23 |
| 8 | $-25$ to $-21$ | $-14$ to $-13$ | 14 to 17 | 23 to 28 |
| 9 | $-35$ to $-29$ | $-12$ to $-11$ | 16 to 18 | 43 to 50 |
| 10 | $-26$ to $-20$ | $-15$ to $-14$ | 14 to 15 | 20 to 27 |
| 11 | $-22$ to $-19$ | $-14$ to $-13$ | 11 to 13 | 18 to 22 |
| 12 | $-21$ to $-13$ | $-12$ to $-11$ | 8 to 12 | 11 to 15 |
| 13 | $-34$ to $-23$ | $-14$ to $-13$ | 13 to 17 | 23 to 32 |
| 14 | $-16$ to $-13$ | $-10$ to $-9$ | 12 to 15 | 20 to 24 |
| 15 | $-12$ to $-8$ | $-8$ to $-7$ | 8 to 10 | 11 to 15 |

[a] Davidson, Suppes, and Siegel, 1957, p. 62.

Coombs' folding model, there exists a scale $\phi$ defined on $G$ such that for each $s$ in $S$ and all $x, y$ in $G$,

$$x \succsim_s y \quad \text{iff} \quad |\phi(x) - \phi[I(s)]| \leqslant |\phi(y) - \phi[I(s)]|. \tag{6}$$

Given the judgments of grade expectations of the individual students, therefore, it should be possible to construct a joint scale of grades (or a J-scale, for short) in accordance with Equation (6).

Since transitivity of $\succsim_s$ was found to hold in all but a very few instances, a rank ordering of the grades (with respect to expectation) was obtained for each subject. Such an ordering was called an individual scale, or an I-scale for short. In order to construct a single numerical J-scale from the various ordinal I-scales, Coombs' unfolding technique was employed. The technique is designed to extract a partial order of grade intervals from a given set of I-scales. To illustrate the logic underlying the procedure, suppose the grade ordering of a given subject is

$$\text{A} - \succ_s \text{A} \succ_s \text{A} + \succ_s \text{B} + \succ_s \text{B} \succ_s \text{B} - \succ \text{C} +.$$

Recalling Equation (6), we note that since the subject's expectation point $I$ is closest to $A-$, it must satisfy $|\phi(A) - \phi(I)| > |\phi(A-) - \phi(I)|$. Moreover, since our subject expects an $A+$ more than he expects a $B+$, and using the natural ordering of the grades (from $A+$ to $C+$) and the former inequality, we obtain

$$|\phi(A+) - \phi(A)| + |\phi(A) - \phi(I)| = |\phi(A+) - \phi(I)|$$
$$< |\phi(B+) - \phi(I)|$$
$$\leqslant |\phi(B+) - \phi(A-)| + |\phi(A-) - \phi(I)|.$$

Subtracting $|\phi(A) - \phi(I)|$ and $|\phi(A-) - \phi(I)|$ from the left- and right-hand side of the above chain of inequalities respectively, yields $|\phi(A+) - \phi(A)| < |\phi(A-) - \phi(B+)|$, and the perceived difference between $A+$ and $A$ is, thus, smaller than that between $A-$ and $B+$. By systematic examination of all I-scales, a partial order of grade intervals is obtained. Typically, no one ordering satisfies all the data, so that a dominant ordering, satisfying the majority of the data, is sought. For further details concerning the unfolding technique, see Coombs (1964, pp. 80–92). In summarizing the analysis of the grade expectation data, Coombs writes:

> The results of this analysis are quite typical, in my experience, of the findings in an unfolding analysis of the data generated by what is presumed to be a unidimensional $J$ scale on a priori grounds—a dominant $J$ scale may usually be found satisfying the majority of cases, 72 % in this instance. ...Whether the cases that do not fit the dominant $J$ scale merely represent unreliability in the data or are significant departures can best be determined by using more powerful methods of collecting the data in the first place, for example, methods with redundancy which permit control of inconsistency. (Coombs, 1964, pp. 94–95.)

The dominant J-scale constructed in the grade expectation study embodied three salient features. (1) A change in a letter grade results in a bigger difference than the addition of a plus or a minus. Indeed, the $(A-, B+)$ and the $(B-, C+)$ intervals were the largest atomic intervals. (2) The addition of a plus results in a bigger difference than the addition of a minus, that is, the $(A+, A)$ and the $(B+, B)$ intervals were larger, respectively than the $(A, A-)$ and the $(B, B-)$ intervals. (3) The spread of the A's was greater than the spread of the B's, e.g., the $(A+, A-)$ interval was larger than the $(B+, B-)$ interval.

Given the partial order of grade intervals, a numerical grade scale was constructed following a method developed by F. Goode (see Coombs, 1964, pp. 96–102). The first step in applying the method consists in expressing larger grade intervals as sums of smaller grade intervals, e.g., the $(A, B)$ interval is expressed as a sum of the three atomic intervals $(A, A-)$, $(A-, B+)$, and $(B+, B)$. Next, one assigns unknown positive quantities to the various intervals, in accordance with their ordering. The smallest interval is expressed as $\Delta_1$, the second smallest interval is then expressed as $\Delta_1 + \Delta_2$, and so on. Thus, successively larger intervals are expressed as sums of additional positive $\Delta$'s, while incomparable intervals are expressed as sums of different $\Delta$'s. When this assignment process is completed, each grade interval is expressible as a linear combination of the $\Delta$'s with integral coefficients. To obtain the scale values of the grades, we set $\phi(A+) = 0$, and let the scale value of each grade be the length of the interval from that grade to $A+$. Hence, each grade $x$ in $G$ is assigned a coefficient vector $v(x)$, with nonnegative integral components, such that for any positive vector $\Delta = (\Delta_1, ..., \Delta_n)$ the function $\phi(x) = v(x) \cdot \Delta$ is a feasible solution of Equation (6). The scale values for the present study can, thus, be described by the following matrix equation:

$$
\begin{bmatrix}
0 & 0 & 0 & 0 & 0 & 0 \\
1 & 1 & 1 & 0 & 1 & 1 \\
2 & 1 & 1 & 0 & 2 & 2 \\
4 & 2 & 2 & 1 & 4 & 3 \\
5 & 3 & 2 & 1 & 5 & 4 \\
5 & 3 & 2 & 1 & 6 & 5 \\
7 & 4 & 3 & 2 & 7 & 6
\end{bmatrix}
\begin{bmatrix}
\Delta_1 \\
\Delta_2 \\
\Delta_3 \\
\Delta_4 \\
\Delta_5 \\
\Delta_6
\end{bmatrix}
=
\begin{bmatrix}
\phi(A+) \\
\phi(A) \\
\phi(A-) \\
\phi(B+) \\
\phi(B) \\
\phi(B-) \\
\phi(C+)
\end{bmatrix}.
$$

Moreover, any positive solution of Equation (6) with $\phi(A+) = 0$ can be obtained from the above equation by an appropriate selection of a positive vector $\Delta$. The dimensionality of $\Delta$, denoted $n$, equals the number of independent unknowns. Hence, the selection of a particular numerical assignment is equivalent to the choice of a particular $\Delta$. The first solution constructed by Coombs was the equal-$\Delta$ solution obtained by letting $\Delta_i = 1$, $i = 1, ..., n$. This solution as well as the maximin $r$ solution are presented in Table 2.

**TABLE 2**

Numerical Solutions for the Grade Expectation Data[a]

| Grade | A+ | A | A− | B+ | B | B− | C+ | min $r$ |
|---|---|---|---|---|---|---|---|---|
| Equal $\Delta$ solution | 0 | 5 | 8 | 16 | 20 | 22 | 29 | .97 |
| Maximin $r$ solution | 0 | 5 | 10 | 29 | 34 | 39 | 58 | .98 |

[a] Coombs, 1964, p. 100.

The table shows that the value of min $r$ for the maximin $r$ solution is 0.98, while the lowest product moment correlation of any solution with the

equal-Δ solution is as high as 0.97. There is no assurance, however, that the
equal-Δ solution will always do so well. The main advantage of the Δ-process
is the simplicity of calculation.

  c.   Our third example is a study of decision making under both risk and
certainty conducted by Tversky (1967b). The experiment was designed
to test expected-utility theory and to construct utility and subjective-probability scales through the use of the additive model. The subjects, 11 inmates
in a state prison, were instructed to state their minimal selling price for
each of several options. That is, they were asked to write down the smallest
amount of money for which they would sell the option to the experimenter.
Three sets of options were employed. Sets I and II consisted of risky options
of the form $(A, n)$, in which one wins $n$ packs of cigarettes (Set I), or bags
of candy (Set II) if event $A$ occurs and nothing if its complement $-A$ occurs.
The chance events were generated by spinning a wheel of fortune. All
combinations of four events and four prizes were used. Set III consisted
of all (riskless) combinations of cigarettes and candy.

  Let $M(A, n)$ denote the cash equivalent (i.e., price) of the risky option
$(A, n)$. Under the subjective expected-utility theory, there exist utility and
subjective-probability functions, denoted $u$ and $P$, respectively, such that

$$M(A, n) \geqslant M(A', n') \quad \text{iff}$$
$$P(A)\, u(n) + P(-A)\, u(0) \geqslant P(A')\, u(n') + P(-A')\, u(0).$$

By letting $u(0) = 0$, $f = \log u$ and $g = \log P$, the above equation reduces to

$$M(A, n) \geqslant M(A', n') \quad \text{iff} \quad g(A) + f(n) \geqslant g(A') + f(n').$$

Hence, expected-utility theory implies the additivity of the price matrices
for the risky options (Sets I and II). The additivity of the price matrices for
Set III is equivalent to the assumption that the subjective values of cigarettes
and candy combine additively.

  Since each option was presented three times, the median price was used
to determine the ordering of the options. (Note that when a magnitude-
estimation procedure, such as pricing, is used, transitivity is automatically
satisfied.) An additive solution that minimizes the number of inversions
was constructed using the algorithm developed by Tversky and Zivian (1966).
  Additivity was strongly supported by the data of all three sets.
  The majority of the bidding matrices (25 out of 33) were perfectly additive,
and the average number of inversions in the matrices that were not perfectly
additive was less than two. The number of inversions, denoted $m$, for all
bidding matrices is shown in Table 3.
  Let us suppose for a moment that the utility of money, denoted $v$, is an

**TABLE 3**

Number of Inversions ($m$) Formed by the Best Additive Solution and the $F$-Ratios for the Interaction Terms for Each Subject in All Sets[a,b]

| | Sets | | | | | |
|---|---|---|---|---|---|---|
| | I | | II | | III | |
| Subjects | $m$ | $F$ | $m$ | $F$ | $m$ | $F$ |
| 1 | 0 | .545 | 0 | .899 | 0 | .762 |
| 2 | 0 | 1.231 | 0 | .979 | 0 | 1.087 |
| 3 | 0 | .730 | 1 | .221 | 1 | .722 |
| 4 | 0 | .180 | 1 | 1.245 | 0 | 1.592 |
| 5 | 3 | 1.755 | 0 | 3.391* | 0 | 1.168 |
| 6 | 0 | .624 | 0 | .706 | 0 | .774 |
| 7 | 3 | 3.910* | 1 | 1.271 | 0 | 1.424 |
| 8 | 0 | .335 | 0 | 1.366 | 0 | .690 |
| 9 | 2 | .591 | 0 | .983 | 2 | .790 |
| 10 | 0 | .032 | 0 | .678 | 0 | 1.020 |
| 11 | 0 | 1.241 | 0 | .337 | 0 | .542 |

[a] Tversky, 1967b, p. 189.

[b] Statistical significance beyond the .10 level is indicated by an asterisk. All $F$-ratios are based on 9 and 32 degrees of freedom, except for Subjects 2 and 10 whose $F$-ratios are based on 9 and 16 degrees of freedom.

increasing power function of money, i.e., $v(x) = x^\theta$ for all $x \geqslant 0$, where $\theta > 0$. [This proposal was first made early in the 18th century by Cramer; see Stevens (1959), p. 58.] Hence, by the definition of $M(A, n)$ and expected utility theory

$$v[M(A, n)] = [M(A, n)]^\theta = P(A) u(n),$$

and, by taking logarithms,

$$\theta \log M(A, n) = \log P(A) + \log u(n) = g(A) + f(n). \tag{7}$$

Consequently, if $v(x) = x^\theta$, then $\log M(A, n)$ is expressible as an additive combination of its probability and value components. Conversely, suppose there exist real-valued functions $f$ and $g$, and a number $\theta$ such that for any $M(A, n) > 0$, $\theta \log M(A, n) = g(A) + f(n)$. By expected-utility theory, on the other hand, and the assumptions that $v$ is an increasing function of money,

$$\log v[M(A, n)] = \log[P(A) u(n)] = \log P(A) + \log u(n).$$

Hence, there exist two scales, $\theta \log M(A, n)$ and $\log v[M(A, n)]$, that are both additive with respect to $A$ and $n$. Assuming that the sets of prizes (e.g., cigarettes) and events are sufficiently rich to satisfy the hypotheses of Theorem 6.2, the two scales are related by a positive linear transformation, and hence, by taking exponentials, $v(x) = \alpha x^\beta$ for all $x \geqslant 0$, where $\alpha, \beta > 0$. Normalizing so that $v(1) = 1$ yields $\alpha = 1$. Therefore, under expected-utility theory together with the above assumptions, the validity of Equation (7) is both necessary and sufficient for the utility of money $v$ to be an increasing (normalized) power function. Note that no such constraints on the utility of prizes $u$ are imposed.

The power-function hypothesis was tested by first applying a logarithmic transformation to all the prices from Sets I and II (but not to those from Set III), and then submitting these to a series of individual two-factor analyses of variance. This analysis provides a statistical test of additivity in the form of a test for the significance of the interaction between the two factors of each price matrix. The analysis showed that while all main effects were highly significant, only two (out of 33) matrices revealed a significant interaction even at the lenient level of $p < 0.10$. The $F$-ratios for the interaction terms are presented in Table 3. The lack of significant interactions in Sets I and II supports the hypothesis that the subjects' prices can be accounted for by a power utility function for money, to the accuracy allowed by the variability of the prices. [When the same analysis was applied to the non-transformed (risky) prices, all but two matrices yielded a significant interaction beyond the 0.05 level, indicating that the proposed test for additivity is not powerless]. Further evidence on this point is presented in Tversky (1967c). Furthermore, the lack of significant interactions in Set III supported the hypothesis that the utilities for cigarettes and candy are approximately linear. This result is less surprising when we note the fact that both commodities are used as currency in the prison.

In further analyses aimed at scaling utility and subjective probability, utility theory was strengthened by imposing the following constraints. (i) Complementarity: subjective probabilities of complementary events sum to one. (ii) Risk invariance: the utility scales derived from risky and from riskless options via additivity should coincide (see Section 8.4.2). It was found, however, that although additivity was satisfied, complementarity and risk invariance could not be satisfied simultaneously. It turned out that the subjects overpriced the risky options. So when complementarity was assumed, the risky utilities exceeded the riskless ones; and when the two utility scales were assumed to coincide, the sum of subjective probabilities of complementary events exceeded unity. Hence, despite the fact that utility theory was supported in each set of data separately, no simultaneous solution satisfying the constraints imposed by the theory was possible. Alternative additive representations, that were not compatible with utility theory but provided a better account of the data, were investigated in Tversky (1967b). Finally, it is of interest to note that the actual attempt to construct measurement scales may highlight some important aspects of the data that are not easily perceived while testing isolated axioms.

d. Most empirical work on additivity in psychology has analyzed numerical, rather than ordinal, measures. Typically, a particular experiment is submitted to an analysis of variance in order to assess the contributions of the various factors and to test the significance of the interactions among them. If the results yield nonsignificant interactions, one concludes that additivity is satisfied and that the particular numerical measure employed can be treated as an additive scale of measurement. Moreover, standard statistical estimation procedures can be used to scale the independent variables additively. It should be noted, however, that the above conclusion is justifiable only if the statistical test is sufficiently powerful; otherwise the lack of significant interactions may be due simply to error variance.

Anderson has employed this approach in a variety of judgment tasks, ranging from impression formation (Anderson, 1962) to psychophysical judgments (Anderson, 1970), in which subjects evaluate multicomponent stimuli, such as lists of adjectives, or combinations of personality traits. In these studies the actual numerical judgments are often well approximated by a simple additive model. This is not always the case, and Anderson has also advocated monotone transformations, although he has not provided formal criteria for their use. When significant interactions are obtained, an ordinal analysis is required to determine whether the interaction is genuine, or whether it is attributable to the particular scale employed, and hence removable by an appropriate order-preserving transformation.

To illustrate the problem, consider the study of Sidowski and Anderson (1967) on the attractiveness of working at a certain occupation in a certain city. The study included four occupations—accountant, teacher, lawyer, and doctor—and four cities, which were preselected by each subject as being: high (A), moderately high (B), neutral (C), and moderately low (D) in attractiveness as places to live. Forty male students judged the attractiveness of each of the 16 city–occupation combinations on a rating scale from 1 to 9. The mean ratings of attractiveness for all combinations are presented in Table 4. Since the data for doctor and lawyer did not differ significantly, they were combined (by the investigators) for clarity.

Table 4 displays a clear interaction effect: whereas the difference between the rating for lawyer–doctor and rating for accountant is approximately constant across cities, the rating for teacher approaches the former in city A and the latter in city D. Indeed, an analysis of variance applied to these data yielded a highly significant interaction ($P < 0.001$) which was interpreted by the investigators in terms of the fact that "a teacher would tend to be

**TABLE 4**

Mean Rating of Attractiveness for City–Occupation Combinations.[a]

|              | City |     |     |        |
| ------------ | ---- | --- | --- | ------ |
| Occupation   | A    | B   | C   | D      |
| Lawyer–doctor | 7.3 | 6.8 | 5.7 | 4.4    |
| Teacher      | 7.3(−) | 6.7 | 5.3 | 3.2(+) |
| Accountant   | 5.9  | 5.4 | 4.3 | 3.2    |

[a] The entries of the table were obtained from the graph in the original report. Although the curve for teacher appears to coincide with the other curves at the endpoints, we have assumed that it lies strictly between them, as was found in Experiment II (Sidowski & Anderson, 1967, Experiment I, p. 279).

in more direct contact with the socioeconomic milieu of a city" and hence be more influenced by the city.

An ordinal analysis of the data, however, reveals that Table 4 satisfies all of the necessary cancellation axioms, and hence it has an order-preserving additive representation. An example of an order-preserving transformation of Table 4 that yields a perfect additive representation is displayed in Table 5.

**TABLE 5**

An Order-Preserving Additive Representation for the Data Presented in Table 4.[a]

|               | City |     |     |     |
| ------------- | ---- | --- | --- | --- |
| Occupation    | A    | B   | C   | D   |
| Lawyer–doctor | 7.3  | 6.8 | 5.7 | 4.4 |
| Teacher       | 6.9  | 6.4 | 5.3 | 4.0 |
| Accountant    | 5.9  | 5.4 | 4.3 | 3.0 |

[a] In constructing an additive representation on the basis of group data, it is assumed that the order of the mean ratings reflects that of the majority of subjects—an assumption that is not necessarily valid.

It is of interest to note that the proposed rescaling preserves not only the order of the entries but also 8 of the 12 original values. In fact, only the two lowest values and two of the four highest values were modified. The interaction between city and occupation, therefore, is attributable to the nature of the rating scale, because it can be eliminated by appropriate rescaling. The above discussion illustrates the range of applicability of the additive model (as exemplified in the work of Anderson and his collaborators), and the importance of the ordinal analysis for an adequate interpretation of the data.


## 9.5 POLYNOMIAL STRUCTURES

In previous sections, we have seen how factorial data together with a proposed measurement model give rise to a set of polynomial inequalities. Generally, for each element $a_1 \cdots a_n$ in $A_1 \times \cdots \times A_n$ we define the corresponding polynomial $p$ in the unknowns corresponding to $a_1, \ldots, a_n$. If the proposed model is decomposable, there will be one unknown for each $a_i$ in $A_i$, and the set $Y$ of unknowns corresponds to $\bigcup_{i=1}^n A_i$. Otherwise, $Y$ may contain several unknowns for some $a_i$ in $A_i$. We define the relation $\gtrsim_I$ between polynomials by specifying that if $p$ corresponds to $a_1 \cdots a_n$ and $q$ to $b_1 \cdots b_n$, then

$$p \gtrsim_I q \quad \text{iff} \quad a_1 \cdots a_n \gtrsim b_1 \cdots b_n.$$

Let $R[Y]$ denote the set of all polynomials in the unknowns of $Y$, with real numbers as coefficients. For $p, q$ in $R[Y]$, we write $p = q$ whenever the equation $p = q$ is an algebraic identity, e.g., $x_1 x_2 - x_1 x_2 = x_3 - x_3$. With respect to the usual addition and multiplication of polynomials, $R[Y]$ is clearly a ring. Throughout, $>$ and hence $>_I$ are assumed nonempty.

THEOREM 4. *A set of polynomial inequalities in the unknowns $Y$ has a solution iff the corresponding relation $\gtrsim_I$ on $R[Y]$ can be extended to a weak order $\gtrsim_{II}$ such that $\langle R[Y], \gtrsim_{II} \rangle$ is an Archimedean weakly ordered ring, i.e., $\gtrsim_{II}$ induces an Archimedean ordered ring structure (p. 58) on $R[Y]/\sim_{II}$.*

We desire to find conditions on a set of inequalities such that the extension to a weakly ordered ring is possible. In the case of linear inequalities considered in Section 9.2, it was sufficient to deal with the semigroup of sums of unknowns. A satisfactory solution to the problem of extending the set of linear inequalities to an Archimedean ordering of the semigroup was achieved. In the case of a ring, the results are not as good. We can find a set of necessary conditions which are quite similar to the necessary and sufficient conditions used in the linear situation. These are given in the Corollary to Theorem 5. The necessary and sufficient conditions for the ring extension, given in Theorem 6, however, do not imply additional consequences that are testable in any simple way. To state these theorems, we introduce the idea of a regular extension of a relation $\gtrsim_I$, closed under addition and multiplication.

DEFINITION 2. *Let* $\succsim_I$ *be a relation on* $R[Y]$. *A pair of relations,* $(\sim_{II}, \succ_{II})$ *is called a* regular extension *of* $\succsim_I$ *iff the following two conditions hold for all* $p, q \in R[Y]$:

(a)  $p \sim_{II} q$ *whenever one of the following holds:*

  (i)  *Extension:*  $p \sim_I q$;

  (ii)  *Polynomial identity:*  $p = q$;

  (iii)  *Closure:   There exist* $p', p'', q', q''$ *with* $p' \sim_{II} q'$ *and* $p'' \sim_{II} q''$ *such that either* $p = p' + p''$, $q = q' + q''$ *or* $p = p'p''$, $q = q'q''$.

(b)  $p \succ_{II} q$ *whenever one of the following holds:*

  (i)  *Extension:*  $p \succ_I q$;

  (ii)  *Additive closure:   There exist* $p', p'', q', q''$ *with* $p' \succ_{II} q'$, $p'' \sim_{II} q''$ *such that* $p = p' + p''$, $q = q' + q''$;

  (iii)  *Multiplicative closure:   There exist* $p', q', r$ *with either* $p' \succ_{II} q'$, $r \succ_{II} 0$ *or* $q' \succ_{II} p'$, $0 \succ_{II} r$, *such that* $p = p'r$, $q = q'r$.

THEOREM 5.   *Any binary relation on* $R[Y]$ *has at least one regular extension (the universal extension) and a unique minimal regular extension.*

We remark that the minimal extension $(\sim^*, \succ^*)$ could be defined recursively. First, Definition 2 (a) can be made a recursive definition of $\sim^*$. Definition 2 (b) assumes that $\sim_{II}$ is defined [2 (b) (ii)] and so it can be made into a recursive definition of $\succ^*$ relative to $\sim^*$. Neither the minimal nor any other regular extension need be an order relation; even if $\succ_I$ is irreflexive, $\succ_{II}$ need not be. The minimal regular and the universal extensions may coincide. Note that $\sim_{II}$ and $\succ_{II}$ are transitive.

If $\succsim'$ is an extension of $\succsim_I$ such that $\langle R[Y], \succsim' \rangle$ is a weakly ordered ring, then clearly, $\sim'$ includes $\sim^*$ and $\succ'$ includes $\succ^*$. The reason for this is that $\sim'$ satisfies part (a) of Definition 4, and $\succ'$ satisfies part (b) relative to $\sim^*$. Hence, if $\succsim'$ is such an extension, then $\succ^*$ is irreflexive. Thus we have established the following.

COROLLARY.   *If there exists an extension of* $\succsim_I$ *such that* $\langle R[Y], \succsim' \rangle$ *is a weakly ordered ring and if* $(\sim^*, \succ^*)$ *is the minimal regular extension of* $\succsim_I$, *then* $\succ^*$ *is irreflexive.*

This corollary is the counterpart of Theorem 1 where irreflexivity was sufficient, as well as necessary, for extension of $\succsim_I$ to an Archimedean weak order of a group. Irreflexivity of the minimal regular extension implies all of the cancellation and independence conditions.

To illustrate, we derive the distributive cancellation condition (Definition 7.5) used in axiomatizing the polynomial $(x_1 + x_2) x_3$. Suppose that this condition is violated in a factorial experiment; then there exist $a_1, a_2, a_3,$ $b_1, b_2, c_1, c_2, c_3, d_1, d_2$ such that

$$a_1 b_2 a_3 \gtrsim d_1 c_2 c_3,$$
$$b_1 a_2 a_3 \gtrsim c_1 d_2 c_3,$$
$$d_1 d_2 c_3 \gtrsim b_1 b_2 a_3,$$
$$c_1 c_2 c_3 > a_1 a_2 a_3.$$

If, for simplicity, we denote the unknowns corresponding to $a_i, b_i$, etc., by $\alpha_i, \beta_i$, etc., then we obtain the polynomial inequalities

$$(\alpha_1 + \beta_2)\, \alpha_3 \gtrsim_{\mathrm{I}} (\delta_1 + \gamma_2)\, \gamma_3,$$
$$(\beta_1 + \alpha_2)\, \alpha_3 \gtrsim_{\mathrm{I}} (\gamma_1 + \delta_2)\, \gamma_3,$$
$$(\delta_1 + \delta_2)\, \gamma_3 \gtrsim_{\mathrm{I}} (\beta_1 + \beta_2)\, \alpha_3,$$
$$(\gamma_1 + \gamma_2)\, \gamma_3 >_{\mathrm{I}} (\alpha_1 + \alpha_2)\, \alpha_3.$$

From (b) (i) and (b) (ii) of Definition 2, we have

$$(\alpha_1 + \alpha_2 + \beta_1 + \beta_2)\, \alpha_3 + (\delta_1 + \delta_2 + \gamma_1 + \gamma_2)\, \gamma_3$$
$$>^* (\alpha_1 + \alpha_2 + \beta_1 + \beta_2)\, \alpha_3 + (\delta_1 + \delta_2 + \gamma_1 + \gamma_2)\, \gamma_3$$

or $p >^* p$, contradicting irreflexivity. Other cancellation or independence conditions, for other polynomial models, may be derived similarly.

Finally, we formulate a necessary and sufficient condition for the existence of an extension of $\gtrsim_{\mathrm{I}}$ which yields a weakly ordered ring.

THEOREM 6. *A set of polynomial inequalities in the unknowns of Y has a solution iff the corresponding relation $\gtrsim_{\mathrm{I}}$ on $R[Y]$ has a regular extension $(\sim_{\mathrm{II}}, >_{\mathrm{II}})$ such that $\gtrsim_{\mathrm{II}}$ is Archimedean and connected and $>_{\mathrm{II}}$ is nonuniversal.*

We conjecture that a finite set of polynomial inequalities has a solution if and only if the relation $>^*$ of the minimal regular extension is irreflexive; i.e., that for finite data, if the minimal extension yields a quasi-ordered ring, then there exists an extension to an Archimedean weak order. This would mean that the conjunction of the necessary cancellation conditions derivable from irreflexivity is also sufficient for solvability of the inequalities. The Corollary to Theorem 5, and Theorem 6, could thus be improved since the assumption of existence of an Archimedean, connected, nonuniversal regular extension does not easily lead to additional cancellation type conditions, other than the consequences of irreflexivity.

The infinite, recursive chain of cancellation conditions, equivalent to irreflexivity of the minimal regular extension, clearly cannot all be tested.

No practical general methods for determination of the solvability of a set of simultaneous polynomial inequalities have been devised; one generally proceeds by trial and error. However, one may hope to find some reasonable limitation on the size and type of cancellation conditions that need to be checked before concluding that a solution exists for a given set of polynomial inequalities. This may depend on the factorial structure that enters into the form of the set of inequalities. This sort of structure was not utilized in the theorems proved in this section. The principle contribution of the present analysis is to indicate the qualitative nature of the conditions underlying general polynomial combination rules, and to set the stage for further research on these questions.

## 9.6 PROOFS

### 9.6.1 Theorem 4 (p. 447)

*A set of polynomial inequalities in the unknowns $Y$ has a solution iff the corresponding relation $\succsim_I$ on $R[Y]$ can be extended to a weak order $\succsim_{II}$ such that $\langle R[Y], \succsim_{II} \rangle$ is an Archimedean weakly ordered ring.*

*PROOF.* First, suppose that a solution to the inequalities is given by a function $\phi$ which assigns some number $\phi(y)$ to any unknown $y \in Y$. Then for any two polynomials $p = p(y_1, ..., y_r)$ and $q = q(y_1, ..., y_s)$, define

$$p \succsim_{II} q \quad \text{iff} \quad p[\phi(y_1), ..., \phi(y_r)] \geqslant q[\phi(y_1), ..., \phi(y_s)].$$

The second inequality involves only real numbers, obtained by substituting $\phi(y_i)$ for $y_i$. Clearly, $\succsim_{II}$ is transitive and connected. By virtue of the fact that $\phi(y)$ gives a solution to $\succsim_I$, $\succsim_{II}$ extends $\succsim_I$: if $p \succsim_I q$, then $p \succsim_{II} q$. Finally, we show that $\langle R[Y], \succsim_{II} \rangle$ is an Archimedean weakly ordered ring. The fact that inequalities are preserved under addition or under multiplication by any polynomial $p \succ_{II} 0$ follows from the corresponding fact for real numbers, e.g., $p[\phi(y_1), ..., \phi(y_r)] \geqslant p'[\phi(y_1), ..., \phi(y_s)]$ implies

$$p[\phi(y_1), ..., \phi(y_r)] + q[\phi(y_1), ..., \phi(y_t)]$$
$$\geqslant p'[\phi(y_1), ..., \phi(y_s)] + q[\phi(y_1), ..., \phi(y_t)].$$

To show the Archimedean property, note that if $p[\phi(y_1), ..., \phi(y_r)] > 0$, then for sufficiently large $n$, $np[\phi(y_1), ..., \phi(y_r)]$ is as large as desired. Thus, for $p \succ_{II} 0$ and for any $q$, there exists $n$ such that $np \succ_{II} q$.

Conversely, suppose that $\langle R[Y], \succsim_{II} \rangle$ is an Archimedean weakly ordered ring and $\succsim_{II}$ is an extension of $\succsim_I$. By the Corollary to Theorem 2.6 there exists an order-preserving homomorphism $\phi$ of $R[Y]$ onto a subring of real numbers. By the properties of a homomorphism, for any $p$,

$$\phi(p) = p[\phi(y_1),..., \phi(y_n)].$$

Thus the numbers $\phi(y)$, $y \in Y$, give a solution to all the inequalities of $\gtrsim_{II}$, in particular, to those of $\gtrsim_I$. $\diamond$

### 9.6.2 Theorem 5 (p. 448)

*Any binary relation on $R[Y]$ has at least one regular extension and a unique minimal regular extension.*

*PROOF.* Note that the universal relation ($pUq$ for all $p$, $q$) satisfies part (a) of Definition 4, and that whenever several relations, $\sim_{II}$, $\sim_{III}$,..., satisfy part (a), then so does their intersection. Thus, there is a unique minimal $\sim^*$ satisfying part (a), namely the intersection of all relations satisfying those conditions. We let $\sim^*$ denote this minimal relation.

Next, note that if $\succ_{II}$ is the universal relation, it satisfies part (b) of Definition 4 with respect to $\sim^*$ defined above, and that the same thing holds for the intersection of several relations, $\succ_{II}$, $\succ_{III}$,..., satisfying part (b) relative to $\sim^*$. Thus, we let the universal extension be $(\sim^*, U)$, where $U$ is the universal relation and $\sim^*$ is minimal, and we let $(\sim^*, \succ^*)$ be the minimal extension, where $\sim^*$ is defined above, and $\succ^*$ is the intersection of all relations satisfying part (b) relative to $\sim^*$. $\diamond$

### 9.6.3 Theorem 6 (p. 449)

*A set of polynomial inequalities in the unknowns $Y$ has a solution iff the corresponding relation $\gtrsim_I$ on $R[Y]$ has a regular extension $(\sim_{II}, \succ_{II})$ such that $\gtrsim_{II}$ is Archimedean and connected and $\succ_{II}$ is nonuniversal.*

*PROOF.* Suppose $(\sim_{II}, \succ_{II})$ is a regular extension, $\succ_{II}$ is nonuniversal, and $\gtrsim_{II} = \succ_{II} \cup \sim_{II}$ is an Archimedean weak order on the additive group of $R[Y]$. Then $\succ_{II}$ is irreflexive; for if $p \succ_{II} p$, then $0 \succ_{II} 0$, hence, for any $q$, $q'$ there exists $n \in I^+$ with $0 = n0 \succ_{II} q - q'$, whence $q \succ_{II} q'$ and $\succ_{II} = U$. It follows immediately that $\succ_{II}$ is asymmetric; in fact, it is the asymmetric part of the weak order $\gtrsim_{II}$. Since the extension is regular, the ring operations are well defined on $R[Y]/\sim_{II}$, which has at least two elements (because $\succ_I$ is assumed nonempty). It follows that $\langle R[Y], \gtrsim_{II} \rangle$ is an Archimedean weakly ordered ring, so the theorem reduces to Theorem 4. The converse is immediate by Theorem 4. $\diamond$

# EXERCISES

**1.** Let $\gtrsim$ be a weak order of $A = A_1 \times A_2$. Show that triple cancellation (Section 6.2.1) does not follow from double cancellation (Definition 6.3) and independence (Definition 6.1).     (9.2)

**2.** Let $A = \{a_0, ..., a_n\}$, $B = \{b_0, b_1\}$. Show that if $\langle A \times B, \gtrsim \rangle$ is an independent weak order in the sense of Definition 6.1, then it has an order-preserving additive representation.     (9.2)

**3.** Show that the transitivity of $>$ does not imply the transitivity of $\gtrsim_I$.     (9.2.1)

**4.** Suppose $\gtrsim$ is connected. Derive triple cancellation (Section 6.2.1) from the irreflexity of $>_I$.     (9.2.1)

**5.** Let $\gtrsim$ be a reflexive and connected binary relation on $A \times B$, and let $>_I$ be defined as in Definition 1. Show that the following three conditions are equivalent:

   (i)   *Irreflexivity*:   For all $x \in A^+$, not $(x >_I x)$

   (ii)  *Generalized cancellation*: For all $x \in A^+, u, v \in A \times B, x + u >_I x + v$ implies $u > v$.

   (iii) *Scott's condition*:   For all sequences $a_0, ..., a_n$ in $A$, $b_0, ..., b_n$ in $B$, and all permutations $\pi, \sigma$ of $\{0, 1, ..., n\}$, if $a_i b_i \gtrsim a_{\pi(i)} b_{\sigma(i)}$ for $i = 0, ..., n - 1$, then $a_{\pi(n)} b_{\sigma(n)} \gtrsim a_n b_n$.     (9.2.1)

**6.** Let $A$ be a finite set, $\gtrsim$ a reflexive binary relation on $A$, and $\circ$ a mapping from $B \subset A \times A$ onto $C \subset A$. Derive a necessary and sufficient condition for the existence of a scale $\phi$ satisfying, for all $a, b \in A$:

   (i)   $a > b$ implies $\phi(a) > \phi(b)$;

   (ii)  $a \sim b$ implies $\phi(a) = \phi(b)$;

   (iii) if $a, b \in A - C, a \circ b \in C$, then $\phi(a \circ b) = \frac{1}{2}[\phi(a) + \phi(b)]$.     (9.2.1)

**7.** Let $A$ be a finite nonempty set, $\gtrsim$ a binary relation on $A \times A$. Find necessary and sufficient conditions for the existence of a real-valued function $\phi$ on $A$ satisfying, for all $a, b, c, d \in A$

$$ab \gtrsim cd \quad \text{iff} \quad \phi(a) - \phi(b) \geqslant \phi(c) - \phi(d).$$     (9.2.1)

**8.** Characterize the set of all positive additive representations of the following structure, where the cell entries indicate the rank order of the elements of $A \times B$.

| $A$ | | | |
|---|---|---|---|
| $B$ | $a_1$ | $a_2$ | $a_3$ |
| $b_3$ | 3 | 5 | 6 |
| $b_2$ | 2 | 4 | 5 |
| $b_1$ | 1 | 2 | 3 |

(9.2.1, 9.4.2)

**9.** Consider the following structure where the cell entries indicate the rank order of the elements of $A \times B$.

| $A$ | | | |
|---|---|---|---|
| $B$ | $a_1$ | $a_2$ | $a_3$ |
| $b_3$ | 6 | 7 | 9 |
| $b_2$ | 2 | 5 | 8 |
| $b_1$ | 1 | 3 | 4 |

(i) Find an additive numerical assignment ($\phi_A$ and $\phi_B$) for $A \times B$ that minimizes the number of inversions.

(ii) Normalize the assignment so that $\phi_A(a_1) = \phi_B(b_1) = 0$ and $\phi_A(a_3) = \phi_B(b_3) = 1$. Find the minimal value of the threshold parameter $\theta$ required for the above numerical assignment.     (9.4.1)

**10.** Let $(\sim_{II}, >_{II})$ be a binary relations on $R[Y]$, defined as in Definition 2. Show that $>_{II}$ is irreflexive iff there are no $p, q \in R[Y]$ such that $p \gtrsim_{II} q$ and $q >_{II} p$.

# Chapter 1O  Dimensional Analysis and Numerical Laws

---

## □ 10.1  INTRODUCTION

Taken together, the numerical measures of physics exhibit a very simple algebraic structure which, although completely familiar and therefore not surprising, tends to be mysterious when given any thought. That is what we do in this chapter—we think about it. We try to describe the structure precisely and, to the extent we can, to account for it.

All of the physical measures with which we shall deal are (or are treated as if they are) ratio scales, i.e., they are completely determined except for an arbitrarily chosen unit. Most, although not all, physical measures have several units in common use; these are usually agreed upon by some international congress or commission. In one system, we find units such as these: second (s) for time, meter (m) for length, newton (N) for force, kilogram (kg) for mass, joule (J) for energy, coulomb (C) for charge, ampere (A) for current, ohm ($\Omega$) for electrical resistance, etc. Each of these is carefully defined in terms of either some natural phenomena (e.g., the period of rotation of the earth about the sun or the wavelength of some monochromatic light source), some standard object (e.g., the length of a certain rod in Paris under specified conditions), or some other units (e.g., kilogram-meter per second squared). For detailed discussions of the choice of units and systems of units, see Kayan (1959).

Since not all of the attributes just listed are independent of each other, some units can be expressed as a numerical factor times certain combinations of other units. For example, when we select charge, temperature, mass, length, time duration, and angle[1] as primary or basic dimensions, then all other known physical attributes that have ratio scale measures can be treated as secondary dimensions that are monomial combinations of these six basic ones. In particular, if we symbolize the dimensions of charge, temperature, mass, length, time, and angle by $Q$, $\Theta$, $M$, $L$, $T$, and $A$, respectively, then density has the dimensions $ML^{-3}$, frequency $T^{-1}A$, force $MLT^{-2}$, current $QT^{-1}$, entropy $\Theta^{-1}ML^2T^{-2}$, etc. This means that, if we wish, we can select arbitrary units for the six basic dimensions and then define all the remaining units in terms of these. When we do, the resulting system of units is called *coherent*. In incoherent systems, the units of (some) measures are selected independently of the primary ones, and so they differ from the corresponding coherent unit by a numerical factor. These dimensional factors must be included in the statement of physical equations; the great advantage of a coherent system of units is that these factors are all 1, and so they can be omitted.

Not only can we express all secondary dimensions as combinations of primary ones, e.g., $Q$, $\Theta$, $M$, $L$, $T$, $A$, but the combinations are always simple monomials, i.e., of the form $Q^\chi\Theta^\theta M^\mu L^\lambda T^\tau A^\alpha$, where $\chi$, $\theta$, $\mu$, $\lambda$, $\tau$, $\alpha$ are small integers. In fact, if we examine Thun's (1960) very complete listing of physical measures, units, and dimensions in the $Q\Theta MLTA$ system, which is reproduced in modified form at the end of the chapter as Table 3, we find that their maximum and minimum values are

|  | $\chi$ | $\theta$ | $\mu$ | $\lambda$ | $\tau$ | $\alpha$ |
|---|---|---|---|---|---|---|
| Maximum: | 2 | 1 | 1 | 4 | 2 | 2 |
| Minimum: | −2 | −1 | −1 | −4 | −3 | −2 |

It is widely agreed that the reason we need only few basic dimensions, even though there are at least 100 distinct useful measures (see Table 3), is because some physical measures are related to others by physical laws.

---

[1] Angle is the bastard quantity of dimensional analysis, about which everyone seems a bit uncomfortable (e.g., Palacios, 1964, p. 43; Sedov, 1959, pp. 2–3). It is said to be dimensionless (because it can be defined as the ratio of two lengths, namely, the arc subtended to the radius of a circle) and also to be extensively measurable (because angles can be concatenated in the obvious way as we saw in Section 3.2.2) and therefore it has a unit. In listing the dimensions of physical quantities, it has to be included in the units but usually it is omitted from the dimensions; so, for example, angular velocity is reported in radians per second, but supposedly it has only the dimension $T^{-1}$. Something is bizarrely wrong; we give what we think is a sensible account of the difficulty in Section 10.6.

It seems to us nowadays a very simple thing to assign dimensions to magnitudes, so simple that we are apt to forget the extremely important implication of the assertions. When we assert that a certain derived magnitude always has certain dimensions, we are in fact asserting the complete accuracy of the law which determines that derived magnitude under all possible conditions. If there is any doubt whatever about the universality of the law, then there is a corresponding doubt about the dimensions of the derived magnitude or even about its existence. (Campbell, p. 416 of the 1957 edition.)

Two familiar examples illustrate the point. First the density of an object is defined to be the ratio of mass to volume, and so it has the dimensions $ML^{-3}$. It is a fact—a law—that for homogeneous materials this ratio is independent of the volume used, and so the notion of the density of a material is a derived measure whose existence depends upon the validity of a law. Second, the displacement $s$ of a body falling in a vacuum for a time $t$ at the surface of a planet is proportional to $t^2$. The constant of proportionality, which depends upon the planet but not on anything else, is interpreted as a measure of the gravitational attraction of the planet, and it has the dimensions of an acceleration, $LT^{-2}$.

Such factors of proportionality are, misleadingly, called dimensional constants. They possess dimensions alright, but mostly they are not constant since they vary from system to system, as in the two examples just cited. They are also called system-dependent or material constants, and they are treated as new measures, somewhat but not entirely on a par with length, mass, time, etc. Campbell (1920) referred to them as *derived measures*, which he contrasted with those *fundamental measures* that result from extensive measurement. Indeed, there are differences. It is hardly feasible to concatenate planets. And substances, although easily concatenated, fail to exhibit the property $a \circ b > a, b$ of extensive quantities, where $\gtrsim$ is the ordering by density. In fact, density of materials exhibits the idempotent property $a \circ a \sim a$, which is characteristic of what are called *intensive attributes*. Temperature is another example of an intensive attribute.

Some dimensional constants—the *universal constants*, listed in Table 1—are truly constant in the sense that, for any fixed system of units, they do not vary.

Universal constants have one disturbing feature. They appear in the expressions for the laws of physics without being previously defined either qualitatively or quantitatively. They are independent of the nature of the bodies to which these laws are applied. Further, they are not physical quantities since they always occur with the same value; and to imagine that in another universe they can take other values is an unjustified and useless metaphysical speculation. As there exists only one example of each, if they were physical quantities they would themselves be their own units and their value would be the number one. Nevertheless, they are not mere numbers

**TABLE 1**

Universal Constants[a]

| | |
|---|---|
| Velocity of light | $c = 2.99793 \times 10^{10}$ cm/s |
| Electron charge | $e = 1.602 \times 10^{-19}$ C |
| Electron mass | $m_e = 9.1091 \times 10^{-28}$ g |
| Proton mass | $m_p = 1.67252 \times 10^{-24}$ g |
| Neutron mass | $m_n = 1.674 \times 10^{-24}$ g |
| Atomic mass unit | amu $= 1.660 \times 10^{-24}$ g |
| Electron-volt | eV $= 1.591 \times 10^{-12}$ erg |
| Gravitational constant | $G = 6.670 \times 10^{-8}$ cm$^3$/g s$^2$ |
| Gas constant | $R = 8.3143 \times 10^7$ erg/mol-g °K |
| Boltzmann's constant | $k = 1.3805 \times 10^{-16}$ erg/°K |
| Planck's constant | $h = 6.626 \times 10^{-27}$ erg s |
| Stefan's constant | $\sigma = 5.670 \times 10^{-5}$ g/s$^3$ °K$^4$ |
| Wien's constant | $b = 0.2897$ cm °K |
| Avogadro's constant | $N_A = 6.0225 \times 10^{23}$/mol-g |

[a] Recommendations of the U. S. National Bureau of Standards, 1963.

since their value depends on the system of units adopted to measure the magnitudes which occur in the equations. Finally, their existence is in some respects precarious since, for example, what has been called the dynamical constant is not mentioned in any physics textbook, whilst the mechanical equivalent of heat, which used to be discussed at great length in textbooks at the beginning of the century, is hardly mentioned in present day works.... (Palacios, 1964, p. 23.)

Other derived measures seem still different. A body of mass $m$ moving at velocity $v$ has momentum $mv$ and kinetic energy $\frac{1}{2}mv^2$, which have the dimensions $MLT^{-1}$ and $ML^2T^{-2}$, respectively. To be sure, these measures arise in the formulation of many laws, but no law seems to be involved in defining them. They appear to be analogous to the definition of the density of objects, but not to the definition of the density of materials, which definitely depends upon the truth of a law. Yet clearly, these two measures are of great importance, whereas most quantities of the form $m^i v^j$, where $i$ and $j$ are integers, apparently are not. We need to understand just how, if at all, laws play a role here.

The laws that seem essential to the definition of some physical measures are very simple—in fact, they seem minimal since they involve only two variables. Most of physics is concerned with laws of greater complexity. Nonetheless, these more complicated laws are all quite compatible with the structure of dimensions constructed in this pairwise fashion. An example illustrates what we mean. Let $\Omega$ denote the angular velocity (dimension $AT^{-1}$) of a viscous fluid, $r$ the radial distance (dimension $L$), and $t$ the

time (dimension $T$); then it is known that the partial differential equation
describing the propagation of vorticity is

$$\frac{\partial \Omega}{\partial t} = \nu \left( \frac{\partial^2 \Omega}{\partial r^2} + \frac{1}{r} \frac{\partial \Omega}{\partial r} \right).$$

The measure $\nu$ is the kinematic viscosity (which equals viscosity/density)
and it has dimension $L^2 T^{-1}$ (see Table 3). Consider the dimensionality of
the three terms of the equation

$$\partial \Omega / \partial t : (A T^{-1})(T^{-1}) = A T^{-2},$$
$$\nu \, \partial^2 \Omega / \partial r^2 : (L^2 T^{-1})(A T^{-1})(L^{-2}) = A T^{-2},$$
$$(\nu/r) \, \partial \Omega / \partial r : (L^2 T^{-1})(L^{-1})(A T^{-1})(L^{-1}) = A T^{-2}.$$

They all have exactly the same dimensions. This illustrates a virtually
universal rule of physical theory; namely, physically meaningful equations
have the property that physical measures which are added or are set equal
to one another have the same dimensions. This rule is not only widely
used to catch blunders in formulating equations, but frequently it helps
greatly in solving problems (see Section 10.5.3).

Some of the goals of the chapter now become clear. We would like to
understand better the distinction between primary and secondary dimensions
and to learn why they are interrelated by (simple) monomial functions.
Perhaps this is related to the distinction Campbell attempted to draw between
fundamental and derived measures, and between extensive and intensive
ones.[2] In any event, some types of simple laws are crucial both to the existence
of some measures and to the monomial character of dimensions. Finally,
we would like to formulate precisely what is involved when we say that
complex physical laws are compatible with the dimensional structure of
the measures, to see how this leads to the method of dimensional analysis,
and to try to explain why the compatibility exists.

The structure of the chapter is this. First we formulate an algebraic
system that describes, but does not account for, the structure of physical
measures including the monomial combining of dimensions. This axiom
system is entirely descriptive in character, but it does make clear why some
dimensions can be expressed as products of others. Within that framework,
we next formulate (Section 10.3) the basic postulate that physical laws are
dimensionally invariant, and from it we derive the classical Pi Theorem,
which is the basis for the method of dimensional analysis. Several examples
of—and cautions about—that method are given in Section 10.5. We then

---

[2] Not all authors agree about the use of these and other words to be introduced later.
In Section 10.9.2 we describe carefully how our usage relates to that of Campbell (1920),
Palacios (1956), Ellis (1966), and Pfanzagl (1968).

attempt to formulate clearly how this algebraic structure of physical measures arises from extensive and additive conjoint measurement theories and we are led to introduce (Section 10.7) two types of elementary, qualitative, trinary laws as the only properties that need be added to the measurement assumptions to get the usual algebraic model (Section 10.9). In 10.10 we introduce the notion of physically similar systems and show why all such families of systems are described by dimensionally invariant laws. In 10.12 we show how poorly these ideas extend to interval scale measurement. Finally, in 10.14 we discuss briefly other concepts of invariance in physics.

## ☐ 10.2 THE ALGEBRA OF PHYSICAL QUANTITIES

It is widely agreed that physical quantities combine additively within a single dimension (at least when that dimension is extensively measurable) and that different dimensions combine multiplicatively; that the multiplicative structure is very much like a finite-dimensional vector space over the rational numbers; that the existence of basic sets of dimensions in terms of which we can express the remaining dimensions corresponds to the existence of finite bases in the vector space; and that numerical physical laws are almost always formulated in terms of a very special class of functions defined on this space. Most texts on dimensional analysis appear to take such an algebra more or less for granted, and they do not attempt to formulate explicitly what is involved (Birkhoff, 1950; Bridgman, 1922, 1931; Duncan, 1953; Focken, 1953; Huntley, 1951; Langhaar, 1951; Palacios, 1956; and Sedov, 1959). Those who have undertaken to make the structure explicit have mainly published in journals (Brand, 1957; Drobot, 1953; Kurth, 1965; Thun, 1960; Whitney, 1968); an exception is Quade (1967). These axiomatizations are all quite similar in spirit; Whitney's strikes us as both the most compact and the most transparent, and so (with some departures[3]) we follow it.

### 10.2.1 The Axiom System

The notion of physical quantity has at least the following features which we must capture. First, in order to specify a particular physical quantity we must state both its dimension and the numerical ratio of the entity

---

[3] The main difference is that we consider only rational powers of dimensions because these are all that seem to arise in physics; Whitney admitted real exponents, which made it necessary for him to include the powers as a separate primitive in the axiom system. The inclusion of real exponents would, however, have simplified slightly the treatment of physically similar systems in Section 10.10.3.

in question to another fixed entity having the same dimension. Specifically, when we say that an object is 5 m long we are saying, first of all, that we are working on the dimension length ($L$) and, second, that the object in question is equivalent to the concatenation of five copies of a rod whose length is called one meter. Implicitly, we take for granted that any physical quantity is measured on a ratio scale, which of course means that the zero of the scale is fixed. Because the zero is fixed, it is trivial to include (formal) negative elements of all physical quantities, and it will be convenient to have them. Second, any two physical quantities can be combined "multiplicatively" to form a new physical quantity. In some cases the combined quantity is easily realized in a physical system. For example, if an object is 5 m long and has a mass of 11 kg, it also has the measure of 55 m-kg, which is a quantity having the dimension $LM$. Such an operation of "multiplication" is assumed to be closed, associative, and commutative. Third, we can also divide one physical quantity by another: the above object also has the measure of 5/11 m/kg, which is a quantity having the dimension $LM^{-1}$. Because we want multiplication of dimensions to be closed and to include division, it is essential to treat the (dimensionless) real numbers also as physical quantities. And finally, we can—at least formally—extract integral roots of positive quantities. In some cases, it is quite meaningful. For example, if we have an area of 25 m², of whatever shape, then we can find a square whose area is also 25 m² and whose side, therefore, is 5 m. Clearly, this is extracting the square root of a physical quantity that represents area. More generally, it will be convenient to suppose that if $x$ is any "positive" physical quantity and $n$ is a positive integer, then there is a physical quantity $x^{1/n}$ that when combined with itself $n$ times yields $x$. If $x$ has dimension $[x]$, then $x^{1/n}$ has dimension $[x]^{1/n}$.

This latter feature probably seems to be the least realistic aspect of the axiomatization to be given. One just never hears physicists speak of dimensions such as $M^{1/113}$. But that objection has little to do with roots as such; the problem is that one does not hear mention of $M^{113}$ either. As we pointed out, only very small integers arise in practice. To our knowledge, no one has yet provided an axiomatization that imposes any limitation on the multiplying and dividing of dimensions, nor do we know of any explanation in physics for these limitations.

In the following definition, we use the symbol $*$ to denote "multiplication" of physical quantities and juxtaposition to denote ordinary multiplication of numbers. It is not strictly necessary to have separate notations, but we find that it helps to avoid confusion about which operations are and are not numerical.

DEFINITION 1. *Suppose that $A$ is a nonempty set, $A^{+}$ is a nonempty* *subset of A, and* $*$ *is a closed operation from* $A \times A$ *into* $A$. *Then* $\langle A, A^+, * \rangle$ *is a structure of physical quantities iff, for all* $a, x, y \in A$,

1.  $*$ *is associative and commutative.*

2.  $\mathrm{Re} \subset A$ *and* $\mathrm{Re} \cap A^+ = \mathrm{Re}^+$.

3.  *If* $\alpha, \beta \in \mathrm{Re}$, *then* $\alpha * \beta = \alpha\beta$.

4.  $1 * a = a$ *and* $0 * a = 0$.

5.  *If* $a \neq 0$, *then exactly one of* $a$ *and* $(-1) * a \in A^+$.

6.  *If* $x, y \in A^+$, *then* $x * y \in A^+$.

7.  *If* $a \neq 0$, *there exists* $a^{-1} \in A$ *such that* $a * a^{-1} = 1$.

8.  *If* $n$ *is an integer* $\neq 0$ *and* $x \in A^+$, *there exists a unique* $x^{1/n} \in A^+$ *such that* $(x^{1/n})^n = x$, *where for* $y \in A^+$, $y^n$ *is defined inductively as*

$$y^n = \begin{cases} y^{n-1} * y, & \text{if } n \geqslant 1, \\ 1, & \text{if } n = 0, \\ y^{n+1} * y^{-1}, & \text{if } n \leqslant -1. \end{cases}$$

A variety of familiar and expected properties flow from these axioms; some are stated as lemmas in Section 10.4.1.

We provide a simple mathematical structure which satisfies the axioms of Definition 1 in a nontrivial way. Not only does this establish their consistency, but it gives some insight into what they mean. Let $A$ be the set of triples $(\alpha, q, r)$, where $\alpha$ is real and not 0 and $q$ and $r$ are rational, together with an element $z$ that will behave as a zero element. Intuitively, we may think of the structure as composed from length and mass, in which case $q$ is interpreted as the length exponent, $r$ the mass one, and $\alpha$ is the ratio of the object to some standard unit object in units $L^q M^r$. Thus, $(1, 1, 0)$ denotes a pure length of unit size, say the standard meter rod, $(\alpha, 1, 0)$ is a rod $\alpha$-meters long, and $(\alpha, 2, 0)$ is an area $\alpha$ m². Similarly, $(\alpha, 0, 1)$ is an object with the mass ratio $\alpha$ relative to the unit mass; $(\alpha, -3, 1)$ refers to a substance with density (which has units $L^{-3}M$) of $\alpha$, etc.

The operation $*$ on $A$ is defined for nonzero elements as:

$$(\alpha, q, r) * (\alpha', q', r') = (\alpha\alpha', q + q', r + r'),$$

and for the zero element by

$$z * a = z = a * z$$

for all $a$ in $A$.

The real numbers are identified with the elements $(\alpha, 0, 0)$ of $A$, which are the dimensionless constants, plus $z$. Let

$$A^+ = \{(\alpha, q, r) \mid \alpha \in \mathrm{Re}^+\}$$
$$(\alpha, q, r)^{-1} = (\alpha^{-1}, -q, -r)$$

and for $(\alpha, q, r) \in A^+$,

$$(\alpha, q, r)^{1/n} = (\alpha^{1/n}, q/n, r/n).$$

It is easy to verify that this structure satisfies Axioms 1–8.

## 10.2.2 General Theorems

An arbitrary structure of physical quantities gives rise to a multiplicative vector space over the rationals. We establish this next.

For any $a \neq 0$, define

$$[a] = \{\alpha * a \mid \alpha \in \mathrm{Re}\}, \qquad [a]^+ = [a] \cap A^+.$$

We can interpret $[a]$ as the dimension of $a$ since it includes every physical quantity that can be obtained from $a$ by multiplying it by a dimensionless real number. This amounts to defining an equivalence relation of "same dimension as" over $A$, for as is shown in Lemma 10, if $a, b \neq 0$, either $[a] = [b]$ or $[a] \cap [b] = \{0\}$. On the set of these dimensions, i.e.,

$$[A] = \{[a] \mid a \in A\},$$

we define the following two notions:

$$[a] * [b] = [a * b], \tag{1}$$

and, for $x \in A^+$ and rational $\rho = i/j$,

$$[x]^\rho = [x^\rho] = [(x^{1/j})^i]. \tag{2}$$

It is shown in Lemma 11 that both definitions are well defined.

THEOREM 1. *Suppose that $\langle A, A^+, * \rangle$ is a structure of physical quantities. Then, the set $[A]$ of dimensions with multiplication and powers defined by Equations* (1) *and* (2) *is a multiplicative vector space over the rationals, where* $[1] = \mathrm{Re}$ *the identity element and* $[a]^{-1} = [a^{-1}]$ *is the inverse of* $[a]$.

For the example given in the previous section, $[(\alpha, q, r)]$ can be identified with the pair $(q, r)$. Thus $[A]$ is the set of rational points in the plane, which is a two-dimensional vector space over Ra.

A *base* for $[A]$ is defined in the usual way: it is a set of elements $[a_1],..., [a_n]$

that *spans* [A] in the sense that, for each [a] ∈ [A], there exist rational numbers $\rho_i$ such that

$$[a] = [a_1]^{\rho_1} * \cdots * [a_n]^{\rho_n},$$

and that are *independent* in the sense that if

$$[a_1]^{\gamma_1} * \cdots * [a_n]^{\gamma_n} = 1,$$

then $\gamma_i = 0$, $i = 1,..., n$.

This suggests the following definitions: a set of elements $a_1,..., a_n \in A^+$ *spans* A, is *independent* in A, and forms a *base* for A if $[a_1],..., [a_n]$, respectively, span [A], are independent in [A], and are a base for [A].

**THEOREM 2.** *Suppose that $\langle A, A^+, * \rangle$ is a structure of physical quantities. Then the elements $a_1,..., a_n \in A^+$ span A iff for every $a \in A$ there exist $\alpha \in$ Re and $\rho_1,..., \rho_n \in$ Ra such that*

$$a = \alpha * a_1^{\rho_1} * \cdots * a_n^{\rho_n}. \tag{3}$$

*They are independent iff*

$$a_1^{\gamma_1} * \cdots * a_n^{\gamma_n} \in \text{Re}$$

*implies $\gamma_i = 0$, $i = 1,..., n$; or equivalently, iff Equation (3) is unique for $a \neq 0$. If they are a base, the $\rho_i$ of Equation (3) depend only on [a].*

The structure $\langle A, A^+, * \rangle$ is said to be of *finite dimension* if it has a base, i.e., if the vector space [A] is of finite dimension. We shall suppose that the structures arising from physical measurements are of finite dimension.

The significance of Theorem 2 is probably clear. We may select a base $a_1,..., a_n$, which constitutes both a choice of basic dimensions and a unit for each, and then every physical quantity $a \neq 0$ is expressed uniquely as in Equation (3), where $a_1^{\rho_1} * \cdots * a_n^{\rho_n}$ describes the resulting secondary dimension and its coherent unit and $\alpha$ is the ratio-scale value of the quantity $a$ relative to that unit.

It is worth noting that we can introduce a formal addition within each dimension. Suppose that $a, b$ are in [c], i.e., $a = \alpha * c$ and $b = \beta * c$, where $\alpha, \beta$ are in Re; then define

$$a \oplus b = (\alpha + \beta) * c.$$

In Lemma 12 it is shown that $\oplus$ is well defined, associative, commutative, and for $\gamma$ in Re, $\gamma * (a \oplus b) = (\gamma * a) \oplus (\gamma * b)$. When a dimension is extensively measurable, this notion of addition agrees with the extensive one. It exists as a formal operation without a corresponding empirical concatenation for nonextensive scales. (See Theorem 11, p. 501.)

## ☐10.3  THE PI THEOREM OF DIMENSIONAL ANALYSIS

In terms of a structure of physical quantities $\langle A, A^+, * \rangle$, a "numerical" law formulates a relationship among positive quantities on two or more dimensions. A typical positive dimension is of the form $P = [a] \cap A^+$, where $a$ is in $A$. Thus, we can view a law as, first, the statement of a function

$$f\colon\ P_1 \times P_2 \times \cdots \times P_s \to \mathrm{Re}, \qquad s \geqslant 2,$$

and second, as the assertion that the physically realizable values of $x_i$ in $P_i$, $i = 1, 2,..., s$, must satisfy the condition

$$f(x_1, x_2,..., x_s) = 0.$$

In practice, such laws are written as if the $x_i$ were real numbers, but in fact they are physical quantities that involve the specification of both the dimension and the unit of measurement in terms of which the numerical dimensionless ratio $x_i$ is given.

With few exceptions,[4] only one type of function is admitted as a possible formulation of a physical law. Such functions are called *dimensionally invariant* or, by some authors, *homogeneous*. The essence of the idea, which we formulate carefully below, is that for $f$ to be a possible candidate for a law it must be of such a form that it is unnecessary to specify the units in terms of which the several physical quantities are reported, provided only that a fixed system of coherent units is used. All well-known physical laws exhibit this property. Why that should be so is not at all obvious, and it has aroused considerable speculative discussion. In Section 10.10 we describe briefly some of the ideas that have been presented to account for this remarkable fact.

### 10.3.1  Similarities

To describe precisely the class of dimensionally invariant functions, we first formulate what is meant by a systematic change of units in $\langle A, A^+, * \rangle$.

---

[4] Palacios (1956) cited certain equations, which describe the behavior of solid bodies not obeying Hooke's law and of non-Newtonian fluids, in which the exponents of the usual quantities such as time, tension, and deformation are constants that vary from one body to another. Such laws are not dimensionally invariant.

Intuitively, when the units of several dimensions are changed, the numerical representation of each quantity that depends upon one or more of these dimensions is altered so that structure within that dimension is preserved. Such changes of unit do not change the dimension of a quantity, but merely the number representing it. Moreover, dimensionless quantities are unchanged. Thus, we see that any change of units determines a mapping of $A$ onto itself which is 1:1 and which in no way disturbs the dimensional structure. This leads us to the following:

DEFINITION 2. *Suppose that* $\langle A, A^+, * \rangle$ *is a structure of physical quantities. A function* $\phi: A \to A$ *is a* similarity *iff it is an isomorphism onto* $A$ *that preserves dimensions, maps* $A^+$ *into* $A^+$, *and leaves* Re *unaltered, i.e., for all* $a, b \in A$,

   (i)   $\phi$ *is 1:1 and onto*;

   (ii)  $\phi(a * b) = \phi(a) * \phi(b)$;

   (iii) $[\phi(a)] = [a]$;

   (iv) *if* $x \in A^+$, *then* $\phi(x) \in A^+$;

   (v) *if* $\alpha \in$ Re, *then* $\phi(\alpha) = \alpha$.

The next theorem shows that a similarity really does correspond to what we think of as changing the units of the base in a finite-dimensional structure.

THEOREM 3. *Suppose that a structure of physical quantities* $\langle A, A^+, * \rangle$ *is of finite dimension and that* $\{a_1, ..., a_n\}$ *is a base. If* $\phi$ *is a similarity, then there exist numbers* $\phi_i > 0$, $i = 1, ..., n$, *such that*

$$\phi(a_i) = \phi_i * a_i,$$

*and for*

$$a = \alpha * a_1^{\rho_1} * \cdots * a_n^{\rho_n}, \qquad \alpha \in \text{Re}, \quad \rho_i \in \text{Ra},$$

*we have*

$$\phi(a) = (\phi_1^{\rho_1} \cdots \phi_n^{\rho_n}) * a. \tag{4}$$

*Conversely, for any* $\phi_i > 0$, $i = 1, ..., n$, *the function defined by Equation* (4) *is a similarity.*

For the previous example

$$A = \{z\} \cup \{(\alpha, q, r) \mid \alpha \in \text{Re}, \quad \alpha \neq 0, \quad q, r \in \text{Ra}\},$$

any similarity has the form

$$\phi(\alpha, q, r) = (\alpha\phi_1{}^q\phi_2{}^r, q, r).$$

## 10.3.2  Dimensionally Invariant Functions

DEFINITION 3.  *Suppose that $\langle A, A^+, * \rangle$ is a structure of physical quantities and that $P_i$, $i = 1,..., s$, are positive dimensions (i.e., of the form $[a_i] \cap A^+$) in that structure. A function $f: \mathsf{X}_{i=1}^s P_i \to \mathrm{Re}$ is dimensionally invariant iff for all $x_i \in P_i$, $i = 1,..., s$,*

$$f(x_1,..., x_s) = 0$$

*is equivalent to: for all similarities $\phi$,*

$$f[\phi(x_1),..., \phi(x_s)] = 0.$$

This formulates precisely a property of most physical laws: they are unaffected by systematic changes of units or, what is the same thing, that only measures having the same dimensions are added and set equal in a physical law. The next theorem establishes the basic mathematical structure of dimensionally invariant functions. It is called the "Pi Theorem" because Buckingham (1914), who first stated it explicitly, used the symbol $\pi$ to stand for the dimensionless constants that arise in its statement. It was used informally much earlier by Fourier (1822) and others.

THEOREM 4.  *Suppose that $\langle A, A^+, * \rangle$ is a finite-dimensional structure of physical quantities, that $P_i$, $i = 1,..., s$, are positive dimensions of the structure which are indexed so that the first $r < s$ form a maximal independent subset of the subspace spanned by all $s$ of them, and that $f: \mathsf{X}_{i=1}^s P_i \to \mathrm{Re}$ is a dimensionally invariant function. Then there exist a function $F: \mathrm{Re}^{s-r} \to \mathrm{Re}$ and $\rho_{ij} \in \mathrm{Ra}$, $i = r + 1,..., s, j = 1,..., r$, such that for all $x_i \in P_i$,*

$$\pi_{i-r} = x_i * x_1^{-\rho_{i1}} * \cdots * x_r^{-\rho_{ir}}, \qquad i = r + 1,..., s, \tag{5}$$

*are real numbers (i.e., dimensionless), and*

$$f(x_1,..., x_s) = 0 \tag{6}$$

*iff*

$$F(\pi_1,..., \pi_{s-r}) = 0. \tag{7}$$

*Conversely, any function of the $\pi$'s given by Equation (5) is dimensionally invariant.*

In practice, physical laws are usually written with a dependent variable expressed as a function of several independent variables, i.e.,

$$x_s = g(x_1, \ldots, x_{s-1}),$$

which, of course, can be put into our form simply by subtracting $x_s$ from both sides. The analogous dimensionless formulation is

$$\pi_{s-r} = G(\pi_1, \ldots, \pi_{s-r-1}), \tag{8}$$

which, when we take Equation (5) into account, can be rewritten as

$$x_s = x_1^{\rho_{s1}} * \cdots * x_r^{\rho_{sr}} * G(\pi_1, \ldots, \pi_{s-r-1}). \tag{9}$$

Equation (9) is of interest only if $x_s$ does not appear in any $\pi$ other than $\pi_{s-r}$. Examples of this form of the theorem are given in Section 10.5.


## 10.4 PROOFS

Throughout this section we suppose that $\langle A, A^+, * \rangle$ is a structure of physical quantities in the sense of Definition 1 (Section 10.2.1). The proofs of several of the lemmas are left as exercises.

### 10.4.1 Preliminary Lemmas

Define

$$-a = (-1) * a$$
$$A^- = \{a \mid -a \in A^+\}.$$

LEMMA 1.   $(-a) * (-b) = a * b$; $(-a) * b = -(a * b)$; $-(-a) = a$.

LEMMA 2.   *If $a \in A$, then either $a = 0$, $a \in A^+$, or $a \in A^-$.*

   *PROOF.*   Axiom 5.                                                            ◇

LEMMA 3.   *If $a \in A^+$ and $b \in A^-$, then $a * b \in A^-$; if $a, b \in A^-$, then $a * b \in A^+$.*

   *PROOF.*   Lemma 1 and Axiom 6.                                               ◇

LEMMA 4.   *If $c \neq 0$ and $a * c = b * c$, then $a = b$.*

LEMMA 5.   *In Axiom 7, $a^{-1}$ is unique.*

**LEMMA 6.** *If $x \in A^+$, then $x^{-1} \in A^+$.*

**LEMMA 7.** *If $x \in A^+$, then $(x^{-1})^n = x^{-n}$.*

*PROOF.* By Lemma 6, $x^{-1} \in A^+$. Suppose that $n \geqslant 1$. The result is true for $n = 1$ by definition. By induction and the definition of $y^n$,

$$(x^{-1})^n = (x^{-1})^{n-1} * (x^{-1})^1 = x^{-(n-1)} * x^{-1} = x^{-n}.$$

For $n = 0$, it is true by definition. For $n \leqslant 1$, it is clear, first, that $(x^{-1})^{-1} = x = x^{-(-1)}$, and so the induction is similar to $n \geqslant 1$. ◇

**LEMMA 8.** *If $n \neq 0$ and $x^n = y^n$, then $x = y$.*

*PROOF.* This simply restates the uniqueness of $x^{1/n}$ in Axiom 8. ◇

For integers $i$ and $j$, define $x^{i/j} = (x^{1/j})^i$.

**LEMMA 9.** *For $\rho$, $\sigma \in Ra$ and $x, y \in A^+$,*

$$(x^\rho)^\sigma = x^{\rho\sigma}; \qquad (x * y)^\rho = x^\rho * y^\rho; \qquad x^{\rho+\sigma} = x^\rho * x^\sigma.$$

*PROOF.* Let $\rho = i/j$ and $\sigma = k/l$. Then,

$$((x^\rho)^\sigma)^{jl} = ((x^{i/j})^{k/l})^{jl} = ((x^{i/j})^{1/l})^{kjl} = (x^{i/j})^{kj} = (x^{1/j})^{ikj} = x^{ik},$$

so by the uniqueness in Axiom 8, $(x^\rho)^\sigma = (x^{i/j})^{k/l} = x^{ik/jl} = x^{\rho\sigma}$. Observe that

$$((x * y)^\rho)^j = ((x * y)^{i/j})^j = (x * y)^i$$
$$= x^i * y^i = (x^{i/j})^j * (y^{i/j})^j = (x^{i/j} * y^{i/j})^j = (x^\rho * y^\rho)^j,$$

so by Lemma 8, $(x * y)^\rho = x^\rho * y^\rho$. Finally,

$$(x^{\rho+\sigma})^{jl} = (x^{i/j+k/l})^{jl}$$
$$= x^{il+jk} = (x^{il/jl})^{jl} * (x^{jk/jl})^{jl}$$
$$= (x^{i/j} * x^{k/l})^{jl} = (x^\rho * x^\sigma)^{jl},$$

and so by Lemma 8, $x^{\rho+\sigma} = x^\rho * x^\sigma$. ◇

For $a \neq 0$, define

$$[a] = \{\alpha * a \mid \alpha \in Re\}.$$

**LEMMA 10.** *For $a, b \neq 0$, either $[a] = [b]$ or $[a] \cap [b] = \{0\}$.*

*PROOF.* Suppose that $c \in [a] \cap [b]$ and $c \neq 0$. By definition, there exist $\alpha, \beta \in \text{Re}$ such that $c = \alpha * a = \beta * b$. Since $a, b \neq 0$, Axioms 3 and 7 imply $a = (1/\alpha) * c$ and $b = (1/\beta) * c$, whence

$$a = (1/\alpha) * \beta * b = (\beta/\alpha) * b \qquad \text{and} \qquad b = (1/\beta) * \alpha * a = (\alpha/\beta) * a,$$

and so, $[a] = [b]$.                                                               $\diamond$

For $a, b \neq 0$, $x \in A^+$, and $\rho \in \text{Ra}$, define

$$[a] * [b] = [a * b],$$
$$[x]^\rho = [x^\rho].$$

LEMMA 11.   $[a] * [b]$ *and* $[x]^\rho$ *are well defined.*

*PROOF.* Suppose that $a, b, c, d \neq 0$ and $[a] = [c]$ and $[b] = [d]$. By definition, there exist $\alpha, \beta \in \text{Re}$ such that $a = \alpha * c$ and $b = \beta * d$, and so by Axioms 1 and 3,

$$a * b = \alpha * c * \beta * d = \alpha * \beta * c * d = (\alpha\beta) * c * d,$$

hence $[a * b] = [c * d]$ by Lemma 10.

If $x, y \in A^+$ and $[x] = [y]$, then $x = \gamma * y$, $\gamma \in \text{Re}$, and by Lemma 9, $x^\rho = (\gamma * y)^\rho = \gamma^\rho * y^\rho$, and so $[x^\rho] = [y^\rho]$.                                  $\diamond$

Suppose that $a, b \in [c]$. Define

$$a \oplus b = (\alpha + \beta) * c,$$

where $a = \alpha * c$ and $b = \beta * c$.

LEMMA 12.   *The operation* $\oplus$ *on* $[c]$ *is well defined, closed, associative, commutative, and* $\gamma * (a \oplus b) = (\gamma * a) \oplus (\gamma * b)$.

### 10.4.2  Theorems 1 and 2   (pp. 462–463)

*The set* $[A] = \{[a] \mid a \in A, a \neq 0\}$, *with multiplication and rational powers defined just prior to Lemma 11, is a multiplicative vector space over the rationals.*

*PROOF.* Since the rationals form a field, it is sufficient to show that $\langle [A], * \rangle$ is a commutative group and that, for $\rho, \sigma \in \text{Ra}$,

$$([a] * [b])^\rho = [a]^\rho * [b]^\rho,$$
$$[a]^{\rho + \sigma} = [a]^\rho * [a]^\sigma,$$
$$([a]^\rho)^\sigma = [a]^{\rho\sigma},$$
$$[a]^1 = [a].$$

The closure, associativity, and commutativity of $\langle [A], * \rangle$ follow from that of $\langle A, * \rangle$. By Axiom 4, $[1] = \text{Re}$ is clearly the identity and $[a^{-1}]$ the inverse of $[a]$. The properties of the powers follow almost immediately from Lemma 9; for example,

$$([a] * [b])^\rho = [a * b]^\rho = [(a * b)^\rho] = [a^\rho * b^\rho] = [a^\rho] * [b^\rho] = [a]^\rho * [b]^\rho. \quad \diamondsuit$$

Given Lemma 9, the proof of Theorem 2 is obvious.

## 10.4.3  Theorem 3  (p. 465)

Let $\{a_1, ..., a_n\}$ be a base of a structure of physical quantities. If $\phi$ is a similarity, then there exist $\phi_i \in \text{Re}^+$, $i = 1, ..., n$, such that

$$\phi(\alpha * a_1^{\rho_1} * \cdots * a_n^{\rho_n}) = \phi_1^{\rho_1} \cdots \phi_n^{\rho_n} * \alpha * a_1^{\rho_1} * \cdots * a_n^{\rho_n},$$

and, conversely, such a function is a similarity.

PROOF.  Suppose, first, that $\phi$ is a similarity. By part (iii) of Definition 2 (Section 10.3.1), there exist $\phi_i$ such that $\phi(a_i) = \phi_i * a_i$. Since $a_i \in A^+$, part (iv) and Lemmas 2 and 3 imply $\phi_i > 0$. By repeated applications of part (ii) to $x^n$ and to $x = (x^{1/n})^n$, $x \in A^+$, we see that for $\rho$ rational, $\phi(x^\rho) = \phi(x)^\rho$. Using this and parts (ii) and (v),

$$\phi(\alpha * a_1^{\rho_1} * \cdots * a_n^{\rho_n}) = \phi(\alpha) * \phi(a_1^{\rho_1}) * \cdots * \phi(a_n^{\rho_n})$$

$$= \alpha * \phi(a_1)^{\rho_1} * \cdots * \phi(a_n)^{\rho_n}$$

$$= \alpha * (\phi_1 * a_1)^{\rho_1} * \cdots * (\phi_n * a_n)^{\rho_n}$$

$$= \phi_1^{\rho_1} \cdots \phi_n^{\rho_n} * \alpha * a_1^{\rho_1} * \cdots * a_n^{\rho_n}.$$

The converse follows almost immediately, using Lemma 4 in proving that $\phi$ is 1:1, Lemma 9 in proving part (ii), and Axiom 6 in proving part (iv). Parts (iii) and (v) hold by the definition of $\phi$.          $\diamondsuit$

## 10.4.4  Theorem 4  (p. 466)

In a finite-dimensional structure of physical quantities, if $f$ is a dimensionally invariant function defined over $s$ positive dimensions $P_i$, the first $r < s$ of which form a maximal independent subset of the space spanned by all $s$ dimensions, then there is a real-valued function $F$ of $s - r$ real variables and rational $\rho_{ij}$, $i = r + 1,\ldots, s$, $j = 1,\ldots, r$, such that for $x_i \in P_i$,

$$\pi_{i-r} = x_i * x_1^{-\rho_{i1}} * \cdots * x_r^{-\rho_{ir}}, \qquad i = r + 1,\ldots, s$$

are real numbers and $f(x_1,\ldots, x_s) = 0$ iff $F(\pi_1,\ldots, \pi_{s-r}) = 0$.

PROOF. Theorem 2 shows that there exist $\rho_{ij} \in \text{Ra}$ such that for $x_i \in P_i$, $i = 1,\ldots, s$, there exists $\pi_{i-r} \in \text{Re}$ with

$$x_i = \pi_{i-r} * x_1^{\rho_{i1}} * \cdots * x_r^{\rho_{ir}}, \qquad i = r + 1,\ldots, s.$$

Let $a_i \in P_i$, $i = 1,\ldots, r$, be arbitrary and define $a_i = a_1^{\rho_{i1}} * \cdots * a_r^{\rho_{ir}}$, $i = r + 1,\ldots, s$. By a well-known result on vector spaces, we may select a base for $A$ that includes the first $r$ of the $a_i$. Since $[x_i] = [a_i]$, there exist $\phi_i \in \text{Re}^+$ such that $x_i = \phi_i * a_i$. By Theorem 3, there exists a similarity $\phi$ such that $\phi(a_i) = \phi_i * a_i = x_i$, $i = 1,\ldots, r$. Observe that, for $i = r + 1,\ldots, s$,

$$\phi(\pi_{i-r} * a_i) = \phi(\pi_{i-r} * a_1^{\rho_{i1}} * \cdots * a_r^{\rho_{ir}})$$

$$= \pi_{i-r} * \phi(a_1)^{\rho_{i1}} * \cdots * \phi(a_r)^{\rho_{ir}}$$

$$= \pi_{i-r} * (\phi_1 * a_1)^{\rho_{i1}} * \cdots * (\phi_r * a_r)^{\rho_{ir}}$$

$$= \pi_{i-r} * x_1^{\rho_{i1}} * \cdots * x_r^{\rho_{ir}}$$

$$= x_i.$$

If we define

$$F(\pi_1,\ldots, \pi_{s-r}) = f(a_1,\ldots, a_r, \pi_1 * a_{r+1},\ldots, \pi_{s-r} * a_s),$$

then by what we have just shown and the definition of a dimensionally invariant function,

$$0 = f(x_1,\ldots, x_s) = f[\phi(a_1),\ldots, \phi(a_r), \phi(\pi_1 * a_{r+1}),\ldots, \phi(\pi_{s-r} * a_s)]$$

iff

$$0 = f(a_1,\ldots, a_r, \pi_1 * a_{r+1},\ldots, \pi_{s-r} * a_s) = F(\pi_1,\ldots, \pi_{s-r}).$$

It is obvious that any function of the $\pi$'s is dimensionally invariant. $\diamond$

## □ 10.5 EXAMPLES OF DIMENSIONAL ANALYSIS

The method of dimensional analysis, which is based on Theorem 4, has evolved during a period of well over 100 years, especially during this century. It has been used to considerable advantage in solving certain problems, mostly in classical physics and engineering. We will simply illustrate the method, and some of its pitfalls, by several examples. The reader seriously interested in learning to use the method should consult several of the books on the subject (Birkhoff, 1950; Bridgman, 1922, 1931; Duncan, 1953; Huntley, 1951; Langhaar, 1951; Palacios, 1956; and Sedov, 1959), which in turn have references to many other books and articles. Palacios's careful discussion of the issues, summary of the history of the subject, and simple examples from various branches of physics make his book an excellent introduction. For examples of really deep and serious applications of the method to mechanics, Sedov's are unsurpassed. The only extended attempt to apply these methods in the social sciences is de Jong (1967).

We will restrict ourselves to examples from mechanics, in which case it is well known that three dimensions suffice as a base; we use $L$, $M$, and $T$ (we are, for the moment, accepting the convention that angle is dimensionless).

### 10.5.1 The Simple Pendulum

Suppose that a rigid pendulum, pinned at the top, is deflected from the vertical and allowed to oscillate in a vacuum. If the length of the pendulum is $l$, its mass is $m$, and we deflect it by an angle $\alpha$ before releasing it, what then is its period $t$ of oscillation? See Figure 1. The only force acting on

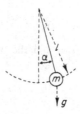

**FIGURE 1.** A simple pendulum.

the pendulum is its weight $mg$, where $g$ is the acceleration due to gravity. It is essential to include $g$ in the analysis of the problem, even though it was not stated explicitly in the formulation. Sedov comments as follows:

> Dimensional analysis enables us to draw conclusions by using arbitrary or special systems of units of measurement to describe physical laws. Consequently, when listing the parameters defining a class of motions, it is necessary to include all the dimensional parameters related to the substance of the phenomena independently of

whether these parameters are constant (in particular, they can be physical constants) or can vary for different motions of the class isolated. It is important that the dimensional parameters should be able to assume various numerical values in different systems of units, although they are possibly identical for all the motions being considered. For example, when considering the motion in which the weight of a body is of importance we must certainly take into account the acceleration due to gravity $g$ as a physical dimensional constant although the value of $g$ is constant under all actual conditions... (Sedov, 1959, p. 21).

The relevant physical quantities and their dimensions for the pendulum problem are shown in the accompanying tabulation. So, in terms of Theorem 4, $r = 3$ and $s = 5$, which means that there are two dimensionless

| Base dimensions | Physical quantities | | | | |
|---|---|---|---|---|---|
| | $t$ | $l$ | $m$ | $g$ | $\alpha$ |
| $L$ | 0 | 1 | 0 | 1 | 0 |
| $M$ | 0 | 0 | 1 | 0 | 0 |
| $T$ | 1 | 0 | 0 | $-2$ | 0 |

parameters $\pi_1$ and $\pi_2$. By inspection, one of them is $\alpha$ (we return to this later); to find the other, let $\rho_x$ be the exponent for quantity $x$, and simply set up the requisite equations for the exponents that result in $L$, $M$, and $T$ each having a zero exponent.

$$L:\ 0\rho_t + 1\rho_l + 0\rho_m + 1\rho_g = 0,$$
$$M:\ 0\rho_t + 0\rho_l + 1\rho_m + 0\rho_g = 0,$$
$$T:\ 1\rho_t + 0\rho_l + 0\rho_m - 2\rho_g = 0,$$

or in matrix form

$$\begin{bmatrix} 0 & 1 & 0 & 1 \\ 0 & 0 & 1 & 0 \\ 1 & 0 & 0 & -2 \end{bmatrix} \begin{bmatrix} \rho_t \\ \rho_l \\ \rho_m \\ \rho_g \end{bmatrix} = \begin{bmatrix} 0 \\ 0 \\ 0 \end{bmatrix}.$$

Since there are three equations and four unknowns, one $\rho$ may be chosen arbitrarily. It is usual to set the exponent of the dependent variable, $\rho_t$ in this case, equal to 1. With that choice, it is easy to see that the solution is: $\rho_l = -\frac{1}{2}$, $\rho_m = 0$, and $\rho_g = \frac{1}{2}$. Thus, $\pi_2 = t(g/l)^{1/2}$. Since the general expression for the dependency is $\pi_2 = G(\pi_1)$ [see Equations (8) and (9)], we conclude that

$$t = (l/g)^{1/2}\,\Phi(\alpha).$$

The exact solution from the laws of mechanics is the same but with $\Phi$ a determined function of $\alpha$. So the correct form has been arrived at, apparently without knowing any mechanics except which variables are relevant and their dimensions.

The last two provisos are really terribly important; one must actually know a good deal of physics before attempting to use dimensional analysis, otherwise the danger of being misled is great. There are four types of error: leaving something out of the base, including a redundant dimension in the base, including as relevant a quantity which is not, and omitting as irrelevant a quantity which really is relevant. We examine by example each of these possibilities in the next section, and, to anticipate, we conclude that the only truly serious errors are those of omission, especially of a relevant quantity. The method can drop out the irrelevant, but it cannot introduce the relevant when the initial formulation omits it and, what is worse, it fails to alert one to the difficulty.

### 10.5.2 Errors of Commission and Omission

*Incomplete bases*: Suppose that, in setting up the basic equations of mechanics, gravitational mass and inertial mass were treated as equal, and so the universal gravitational constant $G$ would be suppressed in the Newtonian gravitational law[5] $F = Gmm'/r^2$. Since we also have Newton's law $F = ma$, it follows that mass has the dimensions $L^3T^{-2}$, and so the base consists of just $L$ and $T$. The pendulum problem is then formulated as

|   | $t$ | $l$ | $m$ | $g$ | $\alpha$ |
|---|-----|-----|-----|-----|----------|
| $L$ | 0 | 1 | 3 | 1 | 0 |
| $T$ | 1 | 0 | $-2$ | $-2$ | 0 |

It is easy to show that $\pi_1 = \alpha$, $\pi_2 = l(g/m)^{1/2}$, and $\pi_3 = t(g/l)^{1/2}$, are dimensionless, and so we obtain

$$t = (l/g)^{1/2}\, \Phi[l(g/m)^{1/2},\, \alpha].$$

Obviously, this is not inconsistent with the previous analysis, but it is a much weaker result which is misleading since it suggests that the period depends on $m$, when in fact it does not.

---

[5] We follow the usual mechanical notation in this section, so here $F$ stands for resultant force, and not the dimension of force. Also note that in Definition 2 we did not introduce separate concepts of gravitational and inertial mass.

A rather more interesting question about incompleteness of the base has been raised by Palacios (1956). Consider a projectile of mass $m$ fired at velocity $v$ and angle $\alpha$ to the horizontal. Assuming a vacuum, at what distance $x$ does it hit the horizontal? See Figure 2. As in the pendulum example,

|   | $x$ | $v$ | $m$ | $g$ | $\alpha$ |
|---|-----|-----|-----|-----|----------|
| $L$ | 1 | 1 | 0 | 1 | 0 |
| $M$ | 0 | 0 | 1 | 0 | 0 |
| $T$ | 0 | −1 | 0 | −2 | 0 |

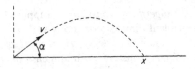

**FIGURE 2.** A projectile problem.

Since $\pi_1 = \alpha$ and $\pi_2 = xg/v^2$ are dimensionless,

$$x = (v^2/g)\,\Phi(\alpha).$$

This analysis fails to distinguish direction, even though it is obviously important. The initial velocity has components both in the $x$ and in the $z$ directions, namely, $v_x = v \cos \alpha$ and $v_z = v \sin \alpha$, the distance traveled has a length component only in the $x$ direction, and gravity has a component only in the $z$ direction. We have ignored all of this information. If we take it into account, not only in the variables, but also in the dimensions by treating length in the $x$ direction as a dimension distinct from length in the $z$ direction, we obtain the results shown in the accompanying tabulation.

|   | $x$ | $v_x$ | $v_z$ | $m$ | $g$ |
|---|-----|-------|-------|-----|-----|
| $L_x$ | 1 | 1 | 0 | 0 | 0 |
| $L_z$ | 0 | 0 | 1 | 0 | 1 |
| $M$ | 0 | 0 | 0 | 1 | 0 |
| $T$ | 0 | −1 | −1 | 0 | −2 |

Since $s - r = 5 - 4$, there is only one dimensionless parameter.

The equations are

$$\rho_x + \rho_{v_z} = 0,$$

$$\rho_{v_z} + \rho_g = 0,$$

$$\rho_m = 0,$$

$$-\rho_{v_x} - \rho_{v_z} - 2\rho_g = 0.$$

Setting $\rho_x = 1$, we find that $\pi_1 = xg/v_x v_z$. Thus,

$$x = (Cv^2/g) \sin \alpha \cos \alpha,$$

which is obviously a more complete result than the previous one. A complete dynamic analysis yields the same equation with $C = 2$.

*Redundant bases and superfluous constants*: Suppose that force is treated as if it were an independent dimension $F$ along with $L$, $M$, and $T$. Then the equation relating force, mass, and acceleration must be written $F = Cma$, where $C$ is a dimensional constant with dimensions $FM^{-1}L^{-1}T^2$. Since, by definition, gravity is the force exerted on mass, $g$ has the dimensions $FM^{-1}$.

|   | $t$ | $l$ | $m$ | $g$ | $C$ | $\alpha$ |
|---|---|---|---|---|---|---|
| $L$ | 0 | 1 | 0 | 0 | $-1$ | 0 |
| $M$ | 0 | 0 | 1 | $-1$ | $-1$ | 0 |
| $T$ | 1 | 0 | 0 | 0 | 2 | 0 |
| $F$ | 0 | 0 | 0 | 1 | 1 | 0 |

So ignoring direction, the simple pendulum has the values shown in the tabulation. Solving, we find

$$t = (Cl/g)^{1/2} \, \Phi(\alpha).$$

This does not differ fundamentally from the first solution, and once it is recognized that $C$ is a universal constant that can be taken to be 1, it reduces to the previous solution.

*Irrelevant quantities*: We already have seen in the pendulum example a case in which mass is irrelevant, and dimensional analysis showed that it is. The addition of irrelevant variables should, however, be avoided whenever possible since, in general, they may introduce an apparent dependency upon dimensionless constants which is not real.

*Omission of relevant quantities*: Beyond any doubt, omitting relevant quantities is the most dangerous error that one can make because it leads to incorrect results, and nothing in the method warns one of the error. Moreover, the matter can be quite subtle, as we now demonstrate. Consider a rectangular, negligibly thick, rigid plate at an angle $\alpha$ to a laminar fluid flow moving at velocity $v$. Suppose that the fluid has density $d$, viscosity $\mu$, and is incompressible. What are the forces perpendicular and parallel to the plane of the plate? The situation is shown in Figure 3. Let the $y$ coordinate

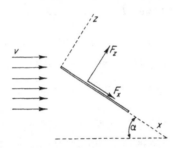

**FIGURE 3.** A plate in an airflow.

lie in the plane of the plate and be perpendicular to the flow of the fluid; let the $x$ coordinate also lie in the plane of the plate and be perpendicular to $y$; and let the $z$ coordinate be perpendicular to the plane of the plate. Thus, the velocity has components $v_x = v \cos \alpha$, $v_y = 0$, and $v_z = v \sin \alpha$. Referring to Table 3 to obtain the dimensions of viscosity, we see that the dimensional structure for the force perpendicular to the plate is as shown in the accompanying tabulation. It is not difficult to show that $\pi_1 = v_z l_x \, d\mu^{-1}$

|       | $F_z$ | $v_x$ | $v_z$ | $l_x$ | $l_y$ | $d$ | $\mu$ |
|-------|-------|-------|-------|-------|-------|-----|-------|
| $L_x$ | 0     | 1     | 0     | 1     | 0     | −1  | 0     |
| $L_y$ | 0     | 0     | 0     | 0     | 1     | −1  | −1    |
| $L_z$ | 1     | 0     | 1     | 0     | 0     | −1  | 0     |
| $M$   | 1     | 0     | 0     | 0     | 0     | 1   | 1     |
| $T$   | −2    | −1    | −1    | 0     | 0     | 0   | −1    |

and $\pi_2 = F_z v_z^{-2} d^{-1} l_x^{-1} l_y^{-1}$ are dimensionless. The quantity $R = vld/\mu$, called the Reynolds number, is of considerable importance in fluid mechanics (see below). Letting $l_x l_y = A =$ the area of the plate, we see then that the force perpendicular to the plate is

$$F_z = v^2 \, dA(\sin \alpha)^2 \, \Phi(R \sin \alpha).$$

This dependence of force agrees with the detailed dynamic solution.

A similar dimensional analysis for $F_x$ yields

$$F_x = v^2 \, dA(\sin \alpha \cos \alpha) \, \Psi(R \sin \alpha).$$

The quantities of interest in aerodynamic theory, for example, are not $F_x$ and $F_z$, but rather the forces perpendicular to and parallel to the direction of flow; they are called the lift $L$ and drag $D$. As is easily seen,

$$L = F_z \cos \alpha - F_x \sin \alpha$$
$$= v^2 \, dA(\sin \alpha)^2 \, (\cos \alpha) \, \Theta(R \sin \alpha),$$

where $\Theta(R \sin \alpha) = \Phi(R \sin \alpha) - \Psi(R \sin \alpha)$, and

$$D = -F_z \sin \alpha + F_x \cos \alpha$$
$$= v^2 \, dA \sin \alpha \, \Theta'(R, \alpha),$$

where $\Theta'(R, \alpha) = -(\sin \alpha)^2 \, \Phi(R \sin \alpha) + (\cos \alpha)^2 \, \Psi(R \sin \alpha)$. Thus,

$$L/D = \sin \alpha \cos \alpha \, \Xi(R, \alpha),$$

where $\Xi(R, \alpha) = \Theta(R \sin \alpha)/\Theta'(R, \alpha)$. This shows one reason why the Reynolds number $R$ is of such great interest: the lift to drag ratio of this model of an airfoil—a quantity which one clearly wants to make as large as possible—depends only on it and on the angle of attack $\alpha$. Thus, if a wind tunnel can be designed to have (approximately) the same value of $R$ as for a full scale aircraft, it is sufficient to estimate $L/D$ from measurements on a model.

Consider now, what happens when we omit one of the variables. We notice that $v_x$ plays no role in the solution of $F_z$, so we omit it on the grounds that it is irrelevant. By omitting it, only one dimensionless constant arises, and so we obtain a more complete solution. Striking out the $v_x$ column, we find that $F_z v_z^{-3/2} l_x^{-1/2} l_y^{-1} d^{-1/2} \mu^{-1/2}$ is dimensionless, and so we conclude that $F_z = C v_z^{3/2} l_x^{1/2} l_y d^{1/2} \mu^{1/2}$. This is wrong. The fact of the matter is that $v_x$ is a relevant factor in the physics of the problem, even if it does not appear explicitly in the equation for $F_z$; it does, of course, appear in the equation for $F_x$. Note that there is an important difference between the actual redundancy of $m$ in the pendulum problem and the apparent redundancy of $v_x$ in this one, namely, $m$ is the only variable involving the dimension $M$, so both of them can be dropped. In this problem, every dimension affects some variable other than $v_x$.

### 10.5.3 Dimensional Analysis as an Aid in Obtaining Exact Solutions

Recall, we stated in Section 1 that the partial differential equation describing the propagation of vorticity is

$$\frac{\partial \Omega}{\partial t} = \nu \left( \frac{\partial^2 \Omega}{\partial r^2} + \frac{1}{r} \frac{\partial \Omega}{\partial r} \right), \tag{10}$$

where $\Omega$ is the angular velocity of a viscous fluid, $r$ is the radial distance, $t$ is time, and $\nu = \mu/d$ is the kinematic viscosity. Suppose that we want to solve this for $\Omega(r, t)$ subject to the initial condition that the circulation around a circle of radius $R$ centered at the origin is a constant, i.e.,

$$4\pi \int_0^R r\Omega(r, 0) \, dr = \Gamma.$$

Following Sedov (1959, pp. 97–98), we first set up a dimensional analysis for $\Omega = \Omega(\Gamma, \nu, r, t)$.

|   | $\Omega$ | $\Gamma$ | $\nu$ | $r$ | $t$ |
|---|---|---|---|---|---|
| $L$ | 0 | 2 | 2 | 1 | 0 |
| $M$ | 0 | 0 | 0 | 0 | 0 |
| $T$ | −1 | −1 | −1 | 0 | 1 |

We find two dimensionless parameters $\pi_1 = r^2 \nu^{-1} t^{-1}$ and $\pi_2 = \Omega \nu t \Gamma^{-1}$, hence

$$\Omega = (\Gamma/\nu t)\, \Phi(\xi),$$

where $\xi = r^2/\nu t$. Now, if we substitute this form for $\Omega$ into the partial differential equation (10), we readily obtain the ordinary differential equation

$$\frac{d}{dt}\, [\xi\Phi(\xi) + 4\xi \, d\Phi(\xi)/d\xi] = 0,$$

and integrating,

$$\xi\Phi(\xi) + 4\xi \, d\Phi(\xi)/d\xi = C.$$

If we assume that $\Phi(0)$ and $d\Phi(0)/d\xi$ are finite, then setting $\xi = 0$ shows that $C = 0$. It follows directly that $\Phi(\xi) = Ae^{-\xi/4}$, where $A$ is a constant. Substituting this into the expression for $\Omega$ and that into the expression for the circulation,

$$4\pi \int_0^R r\Omega(r, t) \, dr = 4\pi \int_0^R (\Gamma A/\nu t) \, re^{-r^2/4\nu t} \, dr$$

$$= \Gamma A 8\pi \int_0^{R^2/4\nu t} e^{-x} \, dx = \Gamma A 8\pi [1 - e^{-R^2/4\nu t}].$$

Since, for $t = 0$, the circulation is assumed to be $\Gamma$, we see that $A = 1/8\pi$ and so the solution to the problem is

$$\Omega = (\Gamma/8\pi\nu t)\, e^{-r^2/4\nu t}.$$

It is impressive how little we need to know about partial differential equations in order to solve this one once we used dimensional analysis to reduce the family of possible solutions.

### 10.5.4  Conclusion

We conclude the discussion of examples by quoting Ellis:

> [Dimensional analysis] is not, as it appears to be, a genuinely a priori way of doing physics. Empirical information comes in at several points. First, there is a judgement based upon experience in the selection of likely factors. (It is not always possible to eliminate wrong selections by dimensional analysis.) Secondly, there is the assumption that we have included all of the relevant dimensional constants. (This is usually based upon a well-informed guess as to the detailed analysis.) Thirdly, there is the information concerning the standard forms of the basic numerical laws contained in the dimensional formulae themselves. With this understanding, dimensional analysis seems much less mysterious. It is, as Bridgman says, "an analysis of an analysis." (Ellis, 1966, pp. 143–144.)

### ☐ 10.6  BINARY LAWS AND UNIVERSAL CONSTANTS

We can select at will the unit for each physical quantity but if we do, equations such as Newton's law $F = ma$ must be written $F = Cma$, where $C$ is a universal constant whose only role is to establish the numerical conversion between the arbitrary force unit and the $MLT^{-2}$ unit. Palacios (1964, p. 67) called these universal constants *superfluous* because they are eliminated simply by using a coherent set of units. In 1935 Giorgi proposed one such system to the International Electrotechnical Commission; later, in 1954, it was adopted by the 10th General Conference on Weights and Measures, which comprised 35 nations (Ruppel, 1959, p. 237). It accepts length, mass, time duration, temperature, and either charge or current (time rate of change of charge) as base, and the units are the meter, kilogram, second, degree absolute (Kelvin), and the coulomb or ampere. By omitting any reference to temperature and by using current rather than charge, this system has come to be referred to as the MKSA—meter, kilogram, second, ampere—system. A collection of papers discussing problems about units is Kayan (1959).

In addition to superfluous universal constants, others, called *indispensable* by Palacios, do not vanish when a coherent set of units is used. They arise whenever two conceptually distinct measures do, in fact, covary without exception. Such laws of covariation seem to be the only binary laws in physics. For example, Einstein showed that any change in the energy of a body is accompanied by a change in its inertial mass, that the two changes are proportional, and that the constant of proportionality is the square of the velocity of light. An historically earlier example is the fact that the inertial mass $m_i$ that enters Newton's law $F = m_i a$ and the gravitational mass $m_g$ that enters his gravitational law $F = m_g m_g'/r^2$ are proportional. The constant $G$ for which $m_g = G^{1/2} m_i$ is called the *universal gravitational constant*. Quantities that always covary are called *inseparable*. Each pair of inseparable quantities introduces an indispensable universal constant. A list of such universal constants was given previously in Table 1 (Section 10.1).

Ellis seems to have taken the position that all universal constants are superfluous:

> We may conclude, then, that simply by adopting different conventions concerning the expression of numerical laws, universal scale-dependent constants may be created or eliminated at will. We cannot, therefore, attach to these constants anything other than a conventional existence; and it is quite absurd to suppose that these constants represent the magnitudes of peculiar invariant properties of space or of matter. (Ellis, 1966, p. 125.)

This view strikes us as too extreme because there do exist logically distinct attributes which do, in fact, invariably covary. Such constants as the velocity of light—which can have any numerical value by adjusting the units of the representations of the attributes—do seem to "represent the magnitudes of peculiar invariant properties of space or of matter."

It should be recognized that the indispensable universal constants are not all independent. In fact, it is clear that the size of the base, which currently is five, sets an upper limit on the number of dimensionally independent universal constants. For both quantum and relativistic mechanics, a base of three is sufficient. Thus, the mass $m_e$ of some elementary particle such as the electron, the velocity of light $c$, and Planck's constant $h$, which are dimensionally independent, will do. Every monomial function of these three is again a universal constant. For example, the universal gravitational constant $G$ is a numerical factor times $hc/m_e^2$. Electromagnetic phenomena add another dimension and another universal constant—the charge $e$ of the electron will do—and thermodynamic phenomena add still another—one possibility is Boltzmann's constant $k$ that relates the average energy of a particle in a large collection of particles to the absolute temperature of the collection.

In carrying out dimensional analyses it is essential to keep the universal constants in mind. We illustrated this in Section 10.5.2 when we discussed incomplete bases. One must either include both $m_l$ and $m_g$ as separate measures or, as is customary, just include $m_l = m$ and in problems where $m_g$ has a bearing also include the universal gravitational constant $G$.

It is curious that in all of the discussions of universal constants that we have read, two spatial universal constants are never mentioned. The fundamental measurement of length in physical three-space involves the selection of a direction along which concatenation takes place. In principle, the measures obtained by concatenating in different directions need not covary, but of course they do. Usually we do not bother to distinguish the three independent dimensions of length; however, we also fail to include the suppressed universal constants which have the numerical value of 1 and, for example, the dimensions $L_x L_y^{-1}$ and $L_x L_z^{-1}$. We suppress them just as we do the superfluous constants, although they are not the same. As we saw by example (Section 10.5.2), the more complete spatial base yields more precise results in dimensional analysis. Thus, taking into account the vectorial character of some physical quantities, the base becomes seven, not five, and the two added universal constants are those we just mentioned.

Another advantage of doing this is that it makes better sense of the measurement of angle. It is obvious that angle can be extensively measured (see Section 3.2.2), and so it has an arbitrary unit; the three most common ones are the radian (the angle subtended when the arc of a circle equals its radius), the degree (1/360 fraction of a circle), and the circle (called a cycle or revolution). Nonetheless, angle is said to be a dimensionless quantity because it is defined to be the ratio of the length of the subtended arc divided by the radius which yields the angle in radian units. Because this is a ratio of two lengths, it is said to be dimensionless. It is oddly uncomfortable to say that angle both is dimensionless and has a unit. Whenever angle plays a role in dimensional analysis, it is usual to treat it as dimensionless and immediately to take it to be one of the $\pi_i$. If, however, we acknowledge that there are three dimensions of length, then angle is no longer dimensionless. But rather than introduce it as a separate variable, it seems better to use trignometric functions of angle to resolve vectors into their three spatial components, as we did in Section 10.5.2.

Ellis appears to have much the same view about angle. He puts it this way:

Even in mechanics, the power of dimensional analysis could be considerably increased. Our laws involving the quantity, *angle*, for example, are always expressed with respect to particular scales of angle (usually the Radian scale). Hence angle is said to be a "dimensionless" quantity, and dimensional analysis is powerless to yield any information at all about the way that this quantity enters into quantitative expressions. But it is only powerless, in this regard, because we choose to express our laws in the way that we do. If we always chose to express our laws with respect to the class of scales of angle similar to our Radian scale, the power of dimensional analysis would be increased (Ellis, 1966, p. 145).

Page (1961) also arrived at the same conclusion; however, he went on to argue that, because

$$\cos \theta = 1 - \frac{\theta^2}{2!} + \frac{\theta^4}{4!} - \cdots,$$

$$\sin \theta = \theta \left[ 1 - \frac{\theta^2}{3!} + \frac{\theta^4}{5!} - \cdots \right],$$

$\cos \theta$ is dimensionless whereas $\sin \theta$ has the dimension of angle. This is in error because among other things it fails to recognize that these expansions hold only for $\theta$ measured in radian units.

## ☐ 10.7 TRINARY LAWS AND DERIVED MEASURES

The algebra of physical quantities that was formulated in Definition 1 is simply an axiomatic attempt to describe how the ratio scale measures of physics combine with one another. One must realize, however, that it does not establish any relationship whatsoever between the physical quantities that it talks about and the numerical measures that arise from the various theories—in particular, those of extensive and conjoint measurement—which have been put forward to account for the measurement of particular attributes. It remains to be shown that these measures form a substructure of an algebra of physical quantities. That this task is not completely trivial can be seen from the following observation. When we apply dimensional analysis, and so Definition 1, to physical problems we implicitly assume that we can treat as physical quantities not only such attributes as mass, length, time, and those others to which the theory of extensive measurement applies, but also intensive attributes such as density and temperature to which we know that it does not apply. The fact is that we have yet to give a systematic measurement theory for many of these attributes. Obviously, this lacuna must be filled before we can show that these measures form a substructure of physical quantities.

It is generally acknowledged that, in addition to those attributes that can be measured directly in terms of extensive measurement procedures, there are others that must be measured indirectly in terms of the extensive measures of other attributes, and this is possible only because certain numerical laws are true. The simplest example is the density of homogeneous materials, which measure exists by virtue of the fact that the ratio of the mass to the

484          10. DIMENSIONAL ANALYSIS AND NUMERICAL LAWS

volume of a homogeneous substance is, under appropriately fixed conditions, independent of the volume used. Other derived quantities, such as momentum and kinetic energy, do not so obviously depend upon the truth of a law for their existence, although clearly they do for their importance.

In this section we attempt to formulate as clearly as we can the nature of such derived measures and the qualitative laws that underlie them, and we use this in Section 10.9 to show how physical measures can be embedded in a structure of physical quantities.

## 10.7.1 Laws of Similitude

Laws such as "$m/V$ is a constant for a given material" or "$F/l$, where $F$ is the applied force and $l$ the length of a spring, is a constant for a given spring" (Hooke's law) have long been interpreted as statements about the similarity of certain classes of physical systems. They can also be looked upon as establishing a (multiplicative) conjoint scale: $m = Vd$, where $d$ is density, and $F = lk$, where $k$ is the spring constant. That is to say, if the set of possible volumes is denoted by $A_1$ and the set of possible materials by $A_2$, then $A_1 \times A_2$ defines a set of objects that vary in mass. Let the qualitative ordering by mass be denoted $\gtrsim$, then if the law $m = Vd$ is true, $\langle A_1 \times A_2, \gtrsim \rangle$ must exhibit all of the necessary conditions of a binary conjoint structure (Chapter 6). Since, in principle, two powdered or laminated substances with different densities can be combined in such proportions so as to generate any intermediate value of density, we will also assume that restricted solvability is satisfied (p. 256). Finally, we assume that each component is essential. Thus, we know that there exist numerical scales $\psi_k$ on $A_k$ such that $\psi = \psi_1 \psi_2$ preserves the qualitative ordering by mass, $\gtrsim$. Since $\log \psi_k$ is an interval scale, each $\psi_k$ is what Stevens (1959) has called a *log-interval scale*, i.e., it is unique up to power transformations of the form $\gamma_k \psi_k^\alpha$, where $\gamma_k, \alpha > 0$, for $k = 1, 2$.

By assuming that the qualitative ordering of mass is a conjoint structure relative to the volume and the material used, we have introduced a scale $\psi_2$ over materials which, presumably, is density or something very like it. There are various troubles, however. The most obvious, though not the most important, is that $\psi_2$ has a free exponent as well as a free unit, whereas density is alleged to have only a free unit. More important, we now have two distinct measures of mass and two of volume—the extensive and the conjoint ones. The conjoint ones arise from independence properties which show that different physical attributes can be combined multiplicatively, as was assumed in the axiomatic structure of physical quantities (Definition 1) and as is true of the physical measures in use. The extensive ones arise because some, though not all, attributes are additive when objects exhibiting the attribute are combined to form others that exhibit it. It is a fact of physics that the two types of scales are not very different from one another; to be specific, in all cases that we know of, when an attribute has both a conjoint scale and an extensive scale, they are power functions of each other. A wholly qualitative theory capable of justifying Definition 1 must, among other things, formulate the nature of the qualitative observations, based only on orderings and concatenation operations, that lead to this power relationship.

To arrive at such a qualitative law, suppose that $\psi$ and $\phi$ are the conjoint and extensive measures on $A_1 \times A_2$, that $\psi_1$ and $\phi_1$ are these two measures on $A_1$, and that $\psi_2$ is the conjoint measure on $A_2$. (We neither assume nor exclude the existence of an extensive measure on $A_2$.) Moreover, let us suppose that, for $a$ in $A_1$ and $u$ in $A_2$,

$$\psi(a, u) = \gamma[\phi(a, u)]^{\beta n},$$
$$\psi_1(a) = \gamma_1[\phi_1(a)]^{\beta m},$$

where $m$ and $n$ are positive integers and $\beta$ is positive. (The free parameter $\beta$ must be introduced because the conjoint measures are log-interval scales.) We choose $m$ and $n$ to be integers simply because physics is that way. For any integer $i$, let $ia$ be $i$ concatenations of $a$ with itself and let $i(a, u)$ be $i$ concatenations of $(a, u)$ with itself. Observe that if we assume unrestricted concatenation, then for all positive integers $i$,

$$
\begin{aligned}
\psi(i^n a, u) &= \psi_1(i^n a)\, \psi_2(u) \\
&= \gamma_1[\phi_1(i^n a)]^{\beta m}\, \psi_2(u) \\
&= \gamma_1[i^n \phi_1(a)]^{\beta m}\, \psi_2(u) \\
&= \gamma_1 i^{\beta mn}[\phi_1(a)]^{\beta m}\, \psi_2(u) \\
&= i^{\beta mn}\psi_1(a)\, \psi_2(u) \\
&= i^{\beta mn}\psi(a, u) \\
&= i^{\beta mn}\gamma[\phi(a, u)]^{\beta n} \\
&= \gamma[i^m \phi(a, u)]^{\beta n} \\
&= \gamma\{\phi[i^m(a, u)]\}^{\beta n} \\
&= \psi[i^m(a, u)],
\end{aligned}
$$

and so we have the necessary condition,

$$(i^n a, u) \sim i^m(a, u).$$

This, it turns out, is the added qualitative property that we need. The only remaining difficulty is that numerical laws are not always of the form

$\phi^n/\phi_1{}^m = \psi_2$, $m, n > 0$; we sometimes have $\phi^n\phi_1{}^m = \psi_2$, where $\phi$ and $\phi_1$ are extensive measures. Thus, the conjoint representation is $\phi^n = \phi_1{}^{-m}\psi_2$, and so the conjoint ordering of the first component is the converse of the extensive one because the exponent is negative. A similar thing can happen on the $\phi$ variable, leading to four possibilities. Let us proceed purely formally and replace $m$ by $-m$ in the above indifference. The problem is that, in general, $i^{-m}$ is not an integer; however, continuing formally, we multiply both sides by $i^m$, thereby obtaining

$$i^m(i^n a, u) \sim (a, u),$$

which is a meaningful equation. A calculation similar to the one above shows that this is, indeed, a necessary condition when one exponent is negative and the other positive. So we are led to the following definition.

DEFINITION 4. *Suppose that* $\langle A_1 \times A_2, \gtrsim \rangle$ *is an additive conjoint structure and that* $\langle A_1 \times A_2, \gtrsim^*, \circ \rangle$ *and* $\langle A_1, \gtrsim_1{}^*, \circ_1 \rangle$, *where* $\gtrsim^*$ *is either* $\gtrsim$ *or* $\lesssim$ *and* $\gtrsim_1{}^*$ *is either* $\gtrsim_1$ *or* $\lesssim_1$, *are extensive structures. A* (*qualitative*) *law of similitude with exponents* $m$ *and* $n$, *where* $m$ *and* $n$ *are positive integers, holds iff one of the following conditions is valid for all* $a \in A_1$, $u \in A_2$, *and* $i \in I^+$ *where the indicated concatenations exist:*

(i) $\gtrsim^* = \gtrsim$ *and* $\gtrsim_1{}^* = \gtrsim_1$ *or* $\gtrsim^* = \lesssim$ *and* $\gtrsim_1{}^* = \lesssim_1$, *and* $i^m(a, u) \sim (i^n a, u)$; *or*

(ii) $\gtrsim^* = \gtrsim$ *and* $\gtrsim_1{}^* = \lesssim_1$ *or* $\gtrsim^* = \lesssim$ *and* $\gtrsim_1{}^* = \gtrsim_1$, *and* $(a, u) \sim i^m(i^n a, u)$.

The main result is that, under plausible conditions, this property is sufficient as well as necessary for the multiplicative conjoint scales to be power functions of the corresponding extensive ones.

Thinking of the image of a conjoint structure as a rectangle with $A_1$ mapped into the vertical side and $A_2$ into the horizontal, we say the structure is *flat* iff for every $a, b \in A_1$ there exists $u, v \in A_2$ such that $(a, u) \sim (b, v)$. It is *tall* iff for every $u, v \in A_2$ there exists $a, b \in A_1$ such that $(a, u) \sim (b, v)$. A structure that is both flat and tall is symmetric in the sense of Definition 6.7 (Section 6.2.3).

THEOREM 5. *Suppose* $\langle A_1 \times A_2, \gtrsim \rangle$ *is a flat conjoint structure that has an additive representation* $\log \psi_1 + \log \psi_2$; *that* $\langle A_1 \times A_2, \gtrsim^*, \circ \rangle$ *and* $\langle A_1, \gtrsim_1{}^*, \circ_1 \rangle$, *where* $\gtrsim^*$ *is either* $\gtrsim$ *or* $\lesssim$ *and* $\gtrsim_1{}^*$ *is either* $\gtrsim_1$ *or* $\lesssim_1$, *are closed extensive structures* (*with no essential maxima*); *that* $\phi$ *and* $\phi_1$ *are,* *respectively, additive extensive scales; and that the range of $\phi_1$ includes the positive rationals. If a law of similitude with exponents m and n holds, then there exist constants $\alpha$, $\gamma$, $\alpha_1$, and $\gamma_1$ such that*

(i)  $\psi_1\psi_2 = \gamma\phi^\alpha$ *and* $\psi_1 = \gamma_1\phi_1^{\alpha_1}$;

(ii)  $\alpha > $ *or* $< 0$ *according as* $\succsim^* = \succsim$ *or* $\precsim$, *and* $\alpha_1 > $ *or* $< 0$ *according as* $\succsim_1^* = \succsim_1$ *or* $\precsim_1$ ;

(iii)  $|\alpha/\alpha_1| = n/m$.

Note that flatness is often an empirically reasonable assumption and that physicists usually assume that extensive measures are onto the positive reals and hence include the positive rationals. It would, however, be desirable to have a version of this theorem in which concatenation is not closed; the assumption about the range of $\phi_i$ would have to be modified accordingly.

If we let sg $x$ denote the sign of $x$, i.e., sg $x$ is 1, 0, or $-1$ when $x$ is greater than, equal to, or less than 0, then it is easy to see from parts (i) and (iii) that

$$\psi_2(u) = (\gamma/\gamma_1)[\phi(a, u)^{(\mathrm{sg}\,\alpha)m}/\phi_1(a)^{(\mathrm{sg}\,\alpha_1)n}]^\beta,$$

where $\beta = |\alpha|/m = |\alpha_1|/n > 0$. Obviously, $\beta$ is a free parameter, as it should be since $\psi_2$ is a log-interval scale. It is customary in physics to reduce $m$ and $n$ so that their greatest common divisor is 1 and then choose $\beta = 1$. The resulting $\psi_2$ is assigned a name and is, thenceforth, treated as if it were a ratio scale. So, for example, when both the volume and the material are varied, the exponents can be chosen to be $m = n = 1$ and[6] density is defined to be $m/V$ not, for example, $(m/V)^{3.83}$ (which is just as satisfactory a measure of density except for those who would continually have to write the exponent 3.83). Presumably, however, if $(m/V)^2$ always appeared in the equations of physics, there would have been a strong tendency to define that to be density. Ellis (1966, pp. 118–121, 125–126) has also pointed out the highly conventional nature of treating these so-called derived measures as ratio scales, when in fact no experiment determines the exponent. This convention is especially apparent when it is realized that such derived measures arise as part of a conjoint structure and they are not extensively measurable; so they must have the two free constants inherent in any conjoint measure. It will be shown in Section 10.12, when we consider interval scales more fully, why this convention does not lead to trouble.

A key to understanding laws of similitude and how they differ from

---

[6] Unfortunately, there is a shortage of symbols. The symbol $m$, which we have just used as an integral exponent in the laws, is also used to denote mass and as an abbreviation for the unit *meter*. The context usually makes clear which meaning is intended.

another type of law discussed below is to realize that only a single physical concatenation underlies the two formal concatenation operations $\circ$ and $\circ_1$. Density illustrates the point clearly. If one has two objects composed of the same substance, physical juxtaposition generates a concatenation of both volume and mass. The law states how these two concatenations relate, i.e., how the single physical concatenation is counted when it is viewed as $\circ$ and as $\circ_1$. The point is that concatenations in the one attribute cannot be chosen independently of those in the other, and the law formulates the nature of the dependence.

## 10.7.2 Laws of Exchange

The other alternative is that there are two extensive attributes that both affect a third and that these two concatenations can be carried out completely independently. It is convenient to assume that components 1 and 2 of a conjoint structure are both extensive. The law in this case formulates how concatenations on one component can be compensated for by concatenations on the other so as to maintain the same level of the conjoint attribute. An example will illustrate what we mean.

Let the conjoint attribute be kinetic energy and assume that velocity is an additive extensive attribute (i.e., ignore relativistic effects). Given a qualitative ordering of energy, it must satisfy the conjoint-measurement axioms because $E = \frac{1}{2}mv^2$; at the same time, both components have extensive measures. For present purposes, we do not care whether $E$ is or is not extensively measurable. Let $\psi_1$ and $\psi_2$ be the (multiplicative) conjoint measures and $\phi_1$ and $\phi_2$ the extensive ones. Our problem is: when can we be sure that $\psi_k = \gamma_k \phi^{\alpha_k}$? As before there are four cases. In the simplest, $\alpha_k > 0$. Suppose that we can write $\alpha_1 = \beta n$ and $\alpha_2 = \beta m$, where $m$ and $n$ are positive integers and $\beta > 0$. Then for any $a \in A_1$ and $u \in A_2$,

$$
\begin{aligned}
\psi_1(i^m a)\, \psi_2(j^n u) &= \gamma_1 [\phi_1(i^m a)]^{\beta n}\, \gamma_2 [\phi_2(j^n u)]^{\beta m} \\
&= \gamma_1 i^{\beta mn} [\phi_1(a)]^{\beta n}\, \gamma_2 j^{\beta nm} [\phi_2(u)]^{\beta m} \\
&= \gamma_1 [j^m \phi_1(a)]^{\beta n}\, \gamma_2 [i^n \phi_2(u)]^{\beta m} \\
&= \psi_1(j^m a)\, \psi_2(i^n u),
\end{aligned}
$$

and so we see that

$$
(i^m a, j^n u) \sim (j^m a, i^n u).
$$

This equivalence formulates that an exchange ratio, characterized by $m$ and $n$, exists between the first and second components in order to maintain the same value of the resulting attribute. Attributes of this type, which differ appreciably from those for which laws of similitude hold, are usually of interest because they enter into conservation laws. When it is asserted that something is conserved in a closed system, e.g., momentum or energy, then the exchange ratio describes the different configurations of the system that are compatible with the unchanging level of the attribute.

As formulated above, the basic equation seems more complex than that for a law of similitude because two, rather than one, arbitrary positive integers are involved. This is illusory since the special case of $j = 1$ is actually equivalent to the general case. Obviously, the only problem is to see that the general follows from the special:

$$(i^m a, j^n u) \sim (a, i^n(j^n u))$$

$$\sim (a, j^n(i^n u))$$

$$\sim (j^m a, i^n u).$$

So we may formulate the definition and theorem as follows.

DEFINITION 5. *Suppose* $\langle A_1 \times A_2, \succsim \rangle$ *is an additive conjoint structure and* $\langle A_k, \succsim_k^*, \circ_k \rangle$, $k = 1, 2$, *where* $\succsim_k^*$ *is either* $\succsim_k$ *or* $\precsim_k$, *are extensive structures. A (qualitative) law of exchange with exponents* $m$ *and* $n$, *where* $m$ *and* $n$ *are positive integers, holds iff one of the following conditions is valid for all* $a \in A_1 u \in A_2$ *and* $i \in I^+$ *where the indicated concatenations exist:*

(i) $\succsim_1^* = \succsim_1$ *and* $\succsim_2^* = \succsim_2$ *or* $\succsim_1 = \precsim_1$ *and* $\succsim_2^* = \precsim_2$, *and* $(i^m a, u) \sim (a, i^n u)$; *or*

(ii) $\succsim_1^* = \succsim_1$ *and* $\succsim_2^* = \precsim_2^*$ *or* $\succsim_1^* = \precsim_1$ *and* $\succsim_2^* = \succsim_2$, *and* $(i^m a, i^n u) \sim (a, u)$.

THEOREM 6. *Suppose* $\langle A_1 \times A_2, \succsim \rangle$ *is a conjoint structure with an additive representation* $\log \psi_1 + \log \psi_2$ *and* $\langle A_k, \succsim_k^*, \circ_k \rangle$, *where for each* $k = 1, 2$, $\succsim_k^* =$ *either* $\succsim_k$ *or* $\precsim_k$, *are both closed positive extensive structures (with no essential maxima) and with additive representations* $\phi_k$. *If a law of exchange with exponents* $m$ *and* $n$ *holds, then there exist constants* $\alpha_k$ *and* $\gamma_k$ *such that, for* $k = 1, 2$,

(i) $\psi_k = \gamma_k \phi_k^{\alpha_k}$;

(ii) $\alpha_k > $ *or* $ < 0$ *according as* $\succsim_k^* = \succsim_k$ *or* $\precsim_k$;

(iii) $|\alpha_1/\alpha_2| = n/m$.

This is Marley's (1968) reformulation of Luce's (1965) version of the result; it eliminates an unnecessary assumption of the earlier theorem at the expense of stating the law in two parts, as in Definition 5. Note that this result is more satisfactory than the corresponding one for laws of similitude (Theorem 5) in that it assumes neither the flatness of the conjoint structure nor that the range of some extensive representation includes all of the positive rational numbers.

It is clear that when Theorems 5 and 6 are applicable, they restrict rather severely the form of laws or axioms in mechanics, or in other parts of science.

### 10.7.3  Compatibility of the Trinary Laws

We must worry about the fact that laws of similitude and exchange may apply simultaneously, thereby overdetermining certain relationships. The question simply is: does compatibility follow from the laws that we have postulated, or must other assumptions be added?

The first potential problem occurs when a conjoint attribute and both of its components have extensive measures. Then two laws of similitude and one of exchange may hold, and we want them to be compatible with a representation of the form $\phi(a, u)^\alpha = \phi_1(a)^{\alpha_1} \phi_2(u)^{\alpha_2}$. It is clear that all six exponents cannot be independent and still have this simple equation true. In fact, any two laws completely determine the third. Suppose, first, that we have the law of exchange with exponents $m$ and $n$ and one law of similitude with exponents $p$ and $q$ on, say, the first component. If the conjoint and extensive ordering are identical, then the other law of similitude holds and the exponents are $mp$ and $nq$.

$$i^{mp}(a, u) \sim (i^{mq}a, u) \sim (i^{qm}a, u) \sim (a, i^{qn}u) \sim (a, i^{nq}u).$$

Observe that $\phi^{nq} = \phi_1^{np}\phi_2^{mp}$ is a representation that is compatible with these three laws. Three other cases arise when the conjoint and extensive ordering are converse; the analysis is similar. Second, suppose that both laws of similitude hold with exponents $m$, $n$ and $p$, $q$ and that the orders agree, then

$$(i^{pn}a, u) \sim i^{pm}(a, u) \sim i^{mp}(a, u) \sim (a, i^{mq}u),$$

which is a law of exchange with exponents $np$ and $mq$. Observe that a representation compatible with all three laws is $\phi^{nq} = \phi_1^{mq}\phi_2^{np}$. The other three cases are similar.

The next problem of compatibility arises when we go to conjoint attributes with more than two components. It is easy to see that it is sufficient to consider the three-dimensional case. Two cases must be considered, the one where there are two laws of exchange, involving two different pairs of components, and the one where there is a law of exchange involving one pair of components and a law of similitude involving the third component.

First, suppose that $\langle A_1 \times A_2, \gtrsim_{12} \rangle$ and $\langle A_2 \times A_3, \gtrsim_{23} \rangle$ satisfy laws of exchange with exponents $m$, $n$ and $p$, $q$, then assuming that the conjoint and extensive orderings are the same and using independence,

$$
\begin{aligned}
(i^{pm}a_1, a_2, a_3) &\sim (a_1, i^{pn}a_2, a_3) \\
&\sim (a_1, i^{np}a_2, a_3) \\
&\sim (a_1, a_2, i^{nq}a_3),
\end{aligned}
$$

and so by independence a law of exchange with exponents $mp$ and $nq$ holds for $\langle A_1 \times A_3, \gtrsim_{13} \rangle$. A representation compatible with all three laws is $\phi_1^{nq} \phi_2^{mq} \phi_3^{mp}$. Similar arguments hold in the other three cases that arise from the second possible form of the law when the conjoint and extensive orders are converse.

Second, suppose there is a law of exchange with exponents $m$, $n$ on the first two components and a law of similitude with exponents $p$, $q$ involving the third one. This is compatible (and, under the conditions of Theorems 5 and 6, it implies) a multiplicative conjoint representation of form

$$
\phi^{\alpha q} = \phi_1^{\beta n} \phi_2^{\beta m} \phi_3^{\alpha p},
$$

If $\alpha/\beta$ is irrational, then no other law of exchange or similitude holds. But if a second law of similitude or of exchange does hold, then by the previous arguments, all three laws of similitude and exchange must hold, with mutually compatible exponents.

### 10.7.4 Some Relations Among Extensive, Difference, and Conjoint Structures

The results just obtained relating extensive and conjoint structures suggest the possibility of similar conditions among other measurement systems. Of those we have examined, the most general ones are the several classes of difference structures. The algebraic one is especially important because the method of magnitude estimation, which is widely used in parts of psychology, can be viewed as a difference structure over numerical scales which result from either extensive measurement, conjoint measurement, or both (see Section 4.6).

As the results are the same for both algebraic- and absolute-difference structures, we state both theorems at once.

In stating the following three theorems, we make the following conventions.

1. By an algebraic- (absolute-) difference structure, we mean one for

which there exists a numerical algebraic- (absolute-) difference representation $\psi$ as in Theorem 4.2 (Theorem 4.6).

2.  By a closed extensive structure, we mean one for which there exists an additive numerical representation $\phi$ of Theorem 3.1 and the range of $\phi$ includes the positive rational numbers.

3.  By an additive conjoint structure, we mean one for which there exists an additive numerical representation $\psi_1 + \psi_2$ of Theorem 6.2 and the range of each $\psi_i$ includes the nonnegative rational numbers.

4.  The relation $\succsim'$ on $A$ is defined in terms of $\succsim$ on $A \times A$ via $A^*$ of Definition 4.4 (Section 4.4.1) [Definition 4.10 (Section 4.10)] by

$$a \succ' b \qquad \text{iff} \quad ab \in A^*,$$
$$a \sim' b \qquad \text{iff} \quad ab \sim aa.$$

5.  We use commas and parentheses in the conjoint structures, but not in the difference ones.

With these conventions, we first establish two relations between difference and extensive structures.

THEOREM 7.  *Suppose* $\langle A \times A, \succsim \rangle$ *is an algebraic- or absolute-difference structure with a representation* $\psi$ *and* $\langle A, \succsim', \circ \rangle$ *is a closed extensive structure with representation* $\phi$. *Then:*

(i)  *there exist* $\alpha \in \mathrm{Re}^+$, $\beta \in \mathrm{Re}$, *with* $\psi = \alpha\phi + \beta$ *iff, for all* $a, b, c, d \in A$,

$$(a \circ c)(b \circ c) \sim (a \circ d)(b \circ d);$$

*and*

(ii)  *there exist* $\alpha \in \mathrm{Re}^+$, $\beta \in \mathrm{Re}$ *with* $\psi = \alpha \log \phi + \beta$ *iff, for all* $a, b, c, d \in A$,

$$(a \circ b)b \sim (c \circ d)d \quad \text{is equivalent to} \quad ab \sim cd.$$

The next theorem relates difference and conjoint structures:

THEOREM 8.  *Suppose* $\langle A \times A, \succsim \rangle$ *is an algebraic- or absolute-difference structure with a representation* $\psi$, $A = A_1 \times A_2$, *and* $\langle A_1 \times A_2, \succsim' \rangle$ *is an additive conjoint structure with representation* $\psi_1 + \psi_2$. *There exist real* $\alpha > 0$ *and* $\beta$ *such that* $\psi = \alpha(\psi_1 + \psi_2) + \beta$ *iff, for all* $a, b \in A_1$ *and* $p, q \in A_2$, $(a, p)(b, p) \sim (a, q)(b, q)$.

Note that the condition in Theorem 8 is analogous to part (i) of Theorem 7 (substitute "," for "$\circ$"). No analog to part (ii) seems possible because interval scales can be negative and so the logarithm is not defined.

Finally, we note that Theorem 4.3 concerning difference-and-ratio structures is similar to Theorems 7 and 8 in that it relates a difference representation to a ratio one for two orderings of $A \times A$. It could have been included here.

Suppose, next, that we have a difference structure for magnitude estimation of a physical attribute that is both extensively and conjointly measurable. A question of compatibility arises: given that a law of similitude holds between the extensive and conjoint structures, how do the conditions of Theorems 7 and 8 relate? A partial answer is formulated in the next result.

THEOREM 9.   Let $A = A_1 \times A_2$. Suppose $\langle A \times A, \gtrsim \rangle$ is an algebraic- or absolute-difference structure, $\langle A_1 \times A_2, \gtrsim' \rangle$ is an additive conjoint structure, $\langle A, \gtrsim', \circ \rangle$ is a closed extensive structure, and $\langle A_1, \gtrsim'^*, \circ_1 \rangle$, where $\gtrsim'^*$ is either $\gtrsim_1'$ or its converse, is a closed extensive structure. If the conjoint and extensive structures are related as in Theorem 5 (law of similitude), then the condition of part (ii) of Theorem 7 is equivalent to the condition in Theorem 8.

It may be useful to remark that the conditions formulated in the laws of similitude and exchange and those in Theorems 7 and 8 are somehow natural and are not easily generalized. In particular, given an arbitrary function relating two scales that preserve the same ordering, it does not appear to be at all easy to find a qualitative condition that is equivalent to that relation. It is as though the power functions of Theorems 5 and 6 and the linear and logarithmic functions of Theorems 7 and 8 have a special natural status; a similar situation was seen in Theorem 4.3. We do not know of any reason why these particular functions should arise rather than any others. Theorems 8 and 9 are left as Exercises 19 and 20.

## 10.8  PROOFS

### 10.8.1  Preliminary Lemma

LEMMA 13.   Suppose that $f$ is a strictly monotonic, positive, real-valued function defined on the positive rationals, that $n$ is a positive integer, and that for all rational $\rho$ and $\sigma$, $f(\rho^n\sigma) = f(\rho^n) f(\sigma)$. Then there exists real $\beta \neq 0$ such that $f(\rho) = \rho^\beta$.

PROOF.   If $f$ is strictly increasing, we extend $f$ to the positive reals by the following definition. For real $\lambda > 0$,

$$f(\lambda) = \sup_{0 < \rho < \lambda} f(\rho).$$

This function is also strictly increasing since when $\lambda < \eta$, there exists a rational $\sigma$ with $\lambda < \sigma < \eta$, and so for $\rho < \lambda$, $f(\rho) < f(\sigma)$; thus, $f(\lambda) < f(\eta)$. Moreover, we show that it satisfies the functional equation $f(\lambda\eta) = f(\lambda) f(\eta)$. Observe, first, that

$$\sup_{\rho < \lambda^{1/n}} f(\rho^n) = \sup_{\sigma < \lambda} f(\sigma).$$

Clearly $\leqslant$ holds. Since there exists a rational $\sigma$ such that $\sigma^{1/n} < \rho < \lambda^{1/n}$, then $\sigma < \rho^n$. By monotonicity, $f(\sigma) < f(\rho^n)$, and so $\geqslant$ also holds. Using this,

$$f(\lambda\eta) = \sup_{\tau < \lambda\eta} f(\tau)$$

$$\geqslant \sup_{\substack{\rho < \lambda^{1/n} \\ \sigma < \eta}} f(\rho^n\sigma)$$

$$= \sup_{\rho < \lambda^{1/n}} f(\rho^n) \sup_{\sigma < \eta} f(\sigma)$$

$$= \sup_{\tau < \lambda} f(\tau) \sup_{\sigma < \eta} f(\sigma)$$

$$= f(\lambda) f(\eta).$$

The inequality also holds in the other direction since when $\tau < \lambda\eta$, we can find $\rho$ and $\sigma$ such that $\tau < \rho^n\sigma$, $\rho < \lambda^{1/n}$, and $\sigma < \eta$. It is well known that $f(\lambda) = \lambda^\beta$ for some $\beta > 0$ (Aczél, 1966, p. 41). If $f$ is decreasing, a similar argument yields the same result with $\beta < 0$.                    $\diamondsuit$

## 10.8.2  Theorem 5  (p. 486)

*Suppose that a flat binary conjoint structure is, with respect to either the same or the converse ordering, also a closed extensive structure (with no essential maximum); that with respect to the induced conjoint ordering or its converse, its first component is also a closed extensive structure (with no essential maximum); and that a law of similitude holds with exponents m and n. If there are extensive measures whose ranges include the positive rationals, then the multiplicative conjoint scales are power functions of the corresponding extensive ones, the exponents are positive or negative according as the extensive orderings are the conjoint ones or their converses, and the absolute value of the ratio of the exponents is n/m.*

PROOF.  Consider, first, the case where both orderings agree and let $\psi_k$, $k = 1, 2$, be the two multiplicative conjoint scales and let $\phi$ and $\phi_1$ be the two extensive scales corresponding to $\psi_1\psi_2$ and $\psi_1$, respectively, where the range of $\phi_1$ includes the positive rationals. Since these functions are order preserving, there exist strictly increasing functions $f$ and $g$ such that for all $a \in A_1$, $u \in A_2$,

$$\psi_1(a)\,\psi_2(u) = f[\phi(a, u)] \quad \text{and} \quad \psi_1(a) = g[\phi_1(a)].$$

By the laws of similitude, for all positive integers,

$$j^m(i^n a, u) \sim (ji)^m (a, u) \sim i^m(j^n a, u),$$

and so by the additivity of $\phi$,

$$j^m \phi(i^n a, u) = i^m \phi(j^n a, u).$$

But

$$j^m \phi(i^n a, u) = j^m f^{-1}[g\phi_1(i^n a)\,\psi_2(u)]$$
$$= j^m f^{-1}\{g[i^n \phi_1(a)]\,x\},$$

where $x$ is in the range of $\psi_2$. Thus,

$$j^m f^{-1}\{g[i^n \phi_1(a)\,x]\} = i^m f^{-1}\{g[j^n \phi_1(a)]\,x\}.$$

Let $\sigma$ be in $\mathrm{Ra^+}$ and $\rho = i/j$, then $\sigma/j^n$ is in the range of $\phi_1$ and so

$$f^{-1}[g(\rho^n \sigma)\,x] = \rho^m f^{-1}[g(\sigma)\,x].$$

By flatness, there exist $x, y$ in range of $\psi_2$ such that $g(1)\,x = g(\sigma)\,y$, so

$$g(\rho^n) = \frac{1}{x} f\{\rho^m f^{-1}[g(1)x]\}$$
$$= \frac{1}{x} f\left\{\rho^m f^{-1}\left[g(\sigma) \cdot \frac{g(1)x}{g(\sigma)}\right]\right\}$$
$$= \frac{y}{x} g(\rho^n \sigma)$$
$$= \frac{g(1)}{g(\sigma)} g(\rho^n \sigma).$$

By Lemma 13, $g(\rho) = \delta \rho^{\alpha m}$, where $\delta = g(1)$ and $\alpha > 0$ because $g$ is increasing. Since the rationals are dense in the real numbers, this holds for all $\rho$ in the range of $\phi_1$.

In the basic functional equation, fix $\sigma$ and $x$, denote $\delta \sigma^{\alpha n} x$ by $\lambda$, and substitute the expression for $g$

$$f^{-1}(\lambda \rho^{\alpha m n}) = \rho^m f^{-1}(\lambda).$$

Writing $x = \rho^m f^{-1}(\lambda)$, we see that $f(x) = \delta_1 x^{\alpha n}$, where $\delta_1 = \lambda/[f^{-1}(\lambda)]^{\alpha n}$. Since the numbers $\rho^m f^{-1}(\lambda)$ are dense in $\mathrm{Re}^+$ as $\rho$ ranges over $\mathrm{Ra}^+$, $f(x) = \delta_1 x^{\alpha n}$ holds for all $x$ in the range of $\phi$.

The case when both orders are converse is similar except that $f$ and $g$ are strictly decreasing.

When one pair of orders is the same and the other pair are converses, we have the law $i^m(i^n a, u) \sim (a, u) \sim j^m(j^n a, u)$. A similar calculation yields the functional equation

$$\rho^m f^{-1}[g(\rho^n \sigma)\, x] = f^{-1}[g(\sigma)\, x].$$

The rest of the proof is substantially the same.                    $\diamondsuit$

### 10.8.3  Theorem 6  (p. 489)

*Suppose that both components of a binary conjoint structure are closed, positive extensive structures (with no essential maxima) with respect to either the induced conjoint order or its converse and that a law of exchange holds with exponents m and n. Then each multiplicative conjoint scale is a power function of the corresponding extensive one, with a positive exponent when the conjoint and extensive orderings are the same and a negative exponent when they are converses, and the absolute value of the ratio of the exponents is n/m.*

PROOF.   Let $\psi_k$ and $\phi_k$ be the conjoint and extensive scales, respectively, with the units of $\psi_k$ selected so that, for each $k$, $\phi_k$ assumes the value 1 for some element. Suppose, first, that the two orderings are the same, then there exist $f_k$ such that $\psi_k = f_k(\phi_k)$ and the $f_k$ are strictly increasing. Recall that the law of exchange implies that, for all positive integers $i$ and $j$, $a \in A_1$, and $u \in A_2$,

$$(i^m a, j^n u) \sim (j^m a, i^n u).$$

Applying the conjoint representation to this,

$$\psi_1(i^m a)/\psi_1(j^m a) = \psi_2(i^n u)/\psi_2(j^n u).$$

Since this ratio does not depend on $a$ and $u$ and since $m$ and $n$ are fixed, it is a function only of $i$ and $j$; denote it $g(i, j)$. If $i/j = k/l$, where $i, j, k$ and $l$ are positive integers, then with no loss of generality there is a positive integer $p$ such that $k = pi$ and $l = pj$, so

$$\begin{aligned}
g(k, l) = g(pi, pj) &= \psi_1[(pi)^m a]/\psi_1[(pj)^m a] \\
&= \psi_1[i^m(p^m a)]/\psi_1[j^m(p^m a)] = g(i, j).
\end{aligned}$$

Thus, we may write $g(i, j)$ as $g(i/j)$.

We show that $g$ is strictly increasing. Suppose that $i/j = \rho > \sigma = k/l$, then $il > kj$, and then since $\phi_1 > 0$,

$$\phi_1[(il)^m\, a] = (il)^m\, \phi_1(a) > (kj)^m\, \phi_1(a) = \phi_1[(kj)^m\, a],$$

whence, by the order-preserving property of $f_1$, $\psi_1[(il)^m\, a] > \psi_1[(kj)^m\, a]$. So,

$$
\begin{aligned}
g(\rho) &= \psi_1(i^m a)/\psi_1(j^m a) \\
&= \psi_1[(il)^m\, a]/\psi_1[(jl)^m\, a] \\
&> \psi_1[(jk)^m\, a]/\psi_1[(jl)^m\, a] \\
&= \psi_1[k^m(j^m a)]/\psi_1[l^m(j^m a)] \\
&= g(\sigma).
\end{aligned}
$$

Next, we show that $g(\rho\sigma) = g(\rho)\, g(\sigma)$, where $\rho = i/j$ and $\sigma = k/l$.

$$
\begin{aligned}
g(\rho\sigma) &= \psi_1[(ik)^m\, a]/\psi_1[(jl)^m\, a] \\
&= \psi_1[i^m(k^m a)]\, \psi_1[k^m(j^m a)]/\psi_1[j^m(k^m a)]\, \psi_1[l^m(j^m a)] \\
&= g(\rho)\, g(\sigma).
\end{aligned}
$$

Thus, by Lemma 13, $g(\rho) = \rho^\beta$, $\beta > 0$. If in the defining equation for $g$ we set $j = 1$, then $\psi_1(i^m a) = i^\beta \psi_1(a)$. For any real $\lambda$ for which there is an $a$ with $\phi_1(a) = \lambda^m$, define $h_1(\lambda) = f_1(\lambda^m)$. Then,

$$
\begin{aligned}
h_1(i\lambda) &= f_1(i^m \lambda^m) \\
&= f_1[\phi_1(i^m a)] \\
&= \psi_1(i^m a) \\
&= i^\beta \psi_1(a) \\
&= i^\beta f_1(\lambda^m) \\
&= i^\beta h_1(\lambda).
\end{aligned}
$$

We now show that $h_1(\lambda) = h_1(1)\, \lambda^\beta$. Suppose not. If for some $\lambda$, $\lambda^\beta > h_1(\lambda)/h_1(1)$, then we can find a rational $i/j$ such that

$$\lambda > i/j > [h_1(\lambda)/h_1(1)]^{1/\beta}.$$

Thus, $(i/j)^\beta > h_1(\lambda)/h_1(1)$ and $i < j\lambda$. Since $h_1$ is strictly increasing, $h_1(1)\, i^\beta = h_1(i) < h_1(j\lambda) = j^\beta h_1(\lambda)$, and so $(i/j)^\beta < h_1(\lambda)/h_1(1)$, a contradiction. The other inequality is similar. Thus,

$$f_1(\lambda) = h_1(\lambda^{1/m}) = h_1(1)\lambda^{\alpha_1},$$

where $\alpha_1 = \beta/m > 0$. The argument for $f_2$ is similar, except that $n$ plays the role of $m$. (Note that we can choose $\phi_1$ so $\phi_1(a) = 1$ for some $a$, so that $h_1(1)$ is defined.)

The arguments in the other three cases are substantially the same except that some functions are strictly decreasing and the corresponding exponents are negative.    $\diamond$

### 10.8.4  Theorem 7  (p. 492)

*Suppose $\langle A \times A, \gtrsim \rangle$ is an algebraic- or absolute-difference structure with the representation $\psi$ and $\langle A, \gtrsim', \circ \rangle$ is a closed extensive structure with a representation $\phi$ whose image includes the positive rationals. Then*

(i)   $\psi$ *and* $\phi$ *are linearly related iff, for all* $a, b, c, d \in A$,

$$(a \circ c)(b \circ c) \sim (a \circ d)(b \circ d),$$

*and*

(ii)   *they are logarithmically related iff, for all* $a, b, c, d \in A$,

$$(a \circ b)b \sim (c \circ d)d \quad \text{is equivalent to} \quad ab \sim cd.$$

PROOF.   The proofs of necessity are left as Exercise 17 and the sufficiency of part (i) as Exercise 18.

To show the sufficiency of part (ii), observe that since $\psi$ and $\phi$ preserve the same order $\gtrsim'$, there exists a strictly increasing function $f$ from, at least, the positive rationals into Re such that $\psi = f\phi$. Denote $w = \phi(a)$, $x = \phi(b)$, $y = \phi(c)$, and $z = \phi(d)$. Consider, first, the case of an algebraic-difference structure. The condition of part (ii) immediately implies

$$f(w + x) - f(x) = f(y + z) - f(z) \quad \text{iff} \quad f(w) - f(x) = f(y) - f(z)$$

for all rational $w, x, y, z > 0$. Let $n$ and $m$ be positive integers, and choose $w = nx$ and $y = nz$. By induction we see that $f(nx) - f(x)$ is independent of $x$. In particular,

$$f(nx/m) - f(x/m) \quad \text{and} \quad f(x) - f(x/m) = f(mx/m) - f(x/m)$$

are both independent of $x$. Therefore, for every rational $r = n/m > 0$,

$$f(rx) - f(x) = f(nx/m) - f(x/m) + f(x/m) - f(x)$$

is independent of $x$. Since $f$ is strictly increasing, this may be extended to all real $k$, $x > 0$, i.e., there exists a function $g$ such that

$$f(kx) - f(x) = g(k).$$

This reduces immediately to Cauchy's functional equation and it is known (Aczél, 1966, p. 41) that the only solution is $f(x) = \alpha \log x + \beta$, $\alpha > 0$.

For an absolute-difference structure, a similar argument shows that $|f(nx) - f(x)|$ is independent of $x$. So, as above, we can show that for all real $k$, $x \geqslant 0$,

$$f(kx) - f(x) = g(k)\,h(x),$$

where $h(x) = 1$ or $-1$. We show $h(x)$ is constant, so we reduce to the preceding case. Suppose not, then there exist $x$ and $y$ such that

$$f(kx) - f(x) = -[f(ky) - f(y)].$$

Suppose $k > 1$, then by the strict monotonicity of $f$,

$$f(x) + f(y) = f(kx) + f(ky) > f(x) + f(y),$$

which is impossible.                                                           $\diamondsuit$

## □10.9 EMBEDDING PHYSICAL ATTRIBUTES IN A STRUCTURE OF PHYSICAL QUANTITIES

### 10.9.1 Assumptions About Physical Attributes

In what follows we shall consider a collection $\mathcal{O}$ of structures $\langle A, \gtrsim \rangle$, where $A$ is a set and $\gtrsim$ a binary relation on $A$, and a subset $\mathcal{E}$ of $\mathcal{O}$ consisting of structures $\langle A, \gtrsim, \circ \rangle$, where $\circ$ is a binary operation on $A$. Intuitively, we think of the elements of $\mathcal{O}$ as *physical attributes* in the sense that $A$ is a set of physical quantities and $\gtrsim$ is determined by some well-specified physical procedure. Those in $\mathcal{E}$ will, of course, be the attributes that are extensively measurable. In practice, if $\langle A, \gtrsim \rangle$ is a physical attribute, so is $\langle A, \lesssim \rangle$; we do not necessarily assume that both are in $\mathcal{O}$.

We axiomatize $\mathcal{O}$ and $\mathcal{E}$, where $\mathcal{E} \subset \mathcal{O}$, as follows:

1. *The set $\mathcal{E}$ is nonempty and $\mathcal{O}$ is finite.*

2. *If $\langle A, \gtrsim, \circ \rangle$ is in $\mathcal{E}$, it is an extensive structure with an additive representation whose range includes all of the positive rational numbers.*

3. *If $\langle A, \gtrsim \rangle$ is in $\mathcal{O}$, it is part of a conjoint structure in the sense that either*

   (i) *$A = A_1 \times A_2$, $\langle A_1 \times A_2, \gtrsim \rangle$ is a symmetric conjoint structure (Definition 6.7) with a multiplicative representation, and $\langle A_i, \gtrsim_i \rangle$, $i = 1, 2$, are in $\mathcal{O}$; or*

   (ii) *there is a symmetric conjoint structure $\langle A_1' \times A_2', \gtrsim' \rangle$ in $\mathcal{O}$, with a multiplicative representation, such that $A_1' = A$, $\gtrsim_1' = \gtrsim$, and $\langle A_2', \gtrsim_2' \rangle$ is in $\mathcal{O}$.*

4.   *If* $\langle A_1 \times A_2 , \gtrsim \rangle$ *is in* $\mathcal{A}$, *then either*
   (i)   *there exist* $\circ_i$ *on* $A_i$, $i = 1, 2$, *such that* $\langle A_i , \gtrsim_i , \circ_i \rangle$ *are both in* $\mathcal{E}$ *and a law of exchange holds; or*
   (ii)   *there exist* $\circ$ *on* $A_1 \times A_2$ *and, for either* $i = 1$ *or* $2$, $\circ_i$ *on* $A_i$ *such that* $\langle A_1 \times A_2 , \gtrsim, \circ \rangle$ *and* $\langle A_i , \gtrsim_i , \circ_i \rangle$ *are both in* $\mathcal{E}$ *and a law of similitude holds.*

5.   *Suppose laws of similitude hold both for* $\langle A_1 \times A_2 , \gtrsim, \circ, \circ_1 \rangle$ *and for* $\langle A_1 \times A_2 , \gtrsim', \circ', \circ_1' \rangle$. *If* $\gtrsim_i' = \gtrsim_i$ *or* $\lesssim_i$, $i = 1, 2$, *then* $\gtrsim' = \gtrsim$ *and* $\circ' = \circ$.

The first two assumptions seem clear. The third simply says that every attribute appears in at least one conjoint triple, either as the basic attribute or as one of the induced component attributes. Moreover, according to the fourth assumption, every such triple is one of just two types: either a law of similitude or of exchange holds, which means that at least two of the three attributes are extensive. The final assumption concerns a situation in which $A_1 \times A_2$ has two distinct orderings whose induced orderings on the two components are identical or converses. Momentum and kinetic energy, with formulas $p = mv$ and $E = \frac{1}{2}mv^2$, show that the possibility is not idle. That case and others like it in which both components are extensive and a law of exchange holds pose no problem since, by Theorem 6, both component measures are power functions of their respective extensive measures and so of each other. But a problem does arise with laws of similitude since then we have the two equations

$$\phi(a, u)^\alpha = \phi_1(a)^{\alpha_1} \psi_2(u),$$

$$\phi^*(a, u)^\beta = \phi_1(a)^{\beta_1} \psi_2^*(u),$$

and nothing forces $\psi_2^*$ to be a power function of $\psi_2$. Assumption 5 simply rules out that possibility from arising. Whether this says something about reality or about physicists is not clear, but it appears to be true of classical physics.

An alternative axiomatic approach which assures that $\psi_2^*$ is a power function of $\psi_2$ was provided by Marley (1970).

Consider any attribute in $\mathcal{A}$. By Assumption 3, it is a part of a conjoint structure. By Assumptions 2–4 and Theorems 5 and 6, either it is extensive or its conjoint scale can be written as a monomial function of extensive representations. By Assumption 5 and the above remarks, this expression is unique.

If either a law of exchange or of similitude holds among the representations of three attributes in $\mathcal{E}$, we may drop one of the three since its representation is a monomial function of the representations of the other two. Continue this until we are forced to stop because no law of exchange or similitude can be found to hold entirely in the remaining set. By Assumption 1, this happens after a finite number of steps. If both an attribute and its converse are in the remaining set, delete one of them. Call the resulting set $\mathcal{B}$. Thus, with the exception of the uniqueness statement, we have proved the following result. The proof of uniqueness is left as Exercise 21.

THEOREM 10. *Suppose that Assumptions 1–5 hold. Then there exists a subset $\mathcal{B}$ of $\mathcal{E}$ that is maximal with respect to the properties*

(i)   *not both an attribute and its converse are in $\mathcal{B}$;*

(ii)  *no law of exchange or similitude holds with all three attributes in $\mathcal{B}$.*

*Further, if $\phi_1,...,\phi_n$ are extensive representations of the n attributes in $\mathcal{B}$ and if $\psi$ is a representation of an attribute in $\mathcal{A}$, then there exist unique real $\alpha > 0$ and unique rational $\rho_i$ such that*

$$\psi = \alpha \prod_{i=1}^{n} \phi_i^{\rho_i}.$$

We now identify any object $a$ in base attribute $A_i$ with that representation $\phi_i$ for which $\phi_i(a) = 1$. Using Theorem 10, an object $a$ in any attribute of $\mathcal{A}$ is identified with some representation $\psi = \prod_{i=1}^{n} \phi_i^{\rho_i}$, with $\psi(a) = 1$. Multiplication of the identified representations corresponds to the operation $*$ on physical quantities. The formal embedding theorem is as follows.

THEOREM 11. *Suppose that the assumptions of Theorem 10 hold and let $\mathcal{B}$ and $\phi_i$, $i = 1,...,n$ be defined as there. Let*

$$A = \left\{ \alpha \prod_{i=1}^{n} \phi_i^{\rho_i} \mid \alpha \in \text{Re}, \quad \rho_i \in \text{Ra} \right\},$$

$$A^+ = \left\{ \alpha \prod_{i=1}^{n} \phi_i^{\rho_i} \mid \alpha \in \text{Re}^+, \quad \rho_i \in \text{Ra} \right\},$$

*and let $*$ denote pointwise multiplication of functions from A. Then*

(i)   $\langle A, A^+, * \rangle$ *is a structure of physical quantities;*

(ii)  $\{\phi_1,...,\phi_n\}$ *is a base of the structure;*

(iii) *if $\psi$ is a representation of an attribute in $\mathcal{A}$, then $\psi \in A^+$.*

The proof of this theorem is routine and is left to the reader (Exercise 22). What the theorem shows is that the axioms of extensive and conjoint measurement plus some assumptions about the occurrence of two types of trinary laws are adequate to construct a structure of physical quantities that satisfies the usual axioms. Moreover, it shows that there is a base composed entirely of extensive representations.

## 10.9.2 Fundamental, Derived, and Quasi-Derived Attributes

As we use the term, an attribute is called *fundamental* if its measurement does not depend on the measurement of anything else. Because of the inherent logical symmetry of conjoint measurement (see Section 6.5.6), this means that all of the attributes in $\mathcal{A}$, and so all of the traditional physical attributes, are fundamental.

Campbell, it will be recalled (p. 7), was aware of a logical distinction between fundamental and extensive (he called them *additive*) attributes, but he claimed that, at the time, they were functionally equivalent. His wording suggests to us that he meant "fundamental" to be the generic term, and this is how we use it. Ellis (1966) drew the opposite conclusion. Other measures that can be expressed in terms of extensive measures Campbell called *derived*. Pfanzagl (1968, p. 31–32) expressed the opinion that no real distinction can be drawn in physics between fundamental and derived measurement. If, as is usual, one places derived in opposition to fundamental, we agree; however, the opposition can be interpreted as against extensive measurement, in which case it is meaningful. Accepting *derived* to mean simply nonextensive, then we may distinguish between two types—e-*derived* for the nonextensive attribute in a law of exchange and s-*derived* for the nonextensive one in a law of similitude.

Whether an attribute is derived, i.e., nonextensive, is to a degree a current fact subject to change by the discovery of a suitable concatenation operation. Moreover, when no very natural method of directly ordering an attribute and/or forming concatenations is known, that attribute is usually treated as derived even though by means of sufficient physical ingenuity it can be shown to be extensive.

Campbell introduced still another notion of derived which appears to be the following. An extensive attribute not in the base $\mathcal{B}$ is called *quasi-derived* relative to $\mathcal{B}$. These are the attributes whose scales are written as monomial powers of the base scales even though they can, themselves, be extensively measured. Closely related is the use of the terms "primary" and "secondary" for, respectively, attributes in $\mathcal{B}$ and those not in $\mathcal{B}$. Thus, the secondary scales consist of the derived and quasi-derived scales. Of course, this distinction is somewhat arbitrary because the choice of a base is not unique. It appears that Palacios (1964) reversed the use of the terms "derived" and "secondary" since, on p. 12, he cautions that "... secondary [in Palacios' sense] quantities must not be confused with those other writers call derived quantities, that is, those which do not occur in the basis of a dimensional system."

Ellis devoted two chapters of his book to what he called *associative* measurement. The basic idea seems to be that two attributes may covary and so the measurement of the one can be reduced to that of the other. Judging by his examples, he is getting at the covariation of a conjoint attribute with one of its components when the other is held fixed. No clear distinction is made between the two types of laws, similitude and exchange, which we have discussed.

Part of the problem in knowing exactly what Campbell and Ellis meant by their terms is that they did not make explicit the role of conjoint measurement and of the laws of similitude and exchange, and so one is forced to infer their meanings from the examples they discuss. Often, it is ambiguous exactly what they intended.

Table 2 provides a summary of how we think the several authors used these terms. The relevant pages are noted after each entry.

**TABLE 2**

A Dictionary of Equivalent Terms[a]

| Krantz *et al.* | Campbell (1920) | Palacios (1956) | Ellis (1966) |
|---|---|---|---|
| *Measurement theory* | | | |
| Fundamental (502) | Fundamental (267–294) | Direct (14) | Direct (55) |
| Derived (502) | Derived (346, 378–379) | Secondary (12) | Derived (118) |
| Extensive (Chapter 3) | Fundamental (267–294) | Primary (12) | Fundamental (56) |
| Conjoint (Chapter 6) | — | — | Associative (56, 90–140) |
| Quasi-derived (502) | Quasi-derived (379–388) | — | — |
| Ordinal (2) | Order (269–270) | — | Elemental (56) |
| *Algebra of physical quantities* | | | |
| Primary or basic (502) | Basic (377–378, 388–393) | Fundamental (12) | Independent (130) |
| Secondary (502) | Derived and quasi-derived (376–393) | Derived (12) | Dependent (130) |

[a] The numbers after each term are the pages on which it is defined and discussed.

## ☐10.10  WHY ARE NUMERICAL LAWS DIMENSIONALLY INVARIANT?

### 10.10.1  Three Points of View[7]

Were we to continue in the spirit of Sections 10.7–9, we should now formulate a general qualitative definition of a physical law, using only orderings and concatenations, and then prove that its numerical representation in terms of the usual physical measures is dimensionally invariant. To the best of our knowledge, no such definition has ever been given, and we have been unable to arrive at one. This failure is somewhat puzzling. Of course, it is not the least surprising that physical laws are usually formulated numerically, for that is certainly the simplest way to communicate them and to work with them, but presumably such equations are only a shorthand summary of what is really a complex set of possible qualitative observations. In principle, then, we should be able to reformulate any numerical law in wholly qualitative terms, much like, but more complex than, the laws of similitude and exchange. What is surprising is that philosophers of science have not only failed to arrive at a general qualitative definition of a physical law, but apparently have failed even to try. Until this is achieved, we very much doubt that there will be any truly satisfactory account of why (most) numerical laws of physics are dimensionally invariant.

It is obvious from the writings of those most concerned about dimensional analysis that the purely descriptive "metaphysical" statement that numerical laws in physics are dimensionally invariant is not considered satisfactory. For example, Bridgman questioned:

> Why is it that an equation which correctly describes a relation between various measurable physical quantities must in its form be independent of the size of the fundamental units? There does not seem to be any necessity for this in the nature of the measuring process itself (Bridgman, 1931, p. 13).

The attempts to account for the fact of dimensionally invariant laws are of three types, which may be tagged as the "it couldn't be otherwise," the "descriptive/deductive," and the "physical similarity" arguments.

An "it couldn't be otherwise" argument proceeds along lines something like these. Since the choice of units is a wholly arbitrary matter—the choice exists because of the way we choose to represent certain qualitative information numerically—any assertion that describes a natural phenomenon cannot depend upon such conventions. This is Sedov's view (1959, p. 16), and it was described as a "principle of theory construction" by Luce (1959) who,

---

[7] In preparing this section, we have drawn heavily upon Causey's (1967) excellent summary of thought about dimensionally invariant laws.

however, backed away from this position (1962) in the face of Rozeboom's (1962) criticisms. Many others have expressed more or less similar arguments. We suspect that many who hold this view are simply saying, as we did earlier, that if we knew how to formulate what we mean by a qualitative physical law, then we would find, as a purely logical consequence of our measurement assumptions, that the numerical representation of the law would be dimensionally invariant. Since this has not been proved, the position that "it couldn't be otherwise" is quite suspect and widely discredited. Perhaps we should add, however, that some of the arguments mounted against this position are no less suspect than the arguments for it. A major one, used by Bridgman for example (1931, p. 41), is stated by Causey as follows:

> ... [a] law does not assert the identity of certain 'quantities'; rather it expresses a *numerical* relation which holds between the numbers obtained when certain phenomena are measured in certain ways. Without further analysis it is not obvious that the same numerical relation will be satisfied by the numbers obtained with all units of measurement (Causey, 1967, p. 26).

We do not agree entirely with this argument because we suspect that any physical law can be formulated (perhaps in an exceedingly complicated way) entirely in terms of qualitative observations, much as we formulated the monomial laws in terms of the axioms of additive extensive and multiplicative conjoint measurement together with laws of similitude and/or exchange. We do, however, agree with the argument to the following extent: until a general qualitative formulation is provided, those who say that "it couldn't be otherwise" are merely asserting their belief in the existence of a theorem—albeit one with an empirically well-supported conclusion.

The descriptive/deductive argument was most clearly stated by Birkhoff (1950); it seems that Bridgman may have also accepted it earlier, but it is difficult to be certain. In essence, the view is that the fundamental equations of physics are, as a matter of fact, all dimensionally invariant; hence, it is claimed, all laws that derive from them must also be dimensionally invariant. Thus, for example, the law relating the period of a pendulum to its length, mass, gravity, and displacement must be dimensionally invariant because Newton's laws of motion are known to be.

Causey (1967, p. 29) rejected the argument on two grounds. First, it accounts only for the dimensional invariance of derived laws; it does not justify using dimensional analysis to obtain new results in theoretical physics, which has sometimes happened. Second, since derived laws depend not only on general laws, but also on initial and boundary conditions, perhaps they are not dimensionally invariant even though the general ones are.

With respect to the second criticism, the problem evaporates if the boundary conditions can also be stated as a dimensionally invariant (although very particular, rather than general) law. From a set of dimensionally invariant laws, some expressing general physical principles and others expressing boundary conditions, it can be shown that only dimensionally invariant consequences follow.

Thus, the descriptive/deductive position justifies applications in which derived laws are obtained when all the general laws and boundary conditions that describe the physics of the situation are known to be dimensionally invariant. This probably covers the bulk of the applications. It fails, however, to account for why all of these general principles and boundary conditions are expressible in a dimensionally invariant way. Certainly, it does not justify the use of the method to obtain new general laws. Of course, the use of dimensional analysis in theoretical physics can be regarded merely as a heuristic principle of theory construction, like many others that are used to invent new theories, which are later tested by other means. Otherwise, it must be justified on other grounds, e.g., by the same arguments that tell us why the already known general laws should be dimensionally invariant. We now turn to the third argument, which at least partly accomplishes this latter purpose.

The third argument is that dimensional analysis has something to do with the existence of physically similar systems and derived measures. This was discussed by Buckingham (1914), Campbell (1920), Lord Rayleigh (1915), and Tolman (1914), but as Causey said:

> Unfortunately, none of these men was able to state his ideas with sufficient clarity to make obvious exactly what empirical assumptions were being made. Consequently these physicists were accused by other physicists of relying on metaphysical principles. It seems to me, on the other hand, that their considerations of similarity represent basically the proper approach to the problem of explaining the empirical basis of dimensional analysis. I think that they were hindered to a great extent by the lack of a good theory of derived measurement, and by the lack of a clear interpretation of dimensional parameters, which we have seen are crucial in dimensional analysis (Causey, 1967, pp. 27–28).

Bridgman (1931, Chapter 7) also discussed physically similar systems, but he did not attempt to justify dimensional invariance in terms of them. Causey (1967, 1969) developed such a theory. In Section 10.10.3 we present an improved version of it, and in the following subsection we describe just how Causey's formulation differs from ours.

## 10.10.2 Physically Similar Systems

It will be well to have a concrete example in mind as we introduce the somewhat abstract definitions. Consider a spring satisfying Hooke's law, i.e., if $F$ is an extensive measure of the applied force and $l$ is an extensive measure of the length of the spring, then there is a dimensional parameter $k$ (with dimensions $MT^{-2}$) characteristic of the spring such that the relation $F = kl$ holds (within limits) for all possible applied forces and lengths. (The measure $k$ defines an ordering over springs which, in this particular case, happens to be an s-derived attribute from Section 10.9. This is not typical, however, since in general a system may have two or more dimensional parameters associated with it.)

In the case of Hooke's law, there are two observables, namely the applied force and the length of the spring. Each such ordered pair (applied force, resulting length) defines a possible configuration of the spring. Keep in mind that by a configuration we mean some realizable combination of physical quantities, not their representation in particular numerical measures. Another way of thinking about it is to imagine that we have a collection of standard forces and another of standard rods (for length); we say that the configuration of the spring consists of that standard force which is equivalent to the applied force together with that standard rod which is equivalent to the corresponding length of the spring. This is not an ordered pair of numbers, but an ordered pair of physical quantities.

The collection of all possible configurations of the spring characterizes its behavior completely, and so we identify a physical system such as a spring with the set of its possible configurations.

Generalizing, suppose that the configuration of a physical system is specified by its value on $r$ positive physical dimensions $P_1, \ldots, P_r$, and let $\langle A, A^+, * \rangle$ be the finite-dimensional structure of physical quantities generated from all possible products of these dimensions (extended to include their formal negative elements). A configuration of this system is some point $p \in \mathsf{X}_{i=1}^r P_i$, where by $\mathsf{X}_{i=1}^r P_i$ we mean the Cartesian product of the $P_i$, not the dimension $P_1 * \cdots * P_r$. As this Cartesian product will recur, it is convenient to call it $\mathscr{P} = \mathsf{X}_{i=1}^r P_i$. The set of all possible configurations of such a system is some subset $S$ of $\mathscr{P}$.

A natural question to raise concerns the relationship between two sets of configurations that correspond to two physical systems of the same type. We leave the empirical notion of "the same type" undefined, but as an example two springs that each satisfy Hooke's law are surely of the same type. If we fix a measure $F$ for force and $l$ for length, then the possible configurations can be represented by

$$\{(F, l) \mid F = kl\} \quad \text{and} \quad \{(F, l) \mid F = k'l\}.$$

Observe that the mapping $(F, l) \to (F, lk'/k)$ carries the first set into the second. This is, of course, a similarity (Definition 2, Section 10.3.1) which, in the *MLT* base, can be described by the real vector $(k/k', k'/k, 1)$ (Theorem 3, Section 10.3.1).

To generalize this, we need some symbolism. If $\phi$ is any similarity of the structure and $p = (p_1,..., p_r)$ is in $\mathscr{P}$, we let $\phi(p)$ denote $(\phi(p_1),..., \phi(p_r))$ in $\mathscr{P}$. If $S \subset \mathscr{P}$, we let

$$\phi(S) = \{p \mid \text{there exists } p' \in S \text{ such that } p = \phi(p')\}.$$

Now, we say that any two subsets $S$ and $S'$ of $\mathscr{P}$ are *similar* if and only if there exists a similarity $\phi$ such that $S' = \phi(S)$. Since the set of all similarities form a commutative group (Exercise 9), being similar is an equivalence relation. One equivalence class, i.e., the set $\mathscr{S}$ of all sets similar to a given set, is called a *family of similar sets*.

Three physical hypotheses underlie the introduction of these definitions. First, the behavior of, at least, some physical systems can be described qualitatively as subsets of some $\mathscr{P}$. Second, two physical systems of the same type can be described as subsets of the same $\mathscr{P}$, and these subsets are similar. And third, if a subset describes the behavior of a physical system and if another subset is similar to it, then there is a physical system of the same type whose behavior is described by the second subset. Because of these assumptions and, more convincingly, because of our later results, we will identify subsets of $\mathscr{P}$ with physical systems and actually substitute the word "system" for "subset."

In addition to the sets of configurations which characterize individual systems, one feature of physical theory is the association of a unique set of dimensional constants, in our example, the spring constant, to each system of a family of physical systems of the same type. These derived system measures have a fixed value for each system in the family, and they enter into the usual statement of the numerical law characterizing the configurations of the several systems. In general, then, some additional positive dimensions $Q_1,..., Q_t$ of $\langle A, A^+, * \rangle$ are singled out and a function $g$ from $\mathscr{S}$ into $\mathsf{X}_{j=1}^{t} Q_j$ associates a $t$-tuple of dimensional constants to each system $S$ in $\mathscr{S}$. Of course, not just any mapping of this sort will do. To get an idea of the property that it must exhibit, consider again springs. Let $S$ be the spring with the spring constant $k = g(S)$ which, in terms of the *MLT* base, has dimensions $MT^{-2}$. Suppose that $\phi$ is the similarity which is equivalent to a real base vector $(\phi_M, \phi_L, \phi_T)$ (see Theorem 3, p. 465). Then, the spring $\phi(S)$ has the spring constant

$$g[\phi(S)] = g(S)\phi_M/\phi_T{}^2 = \phi[g(S)],$$

i.e., $g\phi = \phi g$. This appears to be the basic property of such derived measures (the evidence is Theorem 12), and so we are led to the following definitions.

Suppose that $\mathcal{S}$ is a family of similar systems. A function $g$ from $\mathcal{S}$ into $\mathcal{Q} = \mathsf{X}_{j=1}^{i} Q_j$ is called a *system measure* of $\mathcal{S}$ if and only if, for all similarities $\phi$, $g\phi = \phi g$.

As we mentioned, one key role played by system measures is in formulating numerical laws describing families of systems. We must next state exactly what we mean by such a law. Suppose that $\mathcal{S}$ is a family of similar systems defined in $\mathcal{P}$, that $g$ is a function from $\mathcal{S}$ into $\mathcal{Q}$, and that $f$ is a real-valued function on $\mathcal{P} \times \mathcal{Q}$. We then say that $\mathcal{S}$ *satisfies the law* $(f, g)$ provided that the following condition is satisfied: for all $p$ in $\mathcal{P}$ and $q$ in $\mathcal{Q}$,

$$f(p, q) = 0 \text{ iff there is some } S \in \mathcal{S} \text{ for which } p \in S \text{ and } g(S) = q.$$

The law is said to be *dimensionally invariant* if $f$ is a dimensionally invariant function (Definition 3, Section 10.3), i.e., if for all similarities $\phi$,

$$f(p, q) = 0 \quad \text{iff} \quad f[\phi(p), \phi(q)] = 0.$$

Note that if $f$ is any real-valued function on $\mathcal{P} \times \mathcal{Q}$, then for each $q \in \mathcal{Q}$ we may define a set $S_q$ as

$$S_q = \{p \mid f(p, q) = 0\}.$$

Relative to $\mathcal{P}$ and $\mathcal{Q}$, we denote by $\mathcal{S}_f$ the collection of all nonempty $S_q$. In general $\mathcal{S}_f$ is not a family of similar systems, but when $f$ is dimensionally invariant it is. For if $\phi$ is a similarity, then

$$\begin{aligned}
\phi(S_q) &= \{\phi(p) \mid f(p, q) = 0\} \\
&= \{\phi(p) \mid f[\phi(p), \phi(q)] = 0\} \\
&= S_{\phi(q)},
\end{aligned}$$

and conversely for any $q$ and $q'$, there exists a $\phi$ such that $\phi(q) = q'$, so any two members of $\mathcal{S}_f$ are related by a similarity transformation.

The final notion that we shall need is less obvious than either that of similar systems or of system measures, but it is important. Generally, not all similarities will alter a particular system $S$. Furthermore, different families $\mathcal{S}$ have different sets of similarities that leave their systems unchanged. Intuitively, this set of similarities must have something to do with the structure of $\mathcal{S}$. We define the *stability group* of $\mathcal{S}$ to be

$$SG(\mathcal{S}) = \{\psi \mid \psi(S) = S \quad \text{for all} \quad S \in \mathcal{S}\}.$$

(It is easy to show that $SG(\mathcal{S})$ is a subgroup of the group of similarities; see Exercise 23.) Similarly,

$$SG(\mathcal{Q}) = \{\psi \mid \psi(q) = q \quad \text{for all} \quad q \in \mathcal{Q}\}.$$

To show that $\psi$ is in one of these stability groups, it is sufficient to show that $\psi$ leaves a single element, $S$ or $q$ as the case may be, invariant. For suppose $S, S' \in \mathscr{S}$ and $\psi(S) = S$, then since $S' = \phi(S)$ and since similarities are commutative,

$$\psi(S') = \psi\phi(S) = \phi\psi(S) = \phi(S) = S'.$$

A similar argument holds for $SG(\mathscr{Q})$.

THEOREM 12. *Suppose $\mathscr{S}$ is a family of similar systems defined over $\mathscr{P} = X_{i=1}^{r} P_i$, where $P_1, ..., P_r$ are positive dimensions, and $Q_1, ..., Q_t$ are positive dimensions of the structure spanned by $P_1, ..., P_r$. Let $\mathscr{Q} = X_{j=1}^{t} Q_j$. The following three properties are equivalent:*

   (i)   *There exists a system measure g from $\mathscr{S}$ into $\mathscr{Q}$.*

   (ii)  *There exists a function f from $\mathscr{P} \times \mathscr{Q}$ into Re and a function g from $\mathscr{S}$ into $\mathscr{Q}$ such that $\mathscr{S}$ satisfies the dimensionally invariant law $(f, g)$.*

   (iii) *$SG(\mathscr{S}) \subset SG(\mathscr{Q})$.*

*Assuming the above properties, the following three properties are equivalent:*

   (iv)  *The system measure g is 1:1.*

   (v)   *$\mathscr{S}_f = \mathscr{S}$.*

   (vi)  *$SG(\mathscr{S}) \supset SG(\mathscr{Q})$.*

*Uniqueness. Suppose g is a system measure into $\mathscr{Q}$, then g' is a system measure into $\mathscr{Q}$ iff there exists a similarity $\phi$ such that $g' = \phi g$. If $g' = \phi g$ and $f'(p, q) = f[\phi(p), q]$, then $\mathscr{S}$ satisfies the dimensionally invariant law $(f, g)$ iff $\mathscr{S}$ satisfies the dimensionally invariant law $(f', g')$.*

Theorem 12 raises some questions. Does an arbitrary family of similar systems always have a system measure and hence satisfy a dimensionally invariant law? And does an arbitrary dimensionally invariant function $f$ always lead to the definition of a family of similar systems $\mathscr{S}$ and a system measure $g$ such that $\mathscr{S}$ satisfies the law $(f, g)$? The answer is complicated by the fact that only rational powers of dimensions are permitted in our definition of a structure of physical quantities, and so the families of systems have to be restricted to be compatible with that. Once that is done, then we show in Theorem 13 that the notion of a family of similar systems corresponds to that of a dimensionally invariant function.

If $P_1, ..., P_r$ are positive dimensions, then any similarity on the substructure spanned by the $P_i$ can be regarded as a vector in the multiplicative vector space $(\text{Re}^+)^r$; the similarity $\phi = (\phi_1, ..., \phi_r)$ maps $p_i$ in $P_i$ into $\phi_i * p_i$, $\phi_i$ in $\text{Re}^+$; and

$$\phi(p_1^{\rho_1} * \cdots * p_r^{\rho_r}) = \left( \prod_{i=1}^{r} \phi_i^{\rho_i} \right) * p_1^{\rho_1} * \cdots * p_r^{\rho_r}.$$

If $G$ is any group of similarities, then $G$ is contained in some vector subspace of $(\text{Re}^+)^r$ of minimal dimensionality; if this minimal dimensionality is less than $r$, then there will be nonzero vectors $\alpha = (\alpha_1, ..., \alpha_r)$ orthogonal to all of $G$, in the sense that for all $\phi$ in $G$,

$$\prod_{i=1}^{r} \phi_i^{\alpha_i} = 1.$$

The group $G$ is called *rational* if and only if there exists a nonzero rational vector $\rho$ orthogonal to $G$, i.e., a solution to

$$\prod_{i=1}^{r} \phi_i^{\rho_i} = 1, \qquad \text{all } \phi \text{ in } G,$$

with the $\rho_i$ all rational and some of them nonzero. If such $\rho_i$ exist, then clearly, any quantity

$$p_1^{\rho_1} * \cdots * p_r^{\rho_r}$$

is fixed under all the similarities of $G$.

We say the family $\mathscr{S}$ is *rational* if and only if its stability group is rational. We restrict attention to rational families. Note, however, that this restriction is inessential, since, had we wished, we could have formulated our notion of a structure of physical quantities to include irrational powers of dimensions. Then we would have obtained quantities $p_1^{\alpha_1} * \cdots * p_r^{\alpha_r}$, fixed under $SG(\mathscr{S})$. However, only the rational case seems to arise in physics; and rationality of $\mathscr{S}$ is a necessary condition for the existence of a dimensionally invariant law, as Theorem 13 shows.

THEOREM 13.  *If $\mathscr{S}$ is a rational family of similar systems, then it satisfies a dimensionally invariant law $(f, g)$. Conversely, if $f$ is a dimensionally invariant function, then $\mathscr{S}_f$ is a rational family of similar systems.*

This theorem does not really prove as much as it seems to. In particular, it does not account for the dimensional invariance of laws in which universal constants occur because such laws concern only one system—the universe. No other similar systems are realized except formally. As an example, consider Einstein's famous law $E = mc^2$ in which $c$ is the velocity of light, a universal constant. Treating $E$ and $m$ as physical quantities, the set of possible configurations is

$$S_c = \{(E, m) \mid E = mc^2\}$$

and by what we showed on p. 509, for any similarity $\phi$,

$$\phi(S_c) = S_{\phi(c)} = \{(\phi(E), \phi(m)) \mid \phi(E) = \phi(m)\,\phi(c)^2\}.$$

This is only realized in our universe for $\phi(c) = c$, in which case $\phi(S_c) = S_c$, i.e., the system is unique.

So the best we can say is this. The theory accounts for the fact that families of similar systems can be characterized by dimensionally invariant laws. Moreover, given a dimensionally invariant law, it provides the form of a family of similar systems characterized by it, but it does not tell us when this family is realized physically. Apparently they are not realized when the system measure corresponding to the law is a universal constant. We do not know if the converse is true.

### 10.10.3 Relations to Causey's Theory

The ideas we have just presented, which in this form are due to Luce (1971b), modify those of Causey (1967, 1969); we attempt to sketch the relations. The most obvious difference, but not an essential one, is that Causey formulated the problem in terms of real Euclidean spaces. The key difference is not that, but the restriction on the function $g$. It will be recalled that we have required $g\phi = \phi g$. To state Causey's condition, we must have a little notation. Suppose $q \in \mathcal{Q}$ and $\alpha, \beta \in (\mathrm{Re}^+)^t$, then we define $\alpha q = (\alpha_1 * q_1, ..., \alpha_t * q_t)$, $(\alpha q)(\beta q) = (\alpha\beta)q$ and $(\alpha q)^{-1} = \alpha^{-1}q$. Using these definitions, a 1:1 function $g: \mathcal{S} \to \mathcal{Q}$ is called a *proportional representing measure* provided that, for all $S_i \in \mathcal{S}$, $i = 1, 2, 3, 4$,

$$g(S_1)\,g(S_2)^{-1} = g(S_3)\,g(S_4)^{-1}$$

iff there exists a similarity $\phi$ such that $S_1 = \phi(S_2)$, $S_3 = \phi(S_4)$. Note that $g$ is assumed to be 1:1. A basic analysis of proportionality is given by Büchi and Wright (1955).

Luce (1971b) showed that any 1:1 system measure is a proportional representing measure, but not conversely. Because the concept of a proportional representing measure is strictly weaker than that of a 1:1 system measure and because, as we have seen in Theorem 12, system measure is the appropriate concept, it is clear that in some way Causey's concepts differ from those given here. Luce concluded that Causey's results implicitly require a definition of dimensional invariance that is much weaker than the usual one (Definition 3, Section 10.3.1)—so weak as to be of little interest. As a result, Causey's version of Theorem 13, which he recognized as the goal, is also less satisfactory.

## 10.11  PROOFS

### 10.11.1  Theorem 12  (p. 510)

*Suppose that $\mathscr{S}$ is a family of similar systems over $\mathscr{P}$. The following are equivalent. (i) There exists a system measure g from $\mathscr{S}$ into $\mathscr{Q}$. (ii) There exists a real-valued function f on $\mathscr{P} \times \mathscr{Q}$ and a function g from $\mathscr{S}$ into $\mathscr{Q}$ such that $\mathscr{S}$ satisfies the dimensionally invariant law $(f, g)$. (iii) $SG(\mathscr{S}) \subset SG(\mathscr{Q})$. Assuming these properties, the following are equivalent: (iv) The system measure g is 1:1. (v) $\mathscr{S}_f = \mathscr{S}$. (vi) $SG(\mathscr{S}) \supset SG(\mathscr{Q})$. Uniqueness: The system measures into $\mathscr{Q}$ are unique up to a similarity, and if $g' = \phi g$ and $f'(p, q) = f[\phi(p), q]$, then $\mathscr{S}$ satisfies the dimensionally invariant law $(f, g)$ iff $\mathscr{S}$ satisfies the dimensionally invariant law $(f', g')$.*

PROOF.  (i) implies (ii).  Assume $g$ is a system measure and define $f$ as follows.

$$f(p, q) = \begin{cases} 0, & \text{if there exists } S \in \mathscr{S} \text{ such that } p \in S \text{ and } g(s) = q, \\ 1, & \text{otherwise.} \end{cases}$$

By definition, $\mathscr{S}$ satisfies the law $(f, g)$. We show that $f$ is dimensionally invariant. If $\phi$ is a similarity, then $\phi(p) \in \phi(S) \in \mathscr{S}$ and, by definition, $g\phi(S) = \phi g(S) = \phi(q)$, and so

$$f(p, q) = 0 \quad \text{iff} \quad f[\phi(p), \phi(q)] = 0.$$

(ii) implies (iii).  Suppose $\psi \in SG(\mathscr{S})$ and consider any $S \in \mathscr{S}$. If $p \in S$ and $g(S) = q$, then $f(p, q) = 0$ and, by dimensional invariance, $f[\psi(p), \psi(q)] = 0$. So, by definition, there exists $S'$ such that $\psi(p) \in S'$ and $g(S') = \psi(q)$. Since $\psi(S) = S$, it follows that $S \subset S'$. The converse argument implies $S' = S$ and so

$$\psi(q) = g(S') = g(S) = q,$$

proving $\psi \in SG(\mathscr{Q})$.

(iii) implies (i).  Select any $q \in \mathscr{Q}$ and any $S_0 \in \mathscr{S}$. For $S \in \mathscr{S}$, there exists a similarity $\phi$ such that $S = \phi(S_0)$. Define $g$ by $g(S) = \phi(q)$. First, we show that $g$ is a function.

$$
\begin{array}{lll}
S = S' & \text{implies} & \phi(S_0) = \phi'(S_0), \\
& \text{implies} & \phi^{-1}\phi' \in SG(\mathscr{S}) \subset SG(\mathscr{Q}) \quad \text{(by iii)}, \\
& \text{implies} & \phi(q) = \phi'(q), \\
& \text{implies} & g(S) = g(S').
\end{array}
$$

Second, we show $g\psi = \psi g$.

$$g\psi(S) = g\psi\phi(S_0) = \psi\phi(q) = \psi g\phi(S_0) = \psi g(S).$$

(iv) implies (v). First, suppose $S \in \mathscr{S}$. Since $\mathscr{S}$ satisfies the law $(f, g)$, it follows that $S \subset S_{g(S)}$. Suppose $p \in S_{g(S)} - S$, then there exists $S' \in \mathscr{S}$ such that $p \in S'$ and $g(S') = g(S)$. Since $g$ is 1:1, $S' = S$, so $S = S_{g(S)} \in \mathscr{S}_f$. Conversely, suppose $S_q \in \mathscr{S}_f$, then for $p \in S_q$ there exists $S \in \mathscr{S}$ such that $p \in S$ and $f[p, g(S)] = 0$. Since $g$ is 1:1, this implies $S_q \subset S$. Of course, $S \subset S_q$, and so $S_q \in \mathscr{S}$.

(v) implies (vi). Since $\mathscr{S} = \mathscr{S}_f$, for any $S \in \mathscr{S}$ there exists $q \in \mathscr{Q}$ such that $S = S_q$. If $\psi \in SG(\mathscr{Q})$, then since $f$ is dimensionally invariant,

$$\psi(S) = \psi(S_q) = S_{\psi(q)} = S_q = S,$$

and so $\psi \in SG(\mathscr{S})$.

(vi) implies (iv). Observe,

$$
\begin{array}{lll}
g(S) = g(S') & \text{implies} & g\phi(S_0) = g\phi'(S_0), \\
& \text{implies} & \phi g(S_0) = \phi' g(S_0), \\
& \text{implies} & \phi^{-1}\phi' \in SG(\mathscr{Q}) \subset SG(\mathscr{S}) \quad \text{(by vi)}, \\
& \text{implies} & S = \phi(S_0) = \phi'(S_0) = S',
\end{array}
$$

and so $g$ is 1:1.

Uniqueness: Suppose $g$ is a system measure and $g' = \phi g$. Then

$$g'\psi(S) = \phi g\psi(S) = \phi\psi g(S) = \psi\phi g(S) = \psi g'(S).$$

Next, suppose $g$ and $g'$ are both system measures. Let $q = g(S_0)$ and $q' = g'(S_0)$, and let $\phi$ be any similarity such that $q' = \phi(q)$. Since if $S \in \mathscr{S}$, $S = \psi(S_0)$,

$$
\begin{aligned}
\phi g(S) = \phi g\psi(S_0) &= \phi\psi g(S_0) = \phi\psi(q) \\
&= \psi\phi(q) = \psi(q') = \psi g'(S_0) = g'\psi(S_0) = g'(S).
\end{aligned}
$$

Finally, suppose $g' = \phi g$ and $f'(p, q) = f[\phi(p), q]$, then $\mathscr{S}$ satisfies the dimensionally invariant law $(f, g)$

iff (there exists $S \in \mathscr{S}$ with $p \in \mathscr{S}$ iff $f[p, g(S)] = 0$),

iff (there exists $S \in \mathscr{S}$ with $p \in \mathscr{S}$ iff $f[\phi(p), \phi g(S)] = 0$),

iff (there exists $S \in \mathscr{S}$ with $p \in \mathscr{S}$ iff $f'[p, g'(S)] = 0$),

iff $\mathscr{S}$ satisfies the dimensionally invariant law $(f', g')$. $\Diamond$

### 10.11.2 Theorem 13 (p. 511)

*If $\mathscr{S}$ is a rational family of similar systems, then it satisfies a dimensionally invariant law $(f, g)$. Conversely, if $f$ is a dimensionally invariant function, then $\mathscr{S}_f$ is a rational family of similar systems.*

*PROOF.* Suppose $\mathscr{S}$ is a rational family of similar systems on $\mathsf{X}_{i=1}^r P_i$. Choose a maximal set of linearly independent rational vectors, $(\rho_{j1}, ..., \rho_{jr})$, $j = 1, ..., t$ orthogonal to $SG(\mathscr{S})$. Let

$$Q_j = P_1^{\rho_{j1}} * \cdots * P_r^{\rho_{jr}}.$$

The $Q_j$ are positive dimensions and any $(q_1, ..., q_t)$ in $\mathscr{Q} = \mathsf{X}_{j=1}^t Q_j$ is invariant under all $\phi$ in $SG(\mathscr{S})$. Thus $SG(\mathscr{Q}) \supset SG(\mathscr{S})$, hence, by Theorem 12, $\mathscr{S}$ satisfies a dimensionally invariant law.

Conversely, if $f$ is a dimensionally invariant function on $\mathscr{P} \times \mathscr{Q}$, then, as we showed prior to Theorem 12, $\mathscr{S}_f$ is a family of similar systems. Let $\mathscr{Q} = \mathsf{X}_{j=1}^t Q_j$, $Q_j = P_1^{\rho_{j1}} * \cdots * P_r^{\rho_{rj}}$. Then clearly, any of the vectors $(\rho_{j1}, ..., \rho_{jr})$ is orthogonal to $SG(\mathscr{Q})$, therefore to $SG(\mathscr{S})$. So $\mathscr{S}$ is rational. $\Diamond$

## 10.12 INTERVAL SCALES IN DIMENSIONAL ANALYSIS

The arguments of Section 10.9, which were based upon the axioms of extensive and conjoint measurement and the laws of similitude and exchange, are not the ones usually given to account for the monomial structure of physical quantities. The following is much more typical. Suppose that an attribute with numerical measure $y$ is secondary in the sense that it can be expressed as some explicit function of $n$ primary attributes with measures $x_1, ..., x_n$, i.e., there exists a numerical function $f$ such that

$$y = f(x_1, ..., x_n).$$

Of course, the measurement theory for each of these attributes specifies not just one representation, but a class of them which are related to one another by transformations from some family of transformations. Let these families be $\mathbf{Y}, \mathbf{X}_1, ..., \mathbf{X}_n$, respectively. Since the $x_i$ are measures of primary attributes (i.e., members of a base of the vector space), the $n$ transformations $X_i$ in $\mathbf{X}_i$ can be selected independently and $X_i(x_i)$ is a new number representing the same information as $x_i$. Since $y$ is a secondary measure, when we transform the $x_i$, we automatically transform $y$ by some transformation $Y = Y(X_1, ..., X_n)$ in $\mathbf{Y}$. The claim, then, is that the same function $f$ should apply both before and after the transformation of units, i.e.,

$$
\begin{aligned}
f[X_1(x_1), ..., X_n(x_n)] &= Y(X_1, ..., X_n)(y) \\
&= Y(X_1, ..., X_n)[f(x_1, ..., x_n)].
\end{aligned} \tag{11}
$$

As usual, a measure is called a ratio scale when its family of transformations is the similarity group (i.e., multiplication by a positive constant) and an interval scale when its family of transformation is the affine group (i.e., positive linear transformations). The range of a measure is taken to be $\text{Re}^+$ when it is a ratio scale and $\text{Re}$ when it is an interval scale.

The first theorem, which has been proved in many variants and in many ways, is usually taken as the justification for the monomial dependence among ratio scales.

THEOREM 14.   *Suppose $f$ is continuous in each of its $n$ ratio-scale arguments, $f$ is a ratio scale, and Equation (11) holds. Then there exist constants $\alpha > 0$ and $\beta_i$ such that*

$$
f(x_1, ..., x_n) = \alpha \prod_{i=1}^{n} x_i^{\beta_i}. \tag{12}
$$

As an argument for monomial functions, this theorem has two weaknesses. First, the discussion leading to Equation (11), which is one of the hypotheses of the theorem, is much like the defense of the "it couldn't be otherwise" school for assuming dimensionally invariant laws. In fact, Equation (11) is just dimensional invariance for the case where the number of variables is one more than the number of independent dimensions. The great advantage of the tack taken in Sections 10.7 and 10.9 is that the monomial dependence follows from testable empirical hypotheses—the axioms of extensive and additive conjoint measurement plus the existence of appropriate laws of similitude and exchange—and not from philosophical speculation. Second, the hypothesis that all secondary quantities are ratio scales is simply incorrect for physics; as we pointed out in Section 10.7.1, derived quantities such as density are really log-interval scales. Thus, even for those who are willing to accept Equation (11), we have proved the wrong theorem. The following one is appropriate; it, and Theorem 16, are due to Luce (1959, 1964); also see Osborne (1970). (We have stated them for interval scales, but, of course, an exponential transformation converts any interval scale into a log-interval scale).

THEOREM 15. *Suppose f is continuous in each of its n ratio-scale arguments, f is an interval scale, and Equation (11) holds. Then there exist constants α, β_i, and γ such that either*

$$f(x_1, ..., x_n) = \alpha_i \prod_{i=1}^{n} x_i^{\beta_i} + \gamma \qquad (13)$$

*or*

$$f(x_1, ..., x_n) = \alpha \sum_{i=1}^{n} \beta_i \log x_i + \gamma. \qquad (14)$$

Note that by putting the results in log-interval form, Equation (14) becomes

$$g(x_1, ..., x_n) = \exp f(x_1, ..., x_n) = \delta \left( \prod_{i=1}^{n} x_i^{\beta_i} \right)^{\alpha}, \qquad (15)$$

where $\delta = e^{\gamma} > 0$.

We can now see why treating the derived scales of physics as ratio scales when, in fact, they are log-interval scales has not caused difficulty. We may write Equations (13) and (15) as, respectively,

$$f(x_1, ..., x_n) - \gamma = \alpha \prod_{i=1}^{n} x_i^{\beta_i}$$

and

$$g(x_1, ..., x_n)^{1/\alpha} = \alpha' \prod_{i=1}^{n} x_i^{\beta_i},$$

and these measures are ratio scales. The point is that when the dependent variable is an interval scale, the possibilities are still so limited that, in one case, a choice of the zero and, in the other, a choice of the unit and an exponential transformation results in a monomial function of the $x_i$ that behaves like a ratio scale.

Since it has proved possible to restrict the primary quantities, i.e., the base of the vector space, of physics to extensive ones, it has not been necessary to consider what happens when some of the primary quantities are interval scales. In the behavioral sciences, however, extensive measures are virtually nonexistent, whereas interval scales do arise from various procedures, including conjoint measurement. Thus, it is of interest to know what types of secondary measures can exist when some of the primary quantities are interval scales and to determine the types of laws that can arise in these circumstances, i.e., to work out the analog of the Pi Theorem. The situation is not especially happy.

THEOREM 16. *Suppose f is continuous in and dependent upon each of its n ratio- or interval-scale arguments, at least one of which is an interval scale, and f is either a ratio or an interval scale. Then, with one exception, Equation (11) cannot hold. The exception occurs when $n = 1$ and the dependent variable is an interval scale, in which case*

$$f(x) = \alpha x + \gamma.$$

In summary, then, if we restrict ourselves to ratio and interval scales, we have the following situation. Assuming Equation (11) and excluding the trivial case when the secondary quantity is a linear transformation of just one primary quantity, then secondary quantities can only depend upon ratio scales. In that case, the secondary quantity may be an interval scale, but the dependence given in Equations (13) and (14) is such that it can be treated as if it were a ratio scale. This means, in effect, that the Pi Theorem has no significant generalization to interval scales.[8]

What, then, is the significance of all this for the nonphysical sciences? It suggests that they must either (1) discover their own ratio scales and append them to the existing structure of physical quantities, (2) introduce into that structure new nonbasic quantities that are relevant to the nonphysical sciences, or (3) arrive at lawful formulations having a character different from the dimensionally invariant equations of physics. We comment on each possibility.

First, ratio scales can be obtained in several ways: by extensive measurement (e.g., subjective probability); by polynomial conjoint measurement (e.g., in the distributive combination rule, $\phi_1 + \phi_2$ and $\phi_3$ are ratio scales, Section 7.5); and by taking differences of interval scales such as those obtained from additive conjoint measurement. Extensive measurement is not common in the social sciences (Section 3.14). The use of polynomial conjoint measurement is more promising, since by construction, these scales enter a dimensionally invariant law. As for differences of interval scales, the question is whether they enter dimensionally invariant laws. For example, let $u$ denote an interval scale of utility for money $x$, then $u(x + \Delta x) - u(x)$ suggests itself as a ratio measure of $\Delta x$, but in general, this will not work, because this difference is not independent of $x$. The scale $u$, by its construc-

---

[8] As Causey pointed out, should we choose to treat $y = f(x) = \alpha x + \gamma$ as a secondary quantity different from $x$, then $y/(\alpha x + \gamma)$ is invariant under linear transformations and so it plays the role of one of the $\pi$'s of dimensional analysis.

tion, enters into an additivity or an expected-utility law, but $\Delta u$ may not enter into laws such that dimensional analysis can be applied.

Second, the whole of psychophysical scaling can be looked upon as a major attempt to add nonbasic psychological variables to the existing structure of physical quantities. The well-known experimental and theoretical program of Stevens (1957, 1959a, b, 1960, 1966, and references given there) during the 1950's and 1960's is a recent comprehensive attempt to do this. Although his formulation of the situation is somewhat different, the following remarks seem consistent with his summary of the data. We assume that there is a single psychological attribute, so far unnamed, that we may describe as the total overall subjective intensity of stimulation. It is affected by certain physical attributes such as sound and light energy, vibration, shock intensity, etc. Suppose that there are $m$ relevant physical quantities $z_1, ..., z_m$, that affect subjective intensity, then we assume that it has an interval scale representation[9] of the form $I(z_1, ..., z_m)$. Suppose that $\{x_1, ..., x_n\}$ is a base of the physical structure, then since each $z_j$ is a monomial function of the $x_i$, we can write

$$I(z_1, ..., z_m) = f(x_1, ..., x_n).$$

If we postulate that Equation (11) is valid, then by Theorem 15, subjective intensity of stimulation as a function of the base variables is given by either Equation (13) or Equation (14). In either case, if we fix a reference level $a_1, ..., a_n$ for the base dimensions and we vary any two, say $i$ and $j$, in such a way as to maintain a constant level of subjective intensity, i.e.,

$$f(a_1, ..., x_i, ..., a_j, ..., a_n) = f(a_1, ..., a_i, ..., x_j, ..., a_n),$$

then we see from Equations (13) or (14) that a law of exchange holds which can be written as

$$(x_i/a_i)^{\beta_i} = (x_j/a_j)^{\beta_j}. \tag{16}$$

If we were able to determine the exponents $\beta_i$ in this way, then we would have the dimensions of $I$, and we could then use that knowledge and dimen-

---

[9] Stevens apparently believes that subjective intensity is a ratio scale, although there do not seem to be any strong arguments for this belief aside from those we gave in Section 4.6. Against it we have the fact that his technique of magnitude estimation leads to a function of the physical measures that is fit rather well by expressions of either the form $\alpha(z - \gamma)^\beta$ or $\alpha z^\beta + \gamma$, but not by $\alpha z^\beta$. Of the various possibilities available to us, we can only conclude that either Stevens is not working with a secondary quantity (in the sense of this chapter) or that subjective intensity is an interval scale of the type shown in Equation (14).

sional analysis to determine the dependence of $I$ on various physical quantities. This has yet to be done.

Stevens' work has proceeded somewhat differently, but his results suggest that the program just outlined might be successful. Typically he has selected two physical quantities known to affect subjective intensity, and he has had subjects adjust one so as to match the intensity of the other (Stevens 1959b, 1966). He has concluded that an equation like (16) is valid, but in terms of $z$'s instead of $x$'s. We would, of course, expect this result if the function $I$ were monomial in the $z$'s, at least for those that can be successfully matched. It might be possible to take the various exponents Stevens has obtained and reduce them, via the dimensional formulas of physics, into ones for a base and thereby determine the dimensions of $I$.

Another area in which various secondary measures have been proposed and used is biology; for a survey see Stahl (1962).

Our third alternative is to consider laws of a character rather different from those of physics. Numerical physical laws of the type we have discussed state which combinations of quantities can conceivably be observed at the same time. What we have shown above is that interval scales cannot be incorporated into such laws without violating dimensional invariance (because there are no nonbasic scales to be combined with the basic interval scales in such a way as to cancel out the effects of an affine transformation). In addition to considering the possibility of using laws that violate dimensional invariance, which is not very appealing, we can also entertain laws of a rather different sort. One example is laws that establish relationships at two different times. If $u(a, t)$ is an interval scale measure of some attribute of entity $a$ at time $t$, the quantity

$$\frac{u(a, t) - u(b, t)}{u(a, t') - u(b, t')}$$

is invariant under affine transformations and so asserting that it is a constant is a lawlike statement that is dimensionally invariant. Just what all the possibilities are in this direction has never been worked out.

## 10.13  PROOFS

### 10.13.1  Preliminary Lemma

LEMMA 14. *Suppose that* $f: (\mathrm{Re}^+)^n \to \mathrm{Re}^+$ *is continuous in each of its arguments. If $f$ is homogeneous, i.e., for every $\lambda_i$ , $\eta_i \in \mathrm{Re}^+$, $i = 1,..., n$,*

$$f(\lambda_1 \eta_1 ,..., \lambda_n \eta_n) = f(\lambda_1 ,..., \lambda_n) f(\eta_1 ,..., \eta_n), \tag{17}$$

*then there exist constants $\beta_i$ such that*

$$f(\lambda_1,...,\lambda_n) = \prod_{i=1}^{n} \lambda_i^{\beta_i}.$$

*PROOF.* It will be convenient to use the following abbreviations.

$$f_n(\lambda_i) = f(\lambda_1,...,\lambda_n),$$
$$f_{n-1}(\lambda_i) = f(\lambda_1,...,\lambda_{n-1},1),$$
$$f(\lambda) = f(\lambda,...,\lambda).$$

Using Equation (17) three times and introducing these definitions,

$$f_n(\lambda\eta_i) = f_{n-1}(\eta_i/\eta_n) f(\lambda\eta_n)$$
$$= f_{n-1}(\eta_i/\eta_n) f(\eta_n) f(\lambda).$$

Because the range of $f_n$ is Re$^+$, we may divide out $f_{n-1}(\eta_i/\eta_n)$ to obtain $f(\lambda\eta) = f(\lambda) f(\eta)$. Note that $f$ is continuous because $f_n$ is, and it is well known that the solution to this equation is $f(\lambda) = \lambda^{\beta_n}$. Thus,

$$f_n(\lambda_i) = f_{n-1}(\lambda_i/\lambda_n) \lambda_n^{\beta_n}.$$

To complete the proof, it is sufficient to show that $f_{n-1}$ satisfies Equation (17) since we can then use a finite induction. By Equation (17),

$$f_{n-1}(\lambda_i\eta_i) = f_n(\lambda_1\eta_1,...,\lambda_{n-1}\eta_{n-1},1)$$
$$= f_n(\lambda_1,...,\lambda_{n-1},1) f_n(\eta_1,...,\eta_n,1)$$
$$= f_{n-1}(\lambda_i) f_{n-1}(\eta_i). \qquad \diamond$$

### 10.13.2 Theorem 14 (p. 516)

*If a ratio scale $f$ is continuous in each of its $n$ ratio-scale arguments and Equation (11) holds, then there exist $\alpha > 0$ and $\beta_i$ such that*

$$f_n(x_i) = \alpha \prod_{i=1}^{n} x_i^{\beta_i}.$$

*PROOF.* Let $\lambda_i > 0$ be the transformations on $x_i$ and $\eta_n(\lambda_i) > 0$ on $y$, then Equation (12) reads $f_n(\lambda_i x_i) = \eta_n(\lambda_i) f_n(x_i)$. Setting $x_i = 1$, we see that $\eta_n(\lambda_i) = (1/\alpha) f_n(x_i)$, where $\alpha = f_n(1)$. Thus, $g_n(\lambda_i x_i) = g_n(\lambda_i) g_n(x_i)$, where $g_n = f_n/\alpha$, and the result follows from Lemma 14. $\qquad \diamond$

**10.13.3 Theorem 15** (p. 517)

*If an interval scale f is continuous in each of its n ratio-scale arguments
and Equation (11) holds, then there exist $\alpha$, $\beta_i$, and $\gamma$ such that either*

$$f_n(x_i) = \alpha \prod_{i=1}^{n} x_i^{\beta_i} + \gamma \tag{18}$$

*or*

$$f_n(x_i) = \alpha \sum_{i=1}^{n} \beta_i \log x_i + \gamma. \tag{19}$$

PROOF.  Under these assumptions, Equation (11) takes the form

$$f_n(\lambda_i x_i) = \eta_n(\lambda_i) f_n(x_i) + \zeta_n(\lambda_i). \tag{20}$$

Setting $x_i = 1$ in Equation (20),

$$f_n(\lambda_i) = \delta \eta_n(\lambda_i) + \zeta_n(\lambda_i), \tag{21}$$

where $\delta = f_n(1)$. Using this and writing Equation (20) in two ways,

$$\begin{aligned}
f_n(\lambda_i x_i) &= \eta_n(\lambda_i) f_n(x_i) + \zeta_n(\lambda_i) \\
&= \eta_n(\lambda_i)[\delta \eta_n(x_i) + \zeta_n(x_i)] + \zeta_n(\lambda_i) \\
&= \eta_n(x_i) f_n(\lambda_i) + \zeta_n(x_i) \\
&= \eta_n(x_i)[\delta \eta_n(\lambda_i) + \zeta_n(\lambda_i)] + \zeta_n(x_i).
\end{aligned}$$

Thus,

$$\frac{1 - \eta_n(\lambda_i)}{\zeta_n(\lambda_i)} = \frac{1 - \eta_n(x_i)}{\zeta_n(x_i)},$$

and so either $\eta_n(x_i) \equiv 1$ or there exists $\gamma$ such that $\zeta_n(x_i) = \gamma[1 - \eta_n(x_i)]$.
First, suppose that $\eta_n \equiv 1$, so by Equation (21), $f_n(x_i) = \delta + \zeta_n(x_i)$.
Thus, using Equation (20),

$$\begin{aligned}
\delta + \zeta_n(\lambda_i x_i) &= f_n(\lambda_i x_i) \\
&= f_n(x_i) + \zeta_n(\lambda_i) \\
&= \delta + \zeta_n(x_i) + \zeta_n(\lambda_i).
\end{aligned}$$

Canceling $\delta$ and taking exponentials, yields an equation satisfying
Lemma 14, and so Equation (19) follows with $\gamma = \delta$.

Second, if $\zeta_n = \gamma(1 - \eta_n)$, then by Equation (21), $f_n - \gamma = (\delta - \gamma)\,\eta_n$ and so, by Equation (20),

$$(\delta - \gamma)\,\eta_n(\lambda_i x_i) = f_n(\lambda_i x_i) - \gamma$$
$$= \eta_n(\lambda_i)[f_n(x_i) - \gamma]$$
$$= \eta_n(\lambda_i)(\delta - \gamma)\,\eta_n(x_i).$$

If $\delta = \gamma$, then $f_n(x_i) = \gamma$, which is Equation (18) with $\alpha = 0$, and if $\alpha = \delta - \gamma \neq 0$, then since $\eta_n > 0$, Lemma 14 yields Equation (18) with $\alpha \neq 0$.                                                                      $\diamond$

### 10.13.4  Theorem 16  (p. 518)

*If $f$ is continuous in and dependent upon each of its $n$ arguments, if $f$ and its arguments are either ratio or interval scales and at least one argument is an interval scale, and if Equation (11) holds, then $n = 1$ and $f(x) = \alpha x + \gamma$.*

*PROOF.*  With no loss of generality, suppose that $x_1$ is an interval scale. By restricting ourselves to similarity transformations, we know that $f$ must satisfy either of the two forms in Theorem 15. Suppose, first, that Equation (18) holds, and consider an affine transformation $\lambda_1 x_1 + \rho_1$ on $x_1$ and identity transformations on all other variables, then by Equation (11),

$$f(\lambda_1 x_1 + \rho_1,\, x_2,\,...,\, x_n) = \alpha(\lambda_1 x_1 + \rho_1)^{\beta_1} \prod_{i=2}^{n} x_i^{\beta_i} + \gamma$$

$$= \eta(\lambda_1,\rho_1)\left[\alpha x_1^{\beta_1} \prod_{i=2}^{n} x_i^{\beta_i} + \gamma\right] + \zeta(\lambda_1,\rho_1).$$

Rewriting,

$$\alpha \prod_{i=2}^{n} x_i^{\beta_i}[(\lambda_1 x_1 + \rho_1)^{\beta_1} - \eta(\lambda_1,\rho_1)\,x_1^{\beta_1}] = \zeta(\lambda_1,\rho_1) + \gamma[\eta(\lambda_1,\rho_1) - 1].$$

If $n \geqslant 2$, this equation cannot be satisfied since the $x_i$ are arbitrary. A similar argument shows that Equation (19) cannot hold except for $n = 1$.

With $n = 1$, it is not difficult to see that the above equation can be satisfied if and only if $\beta_1 = 1$, and the one arising from Equation (19) cannot be satisfied at all.                                                                        $\diamond$

## 10.14  PHYSICAL QUANTITIES IN MECHANICS AND GENERALIZATIONS OF DIMENSIONAL INVARIANCE

In this final section, we treat briefly concepts of invariance, distinct from dimensional invariance, which play a role in formulating theories for classical and relativistic particle mechanics. Our axioms for mechanics are based on five major notions: the set $P$ of particles; the position function $s$ such that, for any particle $p$ in $P$ and any instant of time $t$, $s(p, t)$ is the position of $p$ at time $t$; the mass function $m$ such that, for any $p$ in $P$, $m(p)$ is the mass of $p$; the internal force function $f$ such that, for any particles $p$ and $q$ in $P$ and any instant of time $t$, $f(p, q, t)$ is the force exerted on $p$ by $q$ at time $t$; and the external force function $g$ such that, for any particle $p$, any instant of time $t$, and any positive integer $n$, $g(p, t, n)$ is the $n$th external force exerted on $p$ at time $t$.

In physics and in all of the published axiomatizations of mechanics since Newton's *Principia* in 1687, these position, mass, and force functions are treated as numerical, and the problem of physical quantities is not dealt with in an explicit mathematical fashion. For instance, the axioms for particle mechanics given in McKinsey, Sugar, and Suppes (1953) or in Suppes (1957) treat time and mass as real numbers and position and force as triples of real numbers. And so, from a purely mathematical standpoint, it makes sense within such an axiomatic framework to add a time to a mass, even though physically this is considered meaningless. So, to separate questions of measurement from the relationships between physical quantities established by physical law, we formulate the laws of mechanics as constraints among physical quantities, not as constraints holding among the numerical representations of them. This eliminates completely any questions about the units in terms of which different quantities are measured.

Four dimensions play a role in mechanics: length $L$; time duration $T$; mass $M$; and force $F$. In addition, of course, we must specify the location and orientation of the spatial coordinates and the origin of time. By this we might mean, for example, a particular point on the earth with coordinates defined in terms of light rays to certain stars at the time of a certain event described in terms of some unique configuration among the stars. As is customary in physics, we will not explicitly note this choice as a primitive of our systems. It will be implicit in terms of the particular time and position functions chosen.

In the following definitions we shall be dealing with functions from dimensions into dimensions. As Whitney (1968) has shown, it is easy, if somewhat tedious, to translate all of the usual numerical definitions of continuity, differentiability, convergence, vector product, etc. to these functions. The key to doing this is embodied in the following observation. Suppose $h$ is a function from dimension $D$ to dimension $E$ and that $d_0$ and $e_0$ are elements of these dimensions. For any $d \in D$, we know by Theorem 2 that there is a real number $\delta$ such that $d = \delta * d_0$. Similarly,

$e = \epsilon * e_0$. So we may write $e = h(d)$ as

$$\epsilon * e_0 = h(\delta * d_0)$$

which suggests defining the numerical function $h'$ by $h'(\delta) = \epsilon$.

In writing these analogs it is convenient to use analogous notations which, in some fashion, emphasize that we are dealing with physical quantities. Earlier we used $*$ to denote the products of two physical quantities (Definition 1) and $\oplus$ to denote the sum within one dimension (p. 463). We extend these to $n$-tuples or "vectors" of physical quantities as follows:

$\otimes$ is the analog of vector product.

$\odot$ is the analog of inner product, $X^2 = X \odot X$, and $|X| = (X^2)^{1/2} = $ vector length.

$*$ is the analog of multiplying a vector by a scalar or a real matrix.

$\oplus, \sum$ are the analogs of the sum of vectors.

$-X = (-1) * X$, where $X$ is a vector of physical quantities (see p. 461).

DEFINITION 6. *Suppose P is a set and the following are functions*:

$$s: \quad P \times T \to L \times L \times L$$
$$m: \quad P \to M^+$$
$$f: \quad P \times P \times T \to F \times F \times F$$
$$g: \quad P \times T \times I^+ \to F \times F \times F.$$

*Then $\mathscr{P} = \langle P, s, m, f, g \rangle$ is a structure of (classical) particle mechanics iff P is nonempty and finite and, for all $p, q \in P$ and $t \in T$, the following five axioms hold*:

1. *The function s is continuous and piecewise twice differentiable in its second argument.*

2. $f(p, q, t) = -f(q, p, t).$

3. $s(p, t) \otimes f(p, q, t) = -s(q, t) \otimes f(q, p, t).$

4. *The series $\sum_{n=1}^{\infty} g(p, t, n)$ is absolutely convergent.*

5. $m(p) * \partial^2 s(p, t)/\partial t^2 = \sum_{q \in P} f(p, q, t) \oplus \sum_{n=1}^{\infty} g(p, t, n).$

From Axiom 5, which is Newton's Second Law, we can prove at once the standard dimension theorem about force.

THEOREM 17. $L * M * T^{-2} = F.$

Closely related is the following theorem about the number of dimensions—it is to be emphasized that the three dimensions referred to in the theorem have nothing to do with the three-dimensionality of physical space.

THEOREM 18.   *The structure of mechanical quantities spanned by L, T, M, and F is three dimensional.*

We leave the proofs of both of these theorems as Exercise 30.

### 10.14.1  Generalized Galilean Invariance

The first step in analyzing the invariance of a structure of particle mechanics is to define the notion of generalized Galilean transformation of a coordinate system and the corresponding transformation of an entire structure of particle mechanics. For this purpose it is convenient to introduce the set **S** of all structures of particle mechanics and the set $\mathscr{U}$ which is the union of all the first components of members of **S**, that is, the set which is the union of all sets of possible particles. Simply described, $\mathscr{U}$ is the universe of particles. [To allay the fear that the sets **S** and $\mathscr{U}$ may be so large that they might lead to logical contradictions, we remark that there are a number of simple ways to avoid this problem. A discussion may be found in McKinsey and Tarski (1944)]. Using the sets **S** and $\mathscr{U}$ and letting $I^+$ be the set of all positive integers, we may then define a generalized Galilean carrier as follows:

DEFINITION 7.   *A generalized Galilean carrier is an ordered 6-tuple $\langle \phi_1, \phi_2, \mathcal{O}, V, B, r \rangle$ where $\phi_1$ is a 1:1 mapping of $\mathscr{U}$ onto itself, $\phi_2$ is a 1:1 mapping of $I^+$ onto itself, $\mathcal{O}$ is an orthogonal $3 \times 3$ matrix of real numbers, $V \in (L * T^{-1}) \times (L * T^{-1}) \times (L * T^{-1})$, $B \in L \times L \times L$, and $r \in T$.*

The intuitive interpretation of the components of a carrier is this. The function $\phi_1$ replaces the particles of a structure by new particles with similar mechanical properties; this function corresponds to the function establishing an isomorphism between two algebraic objects. The function $\phi_2$ renumbers the given external forces. Multiplying $s[\phi_1^{-1}(p'), t' \ominus r]$ by the orthogonal matrix $\mathcal{O}$ produces a rotation, and possibly a reflection, of the coordinate system. The vector $V$ represents the uniform velocity of the new coordinate system with respect to the old, and $B$ is its translation. The equation

$$t' = r \oplus t$$

shifts the origin of time by an amount $-r$. The parameters $\mathcal{O}, V, B$ and $r$ define a Galilean transformation.

THEOREM 19. *Suppose* $\mathscr{P} = \langle P, s, m, f, g \rangle$ *is a member of* **S** *and* $\Delta = \langle \phi_1, \phi_2, \mathcal{O}\!\mathcal{l}, V, B, r \rangle$ *is a generalized Galilean carrier. Let P' be the set of all elements* $p' \in \mathcal{U}$ *such that for some* $p \in P$

$$p' = \phi_1(p)$$

*and let the functions* $s', m', f'$ *and* $g'$ *be defined by the following equations (for* $p', q' \in P'$, $t' \in T$ *and* $n \in \mathrm{I}^+$).

$$m'(p') = m[\phi_1^{-1}(p')],$$

$$s'(p', t') = \{s[\phi_1^{-1}(p'), t' \ominus r] * \mathcal{O}\!\mathcal{l}\} \oplus (t' * V) \oplus B,$$

$$f'(p', q', t') = f[\phi_1^{-1}(p'), \phi_1^{-1}(q'), t' \ominus r] * \mathcal{O}\!\mathcal{l},$$

$$g'(p', t', n) = g[\phi_1^{-1}(p'), t' \ominus r, \phi_2^{-1}(n)] * \mathcal{O}\!\mathcal{l}.$$

*Then* $\mathscr{P}' = \langle P', s', m', f', g' \rangle$ *is also a member of* **S**.

It is instructive to see how these equations would be written if we had used numerical measures instead of physical quantities. For some particular $l_0$ in $L$, $t_0$ in $T$, $m_0$ in $M$, and $f_0$ in $F$, let $\sigma$, $\mu$, $\tau$, and $\rho$ correspond to, respectively, $s$, $m$, $t$, and $r$. Moreover, let $\phi$ represent the similarity corresponding to a change of units, so that (Theorem 3) $\phi(l_0) = \phi_L * l_0$, $\phi(t_0) = \phi_T * t_0$, etc. Then the first equation translates into

$$\mu'(p') = \phi_M \mu[\phi_1^{-1}(p)]$$

and the second into

$$\sigma'(p', \tau') = \phi_L \left\{ \sigma \left[ \phi_1^{-1}(p'), \frac{\tau' - \rho}{\phi_T} \right] \mathcal{O}\!\mathcal{l} + \frac{\tau'}{\phi_T} V + B \right\}.$$

Usually, in writing such equations, the similarity $\phi$ is suppressed which leads, formally, to equations exactly like those of Theorem 19, but at the expense of making it not very clear how changes in the unit of measurement are to be accommodated.

When the mass of a particle is regarded as a physical quantity, it is an invariant; only its numerical measure changes from one frame of reference to another. To have a similar invariance of time, position, and force would require an absolute reference in space and time, which is precisely what we do not have, either classically or relativistically.

The theorem just stated, whose proof is immediate, suggests a much deeper problem of uniqueness. Is it true that the notion of a generalized Galilean carrier is the broadest notion of a transformation that makes any sense in classical mechanics? The approach to this question is formally just the same as the approach to the uniqueness theorem for measurement structures. The relevant definitions are somewhat more complicated because of the greater complexity of mechanical structures, but the basic idea is the same. The discussion here follows McKinsey and Suppes (1953). What we want to prove is that the transformations embodied in a generalized Galilean carrier are the only ones which take structures of particle mechanics into like structures.

DEFINITION 8. *Let $\phi_1$ be a 1:1 function that maps $T$ onto itself; let $\phi_2$ be a 1:1 function that maps $M^+$ onto itself; let $\phi_3$ be a function from $L \times L \times L \times T$ into $L \times L \times L$; and let $\phi_4$ be a 1:1 function that maps $F \times F \times F$ onto itself. The ordered quadruple $\Phi = \langle \phi_1, \phi_2, \phi_3, \phi_4 \rangle$ is called an* eligible *transformation. If $\Phi = \langle \phi_1, \phi_2, \phi_3, \phi_4 \rangle$ is such an eligible transformation, and $\mathcal{P} = \langle P, s, m, f, g \rangle$ is a structure of particle mechanics, then by $\Phi(\mathcal{P})$ we mean the structure $\langle P, s', m', f', g' \rangle$, where $m', s', f'$, and $g'$ are defined by the following equations (for $p, q \in P$, $t \in T$ and $n \in I^+$).*

$$m'(p) = \phi_2[m(p)],$$

$$s'(p, t) = \phi_3[s(p, \phi_1^{-1}(t)), t],$$

$$f'(p, q, t) = \phi_4[f(p, q, \phi_1^{-1}(t)],$$

$$g'(p, t, n) = \phi_4[g(p, \phi_1^{-1}(t), n)].$$

It is possible to prove a general theorem about the $\Phi$'s that carry structures of particle mechanics into structures of particle mechanics, but the results do not correspond exactly to the notions embodied in that of a generalized Galilean carrier. To make them correspond, one simple general solution is to require that the structures be *ultraclassical* as well as classical. The notion of ultraclassical, which was formalized in McKinsey, Sugar and Suppes (1953), adds the requirement that the forces in a structure of particle mechanics be a function only of distance between the particles. It is sufficient for our purposes here just to impose this requirement on the internal forces, that is, to require the internal forces acting between two particles to be a function of the distance between the particles.

THEOREM 20. *Suppose $\Phi = \langle \phi_1, ..., \phi_4 \rangle$ is an eligible transformation such that for every ultraclassical structure of particle mechanics $\mathcal{P}$, $\Phi(\mathcal{P})$ is again an ultraclassical structure of particle mechanics. Then there exists a vector $V \in (L * T^{-1}) \times (L * T^{-1}) \times (L * T^{-1})$, a vector $B \in L \times L \times L$,* *and an orthogonal* $3 \times 3$ *matrix* $\mathcal{O}$ *of real numbers such that, for every* $t \in T$, $k \in M^+$, $Z \in L \times L \times L$, *and* $W \in F \times F \times F$,

$$\phi_1(t) = r \oplus t,$$
$$\phi_2(k) = k,$$
$$\phi_3(Z, t) = (Z * \mathcal{O}) \oplus (t * V) \oplus B,$$
$$\phi_4(Z) = Z * \mathcal{O}.$$

The proof of this theorem is long and complicated and will not be given here. As is the case in other proofs of uniqueness for measurement structures, its complexity arises from the fact that continuity is not assumed.

Keeping in mind the two theorems about generalized Galilean invariance, we can now extend the discussion of Sections 10.3 and 10.10 on the dimensional invariance of numerical laws to the view that meaningful classical mechanical laws must be invariant not only under measurement similarities, but also under generalized Galilean transformations. Moreover, this is true not only of numerical laws, but of mechanically meaningful numerical functions or concepts as well. The important point to emphasize is that dimensional invariance is not enough, but Galilean invariance is required as well. Indeed, one can conjecture an analog to Theorem 13 in which the existence of families of systems that are related by Galilean carriers implies the existence of laws that exhibit Galilean invariance. This has not been proved.

To make matters explicit and to lead into a generalization and, to some extent, a variant on the earlier definition of dimensionally invariant functions, we first define formally the notion of a generalized Galilean transformation.

DEFINITION 9. *Suppose* $\Delta$ *is a generalized Galilean carrier and* $\Psi$ *is a function from* **S** *into* **S**. *Then* $\Psi$ *is the generalized Galilean transformation corresponding to* $\Delta$ *iff, for every* $\mathcal{P}$ *and* $\mathcal{P}' \in$ **S**, $\Psi(\mathcal{P}) = \mathcal{P}'$ *provided that* $\Delta$, $\mathcal{P}$, *and* $\mathcal{P}'$ *are related as in Theorem 19. Moreover, a function* $\Psi$ *is called simply a* generalized Galilean transformation *if it is the generalized Galilean transformation corresponding to some carrier* $\Delta$.

The following theorem is then easily proved.

THEOREM 21. *The set of all generalized Galilean transformations is a group under the appropriately defined function composition.*

The notion of generalized Galilean invariance of functions is more restricted than Definition 3 (Section 10.3.1) because it is limited to mechanics. On the other hand, the definition has greater potency in the sense that it also considers the problem of Galilean invariance as well as dimensional invariance. Following McKinsey and Suppes (1955), we now make these ideas of invariance precise by defining the notion of a function of a structure of classical particle mechanics which we abbreviate as an SCPM-function. An SCPM-function is one whose first argument has the domain **S** and whose remaining arguments are particles or instants of time. The SCPM-functions which are of interest correspond to various standard mechanical notions, such as mass, acceleration, and angular momentum. It is natural to ask why in defining an SCPM-function it is necessary to include the structure $\mathscr{P}$ of particle mechanics as the first argument. The answer is simply that the same physical quantity can occur as the second component, interpreted as an elapsed time, of many different structures, and similarly for the particles themselves. The requirement that the first argument of an SCPM-function have the set **S** of structures of particle mechanics as its domain, is thus needed in order to make precise which structure we are considering in analyzing a given transformation.

DEFINITION 10. *Let u and v be nonnegative integers. Then a function $\psi$ is an* SCPM-*function (of type $\langle u, v \rangle$) iff the domain of definition of $\psi$ consists of all $(u + v + 1)$-tuples $\langle \mathscr{P}, p_1, ..., p_u, t_1, ..., t_v \rangle$, where $\mathscr{P} = \langle P, s, m, f, g \rangle$ is a structure of classical particle mechanics, $p_i \in P$, $i = 1, ..., u$, and $t_j \in T$, $j = 1, ..., v$.*

In the special case where $u$ (or $v$) is zero, an SCPM-function has no particles (no instants of time) as arguments. Thus, for example, an SCPM-function of type $\langle 0, 1 \rangle$ is one whose domain of definition consists of all couples $\langle \mathscr{P}, t \rangle$, where $\mathscr{P} \in S$ and $t \in T$. As would be expected, we define an SCPM-function as (mechanically) invariant if it is generalized Galilean invariant. The formal definition is as follows:

DEFINITION 11. *An SCPM-function f of type $\langle u, v \rangle$ is (mechanically) invariant iff for every generalized Galilean carrier $\Delta = \langle \phi_1, \phi_2, \mathcal{A}, V, B, r \rangle$, with $\Psi$ the corresponding Galilean transformation, for every*

$$\mathscr{P} = \langle P, s, m, f, g \rangle \in \mathbf{S},$$

*for all $p_1, ..., p_u \in P$, and for all $t_i, ..., t_v \in T$,*

$$f(\mathscr{P}, p_1, ..., p_u, t_1, ..., t_v) = f[\Psi(\mathscr{P}), \phi_1(p_1), ..., \phi_1(p_u), r \oplus t_1, r \oplus t_1, ..., r \oplus t_v].$$

The following are two examples of generalized Galilean invariant functions.

*Ratio of masses* is the SCPM-function $\psi_1$ of type $\langle 2, 0 \rangle$ such that if

$$\mathscr{P} = \langle P, s, m, f, g \rangle \in \mathbf{S}$$

and if $p_1$ and $p_2$ are in $P$, then

$$\psi_1(\mathscr{P}, p_1, p_2) = m(p_1) * m^{-1}(p_2).$$

*Ratio of distances between particles* is the SCPM-function $\psi_2$ of type $\langle 4, 1 \rangle$ such that, if

$$\mathscr{P} = \langle P, s, m, f, g \rangle \in \mathbf{S},$$

and if $p_1$, $p_2$, $p_3$, $p_4$ are in $P$, and $t$ is in $T$, then

$$\psi_2(\mathscr{P}, p_1, p_2, p_3, p_4, t) = |\, s(p_1, t) \ominus s(p_2, t)|/|\, s(p_3, t) \ominus s(p_4, t)|$$

if $|\, s(p_3, t) - s(p_4, t)| \neq 0$, and otherwise

$$\psi_2(\mathscr{P}, p_1, p_2, p_3, p_4, t) = 1.$$

The reader may have noticed that the two examples just given are quite different in character from the functions envisaged in Definition 3. In that case, we were concerned with functions that correspond to physical laws. In the present case, we are concerned more with the sorts of functions that enter into formulation of physical laws. Put another way, the present functions are the sorts that are arguments of physical laws. However, the apparatus being introduced here is quite general, and we can go on to consider functions of the sort envisaged as falling under Definition 3. This arises from the fact that a physical law or any relation between functions can be made to correspond to an SCPM-function. When the relation holds, the value of the function is one; when it does not hold, the value is zero. We may take as a simple example Newton's first law, which says that a particle acted upon by a zero resultant force remains in uniform motion:

*Newton's First Law.* Define the SCPM-function $\psi_3$ of type $\langle 1, 0 \rangle$ such that if

$$\mathscr{P} = \langle P, s, m, f, g \rangle \in \mathbf{S}$$

and $p$ is in $P$, then $\psi_3(\mathscr{P}, p) = 1$ if whenever, for all $t$ in $T$,

$$\sum_{q \in P} f(p, q, t) \oplus \sum_{i=1}^{\infty} g(p, t, i) = 0,$$

then there are vectors $A$, $B$ in $L \times L \times L$ such that for all $t$ in $T$,

$$s(p, t) = A \oplus (t * B),$$

otherwise, $\psi_3(\mathscr{P}, p) = 0$.

It is evident that the SCPM-function used to represent any physical law that is a logical consequence of the axioms of mechanics will be mechanically invariant, and consequently invariant under changes of unit of measurement.

### 10.14.2 Lorentz Invariance and Relativistic Mechanics

We turn now to a brief treatment of relativistic particle mechanics. Much of the analysis parallels what we have already said about classical particle mechanics, and we shall not duplicate that discussion.

To begin with, we formulate axioms for relativistic particle mechanics. The discussion follows Rubin and Suppes (1954). Because of the difficulty of dealing with interacting forces in the special theory of relativity we will eliminate the notion of internal forces and will use just the single notion of external force. With the exception of this change, the axioms are very similar to those given earlier for classical mechanics. It should be noted that Axiom 2 requires that the motion of any particle be slower than the speed of light.

DEFINITION 12. *Suppose $P$, $s$, $m$, and $g$ are as in Definition 6 (Section 10.14) and that $c \in L * T^{-1}$. Then $\mathscr{P} = \langle P, s, m, g, c \rangle$ is a structure of relativistic particle mechanics iff $P$ is nonempty and finite and, for all $p \in P$ and $t \in T$, the following four axioms hold:*

1. *The function $s$ is continuous and piecewise twice differentiable in its second argument.*

2. $| \partial s(p, t)/\partial t | < c.$

3. $\sum_{n=1}^{\infty} g(p, t, n)$ *is absolutely convergent.*

4. $m(p) * \dfrac{\partial}{\partial t} \dfrac{\partial s(p, t)/\partial t}{\left(1 \ominus \dfrac{| \partial s(p, t)/\partial t |^2}{c^2}\right)^{1/2}}$

$$= \left(1 \ominus \frac{| \partial s(p, t)/\partial t |^2}{c^2}\right)^{1/2} * \sum_{n=1}^{\infty} g(p, t, n).$$

We next define the notion of a generalized Lorentz matrix. In this definition $c$ is the physical quantity in $L * T^{-1}$, i.e., the speed of light, as referred to in Definition 12. The number $\delta$ is either 1 or $-1$; when $\delta = -1$, we have a reversal of the direction of time in going from one inertial frame to the new one related to the first by the given Lorentz transformation. The vector $V$ represents the relative velocity of the two inertial frames, and the number $\beta$

is the well-known Lorentz contraction factor. The matrix $\mathcal{E}$ represents a rotation of the spatial coordinates—or a rotation followed by a reflection.

DEFINITION 13. *A matrix $\mathcal{O}$ of order* 4 *is said to be a* generalized Lorentz matrix *iff there exist real numbers $\delta$ and $\beta$, with $\beta > 0$, a vector*

$$V \in (L * T^{-1}) \times (L * T^{-1}) \times (L * T^{-1})$$

*and an orthogonal matrix $\mathcal{E}$ of order 3 such that*[10]

$$\delta^2 = 1, \qquad \beta^2(1 - V^2/c^2) = 1$$

*and*

$$\mathcal{O} = \begin{pmatrix} \mathcal{E} & 0 \\ 0 & \delta \end{pmatrix} \begin{pmatrix} \mathcal{I} + \dfrac{\beta - 1}{V^2} V' * V & \dfrac{-\beta V^*}{c^2} \\ -\beta V & \beta \end{pmatrix}.$$

Corresponding to Theorem 19 (Section 10.14.1), it is easy to prove the following theorem for structures of relativistic particle mechanics.

THEOREM 22. *Suppose $\mathcal{P} = \langle P, s, m, g, c \rangle$ is a structure of relativistic particle mechanics. Let $\mathcal{O}$ be a generalized Lorentz matrix, let*

$$B \in L \times L \times L \times T,$$

*and for each $p \in P$, let the function $h_p$ be defined as follows for all $t \in T$*[11]

$$h_p(t) = \langle \{[s_p(t), t] * \mathcal{O}\} \oplus B \rangle_4 \, .$$

*(The inverse function $h_p^{-1}$ exists.) Let the functions $m'$, $s'$, and $g'$ be defined by the following equations (for $p \in P$, $t \in T$ and $n \in I^+$).*

$$m'(p) = m(p),$$

$$s'(p, t) = \langle \{[(s(p, h_p^{-1}(t')), h_p^{-1}(t')] * \mathcal{O}\} \oplus B \rangle_{1,2,3} \, ,$$

$$g'(p, t'n) = \langle [g(p, h_p^{-1}(t'), n), g(p, h_p^{-1}(t'), n) \odot \partial/\partial t' \, s(p, h_p^{-1}(t')/c^2] * \mathcal{O} \rangle_{1,2,3} \, .$$

*Then $\mathcal{P}' = \langle P, m', s', g', c \rangle$ is a structure of relativistic particle mechanics.*

The proof of this theorem, which is substantially the same as that given in Rubin and Suppes (1954), is more complicated than the proof of Theorem

---

[10] In this definition $V'$ is the transpose of the vector $V$, and so $V' * V$ is a $3 \times 3$ matrix. Also, $\mathcal{I}$ is the $3 \times 3$ identity matrix.

[11] If $Y$ is a 4-vector, $\langle Y \rangle_4$ is the fourth component and $\langle Y \rangle_{1,2,3}$ is the vector composed of the first three components.

19. It is considerably more difficult to prove what amounts to the converse of this theorem; it corresponds in the classical case to Theorem 20 (Section 10.14.1).

DEFINITION 14.   *Let $\phi_1$ be a function that maps $M^+$ into $M^+$, let $\phi_2$ be a 1:1 function that maps $L \times L \times L \times T$ into itself, and let $\phi_3$ be a function that maps $F \times F \times F \times (L * T^{-1}) \times (L * T^{-1}) \times (L * T^{-1})$ into $F \times F \times F$. The ordered triple $\langle \phi_1, \phi_2, \phi_3 \rangle$ is called a* relativistic eligible transformation. *If $\Phi = \langle \phi_1, \phi_2, \phi_3 \rangle$ is such an eligible transformation and $\mathscr{P} = \langle P, s, m, g, c \rangle$ is a structure of relativistic particle mechanics, then by $\Phi(\mathscr{P})$ we mean the structure $\mathscr{P}' = \langle P, s', m', g', c \rangle$, where for each $p \in P$ the function $H_p$ is defined as follows for $t \in T$*

$$H_p(t) = \langle \phi_2[s(p, t), t] \rangle_4,$$

*and $m'$, $s'$ and $g'$ are defined by the following equations for $n \in 1^+$ and $t' \in T$, if the preimage $H^{-1}t'$ of $t'$ under $H_p$ is unique, and otherwise they are undefined*

$$m'(p) = \phi_1[m(p)],$$

$$s'(p, t') = \langle \Phi_2[s(p, H_p^{-1}(t')), H_p^{-1}(t')] \rangle_{1.2.3},$$

$$g'(p, t', n) = \phi_3\left[ f(p, H_p^{-1}(t'), n), \frac{\partial}{\partial t'} s(p, H_p^{-1}(t')) \right].$$

What corresponds then to the uniqueness theorem for generalized Lorentz transformations is the following theorem whose proof is quite long and is not given here. For the statement of the theorem we need the concept of a *c*-line (path of a light ray) and *c*-inertial path (path of a particle with mass under uniform motion). By the *slope* of a line $\mathscr{L}$ in $L \times L \times L \times T$ whose projection on $T$ is a nondegenerate segment, we mean the vector $W$ in $(L * T^{-1}) \times (L * T^{-1}) \times (L * T^{-1})$ such that for any two distinct points $(Z_1, t_1)$ and $(Z_2, t_2)$ of $\mathscr{L}$

$$(Z_1 \ominus Z_2)/(t_1 \ominus t_2) = W.$$

By the *speed* of $\mathscr{L}$ we mean the nonnegative physical quantity $| W | \in L * T^{-1}$. By a *c-line* we mean a line $\mathscr{L}$ whose speed is equal to $c$, and by a *c-inertial path* we mean a line whose speed is less than $c$.

THEOREM 23.   *Suppose $\Phi = \langle \phi_1, \phi_2, \phi_3 \rangle$ is a relativistic eligible transformation such that for every structure of relativistic particle mechanics $\mathscr{P}$, $\Phi(\mathscr{P})$ is again such a structure, and $\phi_2$ carries no c-line into a c-inertial path.*

*Then there exists a generalized Lorentz matrix $\mathcal{A}$ and a vector*

$$B \in L \times L \times L \times T$$

*such that, for any vectors $Z \in L \times L \times L$, $\quad Z_1 \in F \times F \times F$,*

$$Z_2 \in (L * T^{-1}) \times (L * T^{-1}) \times (L * T^{-1}),$$

*and with $|Z_2| < c$, every $t \in T$ and $k \in M^+$*

$$\phi_1(k) = k,$$
$$\phi_2(Z, t) = [(Z, t) * \mathcal{A}] \oplus B,$$
$$\phi_3(Z_1, Z_2) = \langle (Z_1, [Z_1 \odot Z_2]/c^2) * \mathcal{A} \rangle_{1,2,3}.$$

For a detailed study of the structure of the set of relativistic eligible transformations the reader is referred to Rubin and Suppes (1954).

## 10.15  CONCLUDING REMARKS

The contents of the chapter are, properly, a part of the philosophy of science distinct from the theory of measurement. Nonetheless, we elected to include it in a survey of the foundations of measurement because it is such a natural outgrowth of that theory, because it depends so intimately on a fully articulated theory of measurement, and because it has not been the tradition of either philosophers of science or expositors of dimensional analysis to attempt to bridge the gap between theories of (extensive) measurement and the algebra of physical quantities, which is the implicit basis of physics.

As the reader realizes, some of what we have covered is relatively old and has been worked and reworked by many authors. This is true of dimensional analysis, both as an abstract structure and as an applied method. Much of the rest of the chapter is relatively new and has yet to receive the benefit of considerable criticism and revision. This remark applies to the concept of laws of similitude and exchange relating extensive and conjoint measures, the construction of an algebra of physical quantities in terms of these laws together with some additional assumptions, the attempt to state the conceptual basis for the dimensional invariance of physical laws in terms of the concept of physically similar systems, and the axiomatic analysis of Galilean and Lorentz invariance in classical and relativistic particle mechanics. The rawness of some of this material perhaps accounts somewhat for the length of the presentation. Experience surely suggests that clearer

and more concise developments of ideas can be anticipated once others
have had a try at them, and so in all likelihood the bulk of this chapter
will become dated more rapidly than many other parts of this volume.
At least, we hope it will because the construction of a fully axiomatic structure
for physical (and possibly other scientific) measures and their algebra of
combination has been too long neglected.

## EXERCISES

**1.** It is easy to give many different simple mathematical examples of
structures of physical quantities (Definition 1). For one class of examples,
define $*$ as pointwise function multiplication, and let $A^+$ be a nonempty set
of real-valued functions with a common domain and such that 0 is not
included in their range. In order for $\langle A, A^+, *\rangle$ to be a structure of physical
quantities:

   (i)   What closure properties must $A^+$ have?
   (ii)  What must be the definition of $A$?    (10.2.1)

**2.** Suppose that $\langle A, A^+, *\rangle$ is a structure of physical quantities (Definition
1). If $a \in A^+$ and $b \in [a]$, then prove that $b \in A^+$ iff there exists $\alpha \in \text{Re}^+$ such
that $b = \alpha * a$.    (10.2, 10.4.1)

**3.** Suppose that $\langle A, A^+, *\rangle$ is a structure of physical quantities (Definition
1). Define $\succsim_a$ on $[a]$ as follows: if $b = \beta * a$ and $c = \gamma * a$, where $\beta, \gamma \in \text{Re}$,
then $b \succsim_a c$ iff $\beta \geq \gamma$. Prove the following:

   (i)   $\succsim_a$ is a simple ordering of $[a]$.
   (ii)  For $d \neq 0$, $b \succsim_a c$ iff $b * d \succsim_{a*d} c * d$.
   (iii)  If $b, c \in [a]$, $b \in A^+$, and $c \in A^-$, then $b \succ_a c$.    (10.2, 10.4.1)

**4.** Prove Lemma 1.    (10.4.1)

**5.** Prove Lemma 4.    (10.4.1)

**6.** Prove Lemma 5.    (10.4.1)

**7.** Prove Lemma 6.    (10.4.1)

**8.** Prove Lemma 12.    (10.4.1)

**9.** Suppose that $\phi$ and $\psi$ are similarities (Definition 2) of a structure of
physical quantities. Prove in two different ways—using Theorem 3 and not
using it—that $\phi^{-1}$ and $\psi\phi$ are similarities and that $\psi\phi = \phi\psi$.    (10.3, 10.4.3)

**10.** Suppose that $g$ is an arbitrary real-valued function of $m$ real arguments, that $P_i$, $i = 1,..., m$, are positive dimensions of a structure of physical quantities and that $x_i^0 \in P_i$ are fixed elements. Let $f$ be defined as follows: for $x_i^0 \in P_i$, there exist $\alpha_i \in Re$ such that $x_i = \alpha_i * x_i^0$ and

$$f[x_1^* (x_1^0)^{-1}, ..., x_m^* (x_m^0)^{-1}] = g(\alpha_1, ..., \alpha_m).$$

Prove that $f$ is dimensionally invariant (Definition 3).      (10.3)

*In Exercises* 11–16 *use dimensional analysis. The dimensions of the various quantities can be found in Table 3. The symbol $\alpha$ denotes a constant and $\Phi$ an unknown function. See Section 10.5.*

**11.** Two planets subject to their mutual attraction move so that their distance $r$ remains constant. Show that the period $T$ of their rotation is given by

$$T^2 = (r^3/Gm_1)\, \Phi(m_2/m_1),$$

where $m_1$ and $m_2$ are their masses, and $G$ is the universal gravitational constant.

**12.** A body of mass $m$ is rotated in a circle of radius $r$. Show that the force $F$ that must be applied in order to maintain an angular velocity of $\omega$ is given by $F = \alpha r m \omega^2$. (Do not distinguish among lengths in different directions.)

**13.** Consider a spherical drop that is vibrating with period $T$. Assuming that the relevant variables are its radius $r$, density $d$, and surface tension $\sigma$, show that $T^2 = \alpha r^3 d/\sigma$.

**14.** Suppose $R$ is the molar gas constant. Show (without distinguishing the direction of lengths) that the mass $m$, volume $V$, pressure $p$, and temperature $T$ of a gas are related by $pV = \alpha RT$.

**15.** Suppose that a plane condenser has plates of area $A$ separated by the distance $d$. If $\epsilon$ is the permittivity of the dielectric, show that the capacity $C$ of the condenser is given by

$$C = \frac{\epsilon A}{d}\, \Phi\left(\frac{d^2}{A}\right).$$

**16.** Suppose that a uniformly charged sphere of radius $r$ and charge density $\rho$ is in a medium of permittivity $\epsilon$. Show that its potential energy $E$ is given by $E = \alpha r^5 \rho^2/\epsilon$.

**17.** Prove the necessity of the two parts of Theorem 7 for both algebraic- and absolute-difference structures.      (10.7.4)

**18.** Prove the sufficiency of part (i) of Theorem 7 for both algebraic- and absolute-difference structures.    (10.7.4)

**19.** Prove Theorem 8 for both algebraic- and absolute-difference structures.
(10.7.4)

**20.** Prove Theorem 9.    (10.7.4)

**21.** Prove the uniqueness statement of Theorem 10.    (10.9)

**22.** Prove Theorem 11.    (10.9)

**23.** Show that $SG(\mathcal{S})$ is a subgroup of the group of similarities.    (10.10.2)

**24.** In Theorem 12 (Section 10.10.2) give direct proofs that (i) implies (iii), (iii) implies (ii), and (ii) implies (i). In proving the last part, show that if $\mathcal{S}$ satisfies the dimensionally invariant law $(f, g)$ then $g$ is a system measure.
(10.10.2, 10.11.1)

**25.** In Theorem 12 (Section 10.10.2) give direct proofs that in the presence of parts (i)–(iii), (iv) implies (vi), (vi) implies (v), and (v) implies (iv).
(10.10.2, 10.11.1)

**26.** Assuming conditions (i)–(vi) of Theorem 12 (Section 10.10.2) prove that, for all $S_q$, $S_{q'} \in \mathcal{S}_f$, if $S_q = S_{q'}$, then $q = q'$.    (10.10.2)

**27.** Suppose that $f$ is an everywhere differentiable function from $\mathrm{Re}^+$ into $\mathrm{Re}$ and that for all $x, y \in \mathrm{Re}^+$, $f(xy) = f(x)f(y)$. Show directly (using an elementary differential equation) that there exists a constant $\beta$ such that $f(x) = x^\beta, x > 0$.    (10.12)

**28.** For ratio scales, show that solving Equation (11) is equivalent to solving the equation

$$g(x_1,...,x_n)\, g(y_1,...,y_n) = g(x_1 y_1,..., x_n y_n),$$

where $g$ is proportional to $f$. Define $h_i(x)$ to be the value of $g(x_1,...,x_n)$ when $x_j = 1$, $j \neq i$, and $x_i = x$. Show by induction on the preceding result that

$$g(x_1,...,x_n) = \prod_{i=1}^{n} h_i(x_i).    (10.12)$$

**29.** Using Exercises 28 and 29, prove Theorem 13 for functions $f$ that are once differentiable in each argument.    (10.12)

**30.** Prove Theorems 17 and 18.    (10.14)

**31.** Prove Theorem 21.    (10.14.1)

**TABLE 3**

Dimensions and Units of Physical Quantities[a]

| Quantity | Q | θ | M | L | T | A | Symbols | Remarks |
|---|---|---|---|---|---|---|---|---|
| **Base quantities** | | | | | | | | |
| Charge (electric) | 1 | | | | | | $Q$ | Unit: $C$ (coulomb) |
| Temperature | | 1 | | | | | $T, \theta$ | Unit: °K (degree absolute) |
| Mass | | | 1 | | | | $m$ | Unit: kg (kilogram) |
| Length | | | | 1 | | | $l, b, h, s, x, y, z$ | Unit: m (meter) |
| Time | | | | | 1 | | $t$ | Unit: s (second) |
| Plane angle | | | | | | 1 | $\alpha, \beta, \gamma, \theta, \phi, \zeta$ | Unit: rad (radian) = angle of circular arc equal to radius |
| **Kinematic (L, T, A) quantities** | | | | | | | | |
| Curvature | | | | −1 | | | $\rho$ | 1/radius |
| Wave number | | | | −1 | | 1 | $\sigma, \tilde{\nu}$ | 1/wave length |
| Angular acceleration | | | | | −2 | 1 | $\alpha$ | |
| Time constant | | | | | −1 | | $k$ | $x_0 e^{-kt}$ |
| Angular velocity | | | | | −1 | 1 | $\omega, \Omega$ | |
| Frequency | | | | | −1 | 1 | $\nu, f$ | Cycles/s; Unit: Hz (hertz) |
| Plane angle | | | | | | 1 | $\alpha, \beta, \gamma, \theta, \phi, \zeta$ | Unit: rad (radian) |
| Solid angle | | | | | | 2 | $\omega, \Omega$ | Unit: sr(steradian) = $\frac{1}{4}$ surface of sphere of unit radius |
| Period | | | | | 1 | −1 | $T$ | Time per cycle = 1/frequency |
| Time | | | | | 1 | | $t$ | Unit: s (second) |
| Acceleration | | | | 1 | −2 | | $a$ | $d^2x/dt^2$ |
| Acceleration of gravity | | | | 1 | −2 | | $g$ | 9.80665 ms⁻² |
| Velocity, speed | | | | 1 | −1 | | $v$ | $dx/dt$ |
| Velocity of light | | | | 1 | −1 | | $c$ | $2.998 \times 10^8$ ms⁻¹ in vacuo |

*Table continued*

[a] All zero exponents are omitted. Within each subgroup, the organization is lexicographic on the exponents of the basic dimensions. (Adapted from Thun, 1960)

**TABLE 3** *(cont.)*

| Quantity | Q | Θ | M | L | T | A | Symbols | Remarks |
|---|---|---|---|---|---|---|---|---|
| | | | Exponents of | | | | | |
| Wave length | | | | 1 | | −1 | $\lambda$ | Length per cycle = 1/wave number |
| Length | | | | 1 | | | $l, b, h, s, x, y, z$ | Unit: m (meter) |
| Diffusion coefficient | | | | 2 | −1 | | $D$ | $\partial/\partial t = D\,\partial^2/\partial x^2$ |
| Kinematic viscosity | | | | 2 | −1 | | $\nu$ | Viscosity per density |
| Area | | | | 2 | | | $A, s$ | |
| Volume velocity | | | | 3 | −1 | | $U$ | |
| Volume | | | | 3 | | | $V, v$ | |

*Mechanical (M, L, T, A) quantities*

| Quantity | Q | Θ | M | L | T | A | Symbols | Remarks |
|---|---|---|---|---|---|---|---|---|
| Rotational compliance | | | −1 | −2 | 2 | 1 | $C_R$ | Change in angle/torque |
| Rectilinear compliance | | | −1 | | 2 | | $C_M$ | Change in length/force |
| Specific refraction | | | −1 | 3 | | | $r$ | Proportional to 1/density |
| Acoustic capacitance | | | −1 | 4 | 2 | | $C_A$ | Change in volume/pressure |
| Acoustic impedance | | | 1 | −4 | −1 | | $Z_A$ | Pressure/volume velocity |
| Acoustic resistance | | | 1 | −4 | −1 | | $R_A$ | Real component of impedance |
| Acoustic reactance | | | 1 | −4 | −1 | | $X_A$ | Imaginary component of impedance |
| Inertance | | | 1 | −4 | | | $M$ | Pressure/rate of change of volume velocity |
| Density | | | 1 | −3 | | | $d$ | Mass/volume |
| Energy density | | | 1 | −1 | −2 | | $\omega_e$ | Energy/volume |
| Pressure | | | 1 | −1 | −2 | | $p$ | Force/area |
| Stress | | | 1 | −1 | −2 | | $s, \sigma$ | Force/area |
| Modulus of elasticity (Young's) | | | 1 | −1 | −2 | | $E, Y$ | Stress/strain |
| Bulk modulus | | | 1 | −1 | −2 | | $k$ | Compression stress/volume strain |
| Tensile strength | | | 1 | −1 | −2 | | TS | Maximum load/original cross section |

*Table continued*

**TABLE 3** *(cont.)*

| Quantity | Q | Θ | M | L | T | A | Symbols | Remarks |
|---|---|---|---|---|---|---|---|---|
| | | | | Exponents of | | | | |
| Shear modulus | | | 1 | −1 | −2 | | $\mu$ | Shearing stress/shear strain |
| Shear strength | | | 1 | −1 | −2 | | SS | Maximum shear load/cross section |
| Viscosity | | | 1 | −1 | −1 | | $\eta$ | Shearing stress/velocity gradient perpendicular to flow |
| Sound intensity | | | 1 | | −3 | | $I$ | Pressure$^2$/density × velocity of sound |
| Poynting vector | | | 1 | | −3 | | $S$ | Power/area |
| Surface tension | | | 1 | | −2 | | $\gamma, \sigma$ | Energy/area |
| Mechanical rectilinear resistance | | | 1 | | −1 | | $R_M$ | Force/velocity |
| Mass | | | 1 | | | | $m$ | Unit: kg (kilogram) |
| Force | | | 1 | 1 | −2 | | $F$ | Mass × acceleration; Unit: N (newton) |
| Momentum | | | 1 | 1 | −1 | | $p$ | Mass × velocity |
| Impulse | | | 1 | 1 | −1 | | $\mathscr{I}$ | $\int F\,dt = \Delta p$ |
| Radiation intensity | | | 1 | 2 | −3 | −2 | $U$ | Power/solid angle |
| Power | | | 1 | 2 | −3 | | $P$ | Work/time; unit: W (watt) |
| Energy, work | | | 1 | 2 | −2 | | $E, U$ | Unit: $\mathscr{J}$ (joule); volt × coulomb |
| Quantity of heat | | | 1 | 2 | −2 | | $Q, q$ | |
| Moment, torque | | | 1 | 2 | −2 | | $M$ | Radius × force |
| Mechanical rotational resistance | | | 1 | 2 | −1 | −1 | $R_R$ | Torque/angular velocity |
| Action | | | 1 | 2 | −1 | | $A$ | |
| Angular momentum; moment of momentum | | | 1 | 2 | −1 | | $L$ | radius × momentum |
| Moment of inertia | | | 1 | 2 | | | $I, \mathscr{J}$ | $\int r^2\,dm$ |

*Table continued*

**TABLE 3** *(cont.)*

| Quantity | Q | Θ | M | L | T | A | Symbols | Remarks |
|---|---|---|---|---|---|---|---|---|
| | | | Exponents of | | | | | |

*Thermal (θ, M, L, T) quantities*

| Quantity | Q | Θ | M | L | T | A | Symbols | Remarks |
|---|---|---|---|---|---|---|---|---|
| Heat capacity (mass) | | −1 | | 2 | −2 | | $C_m$ | |
| Heat capacity (volume) | | −1 | 1 | −1 | −2 | | $C_V$ | |
| Thermal conductivity | | −1 | 1 | 1 | −3 | | $\lambda$ | Change in quantity of heat × length/ area × time × temperature |
| Entropy | | −1 | 1 | 2 | −2 | | $S$ | |
| Molar gas constant | | −1 | 1 | 2 | −2 | | $R$ | $(M/g)(pV/T) =$ 8.31662 $\mathscr{J}°\mathrm{K}^{-1}$ |
| Temperature gradient | | 1 | | −1 | | | grad $\theta$ | |
| Temperature | | 1 | | | | | $T, \theta$ | Unit: °K (degree absolute) |

*Electrical and magnetic (Q, M, L, T, A) quantities*

| Quantity | Q | Θ | M | L | T | A | Symbols | Remarks |
|---|---|---|---|---|---|---|---|---|
| Permeability | −2 | | 1 | 1 | | | $\mu$ | |
| Impedence (electric) | −2 | | 1 | 2 | −1 | | $Z$ | Potential/current; Unit: $\Omega$ (ohm) |
| Resistance (electric) | −2 | | 1 | 2 | −1 | | $R$ | Real component of impedance |
| Reactance (electric) | −2 | | 1 | 2 | −1 | | $X$ | Imaginary component of impedance |
| Coefficient of inductance | −2 | | 1 | 2 | | | $L, M$ | Unit: H (Henry) |
| Permeance | −2 | | 1 | 2 | | | $\mathscr{P}$ | Magnetic flux/mmf |
| Resistivity | −2 | | 1 | 3 | −1 | | $\rho$ | Resistance × cross section/length = 1/conductivity |
| Magnetic induction; magnetic flux density | −1 | | 1 | | −1 | | $B$ | Permeability × magnetic field intensity |
| Electric field intensity | −1 | | 1 | 1 | −2 | | $E$ | Potential per length |
| Vector potential | −1 | | 1 | 1 | −1 | | $A$ | Current × permeability |

*Table continued*

**TABLE 3** *(cont.)*

| Quantity | Q | Θ | M | L | T | A | Symbols | Remarks |
|---|---|---|---|---|---|---|---|---|
| Potential (electric) | −1 | | 1 | 2 | −2 | | $V, \phi$ | Unit: V (volt) |
| Electromotive force | −1 | | 1 | 2 | −2 | | $V$, emf | $\int E\,dl$ |
| Magnetic flux | −1 | | 1 | 2 | −1 | | $\Phi$ | $\iint B\,ds$, $s$ = area; Unit: Wb (weber) |
| Quantity of magnetism | −1 | | 1 | 2 | −1 | | $m$ | |
| Flux linkage | −1 | | 1 | 2 | −1 | 1 | $\Lambda$ | Magnetic flux × turns |
| Magnetic moment | −1 | | 1 | 3 | −1 | | $M$ | |
| Charge density (volume) | 1 | | | −3 | | | $\rho$ | Charge per volume |
| Current density | 1 | | | −2 | −1 | | $\mathscr{J}$ | Current/area |
| Pole density | 1 | | | −2 | −1 | | $\rho_m$ | Pole strength/volume |
| Electric displacement | 1 | | | −2 | | | $D$ | Charge/area (flux density) |
| Polarization | 1 | | | −2 | | | $P$ | Dipole moment/volume |
| Magnetic field intensity | 1 | | | −1 | −1 | | $H$ | Magnetic induction/permeability |
| Magnetization | 1 | | | −1 | −1 | | $M$ | Magnetic moment/volume |
| Sheet current density | 1 | | | −1 | −1 | | $K$ | Current/length |
| Linear charge density | 1 | | | −1 | | | $L$ | Charge/length |
| Current | 1 | | | | −1 | | $\mathscr{J}, i$ | Unit: A (ampere) |
| Magnetomotive force | 1 | | | | −1 | | mmf, $\mathscr{F}$ | $\int H\,dl$ |
| Charge | 1 | | | | | | $Q$ | Unit: $C$ (coulomb) |
| Flux (electric) | 1 | | | | | | $\psi$ | $\iint D\,ds$, $s$ = area |
| Pole strength | 1 | | | 1 | −1 | | $Q_m, q_m$ | $\iiint \rho_m\,dV$ |
| Dipole moment | 1 | | | 1 | | | $p$ | |

*Table continued*

TABLE 3 *(cont.)*

| Quantity | Exponents of $Q$ | $\Theta$ | $M$ | $L$ | $T$ | $A$ | Symbols | Remarks |
|---|---|---|---|---|---|---|---|---|
| Magnetic (dipole) moment | 1 | | | 2 | −1 | | $m$ | Mechanical moment/magnetic flux density |
| Conductivity | 2 | | −1 | −3 | 1 | | $\gamma$, $\sigma$ | 1/resistivity |
| Permittivity | 2 | | −1 | −3 | 2 | | $\epsilon$ | Electric displacement/electric field intensity |
| Reluctance | 2 | | −1 | −2 | | | $\mathcal{R}$ | mmf/magnetic flux = 1/permeance |
| Admittance (electric) | 2 | | −1 | −2 | 1 | | $Y$ | $1/Z = 1/(R + iX) = G - iB$; Unit: mho |
| Conductance | 2 | | −1 | −2 | 1 | | $G$ | $R/(R^2 + X^2)$ |
| Susceptance | 2 | | −1 | −2 | 1 | | $B$ | $X/(R^2 + X^2)$ |
| Capacitance | 2 | | −1 | −2 | 2 | | $C$ | Charge/potential; Unit: $F$ (Farad) |

# Answers and Hints to Selected Exercises

## Chapter 1

**2.** When weak stochastic transitivity holds, i.e., if $P(x, y) \geqslant \frac{1}{2}$ and $P(y, z) \geqslant \frac{1}{2}$, then $P(x, z) \geqslant \frac{1}{2}$.

**3.** $a \succ d \succ c \succ b$.

**8.** $c_1 \succ_1 b_1 \succ_1 d_1 \succ_1 a_1$ and $d_2 \succ_2 c_2 \succ_2 a_2 \succ_2 b_2$.

**9.** For example, $c_1 J_1 b_1$, $c_2 J_2 a_2$, and $(b_1, c_2) = (c_1, a_2) = 61$.

**10.** *Hint:* Use Exercise 9 and Theorem 2.

**14.** *Hint:* Proceed by induction using

$$(a_1^{(i)}, a_2^{(2)}) \sim (a_1^{(i+1)}, a_2^{(1)}) \quad \text{and} \quad (a_1^{(1)}, a_2^{(i+1)}) \sim (a_1^{(2)}, a_2^{(i)}).$$

## Chapter 2

**3.** First use induction on $n$ to prove the lemma with $m = 1$. Then use induction on $m$, using the fact that the lemma holds for $m = 1$ in the inductive step, e.g.,

$$
\begin{aligned}
(m + 1 + n)a &= (ma) \circ ((n + 1)a) \quad &&\text{(inductive hypothesis)} \\
&= (ma) \circ (a \circ (na)) \quad &&\text{(lemma for } m = 1) \\
&= ((ma) \circ a) \circ (na) \quad &&\text{(Axiom 5)} \\
&= ((m + 1)a) \circ (na).
\end{aligned}
$$

**9.** *Hint*: Use Exercise 8.

**10.** *Hint*: Use Exercise 8.

**13.** Define $\phi$ by $\phi(a) = \inf\{m/n \mid n > 0$ and there exists $b$ such that $a * b,\ a^2 * b \in B^*,\ m \in N_{a*b},\ n \in N_{a^2*b},\ m(a * b) > n(a^2 * b)\}$.

## Chapter 3

**6.** *Hint*: Consider the different values that the expressions on one side of the equation can assume.

**7** *Hint*: Use induction.

**8.** *Hint*: Prove that (1) $W(n,a) \geq W(n,b) + 1$; (2) for $i \in I^+$, $c \in A$, $W(i,c) = iW(1,c) + \sum_{j=1}^{i-1} r(jc, c)$; (3) for $l \in I^+$ there exists $i \in I^+$ such that $\sum_{j=1}^{i-1} r(jc, c) \leq i - l$.

**16.** *Hint*: To show the impossibility of $a, b \in A$ with $(n + 1)a > nb$ and $(n + 1)b > na$ for all $n \in I^+$, suppose $a \gtrsim b$, take $c$ such that $a \gtrsim b \circ c$, and build a standard sequence $c, 2c, 3c, \ldots$ .

**19.** $(x + y + 2 - xy)/(1 - xy)$.

**20.** *Hint*: Use induction.

**21.** *Hint*: To show the impossibility of $a, b \in A$ with $(n + 1)a > nb$ and $(n + 1)b > na$ for all $n \in I^+$, suppose $a > b$ and take $k \in I^+$ and $c \in A$ such that $a \gtrsim (k + 1)c > kc \gtrsim b$, and show that for $n > k$, $(n + 1)b > na$ implies $kc < b$, contrary to choice.

**30.** *Hint*: Use Theorem 13 to obtain $\theta$. *Answer*: 63/101.

## Chapter 4

**2.** All except 5.

**5.** *Hint*: Map Re monotonically into $(0, 1)$ and $w$ into 1 and use the natural algebraic-difference structure on $(0, 1]$.

**6.** Only the weak ordering and the Archimedean axioms hold.

**7.** *Hint*: Suppose $ab \sim bc \sim cd > dd$. Use the representation to show that not $(ac \sim bd)$, violating Axiom 3 of Definition 3.

**8.** *Hint*: Axiomatize $\gtrsim$ where $ab \gtrsim cd$ iff $ad \gtrsim^* cb$, and simply restate the axioms. Note that the resulting system is not very natural: it does not include the transitivity of $\gtrsim^*$. For a more intuitive treatment of algebraic differences from the standpoint of additivity, see Section 6.6.4.

**9.** First let $b'$ solve $cd > b'b > aa$. Then construct a sequence in $\mathscr{L}(cd) \cap \mathscr{G}(b'b)$, e.g., let $b_1 = b$, and solve $cd > b_2 b_1 > b'b$, then solve $cd > b_3 b_2 > b'b$, etc. The procedure cannot continue indefinitely (why not?) yet it stops only when $cd > ab_m$; let $a = a_{m+1}$.

**10.**  (i)  Use Axiom 3 inductively.

(ii)  Take a mesh $cd$ such that $ab$ is between six and seven times $cd$ and $a'b'$ is between four and five times $cd$. (Use part (i); in general, 26 inequalities must be established to obtain this level of accuracy.)

**11.**  (i)  Take a sequence $a_1, ..., a_{m+1}$ in $\mathscr{L}(cd)$, with $m = M(cd, ab)$, $a_1 = b$, $a_{m+1} = a$. Then construct $b_{i+1} b_i$ with $b_1 = b'$ and $cd > b_{i+1} b_i > a_{i+1} a_i$. The construction cannot continue through $i = m$ (Axiom 3), yet, it can stop only for $cd > a'b_i$.

(ii)  Construct the usual sequence $a_1, ..., a_{m+1}$, with $m = M(cd, ab)$; then apply 11 (i) to obtain sequences in $\mathscr{L}(ef)$, of length $\leqslant M(ef, cd) + 1$, that span $a_{i+1} a_i$. The composite sequence spans $ab$ with length $\leqslant M(cd, ab) M(ef, cd) + 1$. This proves the left-hand inequality. For the right-hand inequality, assume the contrary and divide the sequence with mesh $ef$ that spans $ab$ into not more than $M(cd, ab)$ subsequences, each of which has length at most $M(ef, cd)$.

**12.**  Choose any $c_1 d_1$, divide, take the smaller piece, divide, etc. Then show that

$$M(c_{i+j} d_{i+j}, c_i d_i) > 2^j + 1,$$

using 11 (ii). The rest follows Section 2.2.4 closely.

**13.**  Using Exercise 11 (ii), follow the proof of Section 2.2.4.

**14.**  (i)  Use methods similar to 11 (ii).
       (ii)  Follows the proof in Section 2.2.4 closely.

**15.**  Proceed as in Section 2.2.4, using the equation of Exercise 14 (ii).

**16.**  In general, nothing can be said. Two examples are revealing. Let $A_1 = A_2 = \mathrm{Re}^+$. First, define $\gtrsim_1$ and $\gtrsim_2$ by

$$(x_1, y_1) \gtrsim_1 (x_2, y_2) \qquad \text{iff} \qquad x_1 - y_1 \geqslant \log x_2 - \log y_2$$
$$(x_1, y_1) \gtrsim_2 (x_2, y_2) \qquad \text{iff} \qquad x_1/y_1 \geqslant \exp x_2/\exp y_2.$$

Second, define $\gtrsim_1$ and $\gtrsim_2$ by

$$(x_1, y_1) \gtrsim_1 (x_2, y_2) \qquad \text{iff} \qquad x_1 - y_1 \geqslant x_2 - y_2$$
$$(x_1, y_1) \gtrsim_2 (x_2, y_2) \qquad \text{iff} \qquad x_1/y_1 \geqslant (x_2/y_2)^2.$$

We do not know whether the condition is sufficient.

**18.** Axioms 3 (i), 3 (ii), and 4 all fail.

**20.** The triples $(a, c, d)$ and $(e, d, c)$ violate weak monotonicity.

**21.** Yes; No.

## Chapter 5

**1.** (iii) *Hint:* It matters whether $X \cap Y = \varnothing$ or not.

**3.** Axioms 2 and 3 imply neither connectedness nor transitivity; an example can be constructed for either point using a set consisting of two elements.

**4.** $X = \{a, b, c, d\}$, $P(a) = P(b) = 0.19$, $P(c) = 0.22$, $P(d) = 0.40$.

**7.** *Hint:* Set up simultaneous linear equations; use the fact that $P(X) = 1$.

**9.** *Hint:* If $B$ is independent of an atom, $P(B) = 0$ or 1.

**10.** Use Lemma 12 with $C \supset B \supset A \cap B$, $B \supset B \supset A \cap B$.

**12.** *Hint:* Without loss of generality, assume $A \subset C$. For $A \in \mathscr{E} - \mathscr{N}$, show $-B \mid A \sim -B \mid C$ and then use this and the negation of the conclusion to show a contradiction.

## Chapter 6

**8.** Restricted solvability is violated.

**28.** Monotonicity and Archimedean axioms are violated.

**32.** *Hint:* Assume $a \circ x \gtrsim b \circ y$, $f \circ p \gtrsim g \circ q$, and $g \circ y \gtrsim f \circ x$ and prove $a \circ p \gtrsim b \circ q$ by considering $[(u \circ f) \circ v] \circ [(a \circ p) \circ (y \circ w)]$.

## Chapter 7

**1.** *Hint:* Order $(A \times A) \times (S \times S)$ by $[(a, a'), (s, s')] \gtrsim [(a'' \ a'''), (s'', s''')]$ if and only if $d(as, a's') \geqslant d(a''s'', a'''s''')$.

**4.** *Hint:* $[f_1(x_1) + f_2(x_2)]f_3(x_3) = g(x_1 + x_2 + x_3)$, where $f_i$ and $g$ are increasing.

**7.** *Hint:* As $|\phi_1(a_1)| \to \infty$, let $|\phi_1(b_1)|$ and $|\phi_2(a_2)|$ approach 0 suitably, e.g., for $\phi_1(a_1) = -2^n$, let $\phi_1(b_1) = -2^{-(n+3)}$ and $\phi_2(a_2) = -2^{-(n+2)}$.

**9.** *Hint:* The sign-dependence matrix indicates category 1 of Table 1; find an appropriate representation.

**10.** *Hint:* An interesting cancellation condition is the following: If

$$a_1 b_2 a_3 a_4 \sim d_1 d_2 d_3 c_4 ,$$
$$b_1 a_2 a_3 a_4 \sim d_1 d_2 c_3 d_4 ,$$

and

$$b_1 b_2 a_3 a_4 \sim d_1 d_2 c_3 c_4 ,$$

then

$$a_1 a_2 a_3 a_4 \sim d_1 d_2 d_3 d_4 .$$

This leads both to the equality of two different $\oplus$ definitions and to the distributive law for $\oplus$, $*$.

## Chapter 8

**6.** *Hint*: Note that $u(f_X) = u(f_A) P(A) + u(f_{-A}) P(-A)$, where $f_A$ is the restriction of $f_X$ to $A$ and $f_{-A}$ to $-A$. *Answer*: $P(A) = \frac{1}{2}$.

**8.** *Hint*: Prove that if $w$ and $P$ satisfy Equation (6), then there exists a real $\alpha$ such that $(w - \alpha)P$ is a finitely additive measure.

**9.** For example, $a = 4$, $b = 1$, $c = 2.35$.

**10.** *Hint*: Note the sign of $\alpha_2$ in the conclusion of Theorem 5.

## Chapter 9

**1.** *Hint*: Construct an example of a structure where independence and double cancellation are satisfied, but where triple cancellation is violated.

**2.** *Hint*: Arrange the indices so that $a_i b_1 > a_i b_0$ for all $i$, and $a_i b_k \gtrsim a_j b_k$ for $k = 0, 1$ iff $0 \leqslant j \leqslant i \leqslant n$. Define $\phi_B(b_0) = 0 = \phi_A(a_0)$, and $\phi_B(b_1) = 1$. If $a_{i+1} b_0 > a_i b_1$, let $\phi_A(a_{i+1}) = \phi_A(a_i) + 2$. If $a_i b_1 \gtrsim a_{i+1} b_0$, let $\phi_A(a_{i+1}) = \phi_A(a_i) + 2^{(k-i)}$, where $k$ is the smallest integer satisfying $a_k b_1 \gtrsim a_{i+1} b_0$.

**3.** *Hint*: Construct an example of a structure where $>$ is transitive and, for some $x, y, z \in A^+$, $x \gtrsim_1 y$, $y \gtrsim_1 z$, but $x$ and $z$ are incomparable.

**5.** *Hint*: Prove (i) $\to$ (ii) $\to$ (iii) $\to$ (i).

**6.** *Hint*: For $a \in A - C$ define the vectorial representation $v$ as in Section 9.2.1, and for $a, b \in A - C$, $a \circ b \in C$ define $v(a \circ b) = \frac{1}{2}[v(a) + v(b)]$. Define $>_1$ as in Definition 1, and show that the irreflexivity of $>_1$ is the desired condition.

**8.** *Hint*: Find a $4 \times 6$ matrix $M$ of integers such that for any real $\Delta_i > 0$, $i = 1, 2, 3, 4$ the matrix equation

$$(\Delta_1, \Delta_2, \Delta_3, \Delta_4) \cdot M = [\phi_A(a_1), \phi_A(a_2), \phi_A(a_3), \phi_B(b_1), \phi_B(b_2), \phi_B(b_3)]$$

defines a positive additive representation of $A \times B$, and prove that any positive additive representation of $A \times B$ can be obtained from the above equation by an appropriate selection of $\Delta_i$, $i = 1, 2, 3, 4$.

**Chapter 10**

**1.** (i) If $f, g \in A^+$, then $f * g$, $1/f, f^{1/n} \in A^+$, $-f \notin A^+$, and $\text{Re}^+ \subset A^+$.

**18.** *Hint*: Let $\psi = f\phi$. For an algebraic structure, show that condition (i) implies

$$f(x + y) + f(0) = f(x) + f(y).$$

For an absolute structure show

$$|f(x + y) - f(y)| = |f(x) - f(0)|.$$

Define $g(x) = f(x) - f(0)$ and show by induction that $g(nx) = ng(x)$.

**19.** *Hint*: Proceed in much the same way as in part (i) of Theorem 7.

**24.** *Hint*: In proving (iii) implies (ii), construct $g$ as in the proof that (iii) implies (i) and $f$ as in the proof that (i) implies (ii).

**27.** *Hint*: Differentiate the functional equation with respect to $x$, then set $x = 1$, and let $\beta = f'(1)$.

# References

Abelson, R. P., & Tukey, J. W. Efficient conversion of nonmetric information into metric information. *Proc. Amer. Statist. Assoc. Meetings,* Social Statistics Sect., 1959, 226–230.

Abelson, R. P., & Tukey, J. W. Efficient utilization of non-numerical information in quantitative analysis: general theory and the case of simple order. *Ann. Math. Statist.,* 1963, **34,** 1347–1369.

Aczél, J. On mean values. *Bull. Amer. Math. Soc.,* 1948, **54,** 392–400.

Aczél, J. Über die Begründung der Additions-und Multiplikationsformeln von bedingten Wahrscheinlichkeiten. *Magyar Tud. Akad. Mat. Kutató Int. Közl.,* 1961, 6, 110–122.

Aczél, J. *Lectures on functional equations and their applications.* New York: Academic Press, 1966.

Aczél, J. On different characterizations of entropies. In M. Behara, K. Krickeberg, & J. Wolfowitz (Eds.), *Probability and information theory.* Berlin: Springer-Verlag, 1969. Pp. 1–11.

Aczél, J., Belousov, V. D., & Hosszú, M. Generalized associativity and bisymmetry on quasigroups. *Acta Math. Acad. Sci. Hungar.,* 1960, **11,** 127–136.

Aczél, J., & Pfanzagl, J. Remarks on the measurement of subjective probability and information. *Metrika,* 1966, **2,** 91–105.

Aczél, J., Pickert, G., & Radó, F. Nomogramme, Gewebe und Quasigruppen. *Mathematica,* 1960, **2,** 5–24.

Adams, E. W. Survey of Bernoullian utility theory. In H. Solomon (Ed.), *Mathematical thinking in the measurement of behavior.* Glencoe, Illinois: The Free Press, 1960. Pp. 151–268.

Adams, E. W. Elements of a theory of inexact measurement. *Phil. Sci.,* 1965, **32,** 205–228.

Adams, E. W., & Fagot, R. F. A model of riskless choice. Technical Report No. 4, Stanford University, Institute for Mathematical Studies in the Social Sciences, 1956.

Adams, E. W., & Fagot, R. F. A model of riskless choice. *Behav. Sci.,* 1959, **4,** 1–10.

Adams, E. W., Fagot, R. F., & Robinson, R. On the empirical status of axioms in theories of fundamental measurement. *J. Math. Psychol.*, 1970, **7**, 379–409.

Alimov, N. G. On ordered semigroups. *Izv. Akad. Nauk SSSR, Ser. Mat.*, 1950, **14**, 569–576 (in Russian).

Allais, M. Le comportement de l'homme rationnel devant le risque: critique des postulats et axiomes de l'école americaine. *Econometrica*, 1953, **21**, 503–546.

Anderson, N. H. Application of an additive model to impression formation. *Science*, 1962, **138**, 817–818.

Anderson, N. H. Functional measurement and psychophysical judgment. *Psychol. Rev.*, 1970, **77**, 153–170.

Anscombe, F. J., & Aumann, R. J. A definition of subjective probability. *Ann. Math. Statist.*, 1963, **34**, 199–205.

Arrow, K. J. Alternative approaches to the theory of choice in risk-taking situations. *Econometrica*, 1951, **19**, 404–437.

Arrow, K. J. Rational choice functions and orderings. *Economica*, 1959, **26**, 121–127.

Arrow, K. J. Utility and expectation in economic behavior. In S. Koch (Ed.), *Psychology: a study of a science*. Vol. 6. New York: McGraw-Hill, 1963. Pp. 724–752.

Atkinson, J. W. Motivational determinants of risk-taking behavior. *Psychol. Rev.*, 1957, **64**, 359–372.

Aumann, R. J. Utility theory without the completeness axiom. *Econometrica*, 1962, **30**, 445–462.

Aumann, R. J. Utility theory without the completeness axiom: a correction. *Econometrica*, 1964, **32**, 210–212. (a)

Aumann, R. J. Subjective programming. In M. W. Shelly & G. L. Bryan (Eds.), *Human judgments and optimality*. New York: Wiley, 1964. Pp. 217–242. (b)

Baumol, W. The Neumann–Morgenstern utility index—an ordinalist view. *J. Pol. Econ.*, 1951, **59**, 61–66.

Becker, G. M., & McClintock, C. G. Value: behavioral decision theory. In P. Farnsworth (Ed.), *Annual Review of Psychology*, Vol. 18. Palo Alto: Annual Reviews, Inc., 1967. Pp. 239–286.

Behrend, F. A. A system of independent axioms for magnitudes. *J. Proc. Roy. Soc. New South Wales*, 1953, **87**, 27–30.

Behrend, F. A. A contribution to the theory of magnitudes and the foundations of analysis. *Math. Z.*, 1956, **63**, 345–362.

Birkhoff, G. *Lattice theory*. New York: American Mathematical Society Colloquium Publication No. XXV, 1948, 1967.

Birkhoff, G. *Hydrodynamics. A study in logic, fact and similitude.* Princeton, New Jersey: Princeton University Press, 1950, 1960.

Birkhoff, G., & von Neumann, J. The logic of quantum mechanics. *Ann. of Math.*, 1936, **37**, 823–843.

Blackwell, D., & Girshick, M. A. *Theory of games and statistical decisions*. New York: Wiley, 1954.

Blakers, A. L. *Mathematical concepts of elementary measurement.* Vol. 17. *Studies in mathematics.* Stanford: School Mathematics Study Group, 1967.

Blaschke, W., & Bol, G. *Geometrie der Gewebe.* Berlin: Springer, 1938.

Block, H. D., & Marschak, J. Random orderings and stochastic theories of responses. In I. Olkin, S. Ghurye, W. Hoeffding, W. Madow, & H. Mann (Eds.), *Contributions to probability and statistics*. Stanford: Stanford University Press, 1960. Pp. 97–132.

Bock, R. D., & Jones, L. V. *The measurement and prediction of judgment and choice.* San Francisco: Holden Day, 1968.

Bolker, E. D. Functions resembling quotients of measures. *Trans. Amer. Math. Soc.*, 1966, **124**, 292–312.

Bolker, E. D. A simultaneous axiomatization of utility and subjective probability. *Phil. Sci.*, 1967, **34**, 333–340.

Box, G. E. P., & Cox, D. R. An analysis of transformations. *J. Roy. Statist. Soc. Ser. B.*, 1964, **26**, 211–252.

Brand, L. The Pi theorem of dimensional analysis. *Arch. Rational Mech. Anal.*, 1957, **1**, 35–45.

Bridgman, P. *Dimensional analysis.* New Haven: Yale University Press, 1922, 1931.

Buchdahl, H. A. A formal treatment of the consequences of the second law of thermodynamics in Carathéodory's formulation. *Z. Physik*, 1958, **152**, 425–439.

Büchi, J. R., & Wright, J. B. The theory of proportionality as an abstraction of group theory. *Math. Ann.*, 1955, **130**, 102–108.

Buckingham, E. On physically similar systems: illustrations of the use of dimensional equations. *Phys. Rev.*, 1914, **4**, 345–376.

Callen, H. B. *Thermodynamics.* New York: Wiley, 1960.

Campbell, B. A., & Masterson, F. A. Psychophysics of punishment. In B. A. Campbell & R. M. Church (Eds.), *Punishment and aversive behavior.* New York: Appleton-Century-Crofts, 1969. Pp. 3–42.

Campbell, N. R. *Physics: the elements.* Cambridge: Cambridge University Press, 1920. Reprinted as *Foundations of science: the philosophy of theory and experiment.* New York: Dover, 1957.

Campbell, N. R. *An account of the principles of measurement and calculation.* London: Longmans, Green, 1928.

Cantor, G. Beiträge zur Begründung der transfiniten Mengenlehre. *Math. Ann.*, 1895, **46**, 481–512.

Carnap, R. *Logical foundations of probability.* Chicago, Illinois: University of Chicago Press, 1950.

Causey, R. L. Derived measurement and the foundations of dimensional analysis. Technical Report No. 5, University of Oregon, Measurement Theory and Mathematical Models Reports, 1967.

Causey, R. L. Derived measurement, dimensions, and dimensional analysis. *Phil. Sci.*, 1969, **36**, 252–270.

Chisholm, R. M., & Sosa, E. On the logic of "intrinsically better." *Amer. Phil. Quart.*, 1966, **3**, 244–249.

Churchman, C. W. *Prediction and optional decision.* Englewood Cliffs, New Jersey: Prentice-Hall, 1961.

Churchman, C. W., Ackoff, R. L., & Arnoff, E. L. *Introduction to operations research.* New York: Wiley, 1957.

Cliff, N. Adverbs as multipliers. *Psychol. Rev.*, 1959, **66**, 27–44.

Cohen, M. R., & Nagel, E. *An introduction to logic and scientific method.* New York: Harcourt, Brace, 1934.

Coombs, C. H. A theory of psychological scaling. Bulletin No. 34, University of Michigan, Engineering Research Institute, 1952.

Coombs, C. H. *A theory of data.* New York: Wiley, 1964.

Coombs, C. H., Dawes, R. M., & Tversky, A. *Mathematical psychology: an elementary introduction.* Englewood Cliffs, New Jersey: Prentice-Hall, 1970.

Coombs, C. H., & Huang, L. C. Polynomial psychophysics of risk. *J. Math. Psychol.*, 1970, **7**, 317–338.

Coombs, C. H., & Pruitt, D. G. Components of risk and decision making: probability and variance preferences. *J. Exp. Psychol.*, 1960, **60**, 265–277.

Cooper, J. L. B. The foundations of thermodynamics. *J. Math. Anal. Appl.*, 1967, **17**, 172–193.

Copeland, A. H., Sr. Postulates for the theory of probability. *Amer. J. Math.*, 1941, **63**, 741–762.

Copeland, A. H., Sr. Probabilities, observations and predictions. In J. Neyman (Ed.), *Proceedings of the third Berkeley symposium on mathematical statistics and probability*. Vol. 2. Berkeley: University of California Press, 1956. Pp. 41–47.

Cross, D. V. An application of mean value theory to psychophysical measurement. Progress Report No. 6, Report No. 05613-3-P, University of Michigan, The Behavioral Analysis Laboratory, 1965.

Császár, Á. Sur la structure des espaces de probabilité conditionnelle. *Acta. Math. Acad. Sci. Hungar.*, 1955, **6**, 337–361.

Davidson, D., & Suppes, P. A finitistic axiomatization of subjective probability and utility. *Econometrica*, 1956, **24**, 264–275.

Davidson, D., Suppes, P., & Siegel, S. *Decision making: an experimental approach.* Stanford: Stanford University Press, 1957.

Debreu, G. Representation of preference ordering by a numerical function. In R. M. Thrall, C. H. Coombs, & R. L. Davis (Eds.), *Decision processes*. New York: Wiley, 1954. Pp. 159–165.

Debreu, G. Stochastic choice and cardinal utility. *Econometrica*, 1958, **26**, 440–444.

Debreu, G. Cardinal utility for even-chance mixtures of pairs of sure prospects. *Rev. Econ. Studies*, 1959, **71**, 174–177.

Debreu, G. Topological methods in cardinal utility theory. In K. J. Arrow, S. Karlin, & P. Suppes (Eds.), *Mathematical methods in the social sciences, 1959*. Stanford: Stanford University Press, 1960. Pp. 16–26.

deFinetti, B. La prévision: ses lois logiques, ses sources subjectives. *Ann. Inst. H. Poincaré*, 1937, **7**, 1–68. Translated into English in H. E. Kyburg, Jr., & H. E. Smokler (Eds.), *Studies in subjective probability*. New York: Wiley, 1964. Pp. 93–158.

deJong, F. J. *Dimensional analysis for economists.* Amsterdam: North Holland Publ., 1967.

Domotor, Z. Probabilistic relational structures and their applications. Technical Report No. 144, Stanford University, Institute for Mathematical Studies in the Social Sciences, 1969.

Dorsey, N. E. *The velocity of light.* Philadelphia: American Philosophical Society, 1944.

Drobot, S. On the foundations of dimensional analysis. *Studia Math.*, 1953, **14**, 84–99.

Duistermaat, J. J. Energy and entropy as real morphisms for addition and order. *Synthese*, 1968, **18**, 327–393.

Ducamp, A., & Falmagne, J. C. Composite measurement. *J. Math. Psychol.*, 1969, **6**, 359–390.

Duncan, W. J. *Physical similarity and dimensional analysis; an elementary treatise.* London: Arnold, 1953.

Edwards, W. Probability-preferences among bets with differing expected values. *Amer. J. Psychol.*, 1954, **67**, 56–67. (a)

Edwards, W. The reliability of probability preferences. *Amer. J. Psychol.*, 1954, **67**, 68–95. (b)

Edwards, W. Variance preferences in gambling. *Amer. J. Psychol.*, 1954, **67**, 441–452. (c)

Edwards, W. The theory of decision making. *Psychol. Bull.*, 1954, **51**, 380–417. (d)

Edwards, W. Behavioral decision theory. In P. Farnsworth (Ed.), *Annual Review of Psychology*. Vol. 12. Palo Alto: Annual Reviews, Inc., 1961. Pp. 473–498.

Edwards, W. Subjective probabilities inferred from decisions. *Psychol. Rev.*, 1962, **69**, 109–135.

Ellis, B. *Basic concepts of measurement.* London: Cambridge University Press, 1966.

Ellsberg, D. Classic and current notions of "measurable utility." *Econ. J.*, 1954, **64**, 528–556.

Ellsberg, D. Risk, ambiguity, and the Savage axioms. *Quart. J. Econ.*, 1961, **75**, 643–669.

Falk, G., & Jung, H. Axiomatik der Thermodynamik. In S. Flügge (Ed.), *Handbuch der Physik*. Vol. 3/2. *Prinzipien der Thermodynamik und Statistik*. Berlin: Springer, 1959. Pp. 119–175.

Feller, W. *An introduction to probability theory and its applications*. New York: Wiley, Vol. 1: 1950, 1957, 1968; Vol. 2: 1966.

Fine, T. A note on the existence of quantitative probability. *Ann. Math. Statist.*, 1971, **42**, 1182–1186. (a)

Fine, T. *Theories of probability*. New York: Academic Press, 1971. (b)

Fishburn, P. C. *Decision and value theory*. New York: Wiley, 1964.

Fishburn, P. C. Independence in utility theory with whole product sets. *Oper. Res.*, 1965, **13**, 28–45. (a)

Fishburn, P. C. Independence, trade-offs, and transformations in bivariate utility functions. *Management Sci.*, 1965, **11**, 792–801. (b)

Fishburn, P. C. Additivity in utility theory with denumerable product sets. *Econometrica*, 1966, **34**, 500–503. (a)

Fishburn, P. C. A note on recent developments in additive utility theories for multiple-factor situations. *Oper. Res.*, 1966, **14**, 1143–1148. (b)

Fishburn, P. C. Methods of estimating additive utilities. *Management Sci.*, 1967, **13**, 435–453. (a)

Fishburn, P. C. Additive utilities with finite sets: applications in the management sciences. *Naval Res. Logist. Quart.*, 1967, **14**, 1–13. (b)

Fishburn, P. C. Conjoint measurement in utility theory with incomplete product sets. *J. Math. Psychol.*, 1967, **4**, 104–119. (c)

Fishburn, P. C. Additive utilities with incomplete product sets: applications to priorities and assignments. *Oper. Res.*, 1967, **15**, 537–542. (d)

Fishburn, P. C. Bounded expected utility. *Ann. Math. Statist.*, 1967, **38**, 1054–1060. (e)

Fishburn, P. C. Interdependence and additivity in multivariate, unidimensional expected utility theory. *Int. Econ. Rev.*, 1967, **8**, 335–342. (f)

Fishburn, P. C. Preference-based definitions of subjective probability. *Ann. Math. Statist.*, 1967, **38**, 1605–1617. (g)

Fishburn, P. C. Utility theory. *Management Sci.*, 1968, **14**, 335–378.

Fishburn, P. C. A study of independence in multivariate utility theory. *Econometrica*, 1969, **37**, 107–121. (a)

Fishburn, P. C. Weak qualitative probability on finite sets. *Ann. Math. Statist.*, 1969, **40**, 2118–2126. (b)

Fishburn, P. C. *Utility theory for decision making*. New York: Wiley, 1970.

Fisher, I. Mathematical investigations in the theory of value and prices. *Trans. Conn. Acad. Sci.*, 1892, **9**, 1–124.

Fisher, I. A statistical method for measuring "marginal utility" and testing the justice of a progressive income tax. In J. H. Hollander (Ed.), *Economic essays contributed in honor of John Bates Clark*. New York: Macmillan, 1927. Pp. 157–193.

Fletcher, H., & Munson, W. A. Loudness, its definition, measurement and calculation. *J. Acoust. Soc. Amer.*, 1933, **5**, 82–108.

Focken, C. *Dimensional methods and their applications*. London: Arnold, 1953.

Fourier, J. B. J. *Théorie analytique de la chaleur*. Paris: Gauthier-Villars, 1822.

Frisch, R. Sur un problème d'économie pure. *Norsk Matematisk Forenings Skrifter*, 1926, **1**, 1–40.

Frisch, R. New methods of measuring marginal utility. In *Beiträge zur Ökonomischen Theorie*. Tübingen: J. C. B. Mohr (Paul Siebeck), 1932.

Frisch, R. General choice-field theory. In *Report of third annual research conference on economics and statistics*. Cowles Commission for Research in Economics, 1937. Pp. 64–69.

Fuchs, L. *Partially ordered algebraic systems*. Reading, Massachusetts: Addison-Wesley, 1963.

Gale, D. *The theory of linear economic models*. New York: McGraw-Hill, 1960.

Giles, R. *Mathematical foundations of thermodynamics*. New York: Macmillan, 1964.

Gollob, H. F. Impression formation and word combination in sentences. *J. Pers. Soc. Psychol.*, 1968, **10**, 341-353.

Grodal, Brigit, & Mertens, J.-F. Integral representations of utility functions. Discussion Paper 6823, Université Catholique de Louvain, Center for Operations Research and Economics, 1968.

Guilbaud, G. Sur une difficulté de la théorie du risque. *Coll. Int. Centre Nat. Rech. Sci. (Econométrie)*, 1953, **40**, 19-25.

Guild, J. Part III of Quantitative estimation of sensory events (Interim Report). *Rep. British Assoc. Advan. Sci.* 1938, 296-328.

Guttman, L. A basis for scaling qualitative data. *Amer. Sociol. Rev.*, 1944, **9**, 139-150.

Guttman, L. A general nonmetric technique for finding the smallest coordinate space for a configuration of points. *Psychometrika*, 1968, **33**, 469-506.

Halldén, S. *On the logic of "better"*. Lund: Gleerup, 1957.

Hansson, B. Fundamental axioms for preference relations. *Synthese*, 1968, **18**, 423-442. (a)

Hansson, B. Choice structures and preference relations. *Synthese*, 1968, **18**, 443-458. (b)

Harman, H. H. *Modern factor analysis*. Chicago: University of Chicago Press, 1960, 1967.

Hausner, M. Multidimensional utilities. In R. M. Thrall, C. H. Coombs, & R. L. Davis (Eds.), *Decision processes*. New York: Wiley, 1954. Pp. 167-180.

Havel, V. Verallgemeinerte Gewebe. I. *Arch. Math. (Brno)*, 1966, **2**, 63-70.

Helmholtz, H. v. Zählen und Messen erkenntnis-theoretisch betrachet. *Philosophische Aufsätze Eduard Zeller gewidmet*, Leipzig, 1887. Reprinted in *Gesammelte Abhandl.*, Vol. 3, 1895. Pp. 356-391. English translation by C. L. Bryan, *Counting and measuring*. Princeton, New Jersey: van Nostrand, 1930.

Herstein, I. N., & Milnor, J. An axiomatic approach to measurable utility. *Econometrica*, 1953, **21**, 291-297.

Hilbert, D. Grundlagen der Geometrie. In *Festschrift zur Feier der Enthüllung des Gauss–Weber–Denkmals in Göttingen*. Leipzig: Teubner, 1899. English translation by E. J. Townsend, *The foundations of geometry*. LaSalle, Illinois: Open Court, 1947, from the 9th German edition of *Grundlagen der Geometrie*. Stuttgart: Teubner, 1962.

Hofmann, K. H. Zur mathematischen Theorie des Messens. *Rozprawy Mat. (Warsaw)*, 1963, **32**, 1-31.

Hölder, O. Die Axiome der Quantität und die Lehre vom Mass. *Ber. Verh. Kgl. Sächsis. Ges. Wiss. Leipzig, Math.-Phys. Classe*, 1901, **53**, 1-64.

Holman, E. W. Strong and weak extensive measurement. *J. Math. Psychol.*, 1969, **6**, 286-293.

Holman, E. W. A note on additive conjoint measurement. *J. Math. Psychol.*, 1971, **8**, 489-494.

Houthakker, H. S. Revealed preference and the utility function. *Economica*, 1950, **17**, 159-174.

Houthakker, H. S. On the logic of preference and choice. In Anna-Teresa Tymieniecka (Ed.), *Contributions to logic and methodology in honor of J. M. Bochénski*. Amsterdam: North Holland Publ., 1965. Pp. 193-207.

Hull, C. L. *A behavior system*. New Haven, Connecticut: Yale University Press, 1952.

Huntington, E. V. A complete set of postulates for the theory of absolute continuous magnitude. *Trans. Amer. Math. Soc.*, 1902, **3**, 264-279. (a)

Huntington E. V. Complete sets of postulates for the theories of positive integral and of positive rational numbers. *Trans. Amer. Math. Soc.*, 1902, **3**, 280-284. (b)

Huntington, E. V. *The continuum and other types of serial order*. Cambridge, Massachusetts: Harvard University Press, 1917.

Huntley, H. E. *Dimensional analysis*. New York: Rinehart, 1951.

Hurst, P. M., & Siegel, S. Prediction of decision from a higher-ordered metric scale of utility. *J. Exp. Psychol.*, 1956, **52**, 138–144.

Hurvich, L. M., & Jameson, D. Some quantitative aspects of an opponent-colors theory. II. Brightness, saturation, and hue in normal and dichromatic vision. *J. Opt. Soc. Amer.*, 1955, **45**, 602–616.

Jeffrey, R. C. *The logic of decision*. New York: McGraw-Hill, 1965.

Kannai, Y. Existence of a utility in infinite dimensional partially ordered spaces. *Israel J. Math.*, 1963, **1**, 229–234.

Karapetoff, V. Special theory of relativity in hyperbolic functions. *Rev. Mod. Phys.*, 1944, **16**, 33–52.

Kayan, C. F. (Ed.) *Systems of units: national and international aspects*. Washington: American Association for the Advancement of Science, 1959.

Kelley, J. L. *General topology*. Princeton, New Jersey: van Nostrand, 1955.

Kershner, R. B., & Wilcox, L. R. *The anatomy of mathematics*. New York: Ronald Press, 1950.

Keynes, J. M. *A treatise on probability*. New York: Macmillan, 1921, 1929.

Kisch, B. *Scales & weights*. New Haven, Connecticut: Yale University Press, 1965.

Kochen, S., & Specker, E. P. Logical structures arising in quantum theory. In J. W. Addison, L. Henkin, & A. Tarski (Eds.), *Theory of models, proceedings of the 1963 international symposium at Berkeley.* Amsterdam: North-Holland Publ., 1965. Pp. 177–189.

Kolmogorov, A. N. *Grundbegriffe der Wahrscheinlichkeitsrechnung*. Berlin: Springer, 1933. English translation by N. Morrison, *Foundations of the theory of probability*. New York: Chelsea, 1956.

Koopman, B. O. The axioms and algebra of intuitive probability. *Ann. of Math.*, 1940, **41**, 269–292. (a)

Koopman, B. O. The bases of probability. *Bull. Amer. Math. Soc.*, 1940, **46**, 763–774. Reprinted in H. E. Kyburg, Jr., & H. E. Smokler (Eds.), *Studies in subjective probability*. New York: Wiley, 1964. Pp. 159–172. (b)

Koopman, B. O. Intuitive probability and sequences. *Ann. of Math.*, 1941, **42**, 169–187.

Koopmans, T. C. Stationary ordinal utility and impatience. *Econometrica*, 1960, **28**, 287–309.

Koopmans, T. C., Diamond, P. A., & Williamson, R. E. Stationary utility and time perspective. *Econometrica*, 1964, **32**, 82–100.

Kraft, C. H., Pratt, J. W., & Seidenberg, A. Intuitive probability on finite sets. *Ann. Math. Statist.*, 1959, **30**, 408–419.

Krantz, D. H. Conjoint measurement: the Luce-Tukey axiomatization and some extensions. *J. Math. Psychol.*, 1964, **1**, 248–277.

Krantz, D. H. Extensive measurement in semiorders. *Phil. Sci.*, 1967, **34**, 348–362.

Krantz, D. H. A survey of measurement theory. In G. B. Dantzig & A. F. Veinott, Jr. (Eds.), *Mathematics of the decision sciences, part 2*. Vol. 12. Lectures in applied mathematics. Providence, R. I.: American Mathematical Society, 1968. Pp. 314–350.

Krantz, D. H. A theory of magnitude estimation and cross-modality matching. *Science*, 1972, **175**, 1427–1435. (a)

Krantz, D. H. Measurement structures and psychological laws. *J. Math Psychol.*, 1972, **9**, 168–199. (b) MMPP 71-2, University of Michigan, Michigan Mathematical Psychology Program, 1971.

Krantz, D. H., & Tversky, A. A critique of the applicability of cardinal utility theory. Technical Report MMPP 65-4, University of Michigan, Michigan Mathematical Psychology Program, 1965.

Krantz, D. H., & Tversky, A. Conjoint-measurement analysis of composition rules in psychology. *Psychol. Rev.*, 1971, **78**, 151–169.

558

Kristof, W. A foundation of interval scale measurement. Research Bulletin RB 67-22. Princeton, New Jersey, Educational Testing Service, 1967.

Kristof, W. Structural properties and measurement theory of certain sets admitting a concatenation operation. *Brit. J. Math. Statist. Psychol.*, 1968, **21**, 201–229.

Kruskal, J. B. Analysis of factorial experiments by estimating monotone transformations of the data. *J. Roy. Statist. Soc. Ser. B.*, 1965, **27**, 251–263.

Kurth, R. A note on dimensional analysis. *Amer. Math. Monthly*, 1965, **72**, 965–969.

Kyburg, H. E., Jr., & Smokler, H. F. (Eds.) *Studies in subjective probability.* New York: Wiley, 1964.

Lange, O. The determinateness of the utility function. *Rev. Econ. Studies*, 1934, **1**, 218–225.

Langhaar, H. L. *Dimensional analysis and theory of models.* New York: Wiley, 1951.

Lenz, H. Zur Begründung der Winkelmessung. *Math. Nachr.*, 1967, **33**, 363–375.

Levelt, W. J. M., Riemersma, J. B., & Bunt, A. A. Binaural additivity in loudness. *Brit. J. Math. Statist. Psychol.*, 1972, **25**, 51–68.

Levine, M. V. Transformations that render curves parallel. *J. Math. Psychol.*, 1970, **7**, 410–444.

Levine, M. V. Transforming curves into curves with the same shape. *J. Math. Psychol.*, 1972, **9**, 1–16.

Lord, F. M., & Novick, M. R. (with contributions by A. Birnbaum). *Statistical theories of mental test scores.* Reading, Massachusetts: Addison-Wesley, 1968.

Luce, R. D. Semiorders and a theory of utility discrimination. *Econometrica*, 1956, **24**, 178–191.

Luce, R. D. On the possible psychophysical laws. *Psychol. Rev.*, 1959, **66**, 81–95.

Luce, R. D. Comments on Rozeboom's criticisms of "On the possible psychophysical laws." *Psychol. Rev.*, 1962, **69**, 548–551.

Luce, R. D. Detection and recognition. In R. D. Luce, R. R. Bush, & E. Galanter (Eds.), *Handbook of mathematical psychology.* Vol. 1. New York: Wiley, 1963. Pp. 103–189.

Luce, R. D. A generalization of a theorem of dimensional analysis. *J. Math. Psychol.*, 1964, **1**, 278–284.

Luce, R. D. A "fundamental" axiomatization of multiplicative power relations among three variables. *Phil. Sci.*, 1965, **32**, 301–309.

Luce, R. D. Two extensions of conjoint measurement. *J. Math. Psychol.*, 1966, **3**, 348–370.

Luce, R. D. Sufficient conditions for the existence of a finitely additive probability measure. *Ann. Math. Statist.*, 1967, **38**, 780–786.

Luce, R. D. On the numerical representation of qualitative conditional probability. *Ann. Math. Statist.*, 1968, **39**, 481–491.

Luce, R. D. Periodic extensive measurement. *Compositio Math.*, 1971, **23**, 189–198. (a)

Luce, R. D. Similar systems and dimensionally invariant laws. *Phil. Sci.*, 1971, **38**, 157–169. (b)

Luce, R. D., & Krantz, D. H. Conditional expected utility. *Econometrica*, 1971, **39**, 253–271.

Luce, R. D., & Marley, A. A. J. Extensive measurement when concatenation is restricted and maximal elements may exist. In S. Morgenbesser, P. Suppes, & M. G. White (Eds.), *Philosophy, science, and method: essays in honor of Ernest Nagel.* New York: St. Martin's Press, 1969. Pp. 235–249.

Luce, R. D., & Raiffa, H. *Games and decisions: introduction and critical survey.* New York: Wiley, 1957.

Luce, R. D., & Suppes, P. Preference, utility, and subjective probability. In R. D. Luce, R. R. Bush, & E. Galanter (Eds.), *Handbook of mathematical psychology.* Vol. 3. New York: Wiley, 1965. Pp. 249–410.

Luce, R. D., & Tukey, J. W. Simultaneous conjoint measurement: a new type of fundamental measurement. *J. Math. Psychol.*, 1964, **1**, 1–27.

MacCrimmon, K. R. Descriptive and normative implications of the decision theory postulates. In K. Borch and J. Mossin (Eds.), *Risk and uncertainty*. New York: Macmillan, 1968. Pp. 3–32.

MacLane, S., & Birkhoff, G. *Algebra*. New York: Macmillan, 1967.

Malinvaud, E. Note on von Neumann–Morgenstern's strong independence axiom. *Econometrica*, 1952, **20**, 679.

Manne, A. S. The strong independence assumption—gasoline blends and probability mixtures. *Econometrica*, 1952, **20**, 665–669.

Markowitz, H. M. *Portfolio selection*. New York: Wiley, 1959.

Marley, A. A. J. An alternative "fundamental" axiomatization of multiplicative power relations among three variables. *Phil. Sci.*, 1968, **35**, 185–186.

Marley, A. A. J. Additive conjoint measurement with respect to a pair of orderings. *Phil. Sci.*, 1970, **37**, 215–222.

Marschak, J. Rational behavior, uncertain prospects, and measurable utility. *Econometrica*, 1950, **18**, 111–141.

Marschak, J. The payoff-relevant description of states and acts. *Econometrica*, 1963, **31**, 719–725.

McKinsey, J. C. C., Sugar, A. C., & Suppes, P. Axiomatic foundations of classical particle mechanics. *J. Rational Mech. Anal.*, 1953, **2**, 253–272.

McKinsey, J. C. C., & Suppes, P. Transformations of systems of classical particle mechanics. *J. Rational Mech. Anal.*, 1953, **2**, 273–289.

McKinsey, J. C. C., & Suppes, P. On the notion of invariance in classical mechanics. *Brit. J. Phil. Sci.*, 1955, **5**, 290–302.

McKinsey, J. C. C., & Tarski, A. The algebra of topology. *Ann. of Math.* 1944, **45**, 141–191.

Nagel, E. Principles of the theory of probability. *International encyclopedia of unified science*. Vol. 1, Chicago: University of Chicago Press, 1939.

North, J. D. *The measure of the universe; a history of modern cosmology*. Oxford: Clarendon Press, 1965.

Osborne, D. K. Further extensions of a theorem of dimensional analysis. *J. Math. Psychol.*, 1970, **7**, 236–242.

Page, C. H. Physical entities and mathematical representation. *J. Res., Nat. Bur. Standards*, 1961, **65B**, 227–235.

Palacios, J. *Dimensional analysis*. Madrid: Espasa-Calpe, S. A., 1956. English translation by P. Lee & L. Roth, London: Macmillan, 1964.

Pavlik, W. B., & Reynolds, W. F. Effects of deprivation schedule and reward magnitude on acquisition and extinction performance. *J. Comp. Physiol. Psychol.*, 1963, **56**, 452–455.

Pfanzagl, J. *Die axiomatischen Grundlagen einer allgemeinen Theorie des Messens*. Schrift. Stat. Inst. Univ. Wien, Neue Folge Nr. 1. Würzburg: Physica-Verlag, 1959.

Pfanzagl, J. A general theory of measurement—applications to utility. *Naval Res. Logist. Quart.*, 1959, **6**, 283–294. (b)

Pfanzagl, J. Subjective probability derived from the Morgenstern-von Neumann utility concept. In M. Shubik (Ed.), *Essays in mathematical economics in honor of Oskar Morgenstern*. Princeton, New Jersey: Princeton University Press, 1967. Pp. 237–251. (a)

Pfanzagl, J. Characterizations of conditional expectations. *Ann. Math. Statist.*, 1967, **38**, 415–421. (b)

Pfanzagl, J. *Theory of measurement*. New York: Wiley, 1968.

Phelps Brown, E. H., Bernadelli, H., & Lange, O. Notes on the determinateness of the utility function. *Rev. Econ. Studies*, 1935, **2**, 66–77.

Pollatsek, A., & Tversky, A. A theory of risk. *J. Math. Psychol.*, 1970, **7**, 540–553.

Pratt, J. W. Risk aversion in the small and in the large. *Econometrica*, 1964, **32**, 122–136.

Pratt, J. W., Raiffa, H., & Schlaifer, R. The foundations of decisions under uncertainty: an elementary exposition. *J. Amer. Statist. Assoc.*, 1964, **59**, 353–375.

Quade, W. The algebraic structure of dimensional analysis. In F. J. deJong, *Dimensional analysis for economists*. Amsterdam: North Holland Publ., 1967. Pp. 143–199.

Radó, F. Eine Bedingung für die Regularität der Gewebe. *Mathematica (Cluj)*, 1960, **2 (25)**, 325–334.

Radó, F. Einbettung eines Halbgewebes in ein reguläres Gewebe und eines Halbgruppoids in eine Gruppe. *Math. Z.*, 1965, **89**, 395–410.

Raiffa, H. Risk, ambiguity, and the Savage axioms: comment. *Quart. J. Econ.*, 1961, **75**, 690–694.

Raiffa, H. *Decision analysis*. Reading, Massachusetts: Addison-Wesley, 1968.

Raiffa, H., & Schlaifer, R. *Applied statistical decision theory*. Boston, Massachusetts: Harvard University Press, 1961.

Ramsey, F. P. Truth and probability. In F. P. Ramsey, *The foundations of mathematics and other logical essays*. New York: Harcourt, Brace, 1931. Pp. 156–198. Reprinted in H. E. Kyburg, Jr., & H. E. Smokler (Eds.), *Studies in subjective probability*. New York: Wiley, 1964. Pp. 61–92.

Lord Rayleigh. The principle of similitude. *Nature*, 1915, **95**, 66–68, 202, 644.

Reese, T. W. The application of the theory of physical measurement to the measurement of psychological magnitudes, with three experimental examples. *Psychol. Monogr.*, 1943, **55** (Whole no. 251), 1–89.

Reichenbach, H. *Philosophic foundations of quantum mechanics*. Berkeley: University of California Press, 1944.

Rényi, A. On a new axiomatic theory of probability. *Acta Math. Acad. Sci. Hungar.*, 1955, **6**, 285–335.

Rényi, A. On measures of entropy and information. In J. Neyman (Ed.), *Proceedings of the fourth Berkeley symposium on mathematical statistics and probability*. Vol. 1. Berkeley: University of California Press, 1961. Pp. 547–561.

Rescher, N. Semantic foundations for the logic of preference. In N. Rescher (Ed.), *The logic of decision and action*. Pittsburgh: University of Pittsburgh Press, 1967. Pp. 37–79.

Richardson, L. F. Generalized foreign policy. *Brit. J. Psychol. Monogr. Suppl.*, No. 23, 1939.

Richter, M. K. Revealed preference theory. *Econometrica*, 1966, **34**, 635–645.

Roberts, F. S., & Luce, R. D. Axiomatic thermodynamics and extensive measurement. *Synthese*, 1968, **18**, 311–326.

Rosenberg, M. J. Cognitive structure and attitudinal affect. *J. Abn. Soc. Psychol.*, 1956, **53**, 367–372.

Roskies, R. A measurement axiomatization for an essentially multiplicative representation of two factors. *J. Math. Psychol.*, 1965, **2**, 266–276.

Rozeboom, W. W. The untenability of Luce's principle. *Psychol. Rev.*, 1962, **69**, 542–547.

Rubin, H., & Suppes, P. Transformations of systems of relativistic particle mechanics. *Pacific J. Math.*, 1954, **4**, 563–601.

Rudin, W. *Principles of mathematical analysis*. (2nd ed.) New York: McGraw-Hill, 1964.

Ruppel, G. Germany's approach to reconciling system usages. In C. F. Kayan (Ed.), *Systems of units: national and international aspects*. Washington: American Association of the Advancement of Sciences, 1959. Pp. 235–247.

Samuelson, P. A. A note on the pure theory of consumer's behavior. *Economica*, 1938, **5**, 61–71.

Samuelson, P. A. *Foundations of economic analysis*. Cambridge: Harvard University Press, 1947.

Samuelson, P. A., Probability, utility, and the independence axiom. *Econometrica*, 1952, **20**, 670–678.

Savage, L. J. *The foundations of statistics.* New York: Wiley, 1954.

Savage, L. J. The foundations of statistics reconsidered. J. Neyman (Ed.), *Proceedings of the fourth Berkeley symposium on mathematical statistics and probability.* Vol. 1. Berkeley: University of California Press, 1961. Pp. 575–586. Reprinted in H. E. Kyburg, Jr., & H. E. Smokler (Eds.), *Studies in subjective probability.* New York: Wiley, 1964. Pp. 173–188.

Scheffé, H. *The analysis of variance.* New York: Wiley, 1959.

Scott, D. Measurement models and linear inequalities. *J. Math. Psychol.,* 1964, **1,** 233–247.

Scott, D., & Suppes, P. Foundational aspects of theories of measurement. *J. Symbolic Logic,* 1958, **23,** 113–128.

Sedov, L. I. *Similarity and dimensional methods in mechanics.* Moscow, 1943, 1956. English translation of 1956 edition by M. Holt & M. Friedman, New York: Academic Press, 1959.

Seward, J. P., Shea, R. A., & Davenport, R. H. Further evidence for the interaction of drive and reward. *Amer. J. Psychol.,* 1960, **73,** 370–379.

Shepard, R. N. Metric structures in ordinal data. *J. Math. Psychol.,* 1966, **3,** 287–315.

Shuford, E. H., Jr., Albert, A., & Massengill, H. E. Admissible probability measurement procedures. *Psychometrika,* 1966, **31,** 125–145.

Sidowski, J. B., & Anderson, N. H. Judgments of city-occupation combinations. *Psychon. Sci.* 1967, **7,** 279–280.

Siegel, S. A method for obtaining an ordered metric scale. *Psychometrika,* 1956, **21,** 207–216.

Siegel, S. Theoretical models of choice and strategy behavior: stable-state behavior in the two-choice uncertain-outcome situation. *Psychometrika,* 1959, **24,** 303–316.

Siegel, S., Siegel, A. E., & Andrews, J. *Choice, strategy, and utility.* New York: McGraw-Hill, 1964.

Sneed, J. D. Strategy and the logic of decision. *Synthese,* 1966, **16,** 270–283.

Spearman, C. *The abilities of man.* New York: Macmillan, 1927.

Spence, K. W. *Behavior theory and conditioning.* New Haven: Yale University Press, 1956.

Stahl, W. R. Similarity and dimensional methods in biology. *Science,* 1962, **137,** 205–212.

Stevens, J. C., Mack, J. D., & Stevens, S. S. Growth of sensation on seven continua as measured by force of handgrip. *J. Exp. Psychol.,* 1960, **59,** 60–67.

Stevens, S. S. On the theory of scales of measurement. *Science,* 1946, **103,** 677–680.

Stevens, S. S. Mathematics, measurement and psychophysics. In S. S. Stevens (Ed.), *Handbook of experimental psychology.* New York: Wiley, 1951. Pp. 1–49.

Stevens, S. S. On the psychophysical law. *Psychol. Rev.,* 1957, **64,** 153–181.

Stevens, S. S. Measurement, psychophysics, and utility. In C. W. Churchman & P. Ratoosh (Eds.), *Measurement: definitions and theories.* New York: Wiley, 1959. Pp. 18–63. (a)

Stevens, S. S. Cross-modality validation of subjective scales for loudness, vibration, and electric shock. *J. Exp. Psychol.,* 1959, **57,** 201–209. (b)

Stevens, S. S. On the new psychophysics. *Scand. J. Psychol.,* 1960, **1,** 27–35.

Stevens, S. S. Matching functions between loudness and ten other continua. *Percept. & Psychophys.,* 1966, **1,** 5–8.

Stevens, S. S. Measurement, statistics, and the schemapiric view. *Science,* 1968, **161,** 849–856.

Suppes, P. A set of independent axioms for extensive quantities. *Portugal. Math.,* 1951, **10,** 163–172.

Suppes, P. The role of subjective probability and utility in decision-making. J. Neyman (Ed.), *Proceedings of the third Berkeley symposium on mathematical statistics and probability.* Vol. 5. Berkeley: University of California Press, 1956. Pp. 61–73.

Suppes, P. *Introduction to logic.* Princeton, New Jersey: van Nostrand, 1957.

Suppes, P. Probability concepts in quantum mechanics. *Phil. Sci.*, 1961, **28**, 378–389.

Suppes, P. The role of probability in quantum mechanics. In B. Baumrin (Ed.), *Philosophy of science, the Delaware seminar*. New York: Wiley, 1963. Pp. 319–337.

Suppes, P. Logics appropriate to empirical theories. In J. W. Addison, L. Henkin, & A. Tarski (Eds.), *The theory of models, proceedings of the 1963 international symposium at Berkeley*. Amsterdam: North-Holland Publ., 1965. Pp. 364–375.

Suppes, P. The probabilistic argument for a non-classical logic of quantum mechanics. *Phil. Sci.*, 1966, **33**, 14–21.

Suppes, P. *Studies in the methodology and foundations of science*. Dordrecht: Reidel, 1969.

Suppes, P., & Winet, M. An axiomatization of utility based on the notion of utility differences. *Management Sci.*, 1955, **1**, 259–270.

Suppes, P., & Zinnes, J. L. Basic measurement theory. In R. D. Luce, R. R. Bush, & E. Galanter (Eds.), *Handbook of mathematical psychology*. Vol. 1. New York: Wiley, 1963. Pp. 1–76.

Thun, R. E. On dimensional analysis. *IBM J. Res. Develop.*, 1960, **4**, 349–356.

Thurstone, L. L. *Multiple-factor analysis*. Chicago: University of Chicago Press, 1947.

Thurstone, L. L. *The measurement of values*. Chicago: University of Chicago Press, 1959.

Tobin, J. Liquidity preference as behavior towards risk. *Rev. Econ. Studies*, 1958, **25**, 65–86.

Toda, M. Measurement of subjective probability distributions. Report No. 3, Pennsylvania State University, Institute for Research, Division of Mathematical Psychology, 1963.

Toda, M., & Shuford, E. H., Jr. Utility, induced utility, and small worlds. *Behav. Sci.*, 1965, **10**, 238–254.

Tolman, R. C. The principle of similitude. *Phys. Rev.*, 1914, **3**, 244–255.

Torgerson, W. S. *Theory and method of scaling*. New York: Wiley, 1958.

Torgerson, W. S. Distances and ratios in psychological scaling. *Acta Psychol.*, 1961, **19**, 201–205.

Tversky, A. Finite additive structures. Technical Report MMPP 64-6, University of Michigan, Michigan Mathematical Psychology Program, 1964.

Tversky, A. A general theory of polynomial conjoint measurement. *J. Math. Psychol.*, 1967, **4**, 1–20. (a)

Tversky, A. Additivity, utility, and subjective probability. *J. Math. Psychol.*, 1967, **4**, 175–201. (b)

Tversky, A. Utility theory and additivity analysis of risky choices. *J. Exp. Psychol.*, 1967, **75**, 27–36. (c)

Tversky, A. Intransitivity of preferences. *Psychol. Rev.*, 1969, **76**, 31–48.

Tversky, A., & Krantz, D. H. The dimensional representation and the metric structure of similarity data. *J. Math. Psychol.*, 1970, **7**, 572–596.

Tversky, A., & Russo, J. E. Substitutability and similarity in binary choices. *J. Math. Psychol.*, 1969, **6**, 1–12.

Tversky, A., & Zivian, A. A computer program for additivity analysis. *Behav. Sci.*, 1966, **11**, 78–79.

Uzawa, H. Preference and rational choice in the theory of consumption. In K. J. Arrow, S. Karlin, & P. Suppes (Eds.), *Mathematical methods in the social sciences, 1959*. Stanford: Stanford University Press, 1960. Pp. 129–148.

Varadarajan, V. S. Probability in physics and a theorem on simultaneous observability. *Commun. Pure Appl. Math.*, 1962, **15**, 189–217.

Villegas, C. On qualitative probability σ-algebras. *Ann. Math. Statist.*, 1964, **35**, 1787–1796.

Villegas, C. On qualitative probability. *Amer. Math. Monthly*, 1967, **74**, 661–669.

Vinogradov, A. A. Ordered algebraic systems. In R. V. Gamkrelidze (Ed.), *Progress in mathematics*. Vol. 5. New York: Plenum Press, 1969. Pp. 77–126.

von Neumann, J., & Morgenstern, O. *Theory of games and economic behavior*. Princeton: Princeton University Press, 1944, 1947, 1953.

von Wright, G. H. *The logic of preference, an essay*. Edinburgh, Scotland: Edinburgh University Press, 1963.

Wever, E. G. *Theory of hearing*. New York: Wiley, 1949.

Whitney, H. The mathematics of physical quantities. Part I: Mathematical models for measurement. Part II: Quantity structures and dimensional analysis. *Amer. Math. Monthly*, 1968, **75**, 115–138, 227–256.

Williams, A. C., & Nassar, J. I. Financial measurement of capital investments. *Management Sci.*, 1966, **12**, 851–864.

Winkler, R. L. Scoring rules and the evaluation of probability assessors. *J. Amer. Statist. Assoc.*, 1969, **64**, 1073–1078.

Winkler, R. L., & Murphy, A. H. "Good" probability assessors. *J. Appl. Meteorol.*, 1968, **7**, 751–758.

Wold, H. Ordinal preferences or cardinal utility? *Econometrica*, 1952, **20**, 661–664.

Zassenhaus, H. What is an angle? *Amer. Math. Monthly*, 1954, **61**, 369–378.

## SUPPLEMENTARY REFERENCES FOR VOLUME I

The following articles and books are directly relevant to the issues of Volume I but either were overlooked or appeared after 1971. It is a selected list of references, including ones that seem directly relevant to the topics of Volume I. Any entry that might otherwise have been included has not been if it is explicitly treated in Volume II or III. Wherever possible we have included general survey articles that give fuller reference lists. In particular, for Chapter 8 we cover only work specifically concerned with conditional expected utility plus several general surveys and books on the entire area of decision making under risk and uncertainty.

## Chapter 1

Adams, E. W., & Carlstrom, I. F. (1979). Representing approximate ordering and equivalence relations. *J. Math. Psychol.*, **19**, 182–207.

Domotor, Z. (1972). Species of measurement structures. *Theoria*, **38**, 64–81.

Kanger, S. (1972). Measurement: An essay in philosophy of science. *Theoria*, **38**, 1–44.

Krantz, D. H. (1974). Measurement theory and qualitative laws in psychophysics. In D. H. Krantz, R. C. Atkinson, R. D. Luce, & P. Suppes (Eds.), *Contemporary Developments in Mathematical Psychology, Vol. II*. San Francisco: Freeman, pp. 160–199.

Kyburg, H. E., Jr. (1984). *Theory and Measurement*. Cambridge: Cambridge University Press.

Luce, R. D. (1979). Suppes' contribution to the theory of measurement. In R. J. Bogdan (Ed.), *Patrick Suppes*. Dordrecht: Reidel, pp. 93–110.

Luce, R. D., & Narens, L. (1987). Measurement, theory of. In J. Eatwell, M. Milgate, & P. Newman (Eds.), *The New Palgrave: A Dictionary of Economic Theory and Doctrine, Vol. 3*. New York: Macmillan, pp. 428–432.

Luce, R. D., & Suppes, P. (1974). Measurement, theory of. *Encyclopedia Britiannica, 15th Edition*, pp. 739–745.

Ramsey, J. O. (1975). Review of Krantz *et al.*, Foundations of Measurement, Vol. I. *Psychometrika*, **40**, 257–262.

Roberts, F. S. (1979). *Measurement Theory, with Applications to Utility, Decision Making, and the Social Sciences*. Reading, Mass.: Addison-Wesley.

Suppes, P. (1972). Finite equal-interval measurement structures. *Theoria*, **38**, 45–63.

## Chapter 2

Falmagne, J.-C. (1971). Bounded versions of Hölder's theorem with application to extensive measurement. *J. Math. Psychol.*, **8**, 495–507.

Falmagne, J.-C. (1973). Sufficient conditions for an ordered-ring isomorphism onto a positive subinterval: A lemma for polynomial measurement. *J. Math. Psychol.*, **10**, 290–295.

Falmagne, J.-C. (1975). A set of independent axioms for positive Hölder systems. *Phil. Sci.*, **42**, 137–151.

Falmagne, J.-C. (1981). On a recurrent misuse of a classical functional equation result. *J. Math. Psychol.*, **23**, 190–194.

## Chapter 3

Coombs, C. H. (1975). Portfolio theory and the measurement of risk. In M. F. Kaplan & S. Schwartz (Eds.), *Human Judgment and Decision Processes*. New York: Academic Press, pp. 63–85.

Coombs, C. H., & Bowen, J. N. (1971a). Additivity of risk in portfolios. *Perception & Psychophysics*, **10**, 43–46.

Coombs, C. H., & Bowen, J. N. (1971b). A test of VE-theories of risk and the effect of the central limit theorem. *Acta Psychol.*, **35**, 15–28.

Domotor, Z. (1970). Qualitative information and entropy structures. In J. Hintikka & P. Suppes (Eds.), *Information and Inference*. Dordrecht, Holland: Reidel, pp. 148–194.

Fishburn, P. C. (1982). Foundations of risk measurement. II. Effects of gains on risk. *J. Math. Psychol.*, **25**, 226–242.

Fishburn, P. C. (1984). Foundations of risk measurement. I. Risk as probable loss. *Manag. Sci.*, **30**, 396–406.

Landsberg, P. T. (1970). Main ideas in the axiomatics of thermodynamics. *Pure Appl. Chem.*, **22**, 215–227.

Luce, R. D., & Narens, L. (1976). A qualitative equivalent to the relativistic addition law for velocities. *Synthese*, **33**, 483–487.

Norman, R. Z., & Roberts, F. S. (1972). A derivation of a measure of relative balance for social structures and a characterization of extensive ratio systems. *J. Math. Psychol.*, **9**, 66–91.

Segre, B. (1973). Some arithmetical problems on the use of the balance. *Lincei—Rend. Sc. Fis. Mat. e Nat.*, **54**, 912–923.

## Chapter 4

Birnbaum, M. H. (1980). Comparison of two theories of "ratio" and "difference" judgments. *J. Exp. Psychol.: General*, **109**, 304–319.

Doignon, J.-P., & Falmagne, J.-C. (1974). Difference measurement and simple scalability with restricted solvability. *J. Math. Psychol.*, **11**, 473–499.

Falmagne, J.-C. (1974). Difference measurement and simple scalability with restricted solvability. *J. Math. Psychol.*, **11**, 473–499.

Katz, M. (1981). Difference measurement spaces. *J. Math. Psychol.*, **23**, 195–213.

Lehner, P. E. (1981). Folded additive structures: A nonpolynomial model of some factorial interactions. *J. Math. Psychol.*, **23**, 99–114.

Miyamoto, J. M. (1983). An axiomatization of the ratio/difference representation. *J. Math. Psychol.*, **27**, 439–455.

Rule, S. J., Curtis, D. W., & Mullin, L. C. (1981). Subjective ratios and differences in perceived heaviness. *J. Exp. Psychol.: Human Perception and Performance*, 7, 459–466.

Schneider, B., Parker, S., & Stein, D. (1974). The measurement of loudness using direct comparisons of sensory intervals. *J. Math. Psychol.*, 11, 259–273.

Zachow, E.-W. (1978). Positive-difference structures and bilinear utility functions. *J. Math. Psychol.*, 17, 152–164.

## Chapter 5

Chateauneuf, A. (1985). On the existence of a probability measure compatible with a total preorder on a Boolean algebra. *J. Math. Econ.*, 14, 43–52.

Chateauneuf, A., & Jaffray, J. Y. (1984). Archimedean qualitative probabilities. *J. Math. Psychol.*, 28, 191–204.

Chuaqui, R., & Malitz, J. (1983). Preorderings compatible with probability measures. *Trans. Amer. Math. Soc.*, 279, 811–824.

Coletti, G., & Regoli, G. (1983). Probabilità qualitativa nonarchimedea e realizzabilita. *Riv. Mat. Sc. Econ. e Soc.*, 6, 79–99.

Coletti, G., & Regoli, G. (1986). Realizzazioni di probabilità qualitative. *Bolletino della Unione Matematica Funzionale e Applicazioni, Serie VI*, V-C, 95–111.

deFinetti, B. (1970). Logical foundations and measurement of subjective probability. *Acta Psychol.*, 34, 129–145.

Fine, T. L. (1977). An argument for comparative probability. In J. Hintikka (Eds.), *Basic Problems in Methodology and Linguistics*. Dordrecht, Holland: Reidel, pp. 105–119.

Fishburn, P. C. (1975). Weak comparative probability on infinite sets. *The Annals of Probability*, 3, 889–893.

Fishburn, P. C. (1983). A generalization of comparative probability on finite sets. *J. Math. Psychol.*, 27, 298–310.

Fishburn, P. C. (1986a). Interval models for comparative probability on finite sets. *J. Math. Psychol.*, 30, 221–242.

Fishburn, P. C. (1986b). The axioms of subjective probability. *Stat. Sci.*, 1, 347–350.

French, S. (1982). On the axiomatization of subjective probabilities. *Theory and Decision*, 14, 19–34.

Göttinger, H. W. (1974). Subjective qualitative information structures based on orderings. *Theory and Decision*, 5, 69–97.

Kahneman, D., Slovic, P., & Tversky, A. (Eds.) (1982). *Judgment under Uncertainty: Heuristics and Biases*. New York: Cambridge University Press.

Kaplan, M., & Fine, T. L. (1977). Joint orders in comparative probability. *The Annals of Probability*, 5, 161–179.

Luce, R. D., & Narens, L. (1978). Qualitative independence in probability theory. *Theory and Decision*, 9, 225–239.

Niiniluoto, I. (1972). A note on fine and tight qualitative probabilities. *Ann. Math. Stat.*, 43, 1581–1591.

Ochs, W. (1985). Gleason measures and quantum comparative probability. In L. Acardi & W. von Waldenfels (Eds.), *Quantum Probability and Applications*. Berlin: Springer, pp. 388–396.

Roberts, F. S. (1973). A note on Fine's axioms for qualitative probability. *The Annals of Probability*, 1, 484–487.

Savage, L. J. (1971). Elicitation of personal probabilities and expectations. *J. Amer. Stat. Assoc.*, 66, 783–801.

Suppes, P. (1973). New foundations of objective probability: Axioms for propensities. In P. Suppes, L. Henkin, G. C. Noisil, & A. Joja (Eds.), *Logic, Methodology and Philosophy of*

*Science IV: Proceedings of the Fourth International Congress for Logic, Methodology and Philosophy of Science, Bucharest, 1971.* Amsterdam: North-Holland, pp. 515–529.

Suppes, P. (1974). The measurement of belief. *Journal of the Royal Statistical Society (Series B)*, **36**, 160–191.

Suppes, P. (1987). Propensity representations of probability. *Erkenntnis*, **26**, 335–358.

Suppes, P., & Zanotti, M. (1976). Necessary and sufficient conditions for existence of a unique measure strictly agreeing with a qualitative probability ordering. *J. Philosophical Logic*, **5**, 431–438.

Suppes, P., & Zanotti, M. (1982). Necessary and sufficient qualitative axioms for conditional probability. *Zeitschrift Wahrscheinlichkeitstheorie verw. Gebiete*, **60**, 163–169.

Wakker, R. (1981). Agreeing probability measures for comparative probability structures. *Ann. Statistics*, **9**, 658–662.

Wallsten, T. S. (1974). The psychological concept of subjective probability: A measurement theoretical view. In C.A.S. Staël von Holstein (Eds.), *The Concept of Probability in Psychological Experiments*. Dordrecht, The Netherlands: Reidel, pp. 49–72.

## Chapter 6

Adams, E. W., & Fagot, R. F. (1975). On the theory of biased bisection operations and their inverses. *J. Math. Psychol.*, **12**, 35–52.

Dyer, J. S., & Sarin, R. K. (1978). On the relationship between additive conjoint and difference measurement. *J. Math. Psychol.*, **18**, 270–272.

Fagot, R. F., & Stewart, M. R. (1969). Tests of product and additive scaling axioms. *Perception & Psychophysics*, **5**, 117–123.

Fagot, R. F., & Stewart, M. R. (1970). Test of a response bias model of bisection. *Perception & Psychophysics*, **7**, 257–262.

Fishburn, P. C. (1971). Additive representations of real-valued functions on subsets of product sets. *J. Math. Psychol.*, **8**, 382–388.

Irtel, H. (1987). A conjoint Grassmann structure for testing the additivity of binocular color mixtures. *J. Math. Psychol.*, **31**, 192–202.

Jaffray, J. Y. (1974). On the extension of additive utilities to infinite sets. *J. Math. Psychol.*, **11**, 431–452.

Keeney, R. L., & Raiffa, H. (Eds.) (1977). *Conflicting Objectives in Decisions*. New York: Wiley, pp. 148–171.

Krantz, D. H. (1973). Measurement-free tests of linearity in biological systems. *IEEE Transactions on Systems, Man, and Cybernetics*, **3**, 266–271.

Levine, M. V. (1975). Additive measurement with short segments of curves. *J. Math. Psychol.*, **12**, 212–224.

Luce, R. D. (1977). A note on sums of power functions. *J. Math. Psychol.*, **16**, 91–94.

Luce, R. D. (1978). Conjoint measurement: A brief survey. In Hooker, Leach & McClennan (Eds.), *Foundations and Applications of Decision Theory*. Dordrecht: Reidel, pp. 311–336. Also in R. L. Keeney & H. Raiffa (Eds.), *Conflicting Objectives in Decisions: Preferences and Value Tradeoffs*. New York: Wiley, pp. 148–171.

Narens, L. (1974). Minimal conditions for additive conjoint measurement and qualitative probability. *J. Math. Psychol.*, **11**, 404–430.

Ramsey, J. O. (1976). Algebraic representation in the physical and behavioral sciences. *Synthese*, **33**, 417–453.

Suppes, P. (1980). Limitations of the axiomatic method in ancient Greek mathematical science. In J. Hintikka, D. Gruerder, & E. Agazzi (Eds.), *Pisa Conference Proceedings, Vol. 1*. Dordrecht: Reidel, pp. 197–213.

Taylor, M. A. (1971). Classical, cartesian, and solution nets. *Mathematica*, **13**, 151–166.

Taylor, M. A. (1972). Relational systems with a Thomsen or Reidemeister cancellation condition. *J. Math. Psychol.*, **9**, 456–458.

Taylor, M. A. (1973). Cartesian nets and groupoids. *Canadian Mathematical Bulletin*, **16**, 347–362.

von Winterfeldt, D., & Edwards, W. (1986). *Decision Analysis and Behavioral Research.* Cambridge: Cambridge University Press.

Wakker, P. (1984). Cardinal coordinate independence for expected utility. *J. Math. Psychol.*, **28**, 110–117.

Wallsten, T. S. (1972). A conjoint-measurement framework for the study of probabilistic information processing. *Psychol. Rev.*, **79**, 245–260.

Wallsten, T. S. (1976). Using conjoint-measurement models to investigate a theory about probabilistic information processing. *J. Math. Psychol.*, **14**, 144–186.

## Chapter 7

Bell, D. E. (1987). Multilinear representations for ordinal utility functions. *J. Math. Psychol.*, **31**, 44–59.

Coombs, C. H., & Huang, L. C. (1970). Polynomial psychophysics of risk. *J. Math. Psychol.*, **7**, 317–338.

Coombs, C. H., & Lehner, P. E. (1984). Conjoint design and analysis of the bilinear model: An application to judgments of risk. *J. Math. Psychol.*, **28**, 1–42.

Farquhar, P. H. (1977). A survey of multiattribute utility theory and applications. In M. K. Starr & M. Zeleny (Eds.), *Multiple Criteria Decision Making.* Amsterdam: North Holland, pp. 59–89.

Fishburn, P. C. (1978). A survey of multiattribute/multicriterion evaluation theories. In S. Zionts (Ed.), *Multiple Criteria Problem Solving.* New York: Springer-Verlag, pp. 181–224.

Fishburn, P. C., & Keeney, R. L. (1974). Seven independence concepts and continuous multiattribute utility functions. *J. Math. Psychol.*, **11**, 294–327.

Huang, L. C. (1975). A nonsimple conjoint measurement model. *J. Math. Psychol.*, **12**, 437–448.

Miyamoto, J. M. (1983). Measurement foundations for multiattribute psychophysical theories based on first order polynomials. *J. Math. Psychol.*, **27**, 152–182.

Richter, M. K. (1975). Rational choice and polynomial measurement models. *J. Math. Psychol.*, **12**, 99–113.

Young, F. W. (1972). A model for polynomial conjoint analysis algorithms. In R. N. Shepard, A. K. Romney, & S. Nerlove (Eds.), *Multidimensional Scaling: Theory and Application in the Behavioral Sciences, Vol. I*, pp. 69–104.

## Chapter 8

Allais, M., & Hagen, O. (Eds.) (1979). *Expected Utility Hypotheses and the Allais Paradox.* Dordrecht: Reidel.

Balch, M. S., McFadden, D. L., & Wu, S. Y. (Eds.) (1974). *Essays on Economic Behavior under Uncertainty.* Amsterdam: North-Holland.

Bell, D. E., Keeney, R. L., & Raiffa, H. (Eds.) (1977). *Conflicting Objectives in Decisions.* New York: Wiley.

Domotor, Z. (1978). Axiomatization of Jeffrey utilities. *Synthese*, **39**, 165–210.

Farquhar, P. H., & Rao, V. R. (1976). A balance model for evaluating subsets of multiattributed items. *Manag. Sci.*, **22**, 528–539.

Fishburn, P. C. (1972a). *The Mathematics of Decision Making.* The Hague: Mouton.

Fishburn, P. C. (1972b). Subjective expected utility with mixture sets and Boolean algebras. *Ann. Math. Stat.*, **43**, 917–929.

Fishburn, P. C. (1973). A mixture-set axiomatization of conditional subjective expected utility. *Econometrica*, **41**, 1–25.

Fishburn, P. C. (1974). Lexicographic orders, utilities and decision rules: A survey. *Manag. Sci.*, **20**, 1442–1471.

Fishburn, P. C. (1978). Value theory. In J. J. Moder & S. E. Elmaghraby (Eds.), *Handbook of Operations Research*. New York: Van Nostrand Reinhold, pp. 398–422.

Fishburn, P. C. (1982). *The Foundations of Expected Utility*. Dordrecht: Reidel.

Gibbard, A., & Harper, W. L. (1985). Counterfactuals and two kinds of expected utility. In R. Campbell & L. Sowden (Eds.), *Paradoxes of Rationality and Cooperation*. Vancouver: University of British Columbia Press, pp. 133–158.

Hogarth, R. M., & Reder, M. W. (Eds.) (1987). *Rational Choice*. Chicago: University of Chicago Press.

Jeffery, R. C. (1983). *The Logic of Decision*. 2nd Ed. Chicago: University of Chicago Press.

Keeney, R. L., & Raiffa, H. (1976). *Decisions with Multiple Objectives: Preferences and Value Tradeoff*. New York: Wiley.

Luce, R. D. (1972). Conditional expected, extensive utility. *Theory and Decision*, **3**, 101–106.

MacCrimmon, K. R., & Larsson, S. (1979). Utility theory: Axioms versus "paradoxes." In M. Allais & O. Hagen (Eds.), *Expected Utility Hypotheses and the Allais Paradox*. Dordrecht, Holland: Reidel, pp. 333–409.

Machina, M. (1987). Choice under uncertainty: Problems solved and unsolved. *Economic Perspectives*, **1**, 121–154.

Pollak, R. A. (1967). Additive von Neumann-Morgenstern utility functions. *Econometrica*, **35**, 485–494.

Roberts, F. S. (1972). What if utility functions do not exist? *Theory and Decision*, **3**, 126–139.

Skala, H. J. (1975). *Non-Archimedean Utility Theory*. Dordrecht, Holland: Reidel.

Slovic, P., Lichtenstein, S., & Fischhoff, B. (1988). Decision making. In R. C. Atkinson, R. J. Herrnstein, G. Lindzey, & R. D. Luce (Eds.), *Stevens' Handbook of Experimental Psychology, Vol. 2*. New York: Wiley, pp. 673–738.

Spohn, W. (1977). Where Luce and Krantz do really generalize Savage's decision model. *Erkenntnis*, **11**, 113–134.

Stigum, B. P., & Wenstop, F. (Eds.) (1983). *Foundations of Utility and Risk Theory with Applications*. Dordrecht: Reidel.

## Chapter 9

Arbuckle, J., & Larimer, J. (1976). The number of two-way tables satisfying certain additivity axioms. *J. Math. Psychol.*, **13**, 89–100.

de Leeuw, J., Young, F. W., & Takane, Y. (1976). Additive structure in qualitative data: An alternating least squares method with optimal scaling features. *Psychometrika*, **41**, 471–504.

Fishburn, P. C. (1972). Interdependent preferences on finite sets. *J. Math. Psychol.*, **9**, 225–236.

Fishburn, P. C. (1975). Finite additive choice structures. *SIAM J. Appl. Math.*, 29, 263–272.

Sherman, B. F. (1977). An algorithm for finite conjoint additivity. *J. Math. Psychol.*, **16**, 204–218.

Suppes, P. (1975). Approximate probability and expectation of gambles. *Erkenntnis*, **9**, 153–161.

## Chapter 10

Aczél, J., Djokovic, D. Z., & Pfanzagl, J. (1970). On the uniqueness of scales derived from canonical representations. *Metrika*, **16**, 1–8.

Delvendahl, O. (1962–1963). Über die algebraische Struktur des Gr össenkalküls der Physik. *Der Mathematisch-Naturwissenschaftliche Unterricht*, **15**, 446–451.

FitzHugh, R. (1973). Dimensional analysis of nerve models. *J. Theor. Biol.*, **40**, 517–541.

Fleischmann, R. (1959/1960). Einheiteninvariante Gr össengleichungen und Dimensionen. *Der Mathematisch-Naturwissenschaftliche Unterricht*, **12**, 385–443.

Griesel, H. (1969). Algebra und Analysis der Grössensysteme I, II. *Mathematisch Physikalische Semesterberichte*, **16**, 56–93, 189–224.

Narens, L., & Luce, R. D. (1987). Meaningfulness and invariance. In J. Eatwell, M. Milgate, & P. Newman (Eds.), *The New Palgrave: A Dictionary of Economic Theory and Doctrine. Vol. 3.* New York: Macmillan, pp. 417–421.

Roberts, F. S. (1974). Laws of exchange and their applications. *SIAM J. Appl. Math.*, **26**, 260–284.

Roberts, F. S. (1985). Applications of the theory of meaningfulness to psychology. *J. Math. Psychol.*, **29**, 311–332.

Schepartz, B. (1980). *Dimensional Analysis in the Biomedical Sciences.* Springfield, Ill.: Charles C Thomas.

Zebrowski, E., Jr. (1979). *Fundamentals of Physical Measurement.* North Scituate, Mass.: Duxbury Press.

# Author Index

# Subject Index

## S

s-derived attributes, 502
Scalar product, 59
Scale, 283
  interval, 10–11, 516
  log-interval, 10–11, 484
  ordered metric, 431
  ordinal, 11, 33, 38, 43
  rating, 140
  ratio, 10–11, 11n, 85, 516
Scaling, methods of, 33
SCPM-function, 530
Second law of thermodynamics, 111
Secondary attribute, 502, 503
Secondary dimension, 455, 503
Semigroups, ordered local, 44
Semiring, 225
  ordered local, 55
Sensation differences and ratios, 139, 154
Sequence, standard, see Standard sequence
Series, standard, 105, see also Standard sequence
Sets, countable, 40
Sextuple condition, 145
Shock, aversiveness of, 268
σ-algebra, 199
Sign dependent, 329–330
Sign-reversal axiom, 150
Similar sets, 508
Similarity, 465
Similarity group, 516
Similarity transformations, 10, 11n
Similitude, law of, 486, 500
Simple order, 14
Simple polynomial, 328
Simple scalability, 367
Solvability, 83, 146, 165
  restricted, 256, 301, 380
  strong, 276
  unrestricted, 256, 347
Span, 463
Speed, 534
Spring constant, 484, 507
Stability group, 509
Standard sequence, 4, 25, 84, 104, 105, 147, 151, 172, 204, 223, 253, 294, 348, 349, 381
  approximate, 155
  strictly bounded, 84

Standard series, 105, see also Standard sequence
Standard weights, 106
States
  antiequilibrium, 114
  of nature, 412
  of physical system, 111
Stationary relation, 304
Statistical decision theory, 412
Statistical mechanics, entropy of, 113, 114
Strictly bounded, 25, 84
Strong conditional connectedness, 178, 183
Strong solvability, 276
Structure, 8–9
  bisymmetric, 294
  conditional decision, 380
  conjoint, 19, 256, 301
  cross-modality, 165
  decomposable, 247, 317
  difference
    absolute, 172, 491
    algebraic, 151, 168, 491
    positive, 147
    strongly conditional, 183
  empirical relational, 8–9, 11n, 12–13, 33–35
  entropy, 114
  extensive, 92
    closed, 73
    multiples, 103
    periodic, 76
    with no essential maximum, 84
  finite, 15, 19, 168, 217, 297, 414, 430, 433
  flat, 486
  of particle mechanics, 525, 532
  of physical quantities, 461, 501
  of qualitative probability, 204
    conditional, 223
    with equivalent atoms, 216
    with independence, 240
  relational, 8, 12, 14–15, 17–18
    numerical, 8–9, 11n, 12–14, 33
  risk, 125, 127
  symmetric, 256, 486
  tall, 486
  ultraclassical, 528
Structural axioms, 23
Subjective, see Intensity, Probability, Utility
Substitutability, 318

# A CATALOG OF SELECTED
# DOVER BOOKS
## IN SCIENCE AND MATHEMATICS

# Astronomy

CHARIOTS FOR APOLLO: The NASA History of Manned Lunar Spacecraft to 1969, Courtney G. Brooks, James M. Grimwood, and Loyd S. Swenson, Jr. This illustrated history by a trio of experts is the definitive reference on the Apollo spacecraft and lunar modules. It traces the vehicles' design, development, and operation in space. More than 100 photographs and illustrations. 576pp. 6 3/4 x 9 1/4.                    0-486-46756-2

EXPLORING THE MOON THROUGH BINOCULARS AND SMALL TELESCOPES, Ernest H. Cherrington, Jr. Informative, profusely illustrated guide to locating and identifying craters, rills, seas, mountains, other lunar features. Newly revised and updated with special section of new photos. Over 100 photos and diagrams. 240pp. 8 1/4 x 11.                    0-486-24491-1

WHERE NO MAN HAS GONE BEFORE: A History of NASA's Apollo Lunar Expeditions, William David Compton. Introduction by Paul Dickson. This official NASA history traces behind-the-scenes conflicts and cooperation between scientists and engineers. The first half concerns preparations for the Moon landings, and the second half documents the flights that followed Apollo 11. 1989 edition. 432pp. 7 x 10.
0-486-47888-2

APOLLO EXPEDITIONS TO THE MOON: The NASA History, Edited by Edgar M. Cortright. Official NASA publication marks the 40th anniversary of the first lunar landing and features essays by project participants recalling engineering and administrative challenges. Accessible, jargon-free accounts, highlighted by numerous illustrations. 336pp. 8 3/8 x 10 7/8.                    0-486-47175-6

ON MARS: Exploration of the Red Planet, 1958-1978--The NASA History, Edward Clinton Ezell and Linda Neuman Ezell. NASA's official history chronicles the start of our explorations of our planetary neighbor. It recounts cooperation among government, industry, and academia, and it features dozens of photos from Viking cameras. 560pp. 6 3/4 x 9 1/4.                    0-486-46757-0

ARISTARCHUS OF SAMOS: The Ancient Copernicus, Sir Thomas Heath. Heath's history of astronomy ranges from Homer and Hesiod to Aristarchus and includes quotes from numerous thinkers, compilers, and scholasticists from Thales and Anaximander through Pythagoras, Plato, Aristotle, and Heraclides. 34 figures. 448pp. 5 3/8 x 8 1/2.
0-486-43886-4

AN INTRODUCTION TO CELESTIAL MECHANICS, Forest Ray Moulton. Classic text still unsurpassed in presentation of fundamental principles. Covers rectilinear motion, central forces, problems of two and three bodies, much more. Includes over 200 problems, some with answers. 437pp. 5 3/8 x 8 1/2.                    0-486-64687-4

BEYOND THE ATMOSPHERE: Early Years of Space Science, Homer E. Newell. This exciting survey is the work of a top NASA administrator who chronicles technological advances, the relationship of space science to general science, and the space program's social, political, and economic contexts. 528pp. 6 3/4 x 9 1/4.
0-486-47464-X

STAR LORE: Myths, Legends, and Facts, William Tyler Olcott. Captivating retellings of the origins and histories of ancient star groups include Pegasus, Ursa Major, Pleiades, signs of the zodiac, and other constellations. "Classic." -- *Sky & Telescope*. 58 illustrations. 544pp. 5 3/8 x 8 1/2.                    0-486-43581-4

A COMPLETE MANUAL OF AMATEUR ASTRONOMY: Tools and Techniques for Astronomical Observations, P. Clay Sherrod with Thomas L. Koed. Concise, highly readable book discusses the selection, set-up, and maintenance of a telescope; amateur studies of the sun; lunar topography and occultations; and more. 124 figures. 26 halftones. 37 tables. 335pp. 6 1/2 x 9 1/4.                    0-486-42820-6

# Chemistry

MOLECULAR COLLISION THEORY, M. S. Child. This high-level monograph offers an analytical treatment of classical scattering by a central force, quantum scattering by a central force, elastic scattering phase shifts, and semi-classical elastic scattering. 1974 edition. 310pp. 5 3/8 x 8 1/2. 0-486-69437-2

HANDBOOK OF COMPUTATIONAL QUANTUM CHEMISTRY, David B. Cook. This comprehensive text provides upper-level undergraduates and graduate students with an accessible introduction to the implementation of quantum ideas in molecular modeling, exploring practical applications alongside theoretical explanations. 1998 edition. 832pp. 5 3/8 x 8 1/2. 0-486-44307-8

RADIOACTIVE SUBSTANCES, Marie Curie. The celebrated scientist's thesis, which directly preceded her 1903 Nobel Prize, discusses establishing atomic character of radioactivity; extraction from pitchblende of polonium and radium; isolation of pure radium chloride; more. 96pp. 5 3/8 x 8 1/2. 0-486-42550-9

CHEMICAL MAGIC, Leonard A. Ford. Classic guide provides intriguing entertainment while elucidating sound scientific principles, with more than 100 unusual stunts: cold fire, dust explosions, a nylon rope trick, a disappearing beaker, much more. 128pp. 5 3/8 x 8 1/2. 0-486-67628-5

ALCHEMY, E. J. Holmyard. Classic study by noted authority covers 2,000 years of alchemical history: religious, mystical overtones; apparatus; signs, symbols, and secret terms; advent of scientific method, much more. Illustrated. 320pp. 5 3/8 x 8 1/2. 0-486-26298-7

CHEMICAL KINETICS AND REACTION DYNAMICS, Paul L. Houston. This text teaches the principles underlying modern chemical kinetics in a clear, direct fashion, using several examples to enhance basic understanding. Solutions to selected problems. 2001 edition. 352pp. 8 3/8 x 11. 0-486-45334-0

PROBLEMS AND SOLUTIONS IN QUANTUM CHEMISTRY AND PHYSICS, Charles S. Johnson and Lee G. Pedersen. Unusually varied problems, with detailed solutions, cover of quantum mechanics, wave mechanics, angular momentum, molecular spectroscopy, scattering theory, more. 280 problems, plus 139 supplementary exercises. 430pp. 6 1/2 x 9 1/4. 0-486-65236-X

ELEMENTS OF CHEMISTRY, Antoine Lavoisier. Monumental classic by the founder of modern chemistry features first explicit statement of law of conservation of matter in chemical change, and more. Facsimile reprint of original (1790) Kerr translation. 539pp. 5 3/8 x 8 1/2. 0-486-64624-6

MAGNETISM AND TRANSITION METAL COMPLEXES, F. E. Mabbs and D. J. Machin. A detailed view of the calculation methods involved in the magnetic properties of transition metal complexes, this volume offers sufficient background for original work in the field. 1973 edition. 240pp. 5 3/8 x 8 1/2. 0-486-46284-6

GENERAL CHEMISTRY, Linus Pauling. Revised third edition of classic first-year text by Nobel laureate. Atomic and molecular structure, quantum mechanics, statistical mechanics, thermodynamics correlated with descriptive chemistry. Problems. 992pp. 5 3/8 x 8 1/2. 0-486-65622-5

ELECTROLYTE SOLUTIONS: Second Revised Edition, R. A. Robinson and R. H. Stokes. Classic text deals primarily with measurement, interpretation of conductance, chemical potential, and diffusion in electrolyte solutions. Detailed theoretical interpretations, plus extensive tables of thermodynamic and transport properties. 1970 edition. 590pp. 5 3/8 x 8 1/2. 0-486-42225-9

# Engineering

FUNDAMENTALS OF ASTRODYNAMICS, Roger R. Bate, Donald D. Mueller, and Jerry E. White. Teaching text developed by U.S. Air Force Academy develops the basic two-body and n-body equations of motion; orbit determination; classical orbital elements, coordinate transformations; differential correction; more. 1971 edition. 455pp. 5 3/8 x 8 1/2. 0-486-60061-0

INTRODUCTION TO CONTINUUM MECHANICS FOR ENGINEERS: Revised Edition, Ray M. Bowen. This self-contained text introduces classical continuum models within a modern framework. Its numerous exercises illustrate the governing principles, linearizations, and other approximations that constitute classical continuum models. 2007 edition. 320pp. 6 1/8 x 9 1/4. 0-486-47460-7

ENGINEERING MECHANICS FOR STRUCTURES, Louis L. Bucciarelli. This text explores the mechanics of solids and statics as well as the strength of materials and elasticity theory. Its many design exercises encourage creative initiative and systems thinking. 2009 edition. 320pp. 6 1/8 x 9 1/4. 0-486-46855-0

FEEDBACK CONTROL THEORY, John C. Doyle, Bruce A. Francis and Allen R. Tannenbaum. This excellent introduction to feedback control system design offers a theoretical approach that captures the essential issues and can be applied to a wide range of practical problems. 1992 edition. 224pp. 6 1/2 x 9 1/4. 0-486-46933-6

THE FORCES OF MATTER, Michael Faraday. These lectures by a famous inventor offer an easy-to-understand introduction to the interactions of the universe's physical forces. Six essays explore gravitation, cohesion, chemical affinity, heat, magnetism, and electricity. 1993 edition. 96pp. 5 3/8 x 8 1/2. 0-486-47482-8

DYNAMICS, Lawrence E. Goodman and William H. Warner. Beginning engineering text introduces calculus of vectors, particle motion, dynamics of particle systems and plane rigid bodies, technical applications in plane motions, and more. Exercises and answers in every chapter. 619pp. 5 3/8 x 8 1/2. 0-486-42006-X

ADAPTIVE FILTERING PREDICTION AND CONTROL, Graham C. Goodwin and Kwai Sang Sin. This unified survey focuses on linear discrete-time systems and explores natural extensions to nonlinear systems. It emphasizes discrete-time systems, summarizing theoretical and practical aspects of a large class of adaptive algorithms. 1984 edition. 560pp. 6 1/2 x 9 1/4. 0-486-46932-8

INDUCTANCE CALCULATIONS, Frederick W. Grover. This authoritative reference enables the design of virtually every type of inductor. It features a single simple formula for each type of inductor, together with tables containing essential numerical factors. 1946 edition. 304pp. 5 3/8 x 8 1/2. 0-486-47440-2

THERMODYNAMICS: Foundations and Applications, Elias P. Gyftopoulos and Gian Paolo Beretta. Designed by two MIT professors, this authoritative text discusses basic concepts and applications in detail, emphasizing generality, definitions, and logical consistency. More than 300 solved problems cover realistic energy systems and processes. 800pp. 6 1/8 x 9 1/4. 0-486-43932-1

THE FINITE ELEMENT METHOD: Linear Static and Dynamic Finite Element Analysis, Thomas J. R. Hughes. Text for students without in-depth mathematical training, this text includes a comprehensive presentation and analysis of algorithms of time-dependent phenomena plus beam, plate, and shell theories. Solution guide available upon request. 672pp. 6 1/2 x 9 1/4. 0-486-41181-8

HELICOPTER THEORY, Wayne Johnson. Monumental engineering text covers vertical flight, forward flight, performance, mathematics of rotating systems, rotary wing dynamics and aerodynamics, aeroelasticity, stability and control, stall, noise, and more. 189 illustrations. 1980 edition. 1089pp. 5 5/8 x 8 1/4.          0-486-68230-7

MATHEMATICAL HANDBOOK FOR SCIENTISTS AND ENGINEERS: Definitions, Theorems, and Formulas for Reference and Review, Granino A. Korn and Theresa M. Korn. Convenient access to information from every area of mathematics: Fourier transforms, Z transforms, linear and nonlinear programming, calculus of variations, random-process theory, special functions, combinatorial analysis, game theory, much more. 1152pp. 5 3/8 x 8 1/2.          0-486-41147-8

A HEAT TRANSFER TEXTBOOK: Fourth Edition, John H. Lienhard V and John H. Lienhard IV. This introduction to heat and mass transfer for engineering students features worked examples and end-of-chapter exercises. Worked examples and end-of-chapter exercises appear throughout the book, along with well-drawn, illuminating figures. 768pp. 7 x 9 1/4.          0-486-47931-5

BASIC ELECTRICITY, U.S. Bureau of Naval Personnel. Originally a training course; best nontechnical coverage. Topics include batteries, circuits, conductors, AC and DC, inductance and capacitance, generators, motors, transformers, amplifiers, etc. Many questions with answers. 349 illustrations. 1969 edition. 448pp. 6 1/2 x 9 1/4.
          0-486-20973-3

BASIC ELECTRONICS, U.S. Bureau of Naval Personnel. Clear, well-illustrated introduction to electronic equipment covers numerous essential topics: electron tubes, semiconductors, electronic power supplies, tuned circuits, amplifiers, receivers, ranging and navigation systems, computers, antennas, more. 560 illustrations. 567pp. 6 1/2 x 9 1/4.          0-486-21076-6

BASIC WING AND AIRFOIL THEORY, Alan Pope. This self-contained treatment by a pioneer in the study of wind effects covers flow functions, airfoil construction and pressure distribution, finite and monoplane wings, and many other subjects. 1951 edition. 320pp. 5 3/8 x 8 1/2.          0-486-47188-8

SYNTHETIC FUELS, Ronald F. Probstein and R. Edwin Hicks. This unified presentation examines the methods and processes for converting coal, oil, shale, tar sands, and various forms of biomass into liquid, gaseous, and clean solid fuels. 1982 edition. 512pp. 6 1/8 x 9 1/4.          0-486-44977-7

THEORY OF ELASTIC STABILITY, Stephen P. Timoshenko and James M. Gere. Written by world-renowned authorities on mechanics, this classic ranges from theoretical explanations of 2- and 3-D stress and strain to practical applications such as torsion, bending, and thermal stress. 1961 edition. 560pp. 5 3/8 x 8 1/2.          0-486-47207-8

PRINCIPLES OF DIGITAL COMMUNICATION AND CODING, Andrew J. Viterbi and Jim K. Omura. This classic by two digital communications experts is geared toward students of communications theory and to designers of channels, links, terminals, modems, or networks used to transmit and receive digital messages. 1979 edition. 576pp. 6 1/8 x 9 1/4.          0-486-46901-8

LINEAR SYSTEM THEORY: The State Space Approach, Lotfi A. Zadeh and Charles A. Desoer. Written by two pioneers in the field, this exploration of the state space approach focuses on problems of stability and control, plus connections between this approach and classical techniques. 1963 edition. 656pp. 6 1/8 x 9 1/4.
          0-486-46663-9

# Mathematics–Bestsellers

HANDBOOK OF MATHEMATICAL FUNCTIONS: with Formulas, Graphs, and Mathematical Tables, Edited by Milton Abramowitz and Irene A. Stegun. A classic resource for working with special functions, standard trig, and exponential logarithmic definitions and extensions, it features 29 sets of tables, some to as high as 20 places. 1046pp. 8 x 10 1/2. 0-486-61272-4

ABSTRACT AND CONCRETE CATEGORIES: The Joy of Cats, Jiri Adamek, Horst Herrlich, and George E. Strecker. This up-to-date introductory treatment employs category theory to explore the theory of structures. Its unique approach stresses concrete categories and presents a systematic view of factorization structures. Numerous examples. 1990 edition, updated 2004. 528pp. 6 1/8 x 9 1/4. 0-486-46934-4

MATHEMATICS: Its Content, Methods and Meaning, A. D. Aleksandrov, A. N. Kolmogorov, and M. A. Lavrent'ev. Major survey offers comprehensive, coherent discussions of analytic geometry, algebra, differential equations, calculus of variations, functions of a complex variable, prime numbers, linear and non-Euclidean geometry, topology, functional analysis, more. 1963 edition. 1120pp. 5 3/8 x 8 1/2. 0-486-40916-3

INTRODUCTION TO VECTORS AND TENSORS: Second Edition–Two Volumes Bound as One, Ray M. Bowen and C.-C. Wang. Convenient single-volume compilation of two texts offers both introduction and in-depth survey. Geared toward engineering and science students rather than mathematicians, it focuses on physics and engineering applications. 1976 edition. 560pp. 6 1/2 x 9 1/4. 0-486-46914-X

AN INTRODUCTION TO ORTHOGONAL POLYNOMIALS, Theodore S. Chihara. Concise introduction covers general elementary theory, including the representation theorem and distribution functions, continued fractions and chain sequences, the recurrence formula, special functions, and some specific systems. 1978 edition. 272pp. 5 3/8 x 8 1/2. 0-486-47929-3

ADVANCED MATHEMATICS FOR ENGINEERS AND SCIENTISTS, Paul DuChateau. This primary text and supplemental reference focuses on linear algebra, calculus, and ordinary differential equations. Additional topics include partial differential equations and approximation methods. Includes solved problems. 1992 edition. 400pp. 7 1/2 x 9 1/4. 0-486-47930-7

PARTIAL DIFFERENTIAL EQUATIONS FOR SCIENTISTS AND ENGINEERS, Stanley J. Farlow. Practical text shows how to formulate and solve partial differential equations. Coverage of diffusion-type problems, hyperbolic-type problems, elliptic-type problems, numerical and approximate methods. Solution guide available upon request. 1982 edition. 414pp. 6 1/8 x 9 1/4. 0-486-67620-X

VARIATIONAL PRINCIPLES AND FREE-BOUNDARY PROBLEMS, Avner Friedman. Advanced graduate-level text examines variational methods in partial differential equations and illustrates their applications to free-boundary problems. Features detailed statements of standard theory of elliptic and parabolic operators. 1982 edition. 720pp. 6 1/8 x 9 1/4. 0-486-47853-X

LINEAR ANALYSIS AND REPRESENTATION THEORY, Steven A. Gaal. Unified treatment covers topics from the theory of operators and operator algebras on Hilbert spaces; integration and representation theory for topological groups; and the theory of Lie algebras, Lie groups, and transform groups. 1973 edition. 704pp. 6 1/8 x 9 1/4. 0-486-47851-3

A SURVEY OF INDUSTRIAL MATHEMATICS, Charles R. MacCluer. Students learn how to solve problems they'll encounter in their professional lives with this concise single-volume treatment. It employs MATLAB and other strategies to explore typical industrial problems. 2000 edition. 384pp. 5 3/8 x 8 1/2. 0-486-47702-9

NUMBER SYSTEMS AND THE FOUNDATIONS OF ANALYSIS, Elliott Mendelson. Geared toward undergraduate and beginning graduate students, this study explores natural numbers, integers, rational numbers, real numbers, and complex numbers. Numerous exercises and appendixes supplement the text. 1973 edition. 368pp. 5 3/8 x 8 1/2. 0-486-45792-3

A FIRST LOOK AT NUMERICAL FUNCTIONAL ANALYSIS, W. W. Sawyer. Text by renowned educator shows how problems in numerical analysis lead to concepts of functional analysis. Topics include Banach and Hilbert spaces, contraction mappings, convergence, differentiation and integration, and Euclidean space. 1978 edition. 208pp. 5 3/8 x 8 1/2. 0-486-47882-3

FRACTALS, CHAOS, POWER LAWS: Minutes from an Infinite Paradise, Manfred Schroeder. A fascinating exploration of the connections between chaos theory, physics, biology, and mathematics, this book abounds in award-winning computer graphics, optical illusions, and games that clarify memorable insights into self-similarity. 1992 edition. 448pp. 6 1/8 x 9 1/4. 0-486-47204-3

SET THEORY AND THE CONTINUUM PROBLEM, Raymond M. Smullyan and Melvin Fitting. A lucid, elegant, and complete survey of set theory, this three-part treatment explores axiomatic set theory, the consistency of the continuum hypothesis, and forcing and independence results. 1996 edition. 336pp. 6 x 9. 0-486-47484-4

DYNAMICAL SYSTEMS, Shlomo Sternberg. A pioneer in the field of dynamical systems discusses one-dimensional dynamics, differential equations, random walks, iterated function systems, symbolic dynamics, and Markov chains. Supplementary materials include PowerPoint slides and MATLAB exercises. 2010 edition. 272pp. 6 1/8 x 9 1/4. 0-486-47705-3

ORDINARY DIFFERENTIAL EQUATIONS, Morris Tenenbaum and Harry Pollard. Skillfully organized introductory text examines origin of differential equations, then defines basic terms and outlines general solution of a differential equation. Explores integrating factors; dilution and accretion problems; Laplace Transforms; Newton's Interpolation Formulas, more. 818pp. 5 3/8 x 8 1/2. 0-486-64940-7

MATROID THEORY, D. J. A. Welsh. Text by a noted expert describes standard examples and investigation results, using elementary proofs to develop basic matroid properties before advancing to a more sophisticated treatment. Includes numerous exercises. 1976 edition. 448pp. 5 3/8 x 8 1/2. 0-486-47439-9

THE CONCEPT OF A RIEMANN SURFACE, Hermann Weyl. This classic on the general history of functions combines function theory and geometry, forming the basis of the modern approach to analysis, geometry, and topology. 1955 edition. 208pp. 5 3/8 x 8 1/2. 0-486-47004-0

THE LAPLACE TRANSFORM, David Vernon Widder. This volume focuses on the Laplace and Stieltjes transforms, offering a highly theoretical treatment. Topics include fundamental formulas, the moment problem, monotonic functions, and Tauberian theorems. 1941 edition. 416pp. 5 3/8 x 8 1/2. 0-486-47755-X

**Browse over 9,000 books at www.doverpublications.com**

# Mathematics–Logic and Problem Solving

PERPLEXING PUZZLES AND TANTALIZING TEASERS, Martin Gardner. Ninety-three riddles, mazes, illusions, tricky questions, word and picture puzzles, and other challenges offer hours of entertainment for youngsters. Filled with rib-tickling drawings. Solutions. 224pp. 5 3/8 x 8 1/2.                           0-486-25637-5

MY BEST MATHEMATICAL AND LOGIC PUZZLES, Martin Gardner. The noted expert selects 70 of his favorite "short" puzzles. Includes The Returning Explorer, The Mutilated Chessboard, Scrambled Box Tops, and dozens more. Complete solutions included. 96pp. 5 3/8 x 8 1/2.                           0-486-28152-3

THE LADY OR THE TIGER?: and Other Logic Puzzles, Raymond M. Smullyan. Created by a renowned puzzle master, these whimsically themed challenges involve paradoxes about probability, time, and change; metapuzzles; and self-referentiality. Nineteen chapters advance in difficulty from relatively simple to highly complex. 1982 edition. 240pp. 5 3/8 x 8 1/2.                           0-486-47027-X

SATAN, CANTOR AND INFINITY: Mind-Boggling Puzzles, Raymond M. Smullyan. A renowned mathematician tells stories of knights and knaves in an entertaining look at the logical precepts behind infinity, probability, time, and change. Requires a strong background in mathematics. Complete solutions. 288pp. 5 3/8 x 8 1/2.

0-486-47036-9

THE RED BOOK OF MATHEMATICAL PROBLEMS, Kenneth S. Williams and Kenneth Hardy. Handy compilation of 100 practice problems, hints and solutions indispensable for students preparing for the William Lowell Putnam and other mathematical competitions. Preface to the First Edition. Sources. 1988 edition. 192pp. 5 3/8 x 8 1/2.                           0-486-69415-1

KING ARTHUR IN SEARCH OF HIS DOG AND OTHER CURIOUS PUZZLES, Raymond M. Smullyan. This fanciful, original collection for readers of all ages features arithmetic puzzles, logic problems related to crime detection, and logic and arithmetic puzzles involving King Arthur and his Dogs of the Round Table. 160pp. 5 3/8 x 8 1/2.

0-486-47435-6

UNDECIDABLE THEORIES: Studies in Logic and the Foundation of Mathematics, Alfred Tarski in collaboration with Andrzej Mostowski and Raphael M. Robinson. This well-known book by the famed logician consists of three treatises: "A General Method in Proofs of Undecidability," "Undecidability and Essential Undecidability in Mathematics," and "Undecidability of the Elementary Theory of Groups." 1953 edition. 112pp. 5 3/8 x 8 1/2.                           0-486-47703-7

LOGIC FOR MATHEMATICIANS, J. Barkley Rosser. Examination of essential topics and theorems assumes no background in logic. "Undoubtedly a major addition to the literature of mathematical logic." – *Bulletin of the American Mathematical Society*. 1978 edition. 592pp. 6 1/8 x 9 1/4.                           0-486-46898-4

INTRODUCTION TO PROOF IN ABSTRACT MATHEMATICS, Andrew Wohlgemuth. This undergraduate text teaches students what constitutes an acceptable proof, and it develops their ability to do proofs of routine problems as well as those requiring creative insights. 1990 edition. 384pp. 6 1/2 x 9 1/4.                           0-486-47854-8

FIRST COURSE IN MATHEMATICAL LOGIC, Patrick Suppes and Shirley Hill. Rigorous introduction is simple enough in presentation and context for wide range of students. Symbolizing sentences; logical inference; truth and validity; truth tables; terms, predicates, universal quantifiers; universal specification and laws of identity; more. 288pp. 5 3/8 x 8 1/2.                           0-486-42259-3

# Mathematics–Algebra and Calculus

VECTOR CALCULUS, Peter Baxandall and Hans Liebeck. This introductory text offers a rigorous, comprehensive treatment. Classical theorems of vector calculus are amply illustrated with figures, worked examples, physical applications, and exercises with hints and answers. 1986 edition. 560pp. 5 3/8 x 8 1/2.        0-486-46620-5

ADVANCED CALCULUS: An Introduction to Classical Analysis, Louis Brand. A course in analysis that focuses on the functions of a real variable, this text introduces the basic concepts in their simplest setting and illustrates its teachings with numerous examples, theorems, and proofs. 1955 edition. 592pp. 5 3/8 x 8 1/2.        0-486-44548-8

ADVANCED CALCULUS, Avner Friedman. Intended for students who have already completed a one-year course in elementary calculus, this two-part treatment advances from functions of one variable to those of several variables. Solutions. 1971 edition. 432pp. 5 3/8 x 8 1/2.        0-486-45795-8

METHODS OF MATHEMATICS APPLIED TO CALCULUS, PROBABILITY, AND STATISTICS, Richard W. Hamming. This 4-part treatment begins with algebra and analytic geometry and proceeds to an exploration of the calculus of algebraic functions and transcendental functions and applications. 1985 edition. Includes 310 figures and 18 tables. 880pp. 6 1/2 x 9 1/4.        0-486-43945-3

BASIC ALGEBRA I: Second Edition, Nathan Jacobson. A classic text and standard reference for a generation, this volume covers all undergraduate algebra topics, including groups, rings, modules, Galois theory, polynomials, linear algebra, and associative algebra. 1985 edition. 528pp. 6 1/8 x 9 1/4.        0-486-47189-6

BASIC ALGEBRA II: Second Edition, Nathan Jacobson. This classic text and standard reference comprises all subjects of a first-year graduate-level course, including in-depth coverage of groups and polynomials and extensive use of categories and functors. 1989 edition. 704pp. 6 1/8 x 9 1/4.        0-486-47187-X

CALCULUS: An Intuitive and Physical Approach (Second Edition), Morris Kline. Application-oriented introduction relates the subject as closely as possible to science with explorations of the derivative; differentiation and integration of the powers of x; theorems on differentiation, antidifferentiation; the chain rule; trigonometric functions; more. Examples. 1967 edition. 960pp. 6 1/2 x 9 1/4.        0-486-40453-6

ABSTRACT ALGEBRA AND SOLUTION BY RADICALS, John E. Maxfield and Margaret W. Maxfield. Accessible advanced undergraduate-level text starts with groups, rings, fields, and polynomials and advances to Galois theory, radicals and roots of unity, and solution by radicals. Numerous examples, illustrations, exercises, appendixes. 1971 edition. 224pp. 6 1/8 x 9 1/4.        0-486-47723-1

AN INTRODUCTION TO THE THEORY OF LINEAR SPACES, Georgi E. Shilov. Translated by Richard A. Silverman. Introductory treatment offers a clear exposition of algebra, geometry, and analysis as parts of an integrated whole rather than separate subjects. Numerous examples illustrate many different fields, and problems include hints or answers. 1961 edition. 320pp. 5 3/8 x 8 1/2.        0-486-63070-6

LINEAR ALGEBRA, Georgi E. Shilov. Covers determinants, linear spaces, systems of linear equations, linear functions of a vector argument, coordinate transformations, the canonical form of the matrix of a linear operator, bilinear and quadratic forms, and more. 387pp. 5 3/8 x 8 1/2.        0-486-63518-X

# Mathematics–Probability and Statistics

BASIC PROBABILITY THEORY, Robert B. Ash. This text emphasizes the probabilistic way of thinking, rather than measure-theoretic concepts. Geared toward advanced undergraduates and graduate students, it features solutions to some of the problems. 1970 edition. 352pp. 5 3/8 x 8 1/2. 0-486-46628-0

PRINCIPLES OF STATISTICS, M. G. Bulmer. Concise description of classical statistics, from basic dice probabilities to modern regression analysis. Equal stress on theory and applications. Moderate difficulty; only basic calculus required. Includes problems with answers. 252pp. 5 5/8 x 8 1/4. 0-486-63760-3

OUTLINE OF BASIC STATISTICS: Dictionary and Formulas, John E. Freund and Frank J. Williams. Handy guide includes a 70-page outline of essential statistical formulas covering grouped and ungrouped data, finite populations, probability, and more, plus over 1,000 clear, concise definitions of statistical terms. 1966 edition. 208pp. 5 3/8 x 8 1/2. 0-486-47769-X

GOOD THINKING: The Foundations of Probability and Its Applications, Irving J. Good. This in-depth treatment of probability theory by a famous British statistician explores Keynesian principles and surveys such topics as Bayesian rationality, corroboration, hypothesis testing, and mathematical tools for induction and simplicity. 1983 edition. 352pp. 5 3/8 x 8 1/2. 0-486-47438-0

INTRODUCTION TO PROBABILITY THEORY WITH CONTEMPORARY APPLICATIONS, Lester L. Helms. Extensive discussions and clear examples, written in plain language, expose students to the rules and methods of probability. Exercises foster problem-solving skills, and all problems feature step-by-step solutions. 1997 edition. 368pp. 6 1/2 x 9 1/4. 0-486-47418-6

CHANCE, LUCK, AND STATISTICS, Horace C. Levinson. In simple, non-technical language, this volume explores the fundamentals governing chance and applies them to sports, government, and business. "Clear and lively ... remarkably accurate." – Scientific Monthly. 384pp. 5 3/8 x 8 1/2. 0-486-41997-5

FIFTY CHALLENGING PROBLEMS IN PROBABILITY WITH SOLUTIONS, Frederick Mosteller. Remarkable puzzlers, graded in difficulty, illustrate elementary and advanced aspects of probability. These problems were selected for originality, general interest, or because they demonstrate valuable techniques. Also includes detailed solutions. 88pp. 5 3/8 x 8 1/2. 0-486-65355-2

EXPERIMENTAL STATISTICS, Mary Gibbons Natrella. A handbook for those seeking engineering information and quantitative data for designing, developing, constructing, and testing equipment. Covers the planning of experiments, the analyzing of extreme-value data; and more. 1966 edition. Index. Includes 52 figures and 76 tables. 560pp. 8 3/8 x 11. 0-486-43937-2

STOCHASTIC MODELING: Analysis and Simulation, Barry L. Nelson. Coherent introduction to techniques also offers a guide to the mathematical, numerical, and simulation tools of systems analysis. Includes formulation of models, analysis, and interpretation of results. 1995 edition. 336pp. 6 1/8 x 9 1/4. 0-486-47770-3

INTRODUCTION TO BIOSTATISTICS: Second Edition, Robert R. Sokal and F. James Rohlf. Suitable for undergraduates with a minimal background in mathematics, this introduction ranges from descriptive statistics to fundamental distributions and the testing of hypotheses. Includes numerous worked-out problems and examples. 1987 edition. 384pp. 6 1/8 x 9 1/4. 0-486-46961-1

# Mathematics–Geometry and Topology

PROBLEMS AND SOLUTIONS IN EUCLIDEAN GEOMETRY, M. N. Aref and William Wernick. Based on classical principles, this book is intended for a second course in Euclidean geometry and can be used as a refresher. More than 200 problems include hints and solutions. 1968 edition. 272pp. 5 3/8 x 8 1/2. 0-486-47720-7

TOPOLOGY OF 3-MANIFOLDS AND RELATED TOPICS, Edited by M. K. Fort, Jr. With a New Introduction by Daniel Silver. Summaries and full reports from a 1961 conference discuss decompositions and subsets of 3-space; n-manifolds; knot theory; the Poincaré conjecture; and periodic maps and isotopies. Familiarity with algebraic topology required. 1962 edition. 272pp. 6 1/8 x 9 1/4. 0-486-47753-3

POINT SET TOPOLOGY, Steven A. Gaal. Suitable for a complete course in topology, this text also functions as a self-contained treatment for independent study. Additional enrichment materials make it equally valuable as a reference. 1964 edition. 336pp. 5 3/8 x 8 1/2. 0-486-47222-1

INVITATION TO GEOMETRY, Z. A. Melzak. Intended for students of many different backgrounds with only a modest knowledge of mathematics, this text features self-contained chapters that can be adapted to several types of geometry courses. 1983 edition. 240pp. 5 3/8 x 8 1/2. 0-486-46626-4

TOPOLOGY AND GEOMETRY FOR PHYSICISTS, Charles Nash and Siddhartha Sen. Written by physicists for physics students, this text assumes no detailed background in topology or geometry. Topics include differential forms, homotopy, homology, cohomology, fiber bundles, connection and covariant derivatives, and Morse theory. 1983 edition. 320pp. 5 3/8 x 8 1/2. 0-486-47852-1

BEYOND GEOMETRY: Classic Papers from Riemann to Einstein, Edited with an Introduction and Notes by Peter Pesic. This is the only English-language collection of these 8 accessible essays. They trace seminal ideas about the foundations of geometry that led to Einstein's general theory of relativity. 224pp. 6 1/8 x 9 1/4. 0-486-45350-2

GEOMETRY FROM EUCLID TO KNOTS, Saul Stahl. This text provides a historical perspective on plane geometry and covers non-neutral Euclidean geometry, circles and regular polygons, projective geometry, symmetries, inversions, informal topology, and more. Includes 1,000 practice problems. Solutions available. 2003 edition. 480pp. 6 1/8 x 9 1/4. 0-486-47459-3

TOPOLOGICAL VECTOR SPACES, DISTRIBUTIONS AND KERNELS, François Trèves. Extending beyond the boundaries of Hilbert and Banach space theory, this text focuses on key aspects of functional analysis, particularly in regard to solving partial differential equations. 1967 edition. 592pp. 5 3/8 x 8 1/2.
0-486-45352-9

INTRODUCTION TO PROJECTIVE GEOMETRY, C. R. Wylie, Jr. This introductory volume offers strong reinforcement for its teachings, with detailed examples and numerous theorems, proofs, and exercises, plus complete answers to all odd-numbered end-of-chapter problems. 1970 edition. 576pp. 6 1/8 x 9 1/4. 0-486-46895-X

FOUNDATIONS OF GEOMETRY, C. R. Wylie, Jr. Geared toward students preparing to teach high school mathematics, this text explores the principles of Euclidean and non-Euclidean geometry and covers both generalities and specifics of the axiomatic method. 1964 edition. 352pp. 6 x 9. 0-486-47214-0

**Browse over 9,000 books at www.doverpublications.com**

# Mathematics–History

THE WORKS OF ARCHIMEDES, Archimedes. Translated by Sir Thomas Heath. Complete works of ancient geometer feature such topics as the famous problems of the ratio of the areas of a cylinder and an inscribed sphere; the properties of conoids, spheroids, and spirals; more. 326pp. 5 3/8 x 8 1/2. 0-486-42084-1

THE HISTORICAL ROOTS OF ELEMENTARY MATHEMATICS, Lucas N. H. Bunt, Phillip S. Jones, and Jack D. Bedient. Exciting, hands-on approach to understanding fundamental underpinnings of modern arithmetic, algebra, geometry and number systems examines their origins in early Egyptian, Babylonian, and Greek sources. 336pp. 5 3/8 x 8 1/2. 0-486-25563-8

THE THIRTEEN BOOKS OF EUCLID'S ELEMENTS, Euclid. Contains complete English text of all 13 books of the Elements plus critical apparatus analyzing each definition, postulate, and proposition in great detail. Covers textual and linguistic matters; mathematical analyses of Euclid's ideas; classical, medieval, Renaissance and modern commentators; refutations, supports, extrapolations, reinterpretations and historical notes. 995 figures. Total of 1,425pp. All books 5 3/8 x 8 1/2.

Vol. I: 443pp. 0-486-60088-2
Vol. II: 464pp. 0-486-60089-0
Vol. III: 546pp. 0-486-60090-4

A HISTORY OF GREEK MATHEMATICS, Sir Thomas Heath. This authoritative two-volume set that covers the essentials of mathematics and features every landmark innovation and every important figure, including Euclid, Apollonius, and others. 5 3/8 x 8 1/2.

Vol. I: 461pp. 0-486-24073-8
Vol. II: 597pp. 0-486-24074-6

A MANUAL OF GREEK MATHEMATICS, Sir Thomas L. Heath. This concise but thorough history encompasses the enduring contributions of the ancient Greek mathematicians whose works form the basis of most modern mathematics. Discusses Pythagorean arithmetic, Plato, Euclid, more. 1931 edition. 576pp. 5 3/8 x 8 1/2.

0-486-43231-9

CHINESE MATHEMATICS IN THE THIRTEENTH CENTURY, Ulrich Libbrecht. An exploration of the 13th-century mathematician Ch'in, this fascinating book combines what is known of the mathematician's life with a history of his only extant work, the Shu-shu chiu-chang. 1973 edition. 592pp. 5 3/8 x 8 1/2.

0-486-44619-0

PHILOSOPHY OF MATHEMATICS AND DEDUCTIVE STRUCTURE IN EUCLID'S ELEMENTS, Ian Mueller. This text provides an understanding of the classical Greek conception of mathematics as expressed in Euclid's Elements. It focuses on philosophical, foundational, and logical questions and features helpful appendixes. 400pp. 6 1/2 x 9 1/4. 0-486-45300-6

BEYOND GEOMETRY: Classic Papers from Riemann to Einstein, Edited with an Introduction and Notes by Peter Pesic. This is the only English-language collection of these 8 accessible essays. They trace seminal ideas about the foundations of geometry that led to Einstein's general theory of relativity. 224pp. 6 1/8 x 9 1/4. 0-486-45350-2

HISTORY OF MATHEMATICS, David E. Smith. Two-volume history – from Egyptian papyri and medieval maps to modern graphs and diagrams. Non-technical chronological survey with thousands of biographical notes, critical evaluations, and contemporary opinions on over 1,100 mathematicians. 5 3/8 x 8 1/2.

Vol. I: 618pp. 0-486-20429-4
Vol. II: 736pp. 0-486-20430-8

**Browse over 9,000 books at www.doverpublications.com**

# Physics

THEORETICAL NUCLEAR PHYSICS, John M. Blatt and Victor F. Weisskopf. An uncommonly clear and cogent investigation and correlation of key aspects of theoretical nuclear physics by leading experts: the nucleus, nuclear forces, nuclear spectroscopy, two-, three- and four-body problems, nuclear reactions, beta-decay and nuclear shell structure. 896pp. 5 3/8 x 8 1/2. 0-486-66827-4

QUANTUM THEORY, David Bohm. This advanced undergraduate-level text presents the quantum theory in terms of qualitative and imaginative concepts, followed by specific applications worked out in mathematical detail. 655pp. 5 3/8 x 8 1/2.
0-486-65969-0

ATOMIC PHYSICS AND HUMAN KNOWLEDGE, Niels Bohr. Articles and speeches by the Nobel Prize–winning physicist, dating from 1934 to 1958, offer philosophical explorations of the relevance of atomic physics to many areas of human endeavor. 1961 edition. 112pp. 5 3/8 x 8 1/2. 0-486-47928-5

COSMOLOGY, Hermann Bondi. A co-developer of the steady-state theory explores his conception of the expanding universe. This historic book was among the first to present cosmology as a separate branch of physics. 1961 edition. 192pp. 5 3/8 x 8 1/2.
0-486-47483-6

LECTURES ON QUANTUM MECHANICS, Paul A. M. Dirac. Four concise, brilliant lectures on mathematical methods in quantum mechanics from Nobel Prize-winning quantum pioneer build on idea of visualizing quantum theory through the use of classical mechanics. 96pp. 5 3/8 x 8 1/2. 0-486-41713-1

THE PRINCIPLE OF RELATIVITY, Albert Einstein and Frances A. Davis. Eleven papers that forged the general and special theories of relativity include seven papers by Einstein, two by Lorentz, and one each by Minkowski and Weyl. 1923 edition. 240pp. 5 3/8 x 8 1/2. 0-486-60081-5

PHYSICS OF WAVES, William C. Elmore and Mark A. Heald. Ideal as a classroom text or for individual study, this unique one-volume overview of classical wave theory covers wave phenomena of acoustics, optics, electromagnetic radiations, and more. 477pp. 5 3/8 x 8 1/2. 0-486-64926-1

THERMODYNAMICS, Enrico Fermi. In this classic of modern science, the Nobel Laureate presents a clear treatment of systems, the First and Second Laws of Thermodynamics, entropy, thermodynamic potentials, and much more. Calculus required. 160pp. 5 3/8 x 8 1/2. 0-486-60361-X

QUANTUM THEORY OF MANY-PARTICLE SYSTEMS, Alexander L. Fetter and John Dirk Walecka. Self-contained treatment of nonrelativistic many-particle systems discusses both formalism and applications in terms of ground-state (zero-temperature) formalism, finite-temperature formalism, canonical transformations, and applications to physical systems. 1971 edition. 640pp. 5 3/8 x 8 1/2. 0-486-42827-3

QUANTUM MECHANICS AND PATH INTEGRALS: Emended Edition, Richard P. Feynman and Albert R. Hibbs. Emended by Daniel F. Styer. The Nobel Prize–winning physicist presents unique insights into his theory and its applications. Feynman starts with fundamentals and advances to the perturbation method, quantum electrodynamics, and statistical mechanics. 1965 edition, emended in 2005. 384pp. 6 1/8 x 9 1/4. 0-486-47722-3